NUMERICAL METHODS IN FINITE ELEMENT ANALYSIS

NUMERICAL METHODS IN FINITE ELEMENT ANALYSIS

KLAUS-JÜRGEN BATHE

Department of Mechanical Engineering
Massachusetts Institute of Technology

EDWARD L. WILSON

Department of Civil Engineering
University of California, Berkeley

PRENTICE-HALL, INC., *Englewood Cliffs, New Jersey*

Library of Congress Cataloging in Publication Data

BATHE, KLAUS-JÜRGEN.
Numerical methods in finite element analysis.

(Prentice-Hall civil engineering and engineering mechanics series)
Includes bibliographical references.
1. Finite element method. 2. Numerical analysis.
I. Wilson, Edward L., 1931– joint author. II. Title.
TA347.F5B37 624'.171'01515 75–46522
ISBN 0–13–627190–1

10 9 8 7 6 5 4

Printed in the United States of America

PRENTICE-HALL INTERNATIONAL, INC., *London*
PRENTICE-HALL OF AUSTRALIA PTY. LIMITED, *Sydney*
PRENTICE-HALL OF CANADA, LTD., *Toronto*
PRENTICE-HALL OF INDIA PRIVATE LIMITED, *New Delhi*
PRENTICE-HALL OF JAPAN, INC., *Tokyo*
PRENTICE-HALL OF SOUTHEAST ASIA PTE. LTD., *Singapore*

CONTENTS

PREFACES

During the last years, the finite element method of analysis has rapidly become a very popular technique for the computer solution of complex problems in engineering. Basically, the method can be understood as an extension of earlier established analysis techniques, in which a structure is represented as an assemblage of discrete truss and beam elements. The same matrix algebra procedures are used, but instead of truss and beam members, finite elements are employed to represent regions of plane stress, plane strain, axisymmetric, three-dimensional, plate, or shell behavior.

In the earliest developments of finite element analysis, quite naturally almost all emphasis was directed toward the development of effective finite elements for the solution of specific problems. However, the potential of the method when used effectively on a digital computer was rapidly realized, and increasingly larger and more complex finite element systems were considered. This, in turn, enhanced the development of efficient data-handling procedures and effective techniques for the solution of the governing finite element equilibrium equations. At present, computer programs are in use that can handle at reasonable cost very large finite element systems, because the algorithms employed have been developed specifically for finite element analysis. As a result of this development, when referring to finite element analysis, a complete numerical process implemented on the digital computer is implied. This numerical process comprises the formulation of the finite element matrices, the numerical integration to evaluate the matrices, the assemblage of the element matrices into the matrices that correspond to the complete finite element system, and the numerical solution of the system equilibrium equations.

The objective in this book is to present each of the above aspects of finite element analysis and thus to provide a basis for the understanding of the complete solution process. Because finite element analysis is basically a numerical procedure, the numerical aspects of the method are emphasized, but whenever possible physical explanations are given.

According to three basic areas in which knowledge is required, the book is divided into three parts. In the first part, important concepts of matrix and linear algebra are presented. Many readers may be familiar with the elementary rules of matrix algebra, but specific attention should be given to the concepts of linear algebra, because they provide the foundation for a thorough understanding of the numerical procedures presented later.

The second part of the book comprises the formulation of the finite element method and the numerical procedures used to evaluate the element matrices and the matrices of the complete element assemblage. Since the early use of finite element analysis, numerous finite elements have been developed. The objective in this book is not to summarize all the finite element models available, but rather to establish the general principles and describe those elements that are believed to be most effective at present.

In the last part of the book, procedures for the effective solution of the finite element equilibrium equations in static and dynamic analysis are presented. This phase of finite element analysis usually comprises most of the computer effort and, consequently, deserves a great deal of attention. It should be realized that because the numerical procedures employed in this phase largely determine the cost of analysis, they also determine whether an analysis can or cannot be performed in practice.

Throughout the presentation, the aim was to establish the numerical procedures using relatively simple and physical concepts, even at the expense of losing some mathematical rigorousness. The principles and procedures are illustrated using well over 100 worked-out examples, which are an integral part of the presentation. Short computer programs are also included to demonstrate the numerical procedures in a compact manner. These programs can be used directly as subroutines in finite element codes.

The primary aim in the writing of this book is to provide a tool for teaching upper-level undergraduate and graduate courses in engineering. Although the topic is finite element analysis procedures, many of the numerical techniques are quite general and could be employed effectively in any discrete method of analysis. In addition, a large community of engineers using finite element computer programs should find much valuable information in the text.

A very difficult aspect of writing a book is to give references that appropriately acknowledge the work of the various researchers in the field. This is particularly difficult in this book, because the field of finite element analysis has been expanding very rapidly. I would like to apologize here for possibly not having referred at all times to the most appropriate published work.

The endeavor to write this book is a result of the excitement and challenge that I have experienced in working in the field of finite element analysis. I would like to thank my former teacher, Edward L. Wilson, who through his active support made it possible for me to pursue my research and complete most of the work on this book while at Berkeley. Although I have been writing this book and the responsibility for the book rests with me, the name of E. L. Wilson on the title page is justified by the fact that my research and development work was largely based on his earlier achievements.

For some years, I have been associated with Fred Peterson of Engineering/Analysis Corporation, Berkeley, whom I would like to thank for his support. I much enjoyed working with Fred. I also would like to thank Steve Kenney for helping me on the problem solutions; Aileen Frankel, who did an outstanding job in typing the manuscript; Bill Grace for helping me during the final phase of work; and Rosalie Herion, the editor of the book, for her patience and diligence. Finally, I would like to thank, in particular, my wife, Zorka, who with her love and patience supported the writing of this book.

K.J. BATHE
M.I.T.

Researchers in both engineering mechanics and applied mathematics have participated in the development of the finite element method. During the period 1850 to 1860, theories on torsion and bending of beams were unified, and the foundations of the field of structural analysis were established. For approximately 100 years, structural analysis was restricted to the study of systems of one-dimensional beam and truss elements, an important characteristic being that an element was always only connected to two joints. In the mid-1950s, two-dimensional structural elements connected to more than two joints were developed in the aircraft industry in order to improve the modeling of the stiffness of thin membrane elements connected to traditional one-dimensional structural elements. In 1960, Clough first introduced the finite element terminology in his paper, "The Finite Element Method in Plane Stress Analysis." In this paper, the method was presented as an extension of structural analysis techniques to the solution of problems in continuum mechanics.

Ritz, in 1909, developed a very powerful method for the approximate solution of field problems in continuum mechanics. This approach involves the approximation of a potential functional in terms of trial functions of unknown magnitudes. The minimization of the functional with respect to each unknown results in a set of equations that are solved for the unknown magnitudes. One of the initial limitations of the Ritz method was that the trial functions satisfy the boundary conditions of the problem. Courant, in 1943, made a significant extension of the Ritz method by introducing separate linear functions over triangular areas and applied the method to the solution of torsion problems. The unknowns in the problem were selected as the values of the functions at the locations of the interconnecting triangular areas. Hence, the traditional limitation of the global Ritz functions in satisfying boundary conditions was eliminated, since the conditions could be satisfied at a finite number of points along a boundary. The Ritz method as used by Courant was identical to the finite element method independently presented by Clough many years later. Of course, the reason that the finite element method met with almost immediate success in 1960 lies in the fact that the large number of numerical operations inherent in the method could be performed by the recently developed digital computer; whereas, this tool was not available to Courant in 1943.

In the mid-1960s, researchers in both the fields of continuum mechanics and structural analysis recognized that the extended Ritz and the finite element methods are identical, and within the next ten years, the development and application of the method progressed at a very impressive rate. The finite element method has been applied to problems in three dimensions, problems involving both material and geometric nonlinearities, time-dependent problems, and problems in many areas beyond structural analysis, such as fluid flow, heat transfer, and magnetic field analysis. Further comments on the history of the development of the method are presented in Chapter 3.

Because of the speed with which the finite element method has expanded, a number of books have recently been published. Most of these books are concerned with the development of various elements and with examples of their applications. The major purpose of this book is to present the fundamental numerical methods required in all finite element applications. In addition, emphasis is placed on the computer implementation of the numerical procedures. Therefore, I feel that this work is unique in the field when compared to other finite element books.

The successful completion of research involving finite element development generally results in the development of a computer program to be used in the practical solution of engineering problems. In order for the program to be an effective analysis tool, it must be based on theories and techniques from three different disciplines. First, the approximations used to develop the properties of the various finite elements must be based on sound fundamental principles of continuum mechanics. Second, the numerical methods selected for spacial integration, solution of equations, evaluation of eigenvalues, and step-by-step time solutions must be accurate and efficient. Third, the computer implementation of the numerical techniques used must be approached with great care if the number of numerical operations are to be minimized, high speed and low speed storage units are to be used effectively, and the resulting program is to be reasonably machine-independent. The purpose of this book is to present the necessary background in all three of these areas.

The development of new finite elements, effective numerical methods, and practical computer programs have been the areas of my personal research of the past 15 years, and this book emphasizes these research areas. During the past five years, my research in these areas has been greatly enhanced due to my collaboration with Klaus-Jürgen Bathe. Working with Jürgen has been a most enjoyable experience.

E. L. WILSON
U. C., Berkeley

PART I

MATRICES AND LINEAR ALGEBRA

CHAPTER 1

ELEMENTARY CONCEPTS OF MATRICES

1.1 INTRODUCTION

The practical use of finite element analysis is based on matrix algebra and the use of the electronic computer, because it is only in matrix form that the complete solution process can be expressed in a compact and elegant manner. The objective in this chapter is to present briefly the fundamentals of matrix algebra that are needed for an understanding of the solution procedures discussed later.[1-3] In the presentation, emphasis is directed to those aspects of matrix algebra that are important in finite element analysis.

From the practical point of view, matrix algebra can be regarded merely as an effective tool with which to manipulate in an elegant manner large amounts of data, and we use this approach in this chapter. Some concepts of linear algebra that are also required for a thorough understanding of the finite element method are presented in Chapter 2.

Consider that the objective of finite element analysis is to evaluate the displacements at a large number of points of the structure under consideration. Then, once the physical relationships between the required displacements of the structure and the applied loads are known, it is possible, using the concepts of matrix algebra, to express the complete solution process employing a few symbols to identify the various quantities and a few lines to describe the solution process. However, although the solution process is expressed in an elegant manner in terms of matrices and matrix manipulations, it is frequently important to identify the detailed operations that are followed in the matrix solution in order to design more effective solution schemes. Again, this is possible only if the definitions and concepts of matrix algebra are well understood.

1.2 INTRODUCTION TO MATRICES

The effectiveness of using matrices in practical calculations is readily realized by considering the solution of a set of linear simultaneous equations, such as

$$\begin{aligned} 5x_1 - 4x_2 + x_3 \phantom{{}+ x_4} &= 0 \\ -4x_1 + 6x_2 - 4x_3 + x_4 &= 1 \\ x_1 - 4x_2 + 6x_3 - 4x_4 &= 0 \\ x_2 - 4x_3 + 5x_4 &= 0 \end{aligned} \tag{1.1}$$

where the unknowns are x_1, x_2, x_3, and x_4. Using matrix notation, this set of equations is written as

$$\begin{bmatrix} 5 & -4 & 1 & 0 \\ -4 & 6 & -4 & 1 \\ 1 & -4 & 6 & -4 \\ 0 & 1 & -4 & 5 \end{bmatrix} \begin{bmatrix} x_1 \\ x_2 \\ x_3 \\ x_4 \end{bmatrix} = \begin{bmatrix} 0 \\ 1 \\ 0 \\ 0 \end{bmatrix} \tag{1.2}$$

where it is noted that, rather logically, the coefficients of the unknowns (5, −4, 1, etc.) are grouped together in one array; the left-hand-side unknowns (x_1, x_2, x_3, and x_4) and the right-hand-side known quantities are each grouped together in additional arrays. Although written differently, the relation (1.2) still reads the same way as (1.1). However, using matrix symbols to identify the arrays in (1.2), we can now write the set of simultaneous equations as

$$\mathbf{A}\mathbf{x} = \mathbf{b} \tag{1.3}$$

where $\mathbf{A}$ is the matrix of the coefficients in the set of linear equations, $\mathbf{x}$ is the matrix of unknowns, and $\mathbf{b}$ is the matrix of known quantities; i.e.,

$$\mathbf{A} = \begin{bmatrix} 5 & -4 & 1 & 0 \\ -4 & 6 & -4 & 1 \\ 1 & -4 & 6 & -4 \\ 0 & 1 & -4 & 5 \end{bmatrix}; \qquad \mathbf{x} = \begin{bmatrix} x_1 \\ x_2 \\ x_3 \\ x_4 \end{bmatrix}; \qquad \mathbf{b} = \begin{bmatrix} 0 \\ 1 \\ 0 \\ 0 \end{bmatrix} \tag{1.4}$$

The formal definition of a matrix, as given below, seems now apparent.

Definition: A *matrix* is an array of ordered numbers. A *general matrix* consists of mn numbers arranged in m rows and n columns, giving the following array:

$$\mathbf{A} = \begin{bmatrix} a_{11} & a_{12} & \cdots & a_{1n} \\ a_{21} & a_{22} & \cdots & a_{2n} \\ \vdots & \vdots & & \vdots \\ a_{m1} & a_{m2} & \cdots & a_{mn} \end{bmatrix} \tag{1.5}$$

We say that this matrix has order $m \times n$ (m by n). When we have only one row ($m = 1$) or one column ($n = 1$), we call $\mathbf{A}$ also a vector. Matrices are represented in this book using boldface letters, and upper case letters when they are not vectors. On the other hand, vectors can be upper case or lower case boldface.

We therefore see that the following are matrices:

$$\begin{bmatrix} 1 \\ 2 \end{bmatrix}; \quad \begin{bmatrix} 1 & 4 & -5.3 \\ 3 & 2.1 & 6 \end{bmatrix}; \quad [6.1 \quad 2.2 \quad 3] \tag{1.6}$$

where the first and the last matrices are also column and row vectors, respectively.

A typical element in the ith row and jth column of $\mathbf{A}$ is identified as a_{ij}, e.g., in the first matrix in (1.4), $a_{11} = 5$ and $a_{21} = -4$. Considering the elements a_{ij} in (1.5), we note that the subscript i runs from 1 to m and the subscript j runs from 1 to n. A comma between subscripts will be used when there is any risk of confusion, e.g., $a_{1+r,\, j+s}$.

In general, the utility of matrices in practice arises from the fact that we can identify and manipulate an array of many numbers by use of a single symbol. Relationships among large sets of numbers can be stated in a clear and compact way. Furthermore, if the relationships involve properly defined known and unknown quantities, such as in the solution of a set of simultaneous equations, it is possible to solve for the unknown quantitites in a systematic manner. This is where the computer is of so much help, because it can calculate and manipulate the elements of matrices at very large speeds. However, we as the analysts are required to clearly define the variables in a problem, to prescribe the calculation procedures for the evaluation of the elements in the matrices involved, and to design the solution procedures.

1.3 SPECIAL MATRICES

Whenever the elements of a matrix obey a certain law, we can consider the matrix to be of special form. A *real matrix* is a matrix whose elements are all real. A *complex matrix* has elements that may be complex. We shall deal only with real matrices. In addition, the matrix will often be symmetric, which is the property defined next.

Definition: The *transpose* of the $m \times n$ matrix $\mathbf{A}$, written as $\mathbf{A}^T$, is obtained by interchanging the rows and columns in $\mathbf{A}$. If $\mathbf{A} = \mathbf{A}^T$, it follows that the number of rows and columns in $\mathbf{A}$ are equal and that $a_{ij} = a_{ji}$. Because $m = n$, we say that $\mathbf{A}$ is a *square matrix* of order n, and because $a_{ij} = a_{ji}$, we say that $\mathbf{A}$ is a *symmetric matrix*. Note that symmetry implies that $\mathbf{A}$ is square but not vice versa; i.e., a square matrix need not be symmetric.

A symmetric matrix of order 4 was, for example, the coefficient matrix $\mathbf{A}$ in (1.2). We can verify that $\mathbf{A}^T = \mathbf{A}$ by simply checking that $a_{ji} = a_{ij}$ for $i, j = 1, \ldots, 4$.

Another special matrix is the *identity* (*or unit*) *matrix* $\mathbf{I}_n$, which is a square matrix of order n with only zero elements except for its diagonal entries, which are unity. For example, the identity matrix of order 3 is

$$\mathbf{I}_3 = \begin{bmatrix} 1 & 0 & 0 \\ 0 & 1 & 0 \\ 0 & 0 & 1 \end{bmatrix} \tag{1.7}$$

In practical calculations the order of an identity matrix is often implied and the subscript is not written. In analogy with the identity matrix, we also use *identity* (*or unit*) *vectors* of order n, defined as $\mathbf{e}_i$, where the subscript i indicates that the vector is the ith column of an identity matrix.

We shall work abundantly with symmetric banded matrices. Bandedness means that all elements beyond the bandwidth of the matrix are zero. Because $\mathbf{A}$ is symmetric, we can state this condition as

$$a_{ij} = 0 \qquad \text{for } j > i + m_{\mathbf{A}} \tag{1.8}$$

where $2m_{\mathbf{A}} + 1$ *is the bandwidth of* $\mathbf{A}$. As an example, the following matrix is a symmetric banded matrix of order 5. The half-bandwidth $m_{\mathbf{A}}$ is 2:

$$\mathbf{A} = \begin{bmatrix} 3 & 2 & 1 & 0 & 0 \\ 2 & 3 & 4 & 1 & 0 \\ 1 & 4 & 5 & 6 & 1 \\ 0 & 1 & 6 & 7 & 4 \\ 0 & 0 & 1 & 4 & 3 \end{bmatrix} \tag{1.9}$$

If the half-bandwidth of a matrix is zero, we have nonzero elements only on the diagonal of the matrix and denote it as a *diagonal matrix*. For example, the identity matrix is a diagonal matrix.

In computer calculations with matrices, we need to use a scheme of storing the elements of the matrices in high-speed storage. An obvious way of storing the elements of a matrix $\mathbf{A}$ of order $m \times n$ is simply to dimension in the FORTRAN program an array A(M,N), where M $= m$ and N $= n$, and store each matrix element a_{ij} in the storage location A(I,J). However, in many calculations we store in this way unnecessarily many zero elements of $\mathbf{A}$, which are never needed in the calculations. Also, if $\mathbf{A}$ is symmetric, we should probably take advantage of it and store only the upper half of the matrix, including the diagonal elements. In general, only a restricted number of high-speed storage locations will be available, and it is necessary to use an effective storage scheme in order to be able to take into high-speed core the maximum matrix size possible. If the matrix is too large to be contained in high-speed storage, the solution process involves reading and writing on secondary storage, which can add significantly to the solution cost. Fortunately, in finite element analysis, the system matrices are symmetric and banded. Therefore, with an effective storage scheme, rather large-order matrices can be kept in high-speed core.

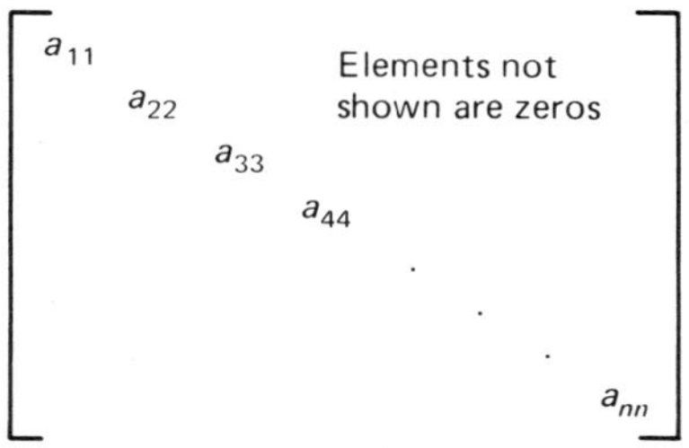

$A(1) = a_{11}$, $A(2) = a_{22}$, $A(3) = a_{33}$
$A(4) = a_{44}$, ..., $A(N) = a_{nn}$

(a) Diagonal matrix

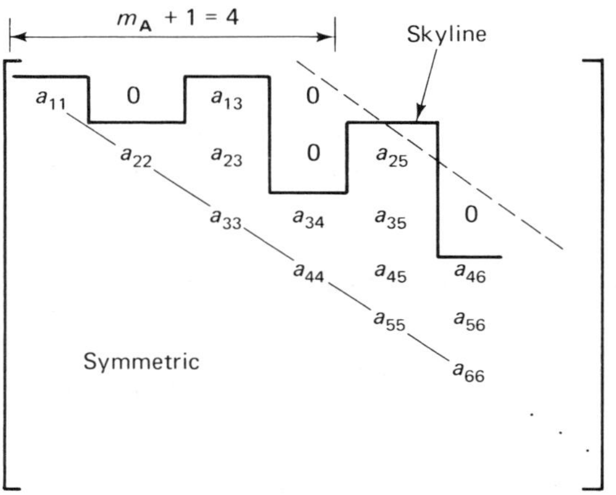

$A(1) = a_{11}$, $A(2) = a_{22}$,
$A(3) = a_{33}$, $A(4) = a_{23}$,
$A(5) = a_{13}$, $A(6) = a_{44}$,
$A(7) = a_{34}$, $A(8) = a_{55}$,
$A(9) = a_{45}$, $A(10) = a_{35}$,
$A(11) = a_{25}$, $A(12) = a_{66}$,
etc.

(b) Banded matrix, $m_A = 3$

Figure 1.1. Storage of matrix **A** in a one-dimensional array.

Let us denote by A(I) the Ith element in the one-dimensional storage array A. A diagonal matrix of order n would simply be stored as shown in Fig. 1.1:

$$A(I) = a_{ii}; \qquad I = i = 1, \ldots, n \tag{1.10}$$

Consider a banded matrix as shown in Fig. 1.1. We will see later that zero elements within the "skyline" of the matrix may be changed to nonzero elements in the solution process; for example, a_{35} may be a zero element but becomes nonzero during the solution process (see Section 7.2.3). Therefore, we need to allocate storage locations to zero elements within the skyline but do not need to store zero elements that are outside the skyline. The storage scheme that will be used in the finite element solution process is indicated in Fig. 1.1 and is explained further in Section 6.2.3.

1.4 MATRIX EQUALITY, ADDITION, AND MULTIPLICATION BY A SCALAR

We have defined matrices to be ordered arrays of numbers and identified them by single symbols. In order to be able to deal with them in a similar way as we deal with ordinary numbers, it is necessary to define rules corresponding to those which govern equality, addition, subtraction, multiplication, and division of ordinary numbers. We shall simply state the matrix rules and not provide motivation for them. The rationale for these rules will appear later, as it will turn out that these are precisely the rules that are needed to use matrices in the solution of practical problems. For matrix equality, matrix addition, and matrix multiplication by a scalar we provide the following definitions.

Definition: The matrices **A** and **B** are *equal* if and only if:

(1) **A** and **B** have the same number of rows and columns.
(2) All corresponding elements are equal; i.e., $a_{ij} = b_{ij}$ for all i and j.

Definition: Two matrices **A** and **B** can be *added* if and only if they have the same number of rows and columns. The addition of the matrices is performed by adding all corresponding elements; i.e., if a_{ij} and b_{ij} denote a general element of **A** and **B**, then $c_{ij} = a_{ij} + b_{ij}$ denotes a general element of **C**, where $\mathbf{C} = \mathbf{A} + \mathbf{B}$. It follows that **C** has the same number of rows and columns as **A** and **B**.

Example 1.1

Calculate $\mathbf{C} = \mathbf{A} + \mathbf{B}$, where

$$\mathbf{A} = \begin{bmatrix} 2 & 1 & 1 \\ 0.5 & 3 & 0 \end{bmatrix}; \qquad \mathbf{B} = \begin{bmatrix} 3 & 1 & 2 \\ 2 & 4 & 1 \end{bmatrix}$$

Here we have

$$\mathbf{C} = \mathbf{A} + \mathbf{B} = \begin{bmatrix} 5 & 2 & 3 \\ 2.5 & 7 & 1 \end{bmatrix}$$

It need be noted that the order in which the matrices are added is not important. The subtraction of matrices is defined in an analogous way.

Example 1.2

Calculate $\mathbf{C} = \mathbf{A} - \mathbf{B}$, where $\mathbf{A}$ and $\mathbf{B}$ are given in Example 1.1.

Here we have

$$\mathbf{C} = \mathbf{A} - \mathbf{B} = \begin{bmatrix} -1 & 0 & -1 \\ -1.5 & -1 & -1 \end{bmatrix}$$

From the definition of the subtraction of matrices, it follows that the subtraction of a matrix from itself results into a matrix with zero elements only. Such a matrix is defined to be a *null matrix* $\mathbf{0}$. We turn now to the multiplication of a matrix by a scalar.

Definition: A matrix is *multiplied* by a scalar by multiplying each matrix element by the scalar; i.e., $\mathbf{C} = k\mathbf{A}$ means that $c_{ij} = ka_{ij}$.

The following example demonstrates this definition.

Example 1.3

Calculate $\mathbf{C} = k\mathbf{A}$, where

$$\mathbf{A} = \begin{bmatrix} 2 & 1 & 1 \\ 0.5 & 3 & 0 \end{bmatrix}; \qquad k = 2$$

We have

$$\mathbf{C} = k\mathbf{A} = \begin{bmatrix} 4 & 2 & 2 \\ 1 & 6 & 0 \end{bmatrix}$$

It should be noted that so far all definitions are completely analogous to those which are used in the calculation with ordinary numbers. Furthermore, to add (or subtract) two general matrices of order $n \times m$ requires nm addition (subtraction) operations, and to multiply a general matrix of order $n \times m$ by a scalar requires nm multiplications. Therefore, when the matrices are of special form, such as symmetric and banded, we should take advantage of the situation by only evaluating the elements below the skyline of the matrix $\mathbf{C}$, because all other elements are zero.

1.5 MULTIPLICATION OF MATRICES

We consider two matrices **A** and **B** and want to find the matrix product $\mathbf{C} = \mathbf{AB}$.

Definition: Two matrices **A** and **B** can be *multiplied* to obtain $\mathbf{C} = \mathbf{AB}$ if and only if the number of columns in **A** is equal to the number of rows in **B**. Assume that **A** is of order $p \times m$ and **B** is of order $m \times q$. Then for each element in **C** we have

$$c_{ij} = \sum_{r=1}^{m} a_{ir}b_{rj} \tag{1.11}$$

where **C** is of order $p \times q$; i.e., the indices i and j in (1.11) vary from 1 to p and 1 to q, respectively.

Therefore, to calculate the (i, j)th element in **C**, we multiply the elements in the ith row of **A** by the elements in the jth column of **B** and add all individual products. By taking the product of each row in **A** with each column in **B**, it follows that **C** must be of order $p \times q$.

Example 1.4

Calculate the matrix product $\mathbf{C} = \mathbf{AB}$, where

$$\mathbf{A} = \begin{bmatrix} 5 & 3 & 1 \\ 4 & 6 & 2 \\ 10 & 3 & 4 \end{bmatrix}; \qquad \mathbf{B} = \begin{bmatrix} 1 & 5 \\ 2 & 4 \\ 3 & 2 \end{bmatrix}$$

We have

$$\begin{aligned} c_{11} &= (5)(1) + (3)(2) + (1)(3) = 14 \\ c_{21} &= (4)(1) + (6)(2) + (2)(3) = 22 \\ c_{31} &= (10)(1) + (3)(2) + (4)(3) = 28 \qquad \text{etc.} \end{aligned}$$

Hence we obtain

$$\mathbf{C} = \begin{bmatrix} 14 & 39 \\ 22 & 48 \\ 28 & 70 \end{bmatrix}$$

As can readily be verified, the number of operations required in this matrix multiplication is $p \times q \times m$. When we deal with matrices in practice, however, we can often reduce the number of operations by taking advantage of zero elements within the matrices.

Example 1.5

Calculate the matrix product $\mathbf{c} = \mathbf{Ab}$, where

$$\mathbf{A} = \begin{bmatrix} 2 & -1 & 0 & 0 \\ & 2 & -1 & 0 \\ & & 2 & -1 \\ \text{symmetric} & & & 1 \end{bmatrix}; \qquad \mathbf{b} = \begin{bmatrix} 4 \\ 1 \\ 2 \\ 3 \end{bmatrix}$$

Here we can take advantage of the fact that the bandwidth of $\mathbf{A}$ is 3; i.e., $m_{\mathbf{A}} = 1$. Thus, taking into account only the elements within the band of $\mathbf{A}$, we have

$$\begin{aligned} c_1 &= (2)(4) + (-1)(1) = 7 \\ c_2 &= (-1)(4) + (2)(1) + (-1)(2) = -4 \\ c_3 &= (-1)(1) + (2)(2) + (-1)(3) = 0 \\ c_4 &= (-1)(2) + (1)(3) = 1 \end{aligned}$$

Hence

$$\mathbf{c} = \begin{bmatrix} 7 \\ -4 \\ 0 \\ 1 \end{bmatrix}$$

As is well known, the multiplication of ordinary numbers is commutative; i.e., $ab = ba$. We need to investigate if the same holds for matrix multiplication. If we consider the matrices

$$\mathbf{A} = \begin{bmatrix} 1 \\ 2 \end{bmatrix}; \qquad \mathbf{B} = [3 \quad 4] \tag{1.12}$$

we have

$$\mathbf{AB} = \begin{bmatrix} 3 & 4 \\ 6 & 8 \end{bmatrix}; \qquad \mathbf{BA} = [11] \tag{1.13}$$

Therefore, the products $\mathbf{AB}$ and $\mathbf{BA}$ are not the same, and it follows that matrix multiplication is not commutative. Indeed, depending on the orders of $\mathbf{A}$ and $\mathbf{B}$, the orders of the two product matrices $\mathbf{AB}$ and $\mathbf{BA}$ can be different, and the product $\mathbf{AB}$ may be defined, whereas the product $\mathbf{BA}$ may not be calculable.

To distinguish the order of multiplication of matrices, we say that in the product $\mathbf{AB}$, the matrix $\mathbf{A}$ premultiplies $\mathbf{B}$, or the matrix $\mathbf{B}$ postmultiplies $\mathbf{A}$. Although $\mathbf{AB} \neq \mathbf{BA}$ in general, it may happen that $\mathbf{AB} = \mathbf{BA}$ for special $\mathbf{A}$ and $\mathbf{B}$, in which case we say that $\mathbf{A}$ and $\mathbf{B}$ commute.

Although the cummutative law does not hold in matrix multiplication, the distributive law and associative law both are valid. The distributive law states that

$$\mathbf{E} = (\mathbf{A} + \mathbf{B})\mathbf{C} = \mathbf{AC} + \mathbf{BC} \tag{1.14}$$

In other words, we may first add **A** and **B** and then do the multiplication by **C**, or we may first multiply **A** and **B** by **C** and then do the addition. Note that considering the number of operations, the evaluation of **E** by adding **A** and **B** first is much more economical, which is important to remember in the design of an analysis program.

The distributive law is proved using (1.11); namely, using

$$e_{ij} = \sum_{r=1}^{m} (a_{ir} + b_{ir})c_{rj} \tag{1.15}$$

we obtain

$$e_{ij} = \sum_{r=1}^{m} a_{ir}c_{rj} + \sum_{r=1}^{m} b_{ir}c_{rj} \tag{1.16}$$

The associative law states that,

$$\mathbf{G} = (\mathbf{AB})\mathbf{C} = \mathbf{A}(\mathbf{BC}) = \mathbf{ABC} \tag{1.17}$$

in other words, that the order of multiplication is immaterial. The proof is carried out by using the definition of matrix multiplication in (1.11) and calculating in either way a general element of **G**.

Since the associative law holds, in practice, a string of matrix multiplications can be carried out in an arbitrary sequence, and by a clever choice of the sequence, many operations can frequently be saved. The only point that must be remembered when manipulating the matrices is that brackets can be removed or inserted, that powers can be combined, but that the order of multiplication must be preserved.

Consider the following examples to demonstrate the use of the associative and distributive laws in order to simplify a string of matrix multiplications.

Example 1.6

Calculate $\mathbf{A}^4$, where

$$\mathbf{A} = \begin{bmatrix} 2 & 1 \\ 1 & 3 \end{bmatrix}$$

One way of evaluating $\mathbf{A}^4$ is to simply calculate

$$\mathbf{A}^2 = \begin{bmatrix} 2 & 1 \\ 1 & 3 \end{bmatrix}\begin{bmatrix} 2 & 1 \\ 1 & 3 \end{bmatrix} = \begin{bmatrix} 5 & 5 \\ 5 & 10 \end{bmatrix}$$

Hence

$$\mathbf{A}^3 = \mathbf{A}^2\mathbf{A} = \begin{bmatrix} 5 & 5 \\ 5 & 10 \end{bmatrix}\begin{bmatrix} 2 & 1 \\ 1 & 3 \end{bmatrix} = \begin{bmatrix} 15 & 20 \\ 20 & 35 \end{bmatrix}$$

and

$$\mathbf{A}^4 = \mathbf{A}^3\mathbf{A} = \begin{bmatrix} 15 & 20 \\ 20 & 35 \end{bmatrix}\begin{bmatrix} 2 & 1 \\ 1 & 3 \end{bmatrix} = \begin{bmatrix} 50 & 75 \\ 75 & 125 \end{bmatrix}$$

Alternatively, we may use

$$\mathbf{A}^4 = \mathbf{A}^2\mathbf{A}^2 = \begin{bmatrix} 5 & 5 \\ 5 & 10 \end{bmatrix}\begin{bmatrix} 5 & 5 \\ 5 & 10 \end{bmatrix} = \begin{bmatrix} 50 & 75 \\ 75 & 125 \end{bmatrix}$$

and save one matrix multiplication.

Example 1.7

Evaluate the product $\mathbf{v}^T\mathbf{A}\mathbf{v}$, where

$$\mathbf{A} = \begin{bmatrix} 3 & 2 & 1 \\ 2 & 4 & 2 \\ 1 & 2 & 6 \end{bmatrix}; \qquad \mathbf{v} = \begin{bmatrix} 1 \\ 2 \\ -1 \end{bmatrix}$$

The formal procedure would be to calculate $\mathbf{x} = \mathbf{A}\mathbf{v}$; i.e.,

$$\mathbf{x} = \mathbf{A}\mathbf{v} = \begin{bmatrix} 3 & 2 & 1 \\ 2 & 4 & 2 \\ 1 & 2 & 6 \end{bmatrix} \begin{bmatrix} 1 \\ 2 \\ -1 \end{bmatrix} = \begin{bmatrix} 6 \\ 8 \\ -1 \end{bmatrix}$$

and then calculate $\mathbf{v}^T\mathbf{x}$ to obtain

$$\mathbf{v}^T\mathbf{A}\mathbf{v} = [1 \quad 2 \quad -1] \begin{bmatrix} 6 \\ 8 \\ -1 \end{bmatrix} = 23$$

However, it is more effective to calculate the required product in the following way. First, we write

$$\mathbf{A} = \mathbf{U} + \mathbf{D} + \mathbf{U}^T$$

where $\mathbf{U}$ is a lower triangular matrix and $\mathbf{D}$ is a diagonal matrix,

$$\mathbf{U} = \begin{bmatrix} 0 & 0 & 0 \\ 2 & 0 & 0 \\ 1 & 2 & 0 \end{bmatrix}; \qquad \mathbf{D} = \begin{bmatrix} 3 & 0 & 0 \\ 0 & 4 & 0 \\ 0 & 0 & 6 \end{bmatrix}$$

Hence we have

$$\mathbf{v}^T\mathbf{A}\mathbf{v} = \mathbf{v}^T(\mathbf{U} + \mathbf{D} + \mathbf{U}^T)\mathbf{v}$$

$$\mathbf{v}^T\mathbf{A}\mathbf{v} = \mathbf{v}^T\mathbf{U}\mathbf{v} + \mathbf{v}^T\mathbf{D}\mathbf{v} + \mathbf{v}^T\mathbf{U}^T\mathbf{v}$$

However, $\mathbf{v}^T\mathbf{U}\mathbf{v}$ is a single number and hence $\mathbf{v}^T\mathbf{U}^T\mathbf{v} = \mathbf{v}^T\mathbf{U}\mathbf{v}$, and it follows that

$$\mathbf{v}^T\mathbf{A}\mathbf{v} = 2\mathbf{v}^T\mathbf{U}\mathbf{v} + \mathbf{v}^T\mathbf{D}\mathbf{v} \tag{a}$$

The higher efficiency in the matrix multiplication is obtained by taking advantage of the fact that $\mathbf{U}$ is a lower triangular and $\mathbf{D}$ is a diagonal matrix. Let $\mathbf{x} = \mathbf{U}\mathbf{v}$; then we have

$$x_1 = 0$$
$$x_2 = (2)(1) = 2$$
$$x_3 = (1)(1) + (2)(2) = 5$$

Hence

$$\mathbf{x} = \begin{bmatrix} 0 \\ 2 \\ 5 \end{bmatrix}$$

Next, we obtain

$$\mathbf{v}^T\mathbf{U}\mathbf{v} = \mathbf{v}^T\mathbf{x} = (2)(2) + (-1)(5) = -1$$

Also

$$\mathbf{v}^T\mathbf{D}\mathbf{v} = (1)(1)(3) + (2)(2)(4) + (-1)(-1)(6)$$
$$= 25$$

Hence using (a) we have $\mathbf{v}^T\mathbf{A}\mathbf{v} = 23$, as before.

Apart from the commutative law, which in general does not hold in matrix multiplications, also the cancellation of matrices in matrix equations cannot be performed, in general, as the cancellation of ordinary numbers. In particular, if $\mathbf{AB} = \mathbf{CB}$, it does not necessarily follow that $\mathbf{A} = \mathbf{C}$. This is easily demonstrated considering a specific case:

$$\begin{bmatrix} 2 & 1 \\ 4 & 0 \end{bmatrix}\begin{bmatrix} 1 \\ 2 \end{bmatrix} = \begin{bmatrix} 4 & 0 \\ 0 & 2 \end{bmatrix}\begin{bmatrix} 1 \\ 2 \end{bmatrix} \tag{1.18}$$

but

$$\begin{bmatrix} 2 & 1 \\ 4 & 0 \end{bmatrix} \neq \begin{bmatrix} 4 & 0 \\ 0 & 2 \end{bmatrix} \tag{1.19}$$

However, it must be noted that $\mathbf{A} = \mathbf{C}$ if the equation $\mathbf{AB} = \mathbf{CB}$ holds for all possible $\mathbf{B}$. Namely, in that case, we simply select $\mathbf{B}$ to be the identity matrix $\mathbf{I}$, and hence $\mathbf{A} = \mathbf{C}$.

It need also be observed that included in the above observation is the fact that if $\mathbf{AB} = \mathbf{0}$, it does not follow that either $\mathbf{A}$ or $\mathbf{B}$ is a null matrix. A specific case demonstrates this observation:

$$\mathbf{A} = \begin{bmatrix} 1 & 0 \\ 2 & 0 \end{bmatrix}; \qquad \mathbf{B} = \begin{bmatrix} 0 & 0 \\ 3 & 4 \end{bmatrix}; \qquad \mathbf{AB} = \begin{bmatrix} 0 & 0 \\ 0 & 0 \end{bmatrix} \tag{1.20}$$

Some special rules concerning the use of transposed matrices in matrix multiplications need be pointed out. It is noted that the transpose of the product of two matrices $\mathbf{A}$ and $\mathbf{B}$ is equal to the product of the transposed matrices in reverse order; i.e.,

$$(\mathbf{AB})^T = \mathbf{B}^T\mathbf{A}^T \tag{1.21}$$

The proof that (1.21) does hold is obtained using the definition for the evaluation of a matrix product given in (1.11).

Considering the matrix products in (1.21), it need be noted that although $\mathbf{A}$ may be symmetric, $\mathbf{AB}$ is, in general, not symmetric. However, if $\mathbf{A}$ is symmetric, the matrix $\mathbf{B}^T\mathbf{AB}$ is always symmetric. The proof follows using (1.21):

$$(\mathbf{B}^T\mathbf{AB})^T = (\mathbf{AB})^T(\mathbf{B}^T)^T \tag{1.22}$$

$$= \mathbf{B}^T\mathbf{A}^T\mathbf{B} \tag{1.23}$$

But, because $\mathbf{A}^T = \mathbf{A}$, we have

$$(\mathbf{B}^T\mathbf{AB})^T = \mathbf{B}^T\mathbf{AB} \tag{1.24}$$

and hence $\mathbf{B}^T\mathbf{AB}$ is symmetric.

1.6 THE INVERSE MATRIX

We have seen that matrix addition and subtraction are carried out using essentially the same laws as those used in the manipulation of ordinary numbers. However, matrix multiplication is quite different and we have to get used to special rules. With regard to matrix division, it strictly does not exist. Instead, an inverse matrix is defined. We shall define and use the inverse of square matrices only.

Definition: The *inverse* of a matrix $\mathbf{A}$ is denoted by $\mathbf{A}^{-1}$. Assume that the inverse exists; then the elements of $\mathbf{A}^{-1}$ are such that $\mathbf{A}^{-1}\mathbf{A} = \mathbf{I}$ and $\mathbf{A}\mathbf{A}^{-1} = \mathbf{I}$. A matrix that possesses an inverse is said to be *nonsingular*. A matrix without an inverse is a *singular* matrix.

As mentioned above, the inverse of a matrix does not need to exist. A trivial example is the null matrix. Assume that the inverse of $\mathbf{A}$ exists. Then we still want to show that either of the conditions $\mathbf{A}^{-1}\mathbf{A} = \mathbf{I}$ or $\mathbf{A}\mathbf{A}^{-1} = \mathbf{I}$ implies the other. Assume that we have evaluated the elements of the matrices $\mathbf{A}_l^{-1}$ and $\mathbf{A}_r^{-1}$ such that $\mathbf{A}_l^{-1}\mathbf{A} = \mathbf{I}$ and $\mathbf{A}\mathbf{A}_r^{-1} = \mathbf{I}$. Then we have

$$\mathbf{A}_l^{-1} = \mathbf{A}_l^{-1}(\mathbf{A}\mathbf{A}_r^{-1}) = (\mathbf{A}_l^{-1}\mathbf{A})\mathbf{A}_r^{-1} = \mathbf{A}_r^{-1} \tag{1.25}$$

and hence $\mathbf{A}_l^{-1} = \mathbf{A}_r^{-1}$.

Example 1.8

Evaluate the inverse of the matrix $\mathbf{A}$, where

$$\mathbf{A} = \begin{bmatrix} 2 & -1 \\ -1 & 3 \end{bmatrix}$$

For the inverse of $\mathbf{A}$ we need $\mathbf{A}\mathbf{A}^{-1} = \mathbf{I}$. By trial and error (or otherwise) we find that

$$\mathbf{A}^{-1} = \begin{bmatrix} \frac{3}{5} & \frac{1}{5} \\ \frac{1}{5} & \frac{2}{5} \end{bmatrix}$$

We check that $\mathbf{A}\mathbf{A}^{-1} = \mathbf{I}$ and $\mathbf{A}^{-1}\mathbf{A} = \mathbf{I}$:

$$\mathbf{A}\mathbf{A}^{-1} = \begin{bmatrix} 2 & -1 \\ -1 & 3 \end{bmatrix}\begin{bmatrix} \frac{3}{5} & \frac{1}{5} \\ \frac{1}{5} & \frac{2}{5} \end{bmatrix} = \begin{bmatrix} 1 & 0 \\ 0 & 1 \end{bmatrix}$$

$$\mathbf{A}^{-1}\mathbf{A} = \begin{bmatrix} \frac{3}{5} & \frac{1}{5} \\ \frac{1}{5} & \frac{2}{5} \end{bmatrix}\begin{bmatrix} 2 & -1 \\ -1 & 3 \end{bmatrix} = \begin{bmatrix} 1 & 0 \\ 0 & 1 \end{bmatrix}$$

To calculate the inverse of a product $\mathbf{AB}$, we proceed as follows. Let $\mathbf{G} = (\mathbf{AB})^{-1}$, where $\mathbf{A}$ and $\mathbf{B}$ are both square matrices. Then

$$\mathbf{GAB} = \mathbf{I} \tag{1.26}$$

and postmultiplying by $\mathbf{B}^{-1}$ and $\mathbf{A}^{-1}$, we obtain

$$\mathbf{GA} = \mathbf{B}^{-1} \tag{1.27}$$

$$\mathbf{G} = \mathbf{B}^{-1}\mathbf{A}^{-1} \tag{1.28}$$

Therefore,

$$(\mathbf{AB})^{-1} = \mathbf{B}^{-1}\mathbf{A}^{-1} \tag{1.29}$$

The reader will notice that the same law of matrix reversal was shown to apply when the transpose of a matrix product is calculated.

Example 1.9

For the matrices **A** and **B** given below, check that $(\mathbf{AB})^{-1} = \mathbf{B}^{-1}\mathbf{A}^{-1}$.

$$\mathbf{A} = \begin{bmatrix} 2 & -1 \\ -1 & 3 \end{bmatrix}; \qquad \mathbf{B} = \begin{bmatrix} 3 & 0 \\ 0 & 4 \end{bmatrix}$$

The inverse of **A** was used in Example 1.8. The inverse of **B** is easy to obtain:

$$\mathbf{B}^{-1} = \begin{bmatrix} \frac{1}{3} & 0 \\ 0 & \frac{1}{4} \end{bmatrix}$$

To check that $(\mathbf{AB})^{-1} = \mathbf{B}^{-1}\mathbf{A}^{-1}$, we need to evaluate $\mathbf{C} = \mathbf{AB}$:

$$\mathbf{C} = \begin{bmatrix} 2 & -1 \\ -1 & 3 \end{bmatrix}\begin{bmatrix} 3 & 0 \\ 0 & 4 \end{bmatrix} = \begin{bmatrix} 6 & -4 \\ -3 & 12 \end{bmatrix}$$

Assume that $\mathbf{C}^{-1} = \mathbf{B}^{-1}\mathbf{A}^{-1}$. Then we would have

$$\mathbf{C}^{-1} = \begin{bmatrix} \frac{1}{3} & 0 \\ 0 & \frac{1}{4} \end{bmatrix}\begin{bmatrix} \frac{3}{5} & \frac{1}{5} \\ \frac{1}{5} & \frac{2}{5} \end{bmatrix} = \begin{bmatrix} \frac{1}{5} & \frac{1}{15} \\ \frac{1}{20} & \frac{1}{10} \end{bmatrix} \tag{a}$$

To check that the matrix given in (a) is indeed the inverse of **C**, we evaluate $\mathbf{C}^{-1}\mathbf{C}$ and find that

$$\mathbf{C}^{-1}\mathbf{C} = \begin{bmatrix} \frac{1}{5} & \frac{1}{15} \\ \frac{1}{20} & \frac{1}{10} \end{bmatrix}\begin{bmatrix} 6 & -4 \\ -3 & 12 \end{bmatrix} = \mathbf{I}$$

But since $\mathbf{C}^{-1}$ is unique and only the correct $\mathbf{C}^{-1}$ satisfies the relation $\mathbf{C}^{-1}\mathbf{C} = \mathbf{I}$, we indeed have found in (a) the inverse of **C**, and the relation $(\mathbf{AB})^{-1} = \mathbf{B}^{-1}\mathbf{A}^{-1}$ is satisfied.

In Examples 1.8 and 1.9, the inverse of **A** and **B** could be found by trial and error. However, to obtain the inverse of a general matrix, we need to have a general algorithm. One way of calculating the inverse of a matrix **A** of order n is to solve the n systems of equations

$$\mathbf{AX} = \mathbf{I} \tag{1.30}$$

where **I** is the identity matrix of order n and we have $\mathbf{X} = \mathbf{A}^{-1}$. For the solution of each system of equations in (1.30), we can use the algorithm presented in Section 7.2.

The above considerations show that a system of equations could be solved by calculating the inverse of the coefficient matrix; i.e., if we have

$$\mathbf{Ay} = \mathbf{c} \tag{1.31}$$

where **A** is of order $n \times n$ and **y** and **c** are of order $n \times 1$, then

$$\mathbf{y} = \mathbf{A}^{-1}\mathbf{c} \tag{1.32}$$

However, the inversion of **A** is very costly and it is much more effective to use Gauss elimination to solve the equations in (1.31) (see Chapter 7).

Indeed, although we may write symbolically that $\mathbf{y} = \mathbf{A}^{-1}\mathbf{c}$, to evaluate $\mathbf{y}$ we actually use Gauss elimination.

1.7 PARTITIONING OF MATRICES

To facilitate matrix manipulations and to take advantage of the special form of matrices, it may be useful to partition a matrix into submatrices. A submatrix is a matrix that is obtained from the original matrix by including only the elements of certain rows and columns. The idea is demonstrated using a specific case in which the dashed lines are the lines of partitioning:

$$\mathbf{A} = \left[\begin{array}{c:ccc:cc} a_{11} & a_{12} & a_{13} & a_{14} & a_{15} & a_{16} \\ a_{21} & a_{22} & a_{23} & a_{24} & a_{25} & a_{26} \\ \hdashline a_{31} & a_{32} & a_{33} & a_{34} & a_{35} & a_{36} \end{array}\right] \tag{1.33}$$

It should be noted that each of the partitioning lines must run all across the original matrix. Using the partitioning, matrix $\mathbf{A}$ is written as

$$\mathbf{A} = \begin{bmatrix} \mathbf{A}_{11} & \mathbf{A}_{12} & \mathbf{A}_{13} \\ \mathbf{A}_{21} & \mathbf{A}_{22} & \mathbf{A}_{23} \end{bmatrix} \tag{1.34}$$

where

$$\mathbf{A}_{11} = \begin{bmatrix} a_{11} \\ a_{21} \end{bmatrix}; \qquad \mathbf{A}_{12} = \begin{bmatrix} a_{12} & a_{13} & a_{14} \\ a_{22} & a_{23} & a_{24} \end{bmatrix}; \qquad \text{etc.} \tag{1.35}$$

The right-hand side of (1.34) could again be partitioned, such as

$$\mathbf{A} = \left[\begin{array}{c:cc} \mathbf{A}_{11} & \mathbf{A}_{12} & \mathbf{A}_{13} \\ \mathbf{A}_{21} & \mathbf{A}_{22} & \mathbf{A}_{23} \end{array}\right] \tag{1.36}$$

and we may write $\mathbf{A}$ as

$$\mathbf{A} = [\bar{\mathbf{A}}_1 \quad \bar{\mathbf{A}}_2]; \qquad \bar{\mathbf{A}}_1 = \begin{bmatrix} \mathbf{A}_{11} \\ \mathbf{A}_{21} \end{bmatrix}; \qquad \bar{\mathbf{A}}_2 = \begin{bmatrix} \mathbf{A}_{12} & \mathbf{A}_{13} \\ \mathbf{A}_{22} & \mathbf{A}_{23} \end{bmatrix} \tag{1.37}$$

The partitioning of matrices can be of advantage in saving computer storage; namely, if submatrices repeat, it would only be necessary to store the submatrix once. The same applies in arithmetic. Using submatrices, we may identify a typical operation that is repeated many times. We would then carry out this operation only once and use the result whenever it is needed.

The rules to be used in calculations with partitioned matrices follow from the definition of matrix addition, subtraction, and multiplication. Using partitioned matrices we can add, subtract, or multiply as if the submatrices were ordinary matrix elements, provided the original matrices have been partitioned in such a way that it is permissible to perform the individual submatrix additions, subtractions, or multiplications.

These rules are easily justified and remembered if we keep in mind that the partitioning of the original matrices is only a device to facilitate the matrix manipulations and does not change any results.

Example 1.10

Evaluate the matrix product $\mathbf{C} = \mathbf{AB}$ of Example 1.4 by using the following partitioning:

$$\mathbf{A} = \left[\begin{array}{cc|c} 5 & 3 & 1 \\ 4 & 6 & 2 \\ \hline 10 & 3 & 4 \end{array}\right]; \qquad \mathbf{B} = \left[\begin{array}{cc} 1 & 5 \\ 2 & 4 \\ \hline 3 & 2 \end{array}\right]$$

Here we have

$$\mathbf{A} = \begin{bmatrix} \mathbf{A}_{11} & \mathbf{A}_{12} \\ \mathbf{A}_{21} & \mathbf{A}_{22} \end{bmatrix}; \qquad \mathbf{B} = \begin{bmatrix} \mathbf{B}_1 \\ \mathbf{B}_2 \end{bmatrix}$$

Therefore,

$$\mathbf{AB} = \begin{bmatrix} \mathbf{A}_{11}\mathbf{B}_1 + \mathbf{A}_{12}\mathbf{B}_2 \\ \mathbf{A}_{21}\mathbf{B}_1 + \mathbf{A}_{22}\mathbf{B}_2 \end{bmatrix} \tag{a}$$

But

$$\mathbf{A}_{11}\mathbf{B}_1 = \begin{bmatrix} 5 & 3 \\ 4 & 6 \end{bmatrix}\begin{bmatrix} 1 & 5 \\ 2 & 4 \end{bmatrix} = \begin{bmatrix} 11 & 37 \\ 16 & 44 \end{bmatrix}$$

$$\mathbf{A}_{12}\mathbf{B}_2 = \begin{bmatrix} 1 \\ 2 \end{bmatrix}[3 \quad 2] = \begin{bmatrix} 3 & 2 \\ 6 & 4 \end{bmatrix}$$

$$\mathbf{A}_{21}\mathbf{B}_1 = [10 \quad 3]\begin{bmatrix} 1 & 5 \\ 2 & 4 \end{bmatrix} = [16 \quad 62]$$

$$\mathbf{A}_{22}\mathbf{B}_2 = [4][3 \quad 2] = [12 \quad 8]$$

Then substituting into (a) we have

$$\mathbf{AB} = \begin{bmatrix} 14 & 39 \\ 22 & 48 \\ 28 & 70 \end{bmatrix}$$

Example 1.11

Taking advantage of partitioning, evaluate $\mathbf{c} = \mathbf{Ab}$, where

$$\mathbf{A} = \left[\begin{array}{cc|cc} 4 & 3 & 1 & 2 \\ 3 & 6 & 2 & 1 \\ \hline 1 & 2 & 8 & 6 \\ 2 & 1 & 6 & 12 \end{array}\right]; \qquad \mathbf{b} = \left[\begin{array}{c} 2 \\ 2 \\ \hline 1 \\ 1 \end{array}\right]$$

The only products that we need to evaluate are

$$\mathbf{w}_1 = \begin{bmatrix} 4 & 3 \\ 3 & 6 \end{bmatrix}\begin{bmatrix} 1 \\ 1 \end{bmatrix} = \begin{bmatrix} 7 \\ 9 \end{bmatrix}$$

and

$$\mathbf{w}_2 = \begin{bmatrix} 1 & 2 \\ 2 & 1 \end{bmatrix}\begin{bmatrix} 1 \\ 1 \end{bmatrix} = \begin{bmatrix} 3 \\ 3 \end{bmatrix}$$

We can now construct **c**:

$$\mathbf{c} = \begin{bmatrix} 2\mathbf{w}_1 + \mathbf{w}_2 \\ 2\mathbf{w}_1 + 2\mathbf{w}_2 \end{bmatrix}$$

or, substituting,

$$\mathbf{c} = \begin{bmatrix} 17 \\ 21 \\ 20 \\ 24 \end{bmatrix}$$

1.8 THE TRACE AND DETERMINANT OF A MATRIX

The trace and determinant of a matrix are only defined if the matrix is square. Both quantities are single numbers, which are evaluated from the elements of the matrix and are therefore functions of the matrix elements.

Definition: The *trace* of the matrix **A** is identified as tr (**A**) and is equal to $\sum_{i=1}^{n} a_{ii}$, where n is of the order of **A**.

Example 1.12

Calculate the trace of the matrix **A** given in Example 1.11.

Here we have

$$\text{tr}(\mathbf{A}) = 4 + 6 + 8 + 12 = 30$$

The determinant of a matrix **A** can be defined in terms of the determinants of submatrices of **A**, and noting that the determinant of a matrix of order 1 is simply the element of the matrix; i.e., if $\mathbf{A} = [a_{11}]$, then $\det \mathbf{A} = a_{11}$.

Definition: The *determinant* of an $n \times n$ matrix **A** is denoted as det **A** and is defined by the recurrence relation

$$\det \mathbf{A} = \sum_{j=1}^{n} (-1)^{1+j} a_{1j} \det \mathbf{A}_{1j} \tag{1.38}$$

where $\mathbf{A}_{1j}$ is the $(n-1) \times (n-1)$ matrix obtained by eliminating the 1st row and jth column from the matrix **A**.

Example 1.13

Evaluate the determinant of **A**, where

$$\mathbf{A} = \begin{bmatrix} a_{11} & a_{12} \\ a_{21} & a_{22} \end{bmatrix}$$

Using the relation in (1.38), we obtain

$$\det \mathbf{A} = (-1)^2 a_{11} \det \mathbf{A}_{11} + (-1)^3 a_{12} \det \mathbf{A}_{12}$$

But

$$\det \mathbf{A}_{11} = a_{22}; \qquad \det \mathbf{A}_{12} = a_{21}$$

Hence

$$\det \mathbf{A} = a_{11}a_{22} - a_{12}a_{21} \tag{a}$$

This relation is the general formula for the determinant of a 2×2 matrix.

It can be shown that to evaluate the determinant of a matrix we may use the recurrence relation given in (1.38) along any row or column, as indicated in Example 1.14.

Example 1.14

Evaluate the determinant of the matrix **A**, where

$$\mathbf{A} = \begin{bmatrix} 2 & 1 & 0 \\ 1 & 3 & 1 \\ 0 & 1 & 2 \end{bmatrix}$$

Using the recurrence relation in (1.38), we obtain

$$\det \mathbf{A} = (-1)^2(2) \det \begin{bmatrix} 3 & 1 \\ 1 & 2 \end{bmatrix} + (-1)^3(1) \det \begin{bmatrix} 1 & 1 \\ 0 & 2 \end{bmatrix} + (-1)^4(0) \det \begin{bmatrix} 1 & 3 \\ 0 & 1 \end{bmatrix}$$

We now employ the formula for the determinant of a 2×2 matrix given in Example 1.13, and have

$$\det \mathbf{A} = (2)\{(3)(2) - (1)(1)\} - \{(1)(2) - (0)(1)\} + 0$$

Hence

$$\det \mathbf{A} = 8$$

Let us check that the same result is obtained by using (1.38) along the second row instead of the first row. In this case we have

$$\det \mathbf{A} = (-1)^3(1) \det \begin{bmatrix} 1 & 0 \\ 1 & 2 \end{bmatrix} + (-1)^4(3) \det \begin{bmatrix} 2 & 0 \\ 0 & 2 \end{bmatrix} + (-1)^5(1) \det \begin{bmatrix} 2 & 1 \\ 0 & 1 \end{bmatrix}$$

Again using the formula given in Example 1.13, we have

$$\det \mathbf{A} = -\{(1)(2) - (0)(1)\} + (3)\{(2)(2) - (0)(0)\} - \{(2)(1) - (1)(0)\}$$

or, as before,

$$\det \mathbf{A} = 8$$

Finally, using (1.38) along the third column, we have

$$\det \mathbf{A} = (-1)^4(0) \det \begin{bmatrix} 1 & 3 \\ 0 & 1 \end{bmatrix} + (-1)^5(1) \det \begin{bmatrix} 2 & 1 \\ 0 & 1 \end{bmatrix} + (-1)^6(2) \det \begin{bmatrix} 2 & 1 \\ 1 & 3 \end{bmatrix}$$

and, as before, obtain $\det \mathbf{A} = 8$.

Many theorems are associated with the use of determinants. Typically, the solution of a set of simultaneous equations can be obtained by a series of determinant evaluations. However, from a modern viewpoint, most of the results that are obtained using determinants can be obtained much more effectively using matrix theory. For example, the solution of simultaneous equations using determinants is very inefficient. As we shall see later, a primary value of using determinants lies in the convenient shorthand notation that we can use in the discussion of certain questions, such as the existence of an inverse of a matrix. We shall use determinants in particular in the solution of eigenvalue problems.

Considering the evaluation of the determinant of a matrix, it may be effective to first factorize the matrix into a product of matrices and then use the following result:

$$\det(\mathbf{BC} \ldots \mathbf{F}) = (\det \mathbf{B})(\det \mathbf{C}) \ldots (\det \mathbf{F}) \tag{1.39}$$

Relation (1.39) states that the determinant of the product of a number of matrices is equal to the product of the determinants of each matrix. The proof of this result is rather lengthy and clumsy [it is obtained using the determinant definition in (1.38)] and therefore we shall not include it here. We shall use the result in (1.39) abundantly in eigenvalue solution when the determinant of a matrix, say matrix $\mathbf{A}$, is required. The specific decomposition used is $\mathbf{A} = \mathbf{LDL}^T$, where $\mathbf{L}$ is a lower unit triangular matrix and $\mathbf{D}$ is a diagonal matrix. In that case,

$$\det \mathbf{A} = \det \mathbf{L} \det \mathbf{D} \det \mathbf{L}^T \tag{1.40}$$

and because $\det \mathbf{L} = 1$, we have

$$\det \mathbf{A} = \prod_{i=1}^{n} d_{ii} \tag{1.41}$$

Example 1.15

Using the $\mathbf{LDL}^T$ decomposition, evaluate the determinant of $\mathbf{A}$, where $\mathbf{A}$ is given in Example 1.14.

The procedure to obtain the $\mathbf{LDL}^T$ decomposition of $\mathbf{A}$ is presented in Section 7.2. Here we simply give $\mathbf{L}$ and $\mathbf{D}$, and it can be verified that $\mathbf{LDL}^T = \mathbf{A}$:

$$\mathbf{L} = \begin{bmatrix} 1 & 0 & 0 \\ \frac{1}{2} & 1 & 0 \\ 0 & \frac{2}{5} & 1 \end{bmatrix}; \qquad \mathbf{D} = \begin{bmatrix} 2 & 0 & 0 \\ 0 & \frac{5}{2} & 0 \\ 0 & 0 & \frac{8}{5} \end{bmatrix}$$

Using (1.41), we obtain

$$\det \mathbf{A} = (2)(\tfrac{5}{2})(\tfrac{8}{5}) = 8$$

This is also the value obtained in Example 1.14.

The determinant and the trace of a matrix are functions of the matrix elements. However, it is important to observe that the off-diagonal elements do not affect the trace of a matrix, whereas the determinant is a function of all the elements in the matrix. Although we can conclude that a large determinant or a large trace means that some matrix elements are large, we cannot conclude that a small determinant or a small trace means that all matrix elements are small.

Example 1.16

Calculate the trace and determinant of **A**, where

$$\mathbf{A} = \begin{bmatrix} 1 & 10{,}000 \\ 10^{-4} & 2 \end{bmatrix}$$

Here we have

$$\text{tr}\,(\mathbf{A}) = 3$$

and

$$\det \mathbf{A} = (1)(2) - (10^{-4})(10{,}000)$$

i.e.,

$$\det \mathbf{A} = 1$$

Hence both the trace and the determinant of **A** are small in relation to the off-diagonal element a_{12}.

REFERENCES

1. C. E. FRÖBERG, *Introduction to Numerical Analysis*, Addison-Wesley Publishing Company, Inc., Reading, Mass., 1969.
2. B. NOBLE, *Applied Linear Algebra*, Prentice-Hall, Inc., Englewood Cliffs, N.J., 1969.
3. R. ZURMÜHL, *Matrizen*, Springer-Verlag, Berlin, 1964.

CHAPTER 2

MATRICES AND VECTOR SPACES

2.1 INTRODUCTION

In Chapter 1 we introduced matrices simply as ordered arrays of numbers that are subjected to specific rules of addition, multiplication, and so on. The basic algebra rules will be used extensively in the next chapters, and it is very important to be thoroughly familiar with them. However, we did not discuss so far how to obtain the elements of a matrix from an actual physical problem, and under what conditions, if any, the rules of matrix algebra that have been presented would actually be applicable. In other words, we did not attempt to give a deeper understanding of what a matrix does represent in the calculations. However, such an understanding is very important—indeed essential—if the numerical techniques to be presented later are to be well understood.

The objective in this chapter is to present all the important concepts of linear algebra that need be known for a thorough understanding of the numerical techniques used in finite element analysis.[1-7] However, in addition to developing important knowledge of matrix and linear algebra, it is also hoped that, by means of the concepts presented in this chapter, the somewhat dry aspects of the use of matrices as merely a "bunch of numbers" will be removed. Namely, it is only with an increased understanding of the use of matrices that the solution techniques can be fully appreciated.

2.2 VECTOR SPACES, SUBSPACES, AND THE SPAN OF A MATRIX

In Chapter 1 we defined a vector of order n to be an array of n numbers written in matrix form. We now want to associate a geometrical interpretation to the elements of a vector. Consider as an example a vector of order 3, such as

$$\mathbf{x} = \begin{bmatrix} x_1 \\ x_2 \\ x_3 \end{bmatrix} = \begin{bmatrix} 2 \\ 4 \\ 3 \end{bmatrix} \tag{2.1}$$

We know from elementary geometry that $\mathbf{x}$ represents a geometrical vector in a chosen coordinate system in three-dimensional space. Figure 2.1 shows assumed coordinate axes and the vector corresponding to (2.1) in this system. We should note that the geometric representation of $\mathbf{x}$ in (2.1) depends completely on the coordinate system chosen; in other words, if (2.1) would give the components of a vector in a different coordinate system, then the geometrical representation of $\mathbf{x}$ would be different from the one in Fig. 2.1. Therefore, the coordinates (or components of a vector) alone do not define the actual geometric quantity, but they need to be given together with the specific coordinate system in which they are measured.

The concepts of three-dimensional geometry generalize to a vector of any finite order n. If $n > 3$, we can no longer obtain a plot of the vector; however, we shall see that mathematically all concepts that pertain to vectors are independent of n. As before, when we considered the specific case $n = 3$, the vector of order n represents a quantity in a specific coordinate system of an n-dimensional space.

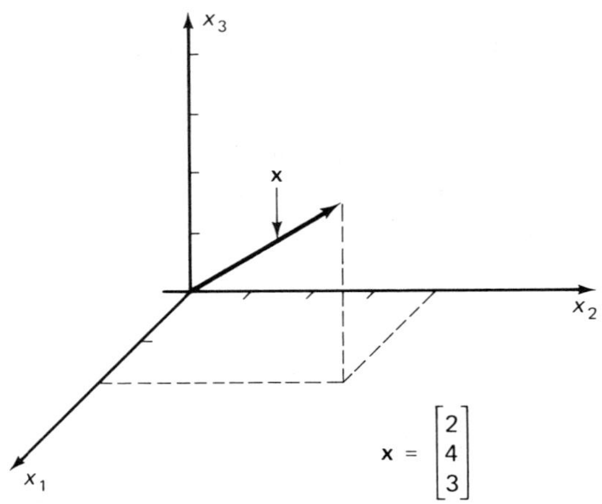

Figure 2.1. Geometric representation of vector $\mathbf{x}$.

Assume that we are dealing with a number of vectors all of order n, which are defined in a fixed coordinate system. Some fundamental concepts that we shall find extremely powerful in the later chapters are summarized in the following definitions and facts.

Definition: A *collection* of vectors $\mathbf{x}_1, \mathbf{x}_2, \ldots, \mathbf{x}_s$ is said to be *linearly dependent* if there exist numbers $\alpha_1, \alpha_2, \ldots, \alpha_s$, which are not all zero, such that

$$\alpha_1\mathbf{x}_1 + \alpha_2\mathbf{x}_2 + \ldots + \alpha_s\mathbf{x}_s = \mathbf{0} \tag{2.2}$$

If the vectors are not linearly dependent, they are called *linearly independent* vectors.

We consider the following examples to clarify the meaning of this definition.

Example 2.1

Let $n = 3$ and determine if the vectors $\mathbf{e}_i$, $i = 1, 2, 3$, are linearly dependent or independent.

According to the definition of linear dependency, we need to check if there are constants α_1, α_2, and α_3, not all zero, to satisfy the equation

$$\alpha_1\begin{bmatrix}1\\0\\0\end{bmatrix} + \alpha_2\begin{bmatrix}0\\1\\0\end{bmatrix} + \alpha_3\begin{bmatrix}0\\0\\1\end{bmatrix} = \begin{bmatrix}0\\0\\0\end{bmatrix} \tag{a}$$

But the equations in (a) read

$$\begin{bmatrix}\alpha_1\\\alpha_2\\\alpha_3\end{bmatrix} = \begin{bmatrix}0\\0\\0\end{bmatrix}$$

which is only satisfied if $\alpha_i = 0$, $i = 1, 2, 3$; therefore, the vectors $\mathbf{e}_i$ are linearly independent.

Example 2.2

With $n = 4$, investigate whether the following vectors are linearly dependent or independent.

$$\mathbf{x}_1 = \begin{bmatrix}1\\1\\0\\0.5\end{bmatrix}; \quad \mathbf{x}_2 = \begin{bmatrix}-1\\0\\1\\0\end{bmatrix}; \quad \mathbf{x}_3 = \begin{bmatrix}0\\-0.5\\-0.5\\-0.25\end{bmatrix}$$

We need to consider the system of equations

$$\alpha_1\begin{bmatrix}1\\1\\0\\0.5\end{bmatrix} + \alpha_2\begin{bmatrix}-1\\0\\1\\0\end{bmatrix} + \alpha_3\begin{bmatrix}0\\-0.5\\-0.5\\-0.25\end{bmatrix} = \begin{bmatrix}0\\0\\0\\0\end{bmatrix}$$

or, considering each row,

$$\begin{aligned}
\alpha_1 - \alpha_2 \qquad\qquad &= 0\\
\alpha_1 \qquad - 0.5\alpha_3 &= 0\\
\alpha_2 - 0.5\alpha_3 &= 0\\
0.5\alpha_1 \qquad - 0.25\alpha_3 &= 0
\end{aligned}$$

where we note that the equations are satisfied for $\alpha_1 = 1$, $\alpha_2 = 1$, and $\alpha_3 = 2$. Therefore, the vectors are linearly dependent.

In the preceding examples, the solution for α_1, α_2, and α_3 could be obtained by inspection. We shall later develop a systematic procedure of checking whether a number of vectors are linearly dependent or independent.

Another way of looking at the problem, which may be more appealing, is to say that the vectors are linearly dependent if any one of them can be expressed in terms of the others. That is, if not all of the α_i in (2.2) are zero, say $\alpha_j \neq 0$, then we can write

$$\mathbf{x}_j = -\sum_{\substack{k=1\\k\neq j}}^{s} \frac{\alpha_k}{\alpha_j}\mathbf{x}_k \tag{2.3}$$

Geometrically, when $n \leq 3$, we could plot the vectors and, if they are linearly dependent, we would be able to plot one vector in terms of multiples of the other vectors. For example, plotting the vectors used in Example 2.1, we immediately observe that none of them can be expressed in terms of multiples of the remaining ones; hence the vectors are linearly independent.

Consider that we are given q vectors of order n, $n \geq q$, which are linearly dependent, but that we only consider any $(q - 1)$ vectors of them. These $(q - 1)$ vectors may still be linearly dependent. However, by continuing to decrease the number of vectors under consideration, we would arrive at p vectors, which are linearly independent, where, in general, $p \leq q$. The other $(q - p)$ vectors can be expressed in terms of the p vectors. We are thus led to the following definition.

Definition: Assume that we have p linearly independent vectors of order n, where $n \geq p$. These p vectors form a *basis* for a *p-dimensional vector space*.

We talk about a vector space of dimension p because any vector in the space can be expressed as a linear combination of the p base vectors. We should note that the base vectors for the specific space considered are not unique; linear combinations of them can give another basis for the same space. Specifically, if $p = n$, then a basis for the space considered is $\mathbf{e}_i$, $i = 1, \ldots, n$ (which we call the *natural basis*), from which it also follows that p cannot be larger than n.

Definition: q vectors, of which p vectors are linearly independent, are said to *span* a p-dimensional vector space.

We therefore realize that all the importance lies in the base vectors, since they are the smallest number of vectors that span the space considered. All q vectors can be expressed in terms of the base vectors, however large q may be (and indeed q could be larger than n)..

Example 2.3

Establish a basis for the space spanned by the three vectors of Example 2.2.

In this case $q = 3$ and $n = 4$. We find by inspection that the two vectors $\mathbf{x}_1$ and $\mathbf{x}_2$ are linearly independent. Hence $\mathbf{x}_1$ and $\mathbf{x}_2$ can be taken as base vectors of the two-dimensional space spanned by $\mathbf{x}_1$, $\mathbf{x}_2$, and $\mathbf{x}_3$. Also, using the result of Example 2.2, we have that $\mathbf{x}_3 = -\frac{1}{2}\mathbf{x}_2 - \frac{1}{2}\mathbf{x}_1$.

Assume that we are given a p-dimensional vector space which we denote as E_p, for which $\mathbf{x}_1, \mathbf{x}_2, \ldots, \mathbf{x}_p$ are chosen base vectors, $p > 1$. Then we may like to consider only all those vectors that can be expressed in terms of $\mathbf{x}_1$ and $\mathbf{x}_2$. But the vectors $\mathbf{x}_1$ and $\mathbf{x}_2$ also form the basis of a vector space that we call E_2. If $p = 2$, we note that E_p and E_2 coincide. We call E_2 a subspace of E_p, the concise meaning of which is defined next.

Definition: A *subspace* of a vector space is a vector space such that any vector in the subspace is also in the original space. If $\mathbf{x}_1, \mathbf{x}_2, \ldots, \mathbf{x}_p$ are the base vectors of the original space, any subset of these vectors forms the *basis* of a subspace; the *dimension* of the subspace is equal to the number of base vectors selected.

Example 2.4

The three vectors $\mathbf{x}_1$, $\mathbf{x}_2$, and $\mathbf{x}_3$ are linearly independent and therefore form the basis of a three-dimensional vector space E_3:

$$\mathbf{x}_1 = \begin{bmatrix} 1 \\ 2 \\ 1 \\ 0 \end{bmatrix}; \qquad \mathbf{x}_2 = \begin{bmatrix} 1 \\ 0 \\ 0 \\ 0 \end{bmatrix}; \qquad \mathbf{x}_3 = \begin{bmatrix} 0 \\ -1 \\ 0 \\ 1 \end{bmatrix} \tag{a}$$

Identify some possible two-dimensional subspaces of E_3.

Using the base vectors in (a), a two-dimensional subspace is formed by any two of the three vectors; e.g., $\mathbf{x}_1$ and $\mathbf{x}_2$ represent a basis for a two-dimensional subspace; $\mathbf{x}_1$ and $\mathbf{x}_3$ are the basis for another two-dimensional subspace; and so on. Indeed, any two linearly independent vectors in E_3 form the basis of a two-dimensional subspace, and it follows that there are an infinite number of two-dimensional subspaces in E_3.

Having considered the concepts of a vector space we may now recognize that the columns of any rectangular matrix $\mathbf{A}$ also span a vector space. We call this space the column space of $\mathbf{A}$. Similarly, the rows of a matrix span a vector space, which we call the row space of **A**. Conversely, we may assemble any q vectors of order n into a matrix of order $n \times q$. The number of linearly independent vectors used is equal to the dimension of the column space of **A**. For example, the three vectors in Example 2.4 form the matrix

$$\mathbf{A} = \begin{bmatrix} 1 & 1 & 0 \\ 2 & 0 & -1 \\ 1 & 0 & 0 \\ 0 & 0 & 1 \end{bmatrix} \tag{2.4}$$

Assume that we are given a matrix **A** and that we need to calculate the dimension of the column space of **A**. In other words, we want to evaluate how many columns in **A** are linearly independent. The number of linearly independent columns in **A** is neither increased nor decreased by taking any linear combinations of them. Therefore, in order to identify the column space of **A**, we may try to transform the matrix by linearly combining its columns to obtain unit vectors $\mathbf{e}_i$. Because unit vectors $\mathbf{e}_i$ with distinct i are linearly independent, the dimension of the column space of **A** would be equal to the number of unit vectors that can be obtained. However, it need be noted that we may not be able to actually obtain unit vectors $\mathbf{e}_i$ (see Example 2.5), but the process followed in the transformation of **A** will always lead to a form that displays the dimension of the column space.

Example 2.5

Calculate the dimension of the column space of the matrix **A** formed by the vectors $\mathbf{x}_1$, $\mathbf{x}_2$, and $\mathbf{x}_3$ considered in Example 2.4.

The matrix considered is

$$\mathbf{A} = \begin{bmatrix} 1 & 1 & 0 \\ 2 & 0 & -1 \\ 1 & 0 & 0 \\ 0 & 0 & 1 \end{bmatrix}$$

Writing the second and third columns as the first and second columns, respectively, we obtain

$$\mathbf{A}_1 = \begin{bmatrix} 1 & 0 & 1 \\ 0 & -1 & 2 \\ 0 & 0 & 1 \\ 0 & 1 & 0 \end{bmatrix}$$

Subtracting the first column from the third column, adding twice the second column to the third column, and finally multiplying the second column by (-1), we obtain

$$\mathbf{A}_2 = \begin{bmatrix} 1 & 0 & 0 \\ 0 & 1 & 0 \\ 0 & 0 & 1 \\ 0 & -1 & 2 \end{bmatrix}$$

But we have now reduced the matrix to a form where we can identify that the three columns are linearly independent; i.e., the columns are linear independent because the first three elements in the vectors are the columns of the identity matrix of order 3. However, since we obtained $\mathbf{A}_2$ from $\mathbf{A}$ by interchanging and linearly combining the original columns of $\mathbf{A}$ and thus have not increased in the solution process the space spanned by the columns of the matrix, we have that the dimension of the column space of $\mathbf{A}$ is 3.

In the above presentation we linearly combined the vectors $\mathbf{x}_1, \ldots, \mathbf{x}_q$, which were the columns of $\mathbf{A}$, in order to identify whether they are linearly independent. Alternatively, to find the dimension of the space spanned by a set of vectors $\mathbf{x}_1, \mathbf{x}_2, \ldots, \mathbf{x}_q$, we could use the definition of vector linear independence in (2.2) and consider the set of simultaneous homogeneous equations

$$\alpha_1 \mathbf{x}_1 + \alpha_2 \mathbf{x}_2 + \ldots + \alpha_q \mathbf{x}_q = \mathbf{0} \tag{2.5}$$

which is, in matrix form,

$$\mathbf{A}\boldsymbol{\alpha} = \mathbf{0} \tag{2.6}$$

where $\boldsymbol{\alpha}$ is a vector with elements $\alpha_1, \ldots, \alpha_q$, and the columns of $\mathbf{A}$ are the vectors $\mathbf{x}_1, \mathbf{x}_2, \ldots, \mathbf{x}_q$. The solution for the unknowns $\alpha_1, \ldots, \alpha_q$ is not changed by linearly combining or multiplying any of the rows in the matrix $\mathbf{A}$. Therefore, we may try to reduce $\mathbf{A}$ by multiplying and combining its rows into a matrix, in which columns consist only of unit vectors. This reduced matrix is called the *row-echelon form* of $\mathbf{A}$. The number of unit column vectors in the row-echelon form of $\mathbf{A}$ is equal to the dimension of the column space of $\mathbf{A}$ and, from the preceding discussion is also equal to the dimension of the row space of $\mathbf{A}$. It follows that the dimension of the column space of $\mathbf{A}$ is equal to the dimension of the row space of $\mathbf{A}$. In other words, the number of linearly independent columns in $\mathbf{A}$ is equal to the number of linearly independent rows in $\mathbf{A}$. This result is summarized in the definition of the rank of a matrix.

Definition: The *rank* of a matrix $\mathbf{A}$ is equal to the dimension of the column space and equal to the dimension of the row space of $\mathbf{A}$.

Example 2.6

Consider the following three vectors:

$$\mathbf{x}_1 = \begin{bmatrix} 1 \\ 2 \\ 1 \\ 3 \\ 4 \\ 3 \end{bmatrix}; \quad \mathbf{x}_2 = \begin{bmatrix} 3 \\ 1 \\ -2 \\ 4 \\ 2 \\ -1 \end{bmatrix}; \quad \mathbf{x}_3 = \begin{bmatrix} 2 \\ 3 \\ 1 \\ 5 \\ 6 \\ 4 \end{bmatrix}$$

Use these vectors as the columns of a matrix **A** and reduce the matrix to row-echelon form.

We have

$$\mathbf{A} = \begin{bmatrix} 1 & 3 & 2 \\ 2 & 1 & 3 \\ 1 & -2 & 1 \\ 3 & 4 & 5 \\ 4 & 2 & 6 \\ 3 & -1 & 4 \end{bmatrix}$$

Subtracting multiples of the first row from the rows below it in order to obtain the unit vector $\mathbf{e}_1$ in the first column, we obtain

$$\mathbf{A}_1 = \begin{bmatrix} 1 & 3 & 2 \\ 0 & -5 & -1 \\ 0 & -5 & -1 \\ 0 & -5 & -1 \\ 0 & -10 & -2 \\ 0 & -10 & -2 \end{bmatrix}$$

Dividing the second row by (-5) and then subtracting multiples of it from the other rows in order to reduce the second column to the unit vector $\mathbf{e}_2$, we obtain

$$\mathbf{A}_2 = \begin{bmatrix} 1 & 0 & \frac{7}{5} \\ 0 & 1 & \frac{1}{5} \\ 0 & 0 & 0 \\ 0 & 0 & 0 \\ 0 & 0 & 0 \\ 0 & 0 & 0 \end{bmatrix}$$

Hence we can give the following equivalent statements:

1. The solution to $\mathbf{A}\boldsymbol{\alpha} = \mathbf{0}$ is

$$\alpha_1 = -\tfrac{7}{5}\alpha_3$$

$$\alpha_2 = -\tfrac{1}{5}\alpha_3$$

2. The three vectors $\mathbf{x}_1$, $\mathbf{x}_2$, and $\mathbf{x}_3$ are linearly dependent. They form a two-dimensional vector space. The vectors $\mathbf{x}_1$ and $\mathbf{x}_2$ are linearly independent and they form a basis of the two-dimensional space in which $\mathbf{x}_1$, $\mathbf{x}_2$, and $\mathbf{x}_3$ lie.
3. The rank of **A** is 2.
4. The dimension of the column space of **A** is 2.
5. The dimension of the row space of **A** is 2.

2.3 MATRIX REPRESENTATION OF LINEAR TRANSFORMATION

In engineering analysis we generally want to relate certain quantities to some other quantities such as stresses to strains, strains to displacements, forces to displacements, temperatures to heat flows, and so on. Each of these relationships can be regarded as a transformation from one entity to another and can be defined concisely and in a compact manner using matrices.

Let us not specify for the moment the specific transformation that we want to consider and use instead a symbolic notation, thus including in the discussion many possible transformations of interest. Let $\mathcal{x}$ be an element of the entity $\mathcal{X}$ and $\mathcal{y}$ be an element of the entity $\mathcal{Y}$. Then the transformation of $\mathcal{x}$ into $\mathcal{y}$ is symbolically written as

$$\mathcal{y} = \mathcal{A}\mathcal{x} \tag{2.7}$$

where $\mathcal{A}$ is the transformation operator. If there is a unique element $\mathcal{y}$ for each element $\mathcal{x}$, and vice versa, we say that the operator is nonsingular; otherwise, the operator is singular. The transformation is linear, if the following two relations hold:

$$\begin{aligned} \mathcal{A}(c\mathcal{x}) &= c\mathcal{A}\mathcal{x} \\ \mathcal{A}(\mathcal{x}_1 + \mathcal{x}_2) &= \mathcal{A}\mathcal{x}_1 + \mathcal{A}\mathcal{x}_2 \end{aligned} \tag{2.8}$$

We shall only be concerned with linear transformations.

Example 2.7

Consider the operator $\mathcal{A} = z^2 p'' + 3p'$, where p represents all polynomials in z of degree ≤ 3. Calculate $\mathcal{y} = \mathcal{A}\mathcal{x}$, where $\mathcal{x} = 3z + 4z^3$. Discuss whether the operator represents a linear transformation and whether it is singular.

Applying the operator $\mathcal{A}$ to $\mathcal{x}$, we obtain

$$\mathcal{x}' = 3 + 12z^2$$

$$\mathcal{x}'' = 24z$$

and hence

$$\mathcal{y} = 9 + 36z^2 + 24z^3$$

The operator $\mathcal{A}$ is linear because differentiation is a linear process. Namely, if $p = c\mathcal{x}$, then $p' = (c\mathcal{x})' = c(\mathcal{x})'$; i.e., the constant c can be taken out of the differentiation, and also if $p = \mathcal{x}_1 + \mathcal{x}_2$, then $p' = \mathcal{x}_1' + \mathcal{x}_2'$.

To identify whether $\mathcal{A}$ is singular, consider a simple element, say $\mathcal{x} = z$. In this case the transformation in (2.7) gives $\mathcal{y} = \mathcal{A}\mathcal{x} = 3$. Now assume that we are given $\mathcal{y} = 3$. Then we need to ask whether there is a unique element $\mathcal{x}$ corresponding to that $\mathcal{y}$. However, there is no unique element $\mathcal{x}$, because $\mathcal{x} = z$, $\mathcal{x} = z + 3$, $\mathcal{x} = z + 10$, etc., would all give $\mathcal{y} = 3$. Hence the operator $\mathcal{A}$ is singular.

Although the specific element $\mathcal{y}$ corresponding to an element $\mathcal{x}$ can always be calculated using the explicit definition of $\mathcal{A}$ as in Example 2.7, a more systematic procedure is needed if the operations are to be performed by the digital computer. This is achieved using the matrix representation of the operator $\mathcal{A}$.

The first step in the matrix representation of a linear transformation is the selection of two sets of base vectors, one set to represent the elements in $\mathcal{X}$ and the other set to represent the elements in $\mathcal{Y}$. Let $\mathcal{v}_1, \ldots, \mathcal{v}_n$ be the basis in $\mathcal{X}$ and $\mathcal{w}_1, \ldots, \mathcal{w}_n$ be the basis in $\mathcal{Y}$. Therefore, each element $\mathcal{x}$ in $\mathcal{X}$ can be represented as

$$\mathcal{x} = x_1\mathcal{v}_1 + \ldots + x_n\mathcal{v}_n \tag{2.9}$$

where $x_1, \ldots, x_n$ are the coordinates of $\mathcal{x}$, and each element $\mathcal{y}$ in $\mathcal{Y}$ can be represented as

$$\mathcal{y} = y_1\mathcal{w}_1 + \ldots + y_n\mathcal{w}_n \tag{2.10}$$

where $y_1, \ldots, y_n$ are the coordinates of $\mathcal{y}$. The matrix representation of the linear transformation $\mathcal{y} = \mathcal{A}\mathcal{x}$ is then

$$\mathbf{y} = \mathbf{A}\mathbf{x} \tag{2.11}$$

where $\mathbf{A}$ is a square matrix of order n and $\mathbf{y}$ and $\mathbf{x}$ are vectors that list the coordinates of $\mathcal{y}$ and $\mathcal{x}$, respectively; i.e.,

$$\mathbf{y} = \begin{bmatrix} y_1 \\ \cdot \\ \cdot \\ \cdot \\ y_n \end{bmatrix}; \qquad \mathbf{x} = \begin{bmatrix} x_1 \\ \cdot \\ \cdot \\ \cdot \\ x_n \end{bmatrix} \tag{2.12}$$

The element a_{ij} of the matrix $\mathbf{A}$ is the coordinate y_i obtained in (2.7) when the operator $\mathcal{A}$ is applied to the base vector $\mathcal{v}_j$. Therefore, to calculate the elements of the jth column of $\mathbf{A}$, we transform $\mathcal{x} = \mathcal{v}_j$; i.e., we calculate $\mathcal{y} = \mathcal{A}\mathcal{v}_j$, and represent $\mathcal{y}$ in terms of the base vectors $\mathcal{w}_i$, $i = 1, \ldots, n$. The coefficients of the $\mathcal{w}_i$ are the elements of the jth column in $\mathbf{A}$.

The effectiveness of using a matrix representation of $\mathcal{A}$ lies in that the transformation $\mathcal{y} = \mathcal{A}\mathcal{x}$ can be systematically carried out using (2.11), and, in particular, can readily be programmed for the digital computer.

It should be noted that the rules of matrix algebra defined in the previous sections can be established using matrices as representations of linear transformations.

Consider as an example the multiplication of two linear transformations $\mathcal{A}$ and $\mathcal{B}$, where

$$\mathcal{y} = \mathcal{A}\mathcal{x} \tag{2.13}$$

and

$$\mathcal{z} = \mathcal{B}\mathcal{y} \tag{2.14}$$

It follows that

$$\mathcal{z} = \mathcal{C}\mathcal{x} \tag{2.15}$$

where, symbolically,

$$\mathcal{C} = \mathcal{B}\mathcal{A} \tag{2.16}$$

The matrix representations of (2.13) and (2.14) are, respectively,

$$\mathbf{y} = \mathbf{A}\mathbf{x} \tag{2.17}$$

and

$$\mathbf{z} = \mathbf{B}\mathbf{y} \tag{2.18}$$

From the definition of the elements in **A** and **B** it follows that the matrix representation of (2.15) is

$$\mathbf{z} = \mathbf{C}\mathbf{x} \tag{2.19}$$

where

$$c_{ij} = \sum_k b_{ik}a_{kj} \tag{2.20}$$

But the relation (2.20) is the matrix multiplication defined earlier in Section 1.5. Similarly, the definitions for matrix addition, subtraction, and the inverse of a matrix given in Chapter 1 would all naturally be arrived at when considering the matrices as representations of linear transformations in chosen bases.

Example 2.8

Represent the linear transformation considered in Example 2.7 as a matrix, and perform the transformation $\mathcal{y} = \mathcal{A}\mathcal{x}$, where $\mathcal{y}$ and $\mathcal{x}$ are given in Example 2.7.

Choose the following two bases:

(1) $\mathcal{v}_1 = 1$, $\mathcal{v}_2 = z$, $\mathcal{v}_3 = z^2$, $\mathcal{v}_4 = z^3$, and $\mathcal{w}_i = \mathcal{v}_i$ for $i = 1, \ldots, 4$.
(2) $\mathcal{v}_1 = 1$, $\mathcal{v}_2 = 1 - z$, $\mathcal{v}_3 = z + 3z^2$, $\mathcal{v}_4 = z^3$, and $\mathcal{w}_i = \mathcal{v}_i$ for $i = 1, \ldots, 4$.

The operator in Example 2.7 was $\mathcal{A} = z^2p'' + 3p'$, where p represents all polynomials in z of degree ≤ 3. To evaluate the elements of the jth column in the matrix representation **A** of the operator $\mathcal{A}$, we form $\mathcal{A}\mathcal{v}_j$ and represent the result in terms of the base vectors $\mathcal{w}_j, j = 1, \ldots, 4$. Considering the basis in (1), we have

$$\begin{aligned}
\mathcal{A}\mathcal{v}_1 &= 0 \\
\mathcal{A}\mathcal{v}_2 &= 3 = 3\mathcal{w}_1 \\
\mathcal{A}\mathcal{v}_3 &= 6z + 2z^2 = 6\mathcal{w}_2 + 2\mathcal{w}_3 \\
\mathcal{A}\mathcal{v}_4 &= 9z^2 + 6z^3 = 9\mathcal{w}_3 + 6\mathcal{w}_4
\end{aligned}$$

Therefore, the matrix representation of $\mathcal{A}$ in the chosen basis is

$$\mathbf{A} = \begin{bmatrix} 0 & 3 & 0 & 0 \\ 0 & 0 & 6 & 0 \\ 0 & 0 & 2 & 9 \\ 0 & 0 & 0 & 6 \end{bmatrix}$$

To transform the element $\mathcal{X}$, where $\mathcal{X} = 3z + 4z^3$, we find the coordinates of $\mathcal{X}$ in the selected basis:

$$\mathbf{x} = \begin{bmatrix} 0 \\ 3 \\ 0 \\ 4 \end{bmatrix}$$

Then, using $\mathbf{y} = \mathbf{A}\mathbf{x}$, we obtain

$$\mathbf{y} = \begin{bmatrix} 9 \\ 0 \\ 36 \\ 24 \end{bmatrix}$$

which gives $\mathcal{Y} = 9 + 36z^2 + 24z^3$, as in Example 2.7.

Using the same procedure to evaluate the matrix representation of $\mathcal{A}$ corresponding to the basis given in (2), we obtain

$$\bar{\mathbf{A}} = \begin{bmatrix} 0 & -3 & 19 & -3 \\ 0 & 0 & -16 & 3 \\ 0 & 0 & 2 & 3 \\ 0 & 0 & 0 & 6 \end{bmatrix}$$

The coordinate vector of $\mathcal{X} = 3z + 4z^3$ in the new basis is

$$\bar{\mathbf{x}} = \begin{bmatrix} 3 \\ -3 \\ 0 \\ 4 \end{bmatrix}$$

To obtain the coordinate vector of $\mathcal{Y}$, we form $\bar{\mathbf{y}} = \mathbf{A}\bar{\mathbf{x}}$ and obtain

$$\bar{\mathbf{y}} = \begin{bmatrix} -3 \\ 12 \\ 12 \\ 24 \end{bmatrix}$$

As a check, we use $\bar{\mathbf{y}}$ to obtain $\mathcal{Y}$ and find

$$\mathcal{Y} = (-3)(1) + (12)(1 - z) + (12)(z + 3z^2) + (24)(z^3)$$

As before,

$$\mathcal{Y} = 9 + 36z^2 + 24z^3$$

2.4 CHANGE OF BASIS

Assume that a transformation has been represented by a matrix $\mathbf{A}$ in a given basis and that we would like to find the matrix $\bar{\mathbf{A}}$, which represents the same transformation in a different basis. One way of obtaining the elements of the matrix $\bar{\mathbf{A}}$ is to simply use the basic procedure of evaluating the matrix of the operator $\mathcal{A}$ in the new basis as we have done in Example 2.8. However, we can also calculate $\bar{\mathbf{A}}$ from $\mathbf{A}$ by relating the new basis to the old basis, which has practical advantages. That is, we can change basis without ever going back to the operator $\mathcal{A}$, and evaluate a basis in which the transformation is, from numerical considerations, represented very effectively. Since we use the matrix $\mathbf{A}$ to relate $\mathbf{x}$ and $\mathbf{y}$ by

$$\mathbf{y} = \mathbf{A}\mathbf{x} \tag{2.21}$$

the representation of the linear transformation will be particularly efficient if $\mathbf{A}$ is diagonal. Fortunately, in finite element analysis, a basis in which the system matrix is diagonal does nearly always exist. However, unfortunately, in most practical calculations we do not know this basis a priori. Therefore, the procedure is to simply represent the transformation first in some convenient basis (i.e., the finite element coordinate basis), and if efficient for the complete solution process, we then transform the matrix into diagonal form. This transformation is carried out in a systematic manner, and as we will show, requires the solution of an eigenproblem. The matrix transformation discussed below is fundamental to the solution of eigenproblems.

Assume that $\mathbf{A}$ is the $n \times n$ matrix representation of an operator $\mathcal{A}$ in the basis $\mathcal{V}_1, \ldots, \mathcal{V}_n$. Assume that the matrix $\mathbf{A}$ has been calculated using one basis only, i.e., $\mathcal{W}_i = \mathcal{V}_i$, $i = 1, \ldots, n$, which is the most important case.

The matrix $\mathbf{A}$ relates $\mathbf{x}$ to $\mathbf{y}$ as given in (2.21), where the ith element of the vectors $\mathbf{x}$ and $\mathbf{y}$ (i.e., x_i and y_i) are the multipliers of the base vectors $\mathcal{V}_i$. Assume that the new base vectors are $\bar{\mathcal{V}}_i$, $i = 1, \ldots, n$, and that the corresponding coordinates are stored in vectors $\bar{\mathbf{x}}$ and $\bar{\mathbf{y}}$. To obtain the matrix $\bar{\mathbf{A}}$, which relates $\bar{\mathbf{x}}$ to $\bar{\mathbf{y}}$, the first step is to express all new base vectors $\bar{\mathcal{V}}_i$, $i = 1, \ldots, n$, in terms of the old base vectors $\mathcal{V}_i$, $i = 1, \ldots, n$; i.e., we evaluate

$$\bar{\mathcal{V}}_i = \sum_{k=1}^{n} p_{ik}\mathcal{V}_k; \qquad i = 1, \ldots, n \tag{2.22}$$

Using this relation we then have

$$\mathbf{x} = \mathbf{P}\bar{\mathbf{x}}; \qquad \mathbf{y} = \mathbf{P}\bar{\mathbf{y}} \tag{2.23}$$

where $\mathbf{P}$ is an $n \times n$ matrix with the p_{ik} of (2.22) being element (k, i) in $\mathbf{P}$. Substituting the relations in (2.23) into (2.21), we obtain

$$\mathbf{P}\bar{\mathbf{y}} = \mathbf{A}\mathbf{P}\bar{\mathbf{x}} \tag{2.24}$$

Solving for $\bar{\mathbf{y}}$, we obtain

$$\bar{\mathbf{y}} = \mathbf{P}^{-1}\mathbf{A}\mathbf{P}\bar{\mathbf{x}} \tag{2.25}$$

and hence

$$\bar{\mathbf{A}} = \mathbf{P}^{-1}\mathbf{A}\mathbf{P} \tag{2.26}$$

In the previous derivation it was assumed that the inverse of $\mathbf{P}$ exists. This is the case provided that a proper basis $\bar{\mathcal{V}}_i$, $i = 1, \ldots, n$, has been chosen; i.e., all $\bar{\mathcal{V}}_i$ are linearly independent.

It should be noted that the existence of the inverse of **P** assures that for any vectors **x** and **y** there are unique vectors $\bar{\mathbf{x}}$ and $\bar{\mathbf{y}}$ as expressed in (2.23), and vice versa.

In an alternative procedure, to obtain $\bar{\mathbf{A}}$ we may use $\mathbf{x} = \mathbf{P}\bar{\mathbf{x}}$ and $\bar{\mathbf{y}} = \mathbf{Q}\mathbf{y}$, where the elements in **Q** are obtained from the relation expressing the base vectors $\mathscr{v}_i$, $i = 1, \ldots, n$, in terms of the base vectors $\bar{\mathscr{v}}_i$, $i = 1, \ldots, n$; i.e., we use

$$\mathscr{v}_i = \sum_{k=1}^{n} q_{ik}\bar{\mathscr{v}}_k \tag{2.27}$$

where q_{ik} is element (k, i) of **Q**. Then, substituting for **x** in (2.21) and evaluating $\bar{\mathbf{y}}$, we obtain

$$\bar{\mathbf{y}} = \mathbf{QAP}\bar{\mathbf{x}} \tag{2.28}$$

and hence

$$\bar{\mathbf{A}} = \mathbf{QAP} \tag{2.29}$$

Because of the uniqueness of the operator $\bar{\mathbf{A}}$, we have

$$\mathbf{Q} = \mathbf{P}^{-1} \tag{2.30}$$

and in (2.27) we essentially evaluated the inverse of **P**.

Example 2.9

Consider the transformation discussed in Example 2.8. Using the basis $\mathscr{v}_1 = 1$, $\mathscr{v}_2 = z$, $\mathscr{v}_3 = z^2$, $\mathscr{v}_4 = z^3$, we obtained

$$\mathbf{A} = \begin{bmatrix} 0 & 3 & 0 & 0 \\ 0 & 0 & 6 & 0 \\ 0 & 0 & 2 & 9 \\ 0 & 0 & 0 & 6 \end{bmatrix}$$

Use a transformation on **A** to change basis to $\bar{\mathscr{v}}_1 = 1$, $\bar{\mathscr{v}}_2 = 1 - z$, $\bar{\mathscr{v}}_3 = z + 3z^2$, and $\bar{\mathscr{v}}_4 = z^3$.

Relation (2.22) for each of the base vectors gives

$$\begin{aligned} \bar{\mathscr{v}}_1 &= \mathscr{v}_1 \\ \bar{\mathscr{v}}_2 &= \mathscr{v}_1 - \mathscr{v}_2 \\ \bar{\mathscr{v}}_3 &= \mathscr{v}_2 + 3\mathscr{v}_3 \\ \bar{\mathscr{v}}_4 &= \mathscr{v}_4 \end{aligned}$$

Hence

$$\mathbf{P} = \begin{bmatrix} 1 & 1 & 0 & 0 \\ 0 & -1 & 1 & 0 \\ 0 & 0 & 3 & 0 \\ 0 & 0 & 0 & 1 \end{bmatrix}$$

Similarly, using the relation (2.27), we have

$$\begin{aligned}\mathbf{v}_1 &= \bar{\mathbf{v}}_1\\ \mathbf{v}_2 &= \bar{\mathbf{v}}_1 - \bar{\mathbf{v}}_2\\ \mathbf{v}_3 &= -\tfrac{1}{3}\bar{\mathbf{v}}_1 + \tfrac{1}{3}\bar{\mathbf{v}}_2 + \tfrac{1}{3}\bar{\mathbf{v}}_3\\ \mathbf{v}_4 &= \bar{\mathbf{v}}_4\end{aligned}$$

and

$$\mathbf{Q} = \begin{bmatrix} 1 & 1 & -\frac{1}{3} & 0 \\ 0 & -1 & \frac{1}{3} & 0 \\ 0 & 0 & \frac{1}{3} & 0 \\ 0 & 0 & 0 & 1 \end{bmatrix}$$

We may check that $\mathbf{QP} = \mathbf{I}$. To obtain $\bar{\mathbf{A}}$ we evaluate (2.29) and obtain

$$\bar{\mathbf{A}} = \begin{bmatrix} 0 & -3 & 19 & -3 \\ 0 & 0 & -16 & 3 \\ 0 & 0 & 2 & 3 \\ 0 & 0 & 0 & 6 \end{bmatrix}$$

This is the same matrix $\bar{\mathbf{A}}$ that we calculated in Example 2.8.

Consider the transformation that corresponds to a change of basis as carried out in Example 2.9; it appears that it requires much numerical effort. First the two matrices **P** and **Q** (or the inverse of **P**) need to be established, and then the matrix product in (2.29) must be evaluated. A solution process requiring many transformations could therefore become very expensive. However, in practice special matrices are employed. The advantage of using these matrices lies in the fact that they can be inverted easily (and accurately) and also that the matrix product in (2.26) is calculated with relatively little effort. We will use most frequently the matrices defined in the following section.

2.5 MATRIX REPRESENTATION OF VARIATIONAL FORMULATION

In the previous sections we discussed the matrix representation of a general linear transformation. The resulting matrix **A** may be a symmetric, nonsymmetric (see Example 2.8), or even complex matrix. However, in finite element analysis we are concerned primarily with the matrix representation of a variational formulation of the problem under consideration and are thus able to establish symmetric matrices in almost all cases. We present the finite element formulation in detail in Chapters 3 to 5, but it is very instructive to briefly focus attention on the formulation here in the light of the procedures given in the previous sections.

As is pointed out later, there are many different finite element formulations. As an example, we consider here the displacement-based formulation, which is most widely used. The basic equation considered is, in this case,

$$\int_{\xi} (\mathcal{L}\,\delta u)^T(\mathcal{S}u)\,d\xi - \int_{\eta} \delta u^T \mathcal{r}\,d\eta = 0 \tag{2.31}$$

where u represents displacement components, $\mathcal{L}$ and $\mathcal{S}$ are operators representing differentiations of u, $\mathcal{r}$ is a forcing function, δ means "variation in," and ξ and η represent appropriate quantities over which is integrated. The matrix representation of (2.31) is

$$\mathbf{A}\mathbf{x} = \mathbf{r} \tag{2.32}$$

where $\mathbf{A}$ is a square symmetric matrix, $\mathbf{x}$ is defined in (2.12), and $\mathbf{r}$ is a forcing vector,

$$\mathbf{r} = \begin{bmatrix} r_1 \\ \cdot \\ \cdot \\ \cdot \\ r_n \end{bmatrix} \tag{2.33}$$

It should be noted that the relation in (2.32) represents a linear transformation in which $\mathbf{x}$ is related to $\mathbf{r}$.

Following the development of Section 2.3, the first step is the selection of base vectors. Here we choose base functions for the displacements u and have

$$u = x_1 u_1 + \ldots + x_n u_n \tag{2.34}$$

The element a_{ij} of the matrix $\mathbf{A}$ is now obtained by substituting into (2.31) for u and δu the jth and ith base functions, i.e., by using in (2.31) $u = u_j$ and $\delta u = \delta u_i$. At the same time, the element r_i is obtained. The only, but important, difference to the procedure used in Section 2.3 is, therefore, that to represent (2.31), we substitute the individual base functions for the displacements and the variations in the displacements, and it is the integral of the resulting product that gives the required element of the matrix $\mathbf{A}$. Since the same base functions are employed for u and δu, a symmetric matrix $\mathbf{A}$ is obtained. The detailed procedures used to evaluate the elements of $\mathbf{A}$ and $\mathbf{r}$ are presented in Chapters 3 to 5. But it is instructive to consider here a small example which demonstrates the important points of the process.

Example 2.10

The variational formulation used to evaluate the displacements and forces in the bar element shown in Fig. 2.2 is

$$\int_L \delta u'(x) EA(x) u'(x)\,dx - \int_V p(x)\,\delta u(x)\,dV = 0 \tag{a}$$

where $u(x)$ is the longitudinal displacement of the bar, $EA(x)$ the stiffness of the bar, $p(x)$ the applied volume loading at coordinate x, and L and V the length and volume of the bar, respectively. Assume as a basis the polynomials x and x^2 and $A = A_0(1 + x/L)$. Evaluate the transformation matrix that relates the coordinates of the selected basis to the applied forces.

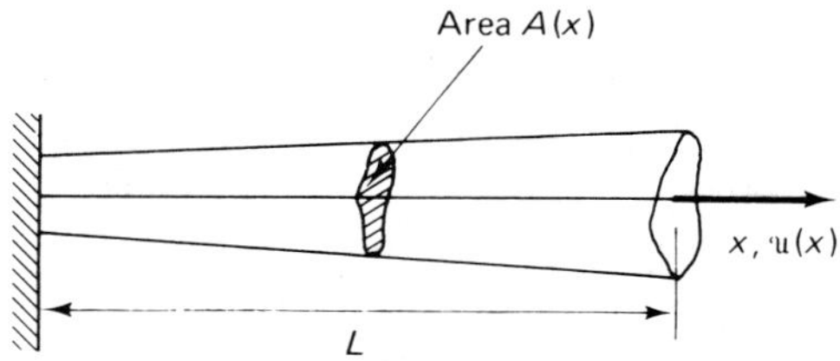

Figure 2.2. Simple bar element.

To calculate the matrix representation of (a) using the specified basis, we have

$$u = \alpha_1 x + \alpha_2 x^2$$

where α_1 and α_2 are the unknown coordinates. Hence we have

$$u' = \alpha_1 + 2\alpha_2 x$$

and

$$\int_L (\delta\alpha_1 + 2x\delta\alpha_2)EA(\alpha_1 + 2x\alpha_2)\,dx - \int_V p(\delta\alpha_1 x + \delta\alpha_2 x^2)\,dV = 0 \tag{b}$$

To evaluate element (i, j) of $\mathbf{A}$, we consider the first integral and let $\delta\alpha_i = 1$, $\alpha_j = 1$, and all other coordinate variables be zero. Thus we obtain

$$a_{11} = \int_L (1)(EA)(1)\,dx$$

$$a_{12} = \int_L (1)(EA)(2x)\,dx$$

$$a_{21} = \int_L (2x)(EA)(1)\,dx$$

$$a_{22} = \int_L (2x)(EA)(2x)\,dx$$

Substituting that $EA = EA_0(1 + x/L)$, we have

$$\mathbf{A} = EA_0\begin{bmatrix} \frac{3}{2}L & \frac{5}{3}L^2 \\ \frac{5}{3}L^2 & \frac{7}{3}L^3 \end{bmatrix}$$

To calculate the ith element of $\mathbf{r}$, we evaluate the second integral in (b) and let $\delta\alpha_i = 1$; thus

$$r_1 = \int_L (p)(A)x\,dx$$

$$r_2 = \int_L (p)(A)(x^2)\,dx$$

where A is the area of the bar at the coordinate x. Substituting $A = A_0(1 + x/L)$ and using p as a constant, we obtain

$$\mathbf{r} = A_0 p\begin{bmatrix} \frac{5}{6}L^2 \\ \frac{7}{12}L^3 \end{bmatrix}$$

Hence the equations to be solved are

$$\begin{bmatrix} \frac{3}{2}L & \frac{5}{3}L^2 \\ \frac{5}{3}L^2 & \frac{7}{3}L^3 \end{bmatrix} \begin{bmatrix} \alpha_1 \\ \alpha_2 \end{bmatrix} = \begin{bmatrix} \frac{5}{6}L^2 \\ \frac{7}{12}L^3 \end{bmatrix} \frac{p}{E}$$

which gives

$$\begin{bmatrix} \alpha_1 \\ \alpha_2 \end{bmatrix} = \begin{bmatrix} \dfrac{105}{78} \\ -\dfrac{37}{52}\dfrac{1}{L} \end{bmatrix} \frac{pL}{E}$$

Hence

$$u = \frac{105}{78}\frac{pL}{E}x - \frac{37}{52}\frac{p}{E}x^2$$

We now turn to the problem of changing basis when **A** has been derived from a variational formulation. Following the development given in Section 2.4, we assume that u is represented in the new basis as

$$u = \bar{x}_1\bar{u}_1 + \dots + \bar{x}_n\bar{u}_n \tag{2.35}$$

where $\bar{u}_i$ is the ith new base function and $\bar{x}_i$ is the corresponding coordinate. In this case we have

$$\mathbf{x} = \mathbf{P}\bar{\mathbf{x}} \tag{2.36}$$

and element (k, i) of **P** is given by p_{ik}, where

$$\bar{u}_i = \sum_{k=1}^{n} p_{ik}u_k \tag{2.37}$$

Noting that we use u^T and u in (2.31), the change of basis now gives

$$\bar{\mathbf{A}}\bar{\mathbf{x}} = \bar{\mathbf{r}} \tag{2.38}$$

where

$$\bar{\mathbf{A}} = \mathbf{P}^T\mathbf{A}\mathbf{P}; \qquad \bar{\mathbf{r}} = \mathbf{P}^T\mathbf{r} \tag{2.39}$$

To demonstrate the procedure, consider the following example.

Example 2.11

Evaluate the matrices **A** and **r** of the problem in Example 2.10 when using as base functions x and $x - x^2$.

1. Calculate **A** and **r** using the basic procedure employed in Example 2.10.
2. Perform the change of basis on the matrices **A** and **r** established in Example 2.10 as given in (2.39).

In the first case we use

$$u = \bar{\alpha}_1 x + \bar{\alpha}_2(x - x^2)$$

and proceed as in Example 2.10 to evaluate $\bar{\mathbf{A}}$ and $\bar{\mathbf{r}}$. Here we have

$$\bar{a}_{11} = EA_0 \int_0^L (1)\left(1 + \frac{x}{L}\right)(1)\,dx = (\tfrac{3}{2}L)EA_0$$

$$\bar{a}_{12} = EA_0 \int_0^L (1)\left(1 + \frac{x}{L}\right)(1 - 2x)\, dx = (\tfrac{3}{2}L - \tfrac{5}{3}L^2)EA_0$$

$$\bar{a}_{21} = EA_0 \int_0^L (1 - 2x)\left(1 + \frac{x}{L}\right)(1)\, dx = \bar{a}_{12} \tag{a}$$

$$\bar{a}_{22} = EA_0 \int_0^L (1 - 2x)\left(1 + \frac{x}{L}\right)(1 - 2x)\, dx = (\tfrac{3}{2}L - \tfrac{10}{3}L^2 + \tfrac{7}{3}L^3)EA_0$$

$$\bar{r}_1 = A_0 p \int_0^L \left(1 + \frac{x}{L}\right)(x)\, dx = A_0 p \tfrac{5}{6}L^2$$

$$\bar{r}_2 = A_0 p \int_0^L \left(1 + \frac{x}{L}\right)(x - x^2)\, dx = A_0 p(\tfrac{5}{6}L^2 - \tfrac{7}{12}L^3) \tag{b}$$

Hence the equations to be solved are

$$\begin{bmatrix} \tfrac{3}{2}L & \tfrac{3}{2}L - \tfrac{5}{3}L^2 \\ \tfrac{3}{2}L - \tfrac{5}{3}L^2 & \tfrac{3}{2}L - \tfrac{10}{3}L^2 + \tfrac{7}{3}L^3 \end{bmatrix} \begin{bmatrix} \bar{\alpha}_1 \\ \bar{\alpha}_2 \end{bmatrix} = \begin{bmatrix} \tfrac{5}{6}L^2 \\ \tfrac{5}{6}L^2 - \tfrac{7}{12}L^3 \end{bmatrix} \frac{p}{E}$$

and we obtain

$$\bar{\alpha}_1 = \frac{105}{78}\frac{pL}{E} - \frac{37}{52}\frac{p}{E}$$

$$\bar{\alpha}_2 = \frac{37}{52}\frac{p}{E}$$

Hence

$$u = \left(\frac{105}{78}\frac{pL}{E} - \frac{37}{52}\frac{p}{E}\right)x + \frac{37}{52}\frac{p}{E}(x - x^2)$$

or

$$u = \frac{105}{78}\frac{pL}{E}x - \frac{37}{52}\frac{p}{E}x^2$$

This is the same solution as we obtained in Example 2.10, because the same space is spanned by the base functions used.

To perform the change of basis directly on the matrices $\mathbf{A}$ and $\mathbf{r}$ derived in Example 2.10, we need to evaluate first the matrix $\mathbf{P}$ defined in (2.37). Here we have, with $u_1 = x$, $u_2 = x^2$, $\bar{u}_1 = x$, and $\bar{u}_2 = x - x^2$,

$$\mathbf{P} = \begin{bmatrix} 1 & 1 \\ 0 & -1 \end{bmatrix}$$

Now, using (2.39) with the matrices derived in Example 2.10, we obtain

$$\bar{\mathbf{A}} = EA_0 \begin{bmatrix} 1 & 0 \\ 1 & -1 \end{bmatrix} \begin{bmatrix} \tfrac{3}{2}L & \tfrac{5}{3}L^2 \\ \tfrac{5}{3}L^2 & \tfrac{7}{3}L^3 \end{bmatrix} \begin{bmatrix} 1 & 1 \\ 0 & -1 \end{bmatrix}$$

or

$$\bar{\mathbf{A}} = EA_0 \begin{bmatrix} \tfrac{3}{2}L & \tfrac{3}{2}L - \tfrac{5}{3}L^2 \\ \tfrac{3}{2}L - \tfrac{5}{3}L^2 & \tfrac{3}{2}L - \tfrac{10}{3}L^2 + \tfrac{7}{3}L^3 \end{bmatrix}$$

and

$$\bar{\mathbf{r}} = A_0 p \begin{bmatrix} 1 & 0 \\ 1 & -1 \end{bmatrix} \begin{bmatrix} \tfrac{5}{6}L^2 \\ \tfrac{7}{12}L^3 \end{bmatrix}$$

or

$$\bar{\mathbf{r}} = A_0 p \begin{bmatrix} \frac{5}{6}L^2 \\ \frac{5}{6}L^2 - \frac{7}{12}L^3 \end{bmatrix}$$

Hence we obtained the same matrices $\bar{\mathbf{A}}$ and $\bar{\mathbf{r}}$ as given in (a) and (b).

The automatic change of basis is of particular importance in eigensolution algorithms, as we will indicate in the next section. However, in the solution of standard eigenproblems a change of basis will only be effective if it is of the form $\mathbf{P}^{-1}\mathbf{A}\mathbf{P}$ (see Section 2.6), and for this reason orthogonal matrices are used very effectively.

Definition: A matrix $\mathbf{P}$ is an *orthogonal matrix* if $\mathbf{P}^T\mathbf{P} = \mathbf{P}\mathbf{P}^T = \mathbf{I}$. Hence for an orthogonal matrix, we have $\mathbf{P}^{-1} = \mathbf{P}^T$.

This definition shows that if an orthogonal matrix is used in the change of basis, we have $\bar{\mathbf{A}} = \mathbf{P}^T\mathbf{A}\mathbf{P} = \mathbf{P}^{-1}\mathbf{A}\mathbf{P}$. For practical use, there exist some orthogonal matrices that can be constructed and employed easily. Two kinds that we shall employ extensively are introduced below.

An orthogonal matrix very frequently used is the *rotation matrix*

$$\mathbf{P} = \begin{bmatrix} 1 & & & & & & \\ & \ddots & & & & & \\ & & \cos\theta & \cdots & -\sin\theta & & \\ & & \vdots & & \vdots & & \\ & & \sin\theta & \cdots & \cos\theta & & \\ & & & & & \ddots & \\ & & & & & & 1 \end{bmatrix} \begin{matrix} \\ \\ i\text{th} \\ \\ j\text{th row} \\ \\ \\ \end{matrix} \tag{2.40}$$

(cos θ in the ith column, −sin θ in the jth column; elements not shown are zeros)

where i and j can be arbitrary but $i \neq j$. The name "rotation matrix" is explained by considering $n = 2$, in which case

$$\mathbf{P} = \begin{bmatrix} \cos\theta & -\sin\theta \\ \sin\theta & \cos\theta \end{bmatrix} \tag{2.41}$$

Referring to Fig. 2.3, we observe that for the base vectors $\mathbf{e}_1$ and $\mathbf{e}_2$, the transformation

$$[\mathbf{v}_1, \mathbf{v}_2] = \mathbf{P}[\mathbf{e}_1, \mathbf{e}_2] \tag{2.42}$$

represents a mere rotation by an angle θ. In the general case, this rotation is carried out in the n-dimensional space.

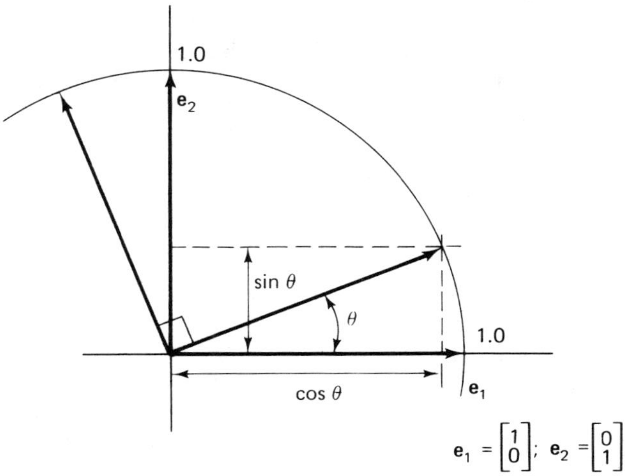

Figure 2.3. Rotation of vectors $\mathbf{e}_1$ and $\mathbf{e}_2$.

Example 2.12

Rotate the vector $\mathbf{v}$ through an angle of 45°, where $\mathbf{v} = \begin{bmatrix} 1 \\ 1 \end{bmatrix}$.

The rotation matrix is in this case with cos 45° = sin 45° = $\sqrt{2}/2$,

$$\mathbf{P} = \begin{bmatrix} \frac{\sqrt{2}}{2} & -\frac{\sqrt{2}}{2} \\ \frac{\sqrt{2}}{2} & \frac{\sqrt{2}}{2} \end{bmatrix}$$

The rotated vector is given by $\mathbf{Pv}$,

$$\mathbf{Pv} = \begin{bmatrix} \frac{\sqrt{2}}{2} & -\frac{\sqrt{2}}{2} \\ \frac{\sqrt{2}}{2} & \frac{\sqrt{2}}{2} \end{bmatrix} \begin{bmatrix} 1 \\ 1 \end{bmatrix} = \begin{bmatrix} 0 \\ \sqrt{2} \end{bmatrix}$$

It should be noted that the length of the vector $\mathbf{Pv}$ is equal to the length of the vector $\mathbf{v}$. The results obtained can easily be interpreted geometrically, as was done for a general vector in two-dimensional space in Fig. 2.3.

Another orthogonal matrix frequently used is the *reflection matrix*,

$$\mathbf{P} = \mathbf{I} - \alpha \mathbf{v}\mathbf{v}^T; \qquad \alpha = \frac{2}{\mathbf{v}^T\mathbf{v}} \tag{2.43}$$

where $\mathbf{v}$ can be arbitrary. This matrix is called a reflection matrix, because the vector $\mathbf{Pw}$ is the reflection of the vector $\mathbf{w}$ in the plane to which $\mathbf{v}$ is orthogonal.

A typical reflection is depicted in Fig. 2.4 for the case $n = 2$ and arbitrary vectors $\mathbf{v}$ and $\mathbf{w}$.

To prove the reflection property when $\mathbf{v}$ has any order n, we use that $\mathbf{w}$ (which is also of order n) has components $\gamma\mathbf{v}$ into the direction $\mathbf{v}$ and $\mathbf{u}$ perpendicular to $\mathbf{v}$. For the component $\gamma\mathbf{v}$, we have

$$
\begin{aligned}
\mathbf{P}\gamma\mathbf{v} &= \gamma\mathbf{v} - \alpha\mathbf{v}\mathbf{v}^T\gamma\mathbf{v} \\
&= \gamma\left[\mathbf{v} - \frac{2}{\mathbf{v}^T\mathbf{v}}\mathbf{v}(\mathbf{v}^T\mathbf{v})\right] \\
&= -\gamma\mathbf{v}
\end{aligned}
\tag{2.44}
$$

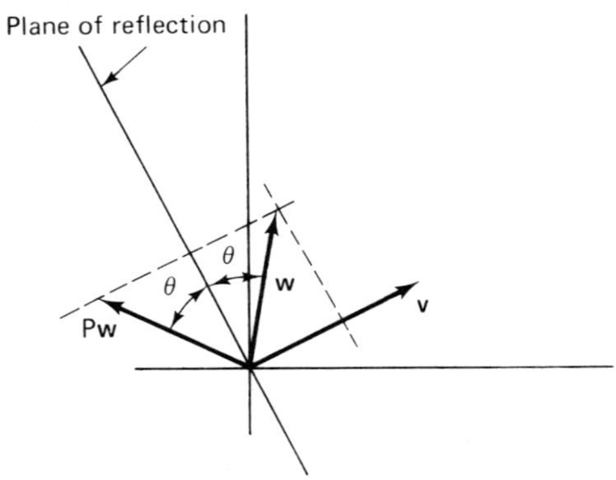

Figure 2.4. Reflection of vector **w**.

Therefore, the component into the direction of $\mathbf{v}$ has its direction reversed. For the component orthogonal to $\mathbf{v}$, we have

$$
\begin{aligned}
\mathbf{P}\mathbf{u} &= \mathbf{u} - \alpha\mathbf{v}\mathbf{v}^T\mathbf{u} \\
&= \mathbf{u}
\end{aligned}
\tag{2.45}
$$

because $\mathbf{v}^T\mathbf{u} = 0$. Therefore, this component is not transformed. The two cases together prove the general reflection property of the matrix $\mathbf{P}$.

Example 2.13

Establish the reflection matrix corresponding to the vector $\mathbf{v} = \begin{bmatrix} 1 \\ 1 \end{bmatrix}$ and reflect the vector $\mathbf{w} = \begin{bmatrix} 0 \\ 1 \end{bmatrix}$

In this case we have

$$
\mathbf{P} = \begin{bmatrix} 1 & 0 \\ 0 & 1 \end{bmatrix} - \frac{2}{2}\begin{bmatrix} 1 & 1 \\ 1 & 1 \end{bmatrix}
$$

Hence

$$\mathbf{P} = \begin{bmatrix} 0 & -1 \\ -1 & 0 \end{bmatrix}$$

The reflection of the vector **w** is **Pw**, where

$$\mathbf{Pw} = \begin{bmatrix} 0 & -1 \\ -1 & 0 \end{bmatrix} \begin{bmatrix} 0 \\ 1 \end{bmatrix} = \begin{bmatrix} -1 \\ 0 \end{bmatrix}$$

This result can easily be interpreted geometrically, as shown for a general reflection in two-dimensional space in Fig. 2.4. Note that the length of the vector **Pw** is equal to the length of the vector **w**.

2.6 THE EIGENPROBLEM $\mathbf{Av} = \lambda\mathbf{v}$, WITH A BEING A SYMMETRIC MATRIX

In previous sections we discussed how a change of basis can be performed in matrix representations of operators. In finite element analysis we are primarily interested in transformations as applied to symmetric matrices that have been obtained from a variational formulation and we shall assume in the discussion to follow that **A** is symmetric. For example, the matrix **A** may represent the stiffness matrix, mass matrix, or heat capacity matrix of an element assemblage.

There are various important applications (see Examples 2.20, 2.21, and 2.22, and Chapter 8), in which for overall solution effectiveness a change of basis is performed using in the transformation matrix the eigenvectors of the eigenproblem

$$\mathbf{Av} = \lambda\mathbf{v} \tag{2.46}$$

The problem in (2.46) is a *standard eigenproblem*. If the solution of (2.46) is considered in order to obtain eigenvalues and eigenvectors, the problem $\mathbf{Av} = \lambda\mathbf{v}$ will be referred to as an *eigenproblem*, whereas if only eigenvalues are to be calculated, $\mathbf{Av} = \lambda\mathbf{v}$ is called an *eigenvalue problem*. The objective in this section is to discuss the various properties that pertain to the solutions of (2.46).

Let n be the order of the matrix **A**. The first important point is that there exist n nontrivial solutions to (2.46). The word "nontrivial" means here that **v** must not be a null vector for which (2.46) is always satisfied. The ith nontrivial solution is given by the *eigenvalue* λ_i and the corresponding *eigenvector* $\mathbf{v}_i$, for which we have

$$\mathbf{Av}_i = \lambda_i \mathbf{v}_i \tag{2.47}$$

Therefore, each solution consists of an eigenpair, and we write the n solutions as $(\lambda_1, \mathbf{v}_1), (\lambda_2, \mathbf{v}_2), \ldots, (\lambda_n, \mathbf{v}_n)$, where

$$\lambda_1 \leq \lambda_2 \leq \ldots \leq \lambda_n \tag{2.48}$$

We also call all n eigenvalues and eigenvectors the *eigensystem* of **A**.

The proof that there must be n eigenvalues and corresponding eigenvectors can conveniently be obtained by writing (2.46) in the form

$$(\mathbf{A} - \lambda\mathbf{I})\mathbf{v} = \mathbf{0} \tag{2.49}$$

But these equation only have a solution provided that

$$\det(\mathbf{A} - \lambda\mathbf{I}) = 0 \tag{2.50}$$

Unfortunately, the necessity for (2.50) to hold can only be explained once the solution of simultaneous equations has been presented. For this reason we postpone until Section 10.2.2 the discussion that (2.50) is indeed required.

Using (2.50), the eigenvalues of $\mathbf{A}$ are thus the roots of the polynomial

$$p(\lambda) = \det(\mathbf{A} - \lambda\mathbf{I}) \tag{2.51}$$

This polynomial is called the *characteristic polynomial* of $\mathbf{A}$. However, since the order of the polynomial is equal to the order of $\mathbf{A}$, we have n eigenvalues, and using (2.49) we obtain n corresponding eigenvectors. It may be noted that the vectors obtained from the solution of (2.49) are only defined within a scalar multiple.

Example 2.14

Consider the matrix

$$\mathbf{A} = \begin{bmatrix} -1 & 2 \\ 2 & 2 \end{bmatrix}$$

Show that the matrix has two eigenvalues. Calculate the eigenvalues and eigenvectors.

The characteristic polynomial of $\mathbf{A}$ is

$$p(\lambda) = \det \begin{bmatrix} -1-\lambda & 2 \\ 2 & 2-\lambda \end{bmatrix}$$

Using the procedure given in Section 1.8 to calculate the determinant of a matrix (see Example 1.13), we obtain

$$\begin{aligned} p(\lambda) &= (-1-\lambda)(2-\lambda) - (2)(2) \\ &= \lambda^2 - \lambda - 6 \\ &= (\lambda + 2)(\lambda - 3) \end{aligned}$$

The order of the polynomial is 2, and hence there are two eigenvalues. In fact, we have

$$\begin{aligned} \lambda_1 &= -2 \\ \lambda_2 &= 3 \end{aligned}$$

The corresponding eigenvectors are obtained by applying (2.49) at the eigenvalues. Thus we have for λ_1,

$$\begin{bmatrix} -1-(-2) & 2 \\ 2 & 2-(-2) \end{bmatrix} \begin{bmatrix} v_1 \\ v_2 \end{bmatrix} = \begin{bmatrix} 0 \\ 0 \end{bmatrix}$$

with the solution

$$\mathbf{v}_1 = \begin{bmatrix} 2 \\ -1 \end{bmatrix}$$

Considering λ_2, we have

$$\begin{bmatrix} -1-3 & 2 \\ 2 & 2-3 \end{bmatrix} \begin{bmatrix} v_1 \\ v_2 \end{bmatrix} = \begin{bmatrix} 0 \\ 0 \end{bmatrix}$$

with the solution

$$\mathbf{v}_2 = \begin{bmatrix} \frac{1}{2} \\ 1 \end{bmatrix}$$

An important observation can be made concerning a change of basis on the matrix **A**. Suppose that **A** represents a linear transformation as discussed in Section 2.4. The change of basis is then carried out using

$$\mathbf{v} = \mathbf{P}\bar{\mathbf{v}} \tag{2.52}$$

to substitute for **v** in the problem $\mathbf{Av} = \lambda\mathbf{v}$. We thus obtain

$$\bar{\mathbf{A}}\bar{\mathbf{v}} = \lambda\bar{\mathbf{v}} \tag{2.53}$$

where

$$\bar{\mathbf{A}} = \mathbf{P}^{-1}\mathbf{AP} \tag{2.54}$$

This transformation is called a *similarity transformation.*

Consider now that **A** is the matrix representation of a variational formulation as discussed in Section 2.5. The change of basis using (2.52) now results in

$$\bar{\mathbf{A}}\bar{\mathbf{v}} = \lambda\mathbf{B}\bar{\mathbf{v}} \tag{2.55}$$

where

$$\bar{\mathbf{A}} = \mathbf{P}^T\mathbf{AP}; \qquad \mathbf{B} = \mathbf{P}^T\mathbf{P} \tag{2.56}$$

The eigenproblem in (2.55) is called a *generalized eigenproblem*. However, since a generalized eigenproblem is more difficult to deal with than a standard problem, the transformation to a generalized problem should be avoided. This is achieved by assuming an orthogonal matrix **P**, which yields $\mathbf{B} = \mathbf{I}$. If **P** is orthogonal, the transformations in (2.53) and (2.55) are referred to as *orthogonal similarity transformations*. We should recall that $\bar{\mathbf{A}} = \mathbf{P}^T\mathbf{AP}$ is a symmetric matrix.

Considering a change of basis, it should be noted that the problems $\bar{\mathbf{A}}\bar{\mathbf{v}} = \lambda\bar{\mathbf{v}}$ in (2.53) and $\bar{\mathbf{A}}\bar{\mathbf{v}} = \lambda\mathbf{B}\bar{\mathbf{v}}$ in (2.55) have the same eigenvalues as the problem $\mathbf{Av} = \lambda\mathbf{v}$, whereas the eigenvectors are related as given in (2.52). To show that the eigenvalues are identical, we consider the characteristic polynomials. For the problem in (2.53) we have

$$\bar{p}(\lambda) = \det(\mathbf{P}^{-1}\mathbf{AP} - \lambda\mathbf{I}) \tag{2.57}$$

which can be written (see Section 1.8)

$$\bar{p}(\lambda) = \det\mathbf{P}^{-1}\det(\mathbf{A} - \lambda\mathbf{I})\det\mathbf{P} \tag{2.58}$$

and hence

$$\bar{p}(\lambda) = \det(\mathbf{A} - \lambda\mathbf{I}) \tag{2.59}$$

Therefore, the characteristic polynomials of the problems $\mathbf{Av} = \lambda\mathbf{v}$ and $\bar{\mathbf{A}}\bar{\mathbf{v}} = \lambda\bar{\mathbf{v}}$ in (2.53) are the same, from which it follows that the eigenvalues are the same.

Considering the problem in (2.55), we have as its characteristic polynomial,

$$\bar{p}(\lambda) = \det(\mathbf{P}^T\mathbf{AP} - \lambda\mathbf{P}^T\mathbf{P}) \tag{2.60}$$

which can be written

$$\bar{p}(\lambda) = \det \mathbf{P}^T \det(\mathbf{A} - \lambda\mathbf{I}) \det \mathbf{P} \tag{2.61}$$

and therefore

$$\bar{p}(\lambda) = \det \mathbf{P}^T \det \mathbf{P}\, p(\lambda) \tag{2.62}$$

where $p(\lambda)$ is given in (2.51). Hence the characteristic polynomials of the problems $\mathbf{Av} = \lambda\mathbf{v}$ and $\bar{\mathbf{A}}\bar{\mathbf{v}} = \lambda\mathbf{B}\bar{\mathbf{v}}$ in (2.55) are the same within a multiplier. This means that the eigenvalues of the two problems are identical.

Example 2.15

Consider the eigenproblem $\mathbf{Av} = \lambda\mathbf{v}$, where $\mathbf{A}$ is the matrix calculated in Example 2.10. Use $L = 6$ and $EA_0 = 1$.

(1) Establish the characteristic polynomial and calculate the eigenvalues and eigenvectors.

(2) Suppose that a change of basis is made using the new basis given in Example 2.11. The eigenproblem is now

$$\mathbf{P}^T\mathbf{AP}\bar{\mathbf{v}} = \lambda\mathbf{P}^T\mathbf{P}\bar{\mathbf{v}} \tag{a}$$

where $\mathbf{P}$ was defined in Example 2.11. Establish the characteristic polynomial and calculate the eigenvalues and eigenvectors.

(3) Assume that a change of basis is performed using an orthogonal matrix $\mathbf{P}$,

$$\mathbf{P} = \begin{bmatrix} \frac{3}{5} & -\frac{4}{5} \\ \frac{4}{5} & \frac{3}{5} \end{bmatrix} \tag{b}$$

Evaluate the characteristic polynomial, the eigenvalues, and eigenvectors of the problem $\mathbf{P}^T\mathbf{AP}\bar{\mathbf{v}} = \lambda\bar{\mathbf{v}}$.

Using $L = 6$ and $EA_0 = 1$ in the matrix $\mathbf{A}$ of Example 2.10, we have

$$\mathbf{A} = \begin{bmatrix} 9 & 60 \\ 60 & 504 \end{bmatrix}$$

Hence

$$p(\lambda) = \lambda^2 - 513\lambda + 936 \tag{c}$$

and

$$\lambda_1 = 1.83; \qquad \lambda_2 = 511.17$$

$$\mathbf{v}_1 = \begin{bmatrix} 0.991 \\ -0.118 \end{bmatrix}; \qquad \mathbf{v}_2 = \begin{bmatrix} 0.118 \\ 0.991 \end{bmatrix} \tag{d}$$

Next we calculate the matrices in (a) and obtain the eigenproblem

$$\begin{bmatrix} 9 & -51 \\ -51 & 393 \end{bmatrix}\bar{\mathbf{v}} = \begin{bmatrix} 1 & 1 \\ 1 & 2 \end{bmatrix}\bar{\mathbf{v}} \tag{e}$$

The characteristic polynomial is

$$p(\lambda) = \det\begin{bmatrix} 9 - \lambda & -51 - \lambda \\ -51 - \lambda & 393 - 2\lambda \end{bmatrix}$$

and evaluated we obtain $p(\lambda)$ in (c).

Hence the eigenvalues of (e) are those given in (d). The eigenvectors are obtained using the fact that

$$(\mathbf{P}^T\mathbf{A}\mathbf{P} - \lambda_i\mathbf{P}^T\mathbf{P})\bar{\mathbf{v}}_i = \mathbf{0} \tag{f}$$

Noting that from (f) the eigenvectors can only be calculated within scalar multiples, we can obtain

$$\bar{\mathbf{v}}_1 = \begin{bmatrix} 0.873 \\ 0.118 \end{bmatrix}; \qquad \bar{\mathbf{v}}_2 = \begin{bmatrix} 1.109 \\ -0.991 \end{bmatrix}$$

and we may check that $\mathbf{v}_i = \mathbf{P}\bar{\mathbf{v}}_i$.

Finally, we use the matrix **P** defined in (b) to perform a change of basis. In this case we obtain the eigenproblem

$$\begin{bmatrix} 383.4 & 220.8 \\ 220.8 & 129.6 \end{bmatrix}\bar{\mathbf{v}} = \lambda\bar{\mathbf{v}}$$

because $\mathbf{P}^T\mathbf{P} = \mathbf{I}$.

This problem has again the same characteristic polynomial as in (c), and we obtain

$$\bar{\mathbf{v}}_1 = \begin{bmatrix} 0.5 \\ -0.864 \end{bmatrix}; \qquad \bar{\mathbf{v}}_2 = \begin{bmatrix} 0.864 \\ 0.5 \end{bmatrix}$$

where we may again check that $\mathbf{v}_i = \mathbf{P}\bar{\mathbf{v}}_i$.

So far we have shown that there are n eigenvalues and corresponding eigenvectors, but we have not yet discussed the properties of the eigenvalues and vectors.

A first observation is that the eigenvalues are real. Consider the ith eigenpair $(\lambda_i, \mathbf{v}_i)$, for which we have

$$\mathbf{A}\mathbf{v}_i = \lambda_i\mathbf{v}_i \tag{2.63}$$

Assume that $\mathbf{v}_i$ and λ_i are complex, which includes the case of real eigenvalues, and let the elements of $\bar{\mathbf{v}}_i$ and $\bar{\lambda}_i$ be the complex conjugates of the elements of $\mathbf{v}_i$ and λ_i. Then premultiplying (2.63) by $\bar{\mathbf{v}}_i^T$, we obtain

$$\bar{\mathbf{v}}_i^T\mathbf{A}\mathbf{v}_i = \lambda_i\bar{\mathbf{v}}_i^T\mathbf{v}_i \tag{2.64}$$

On the other hand, we also obtain from (2.63),

$$\bar{\mathbf{v}}_i^T\mathbf{A} = \bar{\mathbf{v}}_i^T\bar{\lambda}_i \tag{2.65}$$

and postmultiplying by $\mathbf{v}_i$, we have

$$\bar{\mathbf{v}}_i^T \mathbf{A} \mathbf{v}_i = \bar{\lambda}_i \bar{\mathbf{v}}_i^T \mathbf{v}_i \tag{2.66}$$

But the left-hand sides of (2.64) and (2.66) are the same and thus we have

$$(\lambda_i - \bar{\lambda}_i) \bar{\mathbf{v}}_i^T \mathbf{v}_i = 0 \tag{2.67}$$

Since $\mathbf{v}_i$ is nontrivial, it follows that $\lambda_i = \bar{\lambda}_i$, and hence the eigenvalue must be real. However, then it also follows from (2.49) that the eigenvectors can be made real, because the coefficient matrix $\mathbf{A} - \lambda \mathbf{I}$ is real.

Another important point is that the eigenvectors which correspond to distinct eigenvalues are unique (within scalar multipliers) and *orthogonal*, whereas the eigenvectors corresponding to multiple eigenvalues are not unique, but we can always choose an orthogonal set.

Assume first that the eigenvalues are distinct. In this case we have for two eigenpairs,

$$\mathbf{A} \mathbf{v}_i = \lambda_i \mathbf{v}_i \tag{2.68}$$

and

$$\mathbf{A} \mathbf{v}_j = \lambda_j \mathbf{v}_j \tag{2.69}$$

Premultiplying (2.68) by $\mathbf{v}_j^T$ and (2.69) by $\mathbf{v}_i^T$, we obtain

$$\mathbf{v}_j^T \mathbf{A} \mathbf{v}_i = \lambda_i \mathbf{v}_j^T \mathbf{v}_i \tag{2.70}$$

$$\mathbf{v}_i^T \mathbf{A} \mathbf{v}_j = \lambda_j \mathbf{v}_i^T \mathbf{v}_j \tag{2.71}$$

Taking the transpose in (2.71), we have

$$\mathbf{v}_j^T \mathbf{A} \mathbf{v}_i = \lambda_j \mathbf{v}_j^T \mathbf{v}_i \tag{2.72}$$

and thus from (2.72) and (2.70) we obtain

$$(\lambda_i - \lambda_j) \mathbf{v}_j^T \mathbf{v}_i = 0 \tag{2.73}$$

Since we assumed that $\lambda_i \neq \lambda_j$, it follows that $\mathbf{v}_j^T \mathbf{v}_i = 0$, i.e., that $\mathbf{v}_j$ and $\mathbf{v}_i$ are orthogonal. Furthermore, we can scale the elements of the vector $\mathbf{v}_i$ to obtain that

$$\mathbf{v}_i^T \mathbf{v}_j = \delta_{ij} \tag{2.74}$$

where δ_{ij} = Kronecker delta; i.e., $\delta_{ij} = 1$ when $i = j$, and $\delta_{ij} = 0$ when $i \neq j$. If (2.74) is satisfied, we say that the eigenvectors are *orthonormal*.

It should be noted that the solution of (2.49) yields a vector in which only the relative magnitudes of the elements are defined. If all elements are scaled by the same amount, the new vector would still satisfy (2.49). In effect, the solution of (2.49) yields the direction of the eigenvector, and we use the orthonormality condition in (2.74) to fix the magnitudes of the elements in the vector. Therefore, *when we refer to eigenvectors it is from now on implied that the vectors are orthonormal.*

Example 2.16

Check that the vectors calculated in Example 2.14 are orthogonal, and then orthonormalize them.

The orthogonality is checked by forming $\mathbf{v}_1^T\mathbf{v}_2$, which gives

$$\mathbf{v}_1^T\mathbf{v}_2 = (-2)(\tfrac{1}{2}) + (1)(1) = 0$$

Hence the vectors are orthogonal. To orthonormalize the vectors, we need to make the length of the vectors equal to 1. Then we have

$$\mathbf{v}_1 = \frac{1}{\sqrt{5}}\begin{bmatrix} -2 \\ 1 \end{bmatrix}; \qquad \mathbf{v}_2 = \frac{1}{\sqrt{5}}\begin{bmatrix} 1 \\ 2 \end{bmatrix}$$

We now turn to the case in which multiple eigenvalues are also present. The proof of eigenvector orthonormality given in (2.68) to (2.74) is not possible because for a multiple eigenvalue, λ_i is equal to λ_j in (2.73). Assume that $\lambda_i = \lambda_{i+1} = \ldots = \lambda_{i+m-1}$; i.e., λ_i is an m-times multiple root. Then we can show that it is still always possible to choose m orthonormal eigenvectors that correspond to $\lambda_i, \lambda_{i+1}, \ldots, \lambda_{i+m-1}$. This follows because for a symmetric matrix, we can always establish a complete set of n orthonormal eigenvectors. Corresponding to each distinct eigenvalue we have an *eigenspace* with dimension equal to the multiplicity of the eigenvalue. All eigenspaces are unique and are orthogonal to the eigenspaces that correspond to other distinct eigenvalues. The eigenvectors associated with an eigenvalue provide a basis for the eigenspace, and since the basis is not unique if $m > 1$, the eigenvectors corresponding to a multiple eigenvalue are not unique. The formal proofs of the above statements are an application of the principles discussed earlier and are given in the following examples.

Example 2.17

Show that for a symmetric matrix $\mathbf{A}$ of order n, there are always n orthonormal eigenvectors.

Assume that we have calculated an eigenvalue λ_i and corresponding eigenvector $\mathbf{v}_i$. Let us construct an orthonormal matrix $\mathbf{Q}$ whose first column is $\mathbf{v}_i$,

$$\mathbf{Q} = [\mathbf{v}_i \quad \hat{\mathbf{Q}}]; \qquad \mathbf{Q}^T\mathbf{Q} = \mathbf{I}$$

This matrix can always be constructed because the vectors in $\mathbf{Q}$ provide an orthonormal basis for the n-dimensional space in which $\mathbf{A}$ is defined. However, we can now calculate

$$\mathbf{Q}^T\mathbf{A}\mathbf{Q} = \begin{bmatrix} \lambda_i & \mathbf{0} \\ \mathbf{0} & \mathbf{A}_1 \end{bmatrix} \tag{a}$$

where

$$\mathbf{A}_1 = \hat{\mathbf{Q}}^T\mathbf{A}\hat{\mathbf{Q}}$$

and $\mathbf{A}_1$ is a full matrix of order $(n - 1)$. If $n = 2$, we note that $\mathbf{Q}^T\mathbf{A}\mathbf{Q}$ is diagonal. In that case, if we premultiply (a) by $\mathbf{Q}$ and let $a \equiv \mathbf{A}_1$ we obtain

$$\mathbf{A}\mathbf{Q} = \mathbf{Q}\begin{bmatrix} \lambda_i & 0 \\ 0 & a \end{bmatrix}$$

and hence the vector in $\hat{\mathbf{Q}}$ is the other eigenvector, and a is the other eigenvalue regardless of whether λ_i is a multiple eigenvalue or not.

The complete proof is now obtained by induction. Assume that the statement is true for a matrix of order $(n - 1)$; then we will show that it is also true for a matrix of order n. But since we demonstrated that the statement is true for $n = 2$, it follows that it is true for any n.

The assumption that there are n orthonormal eigenvectors for a matrix of order $(n - 1)$ gives

$$\mathbf{Q}_1^T \mathbf{A}_1 \mathbf{Q}_1 = \mathbf{\Lambda} \tag{b}$$

where $\mathbf{Q}_1$ is a matrix of the eigenvectors of $\mathbf{A}_1$ and $\mathbf{\Lambda}$ is a diagonal matrix listing the eigenvalues of $\mathbf{A}_1$. However, if we now define

$$\mathbf{S} = \begin{bmatrix} 1 & \mathbf{0} \\ \mathbf{0} & \mathbf{Q}_1 \end{bmatrix}$$

we have

$$\mathbf{S}^T \mathbf{Q}^T \mathbf{A} \mathbf{Q} \mathbf{S} = \begin{bmatrix} \lambda_i & \\ & \mathbf{\Lambda} \end{bmatrix} \tag{c}$$

Let

$$\mathbf{P} = \mathbf{Q}\mathbf{S}; \qquad \mathbf{P}^T\mathbf{P} = \mathbf{I}$$

Then premultiplying (c) by $\mathbf{P}$ we obtain

$$\mathbf{A}\mathbf{P} = \mathbf{P}\begin{bmatrix} \lambda_i & \\ & \mathbf{\Lambda} \end{bmatrix}$$

Therefore, under the assumption in (b), the statement is also true for a matrix of order n, which completes the proof.

Example 2.18

Show that the eigenvectors corresponding to a multiple eigenvalue of multiplicity m define an m-dimensional space in which each vector is also an eigenvector. This space is called the *eigenspace* corresponding to the eigenvalue considered.

Let λ_i be the eigenvalue of multiplicity m, i.e., we have

$$\lambda_i = \lambda_{i+1} = \ldots = \lambda_{i+m-1}$$

We showed in Example 2.17 that there are m orthonormal eigenvectors $\mathbf{v}_i$, $\mathbf{v}_{i+1}, \ldots, \mathbf{v}_{i+m-1}$ corresponding to λ_i. These vectors provide the basis of an m-dimensional space. Consider any vector $\mathbf{w}$ in this space, such as

$$\mathbf{w} = \alpha_i \mathbf{v}_i + \alpha_{i+1}\mathbf{v}_{i+1} + \cdots + \alpha_{i+m-1}\mathbf{v}_{i+m-1}$$

where the $\alpha_i, \alpha_{i+1}, \ldots$ are constants. The vector $\mathbf{w}$ is also an eigenvector, because we have

$$\mathbf{A}\mathbf{w} = \alpha_i \mathbf{A}\mathbf{v}_i + \alpha_{i+1}\mathbf{A}\mathbf{v}_{i+1} + \ldots + \alpha_{i+m-1}\mathbf{A}\mathbf{v}_{i+m-1}$$

which gives

$$\mathbf{A}\mathbf{w} = \alpha_i \lambda_i \mathbf{v}_i + \alpha_{i+1}\lambda_i \mathbf{v}_{i+1} + \ldots + \alpha_{i+m-1}\lambda_i \mathbf{v}_{i+m-1} = \lambda_i \mathbf{w}$$

Therefore, any vector $\mathbf{w}$ in the space spanned by the m eigenvectors $\mathbf{v}_i, \mathbf{v}_{i+1}, \ldots, \mathbf{v}_{i+m-1}$ is also an eigenvector. It should be noted that the vector $\mathbf{w}$ will be orthogonal to the eigenvectors that correspond to eigenvalues not equal to λ_i. Hence there is one eigenspace that corresponds to each distinct or multiple eigenvalue. The dimension of the eigenspace is equal to the multiplicity of the eigenvalue.

With the main properties of the eigenvalues and eigenvectors of **A** having been presented, we can now write the n solutions to $\mathbf{Av} = \lambda\mathbf{v}$ in various forms. First, we have

$$\mathbf{AV} = \mathbf{V\Lambda} \tag{2.75}$$

where $\mathbf{V}$ is a matrix storing the eigenvectors, $\mathbf{V} = [\mathbf{v}_1, \ldots, \mathbf{v}_n]$, and $\mathbf{\Lambda}$ is a diagonal matrix with the corresponding eigenvalues on its diagonal, $\mathbf{\Lambda} = \text{diag}(\lambda_i)$. Using the orthonormality property of the eigenvectors (i.e., $\mathbf{V}^T\mathbf{V} = \mathbf{I}$), we obtain from (2.75) that

$$\mathbf{V}^T\mathbf{AV} = \mathbf{\Lambda} \tag{2.76}$$

Furthermore, we obtain the *spectral decomposition* of **A**,

$$\mathbf{A} = \mathbf{V\Lambda V}^T \tag{2.77}$$

where it may be convenient to write the spectral decomposition of **A** as

$$\mathbf{A} = \sum_{i=1}^{n} \lambda_i \mathbf{v}_i \mathbf{v}_i^T \tag{2.78}$$

It should be noted that each of the above equations represents the solution to the eigenproblem $\mathbf{Av} = \lambda\mathbf{v}$. Consider the following example.

Example 2.19

Establish the relations given in (2.75) to (2.78) for the matrix **A** used in Example 2.14.

The eigenvalues and eigenvectors of **A** have been calculated in Examples 2.14 and 2.16. Using the information given in those examples, we have, for (2.75),

$$\begin{bmatrix} -1 & 2 \\ 2 & 2 \end{bmatrix} \begin{bmatrix} -\frac{2}{\sqrt{5}} & \frac{1}{\sqrt{5}} \\ \frac{1}{\sqrt{5}} & \frac{2}{\sqrt{5}} \end{bmatrix} = \begin{bmatrix} -\frac{2}{\sqrt{5}} & \frac{1}{\sqrt{5}} \\ \frac{1}{\sqrt{5}} & \frac{2}{\sqrt{5}} \end{bmatrix} \begin{bmatrix} -2 & 0 \\ 0 & 3 \end{bmatrix}$$

for (2.76),

$$\begin{bmatrix} -\frac{2}{\sqrt{5}} & \frac{1}{\sqrt{5}} \\ \frac{1}{\sqrt{5}} & \frac{2}{\sqrt{5}} \end{bmatrix} \begin{bmatrix} -1 & 2 \\ 2 & 2 \end{bmatrix} \begin{bmatrix} -\frac{2}{\sqrt{5}} & \frac{1}{\sqrt{5}} \\ \frac{1}{\sqrt{5}} & \frac{2}{\sqrt{5}} \end{bmatrix} = \begin{bmatrix} -2 & 0 \\ 0 & 3 \end{bmatrix}$$

for (2.77),

$$\begin{bmatrix} -1 & 2 \\ 2 & 2 \end{bmatrix} = \begin{bmatrix} -\frac{2}{\sqrt{5}} & \frac{1}{\sqrt{5}} \\ \frac{1}{\sqrt{5}} & \frac{2}{\sqrt{5}} \end{bmatrix} \begin{bmatrix} -2 & 0 \\ 0 & 3 \end{bmatrix} \begin{bmatrix} -\frac{2}{\sqrt{5}} & \frac{1}{\sqrt{5}} \\ \frac{1}{\sqrt{5}} & \frac{2}{\sqrt{5}} \end{bmatrix}$$

and for (2.78),

$$\mathbf{A} = (-2) \begin{bmatrix} -\frac{2}{\sqrt{5}} \\ \frac{1}{\sqrt{5}} \end{bmatrix} \begin{bmatrix} -\frac{2}{\sqrt{5}} & \frac{1}{\sqrt{5}} \end{bmatrix} + (3) \begin{bmatrix} \frac{1}{\sqrt{5}} \\ \frac{2}{\sqrt{5}} \end{bmatrix} \begin{bmatrix} \frac{2}{\sqrt{5}} & \frac{2}{\sqrt{5}} \end{bmatrix}$$

The relations in (2.76) and (2.77) can be employed effectively in various important applications. The objective in the following examples is to present some solution procedures in which they are used.

Example 2.20

Calculate the kth power of a given matrix $\mathbf{A}$; i.e., evaluate $\mathbf{A}^k$. Demonstrate the result using $\mathbf{A}$ of Example 2.14.

One way of evaluating $\mathbf{A}^k$ is to simply calculate $\mathbf{A}^2 = \mathbf{AA}$, $\mathbf{A}^4 = \mathbf{A}^2\mathbf{A}^2$, etc., as we have done in Example 1.6. However, if k is large, it may be more effective to employ the spectral decomposition of $\mathbf{A}$. Assume that we have calculated the eigenvalues and eigenvectors of $\mathbf{A}$; i.e., we have

$$\mathbf{A} = \mathbf{V}\mathbf{\Lambda}\mathbf{V}^T$$

To calculate $\mathbf{A}^2$, we use

$$\mathbf{A}^2 = \mathbf{V}\mathbf{\Lambda}\mathbf{V}^T\mathbf{V}\mathbf{\Lambda}\mathbf{V}^T$$

but because $\mathbf{V}^T\mathbf{V} = \mathbf{I}$, we have

$$\mathbf{A}^2 = \mathbf{V}\mathbf{\Lambda}^2\mathbf{V}^T$$

Proceeding in the same manner, we thus obtain

$$\mathbf{A}^k = \mathbf{V}\mathbf{\Lambda}^k\mathbf{V}^T$$

As an example, let $\mathbf{A}$ be the matrix considered in Example 2.14. Then we have

$$\mathbf{A}^k = \frac{1}{\sqrt{5}}\begin{bmatrix} -2 & 1 \\ 1 & 2 \end{bmatrix}\begin{bmatrix} -2^k & 0 \\ 0 & 3^k \end{bmatrix}\frac{1}{\sqrt{5}}\begin{bmatrix} -2 & 1 \\ 1 & 2 \end{bmatrix}$$

or

$$\mathbf{A}^k = \frac{1}{5}\left[\begin{array}{c|c} -2^{k+2} + \quad 3^k & 2^{k+1} + (2)(3)^k \\ \hline 2^{k+1} + (2)(3)^k & -2^k + (4)(3)^k \end{array}\right]$$

It is interesting to note that if the largest absolute value of all the eigenvalues of $\mathbf{A}$ is smaller than 1, we have that $\mathbf{A}^k \longrightarrow 0$ as $k \longrightarrow \infty$. Thus defining

$$\rho(\mathbf{A}) = \max_{\text{all } i} |\lambda_i|$$

we have that $\lim_{k\to\infty} \mathbf{A}^k = 0$, provided that $\rho(\mathbf{A}) < 1$.

Example 2.21

Consider the system of differential equations

$$\dot{\mathbf{x}} + \mathbf{A}\mathbf{x} = \mathbf{f}(t) \tag{a}$$

and obtain the solution using the spectral decomposition of $\mathbf{A}$. Demonstrate the result using the matrix $\mathbf{A}$ of Example 2.14 and

$$\mathbf{f}(t) = \begin{bmatrix} e^{-t} \\ 0 \end{bmatrix}$$

Substituting $\mathbf{A} = \mathbf{V}\mathbf{\Lambda}\mathbf{V}^T$ and premultiplying by $\mathbf{V}^T$, we obtain

$$\mathbf{V}^T\dot{\mathbf{x}} + \mathbf{\Lambda}(\mathbf{V}^T\mathbf{x}) = \mathbf{V}^T\mathbf{f}(t)$$

Thus if we define $\mathbf{y} = \mathbf{V}^T\mathbf{x}$, we need to solve the equations

$$\dot{\mathbf{y}} + \mathbf{\Lambda y} = \mathbf{V}^T\mathbf{f}(t)$$

But this is a set of n decoupled differential equations. Consider the rth equation, which is typical:

$$\dot{y}_r + \lambda_r y_r = \mathbf{v}_r^T\mathbf{f}(t)$$

The solution is

$$y_r(t) = y_{r0}e^{-\lambda_r t} + e^{-\lambda_r t}\int_0^t e^{\lambda_r \tau}\mathbf{v}_r^T\mathbf{f}(\tau)\,d\tau$$

where y_{r0} is the value of y_r at time $t = 0$. The complete solution to the system of equations in (a) is

$$\mathbf{x} = \sum_{r=1}^{n} \mathbf{v}_r y_r \tag{b}$$

As an example, we consider the system of differential equations

$$\begin{bmatrix} \dot{x}_1 \\ \dot{x}_2 \end{bmatrix} + \begin{bmatrix} -1 & 2 \\ 2 & 2 \end{bmatrix}\begin{bmatrix} x_1 \\ x_2 \end{bmatrix} = \begin{bmatrix} e^{-t} \\ 0 \end{bmatrix}$$

with initial conditions

$$\mathbf{x}_0 = \begin{pmatrix} 1 \\ 1 \end{pmatrix}$$

In this case we have to solve the two decoupled differential equations

$$\dot{y}_1 + (-2)y_1 = 2e^{-t}$$

$$\dot{y}_2 + 3y_2 = e^{-t}$$

with initial conditions

$$\mathbf{y}_0 = \mathbf{V}^T\mathbf{x}_0 = \frac{1}{\sqrt{5}}\begin{bmatrix} 2 & -1 \\ 1 & 2 \end{bmatrix}\begin{bmatrix} 1 \\ 1 \end{bmatrix} = \frac{1}{\sqrt{5}}\begin{bmatrix} 1 \\ 3 \end{bmatrix}$$

We obtain

$$y_1 = \frac{1}{\sqrt{5}}e^{2t} - \frac{2}{3}e^{-t}$$

$$y_2 = \frac{3}{\sqrt{5}}e^{-3t} + \frac{1}{2}e^{-t}$$

Thus using (b), we have

$$\begin{bmatrix} x_1 \\ x_2 \end{bmatrix} = \frac{1}{\sqrt{5}}\left(\begin{bmatrix} 2 \\ -1 \end{bmatrix}y_1 + \begin{bmatrix} 1 \\ 2 \end{bmatrix}y_2\right)$$

$$= \begin{bmatrix} -\dfrac{\sqrt{5}}{6}e^{-t} + \dfrac{3}{5}e^{-3t} + \dfrac{2}{5}e^{2t} \\ \dfrac{\sqrt{5}}{3}e^{-t} + \dfrac{6}{5}e^{-3t} - \dfrac{1}{5}e^{2t} \end{bmatrix}$$

To conclude the presentation, we may note that by introducing auxiliary variables, higher-order differential equations can be reduced to a system of first-order differential equations. However, the coefficient matrix **A** is in that case nonsymmetric.

Example 2.22

Using the spectral decomposition of an $n \times n$ symmetric matrix $\mathbf{A}$, evaluate the inverse of the matrix. Demonstrate the result using the matrix $\mathbf{A}$ of Example 2.14.

Assume that we have evaluated the eigenvalues λ_i and corresponding eigenvectors $\mathbf{v}_i$, $i = 1, \ldots, n$, of the matrix $\mathbf{A}$; i.e., we have solved the eigenproblem

$$\mathbf{A}\mathbf{v} = \lambda\mathbf{v} \tag{a}$$

Premultiplying both sides of the relation in (a) by $\lambda^{-1}\mathbf{A}^{-1}$, we obtain the eigenproblem

$$\mathbf{A}^{-1}\mathbf{v} = \lambda^{-1}\mathbf{v}$$

But this relation shows that the eigenvalues of $\mathbf{A}^{-1}$ are $1/\lambda_i$ and the eigenvectors are $\mathbf{v}_i$, $i = 1, \ldots, n$. Thus using (2.78) for $\mathbf{A}^{-1}$, we have

$$\mathbf{A}^{-1} = \mathbf{V}\mathbf{\Lambda}^{-1}\mathbf{V}^T$$

or

$$\mathbf{A}^{-1} = \sum_{i=1}^{n} \left(\frac{1}{\lambda_i}\right)\mathbf{v}_i\mathbf{v}_i^T$$

These equations show that we cannot find the inverse of $\mathbf{A}$ if the matrix has a zero eigenvalue.

As an example we evaluate the inverse of the matrix $\mathbf{A}$ considered in Example 2.14. In this case we have

$$\mathbf{A}^{-1} = \frac{1}{5}\begin{bmatrix} 2 & 1 \\ -1 & 2 \end{bmatrix}\begin{bmatrix} -\frac{1}{2} & 0 \\ 0 & \frac{1}{3} \end{bmatrix}\begin{bmatrix} 2 & -1 \\ 1 & 2 \end{bmatrix} = \frac{1}{6}\begin{bmatrix} -2 & 2 \\ 2 & 1 \end{bmatrix}$$

Let us now return to the discussion with which we started this section. Comparing (2.76) with (2.55), we find that we perform in (2.76) a change of basis. The new matrix representation of the operator considered is the diagonal matrix $\mathbf{\Lambda}$. Since the vectors in $\mathbf{V}$ are a new basis, they span the n-dimensional space in which $\mathbf{A}$ and $\mathbf{\Lambda}$ are defined, and any vector $\mathbf{w}$ can be expressed as a linear combination of the eigenvectors $\mathbf{v}_i$; i.e., we have

$$\mathbf{w} = \sum_{i=1}^{n} \alpha_i\mathbf{v}_i \tag{2.79}$$

An important observation is that $\mathbf{\Lambda}$ shows directly whether the operator that $\mathbf{\Lambda}$ represents is singular (see Section 2.3), i.e., whether the matrices $\mathbf{A}$ and $\mathbf{\Lambda}$ are singular. Using the definition given in Section 2.3, we find that $\mathbf{\Lambda}$ and hence $\mathbf{A}$ are singular if and only if an eigenvalue is equal to zero, because in that case $\mathbf{\Lambda}^{-1}$ cannot be calculated. In this context it is useful to define some additional terminology. If all eigenvalues are positive, we say that the matrix and the operator that the matrix represents are *positive definite*. If all eigenvalues are greater than or equal to zero, the matrix is *positive semidefinite;* with negative, zero, or positive eigenvalues, the matrix is *indefinite.*

2.7 THE RAYLEIGH QUOTIENT AND THE MINIMAX CHARACTERIZATION OF EIGENVALUES

In the previous section we defined the eigenproblem $\mathbf{A}\mathbf{v} = \lambda\mathbf{v}$ and discussed the basic properties that pertain to the solutions of the problem. The objective in this section is to complement the information given with some very powerful principles.

A number of important principles are derived using the Rayleigh quotient $\rho(\mathbf{v})$, which is defined as

$$\rho(\mathbf{v}) = \frac{\mathbf{v}^T\mathbf{A}\mathbf{v}}{\mathbf{v}^T\mathbf{v}} \tag{2.80}$$

The first observation is that

$$\lambda_1 \leq \rho(\mathbf{v}) \leq \lambda_n \tag{2.81}$$

and it follows that using the definitions given in Section 2.6, we have for any vector $\mathbf{v}$, if $\mathbf{A}$ is positive definite $\rho(\mathbf{v}) > 0$, if $\mathbf{A}$ is positive semidefinite $\rho(\mathbf{v}) \geq 0$, and for $\mathbf{A}$ indefinite $\rho(\mathbf{v})$ can be negative, zero, or positive. For the proof of (2.81) we use that

$$\mathbf{v} = \sum_{i=1}^{n} \alpha_i \mathbf{v}_i \tag{2.82}$$

where $\mathbf{v}_i$ are the eigenvectors of $\mathbf{A}$. Substituting for $\mathbf{v}$ into (2.80) and using that $\mathbf{A}\mathbf{v}_i = \lambda_i\mathbf{v}_i$, $\mathbf{v}_i^T\mathbf{v}_j = \delta_{ij}$, we obtain

$$\rho(\mathbf{v}) = \frac{\lambda_1\alpha_1^2 + \lambda_2\alpha_2^2 + \ldots + \lambda_n\alpha_n^2}{\alpha_1^2 + \ldots + \alpha_n^2} \tag{2.83}$$

Hence, if $\lambda_1 \neq 0$,

$$\rho(\mathbf{v}) = \lambda_1 \frac{\alpha_1^2 + (\lambda_2/\lambda_1)\alpha_2^2 + \ldots + (\lambda_n/\lambda_1)\alpha_n^2}{\alpha_1^2 + \ldots + \alpha_n^2} \tag{2.84}$$

and if $\lambda_n \neq 0$,

$$\rho(\mathbf{v}) = \lambda_n \frac{(\lambda_1/\lambda_n)\alpha_1^2 + (\lambda_2/\lambda_n)\alpha_2^2 + \ldots + \alpha_n^2}{\alpha_1^2 + \ldots + \alpha_n^2} \tag{2.85}$$

But since $\lambda_1 \leq \lambda_2 \leq \ldots \leq \lambda_n$, the relations in (2.83) to (2.85) show that (2.81) holds. Furthermore, it is seen that if $\mathbf{v} = \mathbf{v}_i$, we have $\rho(\mathbf{v}) = \lambda_i$.

Considering the practical use of the Rayleigh quotient, the following property is of particular value. Assume that $\mathbf{v}$ is an approximation to the eigenvector $\mathbf{v}_i$, say

$$\mathbf{v} = \mathbf{v}_i + \epsilon\mathbf{x} \tag{2.86}$$

Then the Rayleigh quotient of $\mathbf{v}$ will give an approximation to λ_i of order ϵ^2; i.e.,

$$\rho(\mathbf{v}) = \lambda_i + o(\epsilon^2) \tag{2.87}$$

The notation $o(\epsilon^2)$ means "of order ϵ^2" and indicates that if $\delta = o(\epsilon^2)$, then $|\delta| \leq b\epsilon^2$, where b is a constant.

To prove the above property of the Rayleigh quotient, we substitute for $\mathbf{v}$ from (2.82) into the Rayleigh quotient expression to obtain

$$\rho(\mathbf{v}_i + \epsilon\mathbf{x}) = \frac{(\mathbf{v}_i^T + \epsilon\mathbf{x}^T)\mathbf{A}(\mathbf{v}_i + \epsilon\mathbf{x})}{(\mathbf{v}_i^T + \epsilon\mathbf{x}^T)(\mathbf{v}_i + \epsilon\mathbf{x})} \tag{2.88}$$

or

$$\rho(\mathbf{v}_i + \epsilon\mathbf{x}) = \frac{\mathbf{v}_i^T\mathbf{A}\mathbf{v}_i + 2\epsilon\mathbf{v}_i^T\mathbf{A}\mathbf{x} + \epsilon^2\mathbf{x}^T\mathbf{A}\mathbf{x}}{\mathbf{v}_i^T\mathbf{v}_i + 2\epsilon\mathbf{x}^T\mathbf{v}_i + \epsilon^2\mathbf{x}^T\mathbf{x}} \tag{2.89}$$

However, since $\mathbf{x}$ is an error in $\mathbf{v}_i$, we can write

$$\mathbf{x} = \sum_{\substack{j=1\\j\neq i}}^{n} \alpha_j \mathbf{v}_j \tag{2.90}$$

But then using that $\mathbf{v}_i^T\mathbf{v}_j = \delta_{ij}$ and $\mathbf{A}\mathbf{v}_j = \lambda_j\mathbf{v}_j$, we have that $\mathbf{v}_i^T\mathbf{A}\mathbf{x} = 0$ and $\mathbf{x}^T\mathbf{v}_i = 0$, and hence

$$\rho(\mathbf{v}_i + \epsilon\mathbf{x}) = \frac{\lambda_i + \epsilon^2 \sum\limits_{\substack{j=1\\j\neq i}}^{n} \alpha_j^2\lambda_j}{1 + \epsilon^2 \sum\limits_{\substack{j=1\\j\neq i}}^{n} \alpha_j^2} \tag{2.91}$$

However, using the binomial theorem to expand the denominator in (2.91), we have

$$\rho(\mathbf{v}_i + \epsilon\mathbf{x}) = \left(\lambda_i + \epsilon^2 \sum_{\substack{j=1\\j\neq i}}^{n} \alpha_j^2\lambda_j\right)\left[1 - \epsilon^2\left(\sum_{\substack{j=1\\j\neq i}}^{n} \alpha_j^2\right) + \epsilon^4\left(\sum_{\substack{j=1\\j\neq i}}^{n} \alpha_j^2\right)^2 + \cdots\right] \tag{2.92}$$

or

$$\rho(\mathbf{v}_i + \epsilon\mathbf{x}) = \lambda_i + \epsilon^2\left(\sum_{\substack{j=1\\j\neq i}}^{n} \alpha_j^2\lambda_j - \lambda_i \sum_{\substack{j=1\\j\neq i}}^{n} \alpha_j^2\right) + \text{higher-order terms} \tag{2.93}$$

The relation in (2.87) thus follows. We demonstrate the above results in a small example.

Example 2.23

Evaluate the Rayleigh quotients $\rho(\mathbf{v})$ for the matrix $\mathbf{A}$ used in Example 2.14. Using $\mathbf{v}_1$ and $\mathbf{v}_2$ of Example 2.14, consider the following cases:

(1) $\mathbf{v} = \mathbf{v}_1 + 2\mathbf{v}_2$; (2) $\mathbf{v} = \mathbf{v}_1$; (3) $\mathbf{v} = \mathbf{v}_1 + 0.02\mathbf{v}_2$.

In case (1) we have

$$\mathbf{v} = \left[\begin{bmatrix} 2 \\ -1 \end{bmatrix} + \begin{bmatrix} 1 \\ 2 \end{bmatrix}\right] = \begin{bmatrix} 3 \\ 1 \end{bmatrix}$$

and thus

$$\rho(\mathbf{v}) = \frac{[3 \quad 1]\begin{bmatrix} -1 & 2 \\ 2 & 2 \end{bmatrix}\begin{bmatrix} 3 \\ 1 \end{bmatrix}}{[3 \quad 1]\begin{bmatrix} 3 \\ 1 \end{bmatrix}} = \frac{1}{2}$$

Recalling that $\lambda_1 = -2$ and $\lambda_2 = 3$, we have, as expected,

$$\lambda_1 \leq \rho(\mathbf{v}) \leq \lambda_2$$

In case (2) we have

$$\mathbf{v} = \begin{bmatrix} 2 \\ -1 \end{bmatrix}$$

and hence

$$\rho(\mathbf{v}) = \frac{[2 \quad -1]\begin{bmatrix} -1 & 2 \\ 2 & 2 \end{bmatrix}\begin{bmatrix} 2 \\ -1 \end{bmatrix}}{[2 \quad -1]\begin{bmatrix} 2 \\ -1 \end{bmatrix}} = -2$$

and hence, as expected, $\rho(\mathbf{v}) = \lambda_1$.

Finally, in case (3) we use

$$\mathbf{v} = \begin{bmatrix} 2 \\ -1 \end{bmatrix} + \begin{bmatrix} 0.01 \\ 0.02 \end{bmatrix} = \begin{bmatrix} 2.01 \\ -0.98 \end{bmatrix}$$

and hence

$$\rho(\mathbf{v}) = \frac{[2.01 \quad -0.98]\begin{bmatrix} -1 & 2 \\ 2 & 2 \end{bmatrix}\begin{bmatrix} 2.01 \\ -0.98 \end{bmatrix}}{[2.01 \quad -0.98]\begin{bmatrix} 2.01 \\ -0.98 \end{bmatrix}}$$

$$= -1.99950005$$

Here, we note that $\rho(\mathbf{v}) > \lambda_1$ and that $\rho(\mathbf{v})$ approximates λ_1 more closely than $\mathbf{v}$ approximates $\mathbf{v}_1$.

Having introduced the Rayleigh quotient, we can now proceed to a very important principle, the minimax characterization of eigenvalues. We know from Rayleigh's principle that

$$\rho(\mathbf{v}) \geq \lambda_1 \tag{2.94}$$

where $\mathbf{v}$ is any vector. In other words, if we consider the problem of varying $\mathbf{v}$, we will always have $\rho(\mathbf{v}) \geq \lambda_1$, and the minimum is reached when $\mathbf{v} = \mathbf{v}_1$, in which case $\rho(\mathbf{v}_1) = \lambda_1$. Suppose that we now impose a restriction on $\mathbf{v}$, namely, that $\mathbf{v}$ be orthogonal to a specific vector $\mathbf{w}$, and that we consider the problem of minimizing $\rho(\mathbf{v})$ subject to this restriction. After having calculated the minimum of $\rho(\mathbf{v})$ with the condition $\mathbf{v}^T\mathbf{w} = 0$, we could start varying $\mathbf{w}$ and for each new $\mathbf{w}$ evaluate a new minimum of $\rho(\mathbf{v})$. We would find that the maximum value of all the minimum values evaluated is λ_2. This result can be generalized to the following principle, called the *minimax characterization of eigenvalues*,

$$\lambda_r = \max\left\{\min \frac{\mathbf{v}^T\mathbf{A}\mathbf{v}}{\mathbf{v}^T\mathbf{v}}\right\} \qquad r = 1, \ldots, n \tag{2.95}$$

with $\mathbf{v}$ satisfying $\mathbf{v}^T\mathbf{w}_i = 0$ for $i = 1, \ldots, r-1$. In (2.95) we choose vectors $\mathbf{w}_i$, $i = 1, \ldots, r-1$, and then evaluate the minimum of $\rho(\mathbf{v})$ with $\mathbf{v}$ subject to the condition $\mathbf{v}^T\mathbf{w}_i = 0$, $i = 1, \ldots, r-1$. After having calculated this minimum we vary the vectors $\mathbf{w}_i$ and always evaluate a new minimum. The maximum value that the minima reach is λ_r.

The proof of (2.95) is as follows. Let

$$\mathbf{v} = \sum_{i=1}^{n} \alpha_i \mathbf{v}_i \tag{2.96}$$

and evaluate the right-hand side of (2.95), which we call R,

$$R = \max \left\{ \min \frac{\alpha_1^2 \lambda_1 + \ldots + \alpha_r^2 \lambda_r + \alpha_{r+1}^2 \lambda_{r+1} + \ldots + \alpha_n^2 \lambda_n}{\alpha_1^2 + \ldots + \alpha_r^2 + \alpha_{r+1}^2 + \ldots + \alpha_n^2} \right\} \tag{2.97}$$

The coefficients α_i need to satisfy the conditions

$$\mathbf{w}_j^T \sum_{i=1}^{n} \alpha_i \mathbf{v}_i = 0 \qquad j = 1, \ldots, r-1 \tag{2.98}$$

Rewriting (2.97) we obtain

$$R = \max \left\{ \min \left[\lambda_r - \frac{\alpha_1^2(\lambda_r - \lambda_1) + \ldots + \alpha_{r-1}^2(\lambda_r - \lambda_{r-1}) + \alpha_{r+1}^2(\lambda_r - \lambda_{r+1}) + \ldots + \alpha_n^2(\lambda_r - \lambda_n)}{\alpha_1^2 + \ldots + \alpha_r^2 + \alpha_{r+1}^2 + \ldots + \alpha_n^2} \right] \right\} \tag{2.99}$$

But we can now see that for the condition $\alpha_{r+1} = \alpha_{r+2} = \ldots = \alpha_n = 0$, we have

$$R \leq \lambda_r \tag{2.100}$$

and the condition in (2.98) can still be satisfied by a judicious choice for α_r. On the other hand, suppose that we now choose $\mathbf{w}_j = \mathbf{v}_j$ for $j = 1, \ldots, r - 1$. This would require that $\alpha_j = 0$ for $j = 1, \ldots, r-1$, and consequently we would have that $R = \lambda_r$, which completes the proof.

A most important property that can be established using the minimax characterization of eigenvalues is the *eigenvalue separation property*. Suppose that in addition to the problem $\mathbf{A}\mathbf{v} = \lambda\mathbf{v}$, we consider the problems

$$\mathbf{A}^{(m)}\mathbf{v}^{(m)} = \lambda^{(m)}\mathbf{v}^{(m)} \tag{2.101}$$

where $\mathbf{A}^{(m)}$ is obtained by omitting the last m rows and columns of $\mathbf{A}$. Hence $\mathbf{A}^{(m)}$ is a square-symmetric matrix of order $(n - m)$. Using also the notation $\mathbf{A}^{(0)} = \mathbf{A}$, $\lambda^{(0)} = \lambda$, $\mathbf{v}^{(0)} = \mathbf{v}$, the eigenvalue separation property states that the eigenvalues of the problem $\mathbf{A}^{(m+1)}\mathbf{v}^{(m+1)} = \lambda^{(m+1)}\mathbf{v}^{(m+1)}$ separate the eigenvalues of the problem in (2.101); i.e., we have

$$\lambda_1^{(m)} \leq \lambda_1^{(m+1)} \leq \lambda_2^{(m)} \leq \lambda_2^{(m+1)} \ldots \leq \lambda_{n-m-1}^{(m)} \leq \lambda_{n-m-1}^{(m+1)} \leq \lambda_{n-m}^{(m)} \qquad \text{for } m = 0, \ldots, n-2 \tag{2.102}$$

For the proof of (2.102) we consider the problems $\mathbf{A}\mathbf{v} = \lambda\mathbf{v}$ and $\mathbf{A}^{(1)}\mathbf{v}^{(1)} = \lambda^{(1)}\mathbf{v}^{(1)}$. If we can show that the eigenvalue separation property holds for these two problems, it will hold also for $m = 1, 2, \ldots, n-2$. Specifically, we therefore want to prove that

$$\lambda_r \leq \lambda_r^{(1)} \leq \lambda_{r+1}; \qquad r = 1, \ldots, n-1 \tag{2.103}$$

Using the minimax characterization, we have that

$$\left.\begin{aligned} &\lambda_{r+1} = \max\left\{\min \frac{\mathbf{v}^T\mathbf{A}\mathbf{v}}{\mathbf{v}^T\mathbf{v}}\right\} \\ &\mathbf{v}^T\mathbf{w}_i = 0; \qquad i = 1, \ldots, r; \text{ all } \mathbf{w}_i \text{ arbitrary} \end{aligned}\right\} \tag{2.104}$$

Similarly, we have that

$$\left.\begin{aligned} &\lambda_r^{(1)} = \max\left\{\min \frac{\mathbf{v}^T\mathbf{A}\mathbf{v}}{\mathbf{v}^T\mathbf{v}}\right\} \\ &\mathbf{v}^T\mathbf{w}_i = 0; \qquad i = 1, \ldots, r \\ &\mathbf{w}_i \text{ arbitrary for } i = 1, \ldots, r-1 \\ &\mathbf{w}_r = \mathbf{e}_n \end{aligned}\right\} \tag{2.105}$$

where $\mathbf{w}_r$ is constraint to be equal to $\mathbf{e}_n$ to ensure that the last element in $\mathbf{v}$ is zero, because $\mathbf{e}_n$ is the last column of the $n \times n$ identity matrix $\mathbf{I}$. However, since the constraint for λ_{r+1} can be more severe and includes that for $\lambda_r^{(1)}$, we have that

$$\lambda_r^{(1)} \leq \lambda_{r+1} \tag{2.106}$$

To determine λ_r we use

$$\left.\begin{aligned} &\lambda_r = \max\left\{\min \frac{\mathbf{v}^T\mathbf{A}\mathbf{v}}{\mathbf{v}^T\mathbf{v}}\right\} \\ &\mathbf{v}^T\mathbf{w}_i = 0; \qquad i = 1, \ldots, r-1 \\ &\text{all } \mathbf{w}_i \text{ arbitrary} \end{aligned}\right\} \tag{2.107}$$

Comparing the characterizations of $\lambda_r^{(1)}$ and λ_r, i.e., (2.105) with (2.107), we observe that to calculate $\lambda_r^{(1)}$ we have the same constraints as in the calculations of λ_r plus one more (i.e., $\mathbf{v}^T\mathbf{e}_n = 0$), and hence

$$\lambda_r \leq \lambda_r^{(1)} \tag{2.108}$$

But (2.106) and (2.108) together establish the required result given in (2.103).

The eigenvalue separation property now yields the following result. If we write the eigenvalue problems in (2.101) including the problem $\mathbf{A}\mathbf{v} = \lambda\mathbf{v}$ in the form

$$p^{(m)}(\lambda^{(m)}) = \det(\mathbf{A}^{(m)} - \lambda^{(m)}\mathbf{I}); \qquad m = 0, \ldots, n-1 \tag{2.109}$$

where $p^{(0)} = p$, we see that the roots of the polynomials $p(\lambda^{(m+1)})$ separate the roots of the polynomial $p(\lambda^{(m)})$. However, a sequence of polynomials $p_i(x)$, $i = 1, \ldots, q$, form a Sturm sequence if the roots of the polynomial $p_{j+1}(x)$ separate the roots of the polynomial $p_j(x)$. *Hence the eigenvalue separation property states that the characteristic polynomials of the problems* $\mathbf{A}^{(m)}\mathbf{v}^{(m)} = \lambda^{(m)}\mathbf{v}^{(m)}$, $m = 0, 1. \ldots, n-1$, *form a Sturm sequence.* It should be noted that in the presentation we considered all symmetric matrices; i.e., the minimax characterization of eigenvalues and the Sturm sequence property are applicable to positive definite and indefinite matrices. We shall use the Sturm sequence property estensively in later chapters (see Sections 7.2.5, 10.2.2, 11.5, 12.2, and 12.3.4). Consider the following example.

Example 2.24

Consider the eigenvalue problem $\mathbf{A}\mathbf{v} = \lambda\mathbf{v}$, where

$$\mathbf{A} = \begin{bmatrix} 5 & -4 & -7 \\ -4 & 2 & -4 \\ -7 & -4 & 5 \end{bmatrix}$$

Evaluate the eigenvalues of $\mathbf{A}$ and of the matrixes $\mathbf{A}^{(m)}$, $m = 1, 2$. Show that the separation property given in (2.102) holds and sketch the characteristic polynomials $p(\lambda)$, $p^{(1)}(\lambda^{(1)})$, and $p^{(2)}(\lambda^{(2)})$.

We have

$$p(\lambda) = \det(\mathbf{A} - \lambda\mathbf{I}) = (5-\lambda)[(2-\lambda)(5-\lambda) - 16] + 4[-4(5-\lambda) - 28] - 7[16 + 7(2-\lambda)]$$

Hence

$$p(\lambda) = (-6-\lambda)(6-\lambda)(12-\lambda)$$

and the eigenvalues are

$$\lambda_1 = -6, \qquad \lambda_2 = 6, \qquad \lambda_3 = 12$$

Also,

$$p^{(1)}(\lambda^{(1)}) = \det(\mathbf{A}^{(1)} - \lambda^{(1)}\mathbf{I}) = (5-\lambda^{(1)})(2-\lambda^{(1)}) - 16$$

or

$$p^{(1)}(\lambda^{(1)}) = \lambda^{(1)^2} - 7\lambda^{(1)} - 6$$

Hence

$$\lambda_1^{(1)} = \tfrac{7}{2} - \tfrac{1}{2}\sqrt{73} = -0.7720$$

$$\lambda_2^{(1)} = \tfrac{7}{2} + \tfrac{1}{2}\sqrt{73} = 7.772$$

Finally,

$$p^{(2)}(\lambda^{(2)}) = \det(\mathbf{A}^{(2)} - \lambda^{(2)}\mathbf{I}) = 5 - \lambda^{(2)}$$

Hence

$$\lambda_1^{(2)} = 5$$

The separation property holds because

$$\lambda_1 \le \lambda_1^{(1)} \le \lambda_2 \le \lambda_2^{(1)} \le \lambda_3$$

and

$$\lambda_1^{(1)} < \lambda_1^{(2)} < \lambda_2^{(1)}$$

The characteristic polynomials are sketched in Fig. 2.5.

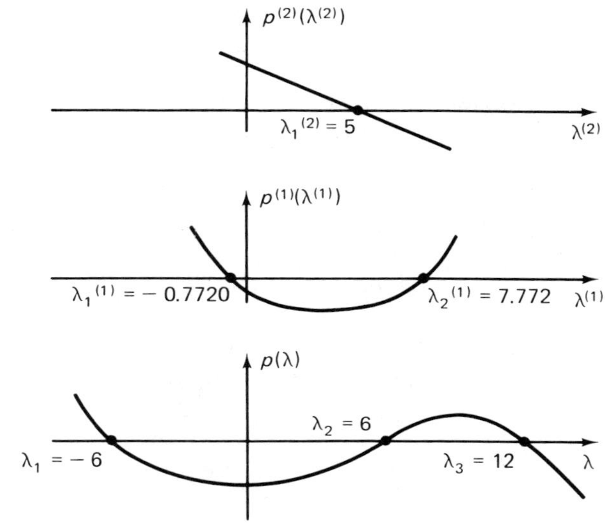

Figure 2.5. Characteristic polynomials.

2.8 VECTOR AND MATRIX NORMS

We have discussed vectors, matrices, eigenvalues, and eigenvectors of symmetric matrices, and have investigated the deeper significance of the elements in these entities. However, one important aspect has not been discussed so far. If we deal with single numbers, we can identify a number as being large or small. Vectors and matrices are functions of many elements, but we shall also need to measure their "size". Specifically, if single numbers are used in iterative processes, the convergence of a series of numbers, say $x_1, x_2, \ldots, x_k$, to a number x is simply measured by

$$\lim_{k \to \infty} |x_k - x| = 0 \tag{2.110}$$

or, said in words, convergence is obtained if the residual $y_k = |x_k - x|$ approaches zero as $k \longrightarrow \infty$. Furthermore, if we can find constants $p \geq 1$ and $c > 0$ such that

$$\lim_{k \to \infty} \frac{|x_{k+1} - x|}{|x_k - x|^p} = c \tag{2.111}$$

we say that convergence is "of order p." If $p = 1$, convergence is linear and the rate of convergence is c, in which case c must be smaller than 1.

In iterative solution processes using vectors and matrices we also need a measure of convergence. Realizing that the "size" of a vector or matrix should depend on the magnitude of all elements in the arrays, we arrive at the definition of vector and matrix norms. A norm is a single number which depends on the magnitude of all elements in the vector or matrix.

Definition: A *norm* of a vector $\mathbf{v}$ of order n written as $||\mathbf{v}||$ is a single number. The norm is a function of the elements of $\mathbf{v}$ and the following conditions are satisfied:

1. $||\mathbf{v}|| \geq 0$ and $||\mathbf{v}|| = 0$ if and only if $\mathbf{v} = 0$. (2.112)
2. $||c\mathbf{v}|| = |c|\,||\mathbf{v}||$ for any scalar c. (2.113)
3. $||\mathbf{v} + \mathbf{w}|| \leq ||\mathbf{v}|| + ||\mathbf{w}||$ for vectors $\mathbf{v}$ and $\mathbf{w}$. (2.114)

The relation (2.114) is the triangle inequality. The following three vector norms are commonly used, and called the infinity, one and two vector norms:

$$||\mathbf{v}||_\infty = \max_i |v_i| \tag{2.115}$$

$$||\mathbf{v}||_1 = \sum_{i=1}^{n} |v_i| \tag{2.116}$$

$$||\mathbf{v}||_2 = \left(\sum_{i=1}^{n} |v_i|^2\right)^{1/2} \tag{2.117}$$

where (2.117) is known as the *Euclidean vector norm*. Geometrically, this norm is equal to the length of the vector $\mathbf{v}$. All three norms are special cases of the vector norm $\sqrt[p]{\sum_i |v_i|^p}$, where for (2.115), (2.116), and (2.117), $p = \infty$, 1, and 2, respectively. It should be noted that each of the norms in (2.115) to (2.117) satisfies the conditions in (2.112) to (2.114).

We can now measure convergence of a sequence of vectors $\mathbf{x}_1, \mathbf{x}_2, \mathbf{x}_3, \ldots, \mathbf{x}_k$ to a vector $\mathbf{x}$. Namely, for the sequence to converge to $\mathbf{x}$ it is sufficient and necessary that

$$\lim_{k\to\infty} ||\mathbf{x}_k - \mathbf{x}|| = 0 \tag{2.118}$$

for any one of the vector norms. The order of convergence p, and in case $p = 1$, the rate of convergence c, are calculated in an analogous manner as in (2.111) but using norms; i.e., we consider

$$\lim_{k\to\infty} \frac{||\mathbf{x}_{k+1} - \mathbf{x}||}{||\mathbf{x}_k - \mathbf{x}||^p} = c \tag{2.119}$$

Considering the relationship among the various vector norms, we find that

$$||\mathbf{v}||_\infty \leq ||\mathbf{v}||_1 \leq n||\mathbf{v}||_\infty; \qquad ||\mathbf{v}||_\infty \leq ||\mathbf{v}||_2 \leq \sqrt{n}\, ||\mathbf{v}||_\infty \tag{2.120}$$

and can now note that although the actual vector length is calculated using the Euclidean norm, by virtue of relation (2.120), the 1 and ∞ norms also provide some measure of the length of the vector.

Example 2.25

Calculate the 1, 2, and ∞ norms of the vector $\mathbf{x}$, and verify the relations in (2.120).

$$\mathbf{x} = \begin{bmatrix} 1 \\ -3 \\ 2 \end{bmatrix}$$

We have

$$\|\mathbf{x}\|_\infty = 3$$

$$\|\mathbf{x}\|_1 = 1 + 3 + 2 = 6$$

$$\|\mathbf{x}\|_2 = \sqrt{1 + 9 + 4} = \sqrt{14}$$

and the relations in (2.120) read

$$3 \leq 6 \leq (3)(3); \qquad 3 \leq \sqrt{14} \leq (\sqrt{3})(3)$$

In analogy to the definition of a vector norm, we also define a matrix norm.

Definition: A *norm* of a matrix $\mathbf{A}$ of order $n \times n$, written as $\|\mathbf{A}\|$, is a single number. The norm is a function of the elements of $\mathbf{A}$, and the following relations hold:

1. $\|\mathbf{A}\| \geq 0$ and $\|\mathbf{A}\| = 0$ if and only if $\mathbf{A} = 0$. (2.121)
2. $\|c\mathbf{A}\| = |c|\,\|\mathbf{A}\|$ for any scalar c. (2.122)
3. $\|\mathbf{A} + \mathbf{B}\| \leq \|\mathbf{A}\| + \|\mathbf{B}\|$ for matrices $\mathbf{A}$ and $\mathbf{B}$. (2.123)
4. $\|\mathbf{AB}\| \leq \|\mathbf{A}\|\,\|\mathbf{B}\|$ for matrices $\mathbf{A}$ and $\mathbf{B}$. (2.124)

The relation in (2.123) is the triangle inequality equivalent to (2.114). The additional condition in (2.124), which was not postulated in the definition of a vector norm, must be satisfied, in order to be able to use matrix norms when matrix products occur.

The following are frequently used matrix norms:

$$\|\mathbf{A}\|_\infty = \max_i \sum_{j=1}^{n} |a_{ij}| \tag{2.125}$$

$$\|\mathbf{A}\|_1 = \max_j \sum_{i=1}^{n} |a_{ij}| \tag{2.126}$$

$$\|\mathbf{A}\|_2 = \sqrt{\tilde{\lambda}_n}; \qquad \tilde{\lambda}_n = \text{maximum eigenvalue of } \mathbf{A}^T\mathbf{A} \tag{2.127}$$

where for a symmetric matrix $\mathbf{A}$ we have $\|\mathbf{A}\|_\infty = \|\mathbf{A}\|_1$. The norm $\|\mathbf{A}\|_2$ is called the *spectral norm* of $\mathbf{A}$. Each of the norms given above satisfies the relations in (2.121) to (2.124). The proof that the relation in (2.124) is satisfied for the infinity norm is given in Example 2.27.

Example 2.26

Calculate the ∞, 1, and 2 norms of the matrix **A**, where **A** was given in Example 2.24.

The matrix **A** considered is

$$\mathbf{A} = \begin{bmatrix} 5 & -4 & -7 \\ -4 & 2 & -4 \\ -7 & -4 & 5 \end{bmatrix}$$

Using the definitions given in (2.125) to (2.127) we have

$$||\mathbf{A}||_1 = 5 + 4 + 7 = 16$$

$$||\mathbf{A}||_\infty = 5 + 4 + 7 = 16$$

To evaluate $||\mathbf{A}||_2$ we first need to calculate $\mathbf{A}^T\mathbf{A}$.

$$\mathbf{A}^T\mathbf{A} = 18\begin{bmatrix} 5 & 0 & -3 \\ 0 & 2 & 0 \\ -3 & 0 & 5 \end{bmatrix}$$

The eigenvalues of $\mathbf{A}^T\mathbf{A}$ are

$$\lambda_1 = 36, \qquad \lambda_2 = 36, \qquad \lambda_3 = 144$$

Hence we have

$$||\mathbf{A}||_2 = \sqrt{144} = 12$$

Example 2.27

Show that for two matrixes **A** and **B**, we have

$$||\mathbf{AB}||_\infty \leq ||\mathbf{A}||_\infty ||\mathbf{B}||_\infty$$

Using the definition of the ∞ matrix norm in (2.125), we have

$$||\mathbf{AB}||_\infty = \max_i \sum_{j=1}^n \left| \sum_{k=1}^n a_{ik}b_{kj} \right|$$

but then

$$||\mathbf{AB}||_\infty \leq \max_i \sum_{j=1}^n \sum_{k=1}^n |a_{ik}|\,|b_{kj}|$$

$$= \max_i \sum_{k=1}^n \left\{ |a_{ik}| \sum_{j=1}^n |b_{kj}| \right\}$$

$$\leq \left\{ \max_i \sum_{k=1}^n |a_{ik}| \right\} \left\{ \max_k \sum_{j=1}^n |b_{kj}| \right\}$$

This proves the desired result.

As in the case of a sequence of vectors, we can now measure the convergence of a sequence of matrices $\mathbf{A}_1, \mathbf{A}_2, \mathbf{A}_3, \ldots, \mathbf{A}_k$ to a matrix **A**. For example, for convergence it is sufficient and necessary that

$$\lim_{k\to\infty} ||\mathbf{A}_k - \mathbf{A}|| = 0 \tag{2.128}$$

for any one of the given matrix norms.

In the definition of a matrix norm we needed relation (2.124) to be able to use norms when we encounter matrix products. Similarly, we also want to use norms when products of matrices with vectors occur. In such a case, in order to obtain useful information by applying norms, we need to employ only specific vector norms with specific matrix norms. Which matrix and vector norms should only be used together is determined by the condition that the following relation hold for any matrix **A** and vector **v**:

$$\|\mathbf{Av}\| \leq \|\mathbf{A}\|\,\|\mathbf{v}\| \tag{2.129}$$

where $\|\mathbf{Av}\|$ and $\|\mathbf{v}\|$ are evaluated using the vector norm and $\|\mathbf{A}\|$ is evaluated using the matrix norm. We may note the close relationship to the condition (2.124), which was required to hold for a matrix norm. If (2.129) holds for a specific vector and matrix norm, the two norms are said to be compatible, and the matrix norm is said to be subordinate to the vector norm. The 1, 2, and ∞ norms of a matrix, as defined above, are subordinate, respectively, to the 1, 2, and ∞ vector norms given in (2.115) to (2.117). In the following example we give the proof that the ∞ norms are compatible and subordinate. The compatibility of the vector and matrix 1 and 2 norms is proved similarly.

Example 2.28

Show that for a matrix **A** and vector **v**, we have

$$\|\mathbf{Av}\|_\infty \leq \|\mathbf{A}\|_\infty \|\mathbf{v}\|_\infty \tag{a}$$

Using the definitions of the infinity norms, we have

$$\begin{aligned}\|\mathbf{Av}\|_\infty &= \max_i \left| \sum_{j=1}^{n} a_{ij} v_j \right| \\ &\leq \max_i \sum_{j=1}^{n} |a_{ij}|\,|v_j| \\ &\leq \left\{\max_i \sum_{j=1}^{n} |a_{ij}|\right\} \left\{\max_j |v_j|\right\}\end{aligned}$$

This proves (a).

To show that the equality sign can be reached, we only need to consider the case where **v** is a full unit vector and $a_{ij} \geq 0$. In this case, $\|\mathbf{v}\|_\infty = 1$ and $\|\mathbf{Av}\|_\infty = \|\mathbf{A}\|_\infty$.

In later chapters we shall encounter various applications of norms. One valuable application arises in the calculation of eigenvalues of a matrix. Namely, if we consider the problem $\mathbf{Av} = \lambda\mathbf{v}$, we obtain, taking norms on both sides,

$$\|\mathbf{Av}\| = \|\lambda\mathbf{v}\| \tag{2.130}$$

and hence using (2.113) and (2.129), we have

$$\|\mathbf{A}\|\,\|\mathbf{v}\| \geq |\lambda|\,\|\mathbf{v}\| \tag{2.131}$$

or

$$|\lambda| \leq ||\mathbf{A}|| \tag{2.132}$$

Therefore, every eigenvalue of **A** is in absolute magnitude smaller than or equal to any norm of **A**. Defining the *spectral radius* $\rho(\mathbf{A})$ as

$$\rho(\mathbf{A}) = \max_i |\lambda_i| \tag{2.133}$$

we have that

$$\rho(\mathbf{A}) \leq ||\mathbf{A}|| \tag{2.134}$$

In practice, the ∞ norm of **A** is calculated most conveniently and thus used effectively to estimate an upper bound on the largest absolute value reached by the eigenvalues.

Example 2.29

Calculate the spectral radius of the matrix **A** considered in Example 2.24. Then show that $\rho(\mathbf{A}) \leq ||\mathbf{A}||_\infty$.

The spectral radius is equal to $\max_i |\lambda_i|$. The eigenvalues of **A** have been calculated in Example 2.24.

$$\lambda_1 = -6; \qquad \lambda_2 = 6; \qquad \lambda_3 = 12$$

Hence

$$\rho(\mathbf{A}) = 12$$

We calculated in Example 2.26 $||\mathbf{A}||_\infty = 16$. Thus the relation $\rho(\mathbf{A}) \leq ||\mathbf{A}||_\infty$ is satisfied.

2.9 REFERENCES

1. K. HOFFMAN and R. KUNZE, *Linear Algebra*, Prentice-Hall, Inc., Englewood Cliffs, N.J., 1961.
2. B. NOBLE, *Applied Linear Algebra*, Prentice-Hall, Inc., Englewood Cliffs, N.J., 1969.
3. C. E. FRÖBERG, *Introduction to Numerical Analysis*, Addison-Wesley Publishing Company, Inc., Reading, Mass., 1969.
4. R. ZURMÜHL, *Matrizen*, Springer-Verlag, Berlin, 1964.
5. S. H. CRANDALL, *Engineering Analysis*, McGraw-Hill Book Company, New York, 1956.
6. R. COURANT and D. HILBERT, *Methods of Mathematical Physics*, Vol. 1, Interscience Publishers, New York, 1953.
7. J. H. WILKINSON, *The Algebraic Eigenvalue Problem*, Clarendon Press, Oxford, 1965.

PART II

THE FINITE ELEMENT METHOD

CHAPTER 3

FORMULATION OF THE FINITE ELEMENT METHOD

3.1 INTRODUCTION

The development of the finite element method as an analysis tool was essentially initiated with the advent of the electronic digital computer. In the numerical solution of a continuum problem it is basically necessary to establish and solve a system of algebraic equations. Using the finite element method on a digital computer it is possible to establish and solve the governing equations for complex systems in a very effective way. It is mainly for the generality of the structure or continuum that can be analyzed, for the relative ease of establishing the governing equations, and for the good numerical properties of the system matrices involved that the finite element method has found wide appeal.[1-14]

The finite element method was initially developed on a physical basis for the analysis of problems in structural mechanics; however, it was soon recognized that the method can be applied equally to the solution of many other classes of problems. The generality of the finite element method will become apparent when the variational formulation of the method is presented. Until this is done, for the introduction and the physical formulation of the method we shall consider the solution of structural mechanics problems only.

As is often the case with original development, it is rather difficult to quote an exact date on which the finite element method was "invented," but the roots of the method can be traced back to three separate groups: applied mathematicians,[15-17] physicists,[18] and engineers. Although in principle published already, the finite element method obtained its real impetus by the independent developments carried out by engineers. Important original contributions have been the papers by Turner et al.[19] and Argyris and Kelsey.[20] The name "finite element" was coined in a paper by Clough, in which the technique was presented for plane stress analysis.[21] Since the early developments of the finite element method, a large amount of research has been devoted to the technique, because the potential of the procedure was quickly realized (see References).

Today, the concept of the finite element method is a very broad one. Even when restricting ourselves to the analysis of structural mechanics problems only, the method can be used in a variety of different ways. This becomes particularly apparent when variational formulations are discussed. A most important formulation, which is widely used for the solution of practical problems, is the displacement-based finite element method. Practically all major general purpose analysis programs have been written using this formulation,[14] because of its simplicity, generality, and good numerical properties. In this chapter we shall concentrate on this formulation; and other formulations, such as equilibrium, hybrid, and mixed methods, will be mentioned only briefly in Section 5.5. Therefore, in the following, the

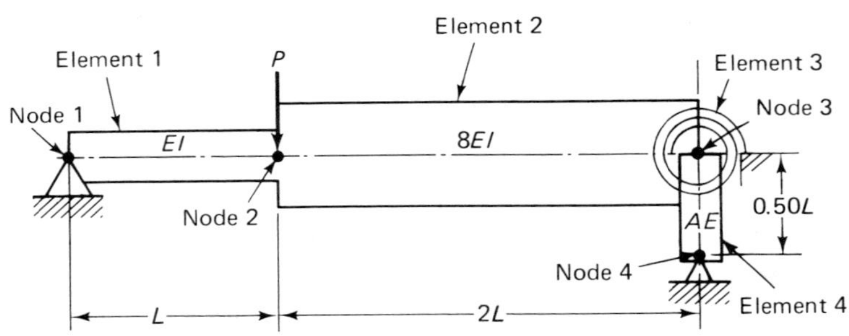

(a) Elements and nodal points

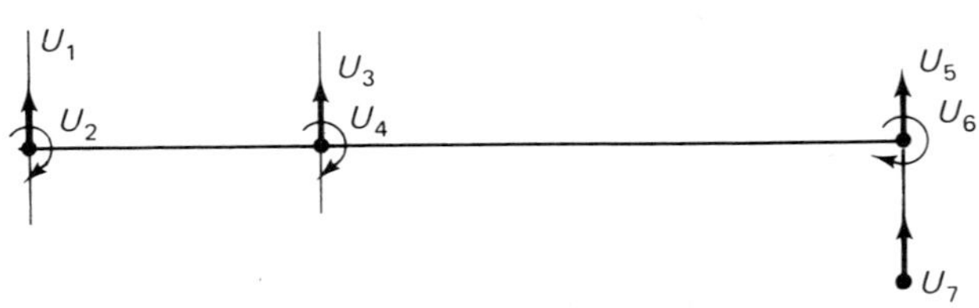

(b) Global degrees of freedom of unrestraint structure

Figure 3.1. Simple beam and truss structure.

name "finite element method" will imply "displacement-based finite element method." However, it should be pointed out that many of the concepts to be presented, and in particular, the numerical techniques for the implementation of the displacement-based finite element method, are in many cases directly applicable to other formulations.

The displacement-based finite element method can be regarded as an extension of the displacement method of analysis, which had been used for many years in the analysis of beam and truss structures.[22,23] It is convenient to recall the displacement method of analysis by means of a small example and at the same time to introduce the basic concepts of the finite element method.

Consider the analysis of the structure in Fig. 3.1. In the displacement method of analysis, we consider the structure as an assemblage of two beams, one truss, and a spring element. The first step is to evaluate the element stiffness matrices that correspond to the global degrees of freedom of the structure. For the beam, spring, and truss elements, respectively, we have in this case

$$\left.\begin{aligned}
\mathbf{K}_1^e &= \frac{EI}{L}\begin{bmatrix} \frac{12}{L^2} & -\frac{6}{L} & -\frac{12}{L^2} & -\frac{6}{L} \\ & 4 & \frac{6}{L} & 2 \\ & & \frac{12}{L^2} & \frac{6}{L} \\ \text{symmetric} & & & 4 \end{bmatrix}; \qquad U_1, U_2, U_3, U_4 \\
\mathbf{K}_2^e &= \frac{EI}{L}\begin{bmatrix} \frac{12}{L^2} & -\frac{12}{L} & -\frac{12}{L^2} & -\frac{12}{L} \\ & 16 & \frac{12}{L} & 8 \\ & & \frac{12}{L^2} & \frac{12}{L} \\ \text{symmetric} & & & 16 \end{bmatrix}; \qquad U_3, U_4, U_5, U_6 \\
\mathbf{K}_3^e &= k_s; \qquad U_6 \\
\mathbf{K}_4^e &= \frac{EA}{L}\begin{bmatrix} 2 & -2 \\ -2 & 2 \end{bmatrix}; \qquad U_5, U_7
\end{aligned}\right\} \tag{3.1}$$

where the subscript on $\mathbf{K}^e$ indicates the element number, and the global degrees of freedom that correspond to the element stiffnesses are written next to the matrices. It should be noted that in this example the element matrices are independent of direction cosines since the centerlines of the elements are aligned with the global axes. If the local axis of an element would not be in the direction of a global axis, the local element stiffness matrix would have to be transformed to obtain the required global element stiffness matrix.

The stiffness matrix of the complete element assemblage is effectively obtained from the stiffness matrices of the individual elements using the *direct stiffness method.* In this procedure the structure stiffness matrix $\mathbf{K}$ is calculated by direct addition of the element stiffness matrices, i.e.,

$$\mathbf{K} = \sum_i \mathbf{K}_i^e \tag{3.2}$$

where the summation includes all elements. To perform the summation, each element matrix $\mathbf{K}_i^e$ is written as a matrix $\mathbf{K}_i$ of the same order as the stiffness matrix $\mathbf{K}$, where all entries in $\mathbf{K}_i$ are zero, except those which correspond to an element degree of freedom. For example, for element 4 we have

$$\begin{array}{c} \\ \mathbf{K}_4 = \end{array} \begin{array}{c} \\ 1 \\ 2 \\ 3 \\ 4 \\ 5 \\ 6 \\ 7 \end{array} \begin{array}{ccccccc} 1 & 2 & 3 & 4 & 5 & 6 & 7 \leftarrow \text{degree of freedom} \end{array}$$

$$\mathbf{K}_4 = \begin{bmatrix} 0 & 0 & 0 & 0 & 0 & 0 & 0 \\ 0 & 0 & 0 & 0 & 0 & 0 & 0 \\ 0 & 0 & 0 & 0 & 0 & 0 & 0 \\ 0 & 0 & 0 & 0 & 0 & 0 & 0 \\ 0 & 0 & 0 & 0 & \frac{2AE}{L} & 0 & -\frac{2EA}{L} \\ 0 & 0 & 0 & 0 & 0 & 0 & 0 \\ 0 & 0 & 0 & 0 & -\frac{2AE}{L} & 0 & \frac{2EA}{L} \end{bmatrix} \tag{3.3}$$

(rows and columns numbered 1–7 by degree of freedom)

Therefore, the stiffness matrix of the structure in Fig. 3.1 is

$$\mathbf{K} = \begin{bmatrix} \frac{12EI}{L^3} & -\frac{6EI}{L^2} & -\frac{12EI}{L^3} & -\frac{6EI}{L^2} & 0 & 0 & 0 \\ & \frac{4EI}{L} & \frac{6EI}{L^2} & \frac{2EI}{L} & 0 & 0 & 0 \\ & & \frac{24EI}{L^3} & -\frac{6EI}{L^2} & -\frac{12EI}{L^3} & -\frac{12EI}{L^2} & 0 \\ & & & \frac{20EI}{L} & \frac{12EI}{L^2} & \frac{8EI}{L} & 0 \\ & \text{symmetric} & & & \frac{12EI}{L^3} + \frac{2AE}{L} & \frac{12EI}{L^2} & -\frac{2AE}{L} \\ & & & & & \frac{16EI}{L} + k_s & 0 \\ & & & & & & \frac{2AE}{L} \end{bmatrix} \tag{3.4}$$

and the equilibrium equations for the system are

$$\mathbf{KU} = \mathbf{R} \tag{3.5}$$

where $\mathbf{U}$ is a vector of the system global displacements and $\mathbf{R}$ is a vector of forces acting into the direction of the structure global displacements:

$$\mathbf{U}^T = [U_1, \ldots, U_7]; \qquad \mathbf{R}^T = [R_1, \ldots, R_7] \tag{3.6}$$

Before solving for the displacements of the structure, we need to impose the boundary conditions that $U_1 = 0$ and $U_7 = 0$. This means that we may consider only five equations in five unknown displacements; i.e.,

$$\bar{\mathbf{K}}\bar{\mathbf{U}} = \bar{\mathbf{R}} \tag{3.7}$$

where $\bar{\mathbf{K}}$ is obtained by eliminating from $\mathbf{K}$ the first and seventh rows and columns, and

$$\bar{\mathbf{U}} = [U_2 \quad U_3 \quad U_4 \quad U_5 \quad U_6]; \qquad \bar{\mathbf{R}}^T = [0 \quad -P \quad 0 \quad 0 \quad 0] \tag{3.8}$$

The solution of (3.7) gives the structure displacements and therefore the element node displacements. The element nodal forces are obtained by multiplying the element stiffness matrices $\mathbf{K}_i^e$ by the element displacements. If the forces at any section of an element are required, we can evaluate them from the element end forces by use of simple statics.

The above discussion demonstrates some important features of the finite element method. The basic process is that the complete structure is considered as an assemblage of individual structural elements. The element stiffness matrices corresponding to the global degrees of freedom of the structural assemblage are calculated and the structure stiffness matrix is formed by addition of the element stiffness matrices. The solution of the equilibrium equations of the assemblage of elements yields the element displacements, which are then used to calculate element stresses. Although, originally, the analysis of beam and truss element assemblages using the displacement method were not considered as finite element analyses, as we shall see later, we actually may regard both the truss and the beam element as two specific finite elements. In this context, the above analysis was a finite element analysis of a beam and truss structure.

A most important step in the analysis, although trivial in this example, is the correct evaluation of the element stiffness properties. It should be noted that whatever element stiffness matrices are used, the element nodal forces of adjacent elements will always be in equilibrium, because this is only a check that the equilibrium equations in (3.7) have been solved correctly. But the accuracy with which the exact displacements and stresses, i.e., "exact" within the context of beam theory, are calculated depends on the correct evaluation of the element stiffness matrices. For example, if, by error, we use in the analysis a modulus of elasticity $E/2$ for element 2, we will solve for an incorrect displacement and stress distribution in the structure.

For the above beam and truss elements, exact stiffness matrices could be calculated, for example, from the solution of the appropriate differential equations of equilibrium,[23] and therefore the exact solution of the problem is obtained. As an example consider the truss element in Fig. 3.2. Assume that the stiffness coefficient k_{22} is required. This coefficient is the force at node 2 of the element when a unit displacement is imposed at the same node.

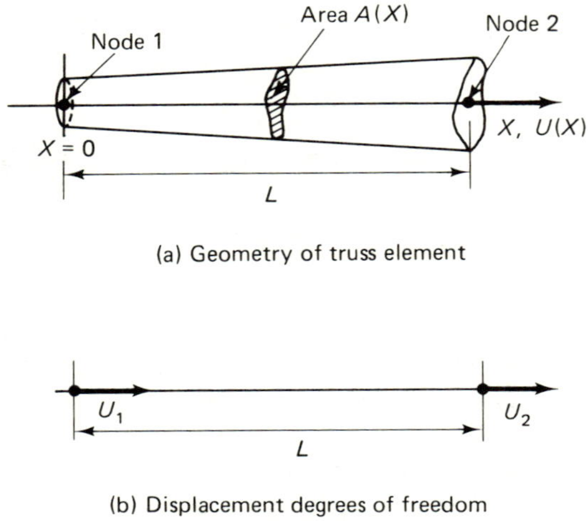

Figure 3.2. Truss element.

The governing differential equation is

$$\frac{dP}{dX} = 0 \tag{3.9}$$

where P is the axial force in the bar [i.e., $P = AE(dU/dX)$], $U(X)$ the displacement, and $AE(X)$ the axial stiffness at coordinate X. The boundary conditions are

$$U|_{X=0} = 0.0; \qquad U|_{X=L} = 1.0 \tag{3.10}$$

For a truss element of constant AE, we obtain by integrating (3.9) and using the boundary conditions in (3.10),

$$U(X) = \frac{X}{L} \tag{3.11}$$

and hence

$$P|_{X=L} = k_{22} = \frac{AE}{L} \tag{3.12}$$

Alternatively, instead of solving the differential equation of equilibrium, we could have used the virtual work principle to calculate the required stiffness coefficient. Before doing so, let us recall the general principle of virtual work (principle of virtual displacements).[23]

Consider the equilibrium of a general three-dimensional body such as in Fig. 3.3. The external forces acting onto the body are surface tractions $\mathbf{f}^S$, body forces $\mathbf{f}^B$, and concentrated forces $\mathbf{F}^i$. These forces include all externally applied forces and reactions and have in general three components corresponding to the three coordinate axes:

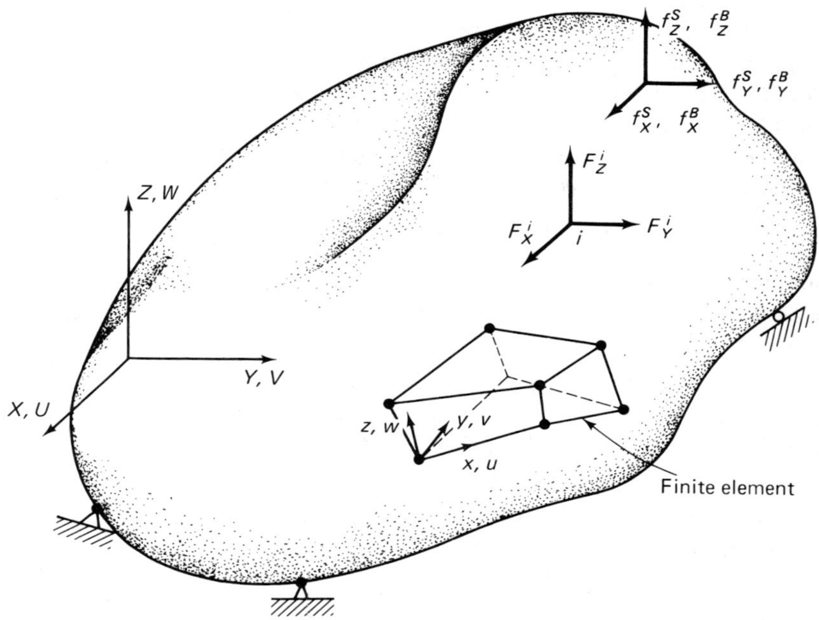

Figure 3.3. General three-dimensional body.

$$\mathbf{f}^B = \begin{bmatrix} f^B_X \\ f^B_Y \\ f^B_Z \end{bmatrix}; \qquad \mathbf{f}^S = \begin{bmatrix} f^S_X \\ f^S_Y \\ f^S_Z \end{bmatrix}; \qquad \mathbf{F}^i = \begin{bmatrix} F^i_X \\ F^i_Y \\ F^i_Z \end{bmatrix} \tag{3.13}$$

The displacements of the body from the unloaded configuration are denoted by $\mathbf{U}$, where

$$\mathbf{U}^T = [U \quad V \quad W] \tag{3.14}$$

The strains corresponding to $\mathbf{U}$ are,

$$\boldsymbol{\epsilon}^T = [\epsilon_{XX} \quad \epsilon_{YY} \quad \epsilon_{ZZ} \quad \gamma_{XY} \quad \gamma_{YZ} \quad \gamma_{ZX}] \tag{3.15}$$

and the stresses corresponding to $\boldsymbol{\epsilon}$ are

$$\boldsymbol{\tau}^T = [\tau_{XX} \quad \tau_{YY} \quad \tau_{ZZ} \quad \tau_{XY} \quad \tau_{YZ} \quad \tau_{ZX}] \tag{3.16}$$

Assume that we require an equation which expresses the equilibrium of the body. Such an equation is needed if the displacements and stresses of the body are to be calculated. One procedure is to focus attention on the forces that act on a typical differential element of the body. Expressing the force equilibrium of the differential element results in the differential equations of equilibrium of the body. As an example, for the truss considered in Fig. 3.2, the differential equation of equilibrium is given in (3.9). On the other hand, considering a two- or three-dimensional body, two and three differential equations of equilibrium, respectively, would be derived.

An equivalent approach to obtain appropriate equations that express the equilibrium of the body is to use the *principle of virtual displacements*. This principle expresses that the equilibrium of the body requires that for any compatible, small virtual displacements imposed onto the body, the total internal virtual work is equal to the total external virtual work; i.e., we have

$$\int_V \bar{\boldsymbol{\epsilon}}^T \boldsymbol{\tau}\, dV = \int_V \bar{\mathbf{U}}^T \mathbf{f}^B\, dV + \int_S \bar{\mathbf{U}}^{S^T} \mathbf{f}^S\, dS + \sum_i \bar{\mathbf{U}}^{i^T} \mathbf{F}^i \qquad (3.17)$$

The internal virtual work is given on the left side of (3.17), and is equal to the actual stresses $\boldsymbol{\tau}$ going through the virtual strains $\bar{\boldsymbol{\epsilon}}$ that correspond to the imposed virtual displacements,

$$\bar{\boldsymbol{\epsilon}}^T = [\bar{\epsilon}_{XX} \quad \bar{\epsilon}_{YY} \quad \bar{\epsilon}_{ZZ} \quad \bar{\gamma}_{XY} \quad \bar{\gamma}_{YZ} \quad \bar{\gamma}_{ZX}] \qquad (3.18)$$

The external work is given on the right side of (3.17) and is equal to the actual forces $\mathbf{f}^B$, $\mathbf{f}^S$, and $\mathbf{F}^i$ going through the virtual displacements $\bar{\mathbf{U}}$, where

$$\bar{\mathbf{U}}^T = [\bar{U} \quad \bar{V} \quad \bar{W}] \qquad (3.19)$$

The superscript S denotes that surface displacements are considered and the superscript i denotes the displacements at the point where the concentrated forces $\mathbf{F}^i$ are applied.

It should be emphasized that the virtual strains used in (3.17) are those corresponding to the imposed body and surface virtual displacements, and that these displacements can be any compatible set of displacements. The equation in (3.17) is an expression of equilibrium, and for different virtual displacements, correspondingly different equations of equilibrium are obtained. The virtual displacement equation has been written in the global coordinate system X, Y, Z of the body, but it is also valid in any other system of coordinates.

We may be more familiar with the differential equations of equilibrium used in the theory of elasticity, such as (3.9) for the truss considered. The equation of virtual work in (3.17) may be derived from these equations. As will be seen many times, the practical use of the principle lies in that it can be used as a mechanism to calculate approximate stress distributions in structural systems. Indeed, *we shall use the principle of virtual displacements as the basis of the displacement-based finite element formulation*. However, it should be noted that the method of weighted residuals and specifically the Galerkin method could also be employed to formulate the displacement-based finite element solution.[5]

We now return to the calculation of the truss element stiffness coefficient k_{22}. In this case the actual displacements due to a unit displacement at node 2 are given in (3.11). Therefore, the actual stresses are

$$\tau_{XX} = \frac{E}{L} \qquad (3.20)$$

The virtual work equation in (3.17) for this specific case reduces to

$$A \int_0^L \bar{\epsilon}_{XX} \tau_{XX} \, dX = \bar{U}_2 P|_{X=L} \tag{3.21}$$

where $\bar{U}_2$ is the imposed virtual displacement at node 2 of the bar and $\bar{\epsilon}_{XX}$ are the corresponding virtual strains. The virtual displacements of the bar due to $\bar{U}_2$ are

$$\bar{U}(X) = \frac{X}{L} \bar{U}_2 \tag{3.22}$$

and corresponding virtual strains are

$$\bar{\epsilon}_{XX} = \frac{1}{L} \bar{U}_2 \tag{3.23}$$

Hence, using the virtual work principle, we obtain

$$P|_{X=L} = k_{22} = A \int_0^L \frac{E}{L^2} \, dX = \frac{AE}{L} \tag{3.24}$$

We need to note that the same stiffness coefficients are obtained as calculated from the solution of the differential equations of equilibrium, if, as in the above truss element case, the exact element displacements are used. Therefore, within the assumptions of truss and beam theory, the exact solution of a truss and beam structure is evaluated when the structure is considered as an assemblage of individual truss and beam elements interconnected at their nodal end points. Basically, the exact stiffness coefficients are obtained because the elements are only connected at their end points, and the exact element displacements for unit end displacements are used in the derivation of the element stiffness matrices. Therefore, in the displacement analysis, all equilibrium and compatibility requirements within each element and at the element boundaries are satisfied.

To demonstrate on a simple example that the exact stiffness coefficients are only obtained provided that the exact displacement distributions for unit nodal point displacements are used, assume that we require the stiffness matrix of a truss element whose axial stiffness varies along its length, such as

$$AE = AE_0\left(1 + \frac{X}{L}\right) \tag{3.25}$$

where L is the length of the element. Let us consider, as before, the calculation of k_{22}. From the solution of the differential equation we obtain

$$k_{22} = \frac{AE_0}{L \log_e 2} = 1.44 \frac{AE_0}{L} \tag{3.26}$$

Alternatively, we may use the virtual work theorem to calculate k_{22}. If the exact displacements of the truss for a unit end displacement are used, we obtain the above exact stiffness coefficient. However, this would necessitate first the calculation of the exact displacements (i.e., the solution of the differential equation of equilibrium), and we may prefer to use as an approximation the corresponding displacements of a uniform truss. We then obtain

$$\begin{aligned} k_{22} &\doteq \int_0^L \frac{1}{L} AE_0\left(1 + \frac{X}{L}\right)\frac{1}{L}\, dX \\ &= 1.5\frac{AE_0}{L} \end{aligned} \tag{3.27}$$

Therefore, the application of the virtual work theorem with an approximate displacement function yields an approximate stiffness coefficient. It should be noted that provided that the functions in the integral have no singularities, we encounter no numerical difficulties in the evaluation of the approximate stiffness coefficient.

It is quite natural to ask at this stage how good the approximation is for a given function and, in particular, how close must the assumed displacement function be to the "exact" function to obtain reasonable results. We shall elaborate on these questions later when the general problems of solution accuracy and convergence of the finite element method are discussed.

Some of the basic ideas having been presented, it is now instructive to briefly outline the finite element analysis of the plane stress problem in Fig. 3.4. It was in the analysis of such problems that the specific term "finite element method" was used for the first time.[21] The unknowns in the problem are considered to be the displacements $U(X, Y)$ and $V(X, Y)$, from which the strain and stress quantities can be calculated.

The basic steps of analysis are as in the above analysis of the truss and beam structure. First, we idealize the plate as an assemblage of discrete elements interconnected at common nodal points, as shown in Fig. 3.4. The solution that is obtained for the displacements $U(X, Y)$ and $V(X, Y)$ from the analysis of the element assemblage is an approximate solution to the actual continuum problem. An important feature of finite element analysis is that the displacements of the total element assemblage are evaluated by focusing attention on the displacements of each individual element. A typical element of the problem considered here is shown in Fig. 3.5. The unknown displacements of the elements are $u(x, y)$ and $v(x, y)$, where the x and y coordinates are local element coordinates. Because the local and global coordinates are measured in this problem in the same directions, the displacements $u(x, y)$ and $v(x, y)$ of an element are also the displacements $U(X, Y)$ and $V(X, Y)$ over the corresponding region in the element assemblage. However, in the general case, the $u(x, y)$ and $v(x, y)$ element displacements need to be transformed using a rotation to obtain the global element assemblage displacements $U(X, Y)$ and $V(X, Y)$.

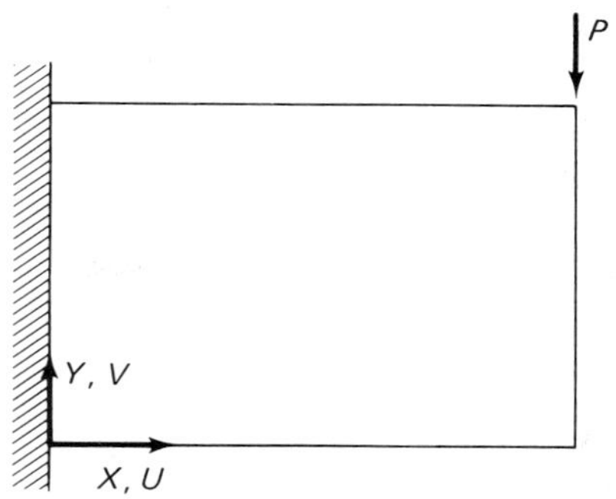

(a) Cantilever plate

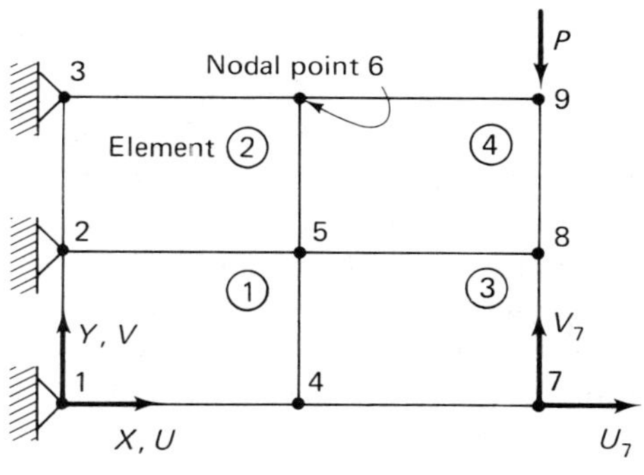

(b) Finite element idealization

Figure 3.4. Finite element plane stress analysis.

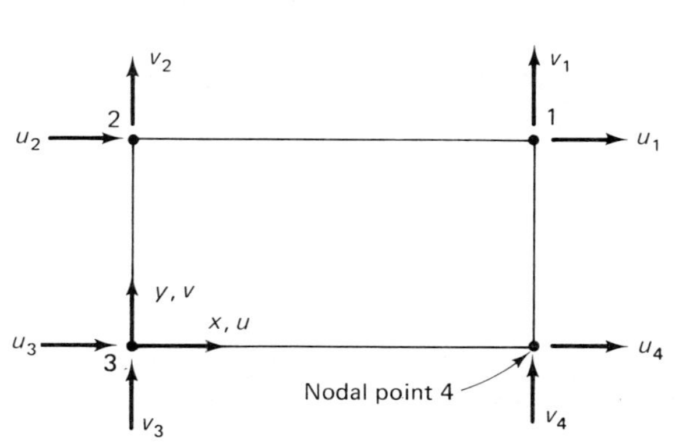

Figure 3.5. Typical two-dimensional four-node finite element defined in local coordinate system.

The solution of the problem proceeds in the following way:

1. Assume at each element nodal point i two unknown displacements u_i and v_i and express $u(x, y)$ and $v(x, y)$ in terms of simple polynomial functions, in which the unknown parameters are the element nodal point displacements. For the element in Fig. 3.5 the unknown displacements are $u_1, v_1, \ldots, u_4, v_4$.
2. Use the virtual displacement theorem to evaluate each element stiffness matrix corresponding to the nodal point degrees of freedom.
3. Calculate the structure stiffness matrix by assembling the element stiffness matrices, impose the boundary conditions, solve the equilibrium equations for the nodal-point displacements, and evaluate the element stresses.

The procedure is conceptually the same as in the analysis of the simple beam and truss structure considered at the beginning of this section. However, there is one important difference. It is important to realize that in the finite element analysis of a continuum, we will obtain only an approximate solution for general loading conditions. The reason is that the polynomial functions assumed as displacement patterns in each finite element can only approximate the exact displacements of the continuum. Therefore, approximate stiffness coefficients are derived using the virtual work theorem, and approximate nodal point displacements and stresses are evaluated. This is quite different from the analysis of the simple beam and truss structure considered at the beginning of this section, in which the polynomial functions for the element displacements did express the displacements of the structure within the region of the elements exactly, and therefore the exact response of the structure is calculated. Indeed, considering general finite element analysis, it is the necessity to approximate the exact response of the continuum which provided the various problems and challenges that led to the development of the field. Although it is considered best to look upon the finite element analysis procedure as a general numerical method (such as the finite difference analysis technique), an advantage of the finite element method is that it can be explained using physical reasoning. Therefore, in this and the next chapters, we aim to present both the physical understanding of the method and the technique as a general numerical procedure.

3.2 FORMULATION OF THE FINITE ELEMENT METHOD USING THE PRINCIPLE OF VIRTUAL DISPLACEMENTS

We have stated in the previous section that the finite element method was originally introduced as an extension of the displacement method of analysis for truss and beam structures. Using the displacement method of analysis, a frame or truss structure is considered as an assemblage of discrete members interconnected at the joints of the structure. The stiffness matrices of the members corresponding to the global structure degrees of freedom are calculated, the structure stiffness matrix is assembled, the load vector is formed, and then the equations of joint equilibrium are solved for the structure joint displacements. The basic process is therefore to consider the complete structure as an assemblage of structural elements and to derive a set of equilibrium equations that correspond to the structure joint displacements, the solution of which yields the displacements at a discrete number of points of the structure. The joint displacements of the structure being known, the displacements at all points of the structure, and therefore the strains and stresses, can be calculated. As was stated in the previous section, the analysis yields the exact displacements and stresses ("exact" within the assumptions of beam theory), provided the beam and truss stiffness matrices have been computed exactly, say from the solution of the appropriate differential equations. However, when element properties vary, we may prefer to calculate approximate matrices using the virtual work theorem, in which case the analysis yields approximate structure displacements, strains, and stresses.

Basically, the same concepts are used in the general finite element analysis of structures. In this analysis the structure is idealized as an assemblage of a discrete number of finite elements. The elements are assumed to be connected at a finite number of nodal points. These nodal points are equivalent to the joints in the analysis of truss or beam structures. It is then assumed that the displacements in each finite element are a function of the element nodal point displacements. Therefore, the displacements at any point in the assemblage of structural elements are expressed in terms of the nodal point displacements of the structural idealization. This enables the use of the virtual displacement theorem, in the same way as in the analysis of truss and beam structures, to calculate the equilibrium equations of the element assemblage corresponding to the nodal point displacements.

3.2.1 Displacement and Strain-Displacement Transformation Matrices for Plane Stress Analysis

Before considering the general finite element formulation, we derive in this section the necessary matrices that pertain to a simple finite element, and that would be required in the analysis employing this element. It is believed that in this way more immediate meaning will be given to the general finite element formulation presented in the next section.

For demonstration purposes, consider again the specific example of the analysis of a cantilever plate subjected to an in-plane load (Fig. 3.4). The structure is in a state of plane stress conditions;[26] therefore, the plate is idealized as an assemblage of two-dimensional plane stress finite elements as shown in Fig. 3.4.

The basic matrices required are a displacement transformation matrix and a strain-displacement transformation matrix for each element of the assemblage. To derive these matrices, we focus attention on a typical element, shown in Fig. 3.5, and assume that the local element displacements u and v are given in the form of polynomials in the local coordinate variables x and y:

$$u(x, y) = \alpha_1 + \alpha_2 x + \alpha_3 y + \alpha_4 xy \tag{3.28}$$

$$v(x, y) = \beta_1 + \beta_2 x + \beta_3 y + \beta_4 xy \tag{3.29}$$

The unknown coefficients $\alpha_1, \ldots, \beta_4$, which are also called the *generalized coordinates*, will be expressed in terms of the unknown element nodal point displacements $u_1, \ldots, u_4$ and $v_1, \ldots, v_4$. Defining

$$\mathbf{u}^T = [u_1 \quad u_2 \quad u_3 \quad u_4 \quad v_1 \quad v_2 \quad v_3 \quad v_4] \tag{3.30}$$

we can write (3.28) and (3.29) in matrix form,

$$\mathbf{u}(x, y) = \mathbf{\Phi}\boldsymbol{\alpha} \tag{3.31}$$

where

$$\mathbf{\Phi} = \begin{bmatrix} \boldsymbol{\phi} & \mathbf{0} \\ \mathbf{0} & \boldsymbol{\phi} \end{bmatrix}; \qquad \boldsymbol{\phi} = [1 \quad x \quad y \quad xy] \tag{3.32}$$

$$\mathbf{u}(x, y) = \begin{bmatrix} u(x, y) \\ v(x, y) \end{bmatrix} \tag{3.33}$$

and

$$\boldsymbol{\alpha}^T = [\alpha_1 \quad \alpha_2 \quad \alpha_3 \quad \alpha_4 \quad \beta_1 \quad \beta_2 \quad \beta_3 \quad \beta_4] \tag{3.34}$$

Equation (3.31) must hold for the nodal points of the element; therefore, using (3.31) for all four nodal points, we have

$$\mathbf{u} = \mathbf{A}\boldsymbol{\alpha} \tag{3.35}$$

in which

$$\mathbf{A} = \begin{bmatrix} \tilde{\mathbf{A}} & \mathbf{0} \\ \mathbf{0} & \tilde{\mathbf{A}} \end{bmatrix} \tag{3.36}$$

and

$$\tilde{\mathbf{A}} = \begin{bmatrix} 1 & x_1 & y_1 & x_1 y_1 \\ 1 & x_2 & y_2 & x_2 y_2 \\ 1 & x_3 & y_3 & x_3 y_3 \\ 1 & x_4 & y_4 & x_4 y_4 \end{bmatrix} \tag{3.37}$$

The generalized coordinates $\boldsymbol{\alpha}$ are therefore

$$\boldsymbol{\alpha} = \mathbf{A}^{-1}\mathbf{u} \tag{3.38}$$

where

$$\mathbf{A}^{-1} = \begin{bmatrix} \tilde{\mathbf{A}}^{-1} & \mathbf{0} \\ \mathbf{0} & \tilde{\mathbf{A}}^{-1} \end{bmatrix} \tag{3.39}$$

and substituting for $\boldsymbol{\alpha}$ into (3.31), we have

$$\mathbf{u}(x, y) = \mathbf{H}(x, y)\mathbf{u} \tag{3.40}$$

where

$$\mathbf{H}(x, y) = \mathbf{\Phi A}^{-1} \tag{3.41}$$

To obtain the element stiffness matrix, we use the virtual work theorem. Assuming plane stress conditions, the element strains are

$$\boldsymbol{\epsilon}^T = [\epsilon_{xx} \quad \epsilon_{yy} \quad \gamma_{xy}] \tag{3.42}$$

where

$$\epsilon_{xx} = \frac{\partial u}{\partial x}; \qquad \epsilon_{yy} = \frac{\partial v}{\partial y}; \qquad \gamma_{xy} = \frac{\partial u}{\partial y} + \frac{\partial v}{\partial x} \tag{3.43}$$

and using (3.40) we have

$$\boldsymbol{\epsilon} = \mathbf{Bu} \tag{3.44}$$

where

$$\mathbf{B} = \mathbf{EA}^{-1} \tag{3.45}$$

and

$$\mathbf{E} = \begin{bmatrix} 0 & 1 & 0 & y & 0 & 0 & 0 & 0 \\ 0 & 0 & 0 & 0 & 0 & 0 & 1 & x \\ 0 & 0 & 1 & x & 0 & 1 & 0 & y \end{bmatrix} \tag{3.46}$$

The element stresses are

$$\boldsymbol{\tau}^T = [\tau_{xx} \quad \tau_{yy} \quad \tau_{xy}] \tag{3.47}$$

Assuming an isotropic linear elastic material, we have for plane stress conditions,

$$\boldsymbol{\tau} = \mathbf{C}\boldsymbol{\epsilon} \tag{3.48}$$

where

$$\mathbf{C} = \frac{E}{1 - \nu^2} \begin{bmatrix} 1 & \nu & 0 \\ \nu & 1 & 0 \\ 0 & 0 & \dfrac{1 - \nu}{2} \end{bmatrix} \tag{3.49}$$

and E and ν are the material constants; i.e., E is Young's modulus and ν is Poisson's ratio.

To perform the actual finite element analysis requires the displacement and strain-displacement transformation matrices given in (3.41) and (3.45), respectively, of each element in the assemblage. The evaluation of these matrices necessitates the calculation of the matrix $\mathbf{A}$ of each element.

So far we considered the displacements of each finite element to be defined in terms of the element local nodal point displacements. However, in the derivation of the equilibrium equations of the complete element assemblage, it is expedient to express the element displacements in terms of the nodal point displacements of the complete element assemblage; i.e., we write for element m,

$$\mathbf{u}^{(m)}(x, y) = \mathbf{H}^{(m)}(x, y)\mathbf{U} \tag{3.50}$$

and similarly for the element strains and stresses:

$$\boldsymbol{\epsilon}^{(m)}(x, y) = \mathbf{B}^{(m)}(x, y)\mathbf{U} \tag{3.51}$$

$$\boldsymbol{\tau}^{(m)}(x, y) = \mathbf{C}^{(m)}\boldsymbol{\epsilon}^{(m)}(x, y) \tag{3.52}$$

where $\mathbf{u}^{(m)}$, $\boldsymbol{\epsilon}^{(m)}$, $\boldsymbol{\tau}^{(m)}$ and $\mathbf{C}^{(m)}$ are the $\mathbf{u}$, $\boldsymbol{\epsilon}$, $\boldsymbol{\tau}$ and $\mathbf{C}$ used above. However, in (3.50) to (3.52) the quantities are related to the vector $\mathbf{U}$, which stores all nodal point displacements in the global coordinate system of the complete finite element assemblage. In order to obtain $\mathbf{H}^{(m)}$ and $\mathbf{B}^{(m)}$ from $\mathbf{H}$ and $\mathbf{B}$, we first relate each local nodal point displacement component to the corresponding global nodal point displacement components, as shown in Fig. 3.6. This corresponds to a transformation from local to global coordinate

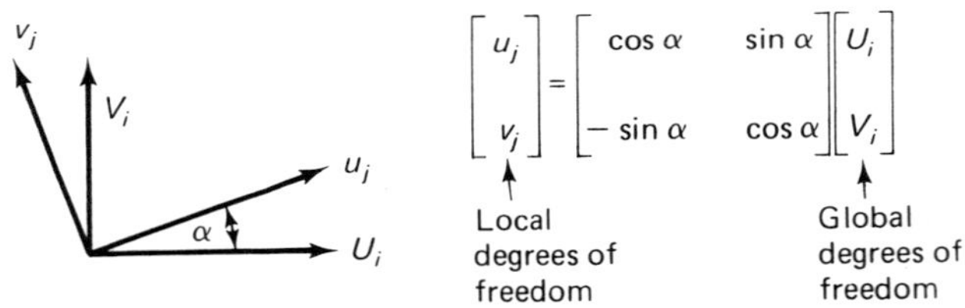

Figure 3.6. Transformation between local and global displacement components at a nodal point.

axes, as carried out in a routine manner in the analysis of conventional truss and beam structures. Next, we expand the displacement and strain-displacement transformation relationships to accommodate all nodal point global displacement components. Thus the columns of $\mathbf{H}^{(m)}$ and $\mathbf{B}^{(m)}$ consist of zero elements if they correspond to a degree of freedom not defined for element m, and they are the appropriate columns of the transformed $\mathbf{H}$ and $\mathbf{B}$, respectively, otherwise.

In the case of the cantilever analysis, the transformation from local to global directions is not required, because the directions of the element degrees of freedom already correspond to the direction of the global structure degrees of freedom (see Fig. 3.4). Hence we obtain the columns of $\mathbf{H}^{(m)}$ and $\mathbf{B}^{(m)}$ directly from the columns of $\mathbf{H}$ and $\mathbf{B}$ of the element under consideration. As an example, consider $\mathbf{H}^{(2)}$, i.e., the matrix defining the displacement variations in element 2 of the cantilever idealization in Fig. 3.4. Using the nodal point displacement definitions of the complete element assemblage given in Fig. 3.4 and of the element given in Fig. 3.5, we have

$$\mathbf{H}^{(2)} = \begin{bmatrix} 0 & 0 & H_{13} & H_{17} & H_{12} & H_{16} & 0 & 0 & H_{14} & H_{18} & H_{11} & H_{15} & 0 & 0 & \ldots \text{zeros} \ldots & 0 \\ 0 & 0 & H_{23} & H_{27} & H_{22} & H_{26} & 0 & 0 & H_{24} & H_{28} & H_{21} & H_{25} & 0 & 0 & \ldots \text{zeros} \ldots & 0 \end{bmatrix} \tag{3.53}$$

Element degrees of freedom (above columns): u_3, v_3 over U_2, V_2; u_2, v_2 over U_3, V_3; u_4, v_4 over U_5, V_5; u_1, v_1 over U_6, V_6 ←element degrees of freedom

Assemblage degrees of freedom: $U_1 \; V_1 \; U_2 \; V_2 \; U_3 \; V_3 \; U_4 \; V_4 \; U_5 \; V_5 \; U_6 \; V_6 \; U_7 \; V_7 \; \ldots \; V_9$ ←assemblage degrees of freedom

where H_{ij} is element (i, j) of $\mathbf{H}$ in (3.41) when element 2 is considered.

It should be noted that in the above derivation of the matrices $\mathbf{H}^{(m)}$ and $\mathbf{B}^{(m)}$, we first focused attention on a typical element to derive displacement and strain-displacement transformation matrices corresponding to local element degrees of freedom. This is the usual procedure when generalized coordinate finite element matrices are derived, which we consider in this chapter. On the other hand, using the isoparametric formulation as presented in Chapter 4, we can establish in many cases the matrices $\mathbf{H}^{(m)}$ and $\mathbf{B}^{(m)}$ directly without taking recourse first to a local element coordinate system.

Whichever procedure will be used to obtain the matrices $\mathbf{H}^{(m)}$ and $\mathbf{B}^{(m)}$ for finite element m in the finite element assemblage, the purpose of the presentation in this section was to develop specific ideas on the evaluation of these matrices. This way more meaning is given to the presentation of the general finite element formulation in the next section.

3.2.2 General Formulation

For a general formulation of the displacement-based finite element method, consider a general three-dimensional body. We will discuss later how the general formulation can be specialized to specific problems. Let the three-dimensional body be subjected to body forces $\mathbf{f}^B$, surface forces $\mathbf{f}^S$, concentrated forces $\mathbf{F}^i$, $i = 1, \ldots, q$, and be subjected to an initial stress $\boldsymbol{\tau}^I$. Identifying virtual quantities by placing a bar over the quantity, the virtual displacement theorem gives the relation

$$\int_V \bar{\boldsymbol{\epsilon}}^T \boldsymbol{\tau} \, dV = \int_V \bar{\mathbf{U}}^T \mathbf{f}^B \, dV + \int_S \bar{\mathbf{U}}^{S^T} \mathbf{f}^S \, dS + \sum_i \bar{\mathbf{U}}^{i^T} \mathbf{F}^i \tag{3.17}$$

where the integrations are performed over the volume V and surface S of the body, respectively, and the summation sign $\sum_i$ includes all points where concentrated forces $\mathbf{F}^i$ are applied. The quantities in the virtual work equation have been defined in (3.13) to (3.19).

The virtual work equation in (3.17) is a statement of the equilibrium of the body. In the finite element analysis we approximate the body as an assemblage of discrete finite elements with the elements being interconnected at nodal points on the element boundaries. Although not necessary, assume that the body has been idealized in such a way that all concentrated forces are applied at the nodal points. The displacements, and hence the strains, measured in a local coordinate system x, y, z within each element are assumed to be a function of the displacements at the n finite element nodal points. Therefore, for element m we have

$$\mathbf{U}^{(m)}(x, y, z) = \mathbf{H}^{(m)}(x, y, z)\mathbf{U} \tag{3.54}$$

$$\boldsymbol{\epsilon}^{(m)}(x, y, z) = \mathbf{B}^{(m)}(x, y, z)\mathbf{U} \tag{3.55}$$

where the superscript m denotes element m and $\mathbf{U}$ is a vector of the three global displacement components U_i, V_i, and W_i at all nodal points, including those at the supports of the element assemblage; i.e., $\mathbf{U}$ is a vector of dimension $3n$,

$$\mathbf{U}^T = [U_1 V_1 W_1 \quad U_2 V_2 W_2 \ldots U_n V_n W_n] \tag{3.56}$$

Although all nodal point displacements are listed in $\mathbf{U}$, it should be realized that considering a specific element, only the displacements at the nodes of the element affect the displacement and strain distribution within the element. In Section 3.2.1 we showed how the matrices $\mathbf{H}^{(m)}$ and $\mathbf{B}^{(m)}$ are derived for a simple plane stress analysis. The purpose of defining the element displacements and strains in terms of the complete array of finite element assemblage nodal point displacements is probably not obvious now. *However, it will turn out that by using (3.54) and (3.55), the assemblage process of the element matrices to the structure matrices, referred to as the direct stiffness method (see Section 3.2.1), is automatically performed in the finite element discretization process.*

The stresses in a finite element are related to the element strains and the element initial stresses, using

$$\boldsymbol{\tau}^{(m)} = \mathbf{C}^{(m)}\boldsymbol{\epsilon}^{(m)} + \boldsymbol{\tau}^{I^{(m)}} \tag{3.57}$$

where $\mathbf{C}^{(m)}$ is the elasticity matrix of element m and $\boldsymbol{\tau}^{I^{(m)}}$ are the element initial stresses. The material law stated in $\mathbf{C}^{(m)}$ for each element can be arbitrary; for example, the coefficients in the matrix could have been established by experiments.

Using the assumption on the displacements within each finite element, as expressed in (3.54), we can now derive equilibrium equations that correspond to the nodal point displacements of the assemblage of finite elements. First, we rewrite (3.17) as a sum of integrations over the volume and areas of all finite elements; i.e.,

$$\begin{aligned}\sum_m \int_{V^{(m)}} \bar{\boldsymbol{\epsilon}}^{(m)T}\boldsymbol{\tau}^{(m)}\, dV^{(m)} &= \sum_m \int_{V^{(m)}} \bar{\mathbf{U}}^{(m)T}\mathbf{f}^{B^{(m)}}\, dV^{(m)} \\ &\quad + \sum_m \int_{S^{(m)}} \bar{\mathbf{U}}^{S^{(m)}T}\mathbf{f}^{S^{(m)}}\, dS^{(m)} + \sum_i \bar{\mathbf{U}}^{iT}\mathbf{F}^i\end{aligned} \tag{3.58}$$

where $m = 1, 2, \ldots, N$ and $N =$ number of elements. It is important to note *that the integrations in (3.58) are performed over the element volumes and surfaces, and that for convenience we may use different element coordinate systems in the evaluations*. Indeed, this is basically the reason why each of the integrals can be evaluated very effectively in general element assemblages. If we substitute into (3.58) for the element displacements, strains, and stresses, using (3.54), (3.55), and (3.57), we obtain

$$\begin{aligned}\bar{\mathbf{U}}^T\left\{\sum_m \int_{V^{(m)}} \mathbf{B}^{(m)T}\mathbf{C}^{(m)}\mathbf{B}^{(m)}\, dV^{(m)}\right\}\mathbf{U} = {} & \bar{\mathbf{U}}^T\left\{\sum_m \int_{V^{(m)}} \mathbf{H}^{(m)T}\mathbf{f}^{B^{(m)}}\, dV^{(m)}\right\} \\ & + \bar{\mathbf{U}}^T\left\{\sum_m \int_{S^{(m)}} \mathbf{H}^{S^{(m)}T}\mathbf{f}^{S^{(m)}}\, dS^{(m)}\right\} \\ & - \bar{\mathbf{U}}^T\left\{\sum_m \int_{V^{(m)}} \mathbf{B}^{(m)T}\boldsymbol{\tau}^{I^{(m)}}\, dV^{(m)}\right\} \\ & + \bar{\mathbf{U}}^T\mathbf{F}\end{aligned} \tag{3.59}$$

where $\mathbf{F}$ is a vector of the externally applied forces to the nodes of the element assemblage. We should note that the ith component in $\mathbf{F}$ is the concentrated nodal force, which corresponds to the ith displacement component in $\mathbf{U}$. In (3.59) the nodal point displacement vector $\mathbf{U}$ of the element assemblage is independent of the element considered and is therefore taken out of the summation signs.

To obtain from (3.59) the equilibrium equations that we want to use for the solution of the nodal point displacements, we invoke the virtual displacement theorem, by imposing unit virtual displacements in turn at all displacement components. In this way we have $\bar{\mathbf{U}}^T = \mathbf{I}$ ($\mathbf{I} =$ the identity matrix), and the equilibrium equations of the element assemblage corresponding to the nodal point displacements are

$$\mathbf{KU} = \mathbf{R} \tag{3.60}$$

where

$$\mathbf{R} = \mathbf{R}_B + \mathbf{R}_S - \mathbf{R}_I + \mathbf{R}_C \tag{3.61}$$

The matrix $\mathbf{K}$ is the stiffness matrix of the element assemblage,

$$\mathbf{K} = \sum_m \int_{V^{(m)}} \mathbf{B}^{(m)T}\mathbf{C}^{(m)}\mathbf{B}^{(m)}\, dV^{(m)} \tag{3.62}$$

The load vector $\mathbf{R}$ includes the effect of the element body forces,

$$\mathbf{R}_B = \sum_m \int_{V^{(m)}} \mathbf{H}^{(m)T}\mathbf{f}^{B^{(m)}}\, dV^{(m)} \tag{3.63}$$

the effect of the element surface forces,

$$\mathbf{R}_S = \sum_m \int_{S^{(m)}} \mathbf{H}^{S^{(m)}T}\mathbf{f}^{S^{(m)}}\, dS^{(m)} \tag{3.64}$$

the effect of the element initial stresses,

$$\mathbf{R}_I = \sum_m \int_{V^{(m)}} \mathbf{B}^{(m)T} \boldsymbol{\tau}^{I^{(m)}} \, dV^{(m)} \tag{3.65}$$

and the concentrated loads,

$$\mathbf{R}_C = \mathbf{F} \tag{3.66}$$

We should note that the summation of the element volume integrals in (3.62) implies the direct addition of the element stiffness matrices to obtain the stiffness matrix of the total element assemblage. In the same way, the assemblage body force vector $\mathbf{R}_B$ is calculated by directly adding the element body force vectors; and similarly $\mathbf{R}_S$, $\mathbf{R}_I$, and $\mathbf{R}_C$ are obtained. Therefore, the formulation of the equilibrium equations in (3.60) in the way carried out above includes the assemblage process to obtain the structure matrices from the element matrices, usually referred to as the direct stiffness method. However, in practice, the finite element matrices are calculated in compact form; i.e., they are of the order of the element degrees of freedom and are assembled using the direct stiffness method as described in Sections 3.1 and 6.2.3.

One interesting and important aspect should be pointed out. It may be of concern that in the derivation of the equilibrium equations, we did not consider rotational degrees of freedom at the finite element nodes. Using beam, plate, and shell elements, rotational degrees of freedom at the element nodal points are defined and included in the finite element solution. However, it should be realized that these rotational degrees of freedom are a consequence of the kinematic assumptions on the behavior of the continuum (as further discussed in Section 3.2.4) and that rotations do not appear explicitly in the general virtual work equation given in (3.17). For this reason we also did not include rotational degrees of freedom at the finite element nodes. On the other hand, if, for example, the general virtual work equation were specialized to a virtual work equation that is applicable to plate analysis only (see Section 3.3.1), rotational degrees of freedom would have to be used in the finite element analysis.

It should be noted that the finite element system equilibrium equations in (3.60) have so far been established for the unsupported element assemblage. The next step, therefore, is to impose the displacement boundary conditions in the same way as in beam and truss structural analysis. Namely, if we write the equilibrium equations in (3.60) in the form

$$\begin{bmatrix} \mathbf{K}_{aa} & \mathbf{K}_{ab} \\ \mathbf{K}_{ba} & \mathbf{K}_{bb} \end{bmatrix} \begin{bmatrix} \mathbf{U}_a \\ \mathbf{U}_b \end{bmatrix} = \begin{bmatrix} \mathbf{R}_a \\ \mathbf{R}_b \end{bmatrix} \tag{3.67}$$

where $\mathbf{U}_a$ are the unknown displacements and $\mathbf{U}_b$ are the known boundary displacements, we have

$$\mathbf{K}_{aa}\mathbf{U}_a = \bar{\mathbf{R}}_a \tag{3.68}$$

where

$$\bar{\mathbf{R}}_a = \mathbf{R}_a - \mathbf{K}_{ab}\mathbf{U}_b \tag{3.69}$$

The unknown nodal point displacements are calculated from (3.68), and the nodal point forces corresponding to $\mathbf{U}_b$ are then

$$\mathbf{R}_b = \mathbf{K}_{ba}\mathbf{U}_a + \mathbf{K}_{bb}\mathbf{U}_b \tag{3.70}$$

If boundary displacements are specified in other than global directions (i.e., we have skew boundary conditions), it is expedient to transform the element stiffness matrices that correspond to a boundary node prior to their assemblage into the structure stiffness matrix (see Fig. 3.7). Namely, by establishing the element stiffness matrices and hence the structure stiffness matrix corresponding to the specified displacement degrees of freedom at the boundary nodes, the given displacements can be directly imposed, as presented in (3.67) to (3.70).

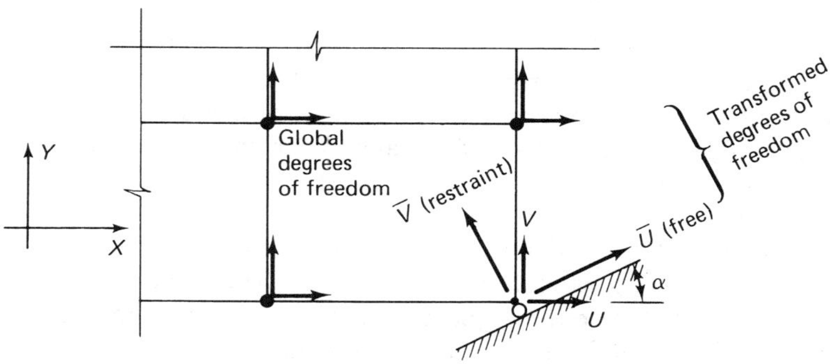

Figure 3.7. Degrees of freedom definition at skew boundary.

As an alternative procedure to the one given in (3.67) to (3.70), a boundary element can be used to impose a given displacement.[24] If a nonzero displacement is to be specified at a degree of freedom i, say $U_i = b$, the equation

$$kU_i = kb \tag{3.71}$$

is added into the equilibrium equations in (3.60), where $k \gg k_{ii}$. Therefore, the solution of the equilibrium equations must now give $U_i = b$. Physically, this can be interpreted as adding at the degree of freedom i a spring of large stiffness k, and specifying a load which because of the relatively flexible element assemblage at this degree of freedom produces the required displacement b. The same procedure can be used to impose skew boundary displacement conditions by choosing the appropriate spring direction, as shown in Fig. 3.8. But in that case, large off-diagonal elements are introduced into the stiffness matrix and care has to be exercised that the accurate solution of the equilibrium equations is still possible (see Section 7.5).

The solution of the equilibrium equations (3.60) yields all element nodal point displacements. The element stresses are then obtained using (3.57)

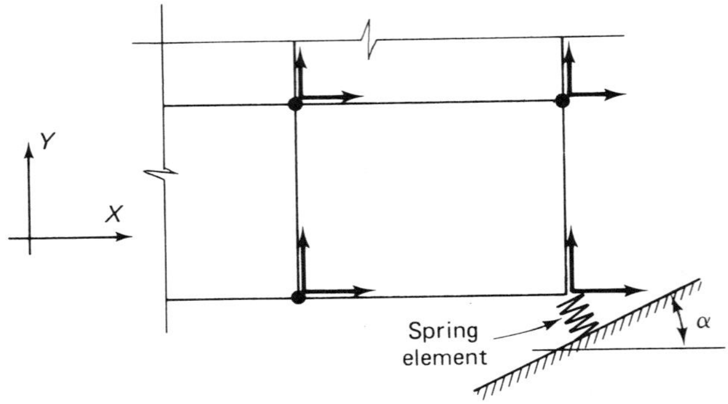

Figure 3.8. Skew boundary condition imposed using spring element.

with (3.55). It should be noted that the final element stresses are given by (3.57); i.e., there are no corrections due to element body or surface forces, as we would apply in the exact analysis of truss or beam structures. In finite element analysis, we do not satisfy local equilibrium in each element. Equilibrium is only satisfied in a global, integrated sense for the complete element assemblage.

Equation (3.60) is a statement of the static equilibrium of the element assemblage. It should be realized, though, that the applied forces may vary with time, in which case the displacements also vary with time and (3.60) is a statement of equilibrium for any specific point in time. However, in most cases when the loads are time dependent, inertia forces need to be considered; i.e., a truly dynamic problem needs to be solved. Using d'Alembert's principle, we can simply include the element inertia forces as part of the body forces.[25] If the element accelerations are approximated in the same way as the element displacements in (3.54), the contribution from the total body forces to the load vector **R** is

$$\mathbf{R}_B = \sum_m \int_{V^{(m)}} \mathbf{H}^{(m)T}[\mathbf{f}^{B(m)} - \rho^{(m)}\mathbf{H}^{(m)}\ddot{\mathbf{U}}]\, dV^{(m)} \tag{3.72}$$

where $\mathbf{f}^{B(m)}$ no longer includes inertia forces, $\ddot{\mathbf{U}}$ lists the nodal point accelerations (i.e., is the second time derivative of **U**), and $\rho^{(m)}$ is the mass density of element m. The equilibrium equations are, in this case,

$$\mathbf{M}\ddot{\mathbf{U}} + \mathbf{K}\mathbf{U} = \mathbf{R} \tag{3.73}$$

where **R** and **U** are time dependent. The matrix **M** is the mass matrix of the structure,

$$\mathbf{M} = \sum_m \int_{V^{(m)}} \rho^{(m)}\mathbf{H}^{(m)T}\mathbf{H}^{(m)}\, dV^{(m)} \tag{3.74}$$

In actually measured dynamic response of structures it is observed that energy is dissipated during vibration, which, in vibration analysis is usually taken account of by introducing velocity dependent damping forces. Introducing the damping forces as additional contributions to the body forces, we obtain corresponding to (3.72),

$$\mathbf{R}_B = \sum_m \int_{V^{(m)}} \mathbf{H}^{(m)T}[\mathbf{f}^{B(m)} - \rho^{(m)}\mathbf{H}^{(m)}\ddot{\mathbf{U}} - \kappa^{(m)}\mathbf{H}^{(m)}\dot{\mathbf{U}}]\, dV^{(m)} \tag{3.75}$$

In this case the vectors $\mathbf{f}^{B(m)}$ no longer include inertia and velocity dependent damping forces, $\dot{\mathbf{U}}$ is a vector of the nodal point velocities (i.e., the first time derivative of $\mathbf{U}$), and $\kappa^{(m)}$ is the damping property parameter of element m. The equilibrium equations are, in this case,

$$\mathbf{M}\ddot{\mathbf{U}} + \mathbf{C}\dot{\mathbf{U}} + \mathbf{K}\mathbf{U} = \mathbf{R} \tag{3.76}$$

where $\mathbf{C}$ is the damping matrix of the structure; i.e., formally,

$$\mathbf{C} = \sum_m \int_{V^{(m)}} \kappa^{(m)}\mathbf{H}^{(m)T}\mathbf{H}^{(m)}\, dV^{(m)} \tag{3.77}$$

In practice it is difficult, if not impossible, to determine for general finite element assemblages the element damping parameters, in particular because the damping properties are frequency dependent. For this reason, the matrix $\mathbf{C}$ is in general not assembled from element damping matrices, but is constructed using the mass matrix and stiffness matrix of the complete element assemblage together with experimental results on the amount of damping. Some formulations used to construct physically significant damping matrices are described in Section 8.3.3.

Example 3.1

Evaluate the matrices **K**, **M**, and **R** for the two-element truss assemblage in Fig. 3.9. Assume a linear displacement distribution between nodal points in each element.

The first step is to construct the matrices $\mathbf{H}^{(m)}$ and $\mathbf{B}^{(m)}$ for $m = 1, 2$. Corresponding to the displacement vector $\mathbf{U}^T = [U_1 U_2 U_3]$, we have

$$\mathbf{H}^{(1)} = \left[\left(1 - \frac{y}{100}\right) \quad \frac{y}{100} \quad 0\right]$$

$$\mathbf{B}^{(1)} = \left[-\frac{1}{100} \quad \frac{1}{100} \quad 0\right]$$

$$\mathbf{H}^{(2)} = \left[0 \quad \left(1 - \frac{y}{80}\right) \quad \frac{y}{80}\right]$$

$$\mathbf{B}^{(2)} = \left[0 \quad -\frac{1}{80} \quad \frac{1}{80}\right]$$

The material property matrices are

$$\mathbf{C}^{(1)} = E; \qquad \mathbf{C}^{(2)} = E$$

where E is Young's modulus of the material. For the volume integrations we need the cross-sectional areas of the elements. We have

$$A^{(1)} = 1 \text{ cm}^2; \qquad A^{(2)} = \left(1 + \frac{y}{40}\right)^2 \text{cm}^2$$

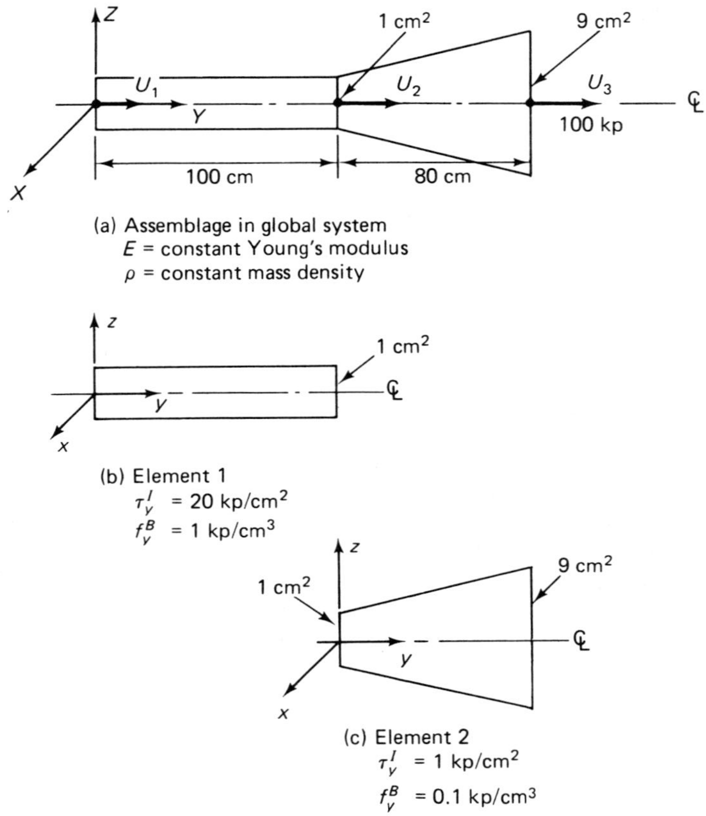

Figure 3.9. Two-element bar assemblage. (1 kp = 1 kgf)

The body forces, initial stresses, and mass densities are given in Fig. 3.9. We therefore have

$$\mathbf{K} = (1)E \int_0^{100} \begin{bmatrix} -\dfrac{1}{100} \\ \dfrac{1}{100} \\ 0 \end{bmatrix} \begin{bmatrix} -\dfrac{1}{100} & \dfrac{1}{100} & 0 \end{bmatrix} dy$$

$$+ E \int_0^{80} \left(1 + \frac{y}{40}\right)^2 \begin{bmatrix} 0 \\ -\dfrac{1}{80} \\ \dfrac{1}{80} \end{bmatrix} \begin{bmatrix} 0 & -\dfrac{1}{80} & \dfrac{1}{80} \end{bmatrix} dy$$

or

$$\mathbf{K} = \frac{E}{100}\begin{bmatrix} 1 & -1 & 0 \\ -1 & 1 & 0 \\ 0 & 0 & 0 \end{bmatrix} + \frac{13E}{240}\begin{bmatrix} 0 & 0 & 0 \\ 0 & 1 & -1 \\ 0 & -1 & 1 \end{bmatrix}$$

$$= \frac{E}{240}\begin{bmatrix} 2.4 & -2.4 & 0 \\ -2.4 & 15.4 & -13 \\ 0 & -13 & 13 \end{bmatrix}$$

Similarly,

$$\mathbf{M} = (1)\rho \int_0^{100} \begin{bmatrix} 1 - \frac{y}{100} \\ \frac{y}{100} \\ 0 \end{bmatrix} \left[\left(1 - \frac{y}{100}\right) \quad \frac{y}{100} \quad 0\right] dy$$

$$+ \rho \int_0^{80} \left(1 + \frac{y}{40}\right)^2 \begin{bmatrix} 0 \\ 1 - \frac{y}{80} \\ \frac{y}{80} \end{bmatrix} \left[0 \quad \left(1 - \frac{y}{80}\right) \quad \frac{y}{80}\right] dy$$

Hence

$$\mathbf{M} = \frac{\rho}{6}\begin{bmatrix} 200 & 100 & 0 \\ 100 & 584 & 336 \\ 0 & 336 & 1024 \end{bmatrix}$$

Also,

$$\mathbf{R}_B = (1)\int_0^{100} \begin{bmatrix} 1 - \frac{y}{100} \\ \frac{y}{100} \\ 0 \end{bmatrix} (1)\, dy + \int_0^{80} \left(1 + \frac{y}{40}\right)^2 \begin{bmatrix} 0 \\ 1 - \frac{y}{80} \\ \frac{y}{80} \end{bmatrix} \left(\frac{1}{10}\right) dy = \frac{1}{3}\begin{bmatrix} 150 \\ 186 \\ 68 \end{bmatrix}$$

$$\mathbf{R}_I = (1)\int_0^{100} \begin{bmatrix} -\frac{1}{100} \\ \frac{1}{100} \\ 0 \end{bmatrix} (20)\, dy + \int_0^{80} \left(1 + \frac{y}{40}\right)^2 \begin{bmatrix} 0 \\ -\frac{1}{80} \\ \frac{1}{80} \end{bmatrix} (1)\, dy = \frac{1}{3}\begin{bmatrix} -60 \\ 47 \\ 13 \end{bmatrix}$$

$$\mathbf{R}_c = \begin{bmatrix} 0 \\ 0 \\ 100 \end{bmatrix}$$

and because $\mathbf{R}_S = \mathbf{0}$, we have

$$\mathbf{R} = \frac{1}{3}\begin{bmatrix} 210 \\ 139 \\ 355 \end{bmatrix}$$

It may be noted that **M** would only be required in a dynamic analysis, in which case nonzero initial conditions on **U** or time varying loads should be specified.

3.2.3 Lumping of Structure Properties and Loads

A physical interpretation of the finite element procedure of analysis as presented in the previous section is that the structure properties—stiffness, mass, and damping—and the loads, internal and external, are lumped at the discrete nodes of the element assemblage using the virtual work principle. *Because the same interpolation functions are employed in the calculation of the load vectors and in the evaluation of the stiffness matrix, we say that "consistent" load vectors and a consistent mass matrix have been evaluated.* We shall discuss in Section 5.3 in what sense the load vectors are consistent. It may now be recognized that instead of performing the integrations leading to the consistent load vector, we may evaluate an approximate load vector by simply adding to the actual applied concentrated node forces $\mathbf{R}_C$ additional forces that are in some sense equivalent to the distributed loads on the elements. A somewhat obvious way of constructing approximate load vectors is to calculate the total body and surface forces corresponding to an

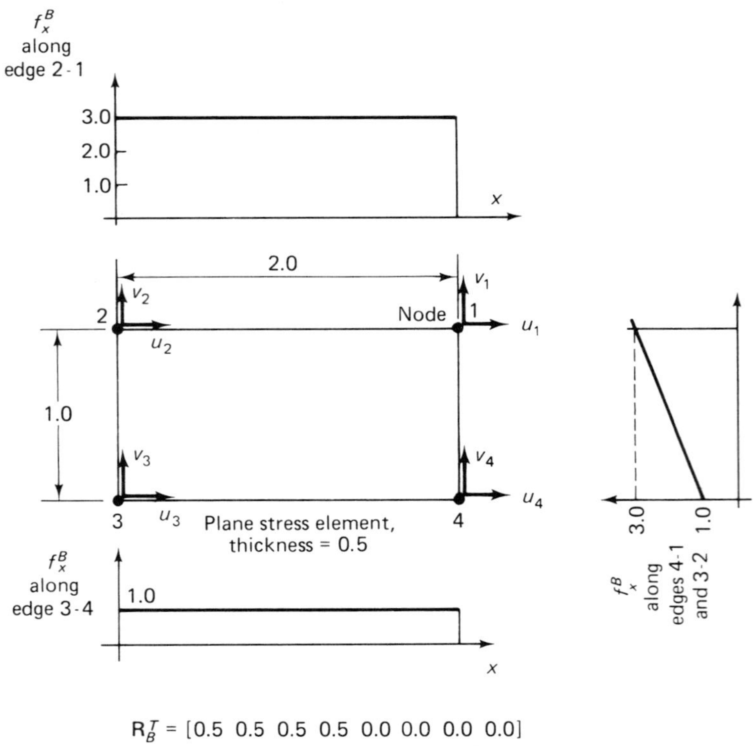

Figure 3.10. Body force distribution and corresponding lumped body force vector $\mathbf{R}_B$ of a rectangular element.

element and to assign equal parts to the appropriate element nodal degrees of freedom. Consider as an example the rectangular plane stress element in Fig. 3.10, with the variation of the body force shown. The total body force is equal to 2.0, and hence we obtain the lumped body force vector given in the figure.

Considering the derivation of an element mass matrix, we recall that the inertia forces have been considered as part of the body forces. Hence we may also establish an approximate mass matrix by lumping equal parts of the total mass to the nodal points. Realizing that each nodal mass essentially corresponds to the mass of an element-contributing-volume around the node, we note that using this procedure of lumping mass, we assume in essence that the accelerations of the contributing volume to a node are constant and equal to the nodal values.

An important advantage of using a lumped mass matrix is that the matrix is diagonal, and, as will be seen later, the numerical operations for the solution of the dynamic equations of equilibrium are reduced.

Example 3.2

Evaluate the lumped body force vector and the lumped mass matrix of the truss element assemblage in Fig. 3.9.

The lumped mass matrix is

$$\mathbf{M} = \rho \int_0^{100} (1) \begin{bmatrix} \frac{1}{2} & 0 & 0 \\ 0 & \frac{1}{2} & 0 \\ 0 & 0 & 0 \end{bmatrix} dy + \rho \int_0^{80} \left(1 + \frac{y}{40}\right)^2 \begin{bmatrix} 0 & 0 & 0 \\ 0 & \frac{1}{2} & 0 \\ 0 & 0 & \frac{1}{2} \end{bmatrix} dy$$

or

$$\mathbf{M} = \frac{\rho}{3} \begin{bmatrix} 150 & 0 & 0 \\ 0 & 670 & 0 \\ 0 & 0 & 520 \end{bmatrix}$$

Similarly, the lumped body force vector is

$$\mathbf{R}_B = \int_0^{100} (1) \begin{bmatrix} \frac{1}{2} \\ \frac{1}{2} \\ 0 \end{bmatrix} (1)\, dy + \int_0^{80} \left(1 + \frac{y}{40}\right)^2 \begin{bmatrix} 0 \\ \frac{1}{2} \\ \frac{1}{2} \end{bmatrix} \left(\frac{1}{10}\right) dy$$

$$= \frac{1}{3} \begin{bmatrix} 150 \\ 202 \\ 52 \end{bmatrix}$$

It may be noted that, as required, the sums of all elements in $\mathbf{M}$ and in $\mathbf{R}_B$ in this example and in Example 3.1 are the same.

3.2.4 Specialization of the General Formulation

In Section 3.2.2 the finite element discretization procedure and derivation of the equilibrium equations was presented in general; i.e., a general three-dimensional body was considered. Although in theory any body may be understood to be three-dimensional, for practical analysis it is in many cases imperative to reduce the dimensionality of the problem. The first step in a finite element analysis is therefore to decide what kind of problem is at hand. This decision is based on the assumptions used in the theory of elasticity approach to the solution of specific problems.[26-28] The classes of problems that are encountered may be summarized as (1) truss, (2) beam, (3) plane stress, (4) plane strain, (5) axisymmetric, (6) plate bending, (7) thin shell, (8) thick shell, and (9) general three-dimensional.

For each of these problem cases, the general formulation is applicable. If a one- or two-dimensional analysis is to be carried out (plane stress, plane strain, or axisymmetric analysis), we must include only those displacement, stress, and strain components that are actually applicable. This has already been done in Section 3.2.1, where we considered a plane stress case, and also in Examples 3.1 and 3.2, in which a bar was analyzed. For the analysis of plate bending and shell problems, an additional consideration is necessary. It should be realized that plates and shells are three-dimensional continua for which static and kinematic assumptions are used to reduce the number of static and kinematic variables.[27,28] For example, in the analysis of a flat plate, using the Kirchhoff plate theory we only have three degrees of freedom: at node k one transverse displacement w_k and two rotations, θ_k^x and θ_k^y, at the center line of the plate, as shown in Fig. 3.11. In shell analysis also the membrane displacements u_k and v_k would be included by simply combining the plate bending and membrane stiffnesses. In addition to neglecting certain stress and strain components in the general formulation, it is also necessary now to calculate the nonvanishing stress and strain quantities using the kinematic assumption for the displacements. It turns out to be convenient in the evaluation of the virtual work equation to work with moments per unit length of plate, curvatures, and a moment-curvature material matrix. This way the integration through the thickness of the plate is taken into account directly. To summarize, we should realize that the general formulation given in (3.54) to (3.66) is the basis of a variety of specific finite element derivations. Some typical elements are presented in Fig. 3.12. The virtual work equation to be used for the derivation of a specific element is obtained by incorporating into the general virtual work equation in (3.17) the assumptions employed in the classical theory of elasticity. We shall discuss the formulation of some specific elements in Section 3.3.

Considering the various finite elements in Fig. 3.12, another important point should be stressed once more. In the general formulation, we could use for each finite element a different local coordinate system; i.e., although a global coordinate system is used in (3.17), the element displacements and strains that are assumed can correspond to a local coordinate system, and

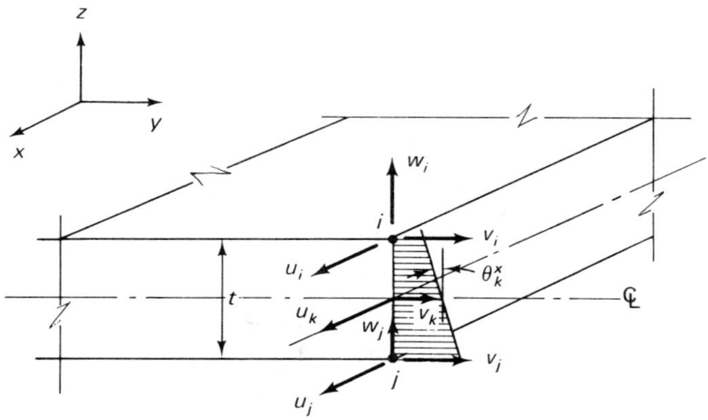

(a) Three-dimensional analysis of plate

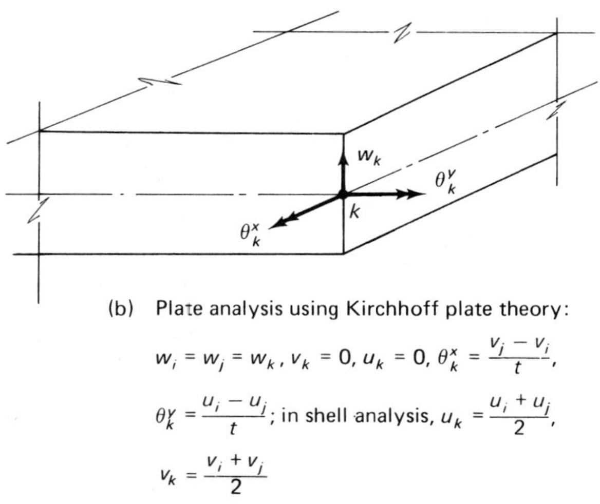

(b) Plate analysis using Kirchhoff plate theory:

$$w_i = w_j = w_k,\ v_k = 0,\ u_k = 0,\ \theta_k^x = \frac{v_j - v_i}{t},$$

$$\theta_k^y = \frac{u_i - u_j}{t};\ \text{in shell analysis, } u_k = \frac{u_i + u_j}{2},$$

$$v_k = \frac{v_i + v_j}{2}$$

Figure 3.11. Kinematic assumptions in plate analysis.

consequently the integrations in the calculation of the element matrices in (3.62) to (3.65) are carried out in the element local coordinate systems. Hence it is possible, without increasing the complexity of the analysis, to have in one element assemblage different kinds of elements. Indeed, it is the possibility that different element types can be combined with relative ease that has made the finite element method very attractive for general applications.

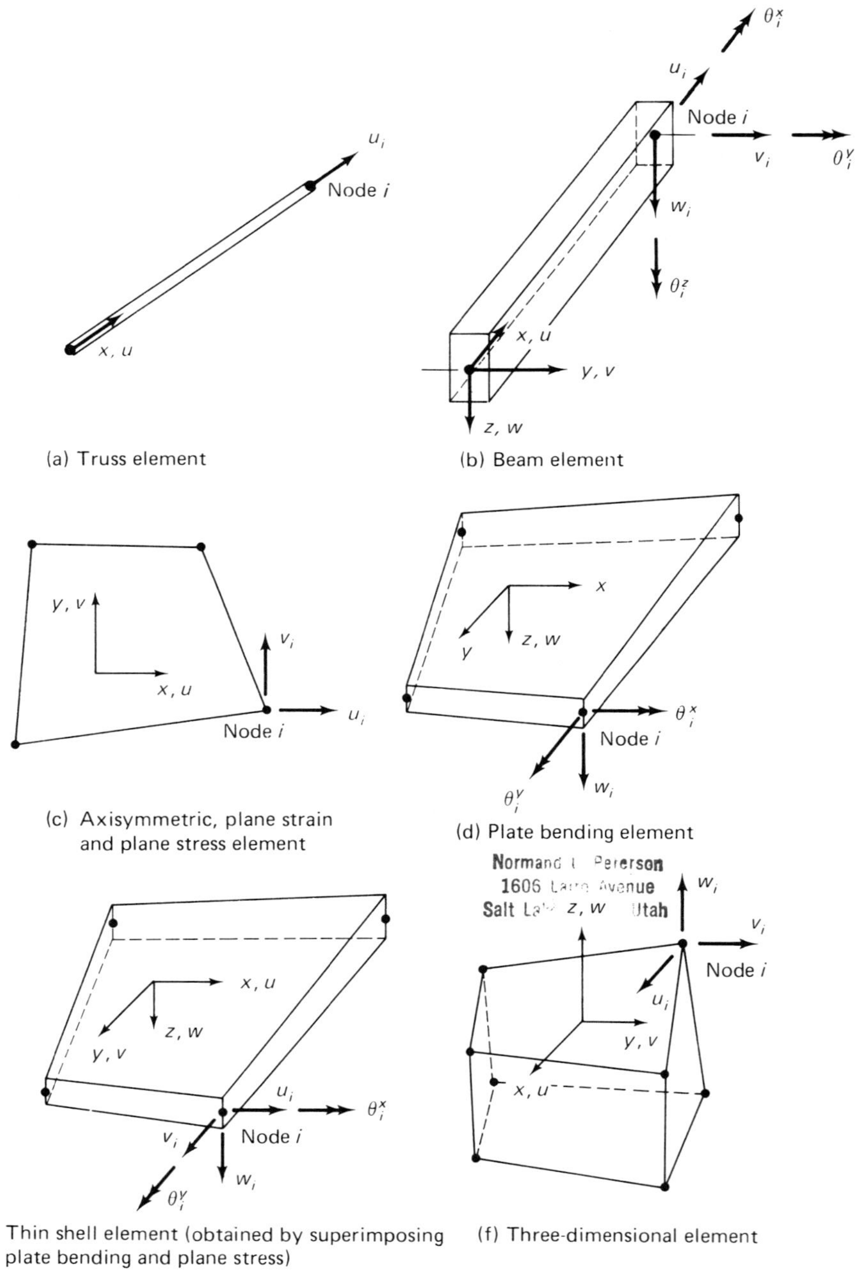

Figure 3.12. Some typical finite elements.

3.2.5 Requirements for Monotonic Convergence

With the idealized structure having been represented as an assemblage of finite elements, the accuracy of the analysis depends mainly on the number of elements used, and on the nature of the assumed displacement functions within the elements.[29,30] In particular, it is important that the accuracy of analysis can be increased by using more elements in the representation of the structure provided that the elements satisfy certain convergence requirements. However, in this context, it should be noted that although convergence in the analysis of the idealized structure might have been obtained (within certain convergence limits), the behavior of the actual structure has been assessed only within the limits of the assumptions used to arrive at the idealized structure.

Considering the convergence requirements, the elements should be complete and compatible, which is also called conforming. If the elements used are both complete and compatible, convergence is monotonic; i.e., the accuracy of the analysis results measured in some norm increases continuously as the number of elements is increased. However, if the elements are only complete and not compatible, the analysis results may still converge in the limit to the "exact" results but in general do not converge monotonically (see Section 5.5).

The requirement of completeness means that the displacement functions of the elements must be able to represent the rigid body displacements and the constant strain states. The rigid body displacements are those displacement modes that the element must be able to undergo without stresses being developed in it. For example, a plane stress element must be able to translate uniformly in either direction of its plane and to rotate without straining.

The necessity for the constant strain states can physically be understood if we imagine that more and more elements are used in the assemblage to represent the structure. Then in the limit as each element approaches a very small size, the strain in each element approaches a constant value, and any complex variation of strain within the structure can be approximated.

The concept of compatibility means that the displacements within the elements and across the element boundaries are continuous. Physically, compatibility assures that no gaps occur between elements when the assemblage is loaded. When only translational degrees of freedom are defined at the element nodes, only continuity in the displacements u, v, and w, whichever is applicable, must be preserved. However, when rotational degrees of freedom are also defined at the element nodes, such as in the plate bending element in Fig. 3.12, it is also necessary to satisfy element continuity in the corresponding first displacement derivatives. This is a consequence of the kinematic assumption on the displacements over the depth of the plate bending element; namely, the continuity in the displacement w and the derivatives

$\partial w/\partial x$ and/or $\partial w/\partial y$ along the respective element edges assures continuity of displacements over the thickness of adjoining elements (see Fig. 3.11).

Compatibility is automatically assured between truss and beam elements because they only join at the nodal points, and compatibility is relatively easy to maintain in two-dimensional plane strain, plane stress, and axisymmetric analysis and in three-dimensional analysis, when only u, v, and w degrees of freedom are used as nodal point variables. However, the requirements of compatibility are difficult to satisfy in plate bending analysis, and in particular in thin shell analysis. A number of nonconforming plate bending and thin shell elements, obtained by combining the plate bending and plane stress stiffness of the element, are presently in use and are known to yield good results.

Considering finally actual practical analyses, in which combinations of various element types are used, compatibility between all elements is not satisfied in many cases. However, for practical purposes, the analysis results are sufficiently accurate because the assumptions used to represent the actual structure by an assemblage of finite elements might not, from a practical point of view, even justify a more refined analysis.

Another important practical point should be noted. Assume that elements are employed that are complete and compatible, i.e., satisfy all criteria for monotonic convergence. *It is important to realize that this does not mean that stresses are continuous across element boundaries.* Element stresses are calculated using (3.55) and (3.57), and the stresses obtained at an element edge (or face) when calculated in adjacent elements may differ substantially if a coarse finite element idealization is used. The stress differences at the element boundaries decrease as the finite element idealization is refined. In practice, acceptable results are usually obtained if element boundary stresses are averaged. Frequently, stress distributions are also based on calculating only stresses in the interior of the elements.

Whether a specific element is complete and compatible depends on the formulation used, and each formulation need be studied individually. Consider the following simple example.

Example 3.3

Investigate if the plane stress element used in Section 3.2.1 is compatible and complete.

We had for the displacements of the element,

$$u(x, y) = \alpha_1 + \alpha_2 x + \alpha_3 y + \alpha_4 xy$$
$$v(x, y) = \beta_1 + \beta_2 x + \beta_3 y + \beta_4 xy$$

Observing that the displacements within an element are continuous, in order to show that the element is compatible we only need to investigate if interelement continuity is also preserved when an element assemblage is loaded. Consider two

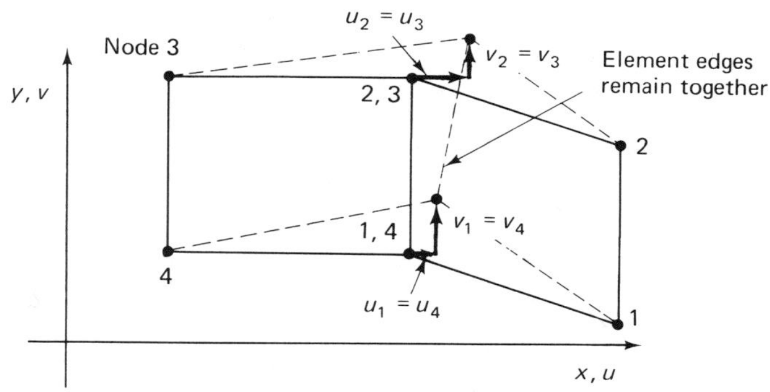

Figure 3.13. Compatibility of plane stress element.

elements interconnected at two node points (Fig. 3.13) onto which we impose two arbitrary displacements. It follows from the displacement assumptions that the points on the adjoining element edges displace linearly, and therefore continuity between the elements is preserved. Hence the element is compatible.

Considering completeness, the displacement functions show that a rigid body translation in the x direction is achieved if only α_1 is nonzero. Similarly, a rigid body displacement in the y direction is imposed by having only β_1 nonzero, and for a rigid body rotation we need only that α_3 and β_2 are nonzero with $\beta_2 = -\alpha_3$. The same conclusion can also be arrived at using the matrix that relates the strains to the generalized coordinates in (3.46). This matrix also shows that constant strain states are possible. Therefore, the element is complete.

In the preceding discussion we considered the conditions for monotonic convergence but did not mention the rate with which convergence occurs. The rate of convergence depends on the polynomial expansions used in the displacement assumptions, and in this context it should be mentioned that the use of *complete polynomials* is most desirable.[31] Figure 3.14 shows the polynomial terms that should be included to have complete polynomials in x and y for two-dimensional analysis. It is seen that all possible terms of the form $x^\alpha\ y^\beta$ are present, where $\alpha + \beta = c$ and c is the degree to which the polynomial is complete. As an example, we may note that the element investigated in Example 3.3 uses a polynomial displacement expansion that is complete to degree 1 only. Considering three-dimensional analysis, a figure analogous to Fig. 3.14 can be drawn, in which the variable z would be included.

It is important to note that when studying the convergence characteristics of an element, the use of an eigenvalue and vector solution is helpful. In Section 5.3 we discuss, in addition to the completeness requirements, that provided that the monotonic convergence conditions are satisfied, the finite

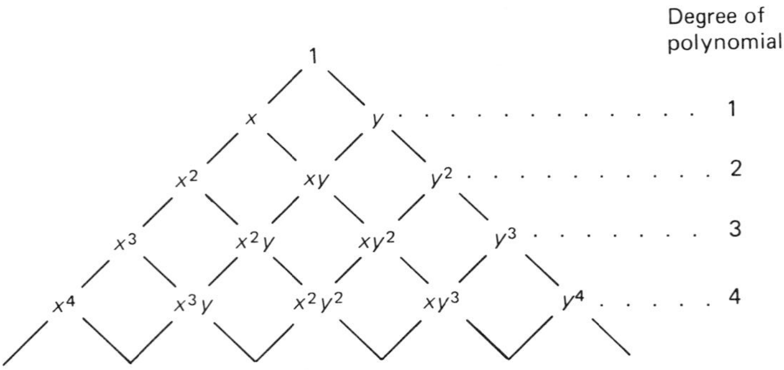

Figure 3.14. Polynomial terms for complete polynomials in two-dimensional analysis.

element analysis overestimates the stiffness of the system; i.e., the calculated displacements are smaller (in some norm) than the exact displacements. Hence, when studying the effectiveness of a finite element, we need to investigate its stiffness characteristics, and an element that is less stiff will be more effective. One way of investigating the stiffness characteristics is to represent the element stiffness matrix in the basis of its eigenvectors. Namely, solving the eigenproblem

$$\mathbf{K}\boldsymbol{\phi} = \lambda\boldsymbol{\phi} \tag{3.78}$$

we have (see Section 2.6)

$$\mathbf{K}\boldsymbol{\Phi} = \boldsymbol{\Phi}\boldsymbol{\Lambda} \tag{3.79}$$

where $\boldsymbol{\Phi}$ is a matrix storing the eigenvectors $\boldsymbol{\phi}_1, \ldots, \boldsymbol{\phi}_n$ and $\boldsymbol{\Lambda}$ is a diagonal matrix storing the corresponding eigenvalues, $\boldsymbol{\Lambda} = \text{diag}(\lambda_i)$. Using the eigenvector orthonormality property we thus have

$$\boldsymbol{\Phi}^T\mathbf{K}\boldsymbol{\Phi} = \boldsymbol{\Lambda} \tag{3.80}$$

We may look at $\boldsymbol{\Lambda}$ as being the stiffness matrix of the element corresponding to the eigenvector displacement modes. The stiffness coefficients $\lambda_1, \ldots, \lambda_n$ directly display how stiff the element is in the corresponding displacement mode. Thus the transformation in (3.80) shows clearly whether the rigid body modes and constant strain states are present and which higher displacement modes are well represented by the element. As an example, the eight eigenvectors and corresponding eigenvalues of a four-node element are shown in Fig. 3.15 (see Section 3.2.1 and Example 4.2). It is seen that the three rigid body modes and constant strain states are present. In addition, the element has two flexural modes.

We shall encounter the procedure of representing the stiffness matrix in the eigenvector basis further in Section 7.2.5 and Chapters 8 and 10.

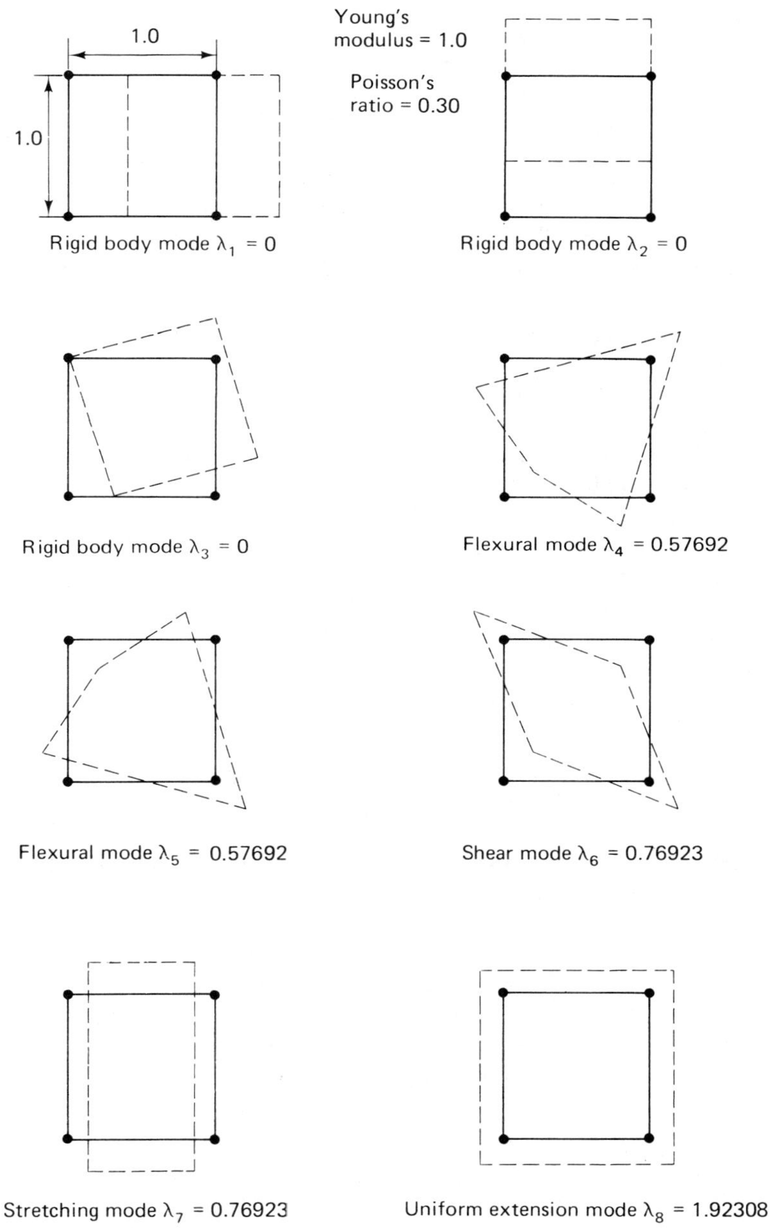

Figure 3.15. Eigenvectors and eigenvalues of four-node plane strain element.

3.3 DERIVATION OF GENERALIZED COORDINATE FINITE ELEMENT MODELS

In the introductory discussion of the plane stress element in Section 3.2, we used for the u and v displacements simple linear polynomial assumptions, where we identified the unknown coefficients in the polynomials as generalized coordinates. The number of unknown coefficients in the polynomials was equal to the number of element nodal point displacements. Expressing the generalized coordinates in terms of the element nodal point displacements, we found that, in general, each polynomial coefficient is not an actual physical displacement but is equal to a linear combination of the element nodal point displacements.

Finite element matrices which are formulated by assuming that the displacements vary in the form of a function whose unknown coefficients are treated as generalized coordinates are referred to as generalized coordinate finite element models. A rather natural class of functions to use for approximating element displacements are polynomials, because polynomials are commonly employed to approximate unknown functions, and the higher the degree of the polynomial, the better is the approximation that we can expect. In addition, polynomials are easy to differentiate; i.e., if the polynomials approximate the displacements of the structure, we can evaluate the strains with relative ease.

Using polynomial displacement assumptions, a very large number of finite elements for practically all problems in structural mechanics have been developed.

As another class, the trigonometric functions can be very effective in specific finite element applications.[32-36] These functions are used in the finite strip method and in the analysis of axisymmetric structures subjected to nonaxisymmetric loading. The advantage of the trigonometric functions lies in their orthogonality properties. Namely, if sine and cosine products are integrated over an appropriate interval, the integral can be zero. But this means that there is no coupling in the equilibrium equations between the generalized coordinates that correspond to the sine and cosine functions, and the equilibrium equations can be solved more effectively. In this context it may be noted that the best functions that we could use in the finite element analysis would be given by the eigenvectors of the problem, because they would give a diagonal stiffness matrix. However, these functions are not known, and for general application, polynomial assumptions for the finite element displacements are most natural.

The objective in this section is to describe the formulation of a variety of generalized coordinate finite element models that use polynomials to approximate the displacement fields. Other functions will, in principle, be used in the same way. In the presentation, emphasis is given to the general formulation rather than to numerically effective finite elements. Therefore, this section serves primarily to enhance general understanding of the finite element method. The currently most effective finite elements for general application are believed to be the isoparametric and related elements that are described in Chapter 4.

3.3.1 General Derivation and Specific Examples

In the following derivations the displacements of the finite elements are always described in the local coordinate systems shown in Fig. 3.12. Also, since we consider one specific element, we shall leave out the superscript (m) used in Section 3.2.2.

For one-dimensional bar elements (truss elements) we have

$$u(x) = \alpha_1 + \alpha_2 x + \alpha_3 x^2 + \dots \tag{3.81}$$

where x varies over the length of the element, u is the local element displacement, and $\alpha_1, \alpha_2, \dots$ are the generalized coordinates. The displacement expansion in (3.81) would also be used for the transverse displacement $w(x)$ of a beam.

For two-dimensional elements (i.e., plane stress, plane strain, and axisymmetric elements), we have for the u and v displacements as a function of the element x and y coordinates,

$$u(x, y) = \alpha_1 + \alpha_2 x + \alpha_3 y + \alpha_4 xy + \alpha_5 x^2 + \dots \tag{3.82}$$

$$v(x, y) = \beta_1 + \beta_2 x + \beta_3 y + \beta_4 xy + \beta_5 x^2 + \dots \tag{3.83}$$

where $\alpha_1, \alpha_2, \dots$ and $\beta_1, \beta_2, \dots$ are the generalized coordinates.

In the case of a plate bending element, the transverse deflection w is assumed as a function of the element coordinates x and y; i.e.,

$$w(x, y) = \gamma_1 + \gamma_2 x + \gamma_3 y + \gamma_4 xy + \gamma_5 x^2 + \dots \tag{3.84}$$

where $\gamma_1, \gamma_2, \dots$ are the generalized coordinates.

Finally, for elements in which the u, v, and w displacements are measured as a function of the element x, y, and z coordinates, we have, in general,

$$u(x, y, z) = \alpha_1 + \alpha_2 x + \alpha_3 y + \alpha_4 z + \alpha_5 xy + \dots \tag{3.85}$$

$$v(x, y, z) = \beta_1 + \beta_2 x + \beta_3 y + \beta_4 z + \beta_5 xy + \dots \tag{3.86}$$

$$w(x, y, z) = \gamma_1 + \gamma_2 x + \gamma_3 y + \gamma_4 z + \gamma_5 xy + \dots \tag{3.87}$$

where $\alpha_1, \alpha_2, \dots, \beta_1, \beta_2, \dots,$ and $\gamma_1, \gamma_2, \dots$ are now the generalized coordinates.

As in the discussion of the plane stress element in Section 3.2.1, relations (3.81) to (3.87) can be written in matrix form,

$$\mathbf{u} = \mathbf{\Phi}\boldsymbol{\alpha} \tag{3.88}$$

where the vector $\mathbf{u}$ corresponds to the displacements used in (3.81) to (3.87), the elements of $\mathbf{\Phi}$ are the corresponding polynomial terms, and $\boldsymbol{\alpha}$ is a vector of the generalized coordinates arranged in the appropriate order.

To evaluate the generalized coordinates in terms of the element nodal point displacements, we need to have as many nodal point displacements as assumed generalized coordinates. Then evaluating (3.88) specifically for the nodal point displacements $\mathbf{u}$ of the element, we obtain

$$\mathbf{u} = \mathbf{A}\boldsymbol{\alpha} \tag{3.89}$$

Assuming that the inverse of $\mathbf{A}$ exists, we have

$$\boldsymbol{\alpha} = \mathbf{A}^{-1}\mathbf{u} \tag{3.90}$$

The element strains to be considered depend on the specific problem to be solved. Denoting by $\boldsymbol{\epsilon}$ a generalized strain vector, whose components are given for specific problems in Table 3.1, we have

$$\boldsymbol{\epsilon} = \mathbf{E}\boldsymbol{\alpha} \tag{3.91}$$

where the matrix $\mathbf{E}$ is established using the displacement assumptions (3.81) to (3.87). A vector of generalized stresses $\boldsymbol{\tau}$ is obtained using the relation

$$\boldsymbol{\tau} = \mathbf{C}\boldsymbol{\epsilon} \tag{3.92}$$

where $\mathbf{C}$ is a generalized elasticity matrix. The quantities $\boldsymbol{\tau}$ and $\mathbf{C}$ are defined for some problems in Tables 3.1 and 3.2. We may note that except in bending problems, the generalized $\boldsymbol{\tau}$, $\boldsymbol{\epsilon}$, and $\mathbf{C}$ matrices are those that are used in the theory of elasticity. The word "generalized" is employed merely to include curvatures and moments as strains and stresses, respectively. The advantage of using curvatures and moments in bending analysis is that in the stiffness evaluation [(3.94) below], integration over the thickness of the corresponding element is not required because this stress and strain variation has already been taken into account (see Example 3.6).

Referring to Table 3.2 it should be noted that all stress–strain matrices can be derived from the general three dimensional stress–strain relationship. The plane strain and axisymmetric stress–strain matrices are obtained simply by deleting in the three-dimensional stress–strain matrix the rows and columns that correspond to the zero-strain components. The stress–strain matrix for plane stress analysis is then obtained from the axisymmetric stress–strain matrix by using the condition that τ_{zz} is zero (see the program QUADS in Section 4.8). To calculate the generalized stress–strain matrix for plate bending analysis, the stress–strain matrix corresponding to plane stress conditions is used, as shown in the following example.

TABLE 3.1 CORRESPONDING KINEMATIC AND STATIC VARIABLES IN VARIOUS PROBLEMS

Problem	Displacement Components	Strain Vector $\boldsymbol{\epsilon}^T$	Stress Vector $\boldsymbol{\tau}^T$
Bar	u	$[\epsilon_{xx}]$	$[\tau_{xx}]$
Beam	w	$[\kappa_{xx}]$	$[M_{xx}]$
Plane stress	u, v	$[\epsilon_{xx} \quad \epsilon_{yy} \quad \gamma_{xy}]$	$[\tau_{xx} \quad \tau_{yy} \quad \tau_{xy}]$
Plane strain	u, v	$[\epsilon_{xx} \quad \epsilon_{yy} \quad \gamma_{xy}]$	$[\tau_{xx} \quad \tau_{yy} \quad \tau_{xy}]$
Axisymmetric	u, v	$[\epsilon_{xx} \quad \epsilon_{yy} \quad \gamma_{xy} \quad \epsilon_{zz}]$	$[\tau_{xx} \quad \tau_{yy} \quad \tau_{xy} \quad \tau_{zz}]$
Three-dimensional	u, v, w	$[\epsilon_{xx} \quad \epsilon_{yy} \quad \epsilon_{zz} \quad \gamma_{xy} \quad \gamma_{yz} \quad \gamma_{zx}]$	$[\tau_{xx} \quad \tau_{yy} \quad \tau_{zz} \quad \tau_{xy} \quad \tau_{yz} \quad \tau_{zx}]$
Plate Bending	w	$[\kappa_{xx} \quad \kappa_{yy} \quad \kappa_{xy}]$	$[M_{xx} \quad M_{yy} \quad M_{xy}]$

Notation: $\epsilon_x = \dfrac{\partial u}{\partial x}, \epsilon_y = \dfrac{\partial v}{\partial y}, \gamma_{xy} = \dfrac{\partial u}{\partial y} + \dfrac{\partial v}{\partial x}, \ldots, \kappa_{xx} = -\dfrac{\partial^2 w}{\partial x^2}, \kappa_{yy} = -\dfrac{\partial^2 w}{\partial y^2}, \kappa_{xy} = 2\dfrac{\partial^2 w}{\partial x\, \partial y}$

TABLE 3.2 GENERALIZED STRESS–STRAIN MATRICES FOR ISOTROPIC MATERIALS AND THE PROBLEMS IN TABLE 3.1

Problem	Material Matrix **C**
Bar	E
Beam	EI
Plane Stress	$\dfrac{E}{1-\nu^2}\begin{bmatrix} 1 & \nu & 0 \\ \nu & 1 & 0 \\ 0 & 0 & \dfrac{1-\nu}{2} \end{bmatrix}$
Plane strain	$\dfrac{E(1-\nu)}{(1+\nu)(1-2\nu)}\begin{bmatrix} 1 & \dfrac{\nu}{1-\nu} & 0 \\ \dfrac{\nu}{1-\nu} & 1 & 0 \\ 0 & 0 & \dfrac{1-2\nu}{2(1-\nu)} \end{bmatrix}$
Axisymmetric	$\dfrac{E(1-\nu)}{(1+\nu)(1-2\nu)}\begin{bmatrix} 1 & \dfrac{\nu}{1-\nu} & 0 & \dfrac{\nu}{1-\nu} \\ \dfrac{\nu}{1-\nu} & 1 & 0 & \dfrac{\nu}{1-\nu} \\ 0 & 0 & \dfrac{1-2\nu}{2(1-\nu)} & 0 \\ \dfrac{\nu}{1-\nu} & \dfrac{\nu}{1-\nu} & 0 & 1 \end{bmatrix}$
Three-dimensional	$\dfrac{E(1-\nu)}{(1+\nu)(1-2\nu)}\begin{bmatrix} 1 & \dfrac{\nu}{1-\nu} & \dfrac{\nu}{1-\nu} & & & \\ \dfrac{\nu}{1-\nu} & 1 & \dfrac{\nu}{1-\nu} & & & \\ \dfrac{\nu}{1-\nu} & \dfrac{\nu}{1-\nu} & 1 & & & \\ & & & \dfrac{1-2\nu}{2(1-\nu)} & & \\ & & & & \dfrac{1-2\nu}{2(1-\nu)} & \\ & & & & & \dfrac{1-2\nu}{2(1-\nu)} \end{bmatrix}$ (elements not shown are zeros)
Plate bending	$\dfrac{Et^3}{12(1-\nu^2)}\begin{bmatrix} 1 & \nu & 0 \\ \nu & 1 & 0 \\ 0 & 0 & \dfrac{1-\nu}{2} \end{bmatrix}$

Notation: E = Young's modulus, ν = Poisson's ratio, t = thickness of plate, I = moment of inertia

Example 3.4

Derive the stress–strain matrix **C** used for plate bending analysis (see Table 3.2).

The strains at a distance z measured downward from the neutral axis of the plate are

$$\left[-z\frac{\partial^2 w}{\partial x^2} \quad -z\frac{\partial^2 w}{\partial y^2} \quad z\frac{2\partial^2 w}{\partial x\,\partial y}\right]$$

In plate bending analysis it is assumed that each layer of the plate acts in plane stress condition. Hence integrating the normal stresses in the plate to obtain moments per unit length, the generalized stress–strain matrix is

$$\mathbf{C} = \int_{-t/2}^{+t/2} z^2 \frac{E}{1-\nu^2}\begin{bmatrix} 1 & \nu & 0 \\ \nu & 1 & 0 \\ 0 & 0 & \dfrac{1-\nu}{2} \end{bmatrix}$$

or

$$\mathbf{C} = \frac{Et^3}{12(1-\nu^2)}\begin{bmatrix} 1 & \nu & 0 \\ \nu & 1 & 0 \\ 0 & 0 & \dfrac{1-\nu}{2} \end{bmatrix}$$

The relation in (3.90) expresses the generalized coordinates in terms of the local element nodal point displacements. For the actual solution of the assemblage of finite elements, we need to express all element nodal point displacement and force quantities in a common coordinate system. To express the element local nodal point displacements in terms of the global assemblage degrees of freedom, we use the transformation

$$\mathbf{u} = \mathbf{T}\mathbf{U} \tag{3.93}$$

where **U** stores all nodal point displacements of the element assemblage.

The element stiffness matrix corresponding to the element assemblage degrees of freedom is now obtained by using the relations (3.90) to (3.93) in the general virtual work expression given in (3.62). Thus we obtain

$$\mathbf{K} = \mathbf{T}^T\mathbf{A}^{-T}\left\{\int_V \mathbf{E}^T\mathbf{C}\mathbf{E}\,dV\right\}\mathbf{A}^{-1}\mathbf{T} \tag{3.94}$$

where we realize that

$$\mathbf{K}_{gc} = \int_V \mathbf{E}^T\mathbf{C}\mathbf{E}\,dV \tag{3.95}$$

is the element stiffness matrix corresponding to the generalized coordinates used, and that $\mathbf{A}^{-T}\mathbf{K}_{gc}\mathbf{A}^{-1}$ is the stiffness matrix of the element corresponding to the element local degrees of freedom **u**. Therefore, in (3.94) we use two congruence transformations to obtain from the stiffness matrix that corresponds to the generalized coordinates the final required stiffness matrix of the element, which is used in the direct addition of the element stiffness matrices to obtain the stiffness matrix of the complete structure.

In general, the elements of $\mathbf{K}_{gc}$ can be calculated explicitly, because the matrices $\mathbf{E}$ and $\mathbf{C}$ are of simple form. However, only in low-order formulations, i.e., only when a few polynomial terms are used in the displacement assumption, is it possible to evaluate the inverse of $\mathbf{A}$ in closed form. For example, in the derivation of the element matrices in Example 3.5 we could find $\mathbf{A}^{-1}$ explicitly. However, in general, the matrix $\mathbf{A}$ is formed and $\mathbf{A}^{-1}$ is evaluated numerically.

The transformation from the local element degrees of freedom to the global structure degrees of freedom is for two- and three-dimensional finite elements similar to that for the one-dimensional truss and beam elements. The ith row in $\mathbf{T}$ simply expresses the ith local element displacement in terms of the structure global displacements. It should be noted that as in matrices $\mathbf{H}^{(m)}$ and $\mathbf{B}^{(m)}$ discussed in Section 3.2, the ith column in $\mathbf{T}$ consists of zero elements if the ith displacement component in $\mathbf{U}$ does not affect any nodal displacement of the element under consideration. Therefore, in practice we can ignore all zero columns in $\mathbf{T}$ and thus calculate compacted element matrices that have order equal to the number of element degrees of freedom. This means that the final element stiffness matrix is defined corresponding to global degrees of freedom, but that for the assemblage of the element it need be identified for which specific global degrees of freedom the element stiffness matrix is defined. In the assemblage that is presented in Section 6.2.3, a connectivity array is used to relate the element global degrees of freedom to the assemblage global degrees of freedom. Hence, for this discussion we can assume that $\mathbf{T}$ stores only the nonzero columns and that the compacted element stiffness matrix is obtained in (3.94).

Once all the matrices that are used in the formulation of the stiffness matrix have been calculated, it is a simple matter to obtain the consistent mass matrix, consistent load vectors due to element body or surface forces, and the loads arising from the element initial stresses. Using (3.54), (3.88), (3.90), and (3.93) we find that

$$\mathbf{H} = \mathbf{\Phi}\mathbf{A}^{-1}\mathbf{T} \tag{3.96}$$

where, as before, the superscript (m) is left out. We thus have for the body force vector of an element,

$$\mathbf{R}_B = \mathbf{T}^T\mathbf{A}^{-T}\left(\int_V \mathbf{\Phi}^T\mathbf{f}^B \, dV\right) \tag{3.97}$$

Similarly, for the surface force vector,

$$\mathbf{R}_S = \mathbf{T}^T\mathbf{A}^{-T}\left(\int_S \mathbf{\Phi}^{S^T}\mathbf{f}^S \, dS\right) \tag{3.98}$$

and for the initial stress load vector,

$$\mathbf{R}_I = \mathbf{T}^T\mathbf{A}^{-T}\left(\int_V \mathbf{E}^T\boldsymbol{\tau}^I \, dV\right) \tag{3.99}$$

The consistent mass matrix is

$$\mathbf{M} = \mathbf{T}^T\mathbf{A}^{-T}\left(\int_V \rho \mathbf{\Phi}^T \mathbf{\Phi} \, dV\right)\mathbf{A}^{-1}\mathbf{T} \tag{3.100}$$

The evaluation of $\mathbf{R}_B$, $\mathbf{R}_S$, and $\mathbf{R}_I$ can be carried out analytically if $\mathbf{f}^B$, $\mathbf{f}^S$, and $\boldsymbol{\tau}^I$ are defined in form of a polynomial; otherwise, we resort to numerical integration of the load integrals, which will be discussed in Chapter 4.

It should be noted that in the derivation of a generalized coordinate finite element model, it is assumed that the inverse of $\mathbf{A}$ can be evaluated. The form of $\mathbf{A}$ depends on the element geometry and the assumed displacement polynomials. Many different element stiffness matrices have been formulated, and we are presenting some element formulations in the following examples. However, in some attempts the matrix $\mathbf{A}$ was found to be ill-conditioned or singular for some element geometries. This means that for the specific displacement assumptions used, there are certain nodal point configurations for which the relationship between the generalized coordinates and the nodal point displacements is ill-defined. The difficulties associated with the inversion of the matrix $\mathbf{A}$ and the numerical effort required in the evaluation of generalized coordinate finite element models largely enhanced the development of the more effective isoparametric finite elements. However, before presenting the isoparametric finite element formulation in Chapter 4, we want to conclude this presentation with some relatively simple generalized coordinate finite element derivations. The purpose in the presentation of these examples is not to describe very effective finite element models but to show how the equations derived in this chapter would actually be applied.

Example 3.5

Discuss the derivation of the stiffness, mass, and load matrices of the axisymmetric three-node finite element in Fig. 3.16.

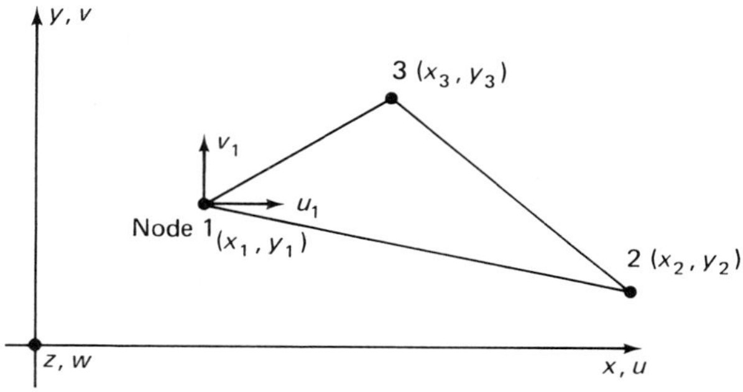

Figure 3.16. Axisymmetric three-node element.

This element was one of the first finite elements developed.[32] For most practical applications, much more effective finite elements are presently available (see Chapter 4), but the element is conveniently used for instructional purposes because the equations to be dealt with are relatively simple.

The displacement assumption used is

$$u(x, y) = \alpha_1 + \alpha_2 x + \alpha_3 y$$
$$v(x, y) = \beta_1 + \beta_2 x + \beta_3 y$$

Therefore, a linear displacement variation is assumed, just as for the derivation of the four-node plane stress element considered in Section 3.2.1, where the fourth node required that the term xy be included in the displacement assumption. Referring to the derivations carried out in Section 3.2.1, we can directly establish the following relationships:

$$\begin{bmatrix} u(x, y) \\ v(x, y) \end{bmatrix} = \begin{bmatrix} 1 & x & y & 0 & 0 & 0 \\ 0 & 0 & 0 & 1 & x & y \end{bmatrix} \begin{bmatrix} \tilde{\mathbf{A}}^{-1} & \mathbf{0} \\ \mathbf{0} & \tilde{\mathbf{A}}^{-1} \end{bmatrix} \begin{bmatrix} u_1 \\ u_2 \\ u_3 \\ v_1 \\ v_2 \\ v_3 \end{bmatrix} \tag{a}$$

in which

$$\tilde{\mathbf{A}} = \begin{bmatrix} 1 & x_1 & y_1 \\ 1 & x_2 & y_2 \\ 1 & x_3 & y_3 \end{bmatrix}$$

Hence

$$\tilde{\mathbf{A}}^{-1} = \frac{1}{\det \tilde{\mathbf{A}}} \begin{bmatrix} x_2 y_3 - x_3 y_2 & x_3 y_1 - x_1 y_3 & x_1 y_2 - x_2 y_1 \\ y_2 - y_3 & y_3 - y_1 & y_1 - y_2 \\ x_3 - x_2 & x_1 - x_3 & x_2 - x_1 \end{bmatrix}$$

where

$$\det \tilde{\mathbf{A}} = x_1(y_2 - y_3) + x_2(y_3 - y_1) + x_3(y_1 - y_2)$$

We may note that $\det \tilde{\mathbf{A}}$ is only zero if the three element nodal points lie on a straight line. The strains are given in Table 3.1; i.e.,

$$\epsilon_{xx} = \frac{\partial u}{\partial x}; \qquad \epsilon_{yy} = \frac{\partial v}{\partial y}; \qquad \gamma_{xy} = \frac{\partial u}{\partial y} + \frac{\partial v}{\partial x}; \qquad \epsilon_{zz} = \frac{\partial w}{\partial z} = \frac{u}{x}$$

Using the assumed displacement polynomials, we obtain

$$\begin{bmatrix} \epsilon_{xx} \\ \epsilon_{yy} \\ \gamma_{xy} \\ \epsilon_{zz} \end{bmatrix} = \begin{bmatrix} 0 & 1 & 0 & 0 & 0 & 0 \\ 0 & 0 & 0 & 0 & 0 & 1 \\ 0 & 0 & 1 & 0 & 1 & 0 \\ \frac{1}{x} & 1 & \frac{y}{x} & 0 & 0 & 0 \end{bmatrix} \begin{bmatrix} \tilde{\mathbf{A}}^{-1} & \mathbf{0} \\ \mathbf{0} & \tilde{\mathbf{A}}^{-1} \end{bmatrix} \begin{bmatrix} u_1 \\ u_2 \\ u_3 \\ v_1 \\ v_2 \\ v_3 \end{bmatrix} \tag{b}$$

The above relations give the matrices $\mathbf{\Phi}(x, y)$, $\mathbf{E}(x, y)$, and $\mathbf{A}^{-1}$ required in the evaluation of the element matrices in (3.94) to (3.100). It may be noted that in plane stress and plane strain analysis the same relation as in (a) would be used, whereas in (b) the last row defining ϵ_{zz} would not be applicable.

Considering the convergence requirements of completeness and compatibility, the discussion in Example 3.3 shows that this three-node element is also compatible and complete.

For the solution of an assemblage of elements, we need in general the transformation matrix **T**. Consider as an example the displacement transformation at one node, say node 1, of the triangular element. Referring to Fig. 3.6 we have

$$\begin{bmatrix} u_1 \\ v_1 \end{bmatrix} = \begin{bmatrix} \cos\alpha & \sin\alpha \\ -\sin\alpha & \cos\alpha \end{bmatrix} \begin{bmatrix} U_i \\ V_i \end{bmatrix}$$

and similarly the displacement components at all nodes can be transformed. For the three-node element in Fig. 3.16 and the displacement components stored as in (a), we have, if we ignore zero columns in the transformation matrix,

$$\mathbf{T} = \left[\begin{array}{ccc|ccc} \cos\alpha & 0 & 0 & \sin\alpha & 0 & 0 \\ 0 & \cos\alpha & 0 & 0 & \sin\alpha & 0 \\ 0 & 0 & \cos\alpha & 0 & 0 & \sin\alpha \\ \hline -\sin\alpha & 0 & 0 & \cos\alpha & 0 & 0 \\ 0 & -\sin\alpha & 0 & 0 & \cos\alpha & 0 \\ 0 & 0 & -\sin\alpha & 0 & 0 & \cos\alpha \end{array}\right]$$

and because **T** is orthogonal, we have $\mathbf{T}^{-1} = \mathbf{T}^T$.

The integrals required in the evaluation of the element matrices (3.94) to (3.100) are, therefore,

$$\int_V \mathbf{E}^T\mathbf{C}\mathbf{E}\, dV =$$

$$\int_A \frac{E(1-\nu)}{(1+\nu)(1-2\nu)} \begin{bmatrix} 0 & 0 & 0 & \frac{1}{x} \\ 1 & 0 & 0 & 1 \\ 0 & 0 & 1 & \frac{y}{x} \\ 0 & 0 & 0 & 0 \\ 0 & 0 & 1 & 0 \\ 0 & 1 & 0 & 0 \end{bmatrix} \begin{bmatrix} 1 & \frac{\nu}{1-\nu} & 0 & \frac{\nu}{1-\nu} \\ \frac{\nu}{1-\nu} & 1 & 0 & \frac{\nu}{1-\nu} \\ 0 & 0 & \frac{1-2\nu}{2(1-2\nu)} & 0 \\ \frac{\nu}{1-\nu} & \frac{\nu}{1-\nu} & 0 & 1 \end{bmatrix} \begin{bmatrix} 0 & 1 & 0 & 0 & 0 & 0 \\ 0 & 0 & 0 & 0 & 0 & 1 \\ 0 & 0 & 1 & 0 & 1 & 0 \\ \frac{1}{x} & 1 & \frac{y}{x} & 0 & 0 & 0 \end{bmatrix} x\, dx\, dy \tag{c}$$

where 1 radian of the axisymmetric element is considered in the volume integration. Similarly, we have

$$\int_V \mathbf{\Phi}^T\mathbf{f}^B dV = \int_A \begin{bmatrix} 1 & 0 \\ x & 0 \\ y & 0 \\ 0 & 1 \\ 0 & x \\ 0 & y \end{bmatrix} \begin{bmatrix} f_x^B \\ f_y^B \end{bmatrix} x\, dx\, dy \tag{d}$$

$$\int_V \mathbf{E}^T \boldsymbol{\tau}^I \, dV = \int_A \begin{bmatrix} 0 & 0 & 0 & \frac{1}{x} \\ 1 & 0 & 0 & 1 \\ 0 & 0 & 1 & \frac{y}{x} \\ 0 & 0 & 0 & 0 \\ 0 & 0 & 1 & 0 \\ 0 & 1 & 0 & 0 \end{bmatrix} \begin{bmatrix} \tau^I_{xx} \\ \tau^I_{yy} \\ \tau^I_{xy} \\ \tau^I_{zz} \end{bmatrix} x \, dx \, dy \tag{e}$$

$$\int \rho \boldsymbol{\Phi}^T \boldsymbol{\Phi} \, dV = \rho \int_A \begin{bmatrix} 1 & 0 \\ x & 0 \\ y & 0 \\ 0 & 1 \\ 0 & x \\ 0 & y \end{bmatrix} \begin{bmatrix} 1 & x & y & 0 & 0 & 0 \\ 0 & 0 & 0 & 1 & x & y \end{bmatrix} x \, dx \, dy \tag{f}$$

where the mass density ρ is assumed to be constant.

For the calculation of the surface load vector $\mathbf{R}_S$, we need to evaluate the integral $\int_S \boldsymbol{\Phi}^{S^T} \mathbf{f}^S \, dS$. In practice, it is expedient to evaluate the integral by introducing auxiliary coordinate systems that are located along the loaded sides of the element. Assume that the side 2–3 of the element is loaded as shown in Fig. 3.17. The load vector $\mathbf{R}_S$ is then evaluated using as the variable s,

$$\mathbf{R}_S = \int_s \begin{bmatrix} 0 & 0 \\ 1 - \frac{s}{L} & 0 \\ \frac{s}{L} & 0 \\ 0 & 0 \\ 0 & 1 - \frac{s}{L} \\ 0 & \frac{s}{L} \end{bmatrix} \begin{bmatrix} f_x^S \\ f_y^S \end{bmatrix} \left[x_2\left(1 - \frac{s}{L}\right) + x_3 \frac{s}{L} \right] ds \tag{g}$$

Considering the integrals in (c) to (g), the following observations may be

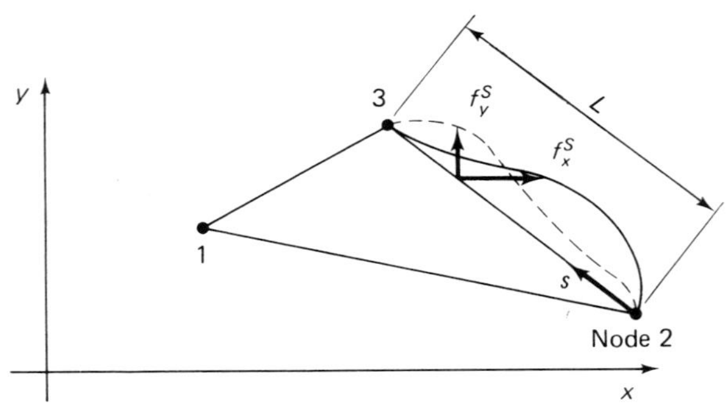

Figure 3.17. Axisymmetric element with surface loading.

made. First, to evaluate the integrals, it is possible to obtain closed-form solutions; alternatively, numerical integration (discussed in Section 4.6) can be used. Second, we find that the stiffness, mass, and load matrices corresponding to plane stress and plane strain finite elements would simply be obtained by (1) not including the fourth row in the strain-displacement matrix **E** used in (c) and (e), (2) employing the appropriate stress–strain matrix **C** in (c), and (3) using as the differential volume element $t\,dx\,dy$ instead of $x\,dx\,dy$, where t is the thickness of the element (conveniently taken equal to 1 in plane strain analysis). Therefore, axisymmetric, plane stress, and plane strain analysis can effectively be implemented in a single computer program. Also, the matrix **E** shows that constant strain conditions ϵ_{xx}, ϵ_{yy}, and γ_{xy} are assumed in either analysis.

The concept of performing axisymmetric, plane strain, and plane stress analysis in an effective manner in one computer program is, in fact, presented in Section 4.8, where we discuss the effective implementation of isoparametric finite element analysis.

Example 3.6

Derive the matrices $\mathbf{\Phi}(x, y)$, $\mathbf{E}(x, y)$, and $\mathbf{A}$ for the rectangular plate bending element in Fig. 3.18.

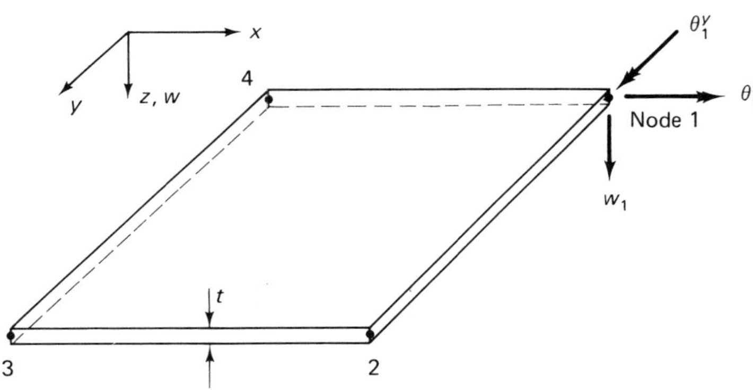

Figure 3.18. Rectangular plate bending element.

This element is one of the first plate bending elements derived.[29,37] As in Example 3.5, the purpose in this presentation is to demonstrate, in a simple example, the general procedure presented in this section. It should be pointed out again that much more effective plate bending elements are already in use (see Chapter 4).[38-41]

As shown in Fig. 3.18, the plate bending element considered has three degrees of freedom per node. Therefore, it is necessary to have 12 unknown generalized coordinates, $\alpha_1, \ldots, \alpha_{12}$, in the displacement assumption for w. The polynomial used is

$$\begin{aligned} w = \alpha_1 + \alpha_2 x + \alpha_3 y + \alpha_4 x^2 + \alpha_5 xy + \alpha_6 y^2 + \alpha_7 x^3 + \alpha_8 x^2 y \\ + \alpha_9 xy^2 + \alpha_{10} y^3 + \alpha_{11} x^3 y + \alpha_{12} xy^3 \end{aligned}$$

Hence

$$\mathbf{\Phi}(x, y) = [1 \quad x \quad y \quad x^2 \quad xy \quad y^2 \quad x^3 \quad x^2y \quad xy^2 \quad y^3 \quad x^3y \quad xy^3] \qquad \text{(a)}$$

We can now calculate $\partial w/\partial x$ and $\partial w/\partial y$; i.e.,

$$\frac{\partial w}{\partial x} = \alpha_2 + 2\alpha_4 x + \alpha_5 y + 3\alpha_7 x^2 + 2\alpha_8 xy + \alpha_9 y^2 + 3\alpha_{11} x^2 y + \alpha_{12} y^3 \qquad \text{(b)}$$

and

$$\frac{\partial w}{\partial y} = \alpha_3 + \alpha_5 x + 2\alpha_6 y + \alpha_8 x^2 + 2\alpha_9 xy + 3\alpha_{10} y^2 + \alpha_{11} x^3 + 3\alpha_{12} xy^2 \qquad \text{(c)}$$

Using the conditions

$$\left.\begin{aligned} w_i = (w)_{x_i, y_i}; \qquad \theta_i^x = \left(\frac{\partial w}{\partial y}\right)_{x_i, y_i} \\ \theta_i^y = \left(-\frac{\partial w}{\partial x}\right)_{x_i, y_i} \qquad\qquad \end{aligned}\right\} i = 1, \ldots, 4$$

we can construct the matrix $\mathbf{A}$; i.e., we obtain

$$\begin{bmatrix} w_1 \\ \cdot \\ \cdot \\ \cdot \\ w_4 \\ \theta_1^x \\ \cdot \\ \cdot \\ \cdot \\ \theta_4^x \\ \theta_1^y \\ \cdot \\ \cdot \\ \cdot \\ \theta_4^y \end{bmatrix} = \mathbf{A} \begin{bmatrix} \alpha_1 \\ \alpha_2 \\ \cdot \\ \cdot \\ \cdot \\ \\ \\ \\ \\ \\ \\ \\ \\ \\ \alpha_{12} \end{bmatrix}$$

where

$$\mathbf{A} = \begin{bmatrix} 1 & x_1 & y_1 & x_1^2 & x_1 y_1 & y_1^2 & x_1^3 & x_1^2 y_1 & x_1 y_1^2 & y_1^3 & x_1^3 y_1 & x_1 y_1^3 \\ \cdot & & & & \cdot & & & & & & & \\ \cdot & & & & \cdot & & & & & & & \\ \cdot & & & & \cdot & & & & & & & \\ 1 & x_4 & y_4 & x_4^2 & x_4 y_4 & y_4^2 & x_4^3 & x_4^2 y_4 & x_4 y_4^2 & y_4^3 & x_4^3 y_4 & x_4 y_4^3 \\ 0 & 0 & 1 & 0 & x_1 & 2y_1 & 0 & x_1^2 & 2x_1 y_1 & 3y_1^2 & x_1^3 & 3x_1 y_1^2 \\ \cdot & & & & \cdot & & & & & & & \\ \cdot & & & & \cdot & & & & & & & \\ \cdot & & & & \cdot & & & & & & & \\ 0 & 0 & 1 & 0 & x_4 & 2y_4 & 0 & x_4^2 & 2x_4 y_4 & 3y_4^2 & x_4^3 & 3x_4 y_4^2 \\ 0 & -1 & 0 & -2x_1 & -y_1 & 0 & -3x_1^2 & -2x_1 y_1 & -y_1^2 & 0 & -3x_1^2 y_1 & -y_1^3 \\ \cdot & & & & \cdot & & & & & & & \\ \cdot & & & & \cdot & & & & & & & \\ \cdot & & & & \cdot & & & & & & & \\ 0 & -1 & 0 & -2x_4 & -y_4 & 0 & -3x_4^2 & -2x_4 y_4 & -y_4^2 & 0 & -3x_4^2 y_4 & -y_4^3 \end{bmatrix} \qquad \text{(d)}$$

which can be shown to be always nonsingular.

To evaluate the matrix **E**, we recall that in plate bending analysis curvatures and moments are used as generalized strains and stresses (see Tables 3.1 and 3.2). Calculating the required derivatives of (b) and (c), we obtain

$$
\begin{aligned}
-\frac{\partial^2 w}{\partial x^2} &= -2\alpha_4 - 6\alpha_7 x - 2\alpha_8 y - 6\alpha_{11} xy \\
-\frac{\partial^2 w}{\partial y^2} &= -2\alpha_6 - 2\alpha_9 x - 6\alpha_{10} y - 6\alpha_{12} xy \qquad \text{(e)} \\
2\frac{\partial^2 w}{\partial x\,\partial y} &= 2\alpha_5 + 4\alpha_8 x + 4\alpha_9 y + 6\alpha_{11} x^2 + 6\alpha_{12} y^2
\end{aligned}
$$

Hence we have

$$
\mathbf{E} = \begin{bmatrix} 0 & 0 & 0 & -2 & 0 & 0 & -6x & -2y & 0 & 0 & -6xy & 0 \\ 0 & 0 & 0 & 0 & 0 & -2 & 0 & 0 & -2x & -6y & 0 & -6xy \\ 0 & 0 & 0 & 0 & 2 & 0 & 0 & 4x & 4y & 0 & 6x^2 & 6y^2 \end{bmatrix} \qquad \text{(f)}
$$

With the matrices $\mathbf{\Phi}$, **A**, and **E** given in (a), (d), and (f) and the material matrix **C** in Table 3.2, the element stiffness matrix, mass matrix, and load vectors in (3.94) to (3.100) can now be calculated.

An important consideration in the evaluation of an element stiffness matrix is whether the element is complete and compatible. The element considered in this example is complete as shown in (e); i.e., the element can represent constant curvature states, but the element is not compatible. The compatibility requirements are violated in a number of plate bending elements, meaning that convergence in the analysis is in general not monotonic.

3.3.2 Spatial Isotropy

Apart from the requirement that there must be a one-to-one correspondence between the generalized coordinates and the element nodal point displacements, another important factor may need to be taken into account. This is the requirement of *spatial isotropy* or *geometric invariance*, which means that the displacements of the element need to be independent of the orientation of the local element coordinate system. Consider as an example a two-dimensional element. With the local element coordinate axes selected, we may assume that

$$u(x, y) = \alpha_1 + \alpha_2 x + \alpha_3 y + \alpha_4 xy \qquad (3.101)$$

$$v(x, y) = \beta_1 + \beta_2 x + \beta_3 y + \beta_4 x^2 \qquad (3.102)$$

We note that the element is complete but only compatible if the orientation of the elements is always the same, i.e., always only element edges along which the displacements vary parabolically and element edges along which the displacements vary linearly are joining. In the analysis of such element assemblages, all convergence requirements are satisfied. However, the element displacements are dependent on the orientation of the coordinate axes, because the polynomial terms associated with α_4 and β_4 are not the same.

For example, assume that the x and y axes are interchanged; then clearly a different displacement field is assumed. Therefore, the stiffness matrix of an element will depend on the orientation of the local element coordinate system.

Although it may seem that such directional behavior is always undesirable, actually, in a variety of element formulations, specific local coordinate axes are assumed, and different displacement variations are associated with each axis. The displacement functions are selected based on the kinematic assumptions, which would be used in the continuum mechanics solution. For example, in the formulation of thick shell elements, a lower order displacement variation is used through the thickness of the elements than over the element surfaces. Similarly, a spar element in bridge analysis may have different displacement expansions in each direction. However, although different displacement functions may be used intentionally for certain directions, for some other directions it may be a requirement that the displacement assumption is independent of the direction considered. Typically, in the analysis of two- and three-dimensional continua with spatially isotropic behavior, geometric invariance is a requirement and results in a restriction on the polynomial terms that may be used in the displacement formulation. In general, a displacement model is geometrically invariant if the same order polynomial terms are used for the displacements u, v, and w if applicable, and in each polynomial term the x, y, and z element coordinates have been appropiately interchanged.

Considering the formulation of generalized coordinate finite element models, it is not always an easy task to satisfy geometric invariance. Indeed, the difficulties of obtaining for some elements spatially isotropic generalized coordinate formulations, for which the inverse of $\mathbf{A}$ is well-defined, have enhanced the development of element formulations in which interpolation functions are used. These developments led to the formulation of isoparametric finite elements. The detailed formulation and calculation of isoparametric elements is presented in Chapter 4, in which other advantages of their use are also pointed out.

REFERENCES

1. O. C. ZIENKIEWICZ and G. S. HOLISTER, *Stress Analysis*, John Wiley & Sons, Inc., New York, 1965.
2. O. C. ZIENKIEWICZ, *The Finite Element Method in Engineering Science*, McGraw-Hill Book Company, New York, 1971.
3. H. C. MARTIN and G. F. CAREY, *Introduction to Finite Element Analysis*, McGraw-Hill Book Company, New York, 1973.
4. C. S. DESAI and J. F. ABEL, *Introduction to the Finite Element Method*, Van Nostrand Reinhold Company, New York, 1972.

5. R. H. GALLAGHER, *Finite Element Analysis Fundamentals*, Prentice-Hall, Inc., Englewood Cliffs, New Jersey, 1975.

6. C. A. FELIPPA, "Refined Finite Element Analysis of Linear and Nonlinear Two-Dimensional Structures," Report UC SESM 66-22, Department of Civil Engineering, University of California, Berkeley, 1966.

7. Proceedings of the 1st, 2nd, and 3rd Conferences on Matrix Methods in Structural Mechanics, Wright-Patterson A.F.B., Ohio, 1965, 1968, 1971.

8. Proceedings of the 1st and 2nd Conferences on Structural Mechanics in Reactor Technology, Berlin, 1971, 1973.

9. I. HOLLAND and K. BELL (eds.), *Finite Element Methods in Stress Analysis*, Tapir, Trondheim, Norway, 1969.

10. R. H. GALLAGHER, Y. YAMADA, and J. T. ODEN (eds), *Recent Advances in Matrix Methods of Structural Analysis and Design*, University of Alabama Press, Huntsville, Ala., 1971.

11. J. T. ODEN, R. W. CLOUGH, and Y. YAMAMOTO, (eds.) *Advances in Computational Methods in Structural Mechanics and Design*, University of Alabama Press, Huntsville, Ala., 1972.

12. J. R. WHITEMAN (ed.), *The Mathematics of Finite Elements and Applications*, Academic Press Inc. Ltd., London, 1973.

13. S. J. FENVES, N. PERRONE, J. ROBINSON, and W. C. SCHNOBRICH (eds.), *Numerical and Computational Methods in Structural Mechanics*, Academic Press, Inc., New York, 1973.

14. W. PILKEY, K. SACZALSKI, and H. SCHAEFFER (eds.), *Structural Mechanics Computer Programs*, University Press of Virginia, Charlottesville, Va., 1974.

15. R. COURANT, "Variational Methods for the Solution of Problems of Equilibriums and Vibrations," *Bulletin of the American Mathematical Society*, Vol. 49, 1943, pp. 1–23.

16. R. COURANT and D. HILBERT, *Methods of Mathematical Physics*, John Wiley & Sons, Inc., New York, 1953.

17. L. COLLATZ, *The Numerical Treatment of Differential Equations*, Springer-Verlag, Berlin, 1st ed., 1950; 3rd ed., English translation, New York, 1966.

18. J. L. SYNGE, *The Hypercircle in Mathematical Physics*, Cambridge University Press, London, 1957.

19. M. J. TURNER, R. W. CLOUGH, H. C. MARTIN, and L. J. TOPP, "Stiffness and Deflection Analysis of Complex Structures," *Journal of Aeronautical Science*, Vol. 23, 1956, pp. 805–823.

20. J. H. ARGYRIS and S. KELSEY, "Energy Theorems and Structural Analysis," *Aircraft Engineering*, Vols. 26 and 27, 1955.

21. R. W. CLOUGH, "The Finite Element in Plane Stress Analysis," Proceedings, 2nd A.S.C.E. Conference on Electronic Computation, Pittsburgh, Pa., Sept. 1960.

22. H. C. MARTIN, *Introduction to Matrix Methods of Structural Analysis*, McGraw-Hill Book Company, New York, 1966.

23. J. S. PRZEMIENIECKI, *Theory of Matrix Structural Analysis*, McGraw-Hill Book Company, New York, 1968.

24. K. J. BATHE, E. L. WILSON, and F. E. PETERSON, "SAP IV—A Structural Analysis Program for Static and Dynamic Response of Linear Systems," Report EERC 73-11, College of Engineering, University of California, Berkeley, June 1973, revised Apr. 1974.

25. W. C. HURTY and M. F. RUBINSTEIN, *Dynamics of Structures*, Prentice-Hall, Inc., Englewood Cliffs, N.J., 1964.

26. S. TIMOSHENKO and J. N. GOODIER, *Theory of Elasticity*, McGraw-Hill Book Company, New York, 1951.

27. S. TIMOSHENKO and S. WOINOWSKI-KRIEGER, *Theory of Plates and Shells*, 2nd ed., McGraw-Hill Book Company, New York, 1959.

28. W. FLÜGGE, *Stresses in Shells*, Springer-Verlag, Berlin, 1960.

29. R. J. MELOSH, "Basis of Derivation of Matrices for the Direct Stiffness Method, *A.I.A.A. Journal*, Vol. 1, 1963, pp. 1631–1637.

30. S. KEY, "A Convergence Investigation of the Direct Stiffness Method," Ph.D. thesis, University of Washington, 1966.

31. R. L. TAYLOR, "On Completeness of Shape Functions for Finite Element Analysis," *International Journal for Numerical Methods in Engineering*, Vol. 4, 1972, pp. 17–22.

32. E. L. WILSON, "Structural Analysis of Axi-symmetric Solids," *A.I.A.A. Journal*, Vol. 3, 1965, pp. 2269–2274.

33. S. GHOSH and E. L. WILSON, "Dynamic Stress Analysis of Axisymmetric Structures under Arbitrary Loading," Report EERC 69-10, College of Engineering, University of California, Berkeley, Sept. 1969.

34. Y. K. CHEUNG, "Finite Strip Method of Analysis of Elastic Slabs," *Proceedings of the American Society of Civil Engineers*, Vol. 94, EM6, 1968, pp. 1365–1378.

35. K. J. WILLAM and A. C. SCORDELIS, "Analysis of Orthotropic Folded Plates with Eccentric Stiffeners," Report UC SESM 70-2, Department of Civil Engineering, University of California, Berkeley, Feb. 1970.

36. C. MEYER, "Analysis and Design of Curved Box Girder Bridges," Report UC SESM 70-22, Department of Civil Engineering, University of California, Berkeley, Dec. 1970.

37. R. W. CLOUGH and J. L. TOCHER, "Finite Element Stiffness Matrices for the Analysis of Plate Bending," Proceedings, Conference on Matrix Methods in Structural Mechanics, Wright-Patterson A.F.B., Ohio, 1965.

38. J. H. ARGYRIS, "Continua and Discontinua," Proceedings, Conference on Matrix Methods in Structural Mechanics, Wright-Patterson A.F.B., Ohio, Oct. 1965.

39. R. W. CLOUGH and C. A. FELIPPA, "A Refined Quadrilateral Element for Analysis of Plate Bending," Proceedings, 2nd Conference on Matrix Methods in Structural Mechanics, Wright-Patterson A.F.B., Ohio, 1968.

40. R. H. GALLAGHER, "Finite Element Analysis of Plate and Shell Structures," Symposium on Applied Finite Element Methods in Civil Engineering, Vanderbilt University, Nashville, Nov. 1969.

41. R. D. COOK, "Some Elements for Analysis of Plate Bending," *A.S.C.E., Journal of Engineering Mechanics Division*, Vol. 98, 1972, pp. 1453–1470.

CHAPTER 4

FORMULATION AND CALCULATION OF ISOPARAMETRIC FINITE ELEMENT MATRICES

4.1 INTRODUCTION

A very important phase of a finite element analysis is the calculation of the finite element matrices. In Chapter 3 we discussed the formulation and calculation of generalized coordinate finite element models. The aim in the presentation of the generalized coordinate finite elements was primarily to enhance the understanding of the finite element method. We already pointed out that in most practical analyses the use of isoparametric finite elements is more effective[1-16]. It is the objective in this chapter to present the formulation of isoparametric finite elements and describe effective implementations.

In the derivation of generalized coordinate finite element models, we used local element coordinate systems x, y, z and assumed the element displacements $u(x, y, z)$, $v(x, y, z)$, and $w(x, y, z)$ in the form of polynomials in x, y, and z with undetermined constant coefficients α_i, β_i, and γ_i, $i = 1, 2, \ldots,$ identified as generalized coordinates. It was not possible to associate a priori a physical meaning to the generalized coordinates; however, on evaluation we found that the generalized coordinates are linear combinations of the element nodal point displacements. We also pointed out that care had to be exercised to be able to find a unique expression of the generalized coordinates in terms of the element nodal point displacements; i.e., the inverse of $\mathbf{A}$ in (3.90) had to be well-defined for any element geometry. The principal idea of the isoparametric finite element formulation is to achieve the relationship between the element displacements at any point and the element nodal point displacements directly through the use of *interpolation functions* (also called *shape functions*). This means that the transformation matrix $\mathbf{A}^{-1}$ and the matrix $\mathbf{T}$ when global coordinates are used [see (3.96)] are not evaluated; instead, the element matrices corresponding to the required degrees of freedom are obtained directly.

4.2 ISOPARAMETRIC DERIVATION OF BAR ELEMENT STIFFNESS MATRIX

Consider the example of a bar element to illustrate the procedure of an isoparametric stiffness formulation. In order to simplify the explanation, assume that the bar lies in the global X-coordinate axis, as shown in Fig. 4.1.

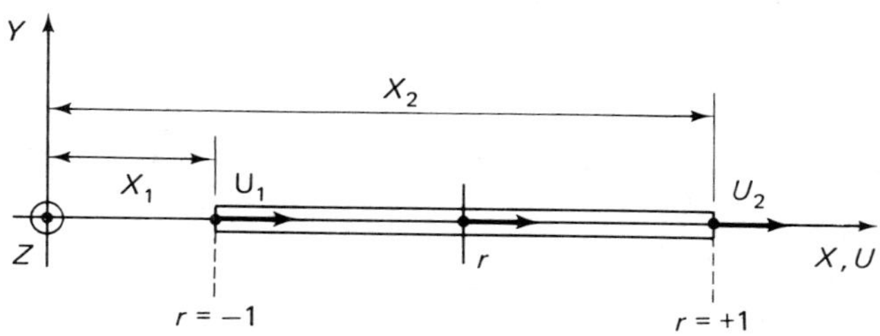

Figure 4.1. Element in global and natural coordinate system.

The first step is to relate the actual global coordinates X to a *natural coordinate system* with variable r, where $-1 \leq r \leq 1$ (Fig. 4.1). This transformation is given by

$$X = \tfrac{1}{2}(1 - r)X_1 + \tfrac{1}{2}(1 + r)X_2 \tag{4.1}$$

or

$$X = \sum_{i=1}^{2} h_i X_i \tag{4.2}$$

where $h_1 = \frac{1}{2}(1 - r)$ and $h_2 = \frac{1}{2}(1 + r)$ are the interpolation or shape functions. Note that (4.2) establishes a unique relationship between the coordinates X and r on the bar.

The bar global displacements are expressed in the same way as the global coordinates:

$$U = \sum_{i=1}^{2} h_i U_i \tag{4.3}$$

where in this case a linear displacement variation is specified. *The interpolation of the element coordinates and element displacements using the same interpolation functions, which are defined in a natural coordinate system, is the basis of the isoparametric finite element formulation.*

For the calculation of the element stiffness matrix we need to find the element strains $\epsilon = dU/dX$. Here we use

$$\epsilon = \frac{dU}{dr}\frac{dr}{dX} \tag{4.4}$$

where, from (4.3),

$$\frac{dU}{dr} = \frac{U_2 - U_1}{2} \tag{4.5}$$

and using (4.2) we obtain

$$\frac{dX}{dr} = \frac{X_2 - X_1}{2} = \frac{L}{2} \tag{4.6}$$

where L is the length of the bar. Hence, as expected, we have

$$\epsilon = \frac{U_2 - U_1}{L} \tag{4.7}$$

The strain-displacement transformation matrix corresponding to (3.55) is therefore

$$\mathbf{B} = \frac{1}{L}[-1 \quad 1] \tag{4.8}$$

In general, the strain-displacement transformation matrix is a function of the natural coordinates and we therefore evaluate the stiffness matrix volume integral in (3.62) by integrating over the natural coordinates. Following this general procedure, although in this example it is not necessary, we have

$$\mathbf{K} = \frac{AE}{L^2}\int_{-1}^{1}\begin{bmatrix}-1\\ 1\end{bmatrix}[-1 \quad 1]\,J\,dr \tag{4.9}$$

where the bar area A and modulus of elasticity E have been assumed constant, and J is the Jacobian relating an element length in the global coordinate system to an element length in the natural coordinate system; i.e.,

$$dX = J\,dr \tag{4.10}$$

From (4.6) we have

$$J = \frac{L}{2} \tag{4.11}$$

Then evaluating (4.9) we obtain the well-known matrix

$$\mathbf{K} = \frac{AE}{L}\begin{bmatrix}1 & -1\\ -1 & 1\end{bmatrix} \tag{4.12}$$

As was stated in the introduction, the isoparametric formulation avoids the construction of the transformation matrices $\mathbf{A}^{-1}$ and $\mathbf{T}$. In order to compare the above formulation with the generalized coordinate formulation, we need to solve from (4.1) for r and then substitute for r into (4.3). We obtain

$$r = \frac{X - [(X_1 + X_2)/2]}{L/2} \tag{4.13}$$

and then

$$U = \alpha_0 + \alpha_1 X \tag{4.14}$$

where

$$\left.\begin{aligned} \alpha_0 &= \tfrac{1}{2}(U_1 + U_2) - \frac{X_1 + X_2}{2L}(U_2 - U_1) \\ \alpha_1 &= \frac{1}{L}(U_2 - U_1) \end{aligned}\right\} \tag{4.15}$$

or

$$\boldsymbol{\alpha} = \begin{bmatrix} \dfrac{1}{2} + \dfrac{X_1 + X_2}{2L} & \dfrac{1}{2} - \dfrac{X_1 + X_2}{2L} \\ -\dfrac{1}{L} & \dfrac{1}{L} \end{bmatrix} \mathbf{U} \tag{4.16}$$

where

$$\boldsymbol{\alpha}^T = [\alpha_0 \quad \alpha_1]; \qquad \mathbf{U}^T = [U_1 \quad U_2] \tag{4.17}$$

and the matrix relating in (4.16) $\boldsymbol{\alpha}$ to $\mathbf{U}$ is $\mathbf{A}^{-1}$. It should be noted that in this example the generalized coordinates α_0 and α_1 relate the global element displacements to the global element coordinates.

4.3 GENERAL ISOPARAMETRIC FORMULATION

Although it is more efficient in most cases to directly calculate the element matrices corresponding to the global degrees of freedom, we shall first present the formulation of the matrices that correspond to the element local degrees of freedom. This is done because additional considerations can be necessary when the element matrices that correspond to the global degrees of freedom are calculated directly (see Section 4.3.2).

4.3.1 Formulation of Isoparametric Finite Element Matrices in Local Coordinate Systems

The basic procedure in the isoparametric finite element formulation is to express the element coordinates and element displacements in the form of interpolations using the natural coordinate system of the element. This coordinate system is one-, two-, or three-dimensional, depending on the dimensionality of the element. The formulation of the element matrices is the same whether we deal with a one-, two-, or three-dimensional element. For this reason we use in the following general presentation the equations of a three-dimensional element. However, the one- and two-dimensional elements are included by simply using only the relevant coordinate axes and the appropriate interpolation functions.

Considering a general three-dimensional element, the coordinate interpolations are

$$\left.\begin{aligned} x &= \sum_{i=1}^{q} h_i x_i \\ y &= \sum_{i=1}^{q} h_i y_i \\ z &= \sum_{i=1}^{q} h_i z_i \end{aligned}\right\} \tag{4.18}$$

where x, y, and z are the coordinates at any point of the element (here local coordinates), and $x_i, y_i, z_i, i = 1, \ldots, q$, are the coordinates of the q element nodes. The interpolation functions h_i are defined in the natural coordinate system of the element, which has variables r, s, and t that each vary from $+1$ to -1. For one- or two- dimensional elements, only the relevant equations in (4.18) would be employed, and the interpolation functions would depend only on the natural coordinate variables r and r, s, respectively.

The unknown quantities in (4.18) are so far the interpolation functions h_i. The fundamental property of the interpolation function h_i is that its value in the natural coordinate system is unity at node i and is zero at all other nodes. Using these conditions the functions h_i corresponding to a specific nodal point layout could be solved for in a systematic manner. However, it is convenient to construct them by inspection, which is demonstrated in the following simple example.

Example 4.1

Construct the interpolation functions corresponding to the three-node truss element in Fig. 4.2.

A first observation is that for the three-node truss element we want interpolation polynomials that involve r^2 as the highest power of r; in other words, the interpolation functions shall be parabolas. The function h_2 can thus be constructed without much effort. Namely, the parabola that satisfies the conditions to be equal to zero at $r = \pm 1$ and equal to 1 at $r = 0$ is given by $(1 - r^2)$. The other two interpolation functions h_1 and h_3 are constructed by superimposing a linear function and a parabola. Consider the interpolation function h_3. Using $\frac{1}{2}(1 + r)$, the conditions that the function shall be zero at $r = -1$ and 1 at $r = +1$ are satisfied. To assure that h_3 is also zero at $r = 0$, we need to use $h_3 = \frac{1}{2}(1 + r) - \frac{1}{2}(1 - r^2)$. The interpolation function h_1 is obtained in a similar manner.

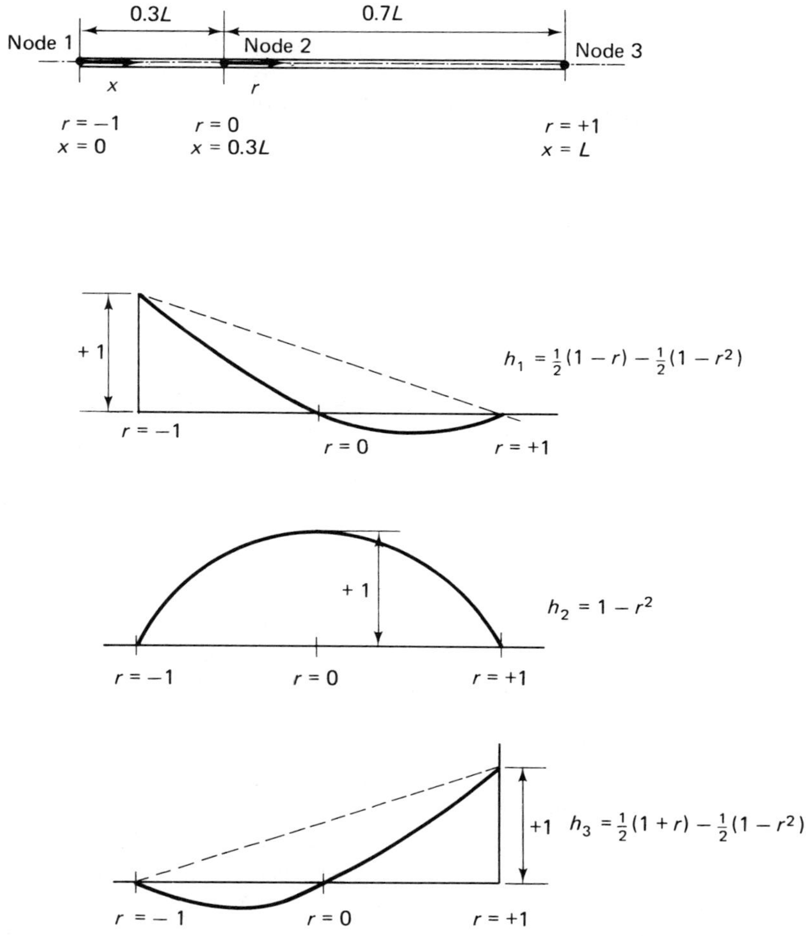

Figure 4.2. One-dimensional interpolation functions of a truss element.

The procedure used in Example 4.1 of constructing the final required interpolation functions suggests an attractive formulation of an element with a variable number of nodes. This formulation is achieved by constructing first the interpolations corresponding to a basic two-node element. The addition of another node then results into an additional interpolation function and a correction to be applied to the already existing interpolation functions. Figure 4.3 gives the interpolation functions of the one-dimensional element considered in Example 4.1, with an additional fourth node possible. As shown in the figure, the element can have from two to four nodes. We

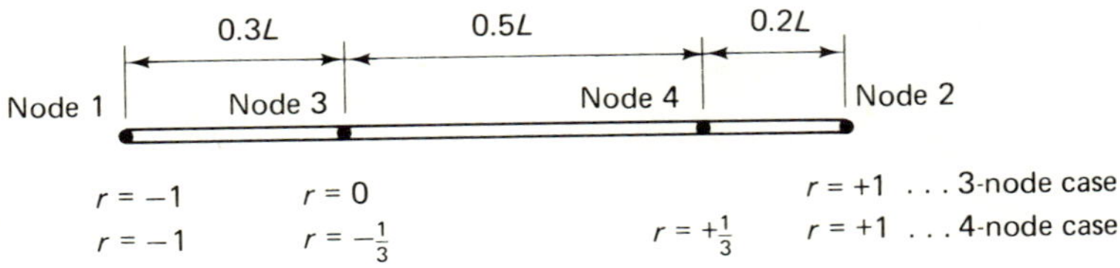

(a) Two to 4 variable-number-nodes truss element

	Include only if node 3 is present	Include only if nodes 3 and 4 are present
$h_1 = \frac{1}{2}(1-r)$	$-\frac{1}{2}(1-r^2)$	$+\frac{1}{16}(-9r^3+r^2+9r-1)$
$h_2 = \frac{1}{2}(1+r)$	$-\frac{1}{2}(1-r^2)$	$+\frac{1}{16}(9r^3+r^2-9r-1)$
$h_3 = (1-r^2)$		$+\frac{1}{16}(27r^3+7r^2-27r-7)$
$h_4 = \frac{1}{16}(-27r^3-9r^2+27r+9)$		

(b) Interpolation functions

Figure 4.3. Interpolation functions of two to four variable-number-nodes one-dimensional element.

should note that nodes 3 and 4 are now intermediate nodes, because nodes 1 and 2 are used to define the two-node element.

This procedure of constructing the element interpolation functions for one-dimensional analysis can be directly generalized for use in two and three dimensions. Figure 4.4 shows the interpolation functions of a four to eight variable-number-nodes two-dimensional element and Fig. 4.5 gives the interpolation functions for three-dimensional eight- to twenty-node elements. The two- and three-dimensional interpolations have been established in a manner analogous to the one-dimensional interpolations, where it should be noted that the basic functions used are, in fact, those already employed in Fig. 4.3. We consider in Figs. 4.4 and 4.5 at most parabolic interpolation, but variable-number-nodes elements with interpolations of higher order could be derived in an analogous way.

The attractiveness of the elements in Figs. 4.3 to 4.5 lies in that the elements can have any number of nodes between the minimum and the maximum. In this context it should be mentioned that a constant strain triangle can be formed from the four-node element in Fig. 4.4 by simply superimposing appropriate element corner nodes. Similarly, wedge elements

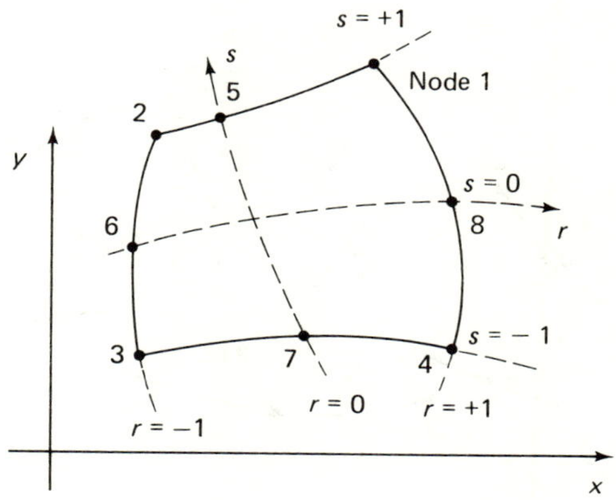

(a) Four to 8 variable-number-nodes two-dimensional element

		Include only if node i is defined			
		$i = 5$	$i = 6$	$i = 7$	$i = 8$
$h_1 =$	$\frac{1}{4}(1+r)(1+s)$	$-\frac{1}{2}h_5$			$-\frac{1}{2}h_8$
$h_2 =$	$\frac{1}{4}(1-r)(1+s)$	$-\frac{1}{2}h_5$	$-\frac{1}{2}h_6$		
$h_3 =$	$\frac{1}{4}(1-r)(1-s)$		$-\frac{1}{2}h_6$	$-\frac{1}{2}h_7$	
$h_4 =$	$\frac{1}{4}(1+r)(1-s)$			$-\frac{1}{2}h_7$	$-\frac{1}{2}h_8$
$h_5 =$	$\frac{1}{2}(1-r^2)(1+s)$				
$h_6 =$	$\frac{1}{2}(1-s^2)(1-r)$				
$h_7 =$	$\frac{1}{2}(1-r^2)(1-s)$				
$h_8 =$	$\frac{1}{2}(1-s^2)(1+r)$				

(b) Interpolation functions

Figure 4.4. Interpolation functions of four to eight variable-number-nodes two-dimensional element.

or a constant strain four-node element can be formed from the eight-node brick element as shown in Fig. 4.6. However, to obtain maximum accuracy, the variable-number-nodes elements should be as nearly rectangular (in three-dimensional analysis, rectangular in each local plane) as possible, and the intermediate side nodes should in general be located at the midpoints between the corner nodes.

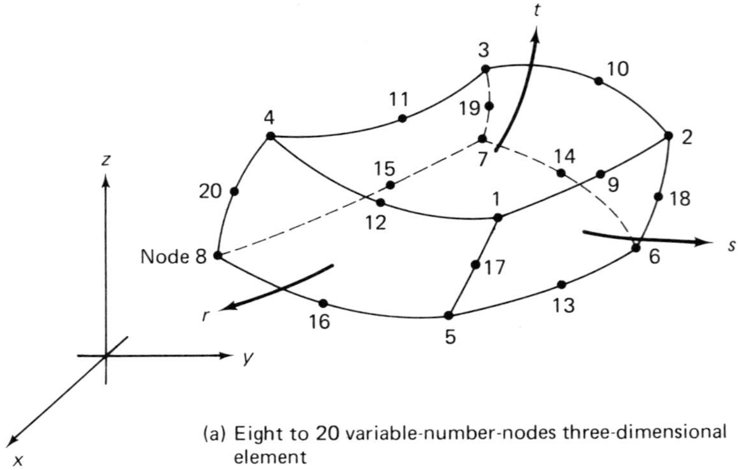

(a) Eight to 20 variable-number-nodes three-dimensional element

$h_1 = g_1 - (g_9 + g_{12} + g_{17})/2$ $\quad h_6 = g_6 - (g_{13} + g_{14} + g_{18})/2$

$h_2 = g_2 - (g_9 + g_{10} + g_{18})/2$ $\quad h_7 = g_7 - (g_{14} + g_{15} + g_{19})/2$

$h_3 = g_3 - (g_{10} + g_{11} + g_{19})/2$ $\quad h_8 = g_8 - (g_{15} + g_{16} + g_{20})/2$

$h_4 = g_4 - (g_{11} + g_{12} + g_{20})/2$ $\quad h_j = g_j$ for $j = 9, \ldots, 20$

$h_5 = g_5 - (g_{13} + g_{16} + g_{17})/2$

$g_i = 0$ if node i is not included; otherwise,

$$g_i = G(r, r_i)\, G(s, s_i)\, G(t, t_i)$$

$$G(\beta, \beta_i) = \tfrac{1}{2}(1 + \beta_i\beta) \quad \text{for } \beta_i = \pm 1$$
$$; \beta = r, s, t$$
$$G(\beta, \beta_i) = (1 - \beta^2) \quad \text{for } \beta_i = 0$$

(b) Interpolation functions

Figure 4.5. Interpolation functions of eight to twenty variable-number-nodes three-dimensional element.

Considering the geometry of the two- and three-dimensional elements in Figs. 4.4 and 4.5 we note that by means of the coordinate interpolation in (4.18), the elements can have, without any difficulties, curved boundaries. This is an important advantage over the generalized coordinate finite element formulation. Another important advantage is the ease with which the element displacement functions can be constructed.

In the isoparametric formulation the element displacements are interpolated in the same way as the geometry; i.e., we use

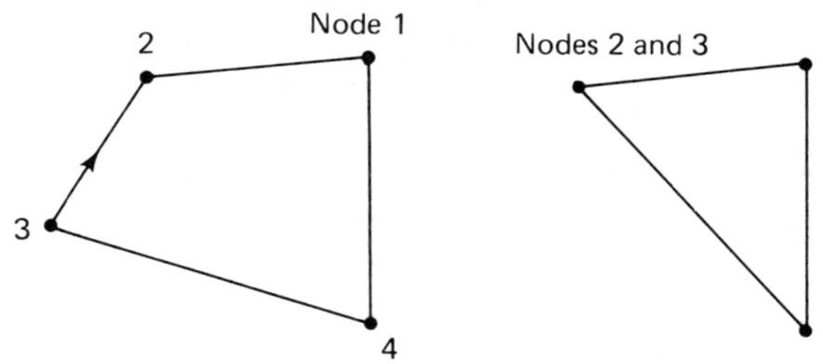

(a) Degeneration of four-node to three-node two-dimensional element

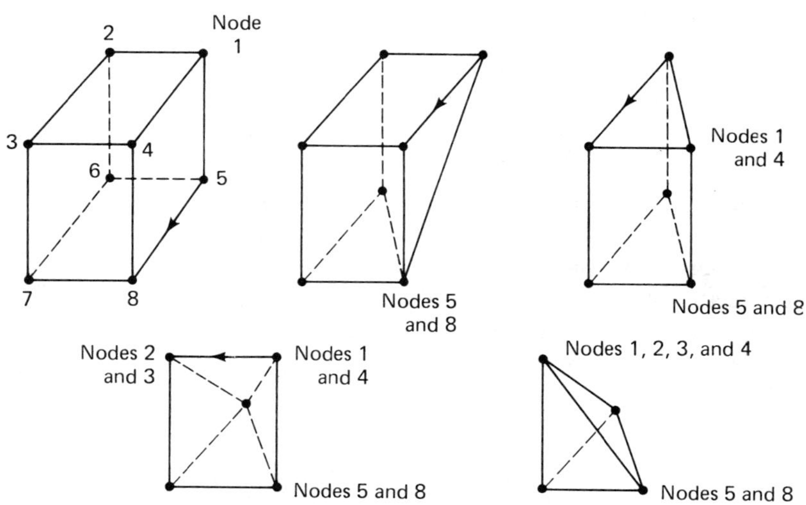

(b) Degenerate forms of eight-node three-dimensional element

Figure 4.6. Degenerate forms of four- and eight-node elements of Figs. 4.4 and 4.5.

$$\left.\begin{aligned} u &= \sum_{i=1}^{q} h_i u_i \\ v &= \sum_{i=1}^{q} h_i v_i \\ w &= \sum_{i=1}^{q} h_i w_i \end{aligned}\right\} \tag{4.19}$$

where u, v, and w are the local element displacements at any point of the element and u_i, v_i, and w_i, $i = 1, \ldots, q$, are the corresponding element displacements at its nodes. Therefore, it is assumed that to each nodal point coordinate necessary to describe the geometry of the element, there corresponds one nodal point displacement.

To be able to evaluate the stiffness matrix of an element, we need to calculate the strain-displacement transformation matrix. The element strains are obtained in terms of derivatives of element displacements with respect to the local coordinates. Because the element displacements are defined in the natural coordinate system using (4.19), we need to relate the x, y, and z derivatives to the r, s, and t derivatives, where we realize that (4.18) is of the form

$$x = f_1(r, s, t); \qquad y = f_2(r, s, t); \qquad z = f_3(r, s, t) \tag{4.20}$$

where f_i denotes "function of." The inverse relationship is

$$r = f_4(x, y, z); \qquad s = f_5(x, y, z); \qquad t = f_6(x, y, z) \tag{4.21}$$

We require the derivatives $\partial/\partial x$, $\partial/\partial y$, and $\partial/\partial z$ and it seems natural to use the chain rule in the following form:

$$\frac{\partial}{\partial x} = \frac{\partial}{\partial r}\frac{\partial r}{\partial x} + \frac{\partial}{\partial s}\frac{\partial s}{\partial x} + \frac{\partial}{\partial t}\frac{\partial t}{\partial x} \tag{4.22}$$

with similar relationships for $\partial/\partial y$ and $\partial/\partial z$. However, to evaluate $\partial/\partial x$ in (4.22), we need to calculate $\partial r/\partial x$, $\partial s/\partial x$, and $\partial t/\partial x$, which means that the explicit inverse relationships in (4.21) would need to be evaluated. These inverse relationships are in general difficult to establish explicitly, and it is necessary to evaluate the required derivatives in the following way. Using the chain rule, we have

$$\begin{bmatrix} \dfrac{\partial}{\partial r} \\[2ex] \dfrac{\partial}{\partial s} \\[2ex] \dfrac{\partial}{\partial t} \end{bmatrix} = \begin{bmatrix} \dfrac{\partial x}{\partial r} & \dfrac{\partial y}{\partial r} & \dfrac{\partial z}{\partial r} \\[2ex] \dfrac{\partial x}{\partial s} & \dfrac{\partial y}{\partial s} & \dfrac{\partial z}{\partial s} \\[2ex] \dfrac{\partial x}{\partial t} & \dfrac{\partial y}{\partial t} & \dfrac{\partial z}{\partial t} \end{bmatrix} \begin{bmatrix} \dfrac{\partial}{\partial x} \\[2ex] \dfrac{\partial}{\partial y} \\[2ex] \dfrac{\partial}{\partial z} \end{bmatrix} \tag{4.23}$$

or, in matrix notation,

$$\frac{\boldsymbol{\partial}}{\boldsymbol{\partial} \mathbf{r}} = \mathbf{J} \frac{\boldsymbol{\partial}}{\boldsymbol{\partial} \mathbf{x}} \tag{4.24}$$

where $\mathbf{J}$ is the *Jacobian operator* relating the natural coordinate derivatives to the local coordinate derivatives. We should note that the Jacobian operator can easily be found using (4.18). We require $\boldsymbol{\partial}/\boldsymbol{\partial}\mathbf{x}$ and use

$$\frac{\boldsymbol{\partial}}{\boldsymbol{\partial} \mathbf{x}} = \mathbf{J}^{-1} \frac{\boldsymbol{\partial}}{\boldsymbol{\partial} \mathbf{r}} \tag{4.25}$$

which requires that the inverse of **J** exists. This inverse exists provided that there is a one-to-one (i.e., unique) correspondence between the natural and the local coordinates of the elements, as expressed in (4.20) and (4.21). In most formulations the one-to-one correspondence between the coordinate systems (i.e., to each r, s, and t, there corresponds only one x, y, and z) is obviously given, such as for the elements in Figs. 4.3 to 4.5. However, in cases where the element is much distorted or folds back upon itself, as in Fig. 4.7, the unique relation between the coordinate systems does not exist.

Using (4.19) and (4.25), we evaluate $\partial u/\partial x, \partial u/\partial y, \partial u/\partial z, \partial v/\partial x, \ldots, \partial w/\partial z$ and can therefore construct the strain-displacement transformation matrix **B**, with

$$\boldsymbol{\epsilon} = \mathbf{B}\mathbf{u} \tag{4.26}$$

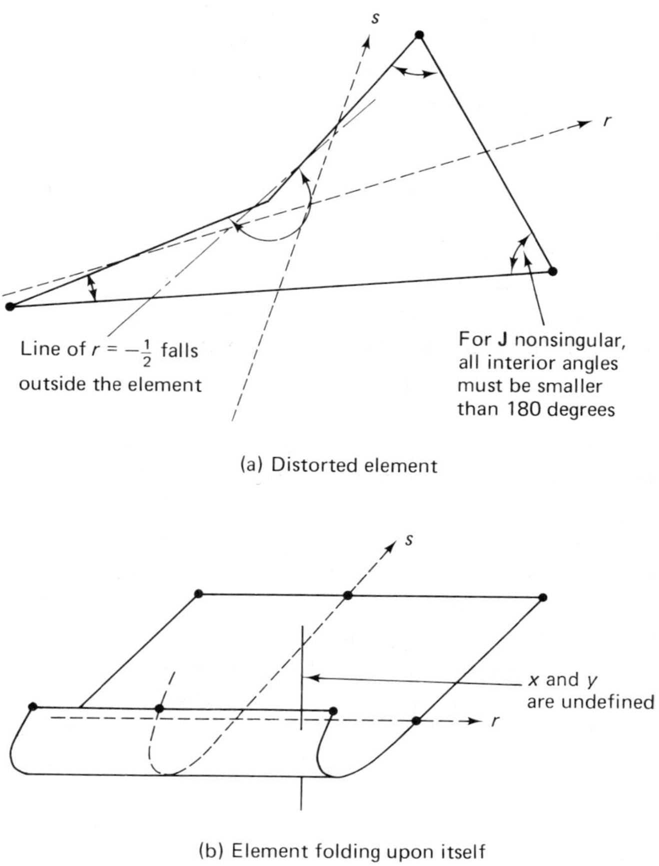

Figure 4.7. Elements with possible singular Jacobian.

where $\mathbf{u}$ is a vector listing the element nodal point displacements of (4.19). The element stiffness matrix corresponding to the local element degrees of freedom is then

$$\mathbf{K} = \int_V \mathbf{B}^T\mathbf{C}\mathbf{B}\, dV \tag{4.27}$$

We should note that the elements of $\mathbf{B}$ are functions of the natural coordinates r, s, and t. Therefore, the volume integration extends over the natural coordinate volume, and the volume differential dV need also be written in terms of the natural coordinates. In general, we have

$$dV = \det \mathbf{J}\, dr\, ds\, dt \tag{4.28}$$

where $\det \mathbf{J}$ is the determinant of the Jacobian operator in (4.24).

An explicit evaluation of the volume integral in (4.27) is generally not possible, and numerical integration must be used. Indeed, numerical integration must be regarded as an integral part of the isoparametric element matrix evaluations. The details of the numerical integration procedure are described in Section 4.6, but the process can briefly be summarized as follows. First, we write (4.27) in the form

$$\mathbf{K} = \int_V \mathbf{F}\, dr\, ds\, dt \tag{4.29}$$

where $\mathbf{F} = \mathbf{B}^T\mathbf{C}\mathbf{B} \det \mathbf{J}$ and the integration is performed in the natural coordinate system of the element. As stated above, the elements of $\mathbf{F}$ depend on r, s, and t, but the actual functional relationship is, in general, unknown. Using numerical integration, the stiffness matrix is now evaluated as

$$\mathbf{K} = \sum_{i,j,k} \alpha_{ijk}\mathbf{F}_{ijk} \tag{4.30}$$

where $\mathbf{F}_{ijk}$ is the matrix $\mathbf{F}$ evaluated at points r_i, s_j, and the t_k, and the α_{ijk} are given constants which depend on the values of r_i, s_j, and t_k. The sampling points r_i, s_j, and t_k of the function and the corresponding weighting factors α_{ijk} are chosen to obtain maximum accuracy in the integration. Naturally, the integration accuracy increases as the number of sampling points is increased.

The purpose of the above brief outline of the numerical integration procedure was to complete the description of the general isoparametric formulation. The relative simplicity of the formulation may already be noted. It is the simplicity of the element formulation and the efficiency with which the element matrices can actually be evaluated in the computer that has drawn much attention to the development of the isoparametric and related elements.

The formulation of the element mass matrix and load vectors is now straightforward. Namely, writing the element displacements in the form

$$\mathbf{u}(r, s, t) = \mathbf{H}\mathbf{u} \tag{4.31}$$

LIVESTOCK-5

Grown, cured and smoked right on the farm—Howard Nail holds some of his homemade orange-juice bacon. He's not exactly liberal with the details of his concoction, but does admit to soaking the slab in orange juice.

soybean meal bags at feed m
mixer/grinder and oversees f

It is in maintaining feed
lenges. Getting corn into ea
planned well in advance. "Al
this year we had to buy Mid
Howard. "We normally get
Florida, but last year was a
What corn there was from Fl

The midwestern corn is ex
River and across the Gulf of
per bushel plus 15¢ for on-
Nail's supply. "Although we
a supply," observes Howard.

where $\mathbf{H}$ is a matrix of the interpolation functions, we have, using (3.74) and (3.63) to (3.65),

$$\mathbf{M} = \int_V \rho \mathbf{H}^T \mathbf{H}\, dV \tag{4.32}$$

$$\mathbf{R}_B = \int_V \mathbf{H}^T \mathbf{f}^B\, dV \tag{4.33}$$

$$\mathbf{R}_S = \int_S \mathbf{H}^{S^T} \mathbf{f}^S\, dS \tag{4.34}$$

$$\mathbf{R}_I = \int_V \mathbf{B}^T \boldsymbol{\tau}^I\, dV \tag{4.35}$$

The above matrices are evaluated using numerical integration as indicated for the stiffness matrix $\mathbf{K}$ in (4.30). In the evaluation we need to use the appropriate function $\mathbf{F}$. To calculate the body force vector $\mathbf{R}_B$ we use $\mathbf{F} = \mathbf{H}^T \mathbf{f}^B \det \mathbf{J}$, for the surface force vector we use $\mathbf{F} = \mathbf{H}^{S^T} \mathbf{f}^S \det \mathbf{J}$, for the initial stress load vector we use $\mathbf{F} = \mathbf{B}^T \boldsymbol{\tau}^I \det \mathbf{J}$, and for the mass matrix we have $\mathbf{F} = \rho \mathbf{H}^T \mathbf{H} \det \mathbf{J}$. The weight coefficients α_{ijk} are the same as in the stiffness matrix evaluation if the same order of numerical integration is used. However, in practice, different integration orders may be employed because the required accuracy in the element matrices varies (see Section 4.7).

The above formulation was for one-, two-, or three-dimensional elements. We shall now consider the specific case of two-dimensional analysis and demonstrate for that case the details of the calculation of element matrices.

Example 4.2

Derive the expressions needed for the evaluation of the stiffness matrix of the isoparametric four-node finite element in Fig. 4.8. Assume plane stress or plane strain conditions.

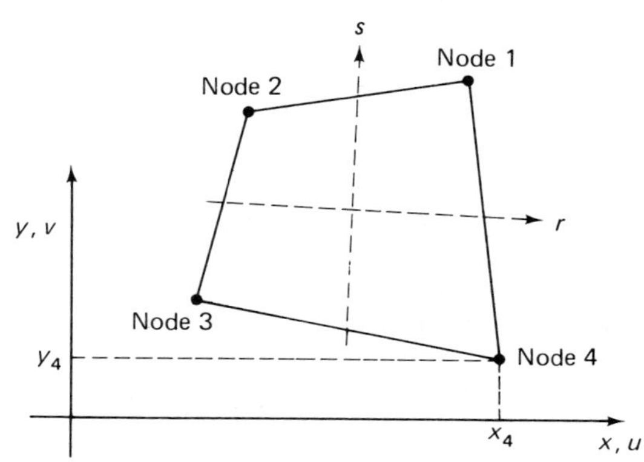

Figure 4.8. Four-node two-dimensional element.

Using the interpolation functions h_1, h_2, h_3, and h_4 defined in Fig. 4.4, the coordinate interpolation given in (4.18) is, for this element,

$$\begin{aligned} x &= \tfrac{1}{4}(1+r)(1+s)x_1 + \tfrac{1}{4}(1-r)(1+s)x_2 \\ &\quad + \tfrac{1}{4}(1-r)(1-s)x_3 + \tfrac{1}{4}(1+r)(1-s)x_4 \\ y &= \tfrac{1}{4}(1+r)(1+s)y_1 + \tfrac{1}{4}(1-r)(1+s)y_2 \\ &\quad + \tfrac{1}{4}(1-r)(1-s)y_3 + \tfrac{1}{4}(1+r)(1-s)y_4 \end{aligned}$$

The displacement interpolation given in (4.19) is

$$\begin{aligned} u &= \tfrac{1}{4}(1+r)(1+s)u_1 + \tfrac{1}{4}(1-r)(1+s)u_2 \\ &\quad + \tfrac{1}{4}(1-r)(1-s)u_3 + \tfrac{1}{4}(1+r)(1-s)u_4 \\ v &= \tfrac{1}{4}(1+r)(1+s)v_1 + \tfrac{1}{4}(1-r)(1+s)v_2 \\ &\quad + \tfrac{1}{4}(1-r)(1-s)v_3 + \tfrac{1}{4}(1+r)(1-s)v_4 \end{aligned}$$

The element strains are given by

$$\boldsymbol{\epsilon}^T = [\epsilon_{xx} \quad \epsilon_{yy} \quad \gamma_{xy}]$$

where

$$\epsilon_{xx} = \frac{\partial u}{\partial x}; \qquad \epsilon_{yy} = \frac{\partial v}{\partial y}; \qquad \gamma_{xy} = \frac{\partial u}{\partial y} + \frac{\partial v}{\partial x}$$

To evaluate the displacement derivatives, we need to evaluate (4.23):

$$\begin{bmatrix} \dfrac{\partial}{\partial r} \\[2ex] \dfrac{\partial}{\partial s} \end{bmatrix} = \begin{bmatrix} \dfrac{\partial x}{\partial r} & \dfrac{\partial y}{\partial r} \\[2ex] \dfrac{\partial x}{\partial s} & \dfrac{\partial y}{\partial s} \end{bmatrix} \begin{bmatrix} \dfrac{\partial}{\partial x} \\[2ex] \dfrac{\partial}{\partial y} \end{bmatrix} \qquad \text{or} \qquad \frac{\boldsymbol{\partial}}{\boldsymbol{\partial} \mathbf{r}} = \mathbf{J} \frac{\boldsymbol{\partial}}{\boldsymbol{\partial} \mathbf{x}}$$

where

$$\frac{\partial x}{\partial r} = \tfrac{1}{4}(1+s)x_1 - \tfrac{1}{4}(1+s)x_2 - \tfrac{1}{4}(1-s)x_3 + \tfrac{1}{4}(1-s)x_4$$

$$\frac{\partial x}{\partial s} = \tfrac{1}{4}(1+r)x_1 + \tfrac{1}{4}(1-r)x_2 - \tfrac{1}{4}(1-r)x_3 - \tfrac{1}{4}(1+r)x_4$$

$$\frac{\partial y}{\partial r} = \tfrac{1}{4}(1+s)y_1 - \tfrac{1}{4}(1+s)y_2 - \tfrac{1}{4}(1-s)y_3 + \tfrac{1}{4}(1-s)y_4$$

$$\frac{\partial y}{\partial s} = \tfrac{1}{4}(1+r)y_1 + \tfrac{1}{4}(1-r)y_2 - \tfrac{1}{4}(1-r)y_3 - \tfrac{1}{4}(1+r)y_4$$

Therefore, for any value r and s, $-1 \leq r \leq +1$ and $-1 \leq s \leq +1$, we can form the Jacobian operator $\mathbf{J}$ by using the above expressions for $\partial x/\partial r$, $\partial x/\partial s$, and $\partial y/\partial r$, $\partial y/\partial s$. Assume that we evaluate $\mathbf{J}$ at $r = r_i$ and $s = s_j$ and denote the operator by $\mathbf{J}_{ij}$ and its determinant by det $\mathbf{J}_{ij}$. Then we have

$$\begin{bmatrix} \dfrac{\partial}{\partial x} \\[2ex] \dfrac{\partial}{\partial y} \end{bmatrix}_{\substack{\text{at } r=r_i \\ s=s_j}} = \mathbf{J}_{ij}^{-1} \begin{bmatrix} \dfrac{\partial}{\partial r} \\[2ex] \dfrac{\partial}{\partial s} \end{bmatrix}_{\substack{\text{at } r=r_i \\ s=s_j}}$$

To evaluate the element strains we use

$$\frac{\partial u}{\partial r} = \tfrac{1}{4}(1+s)u_1 - \tfrac{1}{4}(1+s)u_2 - \tfrac{1}{4}(1-s)u_3 + \tfrac{1}{4}(1-s)u_4$$

$$\frac{\partial u}{\partial s} = \tfrac{1}{4}(1+r)u_1 + \tfrac{1}{4}(1-r)u_2 - \tfrac{1}{4}(1-r)u_3 - \tfrac{1}{4}(1+r)u_4$$

$$\frac{\partial v}{\partial r} = \tfrac{1}{4}(1+s)v_1 - \tfrac{1}{4}(1+s)v_2 - \tfrac{1}{4}(1-s)v_3 + \tfrac{1}{4}(1-s)v_4$$

$$\frac{\partial v}{\partial s} = \tfrac{1}{4}(1+r)v_1 + \tfrac{1}{4}(1-r)v_2 - \tfrac{1}{4}(1-r)v_3 - \tfrac{1}{4}(1+r)v_4$$

Therefore,

$$\begin{bmatrix} \dfrac{\partial u}{\partial x} \\ \dfrac{\partial u}{\partial y} \end{bmatrix}_{\substack{\text{at } r=r_i \\ s=s_j}} = \tfrac{1}{4}\mathbf{J}_{ij}^{-1}\begin{bmatrix} 1+s_j & 0 & -(1+s_j) & 0 & -(1-s_j) & 0 & 1-s_j & 0 \\ 1+r_i & 0 & 1-r_i & 0 & -(1-r_i) & 0 & -(1+r_i) & 0 \end{bmatrix}\mathbf{u} \quad \text{(a)}$$

and

$$\begin{bmatrix} \dfrac{\partial v}{\partial x} \\ \dfrac{\partial v}{\partial y} \end{bmatrix}_{\substack{\text{at } r=r_i \\ s=s_j}} = \tfrac{1}{4}\mathbf{J}_{ij}^{-1}\begin{bmatrix} 0 & 1+s_j & 0 & -(1+s_j) & 0 & -(1-s_j) & 0 & 1-s_j \\ 0 & 1+r_i & 0 & 1-r_i & 0 & -(1-r_i) & 0 & -(1+r_i) \end{bmatrix}\mathbf{u} \quad \text{(b)}$$

where

$$\mathbf{u}^T = [u_1 v_1 \quad u_2 v_2 \quad u_3 v_3 \quad u_4 v_4]$$

Evaluating the relations in (a) and (b), we can establish the strain-displacement transformation matrix at the point (r_i, s_j); i.e., we obtain

$$\boldsymbol{\epsilon}_{ij} = \mathbf{B}_{ij}\mathbf{u}$$

where the subscripts i and j indicate that the strain-displacement transformation is evaluated at the points r_i and s_j. For example, if $x = r$, $y = s$, i.e., the stiffness matrix of a square element is required that has side lengths equal to 2, the Jacobian operator is the identity matrix, and hence

$$\mathbf{B}_{ij} = \tfrac{1}{4}\begin{bmatrix} 1+s_j & 0 & -(1+s_j) & 0 & -(1-s_j) & 0 & 1-s_j & 0 \\ 0 & 1+r_i & 0 & 1-r_i & 0 & -(1-r_i) & 0 & -(1+r_i) \\ 1+r_i & 1+s_j & 1-r_i & -(1+s_j) & -(1-r_i) & -(1-s_j) & -(1+r_i) & 1-s_j \end{bmatrix}$$

The matrix $\mathbf{F}_{ij}$ in (4.30) is now simply

$$\mathbf{F}_{ij} = \mathbf{B}_{ij}^T\mathbf{C}\mathbf{B}_{ij} \det \mathbf{J}_{ij}$$

where the material property matrix $\mathbf{C}$ is given in Table 3.2. In the case of plane stress or plane strain conditions, we integrate in the r, s plane and assume that the function $\mathbf{F}$ is constant through the thickness of the element. The stiffness matrix of the element is therefore

$$\mathbf{K} = \sum_{i,j} t_{ij}\alpha_{ij}\mathbf{F}_{ij}$$

where t_{ij} is the thickness of the element at the sampling point r_i, s_j ($t_{ij} = 1.0$ in plane strain analysis). With the matrices $\mathbf{F}_{ij}$ given as above and the weighing factors α_{ij} being determined, the required stiffness matrix can readily be evaluated.

For the actual implementation it should be noted that in the evaluation of $\mathbf{J}_{ij}$ and of the matrices defining the displacement derivatives in (a) and (b), only the eight possible derivatives of the interpolation functions $h_1, \ldots, h_4$ are required. Therefore, it is expedient to calculate these derivatives corresponding to the point (r_i, s_j) once at the start of the evaluation of $\mathbf{B}_{ij}$ and use them whenever they are required.

It should also be realized that considering the specific point (r_i, s_j), the relations in (a) and (b) may be written, respectively, as

$$\left.\begin{aligned}\frac{\partial u}{\partial x} &= \sum_{i=1}^{4} \frac{\partial h_i}{\partial x} u_i \\ \frac{\partial u}{\partial y} &= \sum_{i=1}^{4} \frac{\partial h_i}{\partial y} u_i\end{aligned}\right\} \tag{c}$$

and

$$\left.\begin{aligned}\frac{\partial v}{\partial x} &= \sum_{i=1}^{4} \frac{\partial h_i}{\partial x} v_i \\ \frac{\partial v}{\partial y} &= \sum_{i=1}^{4} \frac{\partial h_i}{\partial y} v_i\end{aligned}\right\} \tag{d}$$

Hence we have

$$\mathbf{B} = \begin{bmatrix} \frac{\partial h_1}{\partial x} & 0 & \frac{\partial h_2}{\partial x} & 0 & \frac{\partial h_3}{\partial x} & 0 & \frac{\partial h_4}{\partial x} & 0 \\ 0 & \frac{\partial h_1}{\partial y} & 0 & \frac{\partial h_2}{\partial y} & 0 & \frac{\partial h_3}{\partial y} & 0 & \frac{\partial h_4}{\partial y} \\ \frac{\partial h_1}{\partial y} & \frac{\partial h_1}{\partial x} & \frac{\partial h_2}{\partial y} & \frac{\partial h_2}{\partial x} & \frac{\partial h_3}{\partial y} & \frac{\partial h_3}{\partial x} & \frac{\partial h_4}{\partial y} & \frac{\partial h_4}{\partial x} \end{bmatrix} \tag{e}$$

where it is implied that in (c) and (d), the derivatives are evaluated at point (r_i, s_j), and therefore in (e), we have, in fact, the matrix $\mathbf{B}_{ij}$.

Example 4.3

Derive the expressions needed for the evaluation of the mass matrix of the element considered in Example 4.2.

The mass matrix of the element is given by

$$\mathbf{M} = \sum_{i,j} \alpha_{ij} t_{ij} \mathbf{F}_{ij}$$

where

$$\mathbf{F}_{ij} = \rho_{ij} \mathbf{H}_{ij}^T \mathbf{H}_{ij} \det \mathbf{J}_{ij}$$

and $\mathbf{H}_{ij}$ is the displacement interpolation matrix. The displacement interpolation functions for u and v of the four-node element have been given in Example 4.2, and we have

$$\mathbf{H}_{ij} = \frac{1}{4}\begin{bmatrix} (1+r_i)(1+s_j) & 0 & (1-r_i)(1+s_j) & 0 & (1-r_i)(1-s_j) & 0 & (1+r_i)(1-s_j) & 0 \\ 0 & (1+r_i)(1+s_j) & 0 & (1-r_i)(1+s_j) & 0 & (1-r_i)(1-s_j) & 0 & (1+r_i)(1-s_j) \end{bmatrix}$$

The determinant of the Jacobian matrix, det $\mathbf{J}_{ij}$, was given in Example 4.2, and ρ_{ij} is the mass density at the sampling point r_i, s_j. Therefore, all required variables for the evaluation of the mass matrix have been defined.

Example 4.4

Derive the expressions needed for the evaluation of the body force vector $\mathbf{R}_B$ and the initial stress vector $\mathbf{R}_I$ given in (4.33) and (4.35), respectively.

These vectors are obtained using the matrices $\mathbf{H}_{ij}$, $\mathbf{B}_{ij}$, and $\mathbf{J}_{ij}$ defined in Examples 4.2 and 4.3; i.e., we have

$$\mathbf{R}_B = \sum_{i,j} \alpha_{ij} t_{ij} \mathbf{H}_{ij}^T \mathbf{f}_{ij}^B \det \mathbf{J}_{ij}$$

$$\mathbf{R}_I = \sum_{i,j} \alpha_{ij} t_{ij} \mathbf{B}_{ij}^T \boldsymbol{\tau}_{ij}^I \det \mathbf{J}_{ij}$$

where $\mathbf{f}_{ij}^B$ and $\boldsymbol{\tau}_{ij}^I$ are the body force vector and initial stress vector evaluated at the integration sampling points.

Example 4.5

Derive the expressions needed in the calculation of the surface force vector $\mathbf{R}_S$, when the element edge 1-2 of the four-node isoparametric element is loaded as shown in Fig. 4.9.

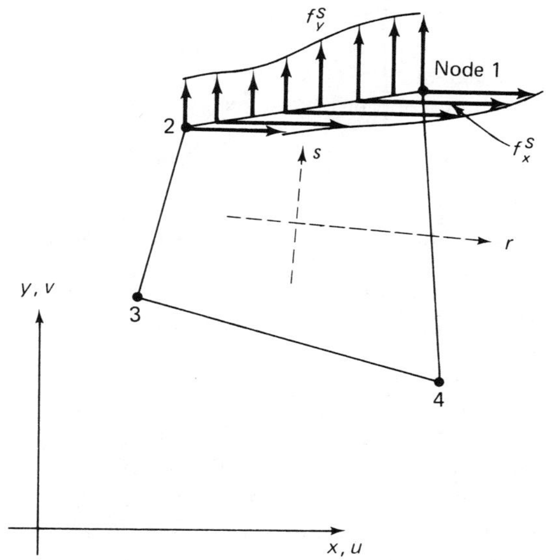

Figure 4.9. Pressure distribution along edge 1-2 of a four-node element.

The first step is to establish the displacement interpolations. Since $s = +1$ at the edge 1-2, we have, using the interpolation functions given in Example 4.2,

$$u^S = \tfrac{1}{2}(1 + r)u_1 + \tfrac{1}{2}(1 - r)u_2$$

$$v^S = \tfrac{1}{2}(1 + r)v_1 + \tfrac{1}{2}(1 - r)v_2$$

Hence to evaluate $\mathbf{R}_S$ in (4.34) we can use

$$\mathbf{H}^S = \begin{bmatrix} \frac{1}{2}(1+r) & 0 & \frac{1}{2}(1-r) & 0 & 0 & 0 & 0 & 0 \\ 0 & \frac{1}{2}(1+r) & 0 & \frac{1}{2}(1-r) & 0 & 0 & 0 & 0 \end{bmatrix}$$

and

$$\mathbf{f}^S = \begin{bmatrix} f_x^S \\ f_y^S \end{bmatrix}$$

where f_x^S and f_y^S are the x and y components of the applied surface force. These components may have been given as a function of r.

For the evaluation of the integral in (4.34), we also need the differential surface area dS expressed in the r, s natural coordinate system. If t_r is the thickness, $dS = t_r\, dl$, where dl is a differential length,

$$dl = \left\{\left(\frac{\partial x}{\partial r}\right)^2 + \left(\frac{\partial y}{\partial r}\right)^2\right\}^{1/2} dr$$

But the derivatives $\partial x/\partial r$ and $\partial y/\partial r$ have been given in Example 4.2. Using $s = +1$, we have, in this case,

$$\frac{\partial x}{\partial r} = \frac{x_1 - x_2}{2}; \qquad \frac{\partial y}{\partial r} = \frac{y_1 - y_2}{2}$$

Although the vector $\mathbf{R}_S$ could in this case be evaluated in a closed form solution (provided that the functions used in $\mathbf{f}^S$ are simple), in order to keep generality in the program that calculates $\mathbf{R}_S$, it is expedient to use numerical integration. This way, variable-number-nodes elements can be implemented in an elegant manner in one program. Thus, using the notation defined in this section, we have

$$\mathbf{R}_S = \sum_i \alpha_i t_i \mathbf{F}_i$$

$$\mathbf{F}_i = \mathbf{H}_i^{S^T} \mathbf{f}_i^S\, dl_i$$

It is noted that in this case only one-dimensional numerical integration is required, because s is not a variable.

4.3.2 Element Matrices in Global Coordinate System

So far we considered the calculation of isoparametric element matrices that correspond to local element degrees of freedom. In the evaluation we used local element coordinates x, y, and z, whichever are applicable, and local element degrees of freedom u_i, v_i, and w_i. However, we may note that for the two-dimensional element considered in Examples 4.2 to 4.5, the element matrices could have been evaluated using the global coordinate variables X and Y, and the global nodal point displacements U_i and V_i. Indeed, in the calculations presented, the x and y local coordinates and u and v local displacement components needed simply to be replaced by the X and Y global coordinates and U and V global displacement components, respectively. In such case the matrices would have corresponded directly to the global displacement components.

In general, the calculation of the element matrices should be carried out in the global coordinate system, using global displacement components if the number of natural coordinate variables is equal to the number of global variables. Typical examples are two-dimensional elements that are defined in a global plane, and the three-dimensional elements in Fig. 4.5. In these cases the Jacobian operator in (4.24) is a square matrix, which can be inverted as required in (4.25), and the element matrices correspond directly to the global displacement components.

In those cases where the order of the global coordinate system is higher than the order of the natural coordinate system, it is usually most expedient to calculate first the element matrices in the local coordinate system and corresponding to local displacement components. Afterward, the matrices must be transformed in the usual manner to the global displacement system. Examples are the truss element or the plane stress element when they are oriented arbitrarily in three-dimensional space. However, alternatively, we may include the transformation to the global displacement components directly in the formulation. This is accomplished by introducing a transformation that expresses in the displacement interpolation the local nodal point displacements in terms of their global components.

Example 4.6

Evaluate the element stiffness matrix of the truss element in Fig. 4.10 using directly global nodal point displacements.

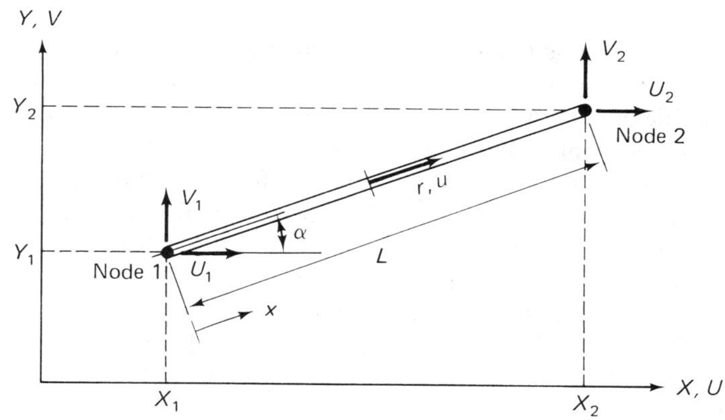

Figure 4.10. Truss element in global coordinate system.

The stiffness matrix of the element is given in (4.27); i.e.,

$$\mathbf{K} = \int_V \mathbf{B}^T \mathbf{C} \mathbf{B} \, dV$$

where **B** is the strain-displacement matrix and **C** is the stress-strain matrix. For the truss element considered we have

$$u = [\cos\alpha \quad \sin\alpha]\begin{bmatrix} \frac{1}{2}(1-r)U_1 + \frac{1}{2}(r+1)U_2 \\ \frac{1}{2}(1-r)V_1 + \frac{1}{2}(r+1)V_2 \end{bmatrix}$$

Then using that the strain $\epsilon = \partial u/\partial x$, which expressed in the natural coordinate system is $\epsilon = (2/L)\,\partial u/\partial r$ (see Section 4.2), we can write the strain-displacement transformation corresponding to the displacement vector $\mathbf{U}^T = [U_1 \quad V_1 \quad U_2 \quad V_2]$ as

$$\mathbf{B} = \frac{1}{L}[\cos\alpha \quad \sin\alpha \quad \cos\alpha \quad \sin\alpha]\begin{bmatrix} -1 & & & \\ & -1 & & \text{zeros} \\ & & 1 & \\ \text{zeros} & & & 1 \end{bmatrix}$$

Also, as given in Section 4.2, we have

$$dV = \frac{AL}{2}\,dr \qquad \text{and} \qquad \mathbf{C} = E$$

Substituting the relations for **B**, **C**, and dV and evaluating the integral, we obtain

$$\mathbf{K} = \frac{AE}{L}\begin{bmatrix} \cos^2\alpha & \cos\alpha\sin\alpha & -\cos^2\alpha & -\cos\alpha\sin\alpha \\ \sin\alpha\cos\alpha & \sin^2\alpha & -\sin\alpha\cos\alpha & -\sin^2\alpha \\ -\cos^2\alpha & -\cos\alpha\sin\alpha & \cos^2\alpha & \cos\alpha\sin\alpha \\ -\sin\alpha\cos\alpha & -\sin^2\alpha & \sin\alpha\cos\alpha & \sin^2\alpha \end{bmatrix}$$

4.4 CONVERGENCE CONSIDERATIONS

In Section 3.2.5 we discussed the requirements for monotonic convergence of the finite element analysis. The two requirements are that the elements must be compatible and complete. We now want to investigate whether the isoparametric element formulation satisfies the convergence criteria.

To investigate the compatibility of an element assemblage, we need to consider each edge, or rather face, between adjacent elements. For compatibility it is necessary that the coordinates and the displacements of the elements at the common face be the same. This is the case if the elements have the same face nodes and the coordinates and displacements along the common face are in each element defined by the same interpolation functions. Examples of adjacent elements that preserve compatibility, and do not, are shown in Fig. 4.11.

Completeness requires that the rigid body displacements and the constant strain states must be possible. Considering a three-dimensional element, this means that the following displacements defined in the local element coordinate system must still be possible when the isoparametric formulation is used:

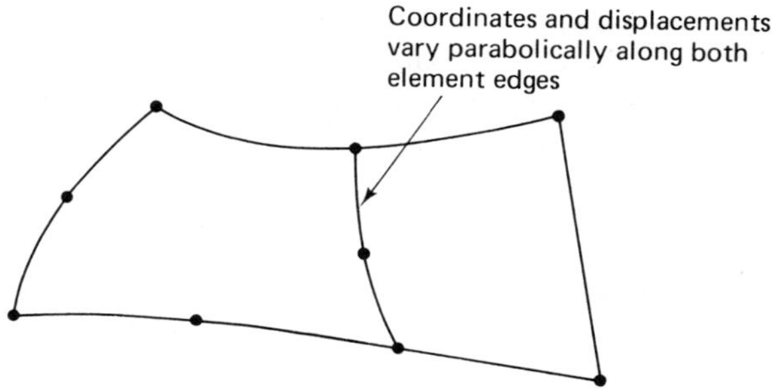

(a) Compatible elements

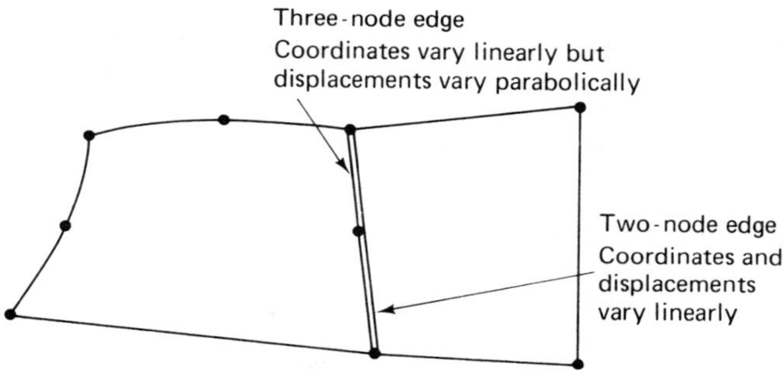

(b) Incompatible elements

Figure 4.11. Compatible and incompatible two-dimensional elements.

$$\left.\begin{aligned} u &= a_1 + b_1x + c_1y + d_1z \\ v &= a_2 + b_2x + c_2y + d_2z \\ w &= a_3 + b_3x + c_3y + d_3z \end{aligned}\right\} \tag{4.36}$$

where a_j, b_j, c_j, and d_j, $j = 1, 2, 3$, are constants. For two- and one-dimensional elements, only the appropriate coordinates and displacements would be included on the right-hand side of (4.36). If a displacement field of the form described in (4.36) could be imposed, the nodal point displacements would be

$$\left.\begin{aligned} u_i &= a_1 + b_1x_i + c_1y_i + d_1z_i \\ v_i &= a_2 + b_2x_i + c_2y_i + d_2z_i \\ w_i &= a_3 + b_3x_i + c_3y_i + d_3z_i \end{aligned}\right\} \tag{4.37}$$

where $i = 1, \ldots, q$ and q = number of nodes. To show that the displacements in (4.36) are possible when the isoparametric formulation is employed, assume that the nodal point displacements of the element are given by (4.37). We should find that, with the nodal point displacements in (4.37), the displacements in the isoparametric formulation are actually those given in (4.36). In the isoparametric formulation we have

$$u = \sum_{i=1}^{q} h_i u_i; \qquad v = \sum_{i=1}^{q} h_i v_i; \qquad w = \sum_{i=1}^{q} h_i w_i \tag{4.38}$$

Substituting from (4.37) we obtain

$$\left.\begin{aligned} u &= a_1 \sum_{i=1}^{q} h_i + b_1 \sum_{i=1}^{q} h_i x_i + c_1 \sum_{i=1}^{q} h_i y_i + d_1 \sum_{i=1}^{q} h_i z_i \\ v &= a_2 \sum_{i=1}^{q} h_i + b_2 \sum_{i=1}^{q} h_i x_i + c_2 \sum_{i=1}^{q} h_i y_i + d_2 \sum_{i=1}^{q} h_i z_i \\ w &= a_3 \sum_{i=1}^{q} h_i + b_3 \sum_{i=1}^{q} h_i x_i + c_3 \sum_{i=1}^{q} h_i y_i + d_3 \sum_{i=1}^{q} h_i z_i \end{aligned}\right\} \tag{4.39}$$

Since in the isoparametric formulation the coordinates are interpolated in the same way as the displacements, we can use (4.18) to obtain from (4.39)

$$\begin{aligned} u &= a_1 \sum_{i=1}^{q} h_i + b_1 x + c_1 y + d_1 z \\ v &= a_2 \sum_{i=1}^{q} h_i + b_2 x + c_2 y + d_2 z \\ w &= a_3 \sum_{i=1}^{q} h_i + b_3 x + c_3 y + d_3 z \end{aligned} \tag{4.40}$$

The displacements defined in (4.40), however, are the same as those given in (4.36), provided that

$$\sum_{i=1}^{q} h_i = 1 \tag{4.41}$$

for any point in the element. Hence the relation in (4.41) is the condition on the interpolation functions in order to satisfy the completeness requirements. We may note that (4.41) is satisfied at the nodes of an element, because the interpolation function h_i has been constructed to be unity at node i with all other interpolation functions, h_j, $j \neq i$, being zero at that node. In fact, although not obvious, the condition in (4.41) will always be satisfied if the isoparametric element is properly constructed. We consider two cases in the following examples.

Example 4.7

Examine whether the plane stress element in Fig. 4.8 satisfies the requirements for monotonic convergence.

For monotonic convergence the element need be compatible and complete. Since along common element edges, the coordinates and displacements vary linearly and are determined only by the coordinates and displacements of the edge nodal points, the element is compatible.

Completeness requires that (4.41) is satisfied. Evaluating $\sum_{i=1}^{4} h_i$ for this element, we obtain

$$\tfrac{1}{4}(1 + r)(1 + s) + \tfrac{1}{4}(1 - r)(1 + s) + \tfrac{1}{4}(1 - r)(1 - s) + \tfrac{1}{4}(1 + r)(1 - s) = 1$$

Hence the element is also complete and all requirements for monotonic convergence are satisfied.

Example 4.8

Investigate whether the requirements for monotonic convergence are satisfied for the variable-number-nodes elements in Figs. 4.4 and 4.5.

Compatibility is maintained between element edges in two-dimensional analysis and element faces in three-dimensional analysis, provided that the same number of nodes are used on connecting element edges and faces. A typical compatible element layout is shown in Fig. 4.11.

Considering the second requirement for monotonic convergence, completeness, (4.41) must be satisfied. However, we showed in Example 4.7 that the basic four-node element is complete, and on studying the interpolation functions given in Fig. 4.4 for the variable-number-nodes element, we find that the total contribution that is added to the basic four-node interpolation functions is always zero for whichever additional node is included. Hence any one of the possible elements defined by the variable number of nodes in Fig. 4.4 is complete. The proof for the three-dimensional elements in Fig. 4.5 is carried out in an analogous manner.

It follows, therefore, that the variable-number-nodes elements satisfy the requirements for monotonic convergence.

4.5 ASSOCIATED ELEMENT FAMILIES

In the isoparametric finite element derivation we assumed that the element coordinates are interpolated in the same way as the element displacements. We may, however, find it more efficient in the solution process to interpolate the coordinates to a lower degree than the displacements. Typically, for the straight-sided eight-node quadrilateral in Fig. 4.12, the displacements u and v are interpolated parabolically, whereas the coordinates x and y of the element need only be interpolated linearly because the element edges are straight. Elements with a lower degree of interpolation on the coordinates than on the displacements are called *subparametric elements.* Such elements satisfy the monotonic convergence requirements, because, firstly, compatibility is satisfied for the same reasoning that is used in the isoparametric formulation. Secondly, a subparametric element is complete because the associated isoparametric element, which is based only on the coordinate interpolations used, is complete. This follows because the addition of higher-order displacement interpolations does not prevent the element from still undergoing the rigid body modes and constant strain states.

In another class of elements, which are called *superparametric elements,*

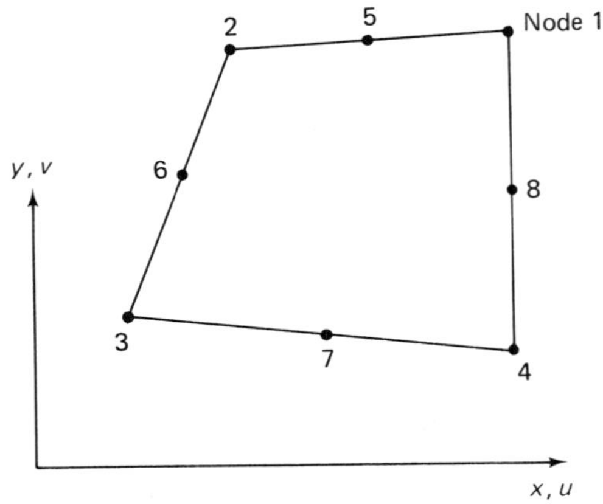

$$x = \sum_{i=1}^{4} h_i x_i; \qquad y = \sum_{i=1}^{4} h_i y_i$$

$$u = \sum_{i=1}^{8} h_i u_i; \qquad v = \sum_{i=1}^{8} h_i v_i$$

Figure 4.12. Two-dimensional subparametric element.

the coordinates are interpolated to a higher degree than the displacements.[17] Superparametric elements have been introduced for thick and thin shell analysis. The basic problems of using isoparametric elements in thick to thin shell analysis are twofold.[17-21, 28-31] Firstly, excessive shear strain energy is stored in the elements and, secondly, as the element becomes thin, the stiffness coefficients corresponding to the transverse displacement degrees of freedom are considerably larger than those corresponding to the longitudinal displacements, which results in numerical ill-conditioning. Considering the first problem, which is important when the parabolic elements in Figs. 4.4 and 4.5 are used, to improve element behavior, selective or reduced numerical integration can be employed (see Section 4.7) or, as discussed below, incompatible displacement modes may be introduced. Next, considering the second problem, to avoid the numerical ill-conditioning, relative displacement degrees of freedom can be used as shown in Fig. 4.13 (see Section 7.5). However, since the relative displacements are small and do not contribute to the overall response of a thin shell, it may be more effective to employ a superparametric element. The geometry of the element is defined by the original interpolation, but the displacement degrees of freedom are assumed to be rotations and translations at midsurface nodes (see Fig. 4.13). Superparametric elements do not necessarily satisfy the convergence criteria, and each class of elements must specifically be examined for compatibility and completeness.

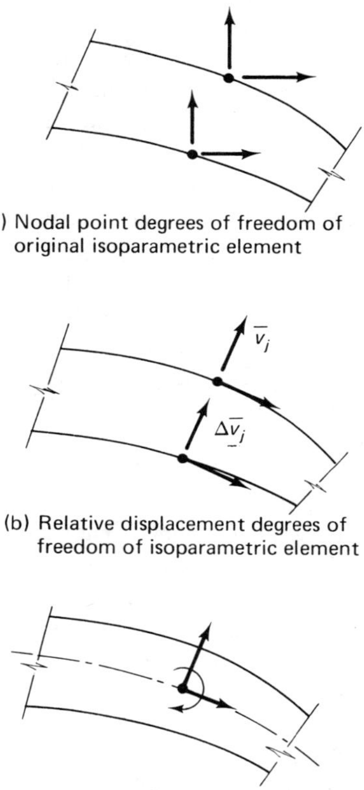

(a) Nodal point degrees of freedom of original isoparametric element

(b) Relative displacement degrees of freedom of isoparametric element

(c) Displacement degrees of freedom of superparametric element

Figure 4.13. Displacement degrees of freedom for thick to thin shell elements.

An important procedure to improve the behavior of isoparametric elements is the introduction of incompatible modes.[18] Consider the simplest case, the addition of incompatible modes to a four-node two-dimensional element. In this case the coordinate and displacement interpolations used are, respectively,

$$x = \sum_{i=1}^{4} h_i x_i; \qquad y = \sum_{i=1}^{4} h_i y_i \tag{4.42}$$

and

$$\left.\begin{aligned} u &= \sum_{i=1}^{4} h_i u_i + \alpha_1(1 - r^2) + \alpha_2(1 - s^2) \\ v &= \sum_{i=1}^{4} h_i v_i + \beta_1(1 - r^2) + \beta_2(1 - s^2) \end{aligned}\right\} \tag{4.43}$$

where the $h_1 \ldots, h_4$ are as defined in Fig. 4.4 or Example 4.2. The $\alpha_1, \alpha_2, \beta_1$, and β_2 are additional degrees of freedom, with a higher-order displacement interpolation than is present in the h_i. If the element considered is rectangular, the additional degrees of freedom allow the element to represent a constant bending moment, and the element will perform much better in many stress analyses.

Considering the actual evaluation of the element stiffness matrix, we note that using the interpolation in (4.43) a matrix of order 12 is obtained, and that the degrees of freedom $\alpha_1, \alpha_2, \beta_1$, and β_2 are not associated with an element nodal point. To reduce the matrix to order 8, the additional degrees of freedom are eliminated using static condensation (see Section 7.2.4), which is equivalent to minimizing the element total potential energy with respect to the variables $\alpha_1, \alpha_2, \beta_1$, and β_2 (see Section 5.3). We shall discuss in Section 4.7 that the same objective is pursued by using reduced order or selective numerical integration.

Since the additional degrees of freedom are not associated with nodal point degrees of freedom, in the analysis using the element, displacement incompatibilities arise between elements. Therefore, the element does not satisfy the requirements for monotonic convergence discussed in Section 3.2.5, and an important question is whether the element does converge in a non-monotonic manner. Referring to Section 5.5 for the test on general convergence of a finite element assemblage, the results are that the four-node incompatible element converges provided that the element is a rectangle (or parallelogram).

Although the use of the variable-number-nodes elements given in Figs. 4.4 and 4.5 is considered to be most effective, it should be noted that in three-dimensional analysis, the use of incompatible modes can in some cases decrease the analysis cost considerably. An element that has been used with much success is the eight-node brick element with incompatible modes analogous to those used in (4.43); i.e., the displacement interpolations employed are

$$\left.\begin{aligned}
u &= \sum_{i=1}^{8} h_i u_i + \alpha_1(1 - r^2) + \alpha_2(1 - s^2) + \alpha_3(1 - t^2) \\
v &= \sum_{i=1}^{8} h_i v_i + \beta_1(1 - r^2) + \beta_2(1 - s^2) + \beta_3(1 - t^2) \\
w &= \sum_{i=1}^{8} h_i w_i + \gamma_1(1 - r^2) + \gamma_2(1 - s^2) + \gamma_3(1 - t^2)
\end{aligned}\right\} \tag{4.44}$$

where the $\alpha_1, \alpha_2, \ldots, \gamma_3$ are displacement components that are not associated with nodal points and are eliminated using static condensation. In analogy to the two-dimensional element discussed above, the three-dimensional brick should be rectangular in all three planes. Observing this rule, the element has been employed successfully in a variety of three-dimensional and shell analyses. Finally, it should be noted that the concept of introducing incompatible modes to increase the effectiveness of an element is quite general, and can also be employed in the formulation of higher-order elements.

4.6 NUMERICAL INTEGRATION

An important aspect of isoparametric and related finite element analysis is the required numerical integration. The required matrix integrals in the finite element calculations have been written in Section 4.3 as

$$\int \mathbf{F}(r)\, dr, \qquad \int \mathbf{F}(r, s)\, dr\, ds, \qquad \int \mathbf{F}(r, s, t)\, dr\, ds\, dt \tag{4.45}$$

in the one-, two-, and three-dimensional cases, respectively. It was stated that these integrals are in practice evaluated numerically using

$$\left.\begin{aligned} \int \mathbf{F}(r)\, dr &= \sum_i \alpha_i \mathbf{F}(r_i) + \mathbf{R}_n \\ \int \mathbf{F}(r, s)\, dr\, ds &= \sum_{i,j} \alpha_{ij} \mathbf{F}(r_i, s_j) + \mathbf{R}_n \\ \int \mathbf{F}(r, s, t)\, dr\, ds\, dt &= \sum_{i,j,k} \alpha_{ijk} \mathbf{F}(r_i, s_j, t_k) + \mathbf{R}_n \end{aligned}\right\} \tag{4.46}$$

where the summations extend over all i, j, and k specified, the α_i, α_{ij}, and α_{ijk} are weighting factors, and $\mathbf{F}(r_i)$, $\mathbf{F}(r_i, s_j)$, and $\mathbf{F}(r_i, s_j, t_k)$ are the matrices $\mathbf{F}(r)$, $\mathbf{F}(r, s)$, and $\mathbf{F}(r, s, t)$ evaluated at the points specified in the arguments. The matrices $\mathbf{R}_n$ are error matrices, which, in practice, are usually not evaluated. Therefore, we use

$$\left.\begin{aligned} \int \mathbf{F}(r)\, dr &= \sum_i \alpha_i \mathbf{F}(r_i) \\ \int \mathbf{F}(r, s)\, dr\, ds &= \sum_{i,j} \alpha_{ij} \mathbf{F}(r_i, s_j) \\ \int \mathbf{F}(r, s, t)\, dr\, ds\, dt &= \sum_{i,j,k} \alpha_{ijk} \mathbf{F}(r_i, s_j, t_k) \end{aligned}\right\} \tag{4.47}$$

The purpose in this section and Section 4.7 is to present the theory and practical implications of numerical integrations.[22-31] An important point is the integration accuracy that is needed, i.e., the number of integration points required in the element formation.

As presented above, in finite element analysis we integrate matrices, which means that each element of the matrix considered is integrated individually. Hence, for the derivation of the numerical integration formulas we can consider a typical element of a matrix which we denote as F.

Consider the one-dimensional case first, i.e., the integration of $\int_a^b F(r)\, dr$. In isoparametric element calculations we would have $a = -1$ and $b = +1$.

Numerical integration of $\int_a^b F(r)\,dr$ is essentially based on passing a polynomial $\psi(r)$ through given values of $F(r)$ and then using $\int_a^b \psi(r)\,dr$ as an approximation to $\int_a^b F(r)\,dr$. The number of evaluations of $F(r)$ and the positions of the sampling points in the interval from a to b determines how well $\psi(r)$ approximates $F(r)$ and hence the error of the numerical integration.

Assume that $F(r)$ has been evaluated at the $n + 1$ distinct points $r_0, r_1, \ldots, r_n$ to obtain $F_0, F_1, \ldots, F_n$, respectively, and that a polynomial $\psi(r)$ is to be passed through these data. Then there is a unique polynomial $\psi(r)$ given as

$$\psi(r) = a_0 + a_1 r + a_2 r^2 + \ldots + a_n r^n \tag{4.48}$$

Using the condition that $\psi(r) = F(r)$ at the $n + 1$ interpolating points, we have

$$\mathbf{F} = \mathbf{V}\mathbf{a} \tag{4.49}$$

where

$$\mathbf{F} = \begin{bmatrix} F_0 \\ F_1 \\ \cdot \\ \cdot \\ \cdot \\ F_n \end{bmatrix}; \qquad \mathbf{a} = \begin{bmatrix} a_0 \\ a_1 \\ \cdot \\ \cdot \\ \cdot \\ a_n \end{bmatrix} \tag{4.50}$$

and $\mathbf{V}$ is the *Vandermonde matrix*,

$$\mathbf{V} = \begin{bmatrix} 1 & r_0 & r_0^2 & \ldots & r_0^n \\ 1 & r_1 & r_1^2 & \ldots & r_1^n \\ \cdot & \cdot & \cdot & & \cdot \\ \cdot & \cdot & \cdot & & \cdot \\ \cdot & \cdot & \cdot & & \cdot \\ 1 & r_n & r_n^2 & \ldots & r_n^n \end{bmatrix} \tag{4.51}$$

Since $\det \mathbf{V} \neq 0$, provided that the r_i are distinct points, we have a unique solution for $\mathbf{a}$.

However, a more convenient way to obtain $\psi(r)$ is to use *Lagrangian interpolation*. First, we recall that the $(n + 1)$ functions $1, r, r^2, \ldots, r^n$ form an $(n + 1)$-dimensional vector space, say V_n, in which $\psi(r)$ is an element (see Section 2.2). Since the coordinates $a_0, a_1, a_2, \ldots, a_n$ of $\psi(r)$ are relatively difficult to evaluate, we seek to use a different basis for the space V_n, in which the coordinates of $\psi(r)$ are more easily evaluated. This basis is provided by the fundamental polynomials of Lagrangian interpolation, given as

$$l_j(r) = \frac{(r - r_0)(r - r_1)\ldots(r - r_{j-1})(r - r_{j+1})\ldots(r - r_n)}{(r_j - r_0)(r_j - r_1)\ldots(r_j - r_{j-1})(r_j - r_{j+1})\ldots(r_j - r_n)} \tag{4.52}$$

where

$$l_j(r_i) = \delta_{ij} \tag{4.53}$$

where δ_{ij} is the Kronecker delta; i.e., $\delta_{ij} = 1$ for $i = j$, and $\delta_{ij} = 0$ for $i \neq j$. Using the property in (4.53), the coordinates of the base vectors are simply the values of $F(r)$, and the polynomial $\psi(r)$ is

$$\psi(r) = F_0 l_0(r) + F_1 l_1(r) + \ldots + F_n l_n(r) \tag{4.54}$$

Example 4.9

Establish the interpolating polynomial $\psi(r)$ for the function $F(r) = 2^r - r$ when the data at the points $r = 0, 1$, and 3 are used. In this case $r_0 = 0, r_1 = 1, r_2 = 3$, and $F_0 = 1$, $F_1 = 1$, $F_2 = 5$.

In the first approach we use the relation in (4.49) to calculate the unknown coefficients a_0, a_1, and a_2 of the polynomial $\psi(r) = a_0 + a_1 r + a_2 r^2$. In this case we have

$$\begin{bmatrix} 1 & 0 & 0 \\ 1 & 1 & 1 \\ 1 & 3 & 9 \end{bmatrix} \begin{pmatrix} a_0 \\ a_1 \\ a_2 \end{pmatrix} = \begin{pmatrix} 1 \\ 1 \\ 5 \end{pmatrix}$$

The solution gives $a_0 = 1$, $a_1 = -\frac{2}{3}$, $a_2 = \frac{2}{3}$, and therefore $\psi(r) = 1 - \frac{2}{3}r + \frac{2}{3}r^2$.

If Lagrangian interpolation is employed, we use the relation in (4.54), which in this case gives

$$\psi(r) = (1)\frac{(r-1)(r-3)}{(-1)(-3)} + (1)\frac{(r)(r-3)}{(1)(-2)} + (5)\frac{(r)(r-1)}{(3)(2)}$$

or, as before,

$$\psi(r) = 1 - \tfrac{2}{3}r + \tfrac{2}{3}r^2$$

Having established an interpolating polynomial $\psi(r)$, we can now obtain an approximation to the integral $\int_a^b F(r)\,dr$. Basically, two classes of procedures can be considered. In the first and simpler method, it is assumed that the sampling points of F are spaced at equal distances, and we define

$$r_0 = a, \qquad r_n = b, \qquad h = \frac{b-a}{n} \tag{4.55}$$

Using Lagrangian interpolation to obtain $\psi(r)$ as an approximation to $F(r)$, we have

$$\int_a^b F(r)\,dr = \sum_{i=0}^{n} \left\{\int_a^b l_i(r)\,dr\right\} F_i + R_n \tag{4.56}$$

or, evaluated,

$$\int_a^b F(r)\,dr = (b-a)\sum_{i=0}^{n} C_i^n F_i + R_n \tag{4.57}$$

where R_n is the remainder and the C_i^n are the *Newton–Cotes constants* for numerical integration with n sampling points.

The Newton–Cotes constants and corresponding remainder terms have been published[23] and are summarized in Table 4.1 for $n = 1$ to 6. The

TABLE 4.1 NEWTON–COTES NUMBERS AND ERROR ESTIMATES

Number of Intervals n	C_0^n	C_1^n	C_2^n	C_3^n	C_4^n	C_5^n	C_6^n	Upper Bound on Error R_n as a Function of the Derivative of F
1	$\frac{1}{2}$	$\frac{1}{2}$						$10^{-1}(b-a)^3F^{\text{II}}(r)$
2	$\frac{1}{6}$	$\frac{4}{6}$	$\frac{1}{6}$					$10^{-3}(b-a)^5F^{\text{IV}}(r)$
3	$\frac{1}{8}$	$\frac{3}{8}$	$\frac{3}{8}$	$\frac{1}{8}$				$10^{-3}(b-a)^5F^{\text{IV}}(r)$
4	$\frac{7}{90}$	$\frac{32}{90}$	$\frac{12}{90}$	$\frac{32}{90}$	$\frac{7}{90}$			$10^{-6}(b-a)^7F^{\text{VI}}(r)$
5	$\frac{19}{288}$	$\frac{75}{288}$	$\frac{50}{288}$	$\frac{50}{288}$	$\frac{75}{288}$	$\frac{19}{288}$		$10^{-6}(b-a)^7F^{\text{VI}}(r)$
6	$\frac{41}{840}$	$\frac{216}{840}$	$\frac{27}{840}$	$\frac{272}{840}$	$\frac{27}{840}$	$\frac{216}{840}$	$\frac{41}{840}$	$10^{-9}(b-a)^9F^{\text{VIII}}(r)$

cases $n = 1$ and $n = 2$ are the well-known *trapezoidal rule* and *Simpson formula*. We note that the formulas for $n = 3$ and $n = 5$ have the same order of accuracy as the formulas for $n = 2$ and $n = 4$, respectively. For this reason, the even formulas with $n = 2$ and 4 are used in practice.

Example 4.10

Evaluate the Newton–Cotes constants when the interpolating polynomial is of order 2; i.e., $\psi(r)$ is a parabola.

In this case we have

$$\int_a^b F(r)\,dr \doteq$$

$$\int_a^b \left[F_0\frac{(r-r_1)(r-r_2)}{(r_0-r_1)(r_0-r_2)} + F_1\frac{(r-r_0)(r-r_2)}{(r_1-r_0)(r_1-r_2)} + F_2\frac{(r-r_0)(r-r_1)}{(r_2-r_0)(r_2-r_1)}\right] dr$$

Using that $r_0 = a$, $r_1 = a + h$, $r_2 = a + 2h$, where $h = (b - a)/2$, the evaluation of the integral gives

$$\int_a^b F(r)\,dr \doteq \frac{b-a}{6}(F_0 + 4F_1 + F_2)$$

Hence the Newton–Cotes constants are as given in Table 4.1 for the case $n = 2$.

Example 4.11

Use Simpson's rule to integrate $\int_0^3 (2^r - r)\,dr$.

In this case $n = 2$ and $h = \frac{3}{2}$. Therefore, $r_0 = 0$, $r_1 = \frac{3}{2}$, $r_2 = 3$, and $F_0 = 1$, $F_1 = 1.328427$, $F_2 = 5$, and we obtain

$$\int_0^3 (2^r - r)\,dr \doteq \frac{3}{6}[(1)(1) + (4)(1.328427) + (1)(5)]$$

or

$$\int_0^3 (2^r - r)\, dr \doteq 5.656854$$

The exact result is

$$\int_0^3 (2^r - r)\, dr = 5.598868$$

Hence the error is

$$R = 0.057986$$

However, using the upper bound value on the error we would have

$$R < \frac{(3-0)^5}{1000}(\ln 2)^4(2^r) = 0.448743$$

To obtain greater accuracy in the integration using the Newton–Cotes formulas we need to employ a smaller interval h; i.e., include more evaluations of the function to be integrated. Then we have the choice between two different strategies: we may use a higher-order Newton-Cotes formula, or, alternatively, employ the lower-order formula in a repeated manner, in which case the integration procedure is referred to as a *composite formula*. Consider the following example.

Example 4.12

Increase the accuracy of the integration in Example 4.11 by using half the interval spacing.

In this case we have $h = \frac{3}{4}$ and the required function values are $F_0 = 1$, $F_1 = 0.931792$, $F_2 = 1.328427$, $F_3 = 2.506828$, and $F_4 = 5$. The choice lies now between using the higher-order Newton–Cotes formula with $n = 4$ or applying the Simpson's rule twice, i.e., to the first two intervals and then to the second two intervals. Using the Newton–Cotes formula with $n = 4$, we obtain

$$\int_0^3 (2^r - r)\, dr \doteq \tfrac{3}{90}(7F_0 + 32F_1 + 12F_2 + 32F_3 + 7F_4)$$

Hence

$$\int_0^3 (2^r - r)\, dr \doteq 5.599232$$

On the other hand, using Simpson's rule twice, we consider

$$\int_0^3 (2^r - r)\, dr \doteq \int_0^{3/2} (2^r - r)\, dr + \int_{3/2}^3 (2^r - r)\, dr$$

The integration is performed using

$$\int_0^{3/2} (2^r - r)\, dr \doteq \frac{\frac{3}{2} - 0}{6}(F_0 + 4F_1 + F_2)$$

where F_0, F_1, and F_2 are the function values at $r = 0$, $r = \frac{3}{4}$, and $r = \frac{3}{2}$, respectively; i.e.,

$$F_0 = 1; \qquad F_1 = 0.931792; \qquad F_2 = 1.328427$$

Hence we use

$$\int_{0}^{3/2} (2^r - r)\, dr \doteq 1.513899 \qquad \text{(a)}$$

Next we need to evaluate

$$\int_{3/2}^{3} (2^r - r)\, dr \doteq \frac{3 - \frac{3}{2}}{6}(F_0 + 4F_1 + F_2)$$

where F_0, F_1, and F_2 are the function values at $r = \frac{3}{2}$, $r = \frac{9}{4}$, and $r = 3$, respectively,

$$F_0 = 1.328427; \qquad F_1 = 2.506828; \qquad F_2 = 5$$

Hence we have

$$\int_{3/2}^{3} (2^r - r)\, dr \doteq 4.088935 \qquad \text{(b)}$$

Adding the results of (a) and (b), we obtain

$$\int_{0}^{3} (2^r - r)\, dr \doteq 5.602834$$

The use of a composite formula has a number of advantages over the application of high-order Newton–Cotes formulas. A composite formula, such as the repetitive use of the Simpson rule, is easy to employ. Convergence is assured as the interval of sampling decreases, and, in practice, a sampling interval could be used that varies from one application of the basic formula to the next. This is particularly advantageous when there are discontinuities in the function to be integrated. For these reasons, in practice, composite formulas are commonly used.

Example 4.13

Use a composite formula that employs Simpson's rule to evaluate the integral $\int_{-1}^{+13} F(r)\, dr$ of the function $F(r)$ in Fig. 4.14.

This function is best integrated by considering three intervals of integration, as follows:

$$\int_{-1}^{13} F\, dr = \int_{-1}^{2} (r^3 + 3)\, dr + \int_{2}^{9} [10 + (r - 1)^{1/3}]\, dr + \int_{9}^{13} [\tfrac{1}{128}(13 - r)^5 + 4]\, dr$$

We evaluate each of the three integrals using Simpson's rule and have

$$\int_{-1}^{2} (r^3 + 3)\, dr \doteq \frac{2 - (-1)}{6}[(1)(2) + (4)(3.125) + (1)(11)]$$

or

$$\int_{-1}^{2} (r^3 + 3)\, dr \doteq 12.75$$

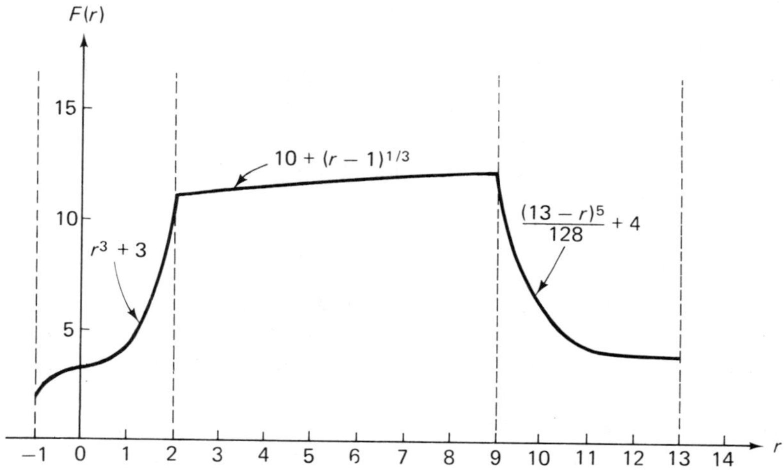

Figure 4.14. Function $F(r)$ integrated in Example 4.13.

$$\int_2^9 [10 + (r-1)^{1/3}]\,dr \doteq \frac{9-2}{6}[(1)(11) + (4)(11.650964) + (1)(12)]$$

or

$$\int_2^9 [10 + (r-1)^{1/3}]\,dr \doteq 81.204498$$

$$\int_9^{13} [\tfrac{1}{128}(13-r)^5 + 4]\,dr \doteq \frac{13-9}{6}[(1)(12) + (4)(4.25) + (1)(4)]$$

or

$$\int_9^{13} [\tfrac{1}{128}(13-r)^5 + 4]\,dr \doteq 22$$

Hence

$$\int_{-1}^{13} F\,dr \doteq 12.75 + 81.204498 + 22$$

or

$$\int_{-1}^{13} F\,dr \doteq 115.954498$$

The basic integration schemes that we considered so far used equally spaced sampling points, although the basic methods could be employed to construct procedures that allow varying the interval of sampling; i.e., the composite formulas have been introduced. The methods considered so far are effective when measurements of an unknown function to be integrated have been taken at equal intervals. However, in the integration of finite element matrices, a subroutine is called to evaluate the unknown function F at given points, and these points may be anywhere on the element. No additional difficulties arise if the sampling points are not equally spaced. Therefore, it seems natural to try to improve the accuracy that can be obtained for a given number of function evaluations by also optimizing the positions of the sampling points. A very important numerical integration procedure in which both the positions of the sampling points and the weights have been optimized is *Gauss quadrature*. The basic assumption in Gauss numerical integration is that

$$\int_a^b F(r)\,dr = \alpha_1 F(r_1) + \alpha_2 F(r_2) + \ldots + \alpha_n F(r_n) + R_n \tag{4.58}$$

where both the weights $\alpha_1, \ldots, \alpha_n$ and the sampling points $r_1, \ldots, r_n$ are variables. It should be recalled that in the derivation of the Newton–Cotes formulas, only the weights were unknown, and they have been determined from the integration of a polynomial $\psi(r)$, which passed through equally spaced sampling points of the function $F(r)$. We now calculate also the positions of the sampling points and, therefore, have $2n$ unknowns to determine a higher-order integration scheme.

In analogy to the derivation of the Newton–Cotes formulas, we use an interpolating polynomial $\psi(r)$ of the form given in (4.54):

$$\psi(r) = \sum_{j=1}^{n} F_j l_j(r) \tag{4.59}$$

where n samplings points are now considered, $r_1, \ldots, r_n$, which are still unknown. For the determination of the values $r_1, \ldots, r_n$, we define a function $P(r)$,

$$P(r) = (r - r_1)(r - r_2)\ldots(r - r_n) \tag{4.60}$$

which is a polynomial of order n. We note that $P(r) = 0$ at the sampling points $r_1, \ldots, r_n$. Therefore, we can write

$$F(r) = \psi(r) + P(r)(\beta_0 + \beta_1 r + \beta_2 r^2 + \ldots) \tag{4.61}$$

Integrating $F(r)$ we obtain

$$\int_a^b F(r)\,dr = \sum_{j=1}^{n} F_j\left[\int_a^b l_j(r)\,dr\right] + \sum_{j=0}^{\infty} \beta_j\left[\int_a^b r^j P(r)\,dr\right] \tag{4.62}$$

where it need be noted that in the first integral on the right of (4.62), functions of order $(n - 1)$ and lower are integrated, and in the second integral the functions that are integrated are of order n and higher. The unknown values $r_j, j = 1, 2, \ldots, n$, can now be determined using the conditions

$$\int_a^b P(r) r^k\,dr = 0 \qquad k = 0, 1, 2, \ldots, n - 1 \tag{4.63}$$

Then, since the polynomial $\psi(r)$ passes through n sampling points of $F(r)$ and $P(r)$ vanishes at these points, the conditions in (4.63) mean that the required integral $\int_a^b F(r)\,dr$ is approximated by integrating a polynomial of order $2n - 1$ instead of $F(r)$.

In summary, using the Newton–Cotes formulas we use $(n + 1)$ equally spaced sampling points and integrate exactly a polynomial of order at most n. On the other hand, in Gauss quadrature we require n unequally spaced sampling points and integrate exactly a polynomial of order at most $(2n - 1)$. Polynomials of order less than n and $(2n - 1)$, respectively, for the two cases would also be integrated exactly.

To determine the sampling points and the integration weights, we realize that they depend on the interval a to b. However, to make the calculations general, we can consider a natural interval from -1 to $+1$ and deduce the sampling points and weights for any interval. Namely, if r_i is a sampling point and α_i is the weight for the interval -1 to $+1$, the corresponding sampling point and weight in the integration from a to b are

$$\frac{a+b}{2} + \frac{b-a}{2} r_i \quad \text{and} \quad \frac{b-a}{2} \alpha_i$$

respectively.

For the above reasons, consider an interval from -1 to $+1$. The sampling points are determined from (4.63) with $a = -1$ and $b = +1$. To calculate the integration weights we substitute for $F(r)$ in (4.58) the interpolating polynomial $\psi(r)$ from (4.59) and perform the integration. It should be noted that because the sampling points have been determined, the polynomial $\psi(r)$ is known, and hence

$$\alpha_j = \int_{-1}^{+1} l_j(r)\, dr; \qquad j = 1, 2, \ldots, n \tag{4.64}$$

The sampling points and weights for the interval -1 to $+1$ have been published and are reproduced in Table 4.2 for values $n = 1$ to 6.[24]

The coefficients in Table 4.2 can be calculated directly using (4.63) and (4.64) (see Example 4.14). However, for larger n the solution becomes cumbersome and it is expedient to use Legendre polynomials to solve for the

TABLE 4.2 Sampling Points and Weights in Gauss–Legendre Numerical Integration

n	r_i	α_i
1	0. (15 zeros)	2. (15 zeros)
2	±0.57735 02691 89626	1.00000 00000 00000
3	±0.77459 66692 41483	0.55555 55555 55556
	0.00000 00000 00000	0.88888 88888 88889
4	±0.86113 63115 94053	0.34785 48451 37454
	±0.33998 10435 84856	0.65214 51548 62546
5	±0.90617 98459 38664	0.23692 68850 56189
	±0.53846 93101 05683	0.47862 86704 99366
	0.00000 00000 00000	0.56888 88888 88889
6	±0.93246 95142 03152	0.17132 44923 79170
	±0.66120 93864 66265	0.36076 15730 48139
	±0.23861 91860 83197	0.46791 39345 72691

coefficients, which are thus referred to as Gauss–Legendre coefficients.

Example 4.14

Derive the sampling points and weights for two-point Gauss quadrature.

In this case $P(r) = (r - r_1)(r - r_2)$ and (4.63) gives the two equations

$$\int_{-1}^{+1} (r - r_1)(r - r_2)\, dr = 0$$

$$\int_{-1}^{+1} (r - r_1)(r - r_2) r\,dr = 0$$

Solving, we obtain

$$r_1 r_2 = -\tfrac{1}{3}$$

and

$$r_1 + r_2 = 0$$

Hence

$$r_1 = -\frac{1}{\sqrt{3}}; \qquad r_2 = +\frac{1}{\sqrt{3}}$$

The corresponding weights are obtained using (4.64), which in this case gives

$$\alpha_1 = \int_{-1}^{+1} \frac{r - r_2}{r_1 - r_2}\, dr$$

$$\alpha_2 = \int_{-1}^{+1} \frac{r - r_1}{r_2 - r_1}\, dr$$

Since $r_2 = -r_1$, we obtain $\alpha_1 = \alpha_2 = 1.0$.

Example 4.15

Use two-point Gauss quadrature to evaluate the integral $\int_0^3 (2^r - r)\, dr$ considered in Examples 4.11 and 4.12.

Using two-point Gauss quadrature, we obtain, from (4.58),

$$\int_0^3 (2^r - r)\, dr \doteq \alpha_1 F(r_1) + \alpha_2 F(r_2) \tag{a}$$

where α_1, α_2 and r_1, r_2 are weights and sampling points, respectively. Since the interval is from 0 to 3, we need to determine the values α_1, α_2, r_1, and r_2 from the values given in Table 4.2,

$$\alpha_1 = \tfrac{3}{2}(1); \qquad \alpha_2 = \tfrac{3}{2}(1)$$

$$r_1 = \tfrac{3}{2}\left(1 - \frac{1}{\sqrt{3}}\right); \qquad r_2 = \tfrac{3}{2}\left(1 + \frac{1}{\sqrt{3}}\right)$$

where $1/\sqrt{3} = 0.5773502692$. Thus

$$F(r_1) = 0.91785978$$

$$F(r_2) = 2.78916389$$

and (a) gives

$$\int_0^3 (2^r - r)\, dr \doteq 5.56053551$$

The Gauss–Legendre integration procedure is commonly used in isoparametric finite element analysis. However, it should be noted that other integration schemes, in which both the weights and sampling positions were varied to obtain maximum accuracy, have been derived also.[21-31] In addition, for the specific objectives in finite element analysis, the basic Gauss quadrature has been applied in modified form as described in Section 4.7.

So far we considered the integration of a one-dimensional function $F(r)$. To evaluate a two- or three-dimensional integral using the methods presented above, the procedure followed is as in the analytical evaluation of multidimensional integrals; i.e., successively, the innermost integral is evaluated by keeping the variables corresponding to the other integrals constant. Therefore, we have for a two-dimensional integral

$$\int F(r, s)\, dr\, ds = \sum_i \alpha_i \int F(r_i, s)\, ds \tag{4.65}$$

or

$$\int F(r, s)\, dr\, ds = \sum_{i,j} \alpha_i \alpha_j F(r_i, s_j) \tag{4.66}$$

and corresponding to (4.47), $\alpha_{ij} = \alpha_i \alpha_j$, where α_i and α_j are the integration weights for one-dimensional integration. Similarly, for a three-dimensional integral,

$$\int F(r, s, t)\, dr\, ds\, dt = \sum_{i,j,k} \alpha_i \alpha_j \alpha_k F(r_i, s_j, t_k) \tag{4.67}$$

and $\alpha_{ijk} = \alpha_i \alpha_j \alpha_k$. We should note that it is not necessary in the numerical integration to use the same quadrature rule in the two or three dimensions; i.e., we may employ different numerical integration schemes in the r, s, and t directions, respectively.

Example 4.16

Evaluate the integral $\int_{-1}^{+1} \int_{-1}^{+1} r^2 s^2\, dr\, ds$ using (1) Simpson's rule in both r and s, (2) Gauss quadrature in both r and s, and (3) Gauss quadrature in r and Simpson's rule in s.

(1) Using Simpson's rule, we have

$$\int_{-1}^{+1} \int_{-1}^{+1} r^2 s^2\, dr\, ds = \int_{-1}^{+1} \tfrac{1}{3}[(1)(1) + (4)(0) + (1)(1)] s^2\, ds$$

$$= \int_{-1}^{+1} \tfrac{2}{3} s^2\, ds = \tfrac{1}{3}[(1)(\tfrac{2}{3}) + (4)(0) + (1)(\tfrac{2}{3})] = \tfrac{4}{9}$$

(2) Using two-point Gauss quadrature, we have

$$\int_{-1}^{+1} \int_{-1}^{+1} r^2 s^2\, dr\, ds = \int_{-1}^{+1} \left[(1)\left(\frac{1}{\sqrt{3}}\right)^2 + (1)\left(\frac{1}{\sqrt{3}}\right)^2\right] s^2\, ds$$

$$= \int_{-1}^{+1} \tfrac{2}{3} s^2\, ds = \tfrac{2}{3}\left[(1)\left(\frac{1}{\sqrt{3}}\right)^2 + (1)\left(\frac{1}{\sqrt{3}}\right)^2\right] = \tfrac{4}{9}$$

(3) Finally, using Gauss quadrature in r and Simpson's rule in s, we have

$$\int_{-1}^{+1}\int_{-1}^{+1}\left[(1)\left(\frac{1}{\sqrt{3}}\right)^2 + (1)\left(\frac{1}{\sqrt{3}}\right)^2\right]s^2\, ds$$

$$= \int_{-1}^{+1} \tfrac{2}{3}s^2\, ds = \tfrac{1}{3}[(1)(\tfrac{2}{3}) + (4)(0) + (1)(\tfrac{2}{3})] = \tfrac{4}{9}$$

We should note that the above numerical integrations are exact because both integration schemes, i.e., Simpson's rule and two-point Gauss quadrature, integrate a parabola exactly.

4.7 PRACTICAL CONSIDERATIONS IN ISOPARAMETRIC ELEMENT CALCULATIONS

In the previous section we described the derivation of important numerical integration methods and the application of the different procedures. We pointed out that using the Newton–Cote formulas, $(n + 1)$ function evaluations are required to integrate without error a polynomial of order n. On the other hand, if Gauss quadrature is used, a polynomial of order $(2n - 1)$ is integrated exactly with only n function evaluations. In finite element analysis a large number of function evaluations directly increases the cost of analysis, and the use of Gauss quadrature is usually most effective.

Having selected a numerical integration scheme, the order of numerical integration to be used in the evaluation of the various finite element integrals needs to be determined. The choice of the order of numerical integration is important in practice, because, firstly, the cost of analysis increases when a higher-order integration is employed, and, secondly, using a different integration order, the results can be affected by a very large amount. These considerations are particularly important in three-dimensional analysis.

The matrices to be evaluated by numerical integration are the stifiness matrix $\mathbf{K}$, the mass matrix $\mathbf{M}$, the body force vector $\mathbf{R}_B$, the initial stress vector $\mathbf{R}_I$, and the surface load vector $\mathbf{R}_S$. In general, the appropriate integration order depends on the matrix that is evaluated and the specific finite element considered. To demonstrate the important aspects, consider the Gauss numerical integration order required to evaluate the matrices of the variable-number-nodes elements in Figs. 4.3 to 4.5.

A first observation in the selection of the order of numerical integration is that using a high-enough order, all matrices are evaluated exactly. On the other hand, using too low an order of integration, the matrices may be evaluated very inaccurately and, in fact, the problem solution may not be possible. For example, considering an element stiffness matrix, if the order of numerical integration is too low, the matrix can have a larger number of zero eigenvalues than the number of physical rigid body modes. Hence, for a successful solution of the equilibrium equations of the element assemblage, it would be necessary that the deformation modes corresponding to all zero eigenvalues of the element be properly restrained in the assemblage of finite elements, because otherwise the structure stiffness matrix would be singular. A simple example would be the evaluation of the stiffness matrix of the one-dimensional element in Fig. 4.2. If one-point Gauss numerical integration is used, the row and column corresponding to the degree of freedom at the midnode of the element would be null vectors, which may result in a structure stiffness matrix that is singular. Therefore, the integration order need in general be higher than a certain limit.

The integration order required to evaluate a specific element matrix without error can be determined by studying the order of the function to be integrated. In the case of the stiffness matrix, we need to evaluate

$$\mathbf{K} = \int_V \mathbf{B}^T \mathbf{C} \mathbf{B} \det \mathbf{J} \, dV \tag{4.68}$$

where $\mathbf{C}$ is a constant material property matrix; $\mathbf{B}$ the strain-displacement matrix in the natural coordinate system r, s, t; $\det \mathbf{J}$ the determinant of the Jacobian transforming local (or global) to natural coordinates (see Section 4.3); and the integration is performed over the element volume in the natural coordinate system. The matrix function $\mathbf{F}$ to be integrated is, therefore,

$$\mathbf{F} = \mathbf{B}^T \mathbf{C} \mathbf{B} \det \mathbf{J} \tag{4.69}$$

The matrices $\mathbf{J}$ and $\mathbf{B}$ have been defined in Section 4.3.

A case for which the order of the variables in $\mathbf{F}$ can be evaluated with relative ease arises when the four-node two-dimensional element studied in Example 4.2 is used as a rectangular or parallelogram element. It is instructive to consider this case in detail because the procedure of evaluating the required integration order is displayed clearly.

Example 4.17

Evaluate the required Gauss numerical integration order for the calculation of the stiffness matrix of a four-node rectangular element.

The integration order to be used depends on the order of the variables r and s in $\mathbf{F}$ defined in (4.69). For a rectangular element with sides $2a$ and $2b$, we can write

$$x = ar$$
$$y = bs$$

and consequently the Jacobian matrix J is

$$\mathbf{J} = \begin{bmatrix} a & 0 \\ 0 & b \end{bmatrix}$$

Since the elements of $\mathbf{J}$ are constant, referring to the information given in Example 4.2, the elements of the strain-displacement matrix $\mathbf{B}$ are therefore functions of r or s only. But the determinant of $\mathbf{J}$ is also constant; hence

$$\mathbf{F} = f(r^2, rs, s^2)$$

where f denotes "function of."

Using two-point Gauss numerical integration in the r and s directions, all functions in r and s involving at most cubic terms are integrated without error; i.e., for integration order n, the order of r and s integrated exactly is $(2n - 1)$. Hence two-point Gauss integration is adequate.

In an analogous manner, the required integration order to evaluate exactly the stiffness matrices, mass matrices, and element load vectors of other elements can be assessed. In this context it should be noted that the Jacobian matrix is not constant for nonparallelogram element shapes, which increases the required integration order.

We discuss in Chapter 5 that provided that the convergence criteria of Section 3.2.5 are satisfied, the displacement formulation of finite element analysis yields a lower bound on the "exact" strain energy of the system considered; i.e., physically, a displacement formulation results in overestimating the system stiffness. Therefore, by not evaluating the element stiffness matrices exactly in the numerical integration, in fact, better results can be obtained provided that the error in the numerical integration compensates appropriately for the overestimation of structural stiffness due to the finite element discretization.[28-31] In other words, a *reduction of the numerical integration order* from the order that is required to evaluate the element stiffness matrix exactly should lead in many cases to improved results. This is indeed the case, and a great deal of research effort has been spent to evaluate the optimum integration order and scheme for isoparametric finite element analysis. In addition to merely using a reduced integration order, it may well be beneficial to integrate the different strain terms with different orders of integration. In certain analyses drastic improvements have been obtained; however, it is instructive to summarize the difficulties encountered in the search for the optimum integration scheme.

The first important point is that the bounding property of the displacement analysis is lost, and if the order of integration is too low, the solution obtained may be meaningless. In particular, as pointed out already, if the integration order for the evaluation of an element stiffness matrix is too low to include all displacement modes, the rank of the element matrix will be smaller than if evaluated exactly. This causes solution difficulties if the element is not provided with sufficient stiffness restraint in the assemblage of elements; i.e., the stiffness matrix of the total element assemblage could, in some rare cases, be singular and is more frequently ill-conditioned. If selective integration is employed, it is also necessary to consider that convergence is assured and that the stiffness matrix remains nondirectional.

It may be argued that the lowest order of numerical integration required for convergence is that order which would evaluate the volume of the element under consideration without error.[12] However, based on the above observations, this rule has to be used with great care, and instead the minimum order of integration given in Table 4.3 using Gauss quadrature is recommended. In Table 4.3 two-dimensional elements are considered; however, the information given can be employed in the calculation of one- and three-dimensional element matrices by deducing the appropriate integration orders for each direction or plane of the element under consideration.

TABLE 4.3 RECOMMENDED ORDER OF GAUSS NUMERICAL INTEGRATION FOR TWO-DIMENSIONAL ISOPARAMETRIC ELEMENTS

Element	Usually Best Integration Order	Maximum Integration Order
4-node rectangle 	2 by 2	2 by 2
4-node distorted 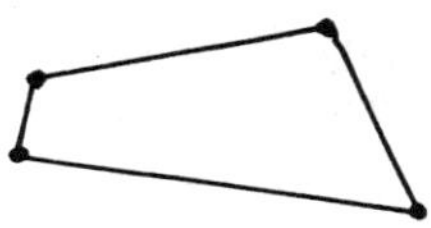	2 by 2	3 by 3
8-node rectangle 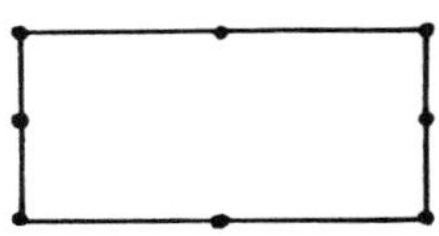	2 by 2	3 by 3
8-node distorted 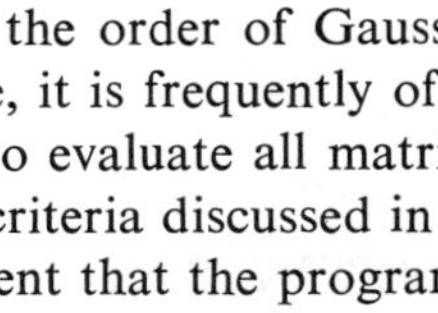	3 by 3	4 by 4

Although the order of Gauss quadrature given in Table 4.3 is usually most effective, it is frequently of value to analyze the finite element system exactly, i.e., to evaluate all matrices without error, because in this way the convergence criteria discussed in Chapter 5 can be satisfied. Therefore, it is most convenient that the program user can specify which integration order should be used.

In the above discussion, attention was focused on the evaluation of element stiffness matrices. Considering the element mass matrices and force vectors, it should be noted that a reduced integration order is usually justified, and again the integration orders given in Table 4.3 are recommended for use.

4.8 COMPUTER PROGRAM IMPLEMENTATION OF ISOPARAMETRIC FINITE ELEMENTS

In Section 4.3 we discussed the isoparametric finite element formulation and showed the specific expressions needed in the calculation of four-node plane stress (or plane strain) elements (see Example 4.2). An important advantage of isoparametric element calculations is the similarity between the calculations of different elements. For example, the calculation of three-dimensional elements is a relatively simple extension from the calculation of two-dimensional elements. Also, in one subroutine, elements with a variety of nodal point configurations can be calculated if an algorithm for selecting the appropriate interpolation functions is used (see Section 4.3).

The purpose of this section is to give an actual computer program for the calculation of the stiffness matrix of four-node isoparametric elements. In essence, SUBROUTINE QUADS given below is the computer program implementation of the procedures presented in Example 4.2. In addition to the plane stress and plane strain analysis capability, axisymmetric conditions can also be analyzed. It is believed that by showing the actual computer implementation of the element, the relative ease of implementing isoparametric elements is demonstrated best. The input and output variables and the flow of the program are described by means of comment cards.

```
      SUBROUTINE QUADS(NEL,ITYPE,NINT,THIC,YM,PR,XX,S,IOUT)
C
C . . . . . . . . . . . . . . . . . . . . . . . . . . . . . . . . . . .
C .                                                                   .
C .   P R O G R A M                                                   .
C .        TO CALCULATE ISOPARAMETRIC QUADRILATERAL ELEMENT STIFFNESS .
C .        MATRIX FOR AXISYMMETRIC, PLANE STRESS AND PLANE STRAIN     .
C .        CONDITIONS                                                 .
C .                                                                   .
C .   - - INPUT VARIABLES - -                                         .
C .        NEL        = NUMBER OF ELEMENT                             .
C .        ITYPE      = ELEMENT TYPE                                  .
C .                         EQ.0 = AXISYMMETRIC                       .
C .                         EQ.1 = PLANE STRAIN                       .
C .                         EQ.2 = PLANE STRESS                       .
C .        NINT       = GAUSS NUMERICAL INTEGRATION ORDER             .
C .        THIC       = THICKNESS OF ELEMENT                          .
C .        YM         = YOUNG*S MODULUS                               .
C .        PR         = POISSON*S RATIO                               .
C .        XX(2,4)    = ELEMENT NODE COORDINATES                      .
C .        S(8,8)     = STIFFNESS MATRIX ON SOLUTION EXIT             .
C .        IOUT       = OUTPUT PRINTING FILE                          .
C .                                                                   .
C .   - - OUTPUT - -                                                  .
C .        S(8,8)     = STIFFNESS MATRIX                              .
C .                                                                   .
C . . . . . . . . . . . . . . . . . . . . . . . . . . . . . . . . . . .
      IMPLICIT REAL*8 (A-H,O-Z)
      SQRT(X)=DSQRT(X)
      ABS(X)=DABS(X)
C . . . . . . . . . . . . . . . . . . . . . . . . . . . . . . . . . . .
C .        THIS PROGRAM IS USED IN SINGLE PRECISION ARITHMETIC ON CDC .
C .        EQUIPMENT AND DOUBLE PRECISION ARITHMETIC ON IBM OR UNIVAC .
C .        MACHINES.FOR SINGLE OR DOUBLE PRECISION ARITHMETIC ACTIVATE.
C .        DEACTIVATE OR ADJUST ABOVE AND DATA XG/WGT CARDS.           .
C . . . . . . . . . . . . . . . . . . . . . . . . . . . . . . . . . . .
C
      DIMENSION D(4,4),B(4,8),XX(2,4),S(8,8),XG(4,4),WGT(4,4),DB(4)
C
C     MATRIX XG STORES GAUSS - LEGENDRE SAMPLING POINTS
C
      DATA XG/   0.D0,     0.D0,     0.D0,     0.D0,    -.5773502691896D0,
     1 .5773502691896D0,    0.D0,     0.D0,    -.7745966692415D0,    0.D0,
     2 .7745966692415D0,    0.D0,    -.8611363115941D0,
```

```
     3 -.3399810435849D0,   .3399810435849D0,   .8611363115941D0 /
C
C     MATRIX WGT STORES GAUSS - LEGENDRE WEIGHTING FACTORS
C
      DATA WGT /  2.D0,    0.D0,    0.D0,    0.D0,    1.D0,    1.D0,
     1 0.D0,    0.D0,    .5555555555556D0,    .8888888888889D0,
     2 .5555555555556D0,    0.D0,    .3478548451375D0,    .6521451548625D0,
     3 .6521451548625D0,    .3478548451375D0 /
C
C     O B T A I N   S T R E S S - S T R A I N   L A W
C
      F=YM/(1. + PR)
      G=F*PR/(1. - 2.*PR)
      H=F + G
C
C     PLANE STRAIN ANALYSIS
C
      D(1,1)=H
      D(1,2)=G
      D(1,3)=0.
      D(2,1)=G
      D(2,2)=H
      D(2,3)=0.
      D(3,1)=0.
      D(3,2)=0.
      D(3,3)=F/2.
      IF (ITYPE.EQ.1) THIC=1.
      IF(ITYPE.EQ.1)GO TO 20
C
C     AXISYMMETRIC ANALYSIS
C
      D(1,4)=G
      D(2,4)=G
      D(3,4)=0.
      D(4,1)=G
      D(4,2)=G
      D(4,3)=0.
      D(4,4)=H
      IF(ITYPE.EQ.0)GO TO 20
C
C     FOR PLANE STRESS ANALYSIS CONDENSE STRESS-STRAIN MATRIX
C
      DO 10 I=1,3
      A=D(I,4)/D(4,4)
      DO 10 J=I,3
      D(I,J)=D(I,J) - D(4,J)*A
   10 D(J,I)=D(I,J)
C
C     C A L C U L A T E   E L E M E N T   S T I F F N E S S
C
   20 DO 30 I=1,8
      DO 30 J=1,8
   30 S(I,J)=0.
      IST=3
      IF (ITYPE.EQ.0) IST=4
      DO 80 LX=1,NINT
      RI=XG(LX,NINT)
      DO 80 LY=1,NINT
      SI=XG(LY,NINT)
C
C     EVALUATE DERIVATIVE OPERATOR B AND THE JACOBIAN DETERMINANT DET
C
      CALL STDM(XX,B,DET,RI,SI,XBAR,NEL,ITYPE,IOUT)
C
C     ADD CONTRIBUTION TO ELEMENT STIFFNESS
C
      IF (ITYPE.GT.0) XBAR=THIC
      WT=WGT(LX,NINT)*WGT(LY,NINT)*XBAR*DET
      DO 70 J=1,8
      DO 40 K=1,IST
      DB(K)=0.0
      DO 40 L=1,IST
   40 DB(K)=DB(K)+D(K,L)*B(L,J)
      DO 60 I=J,8
      STIFF=0.0
      DO 50 L=1,IST
   50 STIFF=STIFF+B(L,I)*DB(L)
   60 S(I,J)=S(I,J)+STIFF*WT
   70 CONTINUE
   80 CONTINUE
C
      DO 90 J=1,8
      DO 90 I=J,8
   90 S(J,I)=S(I,J)
C
      RETURN
C
      END
```

```
      SUBROUTINE STDM(XX,B,DET,R,S,XBAR,NEL,ITYPE,IOUT)
C
C . . . . . . . . . . . . . . . . . . . . . . . . . . . . . . . . . . .
C .                                                                   .
C .   P R O G R A M                                                   .
C .        TO EVALUATE THE STRAIN-DISPLACEMENT TRANSFORMATION MATRIX B.
C .        AT POINT (R,S) FOR A QUADRILATERAL ELEMENT                 .
C .                                                                   .
C . . . . . . . . . . . . . . . . . . . . . . . . . . . . . . . . . . .
      IMPLICIT REAL*8(A-H,O-Z)
      DIMENSION XX(2,4),B(4,8),H(4),P(2,4),XJ(2,2),XJI(2,2)
C
      RP = 1.0 + R
      SP = 1.0 + S
      RM = 1.0 - R
      SM = 1.0 - S
C
C     INTERPOLATION FUNCTIONS
C
      H(1) = 0.25* RP* SP
      H(2) = 0.25* RM* SP
      H(3) = 0.25* RM* SM
      H(4) = 0.25* RP* SM
C
C     NATURAL COORDINATE DERIVATIVES OF THE INTERPOLATION FUNCTIONS
C
C        1. WITH RESPECT TO R
C
      P(1,1) = 0.25* SP
      P(1,2) = - P(1,1)
      P(1,3) = - 0.25* SM
      P(1,4) = - P(1,3)
C
C        2. WITH RESPECT TO S
C
      P(2,1) = 0.25* RP
      P(2,2) = 0.25* RM
      P(2,3) = - P(2,2)
      P(2,4) = - P(2,1)
C
C     EVALUATE THE JACOBIAN MATRIX AT POINT (R,S)
C
   10 DO 30 I=1,2
      DO 30 J=1,2
      DUM = 0.0
      DO 20 K=1,4
   20 DUM=DUM+P(I,K)*XX(J,K)
   30 XJ(I,J)=DUM
C
C     COMPUTE THE DETERMINANT OF THE JACOBIAN MATRIX AT POINT (R,S)
C
      DET = XJ(1,1)* XJ(2,2) - XJ(2,1)* XJ(1,2)
      IF (DET.GT.0.00000001) GO TO 40
      WRITE (IOUT,2000) NEL
      STOP
C
C     COMPUTE INVERSE OF THE JACOBIAN MATRIX
C
   40 DUM=1./DET
      XJI(1,1) = XJ(2,2)* DUM
      XJI(1,2) =-XJ(1,2)* DUM
      XJI(2,1) =-XJ(2,1)* DUM
      XJI(2,2) = XJ(1,1)* DUM
C
C     EVALUATE GLOBAL DERIVATIVE OPERATOR B
C
      K2=0
      DO 60 K=1,4
      K2=K2 + 2
      B(1,K2-1) = 0.
      B(1,K2  ) = 0.
      B(2,K2-1) = 0.
      B(2,K2  ) = 0.
      DO 50 I=1,2
      B(1,K2-1) = B(1,K2-1) + XJI(1,I) * P(I,K)
   50 B(2,K2  ) = B(2,K2  ) + XJI(2,I) * P(I,K)
      B(3,K2  ) = B(1,K2-1)
   60 B(3,K2-1) = B(2,K2  )
C
C     IN CASE OF PLANE STRAIN OR PLANE STRESS ANALYSIS DO NOT INCLUDE
C     THE NORMAL STRAIN COMPONENT
C
      IF (ITYPE.GT.0) RETURN
C
C     COMPUTE THE RADIUS AT POINT (R,S)
C
      XBAR=0.0
```

```
      DO 70 K=1,4
   70 XBAR=XBAR + H(K)*XX(1,K)
C
C     EVALUATE THE HOOP STRAIN-DISPLACEMENT RELATION
C
      IF(XBAR.GT.0.00000001)GC TO 90
C
C     FOR THE CASE OF ZERO RACIUS EQUATE RADIAL TO HOOP STRAIN
C
      DC 80 K=1,8
   80 B(4,K)=B(1,K)
      RETURN
C
C     NCN-ZERO RADIUS
C
   90 DUM=1./XBAR
      K2=0
      DC 100 K=1,4
      K2=K2 + 2
      B(4,K2  ) = 0.
  100 B(4,K2-1)=H(K)*DUM
      RETURN
C
 2000 FORMAT (10HO*** ERROR,
     1       52H ZERO OR NEGATIVE JACOBIAN DETERMINANT FOR ELEMENT (,I4,
     2       1H)  )
C
      END
```

REFERENCES

1. I. C. Taig, "Structural Analysis by the Matrix Displacement Method," English Electric Aviation Report S017, 1961.
2. B. M. Irons, "Engineering Application of Numerical Integration in Stiffness Method," *A.I.A.A. Journal*, Vol. 4, 1966, pp. 2035–2037.
3. B. M. Irons, "Numerical Integration Applied to Finite Element Methods," Conference on the Use of Digital Computers in Structural Engineering, University of Newcastle, England, 1966.
4. B. M. Irons and O. C. Zienkiewicz, "The Isoparametric Finite Element System—a New Concept in Finite Element Analysis," Proceedings, Conference on Recent Advances in Stress Analysis, Royal Aeronautical Society, London, 1968.
5. J. G. Ergatoudis, "Iso-parametric Finite Elements in Two- and Three-Dimensional Analysis," Ph.D. dissertation, University of Wales, Swansea, 1968.
6. J. G. Ergatoudis, B. M. Irons, and O. C. Zienkiewicz, "Curved, Iso-parametric, Quadrilateral Elements for Finite Element Analysis," *International Journal of Solids and Structures*, Vol. 4, 1968, pp. 31–42.
7. J. G. Ergatoudis, B. M. Irons, and O. C. Zienkiewicz, "Three-Dimensional Analysis of Arch Dams and Their Foundations," Symposium on Arch Dams, Institute of Civil Engineering, London, Mar. 1968.
8. R. W. Clough, "Comparison of Three Dimensional Elements," Symposium on Applied Finite Element Methods in Civil Engineering, Vanderbilt University, Nashville, 1969, pp. 1–26.
9. O. C. Zienkiewicz, B. M. Irons, J. Ergatoudis, S. Ahmad, and F. C. Scott, "Isoparametric and Associated Element Families for Two and Three Dimen-

sional Analysis," *Finite Element Methods in Stress Analysis* (I. Holand, and K. Bell, eds.), Tapir, Trondheim, 1969, Chap. 13.

10. S. AHMAD, "Curved Finite Elements in the Analysis of Solid, Shell and Plate Structures," Ph.D. dissertation, University of Wales, Swansea, 1969.

11. S. AHMAD, B. M. IRONS, and O. C. ZIENKIEWICZ, "Analysis of Thick and Thin Shell Structures by Curved Elements," *International Journal for Numerical Methods in Engineering*, Vol. 2, 1970, pp. 419–451.

12. O. C. ZIENKIEWICZ, *The Finite Element Method in Engineering Science*, McGraw-Hill Book Company, Inc., New York, 1971.

13. O. C. ZIENKIEWICZ, D. R. J. OWEN, D. V. PHILLIPS, and G. C. NAYAK, "Finite Element Methods in the Analysis of Reactor Vessels," *Nuclear Engineering and Design*, Vol. 20, 1972, pp. 507–541.

14. E. L. WILSON, "Solid SAP—A Static Analysis Program for Three-Dimensional Solid Structures," Report UC SESM 71-19, Department of Civil Engineering, University of California, Berkeley, 1971.

15. K. J. BATHE, H. OZDEMIR, and E. L. WILSON, "Static and Dynamic Geometric and Material Nonlinear Analysis," Report UC SESM 74-4, Department of Civil Engineering, University of California, Berkeley, Feb. 1974.

16. K. J. BATHE and E. L. WILSON, "Thick Shell Structures," *Structural Mechanics Computer Programs*, (W. Pilkey, K. Saczalski, and H. Schaeffer, eds.), University Press of Virginia, Charlottesville, Va., 1974.

17. S. AHMAD, B. M. IRONS, and O. C. ZIENKIEWICZ, "Curved Thick Shell and Membrane Elements with Particular Reference to Axi-Symmetric Problems," Proceedings, 2nd Conference on Matrix Methods in Structural Mechanics, Wright-Patterson A.F.B., Ohio, 1968.

18. E. L. WILSON, R. L. TAYLOR, W. DOHERTY, and J. GHABOUSSI, "Incompatible Displacement Models," *Numerical and Computer Methods in Structural Mechanics* (S. J. Fenves, N. Perrone, J. Robinson, and W. C. Schnobrich, eds.), Academic Press, Inc., New York, 1973.

19. P. J. BERESFORD, "A Formulation for Non-Conforming Finite Elements," Graduate Student Report 558, SESM Division, Department of Civil Engineering, University of California, Berkeley, 1972.

20. C. K. CHOI and W. C. SCHNOBRICH, "Use of Non-Conforming Modes in Finite Element Analysis of Plates and Shells," Civil Engineering Studies, SRS No. 401, University of Illinois at Urbana–Champaign, 1973.

21. H. H. DOVEY, "Extension of Three-Dimensional Analysis to Shell Structures Using the Finite Element Idealization," Report UC SESM 74-2, Department of Civil Engineering, University of California, Berkeley, January 1974.

22. F. B. HILDEBRAND, *Introduction to Numerical Analysis*, McGraw-Hill Book Company, New York, 1956.

23. C. E. FRÖBERG, *Introduction to Numerical Analysis*, Addison-Wesley Publishing Company, Reading, Mass., 1969.

24. A. N. LOXAN, N. DAVIDS, and A. LEVENSON, "Table of the Zeros of the Legendre Polynomials of Order 1–16 and the Weight Coefficients for Gauss'

Mechanical Quadrature Formula," *Bulletin of the American Mathematical Society*, Vol. 48, 1942, pp. 739–743.

25. B. M. IRONS, "Quadrature Rules for Brick Based Finite Elements," *International Journal for Numerical Methods in Engineering*, Vol. 3, 1971.

26. C. A. FELIPPA, "Refined Finite Element Analysis of Linear and Non-linear Two-Dimensional Structures," Report UC SESM 66-22, Department of Civil Engineering, University of California, Berkeley, Oct. 1966.

27. A. H. STROUD and D. SECREST, "Gaussian Quadrature Formulas," Prentice-Hall, Inc., N.J., 1966.

28. W. P. DOHERTY, E. L. WILSON, and R. L. TAYLOR, "Stress Analysis of Axisymmetric Solids Utilizing Higher Order Quadrilateral Finite Elements," Report UC SESM 69-3, Structural Engineering Laboratory, University of California, Berkeley, 1969.

29. O. C. ZIENKIEWICZ, R. L. TAYLOR, and J. M. TOO, "Reduced Integration Techniques in General Analysis of Plates and Shells," *International Journal for Numerical Methods in Engineering*, Vol. 3, 1971, pp. 275–290.

30. S. F. PAWSEY, "The Analysis of Moderately Thick to Thin Shells by the Finite Element Method," Report UC SESM 70-12, Department of Civil Engineering, University of California, Berkeley, 1970.

31. S. F. PAWSEY and R. W. CLOUGH, "Improved Numerical Integration of Thick Shell Finite Elements," *International Journal for Numerical Methods in Engineering*, Vol. 3, 1971, pp. 575–586.

CHAPTER 5

VARIATIONAL FORMULATION OF THE FINITE ELEMENT METHOD

5.1 INTRODUCTION

The original development of the displacement-based finite element method was carried out using physical considerations and, as discussed, can be regarded as an extention of the matrix displacement analysis of truss and beam structures. However, the important difference between a general finite element analysis and the matrix analysis of a truss or beam structure is that, in general, only approximate results can be obtained. As pointed out in previous sections, truss and beam analyses may be understood to be specific finite element applications, in which "exact" solutions can be calculated. The general finite element analysis process, however, should be understood as a numerical procedure to obtain approximate solutions to problems in continuum mechanics. The approximation achieved was shown to depend on the characteristics and the number of elements that are used to idealize the continuum. The elements have to satisfy convergence requirements, but to obtain accurate results a large enough number of finite elements is also needed. The questions of convergence, accuracy, and improvement in element characteristics in a natural way initiated a theoretical investigation into the fundamentals of finite element analysis. A major advancement was achieved once it was recognized that the finite element analysis procedure is a specific form of Ritz analysis, which had been used many years in the solution of general variational problems. The importance of realizing that the finite element method is a form of Ritz analysis lies in that, automatically, all theorems regarding Ritz solutions are also applicable to finite element solutions. Furthermore, since Ritz analysis can be applied to many different variational problems beyond those encountered in continuum mechanics, the finite element method is also applicable to those problems.

In the following sections we present the variational formulation of continuum mechanics problems and the finite element solution as a Ritz analysis. To demonstrate the immediate applicability of finite element analysis to other variational problems, as an example, the solution of heat transfer problems is presented. Finally, a short survey of other than displacement-based finite elements is given, and the relation of the finite element method to finite difference techniques is described.

5.2 VARIATIONAL FORMULATION OF STRUCTURAL MECHANICS PROBLEMS

In the physical formulation of the finite element analysis procedure, we started, for static and dynamic analysis, from the virtual work principle as stated in (3.17). This equation was

$$\int_V \bar{\boldsymbol{\epsilon}}^T \boldsymbol{\tau}\, dV = \int_V \bar{\mathbf{U}}^T \mathbf{f}^B\, dV + \int_S \bar{\mathbf{U}}^{S^T} \mathbf{f}^S\, dS + \sum_i \bar{\mathbf{U}}^{i^T} \mathbf{F}^i \tag{5.1}$$

where $\boldsymbol{\tau}$ are the actual stresses in the continuous body; $\mathbf{f}^B$, $\mathbf{f}^S$, and $\mathbf{F}^i$ are the body, surface, and concentrated forces acting onto the body; $\bar{\mathbf{U}}$ is an arbitrary virtual displacement field, which satisfies the boundary conditions of the problem; and $\bar{\boldsymbol{\epsilon}}$ are the corresponding virtual strains. It should be realized that there are infinitely many virtual displacement fields for the continuum. If (5.1) is satisfied for all these possible virtual displacement configurations, it would follow that the stresses are those of the exact solution to the problem; in other words, in that case the stresses would be those that are obtained from the exact (analytical) solution to the problem governing differential equations.

In the displacement-based finite element analysis, a finite number of virtual displacement fields are selected, which are defined by the element displacements $\mathbf{U}^{(m)}$,

$$\mathbf{U}^{(m)}(x, y, z) = \mathbf{H}^{(m)}(x, y, z)\mathbf{U}, \qquad m = 1, 2, \ldots \text{ number of elements} \tag{5.2}$$

where the element displacement matrices $\mathbf{H}^{(m)}$ and the vector of nodal point displacements $\mathbf{U}$ have been defined in Chapters 3 and 4. The actual displacements of the body are also assumed to be of the form (5.2), and this enables the derivation of n algebraic equations for the n displacements at the nodal points of the finite element mesh. The displacement distribution, and hence the stress distribution thus solved for, is, in general, approximate since only a finite number of virtual displacement fields are being used and the actual displacement field is constrained to be of the form (5.2).

Consider now the equilibrium of a linear elastic continuum and assume that the virtual displacements are variations in the actual total displacements; then, correspondingly, the virtual strains are variations in the actual total strains. In this case we have

$$\boldsymbol{\tau} = \mathbf{C}\boldsymbol{\epsilon} \tag{5.3}$$

and

$$\bar{\mathbf{U}} = \delta\mathbf{U} \tag{5.4}$$

$$\bar{\boldsymbol{\epsilon}} = \delta\boldsymbol{\epsilon} \tag{5.5}$$

Substituting from (5.3) to (5.5) into (5.1), we have, for the equilibrium of the continuum,

$$\int_V \delta\boldsymbol{\epsilon}^T\mathbf{C}\boldsymbol{\epsilon}\,dV = \int_V \delta\mathbf{U}^T\mathbf{f}^B\,dV + \int_S \delta\mathbf{U}^{S^T}\mathbf{f}^S\,dS + \sum_i \delta\mathbf{U}^{i^T}\mathbf{F}^i \tag{5.6}$$

which, because $\mathbf{C}$ is symmetric, may also be written

$$\delta\left\{\frac{1}{2}\int_V \boldsymbol{\epsilon}^T\mathbf{C}\boldsymbol{\epsilon}\,dV\right\} = \delta\left\{\int_V \mathbf{U}^T\mathbf{f}^B\,dV + \int_S \mathbf{U}^{S^T}\mathbf{f}^S\,dS + \sum_i \mathbf{U}^{i^T}\mathbf{F}^i\right\} \tag{5.7}$$

However, the left-hand side of (5.7) is now recognized as the variation in the strain energy, U, of the elastic continuum, and the right-hand side is the variation in the potential energy of the applied loads, W. The principle of virtual work therefore led to the principle that the variation in the total potential energy, Π, is zero; i.e.,

$$\delta\Pi = 0 \tag{5.8}$$

where

$$\Pi = U - W \tag{5.9}$$

This means that the total potential energy of a system in equilibrium is stationary.[1-17] It can be proved that *the stationary point corresponds to a minimum in the total potential energy*. Furthermore, considering the expressions for the strain energy U and the potential energy of the loads W, we find that $W = 2U$; therefore, $\Pi = -U$ and *the point of equilibrium corresponds to the maximum in the strain energy of the system.*

Example 5.1

Consider a simple spring of stiffness k and applied load P to discuss the above results.

Let u be the displacement of the spring under the load P. We then have

$$U = \tfrac{1}{2}ku^2; \qquad W = Pu$$

and

$$\Pi = \tfrac{1}{2}ku^2 - Pu$$

Since u is the only variable, we have

$$\delta\Pi = (ku - P)\,\delta u$$

which gives the equilibrium equation

$$ku = P \tag{a}$$

Using (a) to evaluate W, we have $W = ku^2$, i.e., $W = 2U$, and $\Pi = -\frac{1}{2}ku^2$. We should note that the exact minimum of the total potential energy and maximum of the strain energy has been obtained, because k is the exact stiffness of the system and corresponds to the solution of the differential equation of equilibrium.

In the description above, the principle of the minimum of the total potential energy of the system was derived from the virtual work principle under the assumption that the continuum is elastic. However, the virtual work principle can be used in much more general cases.

In the finite element analysis of elastic systems, the discretizations of the expressions in (5.1) and (5.7) are equivalent. We should note, however, that to obtain a symmetric system of algebraic equations, either (5.7) must be used or in (5.1) it is necessary to use the same virtual as real displacement field. This was done in the derivation of the governing finite element equations (3.60) in Section 3.2.2.

The stationarity of the total potential energy expressed in (5.8) is the general variational formulation of elastic continuum mechanics problems. For specific problems an expression for the strain energy and the potential energy of the loads, concentrated or distributed, must be established. In mathematical physics the quantity Π is called the functional of the problem considered.

Example 5.2

Consider the cantilever beam in Fig. 5.1. Establish the total potential energy expression of the beam.

The strain energy of the beam is

$$\int_0^L \frac{EI}{2}[w''(x)]^2\,dx$$

and the potential energy of the load is $Rw(x = \frac{3}{4}L)$, where $w(x)$ is the deflection of the beam at a distance x from the left end. Therefore, we have in this case

$$\Pi = \frac{EI}{2}\int_0^L [w''(x)]^2\,dx - Rw(\tfrac{3}{4}L)$$

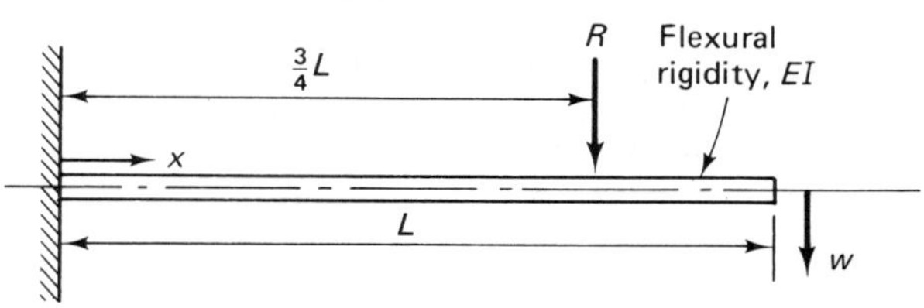

Figure 5.1. Cantilever beam with concentrated load.

In general, the potential of the loads is simply established as the integration of the loads multiplied by the appropriate final displacements. The strain energy depends on the specific strains developed in the system, and the strain vectors for different problems are given in Table 3.1.

5.3 RITZ SOLUTION

Assume that the objective is to perform an analysis of a specific problem. With the functional of the problem being known, the Ritz method is a very powerful tool to carry out the analysis.[6,7,18] In the conventional Ritz analysis the first step is to assume a solution for each unknown variable $u, v, \ldots,$ of the form

$$\left.\begin{aligned} u &= \sum_{i=1}^{q_u} \phi_i^u u_i \\ v &= \sum_{i=1}^{q_v} \phi_i^v v_i \\ &\ldots \end{aligned}\right\} \tag{5.10}$$

where the ϕ_i^u and ϕ_i^v are known functions and the $u_i, v_i, \ldots,$ are the Ritz coefficients (Ritz parameters) to be determined. Therefore, there are $q_u + q_v + \ldots$ unknown coefficients. The Ritz analysis is carried out by substituting the series expressions for the unknown variables into the functional and considering the variation with respect to the Ritz coefficients. This variation yields $q_u + q_v + \ldots$ equations for the unknown Ritz coefficients, the solution of which yields, with (5.10), the approximate solution to the problem.

Example 5.3

Perform a Ritz analysis of the cantilever beam in Fig. 5.1. The governing functional was given in Example 5.2.

For the Ritz solution we assume

$$w = x^2 w_1 + x^3 w_2$$

Therefore, in this case we have

$$\phi_1^w = x^2; \qquad \phi_2^w = x^3$$

Substituting for $w''(x)$ and $w(\frac{3}{4}L)$ into the functional, we obtain

$$\Pi = \frac{EI}{2}\int_0^L \left(4w_1^2 + 24xw_1w_2 + 36x^2w_2^2\right) dx - R[w_1(\tfrac{3}{4}L)^2 + w_2(\tfrac{3}{4}L)^3]$$

or

$$\Pi = \frac{EI}{2}(4w_1^2 L + 12L^2 w_1 w_2 + 12L^3 w_2^2) - R(\tfrac{9}{16}w_1 L^2 + \tfrac{27}{64}w_2 L^3)$$

Since

$$\delta\Pi = \sum_i \frac{\partial \Pi}{\partial w_i} \delta w_i$$

we need that

$$\frac{\partial \Pi}{\partial w_i} = 0; \qquad i = 1, 2$$

This condition gives the following equations for the unknown Ritz coefficients:

$$\frac{EI}{2}(8Lw_1 + 12L^2 w_2) = \tfrac{9}{16}RL^2$$

$$\frac{EI}{2}(12L^2 w_1 + 24L^3 w_2) = \tfrac{27}{64}RL^3$$

The solution is

$$w_1 = \frac{45}{128}\frac{RL}{EI}$$

$$w_2 = -\frac{9}{64}\frac{R}{EI}$$

and hence we obtain

$$w = \frac{45}{128}\frac{RL}{EI}x^2 - \frac{9}{64}\frac{R}{EI}x^3$$

As a comparison the exact result is

$$w = \frac{3}{8}\frac{RL}{EI}x^2 - \frac{1}{6}\frac{R}{EI}x^3 \qquad \text{for } x \leq \frac{3L}{4}$$

$$w = \frac{9}{32}\frac{RL^2}{EI}x - \frac{9}{128}\frac{RL^3}{EI} \qquad \text{for } x \geq \frac{3L}{4}$$

In the general case we have that Π is a function of all Ritz parameters used. Let the Ritz functions be denoted by $f_1, f_2, \ldots, f_n$ and the corresponding Ritz parameters be $a_1, a_2, \ldots, a_n$; then we have, *symbolically*, for the approximate solution,

$$\mathbf{u}^n = \sum_{i=1}^{n} f_i a_i \tag{5.11}$$

and the condition that the functional be a minimum is

$$\frac{\partial}{\partial a_i}\{\Pi(\mathbf{u}^n)\} = 0; \qquad i = 1, \ldots, n \tag{5.12}$$

It should be noted that the f_i include all previously used functions ϕ_i^u, ϕ_i^v, . . . , and the parameters a_i correspond to the values u_i, v_i, ... To evaluate the strain energy in the functional we need to calculate the appropriate strains. Let $\mathcal{L}$ be the operator to be applied to the Ritz displacements in order to obtain strains; then we have symbolically,

$$\boldsymbol{\epsilon} = \mathcal{L}\{\sum_i a_i f_i\} \tag{5.13}$$

and the stresses are

$$\boldsymbol{\tau} = \mathcal{S}\{\sum_i a_i f_i\} \tag{5.14}$$

where the operator $\mathcal{S}$ is obtained by incorporating into $\mathcal{L}$ the constitutive matrix. Thus the strain energy is

$$U = \frac{1}{2}\int \sum_i a_i \mathcal{L} f_i \sum_j a_j \mathcal{S} f_j \tag{5.15}$$

where the integral need be defined over the region of interest. Introducing

$$k_{ij} = \int (\mathcal{L} f_i)(\mathcal{S} f_j) \ ; \qquad \mathbf{a} = \begin{bmatrix} a_1 \\ a_2 \\ \cdot \\ \cdot \\ \cdot \\ a_n \end{bmatrix} \tag{5.16}$$

we have for the strain energy in matrix form,

$$U = \frac{1}{2}\mathbf{a}^T \mathbf{K}\, \mathbf{a} \tag{5.17}$$

where k_{ij} is element (i, j) of $\mathbf{K}$. The potential energy of the loads is given by

$$W = \int r \sum_i a_i f_i \tag{5.18}$$

where r includes body forces, surface forces, and applied concentrated forces, and the integral is defined correspondingly. Let

$$R_i = \int r f_i \ ; \qquad \mathbf{R} = \begin{bmatrix} R_1 \\ R_2 \\ \cdot \\ \cdot \\ \cdot \\ R_n \end{bmatrix} \tag{5.19}$$

Then we can write, in matrix form,

$$W = \mathbf{a}^T \mathbf{R} \tag{5.20}$$

Therefore, we have

$$\Pi = \frac{1}{2}\mathbf{a}^T \mathbf{K}\, \mathbf{a} - \mathbf{a}^T \mathbf{R} \tag{5.21}$$

The n equations for the unknown n Ritz parameters are now obtained using Π in (5.8), which yields

$$\mathbf{K a} = \mathbf{R} \tag{5.22}$$

The solution of (5.22) gives the Ritz coefficients, and using (5.11) we have an approximate solution to the problem considered. Assuming that the Ritz functions satisfy certain continuity conditions, which are discussed below, the total potential energy corresponding to the calculated Ritz coefficients is an upper bound to the exact total potential energy. Also, the strain energy is a lower bound on the exact strain energy. Namely, using (5.22) in (5.21) we obtain $\Pi = -\frac{1}{2}\mathbf{a}^T\mathbf{R}$ and $U = \frac{1}{2}\mathbf{a}^T\mathbf{R}$, and a decrease in Π corresponds to an increase in U. It follows that *in the Ritz analysis the displacements are underestimated and the stiffness of the system is overestimated* (see Section 3.2.5).

We can now identify the finite element analysis procedure as a form of Ritz analysis. The calculation of $\mathbf{K}$ and $\mathbf{R}$ in the conventional Ritz analysis described here is in principle identical to the calculation of a system finite element stiffness matrix and load vector [see the relations in (3.54) and (3.60) to (3.65)]. *In the finite element analysis the Ritz functions are contained in the* $\mathbf{H}^{(m)}$, $m = 1, 2, \ldots$ *(i.e., in the element displacement functions), and the Ritz parameters are the unknown nodal point displacements.*

Although the two analysis procedures are theoretically identical, in practice, the finite element method has most important advantages over a conventional Ritz analysis. One disadvantage in the conventional Ritz analysis is that the functions f_i are defined over the whole region considered. For example, in the analysis of the cantilever in Example 5.3, the Ritz functions spanned from $x = 0$ to $x = L$. Therefore, in the conventional Ritz analysis, the matrix $\mathbf{K}$ is a full matrix. As will be seen in Section 7.2, the numerical operations required for the solution of (5.22) are thus considerable if many functions are used.

A particular difficulty in a conventional Ritz analysis is the selection of appropriate Ritz functions, since the solution is a linear combination of these functions. In order to solve accurately for large displacement or stress gradients, many functions may be needed. However, these functions also unnecessarily span over the regions in which the displacements and stresses vary rather slowly and where not many functions are required.

Another difficulty arises in the conventional Ritz analysis when the total region of interest is made up of subregions with different kinds of strain distributions. As an example, consider a plate that is supported by edge beams and columns. In such a case, the Ritz functions used for one region (e.g., the plate) are not appropriate for the other regions (i.e., the edge beams and columns), and special displacement continuity conditions or interforce relations must be introduced.

The few reasons given already show that the conventional Ritz analysis is, in general, not particularly computer-oriented, except in some cases for the development of special-purpose programs. On the other hand, the finite element method has to a large extent removed the practical difficulties while retaining the advantageous properties of the conventional Ritz method. With regard to the difficulties mentioned above, the selection of Ritz functions is handled by using an adequate element library in the computer program. The use of relatively many functions in regions of high stress and displacement gradients is possible simply by using many elements. The combination of domains with different kinds of strain distributions is possible by using different kinds of elements to idealize the domains. The practical implementation of these possibilities is discussed in Chapter 6.

The main point in recognizing that the finite element analysis is a form of Ritz analysis lies in that the convergence properties and numerical advantages associated with the Ritz method are also applicable to the finite element method. An important numerical property is that the discretization matrix $\mathbf{K}$ in (5.22) is positive definite. In Sections 2.6 and 2.7 we defined a matrix $\mathbf{A}$ as positive definite provided that $\mathbf{u}^T\mathbf{A}\mathbf{u} > 0$ for any vector $\mathbf{u}$. In (5.17) we consider the strain energy of the system, $U = \frac{1}{2}\mathbf{a}^T\mathbf{K}\mathbf{a}$, but since U is positive for any vector $\mathbf{a}$ (assuming as before that the structure is properly supported), we have that $\mathbf{K}$ is a positive definite matrix.

If the structure analyzed is not supported, there will be a number of linearly independent vectors $\mathbf{a}_1, \mathbf{a}_2, \ldots, \mathbf{a}_q$, for which the expression $\mathbf{a}_i^T\mathbf{K}\mathbf{a}_i$ is equal to zero; i.e., zero strain energy is stored in the system when $\mathbf{a}_i$ is the displacement vector. Each vector $\mathbf{a}_i$ is said to represent a *rigid body mode* of the system; hence q is the number of rigid body modes present. Since in this case $\mathbf{a}^T\mathbf{K}\mathbf{a} \geq 0$, $\mathbf{K}$ is said to be positive semidefinite.

A most important aspect is the study of the convergence properties of the finite element method when the procedure is looked upon as a Ritz analysis. To this end we first need to define "convergence"; namely, we may consider convergence of displacements or stresses, pointwise or in a norm. The following definition is mathematically relatively easy to work with. Let $\mathbf{u}$ be the exact solution and $\mathbf{u}^n$ be the solution obtained using n Ritz functions as in (5.11); then the solution converges, provided that

$$\int_V \mathcal{L}(\mathbf{u} - \mathbf{u}^n)\mathcal{S}(\mathbf{u} - \mathbf{u}^n)\, dV \longrightarrow 0 \qquad \text{as} \quad n \longrightarrow \infty \tag{5.23}$$

We say that we obtain "convergence in energy," since the energy norm of the residual displacements $\mathbf{u} - \mathbf{u}^n$ converges to zero.

In the finite element analysis we have functions that extend piecewise continuously over the domain of the finite elements. Let h be the largest dimension of an element; then the statements $n \longrightarrow \infty$ and $h \longrightarrow 0$ are equivalent. Let $\mathbf{u}^h$ denote the finite element solution obtained when h is the largest dimension of any element. Then the definition of convergence in (5.23) can also be written as

$$\int_V \mathcal{L}(\mathbf{u} - \mathbf{u}^h)\mathcal{S}(\mathbf{u} - \mathbf{u}^h)\, dV \longrightarrow 0 \qquad \text{as} \quad h \longrightarrow 0 \tag{5.24}$$

We should note that in the limit, convergence in energy is equivalent to convergence in displacements and stresses. However, for any given finite element mesh, the accuracy of the solution will be different when measured by the energy norm, or by the accuracy in the displacements or the stresses obtained.

Let us briefly summarize the convergence requirements of the conventional Ritz method, from which we can then conclude the already earlier stated convergence requirements of the finite element method (see Section 3.2.5).

Considering the relation in (5.13), we assume that the highest derivative of a Ritz function f_i is of order m; i.e., the operator introduces at most mth-order derivatives. Such a problem we call a C^m variational problem. The question is then, which functions may be used in the Ritz assumption in (5.11) to admit solution convergence. The first criterion is that no singularities must occur in the integration evaluating k_{ij} in (5.16). This means that at least the $(m - 1)$st derivatives of the Ritz trial functions must be continuous; otherwise, the mth derivative would be singular at the discontinuities.

With regard to a finite element analysis, we therefore need to use, in a C^m problem, functions whose $(m - 1)$st derivatives are continuous over the element boundaries. Referring to Table 3.1, we conclude that in the analysis of fully two- and three-dimensional problems, only the displacements between elements must be continuous; and in beam and plate bending problems, we need continuity in the first derivatives of the displacement functions. This is essentially the compatibility condition established in Section 3.2.5.

The second criterion that determines which functions may be used in a Ritz analysis is arrived at by considering the boundary conditions of the problem. There are two classes of boundary conditions, which are referred to as *geometric* and *force boundary conditions*. The geometric boundary conditions are of the translational or rotational displacement type. The order of displacement derivatives in these boundary conditions is, in a C^m problem, at most $m - 1$. Therefore, in fully two- and three-dimensional problems, geometric boundary conditions are boundary deflections, whereas in a bending problem, deflections or rotations can be prescribed. The second class of boundary conditions, i.e., the force conditions also referred to as *natural boundary conditions*, correspond to prescribed boundary forces. The displacement derivatives in these boundary conditions are of order m to $2m - 1$.

It can be proved that for convergence the trial functions in a Ritz analysis need only satisfy the geometric boundary conditions (see Example 5.4), which is a fact that may not be expected because we know that the displacement functions corresponding to the exact solution of the continuum mechanics problem do also satisfy the force boundary conditions. Actually, to reach a better solution, functions satisfying the force boundary conditions should be used. However, to obtain approximate solutions, we may use a larger class of functions—namely all those satisfying the geometric boundary conditions only.

We may note that while in the Ritz analysis of a continuum, the geometric boundary conditions are specified at the boundaries of the continuum, in the finite element analysis they are also the interelement continuity conditions to be satisfied. On the other hand, the natural or force boundary conditions, which are of order m to $2m - 1$, correspond to actual boundary and surface forces. They are taken account of by including their effect into the load vector, as shown in (3.63) and (3.64).

It is an important practical point that the finite element functions need only satisfy geometric boundary conditions, since the simpler the finite element functions that we can use in the analysis, the easier is the implementation. For example, the analysis of fully two- and three-dimensional problems (i.e., $m = 1$) has been found to be much simpler than the analysis of bending problems (i.e., $m = 2$) mainly because in plate and shell bending analysis, the interelement continuity conditions are of higher order. Indeed, because of the difficulties arising in the finite element discretization of the functionals corresponding to plate bending and shell behavior, research has been concentrated on development of techniques to analyze shell structures using three-dimensional elements (see Section 4.5).

Example 5.4

Consider the total potential energy of the cantilever beam of Example 5.2 and examine the requirements that the Ritz functions need only satisfy the geometric boundary conditions.

The total potential energy of the cantilever beam was

$$\Pi = \frac{EI}{2} \int_0^L (w'')^2 \, dx - Rw(\tfrac{3}{4}L)$$

The minimum condition $\delta\Pi = 0$ yields

$$EI \int_0^L w'' \, \delta w'' \, dx - R \, \delta w(\tfrac{3}{4}L) = 0 \tag{a}$$

But $\delta w'' = d(\delta w')/dx$; hence

$$\int_0^L w'' \, \delta w'' \, dx = w'' \, \delta w' \Big|_0^L - \int_0^L w''' \, \delta w' \, dx$$

Similarly, we can substitute for $\int_0^L w''' \, \delta w' \, dx$; hence (a) becomes

$$EI \left\{ w'' \, \delta w' \Big|_0^L - w''' \, \delta w \Big|_0^L + \int_0^L w^{\text{iv}} \, \delta w \, dx \right\} - R \, \delta w(\tfrac{3}{4}L) = 0 \tag{b}$$

The minimum condition of the total potential energy in (a) has therefore been recast to the condition in (b), where the first two terms represent boundary conditions. We should note that the highest order of differentiation of w has increased to $2m$, $m = 2$, which makes (b) more difficult to use than (a), because for the integration to be possible, w''' would now have to be continuous.

In actual analysis we work with (a), where we may recall that $\delta w''(x)$ and $\delta w(\frac{3}{4}L)$ can be understood to represent virtual quantities. We also assumed that the actual displacements and virtual displacements are obtained using the same Ritz displacement assumption,

$$w = \sum_i f_i a_i \qquad \text{and} \qquad \delta w = \sum_i f_i \, \delta a_i$$

where the δa_i are variations in the Ritz coefficients. Therefore, the boundary conditions satisfied by the virtual displacements are also satisfied by the actual real displacements.

Consider the four boundary conditions

$$w''\,\delta w'\Big|_0^L \qquad \text{and} \qquad w'''\,\delta w\Big|_0^L \tag{c}$$

The conditions on w'' and w''' are the natural or force boundary conditions. They correspond here to moment and shear force conditions, respectively. The requirements on w and w' are the geometric boundary conditions. It follows from (c) that if the geometric boundary conditions are zero, we do not need to bother about the force boundary conditions. This is the case at the fixed end of the cantilever (i.e., at $x = 0$). At the free end of the cantilever (i.e., at $x = L$), the deflection and rotation are undetermined and the moment and shear force conditions are prescribed, namely, $w'' = w''' = 0$. In the Ritz or finite element analysis, these boundary conditions are appropriately taken account of by the body and surface load vectors. Therefore, only the geometric boundary conditions at $x = 0$ need be satisfied by the Ritz trial functions.

The basis of the Ritz analysis is the assumption that an approximate solution is obtained using

$$\mathbf{u}^n = \sum_{i=1}^{n} f_i a_i \tag{5.11}$$

This expression for the displacements is substituted into the total potential energy expression, and the minimizing condition is used to calculate the undetermined Ritz parameters a_i, $i = 1, \ldots, n$. Referring to Section 2.5, we may identify the Ritz functions as basis vectors and realize that in the Ritz analysis, we actually find the minimum of the total potential energy within the subspace spanned by the Ritz trial functions. As discussed above, provided that the trial functions satisfy certain continuity conditions, convergence is achieved as $n \longrightarrow \infty$, i.e., as the subspace is enlarged to, finally, an infinite-dimensional space. This appears physically reasonable, since the exact solution should be reached as the space considered increases in dimension. It is also important to note that as the dimension of the subspace considered increases and provided the currently considered subspace contains the previous one, the minimum of the total potential energy reached will monotonically decrease. In practice, this means that starting with a set of trial functions and always adding new functions to those previously used, the solution will continuously be improved.

We should point out once more that so far we have only been concerned with the displacement-based finite element method. All convergence considerations discussed apply to this discretization procedure, where we assume that the conditions of completeness and compatibility are satisfied. Only in that case are we assured that the total potential energy solved for lies above the exact total potential energy and that the strain energy obtained is a lower bound to the exact strain energy of the system. In practical finite element analysis, the condition of compatibility is very frequently not satisfied, and the strain energy integral in (5.15) is also not evaluated exactly, so that the bounding characteristics and the monotonic convergence property are not applicable (see Section 4.7). However, it is important to know about the aspects discussed above, in order to be familiar with the assumptions used in finite element analysis, and, in particular, to be able to apply the techniques to other than structural mechanics problems.

5.4 FORMULATION OF FIELD PROBLEMS— EXAMPLE: HEAT TRANSFER ANALYSIS

So far we have been concerned with stress analysis of structural continua. However, the Ritz analysis procedure, and hence the finite element method, is equally applicable to other types of problems. The solution procedure can always be readily applied when a functional to the physical problem under consideration is available. This is the case in the analysis of a variety of non-structural stituations; for example, in the analysis of heat conduction, seepage, and general flow problems.[19-22] Once the functional for the specific problem under consideration has been defined, the finite element solution can be performed in an analogous manner to the stress analysis described in preceding sections. To demonstrate the finite element solution of a typical field problem, we consider in the following the analysis of heat conduction. It is interesting to note that the functional that governs heat conduction is also applicable (possibly with some small modifications) to other field problems. Therefore, the solution procedure described below can be followed in a variety of field problems.

The functional governing heat conduction in three dimensions is

$$\Pi = \int_V \frac{1}{2}\left\{k_x\left(\frac{\partial\theta}{\partial x}\right)^2 + k_y\left(\frac{\partial\theta}{\partial y}\right)^2 + k_z\left(\frac{\partial\theta}{\partial z}\right)^2\right\}dV - \int_V \theta q^B \, dV - \int_S \theta^S q^S \, dS - \sum_i \theta^i Q^i \tag{5.25}$$

where θ is the temperature; k_x, k_y, and k_z heat conductivity coefficients; q^B the rate of heat generated per unit volume; q^S the rate of heat transfer per unit surface area of the body; and Q^i are concentrated heat flow inputs. Using the condition of stationarity of Π (i.e., $\delta\Pi = 0$), we obtain (because θ is the only variable)

$$\int_V \bar{\boldsymbol{\theta}}'^T \mathbf{k} \boldsymbol{\theta}' \, dV = \int_V \bar{\theta} q^B \, dV + \int_S \bar{\theta}^S q^S \, dS + \sum_i \bar{\theta}^i Q^i \tag{5.26}$$

where

$$\boldsymbol{\theta}'^T = \begin{bmatrix} \frac{\partial\theta}{\partial x} & \frac{\partial\theta}{\partial y} & \frac{\partial\theta}{\partial z} \end{bmatrix} \tag{5.27}$$

$$\mathbf{k} = \begin{bmatrix} k_x & 0 & 0 \\ 0 & k_y & 0 \\ 0 & 0 & k_z \end{bmatrix} \tag{5.28}$$

and a bar denotes "variation in." However, referring to the discussion in Section 5.2, we may consider the variation in θ as a virtual quantity, from which it follows that the equation governing rate of heat transfer in (5.26) is analogous to the virtual work equation given in (5.1) for stress analysis. In terms of the equilibrium considerations used in stress analysis, the relation in (5.26) is a statement of equilibrium of heat flow; i.e., (5.26) states that the rate of heat transfer by conduction shall be equal to the rate of heat generation. It is this physical interpretation that adds insight to the problem, in particular, for comparison of the problems of stress and heat transfer analysis. Because of the complete analogy of the governing equations, all concepts discussed in the finite element solution of (5.1) are therefore directly applicable to the finite element solution of (5.26). In particular, the procedures used to construct the finite element governing equilibrium equations and the considerations of convergence and accuracy can by used directly. This means that in practical analysis, a computer program developed for stress analysis can be modified with relative ease to perform heat transfer analysis. A main difference is that only one unknown variable, the temperature θ, must be calculated.

Assume that the complete body under consideration has been idealized as an assemblage of finite elements; then, in analogy to stress analysis, we have, for element m,

$$\theta^{(m)}(x, y, z) = \mathbf{H}^{(m)}(x, y, z)\boldsymbol{\theta} \tag{5.29}$$

$$\boldsymbol{\theta}'^{(m)}(x, y, z) = \mathbf{B}^{(m)}(x, y, z)\boldsymbol{\theta} \tag{5.30}$$

where the superscript (m) denotes element m and $\boldsymbol{\theta}$ is a vector of the nodal point temperatures at all n nodal points,

$$\boldsymbol{\theta}^T = [\theta_1 \quad \theta_2 \ldots \theta_n] \tag{5.31}$$

The matrices $\mathbf{H}^{(m)}$ and $\mathbf{B}^{(m)}$ now define the temperatures and temperature gradients within element m as a function of the nodal point temperatures. The matrices are constructed using the procedure employed in stress analysis (see Example 5.5).

Using the relations in (5.29) and (5.30) and substituting into (5.26) in the same way as was done in Section 3.2.2, we obtain the equation

$$\mathbf{K}\boldsymbol{\theta} = \mathbf{Q} \tag{5.32}$$

where $\mathbf{K}$ is the conductivity matrix,

$$\mathbf{K} = \sum_m \int_{V^{(m)}} \mathbf{B}^{(m)T}\mathbf{k}^{(m)}\mathbf{B}^{(m)}\, dV^{(m)} \tag{5.33}$$

and $\mathbf{Q}$ is the total nodal point heat flow input,

$$\mathbf{Q} = \mathbf{Q}_B + \mathbf{Q}_S + \mathbf{Q}_C \tag{5.34}$$

where

$$\mathbf{Q}_B = \sum_m \int_{V^{(m)}} \mathbf{H}^{(m)T} q^{B^{(m)}} \, dV^{(m)} \tag{5.35}$$

$$\mathbf{Q}_S = \sum_m \int_{S^{(m)}} \mathbf{H}^{S^{(m)}T} q^{S^{(m)}} \, dS^{(m)} \tag{5.36}$$

and $\mathbf{Q}_C$ is a vector of concentrated nodal point heat flow input.

The heat conduction problem considered above is equivalent to the static stress analysis problem, because time effects on the temperature distribution have not been included; i.e., steady-state conditions are assumed. However, with significant heat flow input changes specified in (5.32), it is important to include a term that takes account of the rate at which heat is stored within the material. This rate of heat absorption is given by

$$q^c = c\dot{\theta} \tag{5.37}$$

where c is the material heat capacity and q^c can be understood to be part of the rate of heat generated, q^B, per unit volume (in the same way as the inertia forces were taken as part of the body forces in stress analysis).

Using for element m,

$$\dot{\theta}^{(m)}(x, y, z) = \mathbf{H}^{(m)}(x, y, z)\dot{\boldsymbol{\theta}}(t) \tag{5.38}$$

and substituting into (5.35), we obtain

$$\mathbf{Q}_B = \sum_m \int_{V^{(m)}} \mathbf{H}^{(m)T}\{q^{B^{(m)}} - c^{(m)}\mathbf{H}^{(m)}\dot{\boldsymbol{\theta}}\} \, dV^{(m)} \tag{5.39}$$

where $q^{B^{(m)}}$ no longer includes the rate at which heat is stored within the material. The heat flow equilibrium equations are now

$$\mathbf{C}\dot{\boldsymbol{\theta}} + \mathbf{K}\boldsymbol{\theta} = \mathbf{Q} \tag{5.40}$$

where $\mathbf{C}$ is the heat capacity matrix,

$$\mathbf{C} = \sum_m \int_{V^{(m)}} \mathbf{H}^{(m)T} c^{(m)} \mathbf{H}^{(m)} \, dV^{(m)} \tag{5.41}$$

and $\mathbf{Q}$ and $\boldsymbol{\theta}$ are time-dependent.

Using the above equations, the heat transfer analysis is carried out effectively with the variable-number-nodes isoparametric elements discussed in Section 4.3. The evaluation of the conductivity and heat capacity matrices is analogous to the calculation of the stiffness and mass matrices, respectively, in stress analysis, where we recall that an important part of the calculations is numerical integration.

Example 5.5

Outline the calculations of the conductivity matrix **K**, heat capacity matrix **C**, and of the nodal point heat flow input vector $\mathbf{Q}_B$ for the two-dimensional isoparametric element in Fig. 4.8.

Referring to the calculation of the element matrices used in stress analysis as given in Examples 4.2 to 4.4, we obtain for heat transfer analysis,

$$\mathbf{K} = \sum_{i,j} \alpha_{ij} t_{ij} \, \mathbf{B}_{ij}^T \mathbf{k}_{ij} \mathbf{B}_{ij} \det \mathbf{J}_{ij}$$

$$\mathbf{C} = \sum_{i,j} \alpha_{ij} t_{ij} \, c_{ij} \mathbf{H}_{ij}^T \mathbf{H}_{ij} \det \mathbf{J}_{ij}$$

$$\mathbf{Q}_B = \sum_{i,j} \alpha_{ij} t_{ij} \, \mathbf{H}_{ij}^T \, q_{ij}^B \det \mathbf{J}_{ij}$$

where α_{ij} is the Gauss weighting factor, t_{ij} the thickness of the element, $\det \mathbf{J}_{ij}$ the determinant of the Jacobian transforming from the global to the natural coordinate system, c_{ij} the heat capacity, and $\mathbf{k}_{ij}$ the heat conductivity matrix,

$$\mathbf{k}_{ij} = \begin{bmatrix} k_x & 0 \\ 0 & k_y \end{bmatrix}_{\text{at } \substack{r=r_i \\ s=s_j}}$$

In the above equations the subscript (i, j) indicates that the quantity is evaluated at the (i, j)th integration point. The evaluation of the Jacobian determinant, $\det \mathbf{J}_{ij}$, was discussed in detail in Example 4.2, and the factors α_{ij} have been given in Section 4.6. However, we still need to describe the temperature interpolation and temperature-gradient interpolation matrices $\mathbf{H}_{ij}$ and $\mathbf{B}_{ij}$, respectively. The matrix $\mathbf{H}_{ij}$ is obtained by assembling the interpolation functions given in Example 4.2:

$$\mathbf{H}_{ij} = \tfrac{1}{4}[(1 + r_i)(1 + s_j) \quad (1 - r_i)(1 + s_j) \quad (1 - r_i)(1 - s_j) \quad (1 + r_i)(1 - s_j)]$$

The temperature-gradient interpolation matrix $\mathbf{B}_{ij}$ is constructed in an analogous manner as the strain interpolation matrix in Example 4.2; i.e., we have

$$\mathbf{B}_{ij} = \tfrac{1}{4}\mathbf{J}_{ij}^{-1} \begin{bmatrix} 1 + s_j & -(1 + s_j) & -(1 - s_j) & 1 - s_j \\ 1 + r_i & 1 - r_i & -(1 - r_i) & -(1 + r_i) \end{bmatrix}$$

The heat conduction analysis using the finite element analysis described above is a Ritz analysis, with monotonic convergence properties, provided that temperature compatibility between element edges is preserved. This follows because the highest temperature derivative in (5.26) is of order 1. (The heat conduction analysis is therefore comparable to fully two- or three-dimensional stress analysis.) However, in practice, advantage may be taken as in stress analysis of reduced or selective integration and of incompatible modes in which case the monotonic convergence properties are lost.

In the above discussion we did not mention how various boundary conditions are imposed. Nodal point temperatures can be specified in the same way as are displacements in stress analysis. A common procedure is to use matrix partitioning (see Section 3.2.2). On the other hand, to specify θ_i

at node i we may modify the corresponding heat flow equilibrium equation by adding a large conductivity value k_c to the ith diagonal element k_{ii} of $\mathbf{K}$ and specifying a heat flow input of $Q_i = k_c\theta_i$. If $k_c \gg k_{ii}$, the calculated temperature at node i is θ_i. This procedure is equivalent to using the boundary element in stress analysis.

Other important boundary conditions are convection and radiation, in which the rate of heat transfer q^S in (5.36) is dependent on the temperatures of the surface of the body and its environment. In a linear convection boundary condition we have

$$q^S = h(\theta_e - \theta) \tag{5.42}$$

where h is a convection constant and θ_e is the prescribed temperature of the environment. Using the surface interpolations to define θ and θ_e, and substituting into (5.36), we obtain

$$\begin{aligned}\mathbf{Q}_S = \sum_m \int_{S^{(m)}} h^{(m)}\mathbf{H}^{S(m)T}\mathbf{H}^{S(m)}\, dS^{(m)}\, \boldsymbol{\theta}_e \\ -\sum_m \int_{S^{(m)}} h^{(m)}\mathbf{H}^{S(m)T}\mathbf{H}^{S(m)}\, dS^{(m)}\, \boldsymbol{\theta}\end{aligned} \tag{5.43}$$

where the second term contributes now to the conductivity matrix of the system. The linear convection boundary condition can therefore be imposed with relative ease. However, in many heat transfer problems, the boundary conditions are nonlinear, and iterative techniques must be used.[22]

5.5 NONCONFORMING, MIXED, AND HYBRID FINITE ELEMENT MODELS; FINITE DIFFERENCE DIFFERENTIAL AND ENERGY METHODS

The conditions imposed so far on the assumed displacement (or field) functions were those of completeness and compatibility. If these conditions are satisfied, the calculated solution converges monotonically (i.e., one-sided) to the exact solution. The completeness condition can, in general, be satisfied with relative ease. The compatibility condition can also be satisfied without major difficulties in C^1 problems; for example, in plane stress and plane strain problems or the analysis of three-dimensional solids such as dams. However, in the analysis of bending problems, and, in particular, in shell analysis, continuity of first displacement derivatives along interelement boundaries is difficult to maintain. Furthermore, considering complex analyses in which completely different finite elements must be used to idealize different regions of the structure, compatibility may be almost impossible to maintain. However, although the compatibility requirements are violated, experience shows that good results can be obtained.

A practical observation to be made in the use of nonconforming (incompatible) displacement-based finite elements is that the continuity requirements on the displacement functions and their derivatives over the complete domain as required in a conventional Ritz analysis are not of the same concern in finite element analysis. The functions and derivatives must be, as discussed in Section 5.3, continuous within each finite element, which can be satisfied easily, but any singularities in the derivatives across element boundaries do not enter into the calculation of the element stiffness matrices. Therefore, although infinite strain energies would be contributed by the singularities to the total potential energy, in practice, these contributions are not calculated and thus are simply ignored.

Since in finite element analysis using nonconforming elements the Ritz analysis requirements are not satisfied, the calculated total potential energy is not necessarily an upper bound to the exact total potential energy of the system, and, consequently, monotonic convergence is not assured. However, having relaxed the objective of monotonic convergence in the analysis, we still need to establish conditions that will assure at least a nonmonotonic convergence.[12-17] Referring to Section 3.2.5, the element completeness condition must always be satisfied, where it may be noted that this condition is not affected by the size of the finite element. On the other hand, the compatibility condition can be relaxed somewhat at the expense of not obtaining a monotonically convergent solution. However, when relaxing the compatibility requirement, the important point is that the essential ingredients of the completeness condition not be lost. We recall that as the finite element mesh is refined (i.e., the size of the elements gets smaller), each element should approach a constant strain condition. Therefore, the second condition on convergence of an assemblage of incompatible finite elements, where the elements may again be of any size, is that the elements together can represent constant strain conditions. We should note that this is not a condition on a single individual element but on an assemblage of elements. That is, although an individual element may be able to represent all constant strain states (i.e., be complete), when the element is used in an assemblage, the incompatibilities between elements may prohibit constant strain states from being represented. We may call this condition the *completeness condition on an element assemblage*.

As a test to investigate whether an assemblage of nonconforming elements is complete, the *patch test* has been proposed.[12,23,24] In this test a patch of elements is subjected to the specific nodal point displacements that in an exact analysis correspond to constant strain conditions. If the element strains do actually represent the constant strain conditions, the patch test is passed; i.e., the completeness condition is satisfied by the element assemblage. We should note that the success or failure of the patch test may depend strongly on the geometry of the elements used, i.e., on the topology of the element layout. For example, an assemblage of the four-node two-dimensional elements with incompatible modes, discussed in Section 4.5, passes the patch test if all elements are rectangular (or parallelograms), but would fail the test otherwise (see Example 5.6).

Although the patch test requires the consideration of an element assemblage, it is possible to perform the test without actually using the computer. To this aim we need to subject an individual element of the element assemblage to the nodal point displacements, which by an exact analysis correspond to element constant strain conditions, and check whether constant strain conditions are indeed represented.

Example 5.6

Consider the four-node two-dimensional element with incompatible modes in Fig. 5.2, and check whether the patch test is passed. The element under consideration was discussed in Section 4.5.

As shown in Fig. 5.2, assume that the displacement field to which the element is subjected in the element assemblage is

$$u^c = ax + b$$

where u^c corresponds to constant strain conditions $\epsilon^c_{xx} = a$, $\epsilon^c_{yy} = 0$, and $\gamma^c_{xy} = 0$.

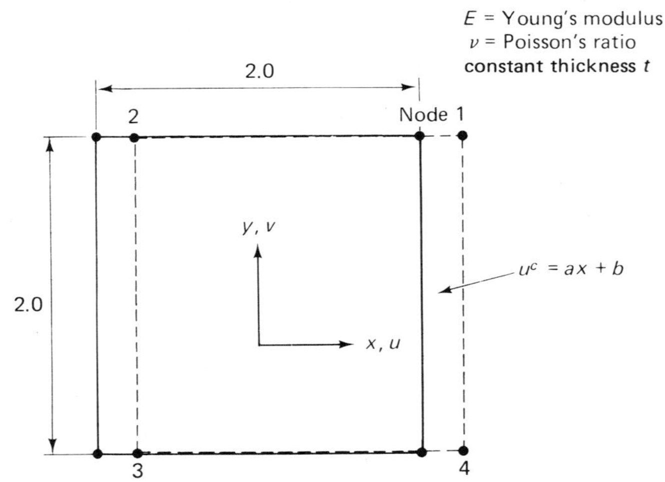

Displacement interpolation functions

$$u = \sum_{i=1}^{4} h_i u_i + \alpha_1 \phi_1 + \alpha_2 \phi_2$$

$$v = \sum_{i=1}^{4} h_i v_i + \alpha_3 \phi_3 + \alpha_4 \phi_4$$

$$\phi_1 = \phi_3 = (1 - x^2); \quad \phi_2 = \phi_4 = (1 - y^2)$$

Figure 5.2. Four-node rectangular element with incompatible modes (plane stress conditions).

In order to complete the patch test we would afterward also consider nonzero constant strain conditions for ϵ_{yy}^c and γ_{xy}^c. To investigate whether on imposing u^c onto the finite element, an incompatible mode ϕ_i is activated, we consider the integral $\int_V (\mathcal{L}\phi_i)(\mathcal{S}u^c)\, dV$. This is equivalent to evaluating whether constraining forces that correspond to the incompatible mode are required when u_c is imposed onto the element. If no constraining forces corresponding to any one of the possible incompatible modes are present, it follows that the incompatible modes have not been activated and the patch test is passed.

For plane stress conditions, we have the material law given in Table 3.2. The evaluation of the integral for the incompatible mode ϕ_1 in Fig. 5.2 is therefore

$$\frac{Et}{1-\nu^2}\int_A (a)(-2x)\, dx\, dy = 0$$

But this is the requirement of the patch test.

Considering ϕ_2, ϕ_3, and ϕ_4 and then displacement fields that correspond to constant strain conditions ϵ_{yy}^c and γ_{xy}^c in a similar manner would complete the patch test. It is found that since the incompatible modes are of the form $(1 - x^2)$ and $(1 - y^2)$, all integrals of the form $\int_V (\mathcal{L}\phi_i)(\mathcal{S}u^c)\, dV$ are zero, and thus the patch test is passed. However, it should be observed that the success of the patch test depends on the geometry of the element under consideration, and indeed, the test is not passed when the element in Fig. 5.2 is not rectangular (or a parallelogram).

We noted that in C^1 problems, in general, the compatibility conditions are relatively easy to satisfy. Yet incompatible elements such as the element investigated in Example 5.6 are used in practice. The reason is that a completely compatible displacement formulation is frequently much too stiff, and to obtain good results many elements or higher-order elements need be used. This may be considered to be expensive, and the use of nonconforming elements to relieve the constraints in the system appears attractive.

The search for improvements in element characteristics and the recognition that in the finite element method we basically aim to minimize a functional brought about the development of other finite element formulations, namely, *hybrid and mixed models*.[25-32] We shall not present the various formulations in detail, but do want to mention briefly some important points about these finite element models, because such discussion will contribute to the understanding and further indicate the generality of the finite element method.

The displacement-based finite element technique was shown to be an application of the theorem of minimum total potential energy. The essential steps in the solution may be recalled as follows:

1. Subdivide the complete domain (structure) into subdomains (finite elements).

2. Express the displacements within each subdomain in terms of displacement functions whose magnitudes are defined by discrete subdomain nodal point displacements.
3. Calculate the total potential energy of each subdomain and assemble the total potential energy Π of the complete domain. The unknowns are the nodal point displacements.
4. Use the minimum condition of the total potential energy of the system as a mechanism to calculate the nodal point displacements.

We note that in the functional of the total potential energy, the only variables are displacements. Once they have been calculated, other quantities of interest, such as strains and stresses, can be obtained as described in Chapters 3 and 4. The essential steps in the solution are that displacement functions are assumed in terms of a discrete set of unknown displacements, which are solved for using the condition that $\delta\Pi$ is equal to zero.

So far the reader may have obtained the impression that the use of the principle of minimum potential energy leads to the only variational formulation of problems in structural mechanics. This is by no means so. A number of different variational principles are in use, which lead to different finite element formulations. Presently, in nearly all practical analyses, the displacement-based finite element method is used; however, it should be noted that the numerical methods discussed in the next chapters can, to a large degree, also be used in the implementation of other finite element procedures.

In the minimum potential energy formulation, the only field variables are the element displacements. One source for additional variational formulations is that, in an analogous manner, element interior and boundary stresses, or combinations of displacements and stresses, could be assumed. The use of different field variables naturally results in different requirements on the continuity of the variables, and this presents another source for the derivation of additional variational formulations. Namely, if the field variables assumed in the formulation do not satisfy the continuity requirements, then additional variables, in the form of Lagrange multipliers, can be introduced to minimize the violations in the required continuities. Hence there are a large number of different variational formulations possible, which, analogous to the minimization of the total potential energy using displacement-based finite elements, yield other finite element discretizations.[2,3,5,25]

The three widely used variational principles are the total potential energy principle, the complementary energy principle, and the Reissner variational principle. The potential energy principle was used in the previous sections as the basis of the displacement-based compatible finite element models. The dual formulation to the potential energy principle is the complementary energy principle, in which the field variables are stresses instead of displacements. Use of the complementary energy principle results in stress-based *equilibrium finite element models.*

A most general variational principle is that of Reissner, in which the primary field variables are both displacements and stresses. The application of this principle results into finite element discretization with nodal displacement and stress variables, referred to as *mixed models.*

The three basic variational principles can be used to derive additional formulations with the aim of improving the accuracy and efficiency of the finite element discretization. As an example, the use of nonconforming finite elements may be understood as the application of a modified total potential energy principle, in which displacement continuity has been relaxed. Hybrid formulations are obtained if in addition to one field variable, which may be displacement or stress, other variables, additional stresses, or displacements are introduced and the parameters that correspond to the additional variables are eliminated at the element level prior to global assembly.

Since a large number of different finite elements could be used, a natural question is: Which models are the most efficient? This question is difficult to answer since many aspects, such as simplicity of formulation, accuracy, numerical stability, and convergence, need be considered. In particular, we should note that the efficiency of an element may to a large degree be problem-dependent. Furthermore, new ideas in the implementation of an element that was believed to be inefficient could, at any time, make the element effective. However, what may be stated is that currently the displacement-based isoparametric and related elements described in Chapter 4 are used a great deal in practical analysis, because of their efficient formulation, general applicability, numerical stability, and theoretical foundation.

A most important aspect is that of convergence. We have shown that using the displacement-based compatible finite element models, the finite element solution always underestimates the strain energy. On the other hand, using the finite element equilibrium model, the strain energy is always overestimated, and in both formulations the error in the energy norm approaches zero monotonically as the number of elements increases. Therefore, if both formulations are implemented, upper and lower bounds to the solution of a problem can be evaluated. It should be noted that for many hybrid and mixed finite element formulations, the questions of convergence have not been resolved completely.

In the convergence investigation of the displacement-based compatible finite elements, it was convenient to use the energy norm (see Section 5.3). However, in practice, we may be more interested in the convergence rates of displacements and stresses. These convergence rates can be calculated for conforming and nonconforming elements by developing the differential equations of equilibrium that correspond to the finite element equilibrium equations. These differential equations should be consistent with the exact differential equations of equilibrium corresponding to the variational problem considered.[33,34] Namely, the leading terms should be those of the exact differential equations of equilibrium, and the additional terms should be associated with powers of the size of the elements. It then follows that as the element size decreases, the differential equations of equilibrium corresponding to the finite element idealization approach the differential equations that we actually want to solve, and the rate of convergence in displacements (and stresses) is displayed by the lowest power of element size occurring with the additional terms. We demonstrate the procedure of convergence analysis by means of a simple example.

Example 5.7

Consider a uniform bar with load $p(x)$ per unit length (Fig. 5.3). Assume that the bar is idealized as an assemblage of bar elements of unequal lengths and that the distributed load application is idealized by the application of lumped loads at

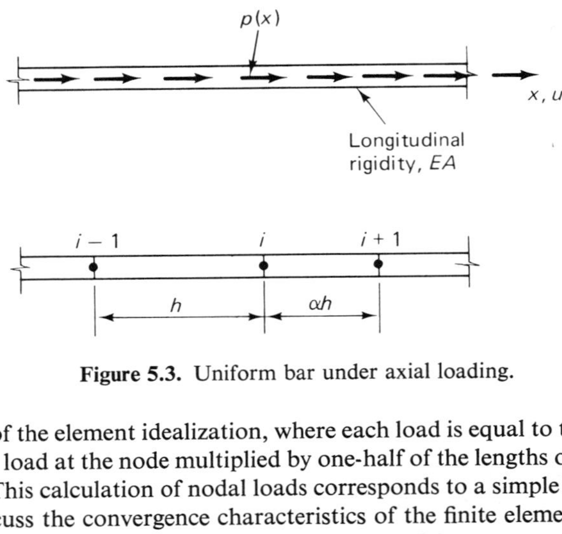

Figure 5.3. Uniform bar under axial loading.

the nodes of the element idealization, where each load is equal to the value of the distributed load at the node multiplied by one-half of the lengths of the adjoining elements. This calculation of nodal loads corresponds to a simple lumping of the loads. Discuss the convergence characteristics of the finite element solution.

The stiffness matrix of a bar element of length h is

$$\mathbf{K} = \frac{EA}{h}\begin{bmatrix} 1 & -1 \\ -1 & 1 \end{bmatrix}$$

and, correspondingly, the equilibrium equation at node i is

$$\frac{EA}{h}[-u_{i-1} + u_i] + \frac{EA}{\alpha h}[u_i - u_{i+1}] = \frac{p_i h(1 + \alpha)}{2} \tag{a}$$

The differential equation of equilibrium corresponding to the finite element idealization at node i is obtained by using the Taylor series expansion of the displacements at nodes $i-1$ and $i+1$ about node i. We have

$$\begin{aligned} u_{i+1} &= u_i + u_i'(\alpha h) + u_i''\frac{(\alpha h)^2}{2} + u_i'''\frac{(\alpha h)^3}{6} + \dots \\ u_{i-1} &= u_i - u_i'h + u_i''\frac{h^2}{2} - u_i'''\frac{h^3}{6} + \dots \end{aligned} \qquad \text{(b)}$$

Substituting (b) into (a), we obtain

$$u_i'' - \frac{h}{3}(1-\alpha)u_i''' + \frac{h^2}{12}\frac{1+\alpha^3}{1+\alpha}u_i^{\text{iv}} + \dots + \frac{p_i}{EA} = 0$$

However, we recall that the exact differential equation of equilibrium at point i is

$$u_i'' + \frac{p_i}{EA} = 0$$

Therefore, the finite element formulation converges to the exact solution as $h \longrightarrow 0$. Also, for $\alpha \neq 1$, convergence is linear, and for $\alpha = 1$, convergence is quadratic.

Convergence calculations as presented in Example 5.7 have been carried out for a number of finite elements. Therefore, in addition to knowing that monotonic convergence occurs in the energy norm, the actual convergence rate in displacements and stresses may also possibly be calculated. It should be realized, as shown in this example, that the rate of convergence depends, in general, on the topology of the finite element mesh.

The example solution of the bar above also indicates another noticeable feature of the finite element method, namely its close relationship to finite difference analysis.[35-41] In conventional finite difference analysis, an approximate solution is obtained by finite differencing the problem governing differential equations of equilibrium. In the case of the bar, the governing differential equation is

$$u'' + \frac{p}{EA} = 0 \qquad (5.44)$$

Using an equal mesh point spacing and the central difference formula, we obtain as the difference equation at point i,

$$\frac{1}{h^2}(u_{i+1} - 2u_i + u_{i-1}) + \frac{p_i}{EA} = 0 \qquad (5.45)$$

which is identical to (a) in Example 5.7. For $\alpha \neq 1$, however, a different finite difference scheme would need to be chosen. It is interesting to note that *finite difference equations can, therefore, also be interpreted as stiffness relations.*[35-37]

A main difficulty in the use of the conventional finite difference method is the incorporation of the boundary conditions. Since in the analysis the differential equations of equilibrium of the system are approximated directly by the difference scheme, it is necessary to satisfy in the differencing both the geometric and the natural boundary conditions. This can be difficult to achieve at arbitrary boundaries, since the topology of the finite difference mesh restricts the form of differencing that can be carried out, and it may be difficult to maintain symmetry properties in the coefficient matrix.[37]

The difficulties associated with conventional finite difference analysis have given new impetus to the development of a finite difference analysis procedure that is based on the principle of minimum total potential energy, referred to as the *finite difference energy method*.[38–42] In this scheme the displacement derivatives in the total potential energy, Π, of the system are approximated by finite differences, and the minimum condition of Π is used to calculate the unknown displacement parameters. Since the variational formulation of the problem under consideration is employed, only the geometric boundary conditions need be satisfied in the differencing. Furthermore, symmetry and positive definiteness in the coefficient matrix of the algebraic equations are assured.

As might well be expected, the finite difference energy method is very closely related to the displacement-based finite element method, and in some cases the same algebraic equations are generated. The specific differences between the finite element method and the finite difference energy method lie essentially in the choice of the generalized displacement components and the locations of the nodes that correspond to the generalized displacements. The solution procedures used are identical in both methods (see Example 5.9). Because of the close relationship between the finite element method and the finite difference energy method, it is also possible to identify equivalent physical models that are actually being analyzed when using the finite difference energy procedure.[40]

An advantage of the finite difference energy method lies in the effectiveness with which the coefficient matrix of the algebraic equations can be generated. This is due to the simple scheme of energy integration employed, which has been shown to be equivalent to using incompatible finite elements. However, when comparing the finite difference energy method with the finite element analysis procedure, it must be noted that the finite difference method has been applied to specific problems (e.g., the analysis of shells of revolution) and that finite element programs could also be written for that class of problems which are unusually effective. The main advantage of the finite element method is that the procedure can be used effectively in general-purpose analysis programs. To demonstrate the finite difference analysis methods, consider the following examples.

Example 5.8

Consider the simply supported beam in Fig. 5.4. Use conventional finite differencing to establish the system equilibrium equations.

The finite difference grid used for the beam analysis is shown in Fig. 5.4. In the conventional finite difference analysis the differential equation of equilibrium and the geometric and natural boundary conditions are considered (see Example 5.4); i.e., we approximate by finite differences at each station,

$$EI\frac{d^4w}{dx^4} = q \tag{a}$$

and use the conditions that $w = 0$ and $w'' = 0$ at $x = 0$ and $x = L$.

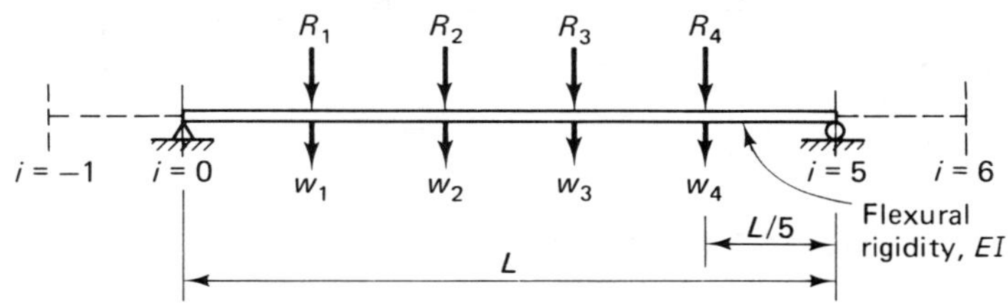

Figure 5.4. Finite difference stations for simply supported beam.

Using central differencing, (a) is approximated at station i by

$$\frac{EI}{\left(\frac{L}{5}\right)^3}\{w_{i-2} - 4w_{i-1} + 6w_i - 4w_{i+1} + w_{i+2}\} = R_i \tag{b}$$

where $R_i = q_iL/5$ and is the concentrated load applied at station i. The condition that w'' is zero at station i is approximated using

$$w_{i-1} - 2w_i + w_{i+1} = 0 \tag{c}$$

Applying (b) at each finite difference station, $i = 1, 2, 3, 4$, and using condition (c) at the support points, we obtain the system of equations

$$\frac{125EI}{L^3}\begin{bmatrix} 5 & -4 & 1 & 0 \\ -4 & 6 & -4 & 1 \\ 1 & -4 & 6 & -4 \\ 0 & 1 & -4 & 5 \end{bmatrix}\begin{bmatrix} w_1 \\ w_2 \\ w_3 \\ w_4 \end{bmatrix} = \begin{bmatrix} R_1 \\ R_2 \\ R_3 \\ R_4 \end{bmatrix}$$

where the coefficient matrix of the displacement vector can be regarded as a stiffness matrix.

Example 5.9

Consider the cantilever beam of Example 5.2. Evaluate the tip deflection using the conventional finite difference method and the finite difference energy method.

The finite difference mesh used is shown in Fig. 5.5. Using the conventional finite difference procedure and central differencing as in Example 5.8, we obtain the equilibrium equations

$$\frac{64EI}{L^3}\begin{bmatrix} 7 & -4 & 1 & 0 \\ -4 & 6 & -4 & 1 \\ 1 & -4 & 5 & -2 \\ 0 & 1 & -2 & 1 \end{bmatrix}\begin{bmatrix} w_2 \\ w_3 \\ w_4 \\ w_5 \end{bmatrix} = \begin{bmatrix} 0 \\ 0 \\ R \\ 0 \end{bmatrix} \tag{a}$$

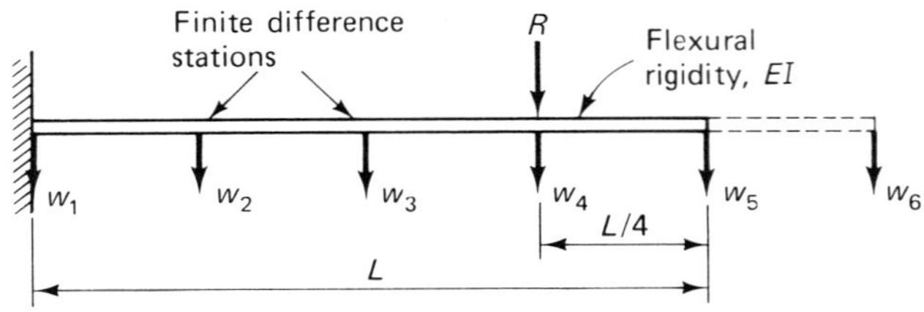

Figure 5.5. Finite difference stations on cantilever beam.

It may be noted that in addition to the equations employed in Example 5.8, the conditions that $w' = 0$ at the fixed end and $w''' = 0$ at the free end are also used. For w' and w''' equal to zero at station i, we employ, respectively,

$$w_{i+1} - w_{i-1} = 0$$

$$w_{i+2} - 2w_{i+1} + 2w_{i-1} - w_{i-2} = 0$$

Using the finite difference energy method, the total potential energy Π given in Example 5.2 is considered

$$\Pi = \frac{EI}{2}\int_0^L [w''(x)]^2\,dx - Rw(\tfrac{3}{4}L)$$

To evaluate the integral we need to approximate $w''(x)$. Using central differencing we obtain for station i,

$$w_i'' = \frac{1}{\left(\frac{L}{4}\right)^2}(w_{i+1} - 2w_i + w_{i-1}) \tag{b}$$

An approximate solution can now be obtained by evaluating Π at the finite difference stations using (b) and replacing the integral by a summation process; i.e., we use the approximation

$$\Pi = \frac{L}{8}\Pi_1 + \frac{L}{4}(\Pi_2 + \Pi_3 + \Pi_4) + \frac{L}{8}\Pi_5 - Rw_4 \tag{c}$$

where

$$\Pi_i = \frac{1}{2}[w_{i-1} \quad w_i \quad w_{i+1}]\begin{bmatrix} 1 \\ -2 \\ 1 \end{bmatrix}\frac{EI}{(L/4)^4}[1 \quad -2 \quad 1]\begin{bmatrix} w_{i-1} \\ w_i \\ w_{i+1} \end{bmatrix}$$

Therefore, we can write, in analogy to the finite element analysis procedure,

$$\Pi_i = \tfrac{1}{2}\mathbf{U}^T\mathbf{B}_i^T\mathbf{C}_i\mathbf{B}_i\mathbf{U}$$

where $\mathbf{B}_i$ is a generalized strain-displacement transformation matrix, $\mathbf{C}_i$ is the stress–strain matrix, and $\mathbf{U}$ is a vector listing all nodal point displacements. Using the direct stiffness method to calculate the total potential energy as given in (c) and employing the condition that the total potential energy is stationary (i.e., $\delta\Pi = 0$), we obtain the equilibrium equations

$$\frac{64EI}{L^3}\begin{bmatrix} 7 & -4 & 1 & & \\ -4 & 6 & -4 & 1 & \\ 1 & -4 & 5.5 & -3 & 0.5 \\ & 1 & -3 & 3 & -1 \\ & & 0.5 & -1 & 0.5 \end{bmatrix}\begin{bmatrix} w_2 \\ w_3 \\ w_4 \\ w_5 \\ w_6 \end{bmatrix} = \begin{bmatrix} 0 \\ 0 \\ R \\ 0 \\ 0 \end{bmatrix} \qquad \text{(d)}$$

where the condition of zero slope at the fixed end has already been used.

The close similarity between the equilibrium equations in (a) and (d) should be noted. Indeed, if we eliminate w_6 from the equations, we obtain the equations in (a). Hence, using the finite difference energy method and the conventional finite difference method we obtain in this case the same equilibrium equations.

As an example, let $R = 1$, $EI = 10^3$, and $L = 10$. Then we obtain, using the equations in (a) or (d) (without w_6),

$$\mathbf{U} = \begin{bmatrix} 0.023437 \\ 0.078125 \\ 0.14843 \\ 0.21875 \end{bmatrix}$$

The exact answer for the tip deflection is $w_5 = 0.2109375$. Hence the finite difference analysis gives a good approximate solution.

REFERENCES

1. J. H. Argyris, *Energy Theorems and Structural Analysis*, Butterworth & Company Ltd., London, 1960.
2. S. G. Mikhlin, *Variational Methods in Mathematical Physics*, Pergamon Press, Inc., Elmsford, N.Y., 1964.
3. K. Washizu, *Variational Methods in Elasticity and Plasticity*, Pergamon Press, Inc., Elmsford, N.Y., 1967.
4. R. Courant, "Variational Methods for the Solution of Problems of Equilibrium and Vibrations," *Bulletin of the American Mathematical Society*, Vol. 49, 1943, pp. 1–23.
5. R. Courant and D. Hilbert, *Methods of Mathematical Physics*, John Wiley & Sons, Inc., New York, 1962.
6. L. Collatz, *The Numerical Treatment of Differential Equations*, Springer-Verlag, New York, 1966.

7. S. H. Crandall, *Engineering Analysis*, McGraw-Hill Book Company, New York, 1956.

8. H. L. Langhaar, *Energy Methods in Applied Mechanics*, John Wiley & Sons, Inc., New York, 1962.

9. S. W. Key, "A Convergence Investigation of the Direct Stiffness Method," Ph.D. dissertation, University of Washington, 1966.

10. C. A. Felippa and R. W. Clough, "The Finite Element Method in Solid Mechanics," Symposium on Numerical Solutions of Field Problems in Continuum Mechanics, Durham, N.C., Apr. 1968.

11. J. T. Oden, *Finite Elements of Nonlinear Continua*, McGraw-Hill Book Company, New York, 1972.

12. G. Strang and G. J. Fix, *An Analysis of the Finite Element Method*, Prentice-Hall, Inc., Englewood Cliffs, N.J., 1973.

13. J. R. Whiteman (ed.), *The Mathematics of Finite Elements and Applications*, Academic Press Inc. Ltd., London, 1973.

14. J. T. Oden, R. W. Clough, and Y. Yamamoto (eds.), "Basic Theory," *Advances in Computational Methods in Structural Mechanics and Design*, University of Alabama Press, Huntsville, Ala., 1972.

15. R. H. Gallagher, Y. Yamada, and J. T. Oden (eds.), "Basic Theory," *Recent Advances in Matrix Methods of Structural Analysis and Design*, University of Alabama Press, Huntsville, Ala., 1971.

16. E. R. de Arantes e Oliveira, "Completeness and Convergence in the Finite Element Method," Proceedings, 2nd Conference on Matrix Methods in Structural Mechanics, Wright-Patterson A.F.B., Ohio, 1968, pp. 1061–1089.

17. E. R. de Arantes e Oliveira, "Theoretical Foundations of the Finite Element Method," *International Journal of Solids and Structures*, Vol. 4, 1968, p. 929.

18. W. Ritz, "Über eine neue Methode zur Lösung gewisser Variationsprobleme der mathematischen Physik," *Zeitschrift für Angewandte Mathematik und Mechanik,* Vol. 135, Heft 1, 1908, pp. 1–61.

19. E. L. Wilson and R. E. Nickell, "Application of the Finite Element Method to Heat Conduction Analysis," *Nuclear Engineering and Design*, Vol. 4, 1966, pp. 276–286.

20. O. C. Zienkiewicz, P. Mayer, and Y. K. Cheung, "Solution of Anistropic Seepage Problem by Finite Elements," *A.S.C.E., Journal of Engineering, Mechanics Division*, Vol. 92, 1966, pp. 111–120.

21. J. H. Argyris, C. Mareczek, and D. W. Scharpf, "Two- and Three-Dimensional Flow Using Finite Elements," *Journal of the Royal Aeronautical Society*, Vol. 73, 1969, pp. 961–964.

22. E. L. Wilson, K. J. Bathe, and F. E. Peterson, "Finite Element Analysis of Linear and Nonlinear Heat Transfer," *Nuclear Engineering and Design*, Vol. 29, 1974.

23. G. P. Bazely, Y. K. Cheung, B. M. Irons, and O. C. Zienkiewicz, "Triangular Elements in Plate Bending Conforming and Non-Conforming Solutions," Proceedings, Conference on Matrix Methods in Structural Mechanics, Wright-Patterson A.F.B., Ohio, 1965.

24. B. M. Irons, O. C. Zienkiewicz, and E. R. de Arantes e Oliveira, "Comments on the Paper: Theoretical Foundations of the Finite Element Method," *International Journal of Solids and Structures*, Vol. 6, 1970, pp. 695–697.

25. T. H. H. Pian, "Formulations of Finite Element Methods for Solid Continua," *Recent Advances in Matrix Methods of Structural Analysis and Design*, University of Alabama Press, Huntsville, Ala., 1971.

26. T. H. H. Pian and P. Tong, "Basis of Finite Element Methods for Solid Continua," *International Journal for Numerical Methods in Engineering*, Vol. 1, 1969, pp. 3–28.

27. T. H. H. Pian, "Variational Formulations of Numerical Methods in Solid Continuum," Proceedings, Symposium on Computer Aided Engineering, University of Waterloo, Waterloo, Ontario, Canada, 1971.

28. E. Reissner, "On a Variational Theorem in Elasticity," *Journal of Mathematics and Physics*, Vol. 29, 1950, pp. 90–95.

29. L. R. Herrmann, "Finite Element Bending Analysis for Plates," *A.S.C.E., Journal of Engineering Mechanics Division*, Vol. 93, 1967, pp. 13–76.

30. W. Wunderlich, "Discretization of Structural Problems by a Generalized Variational Approach," Proceedings, I.A.S.S. Symposium, Honolulu, Hawaii, 1971.

31. R. L. Taylor, K. S. Pister, and L. R. Herrmann, "On a Variational Theorem for Incompressible and Nearly Incompressible Orthotropic Elasticity," *International Journal of Solids and Structures*, Vol. 4, 1968, pp. 875–883.

32. B. F. de Veubeke and G. Sander, "An Equilibrium Model for Plate Bending," *International Journal of Solids and Structures*, Vol. 4, 1968, pp. 447–468.

33. J. E. Walz, R. E. Fulton, and N. J. Cyrus, "Accuracy and Convergence of Finite Element Approximations," Proceedings, 2nd Conference on Matrix Methods in Structural Mechanics, Wright-Patterson A.F.B., Ohio, 1968.

34. J. G. A. Croll and A. C. Walker, "The Finite Difference and Localized Ritz Methods," *International Journal for Numerical Methods in Engineering*, Vol. 3, 1971, pp. 155–160.

35. A. Ghali and A. M. Neville, *Structural Analysis*, Intext Educational Publishers, Scranton, Pa., 1972.

36. A. Ghali and K. J. Bathe, "Analysis of Plates Subjected to In-Plane Forces Using Large Finite Elements," *International Association for Bridge and Structural Engineering Bulletin*, Vol. 30-I, 1970, pp. 61–72.

37. A. Ghali and K. J. Bathe, "Analysis of Plates in Bending Using Large Finite Elements," *International Association for Bridge and Structural Engineering Bulletin*, Vol. 30-II, 1970, pp. 29–40.

38. G. E. Forsythe and W. R. Wasow, *Finite Difference Methods for Partial Differential Equations*, John Wiley & Sons, Inc., New York, 1960.

39. S. W. Key and R. D. Krieg, "Comparison of Finite Element and Finite Difference Methods," *Numerical and Computer Methods in Structural Mechanics*

(S. J. Fenves, N. Perrone, J. Robinson, and W. C. Schnobrich, eds.), Academic Press, Inc., New York, 1973.

40. D. BUSHNELL, "Finite Difference Energy Models Versus Finite Element Models: Two Variational Approaches in One Computer Program," *Numerical and Computer Methods in Structural Mechanics* (S. J. Fenves, N. Perrone, J. Robinson, and W. C. Schnobrich, eds.), Academic Press, Inc., New York, 1973.

41. D. BUSHNELL and B. O. ALMROTH, "Finite Difference Energy Method for Nonlinear Shell Analysis," *Journal of Computers and Structures*, Vol. 1, 1971, p. 361.

42. C. A. FELIPPA, "Finite Element and Finite Difference Technique for the Numerical Solution of Partial Differential Equations," Proceedings, Conference on Computer Simulation, Montreal, July, 1973.

CHAPTER 6

IMPLEMENTATION OF THE FINITE ELEMENT METHOD

6.1 INTRODUCTION

In the previous chapters we presented the formulation and general theory of the finite element method. The objective in this chapter is to discuss some important computational aspects pertaining to the implementation of finite element analysis. Although the implementation of the displacement-based finite element analysis is discussed, it should be noted that most of the concepts presented in this chapter can also be employed in finite element analysis using hybrid and mixed formulations.

The main advantage that the finite element method has over other analysis techniques is its large generality. Normally, as was pointed out, it seems possible, by using many elements, to virtually approximate any continuum with complex boundary and loading conditions to such a degree that an accurate analysis can be carried out. In practice, however, obvious engineering limitations arise, a most important one being the cost of the analysis. As the number of elements used is increased, the manpower required to prepare the relevant data and interpret the results increases, and also a larger amount of computer time is needed for the analysis. Furthermore, the limitations of the program and the computer may prevent the use of a larger number of finite elements to idealize the continuum. The limitations may be due to the high-speed and back-up (low-speed) storage available, or the round-off and truncation errors occurring in the analysis because of finite precision arithmetic. Also, the malfunctioning of a hardware component, if the analysis is carried out using many computer hours, is possible. It is therefore desirable to use efficient finite element programs.

The effectiveness of a program depends essentially on the following factors. Firstly, the use of efficient finite elements is important. We have summarized the isoparametric elements in the preceding chapter. Secondly, efficient programming methods and sophisticated use of the available computer hardware and software are important. Although this aspect of program development is computer-dependent, using standard FORTRAN IV and high- and low-speed storage in a system-independent manner, very effective computer programs can be developed. If such a program is to be permanently installed on a specific computer, its efficiency may normally be increased with relatively little effort by making use of the specific hardware and software options available. In the following, we therefore discuss the design of finite element programs in which computer-independent procedures are used. Also, numerical algorithms are presented using standard FORTRAN IV.

The third very important aspect in the development of a finite element program is the use of appropriate numerical techniques. As an example, if inappropriate techniques for the solution of the frequencies of a system in a dynamic analysis are employed, the cost may be many times larger than with effective techniques, and, even worse, a solution may not be possible at all if an unstable algorithm is employed. In order to implement the finite element method in practice, we need to use the digital computer. However, even with a relatively large capacity computer available, the feasibility of a problem solution and the effectiveness of an analysis depend directly on the numerical procedures employed.

Assume that an actual structure has been idealized as an assemblage of finite elements. The stress analysis process can be understood to consist of essentially three phases:

1. Calculation of structure matrices $\mathbf{K}, \mathbf{M}, \mathbf{C}$, and $\mathbf{R}$, whichever are applicable.
2. Solution of equilibrium equations.
3. Evaluation of element stresses.

In case of the analysis of a field problem, the steps are identical, but the pertinent matrices and solution quantities need be considered.

The objective in this chapter is to describe a program implementation of the first and third phases and to present a small computer program that has all the important features of a general code. The solution of the equilibrium equations is discussed in detail in the next chapters. However, although the total solution may be subdivided into these three phases, it should be realized that the specific implementation of one phase can have a pronounced effect on the efficiency of the other phases, and, indeed, in some programs the first two phases are carried out simultaneously.[1-4]

As might be imagined, there is no unique optimum program organization for the evaluation of the structure matrices, and various different and effective strategies are currently followed.[1-10] In addition, new ideas may be developed, particularly as new computer equipment becomes available. However, although a program design might at first sight be quite different, in effect, the same basic steps are followed. For this reason, it is deemed to be most instructive to discuss in detail all the important features of one specific implementation, namely the one that is basically used in the programs SAP, ADINA, and EASE.[5-10] In the first part of this chapter, the program algorithms used are discussed, and in the second part a small program is presented that uses all the important features of the above general-purpose programs.

6.2 COMPUTER PROGRAM ORGANIZATION FOR CALCULATION OF STRUCTURE MATRICES

The final results of this phase are the required structure matrices for the solution of the system equilibrium equations. In a static analysis the computer program needs to calculate the structure stiffness matrix and the load vectors. In a dynamic analysis, the program must also establish the system mass and damping matrices. In the implementation to be described here, the calculation of the structure matrices is performed as follows:

1. The nodal point and element information are read and/or generated.
2. The element stiffness matrices, mass and damping matrices, and equivalent nodal loads are calculated.
3. The structure matrices **K**, **M**, **C**, and **R**, whichever are applicable, are assembled.

6.2.1 Nodal Point and Element Information Read-In

Consider first the data that correspond to the nodal points. Assume that the program is set up to allow a maximum of six degrees of freedom at each node, three translational and three rotational degrees of freedom, as shown in Fig. 6.1. Corresponding to each nodal point, it must then be identified which of these degrees of freedom shall actually be used in the analysis, i.e., which of the six possible nodal degrees of freedom correspond to degrees of freedom of the finite element assemblage. This is achieved by defining an identification array, the array ID, of dimension 6 times NUMNP, where NUMNP is equal to the number of nodal points in the system. Element (i, j) in the ID matrix corresponds to the ith degree of freedom at the nodal point j. If ID(I, J) = 0, the corresponding degree of freedom is defined in the element system, and if ID(I, J) = 1, the degree of freedom is not defined. It

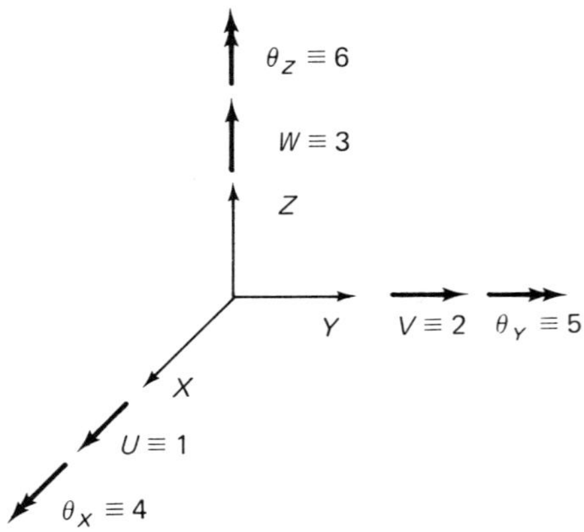

Figure 6.1. Possible degrees of freedom at a nodal point.

should be noted that using the same procedure, an ID array for more (or less) than 6 degrees of freedom per nodal point could be established, and, indeed, the number of degrees of freedom per nodal point could be a variable. Consider the following simple example.

Example 6.1

Establish the ID array for the plane stress element idealization of the cantilever in Fig. 6.2 (and Fig. 3.4), in order to define the active and nonactive degrees of freedom.

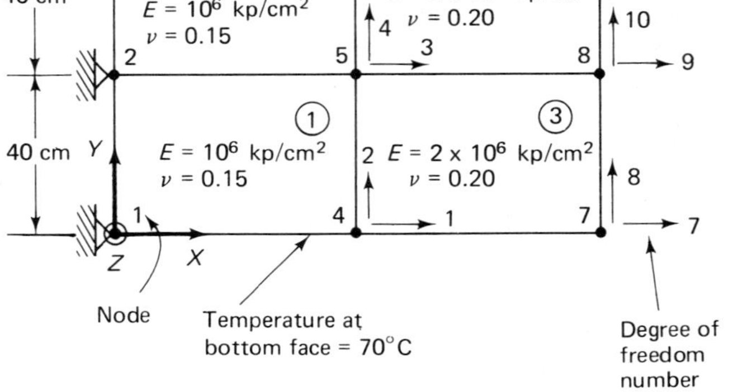

Figure 6.2. Finite element cantilever idealization.

The active degrees of freedom are defined by ID(I, J) = 0, and the nonactive degrees of freedom are defined by ID(I, J) = 1. Since the cantilever is in the X–Y plane, and plane stress elements are used in the idealization, only X and Y translational degrees of freedom are active. By inspection, the ID array is given by

$$\text{ID} = \begin{bmatrix} 1 & 1 & 1 & 0 & 0 & 0 & 0 & 0 & 0 \\ 1 & 1 & 1 & 0 & 0 & 0 & 0 & 0 & 0 \\ 1 & 1 & 1 & 1 & 1 & 1 & 1 & 1 & 1 \\ 1 & 1 & 1 & 1 & 1 & 1 & 1 & 1 & 1 \\ 1 & 1 & 1 & 1 & 1 & 1 & 1 & 1 & 1 \\ 1 & 1 & 1 & 1 & 1 & 1 & 1 & 1 & 1 \end{bmatrix}$$

Once all active degrees of freedom have been defined by zeros in the ID array, the equation numbers corresponding to these degrees of freedom are assigned. The procedure is to simply scan column after column through the ID array and replace each zero by an equation number, which increases successively from 1 to the total number of equations.

Example 6.2

Modify the ID array obtained in Example 6.1 for the analysis of the cantilever plate in Fig. 6.2, to obtain the ID array that defines the equation numbers corresponding to the active degrees of freedom.

As explained above, we simply replace the zeros, column by column, in succession by equation numbers to obtain

$$\text{ID} = \begin{bmatrix} 0 & 0 & 0 & 1 & 3 & 5 & 7 & 9 & 11 \\ 0 & 0 & 0 & 2 & 4 & 6 & 8 & 10 & 12 \\ 0 & 0 & 0 & 0 & 0 & 0 & 0 & 0 & 0 \\ 0 & 0 & 0 & 0 & 0 & 0 & 0 & 0 & 0 \\ 0 & 0 & 0 & 0 & 0 & 0 & 0 & 0 & 0 \\ 0 & 0 & 0 & 0 & 0 & 0 & 0 & 0 & 0 \end{bmatrix}$$

Apart from the definition of all active degrees of freedom, we also need to read the X,Y,Z global coordinates and, if required, the temperature corresponding to each nodal point. For the cantilever beam of Fig. 6.2, the X,Y,Z coordinate arrays and nodal point temperature array T would be as follows:

$$\left.\begin{aligned} \text{X}^T &= [0.0 \quad 0.0 \quad 0.0 \quad 60.0 \quad 60.0 \quad 60.0 \quad 120.0 \quad 120.0 \quad 120.0] \\ \text{Y}^T &= [0.0 \quad 40.0 \quad 80.0 \quad 0.0 \quad 40.0 \quad 80.0 \quad 0.0 \quad 40.0 \quad 80.0] \\ \text{Z}^T &= [0.0 \quad 0.0 \quad 0.0 \quad 0.0 \quad 0.0 \quad 0.0 \quad 0.0 \quad 0.0 \quad 0.0] \\ \text{T}^T &= [70.0 \quad 85.0 \quad 100.0 \quad 70.0 \quad 85.0 \quad 100.0 \quad 70.0 \quad 85.0 \quad 100.0] \end{aligned}\right\} \quad (6.1)$$

At this stage, with all the nodal point data known, the program may read and generate the element information. It is expedient to consider each element type in turn. For example, in the analysis of a container structure, all beam elements, all plane stress elements, and all shell elements are read and generated together. This is efficient, because specific information need be provided for each element of a certain type, which, because of its repetitive nature, can be generated to some degree if all elements of the same type are specified together. Furthermore, the element routine for an element type that reads the element data and calculates the element matrices need only be called once.

The required data corresponding to an element depend on the specific element type. In general, the information required for each element is the element node numbers that correspond to the nodal point numbers of the complete element assemblage, the element material properties, and the surface and body forces applied to the element. Since the element material properties and the element loading are the same for many elements, it is efficient to define material property sets and load sets pertaining to an element type. These sets are specified at the beginning of each group of element data. Therefore, any one of the material property sets and element load sets can be assigned to an element at the same time the element node numbers are read.

Example 6.3

Consider the analysis of the cantilever plate shown in Figs. 3.4 and 6.2 and the local element node numbering defined in Fig. 3.5. For each element give the node numbers that correspond to the nodal point numbers of the complete element assemblage. Also indicate the use of material property sets.

In this analysis we define two material property sets: material property set 1 for $E = 10^6$ kp/cm^2 and $\nu = 0.15$, and material property set 2 for $E = 2 \times 10^6$ kp/cm^2 and $\nu = 0.20$. We then have the following node numbers and material property sets for each element:

Element 1: node numbers: 5, 2, 1, 4; material property set: 1
Element 2: node numbers: 6, 3, 2, 5; material property set: 1
Element 3: node numbers: 8, 5, 4, 7; material property set: 2
Element 4: node numbers: 9, 6, 5, 8; material property set: 2

6.2.2 Calculation of Element Stiffness, Mass, and Equivalent Nodal Loads

The general procedure to calculate element matrices was discussed in Chapter 4, and a computer implementation was presented in Section 4.8. The program organization during this phase consists of calling the appropriate element subroutines for each element. During the element matrix calculations, the element coordinates, properties, and load sets, which have been read and stored in the preceding phase (Section 6.2.1), are needed. After calculation, an element matrix may either be stored on back-up storage, because the assemblage to the structure matrices is carried out later,[7] or the element matrix may be added immediately to the appropriate structure matrix.[8]

6.2.3 Assemblage of Structure Matrices

The assemblage process to obtain the structure stiffness matrix $\mathbf{K}$ is symbolically written

$$\mathbf{K} = \sum_i \mathbf{K}_i \tag{6.2}$$

where the matrix $\mathbf{K}_i$ is the stiffness matrix of the ith element and the summation goes over all elements in the assemblage. In an analogous manner the structure mass matrix and load vectors are assembled from the element mass matrices and element load vectors, respectively. In addition to the element stiffness, mass, and load matrices, concentrated stiffnesses, masses, and loads corresponding to specific degrees of freedom can also be added.

It should be noted that the element stiffness matrices $\mathbf{K}_i$ in (6.2) are of the same order as the structure stiffness matrix $\mathbf{K}$. However, considering the internal structure of the matrices $\mathbf{K}_i$, nonzero elements are only in those rows and columns that correspond to element degrees of freedom (see Section 3.2.2). Therefore, in practice, we only need to store the compacted element stiffness matrix, which is of order equal to the number of element degrees of freedom, together with an array that relates to each element degree of freedom the corresponding assemblage degree of freedom. This vector is conveniently a connectivity array LM, in which entry i gives the equation number that corresponds to the element degree of freedom i.

Example 6.4

Using the convention for the element degrees of freedom shown in Fig. 3.5, establish the vectors defining the assemblage degrees of freedom of the elements in the assemblage of Fig. 6.2 (or 3.4).

Consider element 1 in Fig. 6.2. For this element, nodes 5, 2, 1, and 4 of the element assemblage correspond to the element nodes 1, 2, 3, and 4, respectively (Fig. 3.5). Using the ID array, the equation numbers corresponding to the nodes 5, 2, 1, and 4 of the element assemblage are obtained, and hence the relation between the column (and row) numbers of the compacted, or local, element stiffness matrix and the global stiffness matrix is as follows:

CORRESPONDING COLUMN AND ROW NUMBERS

For compacted matrix	1	2	3	4	5	6	7	8
For $\mathbf{K}_1$	3	4	0	0	0	0	1	2

The array LM, storing the assemblage degrees of freedom of this element, is therefore

$$\text{LM}^T = [3 \quad 4 \quad 0 \quad 0 \quad 0 \quad 0 \quad 1 \quad 2]$$

where a zero means that the corresponding column and row of the compacted element stiffness are ignored and do not enter the global structure stiffness matrix.

Similarly, we can obtain the LM arrays that correspond to the elements 2, 3, and 4. We have for element 2,

$$\mathrm{LM}^T = [5 \quad 6 \quad 0 \quad 0 \quad 0 \quad 0 \quad 3 \quad 4]$$

for element 3,

$$\mathrm{LM}^T = [9 \quad 10 \quad 3 \quad 4 \quad 1 \quad 2 \quad 7 \quad 8]$$

and for element 4,

$$\mathrm{LM}^T = [11 \quad 12 \quad 5 \quad 6 \quad 3 \quad 4 \quad 9 \quad 10]$$

As shown in the example above, the connectivity array of an element is determined from the nodal points to which the element is connected and the equation numbers that have been assigned to these nodal points. Once the array LM has been defined, the corresponding element stiffness matrix can be added in compact form to the structure stiffness matrix **K**, but the process must take due account of the specific storage scheme used for **K**. As already pointed out in Section 1.3, an effective storage scheme for the structure stiffness matrix is to store only the elements below the skyline of the matrix **K** (i.e., the active columns of **K**) in a one-dimensional array A. However, together with the active column storage scheme, we also need a specific procedure of addressing the elements of **K** in A when they are stored as indicated in Section 1.3. Thus, before we are able to proceed with the assemblage of the element stiffness matrices, it is necessary to establish the addresses of the stiffness matrix elements in the one-dimensional array A.

Figure 6.3 shows the element pattern of a typical stiffness matrix. Let us derive the storage scheme and addressing procedure that we propose to use. Since the matrix is symmetric, we naturally only store and work on the part above and including the diagonal. However, in addition, we observe that elements (i, j) of **K** (i.e., k_{ij}) are zero for $j > i + m_{\mathbf{K}}$. The value $m_{\mathbf{K}}$ is known as the *half-bandwidth* of the matrix. Defining by m_i the row number of the first nonzero element in column i (Fig. 6.3), the variables m_i, $i = 1, \ldots, n$, define the *skyline* of the matrix, and the variables $(i - m_i)$ are the *column heights*. Furthermore, the half-bandwidth of the stiffness matrix, $m_{\mathbf{K}}$, equals $\max\{i - m_i\}$, $i = 1, \ldots, n$. In many finite element analyses, the column heights are very different, and it is important that all zero elements outside the skyline not be included in the equation solution (see Section 7.2.3). On the other hand, zero elements within the skyline of the matrix are stored and operated upon, since they will become, in many cases, nonzero elements during the matrix reduction.

The column heights are determined from the connectivity arrays, LM, of the elements; i.e., by evaluating m_i, we also obtain the column height $(i - m_i)$. Consider, as an example, that m_{10} of the stiffness matrix that corresponds

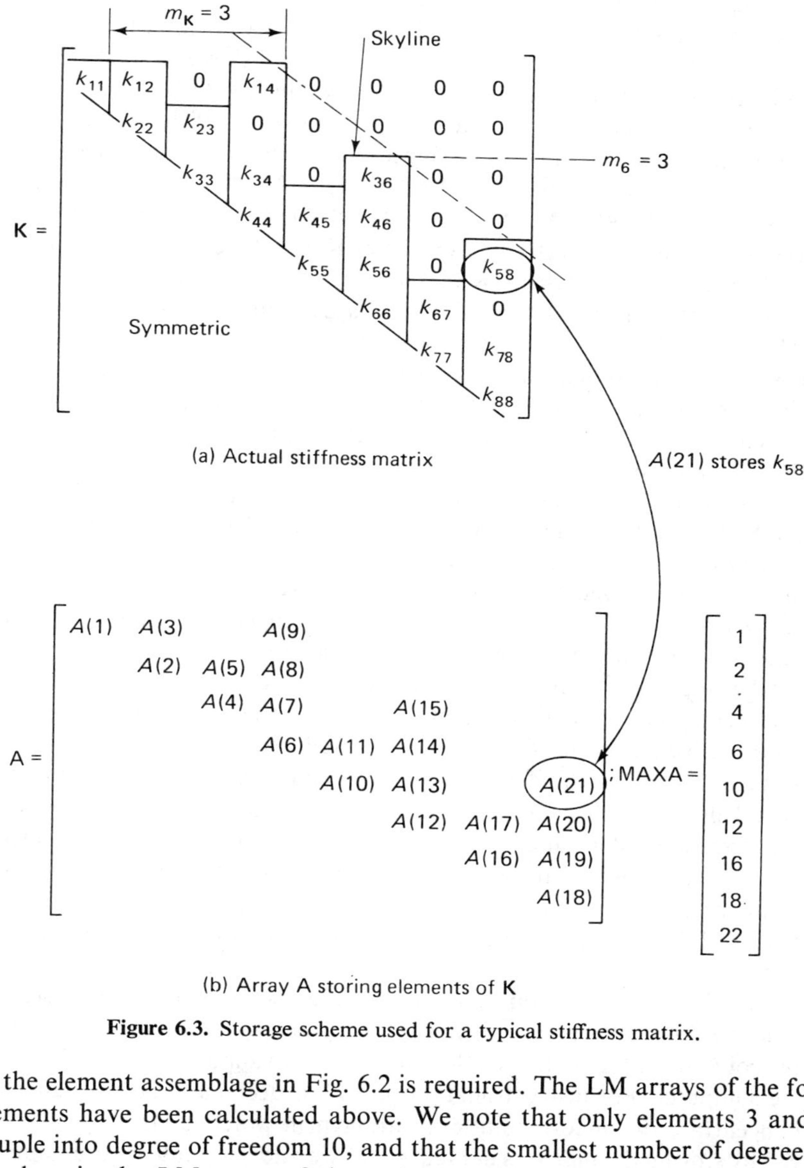

Figure 6.3. Storage scheme used for a typical stiffness matrix.

to the element assemblage in Fig. 6.2 is required. The LM arrays of the four elements have been calculated above. We note that only elements 3 and 4 couple into degree of freedom 10, and that the smallest number of degree of freedom in the LM array of these elements is 1; hence $m_{10} = 1$ and the column height of column 10 is 9.

With the column heights of a stiffness matrix defined, we can now store all elements below the skyline of **K** as a one-dimensional array in A; i.e., the active columns of **K** including the diagonal elements are stored consecutively in A. Figure 6.3 shows which storage locations the elements of the matrix **K** given in the figure would take in A. In addition to A, we also define an array MAXA, which stores the addresses of the diagonal elements of **K** in A; i.e., the address of the ith diagonal element of **K**, k_{ii}, in A is MAXA(I). Referring to Fig. 6.3, it is noted that MAXA(I) is equal to the sum of the column heights up to the $(i - 1)$st column plus I. Hence the number of nonzero elements in the ith column of **K** is equal to MAXA(I+1) − MAXA(I), and the element addresses are MAXA(I), MAXA(I)+1, MAXA(I)+2, . . . , MAXA(I+1)−1. It follows that using this storage scheme of **K** in A together with the address array MAXA each element of **K** in A can be addressed easily.

The storage scheme described above is used in the computer program STAP presented in Section 6.4 and later in the equation and eigenvalue solution subroutines. It may also be noted that this active column storage scheme is employed in the ADINA program.[8] The effectiveness of the scheme lies essentially in that no elements outside the skyline are stored and processed in the calculations.

In the discussion of algorithms for the solution of the equations $\mathbf{KU} = \mathbf{R}$, where **K**, **U**, and **R** are the stiffness matrix, the displacement vector, and the load vector of the element assemblage, respectively, we will show that the solution procedure most commonly used requires about $\frac{1}{2}\, nm_{\mathbf{K}}^2$ operations, where n is the order of the stiffness matrix and $m_{\mathbf{K}}$ is its half-bandwidth assumed to be nearly full; i.e., $(i - m_i) = m_{\mathbf{K}}$ for nearly all i. Therefore, it can be important to minimize $m_{\mathbf{K}}$ both from considerations of storage requirements and number of operations. In case the column heights vary, a mean or "effective" value for $m_{\mathbf{K}}$ must be used (see Section 7.2.3). Since the effective half-bandwidth, if indeed full, is equal to the maximum difference in global degrees of freedom pertaining to any one of the finite elements, we can frequently determine a reasonable nodal point numbering by inspection. However, this nodal point numbering may not be particularly easy to generate, and various automatic schemes are currently used for bandwidth reduction.[11−12] Figure 6.4 shows a typical good and bad nodal point numbering.

It should be pointed out that in the discussion of the above storage scheme, we implicitly assumed that the entire array A (i.e., the sum of all active columns of the matrix **K**) does fit into the available high speed storage of the computer. For instructional purposes it is deemed to be most effective to concentrate on in-core solution, although in practice large systems are solved by storing the matrices in blocks.[5−10] Considering out-of-core solution, in principle, the same storage scheme as for in-core solutions is effective.[8] The main additional problem is one of program logistics; namely, the individual blocks of the matrices must be stored on back-up storage and called into high-speed storage in an effective manner. Various procedures are in use, and

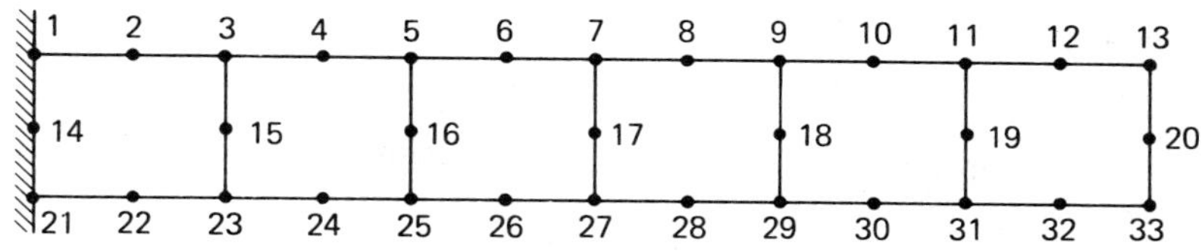

(a) Bad nodal point numbering, $m_K + 1 = 46$

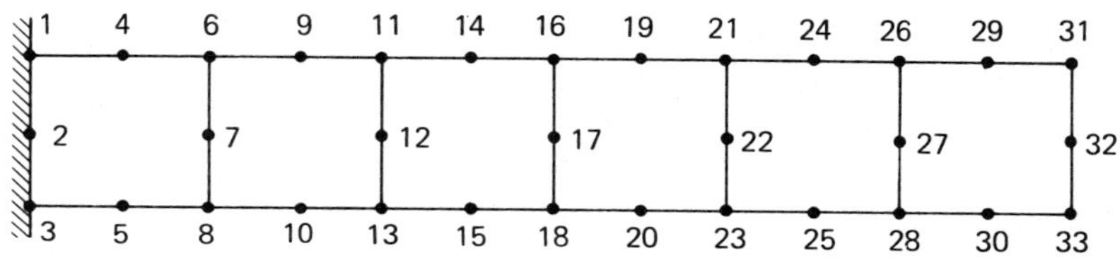

(b) Good nodal point numbering, $m_K + 1 = 16$

Figure 6.4. Bad and good nodal point numbering for finite element assemblage.

in all cases, specific attention has been given to minimize the amount of tape or disc reading required in the assemblage. However, once an effective in-core finite element solution has been studied, little difficulty should be encountered in understanding specific out-of-core implementations.

6.3 CALCULATION OF ELEMENT STRESSES

In the previous section we described the process of assembling the individual finite element matrices into the total structure matrices. The next step is the calculation of nodal point displacements, which is discussed in detail in the following chapters. Once the nodal point displacements have been obtained, element stresses are calculated in the final phase of the analysis.

The equations employed in the element stress calculations are (3.55) and (3.57). However, as in the assemblage of the structure matrices, it is again effective to manipulate compacted finite element matrices, i.e., only deal with the nonzero columns of $\mathbf{B}^{(m)}$ in (3.55). Using the implementation described in the previous section, we calculate the element compacted strain-displacement transformation matrix and then extract the element nodal point displacements from the total displacement vector using the LM array of the element. The procedure is implemented in program STAP described next. It should be noted that finite element stresses can be calculated at any desired location by simply establishing the strain-displacement transformation matrix for the point under consideration in the element. In isoparametric finite element analysis we use the procedures given in Section 4.3.1 (see Example 4.2).

6.4 EXAMPLE PROGRAM STAP

Probably the best way of getting familiar with the implementation of finite element analysis is to study an actual computer program that, although simplified in various areas, shows all the important features of more general codes. The following program, STAP (STatic Analysis Program), is a simple computer program for static linear elastic finite element analysis.

The main objective in the presentation of the program is to show the overall flow of a typical finite element analysis program, and for this reason only a truss element has been made available in STAP. However, the code can generally be used for one-, two- and three-dimensional analysis, and additional elements can be added with relative ease.

Figure 6.5 shows a flow chart of the program, and Figure 6.6 gives the storage allocations used during the various program phases. Before presenting the listing of the program, we give the input manual. It may be noted that the coding and the input manual of program STAP are similar to the procedures used in SAP IV, ADINA, and EASE2.[5-10]

6.4.1 Data Input to Computer Program STAP

I. HEADING CARD (12A6)

Note	Columns	Variable	Entry
(1)	1–80	HED (20)	Enter the master heading information for use in labeling the output

NOTES/

(1) Begin each new data case with a new heading card. Two blank cards must be input after the last data case.

II. CONTROL CARD (4I5)

Note	Columns	Variable	Entry
(1)	1–5	NUMNP	Total number of nodal points; EQ.0, program stop
(2)	6–10	NUMEG	Total number of element groups, GT.0
(3)	11–15	NLCASE	Number of load cases, GT.0
(4)	16–20	MODEX	Flag indicating solution mode; EQ.0, data check only EQ.1, execution

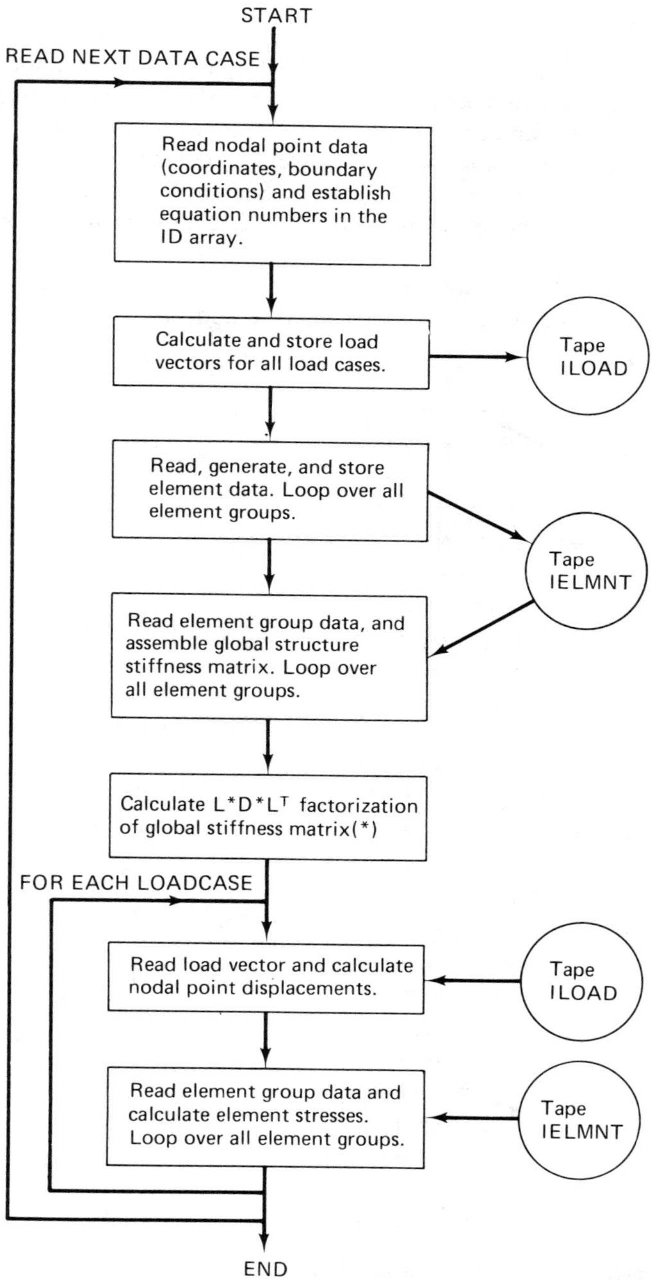

Figure 6.5. Flow chart of program STAP. *See Section 7.2.2.

NOTES/

(1) The total number of nodes (NUMNP) controls the amount of data to be read in Section III. If NUMNP.EQ.0 the program stops.
(2) The total number of elements are dealt with in element groups. An element group consists of a convenient collection of elements. Each element group is input as given in Section V. There must be at least one element per element group, and there must be at least one element group.
(3) The number of load cases (NLCASE) gives the number of load vectors for which the displacement and stress solution is sought.

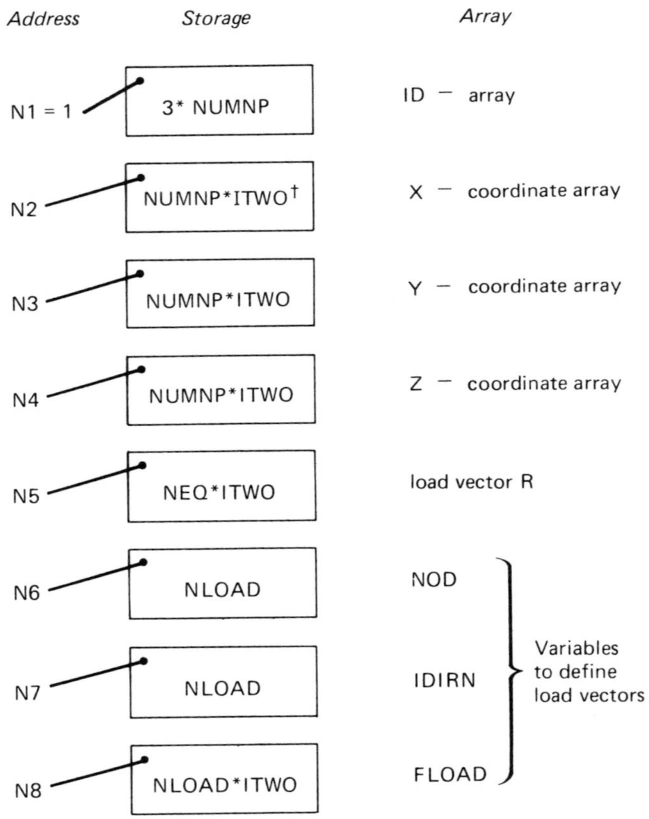

a. Input of ID array, nodal point coordinates, and load vectors.

Figure 6.6. High-speed storage allocation in program STAP. (See user's manual for definition of variables.) †ITWO = 1 in single precision arithmetic. ITWO = 2 in double precision arithmetic.

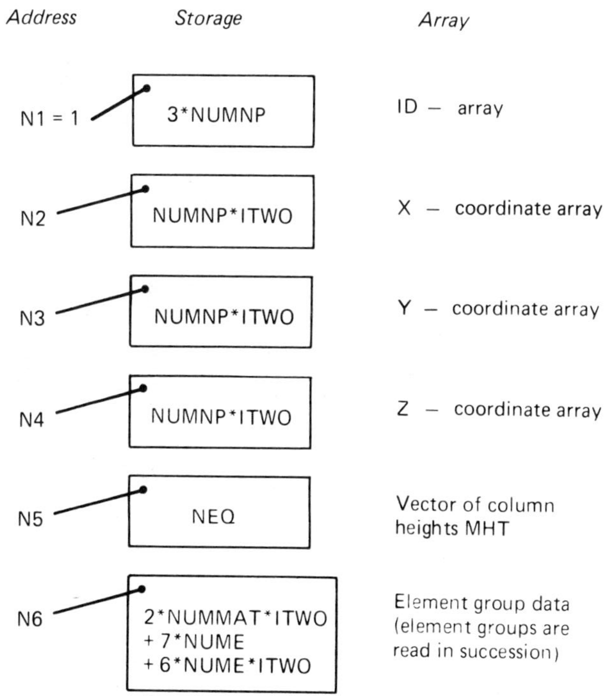

b. Element data input.

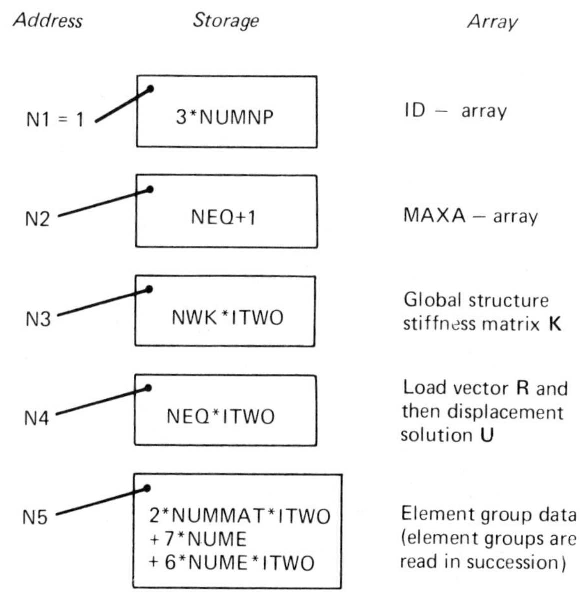

c. Assemblage of global structure stiffness; displacement and stress solution phase.

Figure 6.6. *continued.*

(4) The MODEX parameter determines whether the program is to check the data without executing the analysis (i.e., MODEX. EQ.0) or if the program is to solve the problem (i.e., MODEX. EQ.1). In the data check only mode, the program only reads and prints all data.

III. NODAL POINT DATA CARDS (4I5, 3F10.0, I5)

Note	Columns	Variable	Entry
(1)	1–5	N	Node (joint) number; GE.1 and LE.NUMNP
(2)	6–10	ID (1, N)	X—translation boundary code
	11–15	ID (2, N)	Y—translation boundary code
	16–20	ID (3, N)	Z—translation boundary code
(3)	21–30	X(N)	X—coordinate
	31–40	Y(N)	Y—coordinate
	41–50	Z(N)	Z—coordinate
(4)	51–55	KN	Node number increment for node data generation; EQ.0, no generation

NOTES/

(1) Nodal data must be defined for all (NUMNP) nodes. Node data may be input directly (i.e., each node on its own individual card) or the generation option may be used if applicable (see note 4, p. 219). Admissible node numbers range from 1 to the total number of nodes (NUMNP). The last node that is input must be NUMNP.

(2) Boundary condition codes can only be assigned the following values (M = 1,2,3)

ID(M,N) = 0; unspecified (free) displacement.

ID(M,N) = 1; deleted (fixed) displacement.

An unspecified [ID(M,N) = 0] degree of freedom is free to translate as the solution dictates. Concentrated forces may be applied in this degree of freedom.

One system equilibrium equation is required for each unspecified degree of freedom in the model. The total number of equilibrium equations is defined as NEQ and is always less than three times the total number of nodes in the system.

Deleted [ID(M,N) = 1] degrees of freedom are removed from the final set of equilibrium equations. Deleted degrees of freedom are used to define fixities (points of external reaction), and any loads applied in these degrees of freedom are ignored by the program.

(3) The geometric location of each node is specified by its X, Y, and Z coordinates.

(4) Node cards need not be input in node order sequence; eventually, however, all nodes in the set [1,NUMNP] must be defined. Node data for a series of nodes

$$[N_1, N_1 + 1*KN_1, N_1 + 2*KN_1, \ldots, N_2]$$

may be generated from information given on two cards in sequence—

$$\text{CARD 1} — N_1, ID(1, N_1), \ldots, X(N_1), \ldots, KN_1$$

$$\text{CARD 2} — N_2, ID(1, N_2), \ldots, X(N_2), \ldots, KN_2$$

KN_1 is the node generation parameter given on the first card in the sequence. The first generated node is N_1+1*KN_1; the second generated node is N_1+2*KN_1, etc. Generation continues until node number N_2-KN_1 is established. Note that the node difference N_2-N_1 must be evenly divisible by KN_1.

In the generation the boundary condition codes [ID(L,J) values] of the generated nodes are set equal to those of node N_1. The coordinate values of the generated nodes are interpolated linearly.

IV. LOAD DATA CARDS

Each load case requires the following set of cards. The total number of load cases were defined on the CONTROL CARD (Section II).

CARD 1 (2I5)

Note	Columns	Variable	Entry
(1)	1–5	LL	Enter the load case number
(2)	6–10	NLOAD	Enter the number of concentrated loads applied in this load case

NOTES/

(1) Load cases must be input in ascending sequence beginning with 1.

(2) The variable NLOAD defines the number of cards to be read next for this load case.

NEXT CARDS (2I5, F10.0)

Note	Columns	Variable	Entry
(1)	1–5	NOD	Node number to which this load is applied; GE.1 and LE.NUMNP
(2)	6–10	IDIRN	Degree of freedom number for this load component; EQ.1, X—direction EQ.2, Y—direction EQ.3, Z—direction
	11–20	FLOAD	Magnitude of load

NOTES/

(1) For each concentrated load applied in this load case, one card must be supplied.
(2) All loads must be acting into the global X, Y, or Z direction.

V. TRUSS ELEMENTS

TRUSS elements are two-node members allowed arbitrary orientation in the X, Y, Z system. The truss transmits axial force only, and in general is a six degree of freedom element (i.e., three global translation components at each end of the member). The following sequence of cards is input for each element group. The total number of element groups (NUMEG) was defined on the CONTROL CARD (Section II).

V.1 Element Group Control Card (3I5)

Note	Columns	Variable	Entry
	1–5	NPAR(1)	Enter the number 1
(1)	6–10	NPAR(2)	Number of TRUSS elements in this group; NPAR(2) = NUME GE.1
(2)	11–15	NPAR(3)	Number of different sets of section/material properties; NPAR(3) = NUMMAT GE.1 EQ.0, default set to 1

NOTES/

(1) TRUSS element numbers begin with one (1) and end with the total number of elements in this group, NPAR(2). Element data are input in Section V.3.

(2) The variable NPAR(3) defines the number of sets of material/section properties to be read in Section V.2 below.

V.2 Material/Section Property Cards (I5, 2F10.0)

NUMMAT cards are read in this section.

Note	Columns	Variable	Entry
(1)	1–5	N	Number of property set
	6–15	E(N)	Young's modulus
	16–25	AREA(N)	Section area

NOTES/

(1) Property sets are input in ascending sequence beginning with 1 and ending with NUMMAT. The Young's modulus and section area of each TRUSS element defined below is identified by one of the property sets input here.

V.3 Element Data Cards (5I5)

NUME elements must be input and/or generated in this section in ascending sequence beginning with one (1).

Note	Columns	Variable	Entry
	1–5	M	TRUSS element number; GE.1 and LE.NPAR(2)
	6–10	II	Node number at one end
	11–15	JJ	Node number at other end; GE.1 and LE.NUMNP
(1)	16–20	MTYP	Material property set; GE.1 and LE.NPAR(3)
(2)	21–25	KG	Node generation increment used to compute node numbers for missing elements; EQ.0, default set to 1

NOTES/

(1) The material/section property sets have been defined in Section V.2 above.

(2) Elements must be input in increasing element number order. If cards for elements [M+1, M+2, . . . , M+J] are omitted, these "J" missing elements are generated using MTYP of element "M" and by incrementing the node numbers of successive elements with the value KG; KG is taken from the first card of the element generation sequence (i.e., from the Mth element card).

6.4.2 Listing of Program STAP

```
C . . . . . . . . . . . . . . . . . . . . . . . . . . . . . . . . . . . . .
C .                                                                       .
C .                              S T A P                                  .
C .                                                                       .
C .            AN IN-CORE SOLUTION STATIC ANALYSIS PROGRAM                .
C .                                                                       .
C . . . . . . . . . . . . . . . . . . . . . . . . . . . . . . . . . . . . .
      COMMON /SOL/ NUMNP,NEQ,NWK,NUMEST,MIDEST,MAXEST,MK
      COMMON /DIM/ N1,N2,N3,N4,N5,N6,N7,N8,N9,N10,N11,N12,N13,N14,N15
      COMMON /EL/ IND,NPAR(10),NUMEG,MTOT,NFIRST,NLAST,ITWO
      COMMON /VAR/ NG,MODEX
      COMMON /TAPES/ IELMNT,ILOAD,IIN,IOUT
C
      DIMENSION TIM(5), HED(20)
      DIMENSION IA(1)
      EQUIVALENCE (A(1),IA(1))
C . . . . . . . . . . . . . . . . . . . . . . . . . . . . . . . . . . . . .
C     THE FOLLOWING TWO CARDS ARE USED TO DETERMINE THE MAXIMUM HIGH
C     SPEED STORAGE THAT CAN BE USED FOR SOLUTION. TO CHANGE THE HIGH
C     SPEED STORAGE AVAILABLE FOR EXECUTION CHANGE THE VALUE OF MTOT
C     AND CORRESPONDINGLY COMMON A(MTOT)
C . . . . . . . . . . . . . . . . . . . . . . . . . . . . . . . . . . . . .
      COMMON A(10000)
      MTOT=10000
C
C . . . . . . . . . . . . . . . . . . . . . . . . . . . . . . . . . . . . .
C     DOUBLE PRECISION CARD
C       ITWO = 1 SINGLE PRECISION ARITHMETIC
C       ITWO = 2 DOUBLE PRECISION ARITHMETIC
C . . . . . . . . . . . . . . . . . . . . . . . . . . . . . . . . . . . . .
C
      ITWO=2
C
C     THE FOLLOWING SCRATCH FILES ARE USED
C        IELMNT = TAPE STORING ELEMENT DATA
C        ILOAD  = TAPE STORING LOAD VECTORS
C        IIN    = INPUT TAPE
C        IOUT   = OUTPUT TAPE
C
      IELMNT = 1
      ILOAD = 2
      IIN = 5
      IOUT = 6
C
  200 NUMEST=0
      MAXEST=0
C     * * * * * * * * * * * * * * * * * * * * * * *
C
C     * * *   I N P U T   P H A S E   * * *
C
C     * * * * * * * * * * * * * * * * * * * * * * *
      CALL SECOND (TIM(1))
C
C
C
C     R E A D   C O N T R O L   I N F O R M A T I O N
C
C
      READ (IIN,1000)  HED,NUMNP,NUMEG,NLCASE,MODEX
```

```
      IF (NUMNP.EQ.0) STOP
      WRITE (IOUT,2000) HED,NUMNP,NUMEG,NLCASE,MODEX
C
C
C     R E A D   N O D A L   P O I N T   D A T A
C
C
      N1= 1
      N2=N1 + 3 * NUMNP
      N3=N2 + NUMNP *ITWO
      N4=N3 + NUMNP*ITWO
      N5=N4 + NUMNP*ITWO
      IF (N5.GT.MTOT) CALL ERROR (N5 - MTOT,1)
C
      CALL INPUT (A(N1),A(N2),A(N3),A(N4),NUMNP,NEQ)
C
      NEQ1=NEQ + 1
C
C     C A L C U L A T E   A N D   S T O R E   L O A D   V E C T O R S
C
C
      N6=N5 + NEQ*ITWO
      WRITE(IOUT,2005)
C
      REWIND ILOAD
C
      DO 300 L=1,NLCASE
C
      READ (IIN,1010)   LL,NLOAD
C
      WRITE (IOUT,2010) LL,NLOAD
      IF (LL.EQ.L) GO TO 310
      WRITE (IOUT,2020)
      STOP
  310 CONTINUE
C
      N7=N6 + NLOAD
      N8=N7 + NLOAD
      N9=N8 + NLOAD*ITWO
C
      IF (N9.GT.MTOT) CALL ERROR (N9 - MTOT,2)
C
      CALL LOADS (A(N5),A(N6),A(N7),A(N8),A(N1),NLOAD,NEQ)
C
  300 CONTINUE
C
C
C     R E A D , G E N E R A T E   A N D   S T O R E
C     E L E M E N T   D A T A
C
C     CLEAR STORAGE
C
      N6=N5 + NEQ
      DO 10 I=N5,N6
   10 IA(I)=0
      IND=1
C
      CALL ELCAL
C
      CALL SECOND (TIM(2))
C     * * * * * * * * * * * * * * * * * * * * * * *
C
C     * * *   S O L U T I O N   P H A S E   * * *
C
C     * * * * * * * * * * * * * * * * * * * * * * *
C
C     A S S E M B L E   S T I F F N E S S   M A T R I X
C
C
      CALL ADDRES (A(N2),A(N5))
C
      MM=NWK/NEQ
      N3=N2 + NEQ + 1
      N4=N3 + NWK*ITWO
      N5=N4 + NEQ*ITWO
      N6=N5 + MAXEST
      IF (N6.GT.MTOT) CALL ERROR (N6 - MTOT,4)
C
C     WRITE TOTAL SYSTEM DATA
C
      WRITE(IOUT,2025) NEQ,NWK,MK,MM
C
C     IN DATA CHECK ONLY MODE WE SKIP ALL FURTHER CALCULATIONS
C
      IF (MODEX.GT.0) GO TO 100
      CALL SECOND (TIM(3))
      CALL SECOND (TIM(4))
      CALL SECOND (TIM(5))
      GO TO 120
```

```
C
C     CLEAR STORAGE
C
  100 NNL=NWK + NEQ
      CALL CLEAR(A(N3),NNL)
C
C
      IND=2
C
      CALL ASSEM (A(N5))
C
      CALL SECOND (TIM(3))
C
C     T R I A N G U L A R I Z E   S T I F F N E S S   M A T R I X
C
      KTR=1
      CALL COLSCL (A(N3),A(N4),A(N2),NEQ,NWK,NEQ1,KTR)
C
   35 CALL SECOND (TIM(4))
C
      KTR=2
      IND=3
C
      REWIND ILOAD
      DO 400 L=1,NLCASE
C
      CALL LOADV (A(N4),NEQ)
C
C
C     C A L C U L A T I O N   O F   D I S P L A C E M E N T S
C
C
      CALL COLSCL (A(N3),A(N4),A(N2),NEQ,NWK,NEQ1,KTR)
C
      WRITE(IOUT,2015) L
      CALL WRITE (A(N4),A(N1),NEQ,NUMNP)
C
C
C     C A L C U L A T I O N   O F   S T R E S S E S
C
C
      CALL STRESS (A(N5))
C
  400 CONTINUE
C
      CALL SECOND (TIM(5))
C
C     PRINT SOLUTION TIMES
C
  120 TT=0.
      DO 500 I=1,4
      TIM(I)=TIM(I+1) - TIM(I)
  500 TT=TT + TIM(I)
      WRITE (IOUT,2030) HED,(TIM(I),I=1,4),TT
C
C     READ NEXT ANALYSIS CASE
C
      GO TO 200
C
 1000 FORMAT (20A4/4I5)
 1010 FORMAT (2I5)
C
C
 2000 FORMAT (1H1,20A4 ///
     155H C O N T R O L   I N F O R M A T I O N                 //5X,
     255HNUMBER OF NODAL POINTS  . . . . . . . . . . . (NUMNP)   =,I5//5X,
     355HNUMBER OF ELEMENT GROUPS  . . . . . . . . . . (NUMEG)   =,I5//5X,
     455HNUMBER OF LOAD CASES  . . . . . . . . . . . . (NLCASE)  =,I5//5X,
     555HSOLUTION MODE . . . . . . . . . . . . . . . . (MODEX)   =,I5 /5X,
     655H    EQ.0, DATA CHECK                                     /5X,
     755H    EQ.1, EXECUTION                                          )
 2005 FORMAT (1H1,26H L O A D   C A S E   D A T A )
 2010 FORMAT (////4X,33H LOAD CASE NUMBER . . . . . . . . =,I5//5X,
     132HNUMBER OF CONCENTRATED LOADS . = ,I5)
 2015 FORMAT(1H1,9HLOAD CASE,I3)
 2020 FORMAT (1X,40H*** ERROR  LOAD CASES ARE NOT IN ORDER   )
 2025 FORMAT(1H1,
     155HTOTAL SYSTEM DATA                                       ///5X,
     255HNUMBER OF EQUATIONS . . . . . . . . . . . . . . .(NEQ)  =,I5//5X,
     355HNUMBER OF MATRIX ELEMENTS . . . . . . . . . . . .(NWK)  =,I5//5X,
     455HMAXIMUM HALF BANDWIDTH . . . . . . . . . . . . . (MK )  =,I5//5X,
     555HMEAN HALF BANDWIDTH . . . . . . . . . . . . . . .(MM )  =,I5)
 2030 FORMAT (1H1,48H S O L U T I O N   T I M E   L O G   I N   S E C //
     112X,11HFOR PROBLEM//1X,20A4 ////5X,
     251HTIME FOR INPUT PHASE  . . . . . . . . . . . . . . = ,F12.2//5X,
     351HTIME FOR CALCULATION OF STRUCTURE STIFFNESS MATRIX= ,F12.2//5X,
     451HTRIANGULARIZATION OF STIFFNESS MATRIX . . . . . . =,F12.2//5X,
     551HTIME FOR LOAD CASE SOLUTIONS  . . . . . . . . . . =,F12.2///5X,
     651H T O T A L   S O L U T I O N   T I M E . . . . . = ,F12.2)
C
      END
```

```
      SUBROUTINE ERROR(N,I)
C . . . . . . . . . . . . . . . . . . . . . . . . . . . . . . . . . . . . .
C .                                                                       .
C .   P R O G R A M                                                       .
C .        TO PRINT MESSAGES WHEN HIGH-SPEED STORAGE IS EXCEEDED          .
C .                                                                       .
C . . . . . . . . . . . . . . . . . . . . . . . . . . . . . . . . . . . . .
      COMMON /TAPES/ IELMNT,ILOAD,IIN,IOUT
C
      GO TO (1,2,3,4),I
C
    1 WRITE(IOUT,2000)
      GO TO 6
    2 WRITE(IOUT,2010)
      GO TO 6
    3 WRITE(IOUT,2020)
      GO TO 6
    4 WRITE(IOUT,2030)
C
    6 WRITE (IOUT,2050) N
      STOP
 2000 FORMAT (//48H NOT ENOUGH STORAGE FOR READ-IN OF ID ARRAY AND
     123HNODAL POINT COORDINATES)
 2010 FORMAT (//50H NOT ENOUGH STORAGE FOR DEFINITION OF LOAD VECTORS)
 2020 FORMAT (//42H NOT ENOUGH STORAGE FOR ELEMENT DATA INPUT )
 2030 FORMAT (//53H NOT ENOUGH STORAGE FOR ASSEMBLAGE OF GLOBAL STRUCTUR
     155HE STIFFNESS, AND DISPLACEMENT AND STRESS SOLUTION PHASE )
 2050 FORMAT (// 32H *** ERROR   STORAGE EXCEEDED BY, I9)
      END

      SUBROUTINE INPUT (ID,X,Y,Z,NUMNP,NEQ)
C
C . . . . . . . . . . . . . . . . . . . . . . . . . . . . . . . . . . . . .
C .                                                                       .
C .   P R O G R A M                                                       .
C .      .TO READ, GENERATE, AND PRINT NODAL POINT INPUT DATA             .
C .      . TO CALCULATE EQUATION NUMBERS AND STORE THEM IN ID ARRRAY      .
C .                                                                       .
C .            N=ELEMENT NUMBER                                           .
C .            ID=BOUNDARY CONDITION CODES (0=FREE,1=DELETED)             .
C .            X,Y,Z= COORDINATES                                         .
C .            KN= GENERATION CODE                                        .
C .                     I.E. INCREMENT ON NODAL POINT NUMBER              .
C .                                                                       .
C . . . . . . . . . . . . . . . . . . . . . . . . . . . . . . . . . . . . .
C
C
      IMPLICIT REAL*8(A-H,O-Z)
C . . . . . . . . . . . . . . . . . . . . . . . . . . . . . . . . . . . . .
C .        THIS PROGRAM IS USED IN SINGLE PRECISION ARITHMETIC ON        .
C .        CDC EQUIPMENT AND DOUBLE PRECISION ARITHMETIC ON IBM          .
C .        OR UNIVAC MACHINES .ACTIVATE,DEACTIVATE OR ADJUST ABOVE       .
C .        CARD FOR SINGLE OR DOUBLE PRECISION ARITHMETIC                .
C . . . . . . . . . . . . . . . . . . . . . . . . . . . . . . . . . . . .
      COMMON /TAPES/ IELMNT,ILOAD,IIN,IOUT
      DIMENSION X(1),Y(1),Z(1),ID(3,NUMNP)
C
C
C     READ AND GENERATE NODAL POINT DATA
C
      WRITE (IOUT,2000)
      WRITE (IOUT,2010)
      WRITE(IOUT,2020)
      KNOLD=0
      NOLD=0
C
   10 READ  (IIN,1000)  N,(ID(I,N),I=1,3),X(N),Y(N),Z(N),KN
      WRITE (IOUT,2030) N,(ID(I,N),I=1,3),X(N),Y(N),Z(N),KN
      IF (KNOLD.EQ.0) GO TO 50
      NUM=(N-NOLD) / KNOLD
      NUMN=NUM-1
      IF(NUMN.LT.1) GO TO 50
      XNUM=NUM
      DX=(X(N)-X(NOLD))/XNUM
      DY=(Y(N)-Y(NOLD))/XNUM
      DZ=(Z(N)-Z(NOLD))/XNUM
      K=NOLD
      DO 30 J=1,NUMN
      KK=K
      K=K + KNOLD
      X(K)=X(KK)+DX
      Y(K)=Y(KK)+DY
      Z(K)=Z(KK)+DZ
      DO 30 I=1,3
      ID(I,K)=ID(I,KK)
   30 CONTINUE
C
   50 NOLD=N
      KNOLD=KN
      IF(N.NE.NUMNP) GO TO 10
```

```
C
C     WRITE COMPLETE NODAL DATA
C
      WRITE (IOUT,2015)
      WRITE (IOUT,2020)
      DO 200 N=1,NUMNP
  200 WRITE (IOUT,2030) (N,(ID(I,N),I=1,3),X(N),Y(N),Z(N),KN)
C
C     NUMBER UNKNOWNS
C
      NEQ=0
      DO 100 N=1,NUMNP
      DO 100 I=1,3
      IF (ID(I,N)) 110,120,110
  120 NEQ=NEQ + 1
      ID(I,N)=NEQ
      GO TO 100
  110 ID(I,N)=0
  100 CONTINUE
C
C     WRITE EQUATION NUMBERS
C
      WRITE (IOUT,2040) (N,(ID(I,N),I=1,3),N=1,NUMNP)
      RETURN
C
 1000 FORMAT (4I5,3F10.0,I5)
 2000 FORMAT(1H1,33H N O D A L   P O I N T   D A T A   //)
 2010 FORMAT(18H INPUT NODAL DATA //)
 2015 FORMAT(///22H GENERATED NODAL DATA //)
 2020 FORMAT(7H  NODE ,9X,8HBOUNDARY,25X,11HNODAL POINT,17X,
     1 4HMESH/7H NUMBER,5X,16HCONDITION  CODES,21X,11HCOORDINATES,14X,
     2 10HGENERATING/77X,4HCODE//15X,1HX,4X,1HY,4X,1HZ,15X,1HX,12X,1HY,
     3 12X,1HZ,10X,2HKN)
 2030 FORMAT (I5,6X,3I5,6X,3F13.3,3X,I6)
 2040 FORMAT(//17H EQUATION NUMBERS//,4X,4HNODE,9X,
     1 17HDEGREE OF FREEDOM/3X,6HNUMBER//,
     2 5X,1HN,13X,1HX,4X,1HY,4X,1HZ/(1X,I5,9X,3I5))
C
      END

      SUBROUTINE LOADS (R,NOD,IDIRN,FLOAD,ID,NLOAD,NEQ)
C
C
C . . . . . . . . . . . . . . . . . . . . . . . . . . . . . . . . . . .
C .                                                                   .
C .   P R O G R A M                                                   .
C .       . TO READ NODAL LOAD DATA                                   .
C .       . TO CALCULATE THE LOAD VECTOR R FOR EACH LOAD CASE AND     .
C .         WRITE ONTO TAPE ILOAD                                     .
C .                                                                   .
C . . . . . . . . . . . . . . . . . . . . . . . . . . . . . . . . . . .
C
C
      IMPLICIT REAL*8(A-H,O-Z)
C . . . . . . . . . . . . . . . . . . . . . . . . . . . . . . . . . . .
C .       THIS PROGRAM IS USED IN SINGLE PRECISION ARITHMETIC ON      .
C .       CDC EQUIPMENT AND DOUBLE PRECISION ARITHMETIC ON IBM        .
C .       OR UNIVAC MACHINES .ACTIVATE,DEACTIVATE OR ADJUST ABOVE     .
C .       CARD FOR SINGLE OR DOUBLE PRECISION ARITHMETIC              .
C . . . . . . . . . . . . . . . . . . . . . . . . . . . . . . . . . . .
      COMMON /VAR/ NG,MODEX
      COMMON /TAPES/ IELMNT,ILOAD,IIN,IOUT
      DIMENSION R(NEQ),NOD(1),IDIRN(1),FLOAD(1)
      DIMENSION ID(3,1)
C
      WRITE (IOUT,2000)
      READ (IIN,1000) (NOD(I),IDIRN(I),FLOAD(I),I=1,NLOAD)
      WRITE (IOUT,2010) (NOD(I),IDIRN(I),FLOAD(I),I=1,NLOAD)
      IF (MODEX.EQ.0) RETURN
C
      DO 210 I=1,NEQ
  210 R(I)=0.
C
      DO 220 L=1,NLOAD
      LN=NOD(L)
      LI=IDIRN(L)
      II=ID(LI,LN)
      IF (II) 220,220,240
  240 R(II)=R(II) + FLOAD(L)
C
  220 CONTINUE
C
      WRITE (ILOAD) R
C
  200 CONTINUE
C
 1000 FORMAT (2I5,F10.0)
 2000 FORMAT (////4X,30HNODE       DIRECTION       LOAD/
     1 3X,6HNUMBER,19X,9HMAGNITUDE )
 2010 FORMAT (1H0,I6,9X,I4,7X,E12.5)
      RETURN
      END
```

```
      SUBROUTINE ELCAL
C
C . . . . . . . . . . . . . . . . . . . . . . . . . . . . . . . . . . . .
C .                                                                    .
C .   P R O G R A M                                                    .
C .        TO LOOP OVER ALL ELEMENT GROUPS FOR READING,                .
C .        GENERATING AND STORING THE ELEMENT DATA                     .
C .                                                                    .
C . . . . . . . . . . . . . . . . . . . . . . . . . . . . . . . . . . . .
C
      COMMON /SOL/ NUMNP,NEQ,NWK,NUMEST,MIDEST,MAXEST,MK
      COMMON /EL/ IND,NPAR(10),NUMEG,MTOT,NFIRST,NLAST,ITWO
      COMMON /TAPES/ IELMNT,ILOAD,IIN,IOUT
      COMMON A(1)
C
C
      REWIND IELMNT
      WRITE (IOUT,2000)
C
C
C     LOOP OVER ALL ELEMENT GROUPS
C
C
      DO 100 N=1,NUMEG
      IF (N.NE.1) WRITE (IOUT,2010)
C
      READ (IIN,1000) NPAR
C
C
      CALL ELEMNT
C
      IF (MIDEST.GT.MAXEST) MAXEST=MIDEST
C
      WRITE (IELMNT) MIDEST,NPAR,(A(I),I=NFIRST,NLAST)
C
C
  100 CONTINUE
C
      RETURN
C
 1000 FORMAT (10I5)
 2000 FORMAT (1H1,36HE L E M E N T   G R O U P   D A T A   ///)
 2010 FORMAT (1H1)
C
      END
```

```
      SUBROUTINE ELEMNT
C
C . . . . . . . . . . . . . . . . . . . . . . . . . . . . . . . . . . . .
C .                                                                    .
C .   P R O G R A M                                                    .
C .        TO CALL THE APPROPRIATE ELEMENT SUBROUTINE                  .
C .                                                                    .
C . . . . . . . . . . . . . . . . . . . . . . . . . . . . . . . . . . . .
C
      COMMON /EL/ IND,NPAR(10),NUMEG,MTOT,NFIRST,NLAST,ITWO
C
      NPAR1=NPAR(1)
C
      GO TO (1,2,3),NPAR1
C
    1 CALL TRUSS
      RETURN
C
C     OTHER ELEMENT TYPES WOULD BE CALLED HERE, IDENTIFYING EACH
C     ELEMENT TYPE BY A DIFFERENT NPAR(1) PARAMETER
C
    2 RETURN
C
    3 RETURN
C
      END
```

```
      SUBROUTINE COLHT (MHT,ND,LM)
C . . . . . . . . . . . . . . . . . . . . . . . . . . . . . . . . . . . .
C .                                                                    .
C     P R O G R A M                                                    .
C .        TO CALCULATE COLUMN HEIGHTS                                 .
C .                                                                    .
C . . . . . . . . . . . . . . . . . . . . . . . . . . . . . . . . . . . .
C
      COMMON /SOL/ NUMNP,NEQ,NWK,NUMEST,MIDEST,MAXEST,MK
      DIMENSION LM(1),MHT(1)
C
      LS=100000
      DO 100 I=1,ND
      IF (LM(I)) 110,100,110
  110 IF (LM(I)-LS) 120,100,100
  120 LS=LM(I)
  100 CONTINUE
C
      DO 200 I=1,ND
```

```
      II=LM(I)
      IF (II.EQ.0) GO TO 200
      ME=II - LS
      IF (ME.GT.MHT(II)) MHT(II)=ME
  200 CONTINUE
C
      RETURN
      END


      SUBROUTINE ADDRES(MAXA,MHT)
C
C . . . . . . . . . . . . . . . . . . . . . . . . . . . . . . . . . . . .
C .                                                                     .
C .   P R O G R A M                                                     .
C .        TO CALCULATE ADDRESSES OF DIAGONAL ELEMENTS IN BANDED        .
C .        MATRIX WHOSE COLUMN HEIGHTS ARE KNOWN                        .
C .                                                                     .
C .        MHT  = ACTIVE COLUMN HEIGHTS                                 .
C .        MAXA = ADDRESSES OF DIAGONAL ELEMENTS                        .
C .                                                                     .
C . . . . . . . . . . . . . . . . . . . . . . . . . . . . . . . . . . . .
C
      COMMON /SOL/ NUMNP,NEQ,NWK,NUMEST,MIDEST,MAXEST,MK
      DIMENSION MAXA(1),MHT(1)
C
C     CLEAR ARRAY MAXA
C
      NN=NEQ + 1
      DO 20 I=1,NN
   20 MAXA(I)=0.0
C
      MAXA(1)=1
      MAXA(2)=2
      MK=0
      IF (NEQ.EQ.1) GO TO 100
      DO 10 I=2,NEQ
      IF (MHT(I).GT.MK) MK=MHT(I)
   10 MAXA(I+1)=MAXA(I) + MHT(I) + 1
  100 MK=MK + 1
      NWK=MAXA(NEQ+1) - MAXA(1)
C
      RETURN
      END


      SUBROUTINE CLEAR(A,N)
C . . . . . . . . . . . . . . . . . . . . . . . . . . . . . . . . . . . .
C .                                                                     .
C .   P R O G R A M                                                     .
C .        TO CLEAR ARRAY A                                             .
C .                                                                     .
C . . . . . . . . . . . . . . . . . . . . . . . . . . . . . . . . . . . .
      IMPLICIT REAL*8(A-H,O-Z)
C . . . . . . . . . . . . . . . . . . . . . . . . . . . . . . . . . . . .
C .        THIS PROGRAM IS USED IN SINGLE PRECISION ARITHMETIC ON       .
C .        CDC EQUIPMENT AND DOUBLE PRECISION ARITHMETIC ON IBM         .
C .        OR UNIVAC MACHINES .ACTIVATE,DEACTIVATE OR ADJUST ABOVE      .
C .        CARD FOR SINGLE OR DOUBLE PRECISION ARITHMETIC               .
C . . . . . . . . . . . . . . . . . . . . . . . . . . . . . . . . . . . .
      DIMENSION A(1)
      DO 10 I=1,N
   10 A(I)=0.
      RETURN
      END


      SUBROUTINE ASSEM (AA)
C
C . . . . . . . . . . . . . . . . . . . . . . . . . . . . . . . . . . . .
C .                                                                     .
C .   P R O G R A M                                                     .
C .        TO CALL ELEMENT SUBROUTINES FOR ASSEMBAGE OF THE             .
C .        STRUCTURE STIFFNESS MATRIX                                   .
C .                                                                     .
C . . . . . . . . . . . . . . . . . . . . . . . . . . . . . . . . . . . .
C
      COMMON /EL/ IND,NPAR(10),NUMEG,MTOT,NFIRST,NLAST,ITWO
      COMMON /TAPES/ IELMNT,ILOAD,IIN,IOUT
      DIMENSION AA(1)
C
      REWIND IELMNT
C
      DO 200 N=1,NUMEG
      READ (IELMNT)NUMEST,NPAR,(AA(I),I=1,NUMEST)
C
      CALL ELEMNT
C
  200 CONTINUE
      RETURN
C
C
      END
```

```
      SUBROUTINE ADDBAN (A,MAXA,S,LM,ND)
C
C . . . . . . . . . . . . . . . . . . . . . . . . . . . . . . . . . . .
C .                                                                   .
C .   P R O G R A M                                                   .
C .        TO ASSEMBLE UPPER TRIANGULAR ELEMENT STIFFNESS INTO        .
C .        COMPACTED GLOBAL STIFFNESS                                 .
C .                                                                   .
C .         A = GLOBAL STIFFNESS                                      .
C .         S = ELEMENT STIFFNESS                                     .
C .         ND = DEGREES OF FREEDOM IN ELEMENT STIFFNESS              .
C .                                                                   .
C .                   S(1)          S(2)          S(3)      . . .     .
C .         S    =                  S(ND+1)       S(ND+2)   . . .     .
C .                                               S(2*ND)   . . .     .
C .                                                         . . .     .
C .                                                                   .
C .                                                                   .
C .                   A(1)          A(3)          A(6)      . . .     .
C .         A    =                  A(2)          A(5)      . . .     .
C .                                               A(4)      . . .     .
C .                                                         . . .     .
C .                                                                   .
C .                                                                   .
C . . . . . . . . . . . . . . . . . . . . . . . . . . . . . . . . . . .
C
      IMPLICIT REAL*8(A-H,O-Z)
C . . . . . . . . . . . . . . . . . . . . . . . . . . . . . . . . . . .
C .        THIS PROGRAM IS USED IN SINGLE PRECISION ARITHMETIC ON     .
C .        CDC EQUIPMENT AND DOUBLE PRECISION ARITHMETIC ON IBM       .
C .        OR UNIVAC MACHINES .ACTIVATE,DEACTIVATE OR ADJUST ABOVE    .
C .        CARD FOR SINGLE OR DOUBLE PRECISION ARITHMETIC             .
C . . . . . . . . . . . . . . . . . . . . . . . . . . . . . . . . . . .
C
      DIMENSION A(1),MAXA(1),S(1),LM(1)
C
      NDI=0
      DO 200 I=1,ND
      II=LM(I)
      IF (II) 200,200,100
  100 MI=MAXA(II)
      KS=I
      DO 220 J=1,ND
      JJ=LM(J)
      IF (JJ) 220,220,110
  110 IJ=II - JJ
      IF (IJ) 220,210,210
  210 KK=MI + IJ
      KSS=KS
      IF (J.GE.I) KSS=J + NDI
      A(KK)=A(KK) + S(KSS)
  220 KS=KS + ND - J
  200 NDI=NDI + ND - I
C
      RETURN
      END
```

```
      SUBROUTINE COLSOL(A,V,MAXA,NN,NWK,NNM,KKK)
C . . . . . . . . . . . . . . . . . . . . . . . . . . . . . . . . . . .
C .                                                                   .
C .   P R O G R A M                                                   .
C .        TO SOLVE FINITE ELEMENT STATIC EQUILIBRIUM EQUATIONS IN    .
C .        CORE, USING COMPACTED STORAGE AND COLUMN REDUCTION SCHEME  .
C .                                                                   .
C .  - - INPUT VARIABLES - -                                           .
C .        A(NWK)     = STIFFNESS MATRIX STORED IN COMPACTED FORM     .
C .        V(NN)      = RIGHT-HAND-SIDE LOAD VECTOR                   .
C .        MAXA(NNM)  = VECTOR CONTAINING ADDRESSES OF DIAGONAL       .
C .                     ELEMENTS OF STIFFNESS MATRIX IN A             .
C .        NN         = NUMBER OF EQUATIONS                           .
C .        NWK        = NUMBER OF ELEMENTS BELOW SKYLINE OF MATRIX    .
C .        NNM        = NN + 1                                        .
C .        KKK        = INPUT FLAG                                    .
C .            EQ. 1 TRIANGULARIZATION OF STIFFNESS MATRIX            .
C .            EQ. 2 REDUCTION AND BACK-SUBSTITUTION OF LOAD VECTOR   .
C .        IOUT       = NUMBER OF OUTPUT DEVICE                       .
C .                                                                   .
C .  - - OUTPUT - -                                                   .
C .        A(NWK)     = D AND L - FACTORS OF STIFFNESS MATRIX         .
C .        V(NN)      = DISPLACEMENT VECTOR                           .
C .                                                                   .
C . . . . . . . . . . . . . . . . . . . . . . . . . . . . . . . . . . .
      IMPLICIT REAL*8(A-H,O-Z)
C . . . . . . . . . . . . . . . . . . . . . . . . . . . . . . . . . . .
C .        THIS PROGRAM IS USED IN SINGLE PRECISION ARITHMETIC ON     .
C .        CDC EQUIPMENT AND DOUBLE PRECISION ARITHMETIC ON IBM       .
C .        OR UNIVAC MACHINES .ACTIVATE,DEACTIVATE OR ADJUST ABOVE    .
C .        CARD FOR SINGLE OR DOUBLE PRECISION ARITHMETIC             .
C . . . . . . . . . . . . . . . . . . . . . . . . . . . . . . . . . . .
      COMMON /TAPES/ IELMNT,ILOAD,IIN,IOUT
      DIMENSION A(NWK),V(1),MAXA(1)
C
```

```
C      PERFORM L*D*L(T) FACTORIZATION OF STIFFNESS MATRIX
C
       IF(KKK-2)40,150,150
   40 DO 140 N=1,NN
       KN=MAXA(N)
       KL=KN+1
       KU=MAXA(N+1) - 1
       KH=KU - KL
       IF(KH)110,90,50
   50 K=N-KH
       IC=0
       KLT=KU
       DO 80 J=1,KH
       IC=IC + 1
       KLT=KLT - 1
       KI=MAXA(K)
       ND=MAXA(K+1) - KI - 1
       IF(ND)80,80,60
   60 KK=MINO(IC,ND)
       C=0.
       DO 70 L=1,KK
   70 C=C+A(KI+L)*A(KLT+L)
       A(KLT)=A(KLT) - C
   80 K=K+1
   90 K=N
       B=0.
       DO 100 KK=KL,KU
       K=K - 1
       KI=MAXA(K)
       C=A(KK)/A(KI)
       B=B + C*A(KK)
  100 A(KK)=C
       A(KN)=A(KN) - B
  110 IF (A(KN))120,120,140
  120 WRITE(IOUT,2000) N,A(KN)
       STOP
  140 CONTINUE
       RETURN
C
C      REDUCE RIGHT-HAND-SIDE LOAD VECTOR
C
  150 DO 180 N=1,NN
       KL=MAXA(N) + 1
       KU=MAXA(N+1) - 1
       IF(KU-KL)180,160,160
  160 K=N
       C=0.
       DO 170 KK=KL,KU
       K=K - 1
  170 C=C+A(KK)*V(K)
       V(N)=V(N) - C
  180 CONTINUE
C
C      BACK-SUBSTITUTE
C
       DO 200 N=1,NN
       K=MAXA(N)
  200 V(N)=V(N)/A(K)
       IF (NN.EQ.1) RETURN
       N=NN
       DO 230 L=2,NN
       KL=MAXA(N) + 1
       KU=MAXA(N+1) - 1
       IF(KU-KL)230,210,210
  210 K=N
       DO 220 KK=KL,KU
       K=K - 1
  220 V(K)=V(K)-A(KK)*V(N)
  230 N=N-1
       RETURN
 2000 FORMAT(//48H STOP - STIFFNESS MATRIX NOT POSITIVE DEFINITE   ,//
     1             32H NONPOSITIVE PIVOT FOR EQUATION  ,I4,//
     2             10H PIVOT =   ,E20.12 )
       END


       SUBROUTINE LOADV (R,NEQ)
C
C
C . . . . . . . . . . . . . . . . . . . . . . . . . . . . . . . . . . .
C .                                                                   .
C .    P R O G R A M                                                  .
C .         TO OBTAIN THE LOAD VECTOR                                 .
C . . . . . . . . . . . . . . . . . . . . . . . . . . . . . . . . . . .
       IMPLICIT REAL*8(A-H,O-Z)
C . . . . . . . . . . . . . . . . . . . . . . . . . . . . . . . . . . .
C .         THIS PROGRAM IS USED IN SINGLE PRECISION ARITHMETIC ON    .
C .         CDC EQUIPMENT AND DOUBLE PRECISION ARITHMETIC ON IBM      .
C .         OR UNIVAC MACHINES .ACTIVATE,DEACTIVATE OR ADJUST ABOVE   .
C .         CARD FOR SINGLE OR DOUBLE PRECISION ARITHMETIC            .
C . . . . . . . . . . . . . . . . . . . . . . . . . . . . . . . . . . .
C
```

```
      COMMON /TAPES/ IELMNT,ILCAD,IIN,IOUT
      DIMENSION R(NEQ)
C
      READ (ILOAD) R
C
      RETURN
      END


      SUBROUTINE WRITE (DISP,ID,NEQ,NUMNP)
C
C . . . . . . . . . . . . . . . . . . . . . . . . . . . . . . . . . . . .
C .                                                                    .
C .   P R O G R A M                                                    .
C .      TO PRINT DISPLACEMENTS                                        .
C . . . . . . . . . . . . . . . . . . . . . . . . . . . . . . . . . . . .
      IMPLICIT REAL*8(A-H,O-Z)
C . . . . . . . . . . . . . . . . . . . . . . . . . . . . . . . . . . .
C .         THIS PROGRAM IS USED IN SINGLE PRECISION ARITHMETIC ON       .
C .         CDC EQUIPMENT AND DOUBLE PRECISION ARITHMETIC ON IBM         .
C .         OR UNIVAC MACHINES .ACTIVATE,DEACTIVATE OR ADJUST ABOVE      .
C .         CARD FOR SINGLE OR DOUBLE PRECISION ARITHMETIC               .
C . . . . . . . . . . . . . . . . . . . . . . . . . . . . . . . . . . .
C
      COMMON /TAPES/ IELMNT,ILOAD,IIN,IOUT
      DIMENSION DISP(NEQ),ID(3,NUMNP)
      DIMENSION D(3)
C
C     PRINT DISPLACEMENTS
C
      WRITE (IOUT,2000)
      IC=4
C
      DO 100 II=1,NUMNP
      IC=IC + 1
      IF (IC.LT.56) GO TO 105
      WRITE (IOUT,2000)
      IC=4
  105 DO 110 I=1,3
  110 D(I)=0.
C
      DO 120 I=1,3
      KK=ID(I,II)
      IL=I
  120 IF (KK.NE.0) D(IL)=DISP(KK)
C
  100 WRITE (IOUT,2010) II,D
C
C
      RETURN
C
 2000 FORMAT (///, 26H D I S P L A C E M E N T S  // 7H  NODE  9X
     114HX-DISPLACEMENT 4X 14HY-DISPLACEMENT 4X 14HZ-DISPLACEMENT)
 2010 FORMAT (1X,I3, 8X,3E18.6)
C
      END


      SUBROUTINE STRESS (AA)
C
C . . . . . . . . . . . . . . . . . . . . . . . . . . . . . . . . . . . .
C .                                                                    .
C .   P R O G R A M                                                    .
C .        TO CALL THE ELEMENT SUBROUTINE FOR THE CALCULATION OF       .
C .        STRESSES                                                    .
C .                                                                    .
C . . . . . . . . . . . . . . . . . . . . . . . . . . . . . . . . . . . .
C
      COMMON /VAR/ NG,MODEX
      COMMON /EL/ IND,NPAR(10),NUMEG,MTOT,NFIRST,NLAST,ITWO
      COMMON /TAPES/ IELMNT,ILOAD,IIN,IOUT
      DIMENSION AA(1)
C
C
C     LOOP OVER ALL ELEMENT GROUPS
C
C
      REWIND IELMNT
C
      DO 100 N=1,NUMEG
      NG=N
C
      READ (IELMNT) NUMEST,NPAR,(AA(I),I=1,NUMEST)
C
      CALL ELEMNT
C
  100 CONTINUE
C
      RETURN
      END
```

```
      SUBROUTINE TRUSS
C . . . . . . . . . . . . . . . . . . . . . . . . . . . . . . . . . . . .
C .                                                                     .
C .   P R O G R A M                                                     .
C .        TO SET UP STORAGE AND CALL THE TRUSS ELEMENT SUBROUTINE      .
C .                                                                     .
C . . . . . . . . . . . . . . . . . . . . . . . . . . . . . . . . . . . .
C
      COMMON /SOL/ NUMNP,NEQ,NWK,NUMEST,MIDEST,MAXEST,MK
      COMMON /DIM/ N1,N2,N3,N4,N5,N6,N7,N8,N9,N10,N11,N12,N13,N14,N15
      COMMON /EL/ IND,NPAR(10),NUMEG,MTOT,NFIRST,NLAST,ITWO
      CCMMON /TAPES/ IELMNT,ILOAD,IIN,IOUT
      COMMON A(1)
C
      EQUIVALENCE (NPAR(2),NUME),(NPAR(3),NUMMAT)
C
      NFIRST=N6
      IF(IND.GT.1) NFIRST=N5
      N101=NFIRST
      N102=N101 + NUMMAT*ITWO
      N103=N102 + NUMMAT*ITWO
      N104=N103 + 6*NUME
      N105=N104 + 6*NUME*ITWO
      N106=N105 + NUME
      NLAST=N106
C
      IF (IND.GT.1) GO TO 100
      IF(NLAST.GT.MTOT) CALL ERROR(NLAST-MTOT,3)
      GC TO 200
  100 IF (NLAST.GT.MTOT) CALL ERROR(NLAST-MTOT,4)
C
  200 MIDEST=NLAST - NFIRST
C
      CALL RUSS (A(N1),A(N2),A(N3),A(N4),A(N4),A(N5),A(N101),A(N10
     1 2),A(N103),A(N104),A(N105))
C
      RETURN
C
      END
```

```
      SUBROUTINE RUSS (ID,X,Y,Z,U,MHT,E,AREA,LM,XYZ,MATP)
C . . . . . . . . . . . . . . . . . . . . . . . . . . . . . . . . . . . .
C .                                                                     .
C .        TRUSS ELEMENT SUBRCUTINE                                     .
C .                                                                     .
C . . . . . . . . . . . . . . . . . . . . . . . . . . . . . . . . . . . .
C
      IMPLICIT REAL*8(A-H,O-Z)
      SQRT(X)=DSQRT(X)
C . . . . . . . . . . . . . . . . . . . . . . . . . . . . . . . . . . . .
C .        THIS PROGRAM IS USED IN SINGLE PRECISION ARITHMETIC ON       .
C .        CDC EQUIPMENT AND DOUBLE PRECISION ARITHMETIC ON IBM         .
C .        OR UNIVAC MACHINES .ACTIVATE,DEACTIVATE OR ADJUST ABOVE      .
C .        CARDS FOR SINGLE OR DOUBLE PRECISION ARITHMETIC              .
C . . . . . . . . . . . . . . . . . . . . . . . . . . . . . . . . . . . .
      COMMON /SOL/ NUMNP,NEQ,NWK,NUMEST,MIDEST,MAXEST,MK
      CCMMON /DIM/ N1,N2,N3,N4,N5,N6,N7,N8,N9,N10,N11,N12,N13,N14,N15
      COMMON /EL/ IND,NPAR(10),NUMEG,MTOT,NFIRST,NLAST,ITWO
      COMMON /VAR/ NG,MODEX
      COMMON /TAPES/ IELMNT,ILOAD,IIN,IOUT
      COMMON A(1)
C
      REAL A
C
      DIMENSION X(1),Y(1),Z(1),ID(3,1),E(1),AREA(1),LM(6,1),
     1          XYZ(6,1),MATP(1),U(1),MHT(1)
      DIMENSICN DR(3),IPS(1)
      DIMENSION S(21),ST(6),D(3)
C
      EQUIVALENCE (NPAR(1),NPAR1),(NPAR(2),NUME),(NPAR(3),NUMMAT)
      ND=6
C
      GO TO (300,610,900),IND
C
C
C     R E A D   A N D   G E N E R A T E   E L E M E N T
C     I N F O R M A T I O N
C
C     READ MATERIAL INFORMATICN
C
  300 WRITE (IOUT,2000) NPAR1,NUME
      IF (NUMMAT.EQ.0) NUMMAT=1
      WRITE (IOUT,2010) NUMMAT
C
      WRITE (IOUT,2020)
      DO 10 I=1,NUMMAT
      READ (IIN,1000) N,E(N),AREA(N)
   10 WRITE (IOUT,2030) N,E(N),AREA(N)
C
C     READ ELEMENT INFORMATION
C
```

```
      WRITE (ICUT,2040)
      N=1
  100 READ (IIN,1020) M,II,JJ,MTYP,KG
      IF (KG.EQ.0) KG=1
  120 IF(M.NE.N) GO TO 200
      I=II
      J=JJ
      MTYPE=MTYP
      KKK=KG
C
C     SAVE ELEMENT INFORMATION
C
  200 XYZ(1,N)=X(I)
      XYZ(2,N)=Y(I)
      XYZ(3,N)=Z(I)
C
      XYZ(4,N)=X(J)
      XYZ(5,N)=Y(J)
      XYZ(6,N)=Z(J)
C
C
      MATP(N)=MTYPE
C
      DO 390 L=1,6
  390 LM(L,N)=0.
      DO 400 L=1,3
      LM(L,N)=ID(L,I)
  400 LM(L+3,N)=ID(L,J)
C
C     UPDATE COLUMN HEIGHTS AND BANDWIDTH
C
      CALL COLHT (MHT,ND,LM(1,N))
C
      WRITE (ICUT,2050) N,I,J,MTYPE
      IF (N.EQ.NUME) RETURN
      N=N+1
      I=I+KKK
      J=J+KKK
      IF(N.GT.M) GO TO 100
      GO TO 120
C
C
C
C     A S S E M B L E   S T U C T U R E   S T I F F N E S S   M A T R I X
C
C
C
  610 DO 500 N=1,NUME
      MTYPE=MATP(N)
      XL2=0.
      DO 505 L=1,3
      D(L)=XYZ(L,N) - XYZ(L+3,N)
  505 XL2=XL2 + D(L)*D(L)
      XL=SQRT(XL2)
      XX=E(MTYPE)*AREA(MTYPE)*XL
      DO 510 L=1,3
      ST(L)=D(L)/XL2
  510 ST(L+3)=-ST(L)
C
      KL=0
      DO 600 L=1,6
      YY=ST(L)*XX
      DO 600 K=L,6
      KL=KL + 1
  600 S(KL)=ST(K)*YY
      CALL ADCBAN (A(N3),A(N2),S,LM(1,N),ND)
  500 CONTINUE
      RETURN
C
C
C     S T R E S S   C A L C U L A T I O N S
C
C
900   IPRINT=0
      DO 830 N=1,NUME
      IPRINT=IPRINT + 1
      IF (IPRINT.GT.50) IPRINT=1
      IF (IPRINT.EQ.1)
     *WRITE (IOUT,2060) NG
      MTYPE=MATP(N)
      XL2=0.
      DO 820 L=1,3
      D(L) = XYZ(L,N) - XYZ(L+3,N)
  820 XL2=XL2+D(L)*D(L)
      DO 814 L=1,3
      ST(L)=(D(L)/XL2)*E(MTYPE)
  814 ST(L+3)=-ST(L)
      STR=0.0
      DO 806 L=1,3
      I=LM(L,N)
      IF (I.LE.0) GO TO 807
      STR=STR+ST(L)*U(I)
```

```
  807 J=LM(L+3,N)
      IF (J.LE.0) GO TO 806
      STR=STR+ST(L+3)*U(J)
  806 CONTINUE
      P=STR*AREA(MTYPE)
      WRITE (IOUT,2070) N,P,STR
  830 CONTINUE
C
      RETURN
C
 1000 FORMAT (I5,2F10.0)
 1010 FORMAT (2F10.0)
 1020 FORMAT (5I5)
 2000 FORMAT (36H E L E M E N T   D E F I N I T I O N ///,
     1          14H ELEMENT TYPE ,13(2H .),17H( NPAR(1) ) . . =,I5/,
     2          25H       EQ.1, TRUSS ELEMENTS/,
     3          29H       EQ.2, ELEMENTS CURRENTLY/,
     4          25H       EQ.3, NOT AVAILABLE /,
     5          20H NUMBER OF ELEMENTS.,10(2H .),17H( NPAR(2) ) . . =,
     6          I5//)
 2010 FORMAT (42H M A T E R I A L   D E F I N I T I O N      ///,
     1          37H NUMBER OF DIFFERENT SETS OF MATERIAL
     2             /32H AND CROSS-SECTIONAL  CONSTANTS ,
     3                               4(2H .),17H( NPAR(3) ) . . =,I5//)
 2020 FORMAT (///2X,3HSET,7X,6HYOUNGS,6X,15HCROSS-SECTIONAL/
     1 1X,6HNUMBER,5X,7HMODULUS,10X,4HAREA,/15X,1HE,14X,1HA)
 2030 FORMAT (/I5,4X,E12.5,2X,E14.6)
 2040 FORMAT (1H1,40H E L E M E N T   I N F O R M A T I O N   ///,
     1 8H ELEMENT ,5X,4HNODE,5X,4HNODE,7X,8HMATERIAL/,
     2 9H NUMBER-N,6X,1HI,8X,1HJ,7X,10HSET NUMBER/)
 2050 FORMAT (I5,6X,I5,4X,I5,7X,I5)
 2060 FORMAT (////,46H S T R E S S   C A L C U L A T I O N S   F O R   ,
     1 24HE L E M E N T   G R O U P,I4//,2X,
     27HELEMENT,12X,5HFORCE,12X,6HSTRESS /,2X,6HNUMBER/)
 2070 FORMAT (1X,I5,11X,E13.6,4X,E13.6)
C
      END

      SUBROUTINE SECOND(TIM)
C
C     SUBROUTINE TO OBTAIN TIME
C     THIS SUBROUTINE HAS BEEN USED ON THE IBM 370 AT M.I.T.
C
      CALL TIMING(II)
      TIM=FLOAT(II)/100.
C
      RETURN
      END
```

REFERENCES

1. P. V. MARCAL (ed.), *General Purpose Finite Element Computer Programs*, American Society of Mechanical Engineers, New York, N.Y., 1970.

2. J. T. ODEN, R. W. CLOUGH, and Y. YAMAMOTO (eds.), *Advances in Computational Methods in Structural Mechanics and Design*, University of Alabama Press, Huntsville, Ala., 1972.

3. S. T. FENVES, N. PERRONE, J. ROBINSON, and W. C. SCHNOBRICH, (eds.), *Numerical and Computer Methods in Structural Mechanics*, Academic Press, Inc., New York, N.Y., 1973.

4. W. PILKEY, K. SACZALSKI, and H. SCHAEFFER (eds.), *Structural Mechanics Computer Programs*, University Press of Virginia, Charlottesville, Va., 1974.

5. E. L. WILSON, "SAP—A Structural Analysis Program," Report UC SESM 70-20, Department of Civil Engineering, University of California, Berkeley, 1970.

6. E. L. WILSON, "Solid SAP—A Static Analysis Program for Three-Dimensional Solid Structures," Report UC SESM 71-19, Department of Civil Engineering, University of California, Berkeley, 1971.

7. K. J. BATHE, E. L. WILSON, and F. E. PETERSON, "SAP IV—A Structural Analysis Program for Static and Dynamic Response of Linear Systems," Report EERC 73-11, College of Engineering, University of California, Berkeley, June 1973, revised Apr. 1974.

8. K. J. BATHE, "ADINA—A Finite Element Program for Automatic Dynamic Incremental Nonlinear Analysis," Report 82448-1, Acoustics and Vibration Laboratory, Department of Mechanical Engineering, Massachusetts Institute of Technology, Cambridge, Mass., 1975.

9. EAC/EASE2—User Information Manual/Theoretical, Control Data Corporation Publication No. 84002700, Minneapolis, Minn., 1973.

10. EAC/EASE2 Dynamics—User Information Manual/Theoretical, Control Data Corporation (in press).

11. R. ROSEN, "Matrix Bandwidth Minimization," Proceedings, National Conference A.C.M., 1968, pp. 585–595.

12. E. H. CUTHILL and J. M. MCKEE, "Reducing the Bandwidth of Sparse Symmetric Matrices," Proceedings, National Conference A.C.M., 1969, pp. 151–172.

PART III

SOLUTION OF FINITE ELEMENT EQUILIBRIUM EQUATIONS

CHAPTER 7

SOLUTION OF EQUILIBRIUM EQUATIONS IN STATIC ANALYSIS

7.1 INTRODUCTION

So far we have considered the derivation and calculation of the equilibrium equations of a finite element system. This included the selection and calculation of efficient elements, and the efficient assemblage of the element matrices to the global finite element system matrices. However, the overall effectiveness of an analysis depends to a large degree on the numerical procedures used for the solution of the system equilibrium equations. As discussed in Section 5.3, the accuracy of the analysis can, in general, be improved, if a more refined finite element mesh is used. Therefore, in practice, an analyst tends to employ larger and larger finite element systems to approximate the actual structure. However, this means that the cost of an analysis and, in fact, its practical feasibility depend to a considerable degree on the algorithms available for the solution of the resulting systems of equations. Because of the requirement that large systems be solved, much research effort has gone into optimizing the equation solution algorithms. During the early use of the finite element method, equations of the order 100 were in many cases considered of large order. Currently, equations of the order 10,000 are solved without much difficulty.

Depending on the kind and number of elements used in the assemblage and on the topology of the finite element mesh, in a linear static analysis the time for solution of the equilibrium equations can be a large percentage of the total solution time, whereas in dynamic analysis or in nonlinear analysis, this percentage may be still higher. Therefore, if inappropriate techniques for the solution of the equilibrium equations are used, the total cost of analysis is affected a great deal, and indeed the cost may be many times, say 100 times, larger than is necessary.

In addition to considering the actual computer effort that is spent on the solution of the equilibrium equations, it is important to realize that an analysis may, in fact, not be possible if inappropriate numerical procedures are used. This may be the case because the analysis is simply too costly using the slow solution methods. But, more seriously, the analysis may not be possible because the solution procedures are unstable. We will observe that the stability of the solution procedures is particularly important in dynamic analysis.

In this chapter we are concerned with the solution of the simultaneous linear equations that arise in static analysis (see Section 3.2.2), which are given by

$$\mathbf{K}\mathbf{U} = \mathbf{R} \tag{7.1}$$

where $\mathbf{K}$ is the stiffness matrix, $\mathbf{U}$ is the displacement vector, and $\mathbf{R}$ is the load vector of the finite element system. Since $\mathbf{R}$ and $\mathbf{U}$ may be functions of time t, we may also consider (7.1) as the dynamic equilibrium equations of a finite element system in which inertia and velocity-dependent damping forces have been neglected. It should be realized that since velocities and accelerations do not enter (7.1), we can evaluate the displacements at any time t independent of the displacement history, which is not the case in dynamic analysis (see Chapter 8). However, the relation between the solution of (7.1) and an actual dynamic analysis suggests that the algorithms used for the evaluation of $\mathbf{U}$ in (7.1) may also be employed as part of the solution algorithms used in dynamic analysis. This is indeed the case; *we will see in the following chapters that the procedures discussed here will be the basis of the algorithms employed for eigensolutions and direct step-by-step integration.* Therefore, a detailed study of the procedures used to solve (7.1) is very important.

Essentially, there are two different classes of methods for the solution of simultaneous equations: direct solution techniques and iterative solution methods. In a direct solution the relation in (7.1) is solved using a number of steps and operations that are predetermined in an exact manner, whereas iteration is used when an iterative solution method is employed. Either solution scheme will be seen to have certain advantages; however, in almost all applications the direct methods are currently most effective. For this reason, the larger part of this chapter is devoted to effective direct solution techniques.

7.2 DIRECT SOLUTIONS USING ALGORITHMS BASED ON GAUSS ELIMINATION

The most effective direct solution techniques currently used are basically applications of Gauss elimination, which Gauss proposed over a century ago.[1-11] However, although the basic Gauss solution scheme can be applied to almost any set of simultaneous linear equations, the effectiveness in finite element analysis depends on the specific properties of the finite element stiffness matrix: symmetry, positive definiteness, and bandedness.

In the following we consider first the Gauss elimination procedure as it is used in the solution of positive definite, symmetric, and banded systems. We briefly consider the solution of symmetric, indefinite systems in Section 7.2.5.

7.2.1 Introduction to Gauss Elimination

We propose to introduce the Gauss solution procedure by studying the solution of the equations $\mathbf{KU} = \mathbf{R}$ derived in Example 5.8 with the parameters $L = 5$, $EI = 1$, i.e.,

$$\begin{bmatrix} 5 & -4 & 1 & 0 \\ -4 & 6 & -4 & 1 \\ 1 & -4 & 6 & -4 \\ 0 & 1 & -4 & 5 \end{bmatrix} \begin{bmatrix} U_1 \\ U_2 \\ U_3 \\ U_4 \end{bmatrix} = \begin{bmatrix} 0 \\ 1 \\ 0 \\ 0 \end{bmatrix} \tag{7.2}$$

In this case the stiffness matrix $\mathbf{K}$ corresponds to a simply supported beam with four translational degrees of freedom, as shown in Figs. 5.4 and 7.1. (We should recall that the equilibrium equations have been derived by finite differences; but, in this case, they have the same properties as in finite element analysis.)

To solve the equations in (7.2) using Gauss elimination, we proceed in the following systematic steps:

Step 1: Subtract a multiple of equation 1 from equations 2 and 3 to obtain zero elements in the first column of $\mathbf{K}$. This means that $(-\frac{4}{5})$ times the first row is subtracted from the second row, and $(\frac{1}{5})$ times the first row is subtracted from the third row. The resulting equations are

$$\begin{bmatrix} 5 & -4 & 1 & 0 \\ 0 & \frac{14}{5} & -\frac{16}{5} & 1 \\ 0 & -\frac{16}{5} & \frac{29}{5} & -4 \\ 0 & 1 & -4 & 5 \end{bmatrix} \begin{bmatrix} U_1 \\ U_2 \\ U_3 \\ U_4 \end{bmatrix} = \begin{bmatrix} 0 \\ 1 \\ 0 \\ 0 \end{bmatrix} \tag{7.3}$$

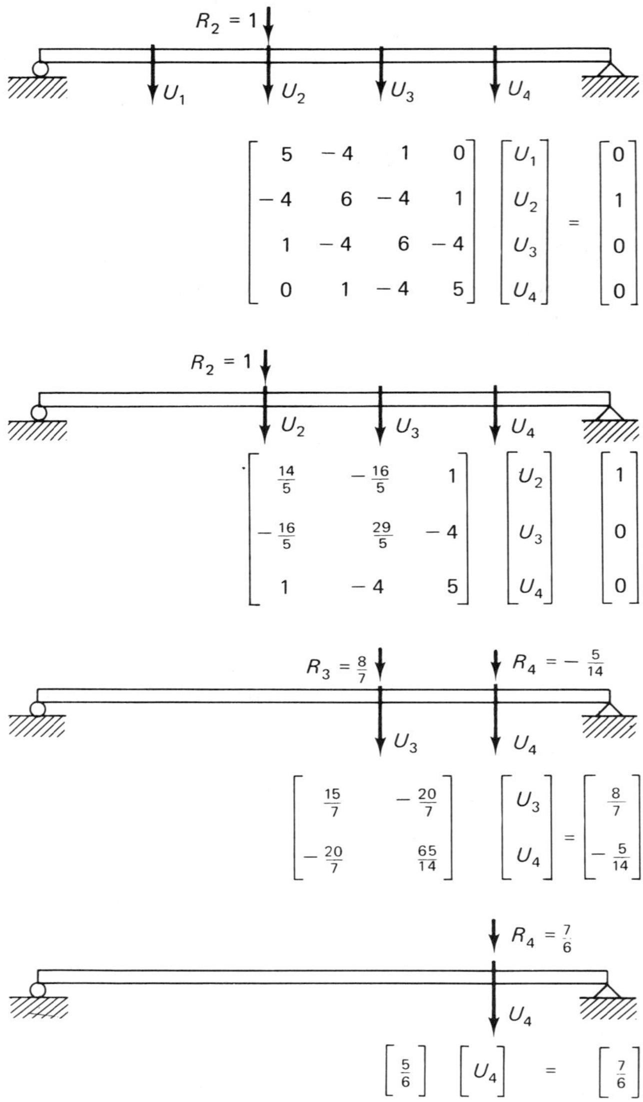

Figure 7.1. Physical systems considered in the Gauss elimination solution of the simply supported beam.

Step 2: Considering next the equations in (7.3), subtract $(-\frac{16}{14})$ times the second equation from the third equation and $(\frac{5}{14})$ times the second equation from the fourth equation. The resulting equations are

$$\begin{bmatrix} 5 & -4 & 1 & 0 \\ 0 & \frac{14}{5} & -\frac{16}{5} & 1 \\ 0 & 0 & \frac{15}{7} & -\frac{20}{7} \\ 0 & 0 & -\frac{20}{7} & \frac{65}{14} \end{bmatrix} \begin{bmatrix} U_1 \\ U_2 \\ U_3 \\ U_4 \end{bmatrix} = \begin{bmatrix} 0 \\ 1 \\ \frac{8}{7} \\ -\frac{5}{14} \end{bmatrix} \tag{7.4}$$

Step 3: Subtract $(-\frac{20}{15})$ times the third equation from the fourth equation in (7.4). This gives

$$\begin{bmatrix} 5 & -4 & 1 & 0 \\ 0 & \frac{14}{5} & -\frac{16}{5} & 1 \\ 0 & 0 & \frac{15}{7} & -\frac{20}{7} \\ 0 & 0 & 0 & \frac{5}{6} \end{bmatrix} \begin{bmatrix} U_1 \\ U_2 \\ U_3 \\ U_4 \end{bmatrix} = \begin{bmatrix} 0 \\ 1 \\ \frac{8}{7} \\ \frac{7}{6} \end{bmatrix} \tag{7.5}$$

Using (7.5) we can now simply solve for the unknowns U_4, U_3, U_2, and U_1:

$$\begin{aligned} U_4 &= \frac{\frac{7}{6}}{\frac{5}{6}} = \frac{7}{5}; \qquad U_3 = \frac{\frac{8}{7} - (-\frac{20}{7})U_4}{\frac{15}{7}} = \frac{12}{5} \\ U_2 &= \frac{1 - (-\frac{16}{5})U_3 - (1)U_4}{\frac{14}{5}} = \frac{13}{5} \\ U_1 &= \frac{0 - (-4)\frac{19}{35} - (1)\frac{36}{15} - (0)\frac{7}{5}}{5} = \frac{8}{5} \end{aligned} \tag{7.6}$$

The procedure in the solution is therefore to subtract in step number i in succession multiples of equation i from equations $i + 1, i + 2, \ldots, n$, where $i = 1, 2, \ldots, n - 1$. In this way the coefficient matrix $\mathbf{K}$ of the equations is reduced to upper triangular form, i.e., a form in which all elements below the diagonal elements are zero. Starting with the last equation, it is then possible to solve for all unknowns in the order $U_n, U_{n-1}, \ldots, U_1$.

It is important to note that at the end of step i the lower right submatrix of order $n - i$ [indicated by dotted lines in (7.3), (7.4), and (7.5)] is symmetric. Therefore, the elements above and including the diagonal can give all elements of the coefficient matrix at all times of the solution. We will see in Section (7.2.3) that in the computer implementation we only work with the upper triangular part of the matrix.

Another important observation is that the above solution assumes in step i a nonzero ith diagonal element in the current coefficient matrix. Namely, it is the nonzero value of the ith diagonal element in the coefficient matrix that makes it possible to reduce the elements below it to zero. Also, in the back-substitution for the solution of the displacements we again divide by the diagonal elements of the coefficient matrix. Fortunately, in the analysis of displacement-based finite element systems, all diagonal elements of the coefficient matrix are positive at all times of the solution, which is another property that makes the application of the Gauss elimination procedure very effective. [This property is not necessarily preserved when the stiffness matrices are derived using hybrid or mixed formulations, or by finite differences (see Section 5.5).] We will prove in Section 7.2.5 that the diagonal elements must remain larger than zero, but this property is also observed by identifying the physical process that is carried out in the Gauss elimination.

In order to identify the physical process corresponding to the mathematical operations in Gauss elimination, we note first that the operations on the coefficient matrix $\mathbf{K}$ are independent of the elements in the load vector $\mathbf{R}$. Therefore, we consider only the operations on the coefficient matrix $\mathbf{K}$. For ease of explanation, let us again use the above example. In this case, $\mathbf{K}$ in (7.2) is the stiffness matrix of the beam element that corresponds to the four degrees of freedom in Fig. 7.1. Since no loads are applied, we now have

$$\begin{bmatrix} 5 & -4 & 1 & 0 \\ -4 & 6 & -4 & 1 \\ 1 & -4 & 6 & -4 \\ 0 & 1 & -4 & 5 \end{bmatrix} \begin{bmatrix} U_1 \\ U_2 \\ U_3 \\ U_4 \end{bmatrix} = \begin{bmatrix} 0 \\ 0 \\ 0 \\ 0 \end{bmatrix} \tag{7.7}$$

Using the condition given by the first equation, i.e.,

$$5U_1 - 4U_2 + U_3 = 0$$

we can write

$$U_1 = \tfrac{4}{5}U_2 - \tfrac{1}{5}U_3 \tag{7.8}$$

and eliminate U_1 from the three equations remaining in (7.7). We thus obtain

$$\begin{aligned} -4(\tfrac{4}{5}U_2 - \tfrac{1}{5}U_3) + 6U_2 - 4U_3 + U_4 &= 0 \\ (\tfrac{4}{5}U_2 - \tfrac{1}{5}U_3) - 4U_2 + 6U_3 - 4U_4 &= 0 \\ U_2 - 4U_3 + 5U_4 &= 0 \end{aligned}$$

or, in matrix form,

$$\begin{bmatrix} \frac{14}{5} & -\frac{16}{5} & 1 \\ -\frac{16}{5} & \frac{29}{5} & -4 \\ 1 & -4 & 5 \end{bmatrix} \begin{bmatrix} U_2 \\ U_3 \\ U_4 \end{bmatrix} = \begin{bmatrix} 0 \\ 0 \\ 0 \end{bmatrix} \tag{7.9}$$

Comparing (7.9) with (7.3), we observe that the coefficient matrix in (7.9) is actually the lower right 3×3 submatrix of the coefficient matrix in (7.3). However, we obtained the coefficient matrix of (7.9) by using (7.7) and the condition in (7.8), which expresses that no force is applied at the degree of freedom 1 of the beam. It follows that the coefficient matrix in (7.9) is, in fact, the stiffness matrix of the beam that corresponds to the degrees of freedom 2, 3, and 4 when no force is applied at the degree of freedom 1, i.e., when the degree of freedom 1 has been released. By the same reasoning, we have obtained in (7.4) the stiffness matrix of the beam when the first two degrees of freedom have been released; and in (7.5), the element (4, 4) of the coefficient matrix represents the stiffness matrix of the beam corresponding to degree of freedom 4 when the degrees of freedom 1, 2, and 3 have all been released.

So far we have assumed that no loads are applied to the structure, since the operations on the stiffness matrix are independent of the operations on the load vector. In case the load vector is not a null vector, the elimination of degrees of freedom proceeds as above, but in this case the equation used to eliminate a displacement variable from the remaining equations involves a load term. In the elimination the effect of this load is carried over to the other degrees of freedom. We may identify that the elimination including the load terms is precisely the Gauss elimination procedure. Therefore, in summary, the physical significance of the Gauss elimination procedure is that n stiffness matrices of order $n, n-1, \ldots, 1$, corresponding to the $(n-i)$ last degrees of freedom, $i = 0, 1, 2, \ldots, n-1$, respectively, are established. In addition, the appropriate load vectors corresponding to the n stiffness matrices are calculated. The unknown displacements are then obtained by considering in succession the systems with only one, two, . . . , degrees of freedom (these correspond to the last, two last, . . . , of the original number of degrees of freedom). Figure 7.1 shows the physical systems considered in the Gauss elimination solution of the simply supported beam.

We can now explain why, because of physical reasons, all diagonal elements in the Gauss elimination procedure must remain positive. This follows because the ith diagonal element is the stiffness at degree of freedom i when the first $(i-1)$ *degrees of freedom of the structure have been released, and this stiffness should be positive. If a zero* (*or negative*) *diagonal element occurs in the Gauss elimination, the structure is not stable.* An example of such a case is shown in Fig. 7.2, where after the release of degrees of freedom U_1, U_2, and U_3, the last diagonal element is zero.

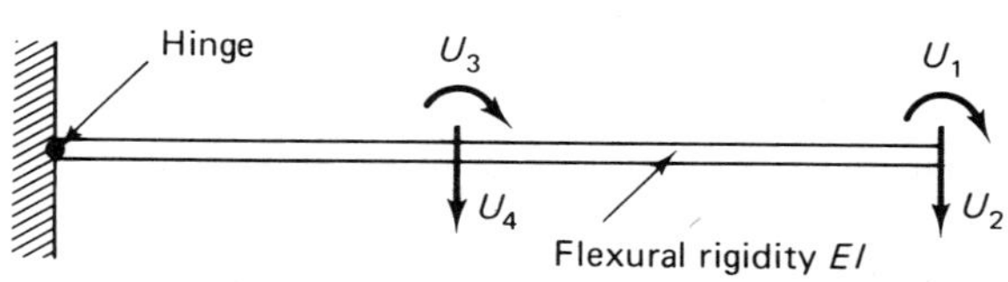

Figure 7.2. Example of an unstable structure.

So far we implied that the elimination proceeds in succession from the first to the $(n-1)$st degree of freedom. However, we may in the same way perform the elimination backward (i.e., from the last to the second degree of freedom) or we may choose any desirable order.

Example 7.1

Obtain the solution to the equilibrium equations of the beam in Fig. 7.1 by eliminating the displacement variables in the order U_3, U_2, U_4.

We may either write down the individual equations during the elimination process, or directly perform Gauss elimination in the prescribed order. In the latter case we obtain, eliminating U_3,

$$\begin{bmatrix} \frac{29}{6} & -\frac{10}{3} & 0 & \frac{2}{3} \\ -\frac{10}{3} & \frac{10}{3} & 0 & -\frac{5}{3} \\ 1 & -4 & 6 & -4 \\ \frac{2}{3} & -\frac{5}{3} & 0 & \frac{7}{3} \end{bmatrix} \begin{bmatrix} U_1 \\ U_2 \\ U_3 \\ U_4 \end{bmatrix} = \begin{bmatrix} 0 \\ 1 \\ 0 \\ 0 \end{bmatrix}$$

Next, we eliminate U_2 and obtain

$$\begin{bmatrix} \frac{3}{2} & 0 & 0 & -1 \\ -\frac{10}{3} & \frac{10}{3} & 0 & -\frac{5}{3} \\ 1 & -4 & 6 & -4 \\ -1 & 0 & 0 & \frac{3}{2} \end{bmatrix} \begin{bmatrix} U_1 \\ U_2 \\ U_3 \\ U_4 \end{bmatrix} = \begin{bmatrix} 1 \\ 1 \\ 0 \\ \frac{1}{2} \end{bmatrix}$$

and finally we eliminate U_4,

$$\begin{bmatrix} \frac{5}{6} & 0 & 0 & 0 \\ -\frac{10}{3} & \frac{10}{3} & 0 & -\frac{5}{3} \\ 1 & -4 & 6 & -4 \\ -1 & 0 & 0 & \frac{3}{2} \end{bmatrix} \begin{bmatrix} U_1 \\ U_2 \\ U_3 \\ U_4 \end{bmatrix} = \begin{bmatrix} \frac{4}{3} \\ 1 \\ 0 \\ \frac{1}{2} \end{bmatrix}$$

The solution for the displacements is now obtained as follows:

$$U_1 = \frac{8}{5}$$

$$U_4 = \frac{\frac{1}{2} - (-1)\frac{8}{5}}{\frac{3}{2}} = \frac{7}{5}$$

$$U_2 = \frac{1 - (-\frac{10}{3})\frac{8}{5} - (-\frac{5}{3})\frac{7}{5}}{\frac{10}{3}} = \frac{13}{5}$$

$$U_3 = \frac{0 - (1)\frac{8}{5} - (-4)\frac{7}{5} - (-4)\frac{13}{5}}{6} = \frac{12}{5}$$

which is the solution already obtained earlier.

7.2.2 The Gauss Elimination Solution

We have seen in the preceding section that the basic procedure of the Gauss elimination solution is to reduce the coefficient matrix of the equations to an upper triangular matrix from which the unknown displacements $\mathbf{U}$ can be calculated by a back-substitution. We now want to formalize the solution procedure using appropriate matrix operations. An additional important purpose of the discussion is to introduce a notation, which can be used throughout the following presentations. The actual computer implementation is given in the next section.

Considering the operations performed in the Gauss elimination solution presented in the preceding section, the reduction of the stiffness matrix $\mathbf{K}$ to upper triangular form can be written

$$\mathbf{L}_{n-1}^{-1} \ldots \mathbf{L}_2^{-1}\mathbf{L}_1^{-1}\mathbf{K} = \mathbf{S} \tag{7.10}$$

where $\mathbf{S}$ is the final upper triangular matrix, and

$$\mathbf{L}_i^{-1} = \begin{bmatrix} 1 & & & & & \\ & \ddots & & & \text{elements not shown are zeros} & \\ & & 1 & & & \\ & & -l_{i+1,i} & \ddots & & \\ & & -l_{i+2,i} & & \ddots & \\ & & \vdots & & & \\ & & -l_{ni} & & & 1 \end{bmatrix}; \qquad l_{i+j,i} = \frac{k_{i+j,i}^{(i)}}{k_{ii}^{(i)}} \tag{7.11}$$

The elements $l_{i+j,i}$ are the Gauss multiplying factors, and the right superscript (i) indicates that an element of the matrix $\mathbf{L}_{i-1}^{-1} \ldots \mathbf{L}_2^{-1}\mathbf{L}_1^{-1}\mathbf{K}$ is used.

We now note that $\mathbf{L}_i$ is obtained by simply reversing the signs of the off-diagonal elements in $\mathbf{L}_i^{-1}$. Therefore, we obtain

$$\mathbf{K} = \mathbf{L}_1\mathbf{L}_2 \ldots \mathbf{L}_{n-1}\mathbf{S} \tag{7.12}$$

where

$$\mathbf{L}_i = \begin{bmatrix} 1 & & & & \\ & \ddots & & & \\ & & 1 & & \\ & & l_{i+1,i} & \ddots & \\ & & l_{i+2,i} & & \ddots \\ & & \vdots & & \\ & & l_{ni} & & 1 \end{bmatrix} \tag{7.13}$$

Hence we can write

$$\mathbf{K} = \mathbf{LS} \tag{7.14}$$

where $\mathbf{L} = \mathbf{L}_1\mathbf{L}_2 \ldots \mathbf{L}_{n-1}$, or

$$\mathbf{L} = \begin{bmatrix} 1 & & & & & \\ l_{21} & 1 & & & & \\ l_{31} & l_{32} & & & & \\ l_{41} & l_{42} & 1 & & & \\ \vdots & \vdots & & \ddots & & \\ & & & & 1 & \\ l_{n1} & \cdots & & & l_{n,n-1} & 1 \end{bmatrix} \tag{7.15}$$

Since **S** is an upper diagonal matrix and the diagonal elements are the pivots in the Gauss elimination, we can write $\mathbf{S} = \mathbf{D}\tilde{\mathbf{S}}$ where **D** is a diagonal matrix storing the diagonal elements of **S**; i.e., $d_{ii} = s_{ii}$. Substituting for **S** into (7.14) and noting that **K** is symmetric and the decomposition is unique, we obtain $\tilde{\mathbf{S}} = \mathbf{L}^T$, and hence

$$\mathbf{K} = \mathbf{LDL}^T \tag{7.16}$$

It is this $\mathbf{LDL}^T$ decomposition of **K** that can be used effectively to obtain the solution to the equations in (7.1) in the following two steps:

$$\mathbf{LV} = \mathbf{R} \tag{7.17}$$

$$\mathbf{DL}^T\mathbf{U} = \mathbf{V} \tag{7.18}$$

where in (7.17) the load vector **R** is reduced to obtain **V**,

$$\mathbf{V} = \mathbf{L}_{n-1}^{-1} \ldots \mathbf{L}_2^{-1}\mathbf{L}_1^{-1}\mathbf{R} \tag{7.19}$$

and in (7.18) the solution **U** is obtained by a back-substitution,

$$\mathbf{L}^T\mathbf{U} = \mathbf{D}^{-1}\mathbf{V} \tag{7.20}$$

In the implementation the vector **V** is very effectively calculated at the same time as the matrices $\mathbf{L}_i^{-1}$ are established. This was done in the example solution of the simply supported beam in Section 7.2.1. However, one point of inefficiency in the solution of the beam should be noted; in (7.6) the matrix **S** was used instead of (7.20), which meant that the multipliers l_{ij} had to be calculated once more.

It should be noted that in practice the matrix multiplications to obtain **L** in (7.15) and **V** in (7.19) are not formally carried out, but that **L** and **V** are established by directly modifying **K** and **R**. This is discussed further in the next section, in which the computer implementation of the solution procedure is presented. However, before proceeding, consider the example of Section 7.2.1 for the derivation of the matrices defined above.

Example 7.2

Establish the matrices $\mathbf{L}_i^{-1}$, $i = 1, 2, 3$, **S**, **L**, and **D** and the vector **V** corresponding to the stiffness matrix and the load vector of the simply supported beam treated in Section 7.2.1.

Using the information given in Section 7.2.1, we can directly write down the required matrices:

$$\mathbf{L}_1^{-1} = \begin{bmatrix} 1 & & & \\ \frac{4}{5} & 1 & & \\ -\frac{1}{5} & 0 & 1 & \\ 0 & 0 & 0 & 1 \end{bmatrix}; \qquad \mathbf{L}_2^{-1} = \begin{bmatrix} 1 & & & \\ 0 & 1 & & \\ 0 & \frac{8}{7} & 1 & \\ 0 & -\frac{5}{14} & 0 & 1 \end{bmatrix}$$

$$\mathbf{L}_3^{-1} = \begin{bmatrix} 1 & & & \\ 0 & 1 & & \\ 0 & 0 & 1 & \\ 0 & 0 & \frac{4}{3} & 1 \end{bmatrix}; \qquad \mathbf{S} = \begin{bmatrix} 5 & -4 & 1 & 0 \\ & \frac{14}{5} & -\frac{16}{5} & 1 \\ & & \frac{15}{7} & -\frac{20}{7} \\ & & & \frac{5}{6} \end{bmatrix}$$

where the matrix $\mathbf{L}_i^{-1}$ stores in the ith column the multipliers that were used in the elimination of the ith equation, and the matrix $\mathbf{S}$ is the upper triangular matrix obtained in (7.5). The matrix $\mathbf{D}$ is a diagonal matrix with the pivot elements on its diagonal. In this case

$$\mathbf{D} = \begin{bmatrix} 5 & & & \\ & \frac{14}{5} & & \\ & & \frac{15}{7} & \\ & & & \frac{5}{6} \end{bmatrix}$$

To obtain $\mathbf{L}$, we use (7.15); hence

$$\mathbf{L} = \begin{bmatrix} 1 & & & \\ -\frac{4}{5} & 1 & & \\ \frac{1}{5} & -\frac{8}{7} & 1 & \\ 0 & \frac{5}{14} & -\frac{4}{3} & 1 \end{bmatrix}$$

and we can check that $\mathbf{S} = \mathbf{D}\mathbf{L}^T$.

The vector $\mathbf{V}$ was obtained in (7.5):

$$\mathbf{V} = \begin{bmatrix} 0 \\ 1 \\ \frac{8}{7} \\ \frac{7}{6} \end{bmatrix}$$

7.2.3 Computer Implementation of Gauss Elimination

A very important aspect in the computer implementation of the Gauss solution procedure is that a minimum solution time should be used. In addition, the high-speed storage requirements should be as small as possible to avoid the use of back-up storage. However, for large systems it will nevertheless be necessary to use back-up storage,[12] and for this reason it should also be possible to modify the solution algorithm for effective out-of-core solution.

As was pointed out in Section 6.2.3, an advantage of finite element analysis is that the stiffness matrix of the element assemblage is not only symmetric and positive definite but is also banded; i.e., $k_{ij} = 0$ for $j > i + m_{\mathbf{K}}$, where $m_{\mathbf{K}}$ is the half-bandwidth of the system. The fact that in finite element analysis all nonzero elements are clustered around the diagonal of the system matrices greatly reduces the total number of operations and the high-speed storage required in the equation solution. However, this property depends on the nodal point numbering of the finite element mesh, and the analyst must take care to obtain an effective nodal point numbering (see Section 6.2.3).

Assume that for a given finite element assemblage a specific nodal point numbering has been determined and the corresponding column heights and the stiffness matrix **K** have been calculated, e.g., as discussed in Section 6.2.3. The $\mathbf{LDL}^T$ decomposition of **K** can be obtained effectively by considering each column in turn; i.e., although the Gauss elimination is carried out by rows, the final elements of **D** and **L** are calculated by columns. Using $d_{11} = k_{11}$, the algorithm for the calculation of the elements l_{ij} and d_{jj} in the jth column is, for $j = 2, \ldots, n$,

$$\left.\begin{aligned} g_{m_j,j} &= k_{m_j,j} \\ g_{ij} &= k_{ij} - \sum_{r=m_m}^{i-1} l_{ri} g_{rj} \qquad i = m_j + 1, \ldots, j-1 \end{aligned}\right\} \tag{7.21}$$

where the elements g_{ij} are only defined as intermediate quantitites, and the calculation is completed using

$$l_{ij} = \frac{g_{ij}}{d_{ii}} \qquad i = m_j, \ldots, j-1 \tag{7.22}$$

$$d_{jj} = k_{jj} - \sum_{r=m_j}^{j-1} l_{rj} g_{rj} \tag{7.23}$$

where $m_m = \max\{m_i, m_j\}$. It should be noted that the summations in (7.21) and (7.23) do not involve multiplications with zero elements outside the skyline of the matrix, and that the l_{ij} are elements of the matrix $\mathbf{L}^T$ rather than of **L**.

Considering the storage arrangements in the reduction, the element l_{ij} when calculated for use in (7.23) immediately replaces g_{ij}, and d_{jj} replaces k_{jj}. Therefore, at the end of the reduction we have elements d_{jj} in the storage locations previously used by k_{jj}, and l_{rj} is stored in the locations of k_{rj}, $j > r$.

In order to get familiar with the solution algorithm, we consider the following examples.

Example 7.3

Use the solution algorithm given in (7.21) to (7.23) to calculate the triangular factors **D** and $\mathbf{L}^T$ of the stiffness matrix of the beam considered in Example 7.2.

The initial elements considered are, when written in their respective matrix locations,

$$\begin{bmatrix} 5 & -4 & 1 & \\ & 6 & -4 & 1 \\ & & 6 & -4 \\ & & & 5 \end{bmatrix}$$

with $m_1 = 1$, $m_2 = 1$, $m_3 = 1$, and $m_4 = 2$. Using (7.21) to (7.23) we obtain, for $j = 2$,

$$d_{11} = k_{11} = 5$$
$$g_{12} = k_{12} = -4$$
$$l_{12} = \frac{g_{12}}{d_{11}} = -\tfrac{4}{5}$$
$$d_{22} = k_{22} - l_{12}g_{12} = 6 - (-4)(-\tfrac{4}{5}) = \tfrac{14}{5}$$

and thus the resulting matrix elements are now, using a dotted line to separate the reduced from the unreduced columns,

$$\begin{bmatrix} 5 & -\frac{4}{5} & \vdots & 1 & \\ & \frac{14}{5} & \vdots & -4 & 1 \\ & & \vdots & 6 & -4 \\ & & \vdots & & 5 \end{bmatrix}$$

Next, we obtain, for $j = 3$,

$$g_{13} = k_{13} = 1$$
$$g_{23} = k_{23} - l_{12}g_{13} = -4 - (-\tfrac{4}{5})(1) = -\tfrac{16}{5}$$
$$l_{13} = \frac{g_{13}}{d_{11}} = \tfrac{1}{5}$$
$$l_{23} = \frac{g_{23}}{d_{22}} = \frac{-\frac{16}{5}}{\frac{14}{5}} = -\tfrac{8}{7}$$
$$d_{33} = k_{33} - l_{13}g_{13} - l_{23}g_{23} = 6 - (\tfrac{1}{5})(1) - (-\tfrac{8}{7})(-\tfrac{16}{5}) = \tfrac{15}{7}$$

and the resulting matrix elements are

$$\begin{bmatrix} 5 & -\frac{4}{5} & \frac{1}{5} & \vdots & \\ & \frac{14}{5} & -\frac{8}{7} & \vdots & 1 \\ & & \frac{15}{7} & \vdots & -4 \\ & & & \vdots & 5 \end{bmatrix}$$

Finally, we have, for $j = 4$,

$$g_{24} = k_{24} = 1$$
$$g_{34} = k_{34} - l_{23}g_{24} = -4 - (-\tfrac{8}{7})(1) = -\tfrac{20}{7}$$
$$l_{24} = \frac{g_{24}}{d_{22}} = \frac{1}{\frac{14}{5}} = \tfrac{5}{14}$$
$$l_{34} = \frac{g_{34}}{d_{33}} = \frac{-\frac{20}{7}}{\frac{15}{7}} = -\tfrac{4}{3}$$
$$d_{44} = k_{44} - l_{24}g_{24} - l_{34}g_{34} = 5 - (\tfrac{5}{14})(1) - (-\tfrac{4}{3})(-\tfrac{20}{7}) = \tfrac{5}{6}$$

and the final elements stored are

$$\begin{bmatrix} 5 & -\frac{4}{5} & \frac{1}{5} & \\ & \frac{14}{5} & -\frac{8}{7} & \frac{5}{14} \\ & & \frac{15}{7} & -\frac{4}{3} \\ & & & \frac{5}{6} \end{bmatrix}$$

We should note that the elements of **D** are stored on the diagonal and the elements l_{ij} have replaced the elements k_{ij}, $j > i$.

Although the solution procedure could already be well demonstrated in Example 7.3, the importance of using the column reduction scheme could not be shown, since the skyline coincides with the band. The effectiveness of the skyline reduction scheme is more apparent in the factorization of the following matrix.

Example 7.4

Use the solution algorithm given in (7.21) to (7.23) to evaluate the triangular factors **D** and $\mathbf{L}^T$ of the stiffness matrix **K**, where

$$\mathbf{K} = \begin{bmatrix} 2 & -2 & & & -1 \\ & 3 & -2 & & 0 \\ & & 5 & -3 & 0 \\ & \text{symmetric} & & 10 & 4 \\ & & & & 10 \end{bmatrix}$$

For this matrix we have $m_1 = 1$, $m_2 = 1$, $m_3 = 2$, $m_4 = 3$, and $m_5 = 1$.
The algorithm gives, in this case with $d_{11} = 2$, for $j = 2$,

$$g_{12} = k_{12} = -2$$

$$l_{12} = \frac{g_{12}}{d_{11}} = \frac{-2}{2} = -1$$

$$d_{22} = k_{22} - l_{12}g_{12} = 3 - (-1)(-2) = 1$$

and thus the resulting matrix elements are

$$\left[\begin{array}{cc:ccc} 2 & -1 & & & -1 \\ & 1 & -2 & & 0 \\ & & 5 & -3 & 0 \\ & & & 10 & 4 \\ & & & & 10 \end{array}\right]$$

For $j = 3$,

$$g_{23} = k_{23} = -2$$

$$l_{23} = \frac{g_{23}}{d_{22}} = \frac{-2}{1} = -2$$

$$d_{33} = k_{33} - l_{23}g_{23} = 5 - (-2)(-2) = 1$$

and the coefficient array is now

$$\left[\begin{array}{ccc:cc} 2 & -1 & & & -1 \\ & 1 & -2 & & 0 \\ & & 1 & -3 & 0 \\ & & & 10 & 4 \\ & & & & 10 \end{array}\right]$$

For $j = 4$,

$$g_{34} = k_{34} = -3$$

$$l_{34} = \frac{g_{34}}{d_{33}} = \frac{-3}{1} = -3$$

$$d_{44} = k_{44} - l_{34}g_{34} = 10 - (-3)(-3) = 1$$

and the resulting matrix elements are

$$\left[\begin{array}{cccc|c} 2 & -1 & & & -1 \\ & 1 & -2 & & 0 \\ & & 1 & -3 & 0 \\ & & & 1 & 4 \\ & & & & 10 \end{array}\right]$$

Finally we have, for $j = 5$,

$$g_{15} = k_{15} = -1$$

$$g_{25} = k_{25} - l_{12}g_{15} = 0 - (-1)(-1) = -1$$

$$g_{35} = k_{35} - l_{23}g_{25} = 0 - (-2)(-1) = -2$$

$$g_{45} = k_{45} - l_{34}g_{35} = +4 - (-3)(-2) = -2$$

$$l_{15} = \frac{g_{15}}{d_{11}} = \frac{-1}{2} = -\frac{1}{2}$$

$$l_{25} = \frac{g_{25}}{d_{22}} = \frac{-1}{1} = -1$$

$$l_{35} = \frac{g_{35}}{d_{33}} = \frac{-2}{1} = -2$$

$$l_{45} = \frac{g_{45}}{d_{44}} = \frac{-2}{1} = -2$$

$$\begin{aligned} d_{55} &= k_{55} - l_{15}g_{15} - l_{25}g_{25} - l_{35}g_{35} - l_{45}g_{45} \\ &= 10 - (-\tfrac{1}{2})(-1) - (-1)(-1) - (-2)(-2) - (-2)(-2) = \tfrac{1}{2} \end{aligned}$$

and the final matrix elements are

$$\begin{bmatrix} 2 & -1 & & & -\frac{1}{2} \\ & 1 & -2 & & -1 \\ & & 1 & -3 & -2 \\ & & & 1 & -2 \\ & & & & \frac{1}{2} \end{bmatrix}$$

As in Example 7.3, we have the elements of $\mathbf{D}$ and $\mathbf{L}^T$ replacing the elements k_{ii} and k_{ij}, $j > i$, of the original matrix $\mathbf{K}$, respectively.

In the preceding discussion we considered only the decomposition of the stiffness matrix **K**, which constitutes the main part of the equation solution. Once the **L**, **D** factors of **K** have been obtained, the solution for **U** is calculated using (7.19) and (7.20), where it may be noted that the reduction of $\mathbf{R}$ in (7.19) can be performed at the same time as the stiffness matrix $\mathbf{K}$ is decomposed or may be carried out separately afterward. The equation to be used is similar to (7.23); i.e., we have $V_1 = R_1$ and calculate for $i = 2, \ldots, n$,

$$V_i = R_i - \sum_{r=m_i}^{i-1} l_{ri} V_r \tag{7.24}$$

where R_i and V_i are the ith elements of $\mathbf{R}$ and $\mathbf{V}$. Considering the storage arrangements, the element V_i replaces R_i.

The back-substitution in (7.20) is performed by evaluating successively $U_n, U_{n-1}, \ldots, U_1$. This is achieved by first calculating $\bar{\mathbf{V}}$, where $\bar{\mathbf{V}} = \mathbf{D}^{-1}\mathbf{V}$. Then using that $\bar{\mathbf{V}}^{(n)} = \bar{\mathbf{V}}$, we have $U_n = \bar{V}_n^{(n)}$, and we calculate for $i = n, \ldots, 2$,

$$\left.\begin{aligned} \bar{V}_r^{(i-1)} &= \bar{V}_r^{(i)} - l_{ri} U_i; \qquad r = m_i, \ldots, i-1 \\ U_{i-1} &= \bar{V}_{i-1}^{(i-1)} \end{aligned}\right\} \tag{7.25}$$

where the superscript $(i-1)$ indicates that the element is calculated in the evaluation of U_{i-1}. It should be noted that $\bar{V}_k^{(j)}$ for all j is stored in the storage location of V_k, i.e., the original storage location of R_k.

Example 7.5

Use the algorithm given in (7.24) and (7.25) to calculate the solution to the problem $\mathbf{KU} = \mathbf{R}$, when $\mathbf{K}$ is the stiffness matrix considered in Example 7.4 and

$$\mathbf{R} = \begin{bmatrix} 0 \\ 1 \\ 0 \\ 0 \\ 0 \end{bmatrix}$$

In the solution we employ the $\mathbf{D}$, $\mathbf{L}^T$ factors of $\mathbf{K}$ calculated in Example 7.4. Using (7.24) for the forward reduction we obtain

$$\begin{aligned} V_1 &= R_1 = 0 \\ V_2 &= R_2 - l_{12}V_1 = 1 - 0 = 1 \\ V_3 &= R_3 - l_{23}V_2 = 0 - (-2)(1) = 2 \\ V_4 &= R_4 - l_{34}V_3 = 0 - (-3)(2) = 6 \\ V_5 &= R_5 - l_{15}V_1 - l_{25}V_2 - l_{35}V_3 - l_{45}V_4 \\ &= 0 - 0 - (-1)(1) - (-2)(2) - (-2)(6) = 17 \end{aligned}$$

Immediately after calculation of V_i, the element replaces R_i. Thus we now have in the vector that initially stored the loads,

$$\mathbf{V} = \begin{bmatrix} 0 \\ 1 \\ 2 \\ 6 \\ 17 \end{bmatrix}$$

The first step in the back-substitution is to evaluate $\bar{\mathbf{V}}$, where $\bar{\mathbf{V}} = \mathbf{D}^{-1}\mathbf{V}$. Here we obtain

$$\bar{\mathbf{V}} = \begin{bmatrix} 0 \\ 1 \\ 2 \\ 6 \\ 34 \end{bmatrix} \tag{a}$$

and thus

$$U_5 = \bar{V}_5 = 34$$

Now we use (7.25) with $\bar{\mathbf{V}}^{(5)} = \bar{\mathbf{V}}$ of (a). Hence we obtain, for $i = 5$,

$$\begin{aligned}
\bar{V}_1^{(4)} &= \bar{V}_1^{(5)} - l_{15}U_5 = 0 - (-\tfrac{1}{2})(34) = 17 \\
\bar{V}_2^{(4)} &= \bar{V}_2^{(5)} - l_{25}U_5 = 1 - (-1)(34) = 35 \\
\bar{V}_3^{(4)} &= \bar{V}_3^{(5)} - l_{35}U_5 = 2 - (-2)(34) = 70 \\
\bar{V}_4^{(4)} &= \bar{V}_4^{(5)} - l_{45}U_5 = 6 - (-2)(34) = 74
\end{aligned}$$

and

$$U_4 = \bar{V}_4^{(4)} = 74$$

For $i = 4$,

$$\bar{V}_3^{(3)} = \bar{V}_3^{(4)} - l_{34}U_4 = 70 - (-3)(74) = 292$$

and

$$U_3 = \bar{V}_3^{(3)} = 292$$

For $i = 3$,

$$\bar{V}_2^{(2)} = \bar{V}_2^{(3)} - l_{23}U_3 = 35 - (-2)(292) = 619$$

and

$$U_2 = \bar{V}_2^{(2)} = 619$$

For $i = 2$,

$$\bar{V}_1^{(1)} = \bar{V}_1^{(2)} - l_{12}U_2 = 17 - (-1)(619) = 636$$

and

$$U_1 = \bar{V}_1^{(1)} = 636$$

The elements stored in the vector that initially stored the loads are after step $i = 5, 4, 3, 2$, respectively:

$$\begin{bmatrix} 17 \\ 35 \\ 70 \\ 74 \\ 34 \end{bmatrix}; \quad \begin{bmatrix} 17 \\ 35 \\ 292 \\ 74 \\ 34 \end{bmatrix}; \quad \begin{bmatrix} 17 \\ 619 \\ 292 \\ 74 \\ 34 \end{bmatrix}; \quad \begin{bmatrix} 636 \\ 619 \\ 292 \\ 74 \\ 34 \end{bmatrix}$$

where the last vector gives the solution $\mathbf{U}$.

Considering the effectiveness of the solution algorithm in (7.21) to (7.25), it should be noted that for a specific matrix $\mathbf{K}$, the algorithm gives an efficient solution, because no operations are performed on zero elements outside the skyline, which also implies that only the elements below the skyline need be stored. However, the total number of operations performed is not an absolute minimum, because, in addition, all those multiplications could be skipped in (7.21) to (7.25) for which l_{ri} or g_{rj} is zero. This skipping, though, would require time-consuming, additional testing and logic in the equation solution. In this context, it should also be realized that we consider a specific matrix **K**, and that by a different nodal point numbering a different stiffness matrix is obtained, which might yield a more effective solution (see Section 6.2.3).

In practice, it is important to be able to estimate the number of operations that are carried out in the solution, because this enables the analyst to estimate the computer cost for a specific analysis. Define one operation to consist of one multiplication (or division), which is nearly always followed by an addition. For an operation count estimate we consider a system with a half-bandwidth $m_{\mathbf{K}}$ that is full; i.e., $m_{\mathbf{K}} = i - m_i$ for all i. In this case the number of operations required for the $\mathbf{LDL}^T$ decomposition of **K** are approximately $n\{m_{\mathbf{K}} + (m_{\mathbf{K}} - 1) + \ldots + 1\} = \frac{1}{2}nm_{\mathbf{K}}^2$, and for the reduction and back-substitution of a load vector, an additional number of approximately $2nm_{\mathbf{K}}$ operations are needed. In practice, a constant bandwidth is encountered rather seldom. However, the above values can still be employed to estimate the approximate number of operations required, if a "mean" half-bandwidth is used for $m_{\mathbf{K}}$. Also, an additional purpose of performing operation counts is to compare the effectiveness of different solution schemes, and for this purpose, systems with full half-bandwidths can be considered.

The solution algorithm in (7.21) to (7.25) has been presented in two-dimensional matrix notation; e.g., element (r, j) of **K** has been denoted by k_{rj}. Also, to demonstrate the working of the algorithm, in Examples 7.3 and 7.4 the elements considered in the reduction have been displayed in their corresponding matrix locations. However, in actual computer solution, the active columns of the matrix **K** are stored effectively in a one-dimensional array. Assume that the storage scheme discussed in Section 6.2.3 is used; i.e., the pertinent elements of **K** are stored in the one-dimensional array A of length NWK and the addresses of the diagonal elements of **K** are stored in MAXA. An effective subroutine that uses the algorithm presented above [i.e., the relations in (7.21) to (7.25)] but operates on the stiffness matrix using the storage scheme described in Section 6.2.3 is given next.

Subroutine COLSOL

Program COLSOL is used to obtain the $\mathbf{LDL}^T$ factorization of a stiffness matrix or reduce and back-substitute the load vector. The complete process gives the solution of the finite element equilibrium equations. The argument variables and use of the subroutine are defined by means of comment cards in the program.

```
      SUBROUTINE COLSOL(A,V,MAXA,NN,NWK,NNM,KKK,IOUT)
C . . . . . . . . . . . . . . . . . . . . . . . . . . . . . . . . . .
C .                                                                  .
C .   P R O G R A M                                                  .
C .        TO SOLVE FINITE ELEMENT STATIC EQUILIBRIUM EQUATIONS IN   .
C .        CORE, USING COMPACTED STORAGE AND COLUMN REDUCTION SCHEME .
C .                                                                  .
C .  - - INPUT VARIABLES - -                                         .
C .        A(NWK)     = STIFFNESS MATRIX STORED IN COMPACTED FORM    .
C .        V(NN)      = RIGHT-HAND-SIDE LOAD VECTOR                  .
C .        MAXA(NNM)   = VECTOR CONTAINING ADDRESSES OF DIAGONAL     .
C .                      ELEMENTS OF STIFFNESS MATRIX IN A           .
C .        NN         = NUMBER OF EQUATIONS                          .
C .        NWK        = NUMBER OF ELEMENTS BELOW SKYLINE OF MATRIX   .
C .        NNM        = NN + 1                                       .
C .        KKK        = INPUT FLAG                                   .
C .            EQ. 1 TRIANGULARIZATION OF STIFFNESS MATRIX           .
C .            EQ. 2 REDUCTION AND BACK-SUBSTITUTION OF LOAD VECTOR  .
C .        IOUT       = NUMBER OF OUTPUT DEVICE                      .
C .                                                                  .
C .  - - OUTPUT - -                                                  .
C .        A(NWK)     = D AND L - FACTORS OF STIFFNESS MATRIX        .
C .        V(NN)      = DISPLACEMENT VECTOR                          .
C .                                                                  .
C . . . . . . . . . . . . . . . . . . . . . . . . . . . . . . . . . .
      IMPLICIT REAL*8(A-H,O-Z)
C . . . . . . . . . . . . . . . . . . . . . . . . . . . . . . . . . .
C .        THIS PROGRAM IS USED IN SINGLE PRECISION ARITHMETIC ON    .
C .        CDC EQUIPMENT AND DOUBLE PRECISION ARITHMETIC ON IBM      .
C .        OR UNIVAC MACHINES .ACTIVATE,DEACTIVATE OR ADJUST ABOVE   .
C .        CARD FOR SINGLE OR DOUBLE PRECISION ARITHMETIC            .
C . . . . . . . . . . . . . . . . . . . . . . . . . . . . . . . . . .
      DIMENSION A(NWK),V(1),MAXA(1)
C
C     PERFORM L*D*L(T) FACTORIZATION OF STIFFNESS MATRIX
C
      IF(KKK-2) 40,150,150
   40 DO 140 N=1,NN
      KN=MAXA(N)
      KL=KN+1
      KU=MAXA(N+1) - 1
      KH=KU - KL
      IF(KH)110,90,50
   50 K=N-KH
      IC=0
      KLT=KU
      DO 80 J=1,KH
      IC=IC + 1
      KLT=KLT - 1
      KI=MAXA(K)
      ND=MAXA(K+1) - KI - 1
      IF(ND)80,80,60
   60 KK=MIN0(IC,ND)
      C=0.
      DO 70 L=1,KK
   70 C=C+A(KI+L)*A(KLT+L)
      A(KLT)=A(KLT) - C
   80 K=K+1
   90 K=N
      B=0.
      DO 100 KK=KL,KU
      K=K - 1
      KI=MAXA(K)
      C=A(KK)/A(KI)
      B=B + C*A(KK)
  100 A(KK)=C
      A(KN)=A(KN) - B
  110 IF (A(KN)) 120,120,140
  120 WRITE(IOUT,2000) N,A(KN)
      STOP
  140 CONTINUE
      RETURN
C
C     REDUCE RIGHT-HAND-SIDE LOAD VECTOR
C
  150 DO 180 N=1,NN
      KL=MAXA(N) + 1
      KU=MAXA(N+1) - 1
      IF(KU-KL) 180,160,160
  160 K=N
      C=0.
      DO 170 KK=KL,KU
      K=K - 1
  170 C=C+A(KK)*V(K)
      V(N)=V(N) - C
  180 CONTINUE
C
C     BACK-SUBSTITUTE
C
      DO 200 N=1,NN
      K=MAXA(N)
  200 V(N)=V(N)/A(K)
```

```
      IF (NN.EQ.1) RETURN
      N=NN
      DO 230 L=2,NN
      KL=MAXA(N) + 1
      KU=MAXA(N+1) - 1
      IF (KU-KL) 230,210,210
  210 K=N
      DO 220 KK=KL,KU
      K=K - 1
  220 V(K)=V(K)-A(KK)*V(N)
  230 N=N-1
      RETURN
 2000 FORMAT(//48H STOP - STIFFNESS MATRIX NCT POSITIVE DEFINITE   ,//
     1          32H NONPOSITIVE PIVOT FOR EQUATICN  ,I4,//
     2          10H PIVOT =  ,E20.12 )
      END
```

7.2.4 Cholesky Factorization, Static Condensation, Substructures, and Frontal Solution

In addition to the $\mathbf{LDL}^T$ decomposition described in the preceding sections, various other schemes are used that are closely related. All methods are applications of the basic Gauss elimination procedure.

In the *Cholesky factorization* the stiffness matrix is decomposed as follows:[7, 8]

$$\mathbf{K} = \tilde{\mathbf{L}}\tilde{\mathbf{L}}^T \tag{7.26}$$

where

$$\tilde{\mathbf{L}} = \mathbf{L}\mathbf{D}^{1/2} \tag{7.27}$$

Therefore, the Cholesky factors could be calculated from the $\mathbf{D}$ and $\mathbf{L}$ factors, but, more generally, the elements of $\tilde{\mathbf{L}}$ are calculated directly. An operation count shows that slightly more operations are required in the equation solution if the Cholesky factorization is used rather than the $\mathbf{LDL}^T$ decomposition. In addition, the Cholesky factorization is only suitable for the solution of positive definite systems, for which all diagonal elements d_{ii} are positive, because, otherwise, complex arithmetic would be required. On the other hand, the $\mathbf{LDL}^T$ decomposition can also be used effectively on indefinite systems (see Section 7.2.5).

Considering a main use of the Cholesky factorization, it should be noted that the decomposition is used effectively in the transformation of a generalized eigenproblem to the standard form (see Section 10.2.5).

Example 7.6

Calculate the Cholesky factor $\tilde{\mathbf{L}}$ of the stiffness matrix $\mathbf{K}$ of the simply supported beam treated in Section 7.2.1 and in Examples 7.1, 7.2, and 7.3.

The $\mathbf{L}$ and $\mathbf{D}$ factors of the beam stiffness matrix have been given in Example 7.2. Rounding to three significant decimals, we have

$$\mathbf{L} = \begin{bmatrix} 1.000 & & & 0 \\ -0.800 & 1.000 & & \\ 0.200 & -1.143 & 1.000 & \\ 0.000 & 0.357 & -1.333 & 1.000 \end{bmatrix}; \quad \mathbf{D} = \begin{bmatrix} 5.000 & & & 0 \\ & 2.800 & & \\ & 0 & 2.143 & \\ & & & 0.833 \end{bmatrix}$$

Hence

$$\tilde{\mathbf{L}} = \begin{bmatrix} 1.000 & & & \\ -0.800 & 1.000 & & \\ 0.200 & -1.143 & 1.000 & \\ 0.000 & 0.357 & -1.333 & 1.000 \end{bmatrix} \begin{bmatrix} 2.236 & & & \\ & 1.673 & & \\ & & 1.464 & \\ & & & 0.913 \end{bmatrix}$$

or

$$\tilde{\mathbf{L}} = \begin{bmatrix} 2.236 & & & \\ -1.789 & 1.673 & & \\ 0.447 & -1.912 & 1.464 & \\ 0 & 0.597 & -1.952 & 0.913 \end{bmatrix}$$

An algorithm that in some cases can effectively be used in the solution of the equilibrium equations is *static condensation.*[13-15] The name "static condensation" was coined in dynamic analysis, for which the solution technique is demonstrated in Section 10.3.1. *Static condensation is employed to reduce the number of element degrees of freedom and thus, in effect, to perform part of the solution of the total finite element system equilibrium equations prior to assembling the structure matrices* **K** *and* **R**. Consider the three-node truss element in Fig. 7.3. Since the degree of freedom at the midnode does not correspond to a degree of freedom of any other element, we can eliminate it to obtain the element stiffness matrix that corresponds to the degrees of freedom 1 and 3 only. The elimination of the degree of freedom 2 is carried out using, in essence, Gauss elimination, as presented in Section 7.2.1 (see Example 7.1).

In order to establish the equations used in static condensation, we assume that the stiffness matrix and corresponding displacement and force vectors of the element under consideration are partitioned into the form

$$\begin{bmatrix} \mathbf{K}_{aa} & \mathbf{K}_{ac} \\ \mathbf{K}_{ca} & \mathbf{K}_{cc} \end{bmatrix} \begin{bmatrix} \mathbf{U}_a \\ \mathbf{U}_c \end{bmatrix} = \begin{bmatrix} \mathbf{R}_a \\ \mathbf{R}_c \end{bmatrix} \tag{7.28}$$

where $\mathbf{U}_a$ and $\mathbf{U}_c$ are the vectors of displacements to be retained and condensed out, respectively. The matrices $\mathbf{K}_{aa}$, $\mathbf{K}_{ac}$, and $\mathbf{K}_{cc}$, and vectors $\mathbf{R}_a$ and $\mathbf{R}_c$ correspond to the displacement vectors $\mathbf{U}_a$ and $\mathbf{U}_c$.

Using the second matrix equation in (7.28), we obtain

$$\mathbf{U}_c = \mathbf{K}_{cc}^{-1}(\mathbf{R}_c - \mathbf{K}_{ca}\mathbf{U}_a) \tag{7.29}$$

The relation in (7.29) is used to substitute for $\mathbf{U}_c$ into the first matrix equation in (7.28) to obtain the condensed equations

$$(\mathbf{K}_{aa} - \mathbf{K}_{ac}\mathbf{K}_{cc}^{-1}\mathbf{K}_{ca})\mathbf{U}_a = \mathbf{R}_a - \mathbf{K}_{ac}\mathbf{K}_{cc}^{-1}\mathbf{R}_c \tag{7.30}$$

Comparing (7.30) with the Gauss solution scheme introduced in Section 7.2.1, it is seen that static condensation is, in fact, Gauss elimination on the degrees of freedom $\mathbf{U}_c$ (see Example 7.7). In practice, therefore, static condensation is carried out effectively by using Gauss elimination sequentially on each degree of freedom to be condensed out, instead of following through the formal matrix procedure given in (7.28) to (7.30). The physical meaning of Gauss elimination was discussed in Section 7.2.1. Since the system stiffness matrix is obtained by direct addition of the element stiffness matrices, we realize that when condensing out element degrees of freedom, in fact, part of the total Gauss solution is already carried out on the element level.

The advantage of using static condensation on the element level is that the order of the system matrices is reduced, which may mean that use of back-up storage is prevented. In addition, if subsequent elements are identical, the stiffness matrix of only the first element need be derived, and performing static condensation on the element internal degrees of freedom also reduces the computer effort required. It should be noted, though, that if static condensation is actually carried out for each element (and no advantage is taken of possible identical finite elements), the total effort involved in the static condensation on all element stiffness matrices plus the Gauss elimination solution of the resulting assembled equilibrium equations is, in fact, the same as using Gauss elimination on the system equations established from the uncondensed element stiffness matrices.

Example 7.7

The stiffness matrix of the truss element in Fig. 7.3 is given below. Use static condensation as given in (7.28) to (7.30) to condense out the internal element degree of freedom. Then use Gauss elimination directly on the internal degree of freedom.

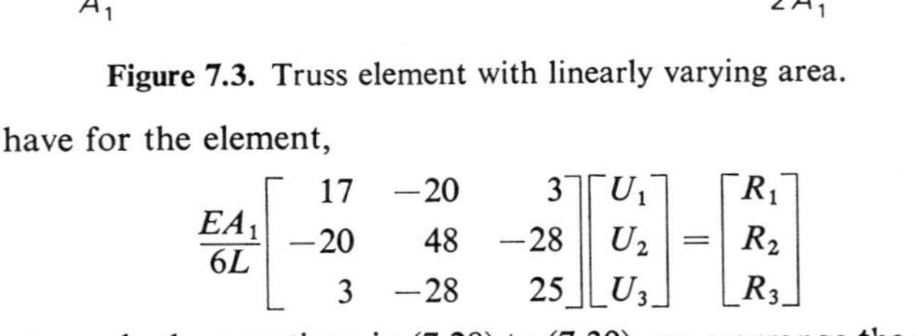

Figure 7.3. Truss element with linearly varying area.

We have for the element,

$$\frac{EA_1}{6L}\begin{bmatrix} 17 & -20 & 3 \\ -20 & 48 & -28 \\ 3 & -28 & 25 \end{bmatrix}\begin{bmatrix} U_1 \\ U_2 \\ U_3 \end{bmatrix} = \begin{bmatrix} R_1 \\ R_2 \\ R_3 \end{bmatrix} \tag{a}$$

In order to apply the equations in (7.28) to (7.30), we rearrange the equations in (a) to obtain

$$\frac{EA_1}{6L}\begin{bmatrix} 17 & 3 & -20 \\ 3 & 25 & -28 \\ -20 & -28 & 48 \end{bmatrix}\begin{bmatrix} U_1 \\ U_3 \\ U_2 \end{bmatrix} = \begin{bmatrix} R_1 \\ R_3 \\ R_2 \end{bmatrix}$$

The relation in (7.30) now gives

$$\frac{EA_1}{6L}\left\{\begin{bmatrix}17 & 3\\ 3 & 25\end{bmatrix} - \begin{bmatrix}-20\\ -28\end{bmatrix}[\tfrac{1}{48}][-20 \quad -28]\right\}\begin{bmatrix}U_1\\ U_3\end{bmatrix} = \begin{bmatrix}R_1 + \frac{20}{48}R_2\\ R_3 + \frac{28}{48}R_2\end{bmatrix}$$

or

$$\frac{13}{9}\frac{EA_1}{L}\begin{bmatrix}1 & -1\\ -1 & 1\end{bmatrix}\begin{bmatrix}U_1\\ U_3\end{bmatrix} = \begin{bmatrix}R_1 + \frac{5}{12}R_2\\ R_3 + \frac{7}{12}R_2\end{bmatrix}$$

Also, (7.29) yields

$$U_2 = \frac{1}{24}\left(\frac{3L}{EA_1}R_2 + 10U_1 + 14U_3\right) \tag{b}$$

Using Gauss elimination directly on (a) for U_2, we obtain

$$\frac{EA_1}{6L}\begin{bmatrix}17 - \frac{(20)(20)}{48} & 0 & 3 - \frac{(20)(28)}{48}\\ -20 & 48 & -28\\ 3 - \frac{(20)(28)}{48} & 0 & 25 - \frac{(28)(28)}{48}\end{bmatrix}\begin{bmatrix}U_1\\ U_2\\ U_3\end{bmatrix} = \begin{bmatrix}R_1 + \frac{20}{48}R_2\\ R_2\\ R_3 + \frac{28}{48}R_2\end{bmatrix} \tag{c}$$

But separating the equations for U_1 and U_3 from the equation for U_2, we can rewrite the relation in (c) as

$$\frac{13}{9}\frac{EA_1}{L}\begin{bmatrix}1 & -1\\ -1 & 1\end{bmatrix}\begin{bmatrix}U_1\\ U_3\end{bmatrix} = \begin{bmatrix}R_1 + \frac{5}{12}R_2\\ R_3 + \frac{7}{12}R_2\end{bmatrix}$$

and

$$U_2 = \frac{1}{24}\left[\frac{3L}{EA_1}R_2 + 10U_1 + 14U_3\right]$$

which are the relations obtained using the formal static condensation procedure.

Example 7.8

Use the stiffness matrix for the three degree of freedom truss element in Example 7.7 to establish the equilibrium equations of the structure shown in Fig. 7.4. Use Gauss elimination directly on degrees of freedom U_2 and U_4. Show that the

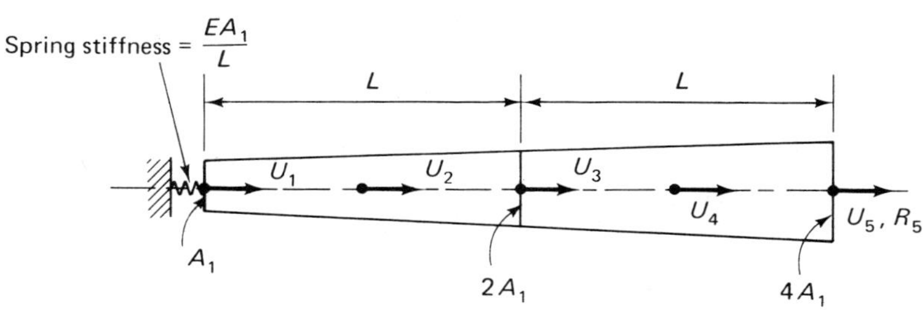

Figure 7.4. Structure composed of two truss elements of Fig. 7.3 and a spring support.

resulting equilibrium equations are identical to those obtained when the two degree of freedom truss element stiffness matrix derived in Example 7.7 (the internal degree of freedom has been condensed out) is used to assemble the stiffness matrix corresponding to U_1, U_3, and U_5.

The stiffness matrix of the three element assemblage in Fig. 7.3 is obtained using the direct stiffness method (see Section 3.1); i.e., we calculate

$$\mathbf{K} = \sum_{m=1}^{3} \mathbf{K}_m \tag{a}$$

where

$$\mathbf{K}_1 = \frac{EA_1}{6L}\begin{bmatrix} 6 & 0 & 0 & 0 & 0 \\ 0 & 0 & 0 & 0 & 0 \\ 0 & 0 & 0 & 0 & 0 \\ 0 & 0 & 0 & 0 & 0 \\ 0 & 0 & 0 & 0 & 0 \end{bmatrix}$$

$$\mathbf{K}_2 = \frac{EA_1}{6L}\begin{bmatrix} 17 & -20 & 3 & 0 & 0 \\ -20 & 48 & -28 & 0 & 0 \\ 3 & -28 & 25 & 0 & 0 \\ 0 & 0 & 0 & 0 & 0 \\ 0 & 0 & 0 & 0 & 0 \end{bmatrix}$$

$$\mathbf{K}_3 = \frac{EA_1}{6L}\begin{bmatrix} 0 & 0 & 0 & 0 & 0 \\ 0 & 0 & 0 & 0 & 0 \\ 0 & 0 & 34 & -40 & 6 \\ 0 & 0 & -40 & 96 & -56 \\ 0 & 0 & 6 & -56 & 50 \end{bmatrix}$$

Hence the equilibrium equations of the structure are

$$\frac{EA_1}{6L}\begin{bmatrix} 23 & -20 & 3 & 0 & 0 \\ -20 & 48 & -28 & 0 & 0 \\ 3 & -28 & 59 & -40 & 6 \\ 0 & 0 & -40 & 96 & -56 \\ 0 & 0 & 6 & -56 & 50 \end{bmatrix}\begin{bmatrix} U_1 \\ U_2 \\ U_3 \\ U_4 \\ U_5 \end{bmatrix} = \begin{bmatrix} 0 \\ 0 \\ 0 \\ 0 \\ R_5 \end{bmatrix}$$

Using Gauss elimination on degrees of freedom U_2 and U_4, we obtain

$$\frac{EA_1}{6L}\begin{bmatrix} 23 - \frac{(20)(20)}{48} & 0 & 3 - \frac{(20)(28)}{48} & 0 & 0 \\ -20 & 48 & -28 & 0 & 0 \\ 3 - \frac{(20)(28)}{48} & 0 & 59 - \frac{(28)(28)}{48} - \frac{(40)(40)}{96} & 0 & 6 - \frac{(40)(56)}{96} \\ 0 & 0 & -40 & 96 & -56 \\ 0 & 0 & 6 - \frac{(40)(56)}{96} & 0 & 50 - \frac{(56)(56)}{96} \end{bmatrix}\begin{bmatrix} U_1 \\ U_2 \\ U_3 \\ U_4 \\ U_5 \end{bmatrix} = \begin{bmatrix} 0 \\ 0 \\ 0 \\ 0 \\ R_5 \end{bmatrix} \tag{b}$$

Now, extracting the equilibrium equations corresponding to degrees of freedom 1, 3, and 5 and degrees of freedom 2 and 4 separately, we have

$$\frac{13}{9}\frac{EA_1}{L}\begin{bmatrix} \frac{22}{13} & -1 & 0 \\ -1 & 3 & -2 \\ 0 & -2 & 2 \end{bmatrix}\begin{bmatrix} U_1 \\ U_3 \\ U_5 \end{bmatrix} = \begin{bmatrix} 0 \\ 0 \\ R_5 \end{bmatrix} \tag{c}$$

and

$$\begin{aligned} U_2 &= \tfrac{1}{12}[5U_1 + 7U_3] \\ U_4 &= \tfrac{1}{12}[5U_3 + 7U_5] \end{aligned} \tag{d}$$

However, using the two degree of freedom truss element stiffness matrix derived in Example 7.7 to directly assemble the structure stiffness matrix corresponding to degrees of freedom 1, 3, and 5, we use as element stiffness matrices in (a),

$$\mathbf{K}_1 = \frac{13}{9}\frac{EA_1}{L}\begin{bmatrix} \frac{9}{13} & 0 & 0 \\ 0 & 0 & 0 \\ 0 & 0 & 0 \end{bmatrix}$$

$$\mathbf{K}_2 = \frac{13}{9}\frac{EA_1}{L}\begin{bmatrix} 1 & -1 & 0 \\ -1 & 1 & 0 \\ 0 & 0 & 0 \end{bmatrix} \tag{e}$$

$$\mathbf{K}_3 = \frac{13}{9}\frac{EA_1}{L}\begin{bmatrix} 0 & 0 & 0 \\ 0 & 2 & -2 \\ 0 & -2 & 2 \end{bmatrix} \tag{f}$$

and obtain the stiffness matrix in (c). Also, the relation (b) of Example 7.7 corresponds to relations (d) in this example. It should be noted that the total effort to solve the equilibrium equations using the condensed truss element stiffness matrix is less than when the original three degree of freedom element stiffness matrix is used, because in the first case the internal degree of freedom was statically condensed out only once, whereas in (b) the element internal degree of freedom is, in fact, statically condensed out twice. The direct solution using the condensed element stiffness matrices in (e) and (f) is, however, only possible because the stiffness matrices are multiples of each other.

As indicated in Example 7.8, it can be particularly effective to employ static condensation when the same element is used many times. An application of this concept is employed in *substructure analysis*, in which the total structure is considered to be an assemblage of substructures.[16-22] Each substructure, in turn, is idealized as an assemblage of finite elements, and all internal degrees of freedom are statically condensed out. The total structure stiffness is formed by assembling the condensed substructure stiffness matrices. Therefore, *in effect, a substructure is used in the same way as an individual finite element with internal degrees of freedom that are statically condensed out prior to the element assemblage process.* If many substructures are identical, it is effective to establish a library of substructure, from which a condensed total structure stiffness matrix is formed.

It should be noted that the unreduced complete structure stiffness matrix is never calculated in substructure analysis, and input data are required only for each substructure in the library plus information on the assemblage of the substructures to make up the complete structure. Typical applications of finite element analysis using substructuring are found in the analysis of buildings and ship hulls, where the substructure technique has allowed economical analysis of very large finite element systems.

As a simple example of substructure analysis, we refer to Example 7.8, in which each substructure is simply composed of one element, and the uncondensed and condensed stiffness matrix of a typical substructure was given in Example 7.7.

The effectiveness of analysis using the basic substructure concept described above can in many cases still be improved upon by defining different levels of substructures; i.e., since each substructure can be looked upon as a "super-finite element," it is possible to define second, third, etc., levels of substructuring. In a similar procedure, two substructures are always combined to define the next-higher-level substructure until the final substructure is, in fact, the actual structure under consideration. The procedure may be employed in one-, two-, or three-dimensional analysis and, as pointed out earlier, is indeed only an effective application of Gauss elimination, in which advantage is taken of the repetition of submatrices, which are the stiffness matrices that correspond to the identical substructures. The possibility of using the solution procedure effectively therefore depends on whether the structure is made up of repetitive substructures, and this is the reason the procedure is most effective in special-purpose programs.

Example 7.9

Use substructuring to evaluate the stiffness matrix and the load vector corresponding to the end nodal point degrees of freedom U_1 and U_9 of the bar in Fig. 7.5.

The basic element of which the bar is composed is the three degree of freedom truss element considered in Example 7.7. The equilibrium equations of the element corresponding to the two degrees of freedom U_1 and U_3 as shown in Fig. 7.5 are

$$\frac{13}{9}\frac{A_1E}{L}\begin{bmatrix} 1 & -1 \\ -1 & 1 \end{bmatrix}\begin{bmatrix} U_1 \\ U_3 \end{bmatrix} = \begin{bmatrix} R_1 + \frac{5}{12}R_2 \\ R_3 + \frac{7}{12}R_2 \end{bmatrix} \tag{a}$$

Since the internal degree of freedom U_2 has been statically condensed out to obtain the equilibrium relations in (a), we may regard the two degree of freedom element as a first-level substructure. We should recall that once U_1 and U_3 have been calculated, we can evaluate U_2 using the relation (b) in Example 7.7:

$$U_2 = \frac{1}{24}\left[\frac{3L}{EA_1}R_2 + 10U_1 + 14U_3\right] \tag{b}$$

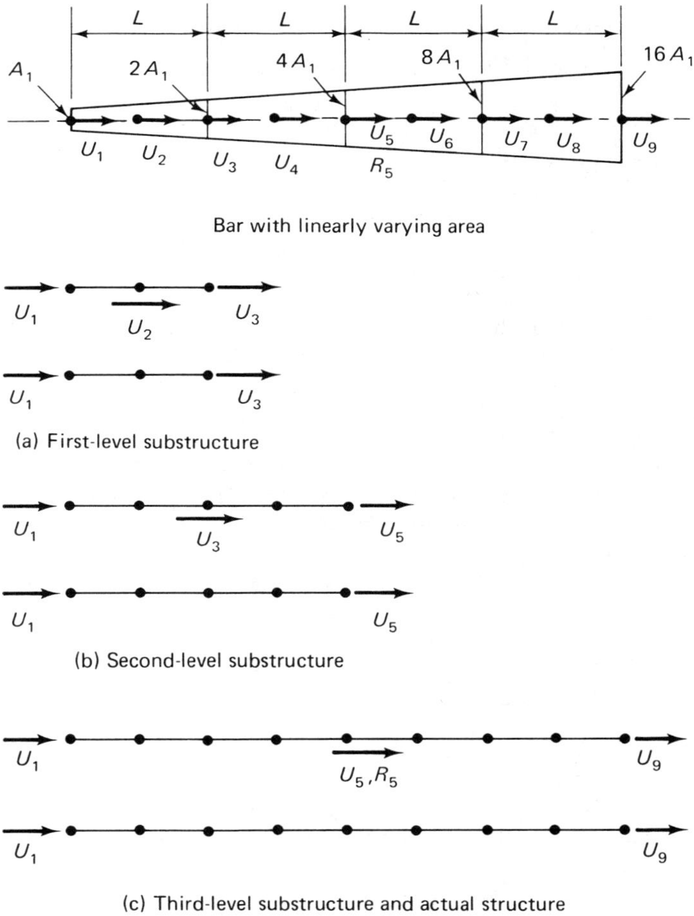

Figure 7.5. Analysis of bar using substructuring.

It is now effective to evaluate a second-level substructure corresponding to degrees of freedom U_1 and U_5 of the bar. For this purpose we use the stiffness matrix and load vector in (a) to evaluate the equilibrium relations corresponding to U_1, U_3, and U_5:

$$\frac{13}{9}\frac{A_1E}{L}\begin{bmatrix} 1 & -1 & 0 \\ -1 & 3 & -2 \\ 0 & -2 & 2 \end{bmatrix}\begin{bmatrix} U_1 \\ U_3 \\ U_5 \end{bmatrix} = \begin{bmatrix} R_1 + \frac{5}{12}R_2 \\ R_3 + \frac{7}{12}R_2 + \frac{5}{12}R_4 \\ R_5 + \frac{7}{12}R_4 \end{bmatrix} \tag{c}$$

The relation to calculate U_4 is similar to the one in (b):

$$U_4 = \frac{1}{24}\left[\frac{3L}{2EA_1}R_4 + 10U_3 + 14U_5\right]$$

Using Gauss elimination on the equations in (c) to condense out U_3, we obtain

$$\frac{13}{9}\frac{A_1E}{L}\begin{bmatrix} \frac{2}{3} & 0 & -\frac{2}{3} \\ -1 & 3 & -2 \\ -\frac{2}{3} & 0 & \frac{2}{3} \end{bmatrix}\begin{bmatrix} U_1 \\ U_3 \\ U_5 \end{bmatrix} = \begin{bmatrix} R_1 + \frac{22}{36}R_2 + \frac{1}{3}R_3 + \frac{5}{36}R_4 \\ R_3 + \frac{7}{12}R_2 + \frac{5}{12}R_4 \\ \frac{14}{36}R_2 + \frac{2}{3}R_3 + \frac{31}{36}R_4 + R_5 \end{bmatrix}$$

or

$$\left(\frac{2}{3}\right)\left(\frac{13}{9}\right)\frac{A_1E}{L}\begin{bmatrix} 1 & -1 \\ -1 & 1 \end{bmatrix}\begin{bmatrix} U_1 \\ U_5 \end{bmatrix} = \begin{bmatrix} R_1 + \frac{22}{36}R_2 + \frac{1}{3}R_3 + \frac{5}{36}R_4 \\ \frac{14}{36}R_2 + \frac{2}{3}R_3 + \frac{31}{36}R_4 + R_5 \end{bmatrix} \tag{d}$$

and

$$U_3 = \frac{1}{3}\left[\frac{9}{13}\frac{L}{A_1E}(R_3 + \tfrac{7}{12}R_2 + \tfrac{5}{12}R_4) + U_1 + 2U_5\right]$$

We should note that the stiffness matrix of the second-level substructure in (d) is simply $\frac{2}{3}$ times the stiffness matrix of the first-level substructure in (a). Therefore, we could continue to build up even higher-level substructures in an analogous manner; i.e., the stiffness matrix of the nth-level substructure would simply be a factor times the stiffness matrix given in (a).

In most cases loads are only applied at the boundary degrees of freedom between substructures, such as in this example. Using the stiffness matrix of the second-level substructure to assemble the stiffness matrix of the complete bar, and assembling the actual load vector for this example, we obtain

$$\frac{2}{3}\left(\frac{13}{9}\right)\frac{A_1E}{L}\begin{bmatrix} 1 & -1 & 0 \\ -1 & 5 & -4 \\ 0 & -4 & 4 \end{bmatrix}\begin{bmatrix} U_1 \\ U_5 \\ U_9 \end{bmatrix} = \begin{bmatrix} 0 \\ R_5 \\ 0 \end{bmatrix}$$

Eliminating U_5 we have

$$\left(\frac{4}{5}\right)\left(\frac{2}{3}\right)\left(\frac{13}{9}\right)\frac{A_1E}{L}\begin{bmatrix} 1 & -1 \\ -1 & 1 \end{bmatrix}\begin{bmatrix} U_1 \\ U_9 \end{bmatrix} = \begin{bmatrix} \frac{1}{5}R_5 \\ \frac{4}{5}R_5 \end{bmatrix}$$

where the stiffness matrix is simply the third-level substructure stiffness matrix corresponding to the algorithm given above. We also have

$$U_5 = \frac{1}{5}\left[\frac{27}{26}\frac{L}{A_1E}R_5 + U_1 + 4U_9\right]$$

To solve for specific displacements, it is necessary to impose boundary conditions on the bar, calculate U_1 and U_9, and then obtain the internal bar displacements using previously derived relations. It should be noted that corresponding relations must also be employed to evaluate U_6 to U_8.

So far, we have not mentioned how to proceed in the solution if the total system matrix cannot be contained in high-speed storage.[12,23-34] If substructuring is used, it is effective to keep the size of each uncondensed substructure stiffness matrix small enough so that the static condensation of the internal degrees of freedom can be carried out in high-speed core. Therefore, back-up storage would mainly be required to store the required information for the calculation of the displacements of the substructure internal nodes as expressed in (7.29).

However, it may be necessary to use multilevel substructuring (i.e., to define substructures of substructures) in order that the final equations to be solved can be taken into high-speed storage.

In general, it is important to use back-up storage effectively, because a great deal of reading and writing can be very expensive and indeed may limit the system size that can be solved, because not enough back-up storage may be available. In out-of-core solutions the particular scheme used for solving the system equilibrium equations is largely coupled with the specific procedure employed to assemble the element stiffness matrices to the global structure stiffness matrix. In many programs the structure stiffness matrix is assembled prior to performing the Gauss solution. In programs SAP, ADINA, and EASE, the equations are essentially assembled as described in Chapter 6, but in blocks that can be taken into high-speed core.[25-27] The solution of the system equations is obtained by first reducing the blocks of the stiffness matrix and load vectors consecutively, and then performing the back-substitution. Similar and effective procedures are presently used in many different structural analysis programs.[28-31]

Instead of first assembling the complete structure stiffness matrix, we may assemble and reduce the equations at the same time. In this case back-up storage for the total unreduced stiffness matrix is not required. A specific solution scheme called the *frontal solution method* has been used effectively.[32-34] In the solution procedure only those equations that are actually required for the elimination of a specific degree of freedom are assembled, the degree of freedom considered is statically condensed out, and so on.

As an example, consider the analysis of the plane stress finite element idealization of the sheet in Fig. 7.6. There are two equations associated with each node of the finite element mesh, namely, the equations corresponding to the U and V displacements, respectively. In the frontal solution scheme the equations are statically condensed out in the order of the elements; i.e., the first equations considered would be those corresponding to nodes 1, 2, m, and $m + 1$, the next equations considered would include those for nodes 3 and $m + 2$, and so on. To eliminate the degrees of freedom of node 1 it is necessary to assemble the equations that correspond to nodes $1, 2, m$, and $m + 1$, because the equations associated with nodes $2, m$, and $m + 1$ couple with those corresponding to node 1. This means that the stiffness matrices of elements $1, 2, q$, and $q + 1$ need first be calculated and assembled, and then the degrees of freedom corresponding to node 1 are statically condensed out. Next, for the elimination of the equations corresponding to node 2, the final equations corresponding to the degrees of freedom at nodes 3 and $m + 2$ are required also, meaning that the stiffness matrices of elements 3 and $q + 2$ must first be calculated and assembled; and so on.

It may now be realized that the complete procedure, in effect, consists of statically condensing out one degree of freedom after the other, and always

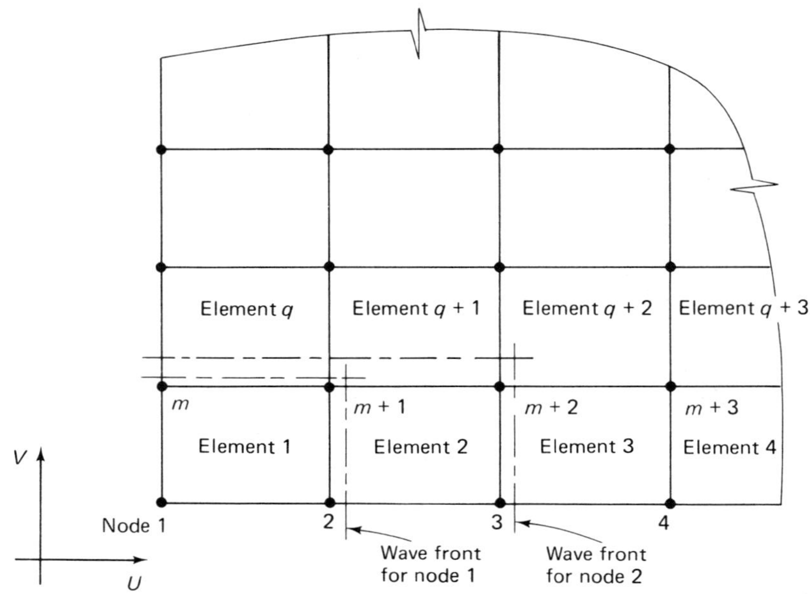

Figure 7.6. Frontal solution of plane stress finite element idealization.

assembling only those equations (or, rather, element stiffness matrices) that are actually required during the specific condensation to be performed. The finite elements that must be considered in the static condensation of the equations corresponding to one specific node define the wave front at that time, as shown in Fig. 7.6.

In principle, the frontal solution is Gauss elimination and the important aspect is the specific computer implementation. Since the equations are assembled in the order of the elements, the length of the wave front and therefore the half-bandwidth dealt with are determined by the element numbering. Therefore, an effective ordering of the elements is necessary.

An advantage of the frontal solution technique is that at any one time only the equations that are currently needed are assembled in high-speed storage. However, if the wave front is large, the total high-speed storage required may well exceed the storage that is available, in which case additional out-of-core operations are required, that decrease the effectiveness of the method considerably. It should also be noted that elements can be added with relative ease, because no nodal point renumbering is required.

7.2.5 Solution of Equations with Symmetric Nonpositive Definite Coefficient Matrices

In the Gauss elimination discussed so far, we implicitly assumed that the stiffness matrix **K** is positive definite, namely that the structure considered is properly restrained and stable. As we discussed in Sections 2.6 and 2.7, positive definiteness of the stiffness matrix means that for any displacement vector $\mathbf{U}$, we have

$$\mathbf{U}^T\mathbf{K}\mathbf{U} > 0 \tag{7.31}$$

Since $\frac{1}{2}\mathbf{U}^T\mathbf{K}\mathbf{U}$ is the strain energy stored in the system for the displacement vector $\mathbf{U}$ (see Section 5.3), (7.31) expresses that for any displacement vector $\mathbf{U}$, the strain energy of a system with a positive definite stiffness matrix must be positive.

It should be noted that the stiffness matrix of a finite element is not positive definite unless the element has been properly restrained; i.e., the rigid body motions have been suppressed (see Section 3.2.2). Instead, the stiffness matrix of an unrestrained finite element is positive semidefinite,

$$\mathbf{U}^T\mathbf{K}\mathbf{U} \geq 0 \tag{7.32}$$

where $\mathbf{U}^T\mathbf{K}\mathbf{U} = 0$ when $\mathbf{U}$ corresponds to a rigid body mode. Considering the finite element assemblage process, it should be realized that positive semidefinite element matrices are added to obtain the positive semidefinite stiffness matrix corresponding to the complete structure. The stiffness matrix of the structure is then rendered positive definite by eliminating the rows and columns that correspond to appropriate restraint degrees of freedom, i.e., by eliminating the possibility for the structure to undergo rigid body motions.

It is instructive to consider in more detail the meaning of positive definiteness of the structure stiffness matrix. In Section 2.6 we discussed the representation of a matrix by its eigenvalues and eigenvectors. Following the development given in Section 2.6, the eigenproblem for the stiffness matrix $\mathbf{K}$ can be written

$$\mathbf{K}\boldsymbol{\phi} = \lambda\boldsymbol{\phi} \tag{7.33}$$

The solution to (7.33) are the eigenpairs $(\lambda_i, \mathbf{v}_i)$, $i = 1, \ldots, n$, and the complete solution can be written

$$\mathbf{K}\boldsymbol{\Phi} = \boldsymbol{\Phi}\boldsymbol{\Lambda}$$

where $\boldsymbol{\Phi}$ is a matrix of the orthonormalized eigenvectors, $\boldsymbol{\Phi} = [\boldsymbol{\phi}_1, \ldots, \boldsymbol{\phi}_n]$, and $\boldsymbol{\Lambda}$ is a diagonal matrix of the corresponding eigenvalues, $\boldsymbol{\Lambda} = \text{diag}(\lambda_i)$. Since $\boldsymbol{\Phi}^T\boldsymbol{\Phi} = \boldsymbol{\Phi}\boldsymbol{\Phi}^T = \mathbf{I}$, we also have

$$\boldsymbol{\Phi}^T\mathbf{K}\boldsymbol{\Phi} = \boldsymbol{\Lambda} \tag{7.34}$$

and

$$\mathbf{K} = \boldsymbol{\Phi}\boldsymbol{\Lambda}\boldsymbol{\Phi}^T \tag{7.35}$$

Referring to Section 2.7, we recall that λ_i of $\mathbf{K}$ represents the minimum that can be reached by the Rayleigh quotient when an orthonormality constraint is satisfied on the eigenvectors $\boldsymbol{\phi}_1, \ldots, \boldsymbol{\phi}_{i-1}$,

$$\left.\begin{aligned} \lambda_i &= \min\left\{\frac{\boldsymbol{\phi}^T\mathbf{K}\boldsymbol{\phi}}{\boldsymbol{\phi}^T\boldsymbol{\phi}}\right\} \\ \text{with} \qquad & \\ \boldsymbol{\phi}^T\boldsymbol{\phi}_r &= 0; \qquad \text{for } r = 1, 2, \ldots, i-1 \end{aligned}\right\} \tag{7.36}$$

Therefore, $\frac{1}{2}\lambda_1$ is the minimum strain energy that can be stored in the element assemblage and the corresponding displacement vector is $\boldsymbol{\phi}_1$. For a positive definite system stiffness matrix, we therefore have $\lambda_1 > 0$. On the other hand, for the stiffness matrix of an unrestrained system, we have $\lambda_1 = \lambda_2 = \ldots = \lambda_m = 0$, where m is the number of rigid body modes present, $m < n$. As the system is restrained, the number of eigenvalues of $\mathbf{K}$ is decreased by one for each degree of freedom that is eliminated, and a zero eigenvalue is lost if the restraint results in the elimination of a rigid body mode.

Example 7.10

Identify whether the deletion of the four degrees of freedom of the plane stress element in Fig. 7.7 results in the elimination of the rigid body modes.

The plane stress element has three rigid body modes: (1) uniform horizontal translation, (2) uniform vertical translation, and (3) in-plane rotation. Consider

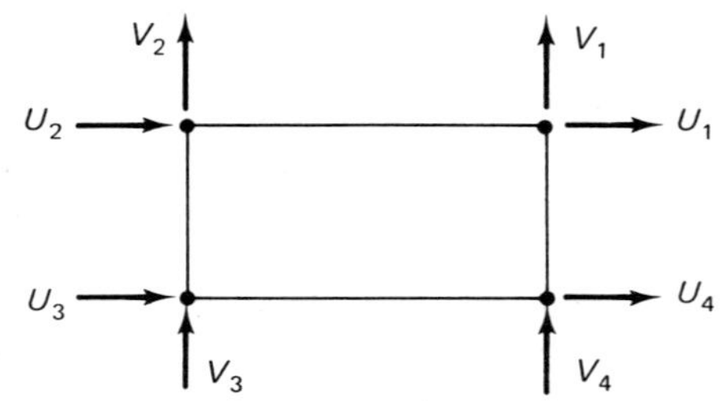

Degrees of freedom of plane stress element

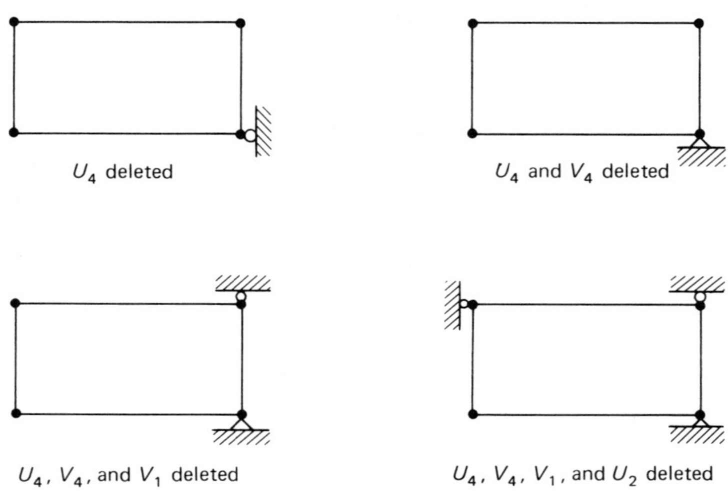

Figure 7.7. Deletion of degrees of freedom of plane stress element.

the sequential deletion of the degrees of freedom, as shown in Fig. 7.7. The deletion of U_4 results into eliminating the horizontal translation rigid body mode. Similarly, the deletion of V_4 results into the deletion of the vertical translation rigid body mode. However, deleting V_1 in addition does not result into the elimination of the last rigid body mode; i.e., the in-plane rotation rigid body mode is only eliminated with the additional deletion of U_2. Therefore, the deletion of U_4 and V_4, and V_1 and U_2 eliminates all rigid body modes of the element, although we should note that, in fact, by the deletion of U_4, V_4, and U_2 alone we would achieve the same result.

The transformation performed in (7.34) has important meaning. Considering the relation and referring to Section 2.5, we realize that in (7.34) a change of basis is performed. The new basis vectors are the finite element interpolations corresponding to the eigenvectors of $\mathbf{K}$, and in this basis the operator is represented by a diagonal matrix with the eigenvalues of $\mathbf{K}$ on its diagonal. We may therefore look at $\mathbf{\Lambda}$ as being the stiffness matrix of the system when the finite element displacement functions used in the virtual work theorem in (3.17) are those corresponding to nodal point displacements $\boldsymbol{\phi}_i$, $i = 1, \ldots, n$, instead of unit nodal point displacements U_i, $i = 1, \ldots, n$ (see Section 3.2.2). The relation in (7.34) is therefore a statement of virtual work resulting into a diagonal stiffness matrix. If the system considered is properly restrained, all stiffness coefficients in $\mathbf{\Lambda}$ are positive; i.e., the stiffness matrix $\mathbf{\Lambda}$ (and hence $\mathbf{K}$) is positive definite, whereas for an unrestrained system some diagonal elements in $\mathbf{\Lambda}$ are zero.

Before studying the solution of nonpositive definite systems of equations, another most important observation should be discussed. In Section 2.7 we introduced the Sturm sequence property of the leading principal minors of a matrix. We should note here the physical meaning of the Sturm sequence. Let $\mathbf{K}^{(r)}$ be the matrix of order $n - r$ obtained by deleting from $\mathbf{K}$ the last r rows and columns, and consider the eigenproblem

$$\mathbf{K}^{(r)}\boldsymbol{\phi}^{(r)} = \lambda^{(r)}\boldsymbol{\phi}^{(r)} \tag{7.37}$$

where $\boldsymbol{\phi}^{(r)}$ is a vector of order $n - r$. *We say that* (*7.37*) *is the eigenproblem of the rth associated constraint problem of the problem* $\mathbf{K}\boldsymbol{\phi} = \lambda\boldsymbol{\phi}$. Then we have shown in Section 2.7 that the eigenvalues of the $(r + 1)$st constraint problem separate those of the rth constraint problem,

$$\lambda_1^{(r)} \leq \lambda_1^{(r+1)} \leq \lambda_2^{(r)} \leq \lambda_2^{(r+1)} \ldots \leq \lambda_{n-r-1}^{(r)} \leq \lambda_{n-r-1}^{(r+1)} \leq \lambda_{n-r}^{(r)} \tag{7.38}$$

As an example, the eigenproblems of the simply supported beam discussed in Section 7.2.1 and of its associated constraint problems can be considered. Figure 7.8 shows the eigenvalues calculated and, in particular, displays their separation property. We should note that as we proceed from the $(r + 1)$st constraint problem to the rth constraint problem by including the $(n - r)$th degree of freedom, the new system has an eigenvalue smaller than (or equal

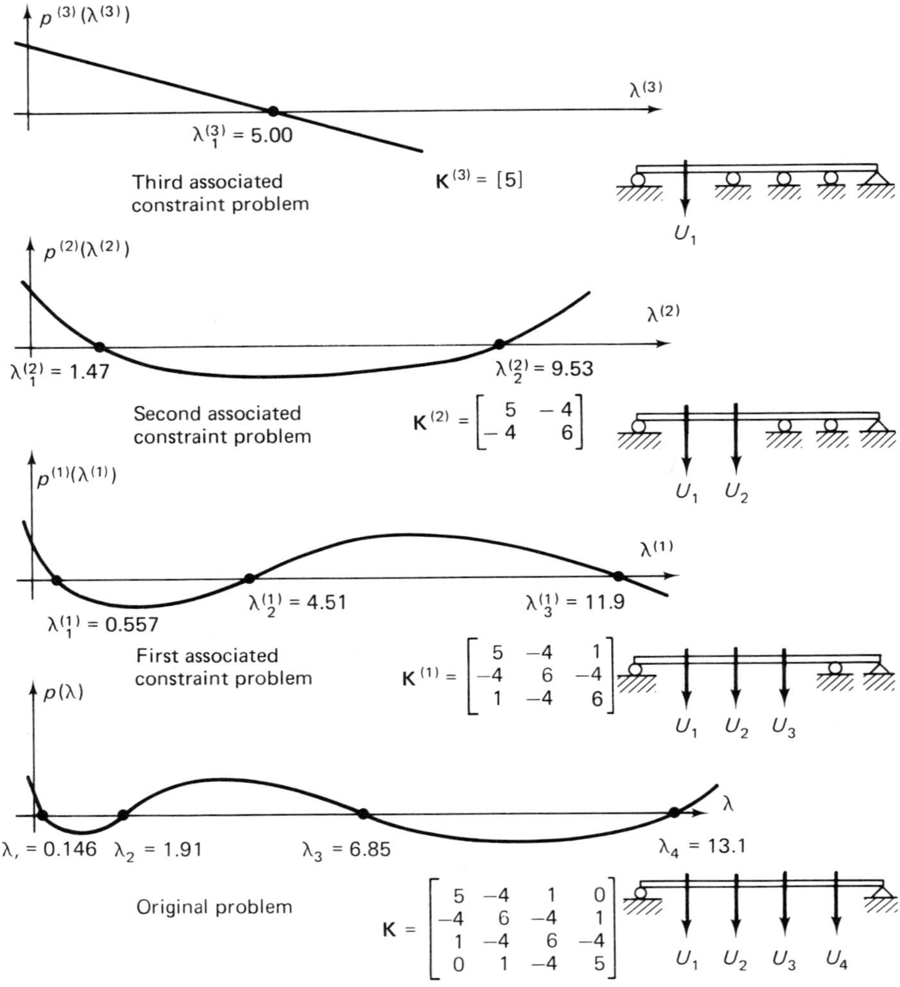

$$\mathbf{K}^{(2)} = \begin{bmatrix} 5 & -4 \\ -4 & 6 \end{bmatrix} \qquad \mathbf{K}^{(1)} = \begin{bmatrix} 5 & -4 & 1 \\ -4 & 6 & -4 \\ 1 & -4 & 6 \end{bmatrix}$$

$$\mathbf{K} = \begin{bmatrix} 5 & -4 & 1 & 0 \\ -4 & 6 & -4 & 1 \\ 1 & -4 & 6 & -4 \\ 0 & 1 & -4 & 5 \end{bmatrix}$$

Figure 7.8. Eigenvalue solutions of simply supported beam and of associated constraint problems.

to) the smallest eigenvalue of the $(r + 1)$st constraint problem, and also an eigenvalue larger than (or equal to) the largest eigenvalue of the $(r + 1)$st constraint problem.

Using the separation property of the eigenvalues, and realizing that any rows and columns may be interchanged at convenience to become the last rows and columns in the matrix $\mathbf{K}$, it follows that if the stiffness matrix corresponding to the n degrees of freedom is positive definite (i.e., $\lambda_1 > 0$), then any stiffness matrix obtained by deleting any rows and corresponding columns is also positive definite. Furthermore, the smallest eigenvalue of the new matrix can only have increased, and the largest eigenvalue can only have decreased. This conclusion applies also if the matrix $\mathbf{K}$ is positive semidefinite and would apply if it were indefinite since we showed the eigenvalue separation theorem to be applicable to all symmetric matrices.

We shall encounter the use of the Sturm sequence property of the leading principal minors more extensively in the design of eigenvalue solution algorithms (see Section 11.5). However, in the following we use the property to yield more insight into the solution of a set of simultaneous equations with a symmetric positive definite, positive semidefinite, and indefinite coefficient matrix. A symmetric indefinite coefficient matrix will be encountered in the solution of eigenproblems (see Section 11.4.2).

Using physical reasoning, we showed in Sections 7.2.1 and 7.2.2 that if $\mathbf{K}$ is the stiffness matrix of a properly restrained structure, we can factorize $\mathbf{K}$ into the form

$$\mathbf{K} = \mathbf{L}\mathbf{D}\mathbf{L}^T \tag{7.39}$$

where $\mathbf{L}$ is a lower unit triangular matrix and $\mathbf{D}$ is a diagonal matrix, with $d_{ii} > 0$. It follows (see Section 1.8) that

$$\begin{aligned} \det \mathbf{K} &= \det \mathbf{L} \det \mathbf{D} \det \mathbf{L}^T \\ &= \prod_{i=1}^{n} d_{ii} > 0 \end{aligned} \tag{7.40}$$

This result can also be obtained by considering the characteristic polynomial of $\mathbf{K}$, defined as

$$p(\lambda) = \det (\mathbf{K} - \lambda \mathbf{I}) \tag{7.41}$$

Since λ_1 is the smallest root of $p(\lambda)$ and $\lambda_1 > 0$ for $\mathbf{K}$ being positive definite, it follows that $\det \mathbf{K} > 0$. However, it does not yet follow that $d_{ii} > 0$ for all i.

In order to formally prove that $d_{ii} > 0$, $i = 1, \ldots, n$, when $\mathbf{K}$ is positive definite, and to identify what happens during the factorization of $\mathbf{K}$, it is expedient to compare the triangular factors of $\mathbf{K}$ with those of $\mathbf{K}^{(i)}$, where $\mathbf{K}^{(i)}$ is the stiffness matrix of the ith associated constraint problem. Assuming that the factors $\mathbf{L}$ and $\mathbf{D}$ of $\mathbf{K}$ have been calculated, we have, for the associated constraint problems,

$$\mathbf{K}^{(i)} = \mathbf{L}^{(i)}\mathbf{D}^{(i)}\mathbf{L}^{(i)T} \qquad i = 1, \ldots, n-1 \tag{7.42}$$

where $\mathbf{L}^{(i)}$ and $\mathbf{D}^{(i)}$ are analogously the factors of $\mathbf{K}^{(i)}$. Since $\mathbf{L}$ is a lower-unit triangular matrix and $\mathbf{D}$ is a diagonal matrix, the factors $\mathbf{L}^{(i)}$ and $\mathbf{D}^{(i)}$ are obtained from $\mathbf{L}$ and $\mathbf{D}$, respectively, by eliminating the last i rows and columns. Therefore, $\mathbf{L}^{(i)}$ and $\mathbf{D}^{(i)}$ are the leading principal submatrices of $\mathbf{L}$ and $\mathbf{D}$, respectively, and they are actually evaluated in the factorization of

K. However, because $\lambda_1^{(i)} > 0$, it now follows that we can use the argument in (7.39) and (7.40) starting with $i = n - 1$ to show that $d_{ii} > 0$ for all i. Therefore, the factorization of **K** into $\mathbf{LDL}^T$ is indeed possible if **K** is positive definite. We demonstrate the analysis in the following example.

Example 7.11

Consider the simply supported beam in Fig. 7.8 and the associated constraint problems. The same beam was used in Section 7.2.1. Establish the $\mathbf{L}^{(i)}$ and $\mathbf{D}^{(i)}$ factors of the matrices $\mathbf{K}^{(i)}$, $i = 1, 2, 3$, and show that d_{ii} must be greater than zero because $\lambda_1 > 0$.

The required triangular factorizations are

$$[5] = [1][5][1] \tag{a}$$

$$\begin{bmatrix} 5 & -4 \\ -4 & 6 \end{bmatrix} = \begin{bmatrix} 1 & 0 \\ -\frac{4}{5} & 1 \end{bmatrix} \begin{bmatrix} 5 & 0 \\ 0 & \frac{14}{5} \end{bmatrix} \begin{bmatrix} 1 & -\frac{4}{5} \\ 0 & 1 \end{bmatrix} \tag{b}$$

$$\begin{bmatrix} 5 & -4 & 1 \\ -4 & 6 & -4 \\ 1 & -4 & 6 \end{bmatrix} = \begin{bmatrix} 1 & 0 & 0 \\ -\frac{4}{5} & 1 & 0 \\ \frac{1}{5} & -\frac{8}{7} & 1 \end{bmatrix} \begin{bmatrix} 5 & 0 & 0 \\ 0 & \frac{14}{5} & 0 \\ 0 & 0 & \frac{15}{7} \end{bmatrix} \begin{bmatrix} 1 & -\frac{4}{5} & \frac{1}{5} \\ 0 & 1 & -\frac{8}{7} \\ 0 & 0 & 1 \end{bmatrix} \tag{c}$$

where the matrices $\mathbf{L}^{(i)}$ and $\mathbf{D}^{(i)}$ are obtained from the **L** and **D** factors given in Example 7.2 by eliminating the last i rows and columns. [As a check, we may want to calculate the product of the matrices on the right sides of the relations in (a), (b), and (c) to obtain the matrices on the left sides.]

Considering the elements d_{ii}, we have that $\lambda_1^{(3)} \geq \lambda_1^{(2)} \geq \lambda_1^{(1)} \geq \lambda_1 > 0$. But using the relation (a), we have $\lambda_1^{(3)} = d_{11}$; hence $d_{11} > 0$. Next we consider $\mathbf{K}^{(2)}$. Since $\lambda_1^{(2)} > 0$, we have, using (7.39) and (7.40), that $d_{11}d_{22} > 0$, which means that $d_{22} > 0$. Similarly, considering $\mathbf{K}^{(1)}$, we have $\lambda_1^{(1)} > 0$, and hence $d_{11}d_{22}d_{33} > 0$, from which it follows that $d_{33} > 0$. Finally, considering **K**, we have $\lambda_1 > 0$, and hence $d_{11}d_{22}d_{33}d_{44} > 0$. Therefore, $d_{44} > 0$ also.

Assume next that the matrix **K** is the stiffness matrix of a finite element assemblage that is unrestrained. In this case, **K** is positive semidefinite, $\lambda = 0.0$ is a root, and det **K** is zero, which, using (7.40), means that d_{ii} for some i must be zero. Therefore, the factorization of **K** as shown in the preceding sections is, in general, not possible because a pivot element will be zero. It is again instructive to consider the associated constraint problems. When **K** is positive semidefinite, we note that the characteristic polynomial corresponding to $\mathbf{K}^{(i)}$ will have a zero eigenvalue, and this zero eigenvalue will be retained in all matrices $\mathbf{K}^{(i-1)}, \ldots, \mathbf{K}$. This follows because, firstly, the Sturm sequence property assures that the smallest eigenvalue of $\mathbf{K}^{(i-1)}$ is smaller or equal to the smallest eigenvalue of $\mathbf{K}^{(i)}$, and secondly, **K** has no negative eigenvalues. For the ith associated constraint problem, we will therefore have

$$\det(\mathbf{L}^{(i)}\mathbf{D}^{(i)}\mathbf{L}^{(i)T}) = 0 \tag{7.43}$$

from which it follows that an element of $\mathbf{D}^{(i)}$ is zero. However, assuming that the zero root only occurs in the ith associated constraint problem [i.e., det $(\mathbf{L}^{(r)}\mathbf{D}^{(r)}\mathbf{L}^{(r)T}) > 0$ for $r > i$], it follows that $d_{n-i,n-i}$ is zero. In summary, therefore, if $\mathbf{K}$ is positive semidefinite, the factorization of $\mathbf{K}$ into $\mathbf{LDL}^T$ (i.e., the Gauss elimination process) will break down at the time a zero diagonal element d_{kk} is encountered, which means that the $(n - k)$th associated constraint problem with a zero eigenvalue prevents the continuation of the factorization process.

In the case of a positive semidefinite matrix, a zero diagonal element must be encountered at some stage of the factorization. However, considering the decomposition of an indefinite matrix (i.e., some of the eigenvalues of the matrix are negative and some are positive), a zero diagonal element is encountered only if one of the associated constraint problems has a zero eigenvalue. Namely, as in the case of a positive semidefinite matrix, d_{kk} is zero if the $(n - k)$th associated constraint problem has a zero eigenvalue. However, if none of the associated constraint problems has a zero eigenvalue, all elements d_{ii} are nonzero and, in exact arithmetic, no difficulties are encountered in the factorization. We shall discuss the decomposition of indefinite coefficient matrices further in the solution of eigenproblems (see Sections 11.4.2 and 12.2). Figure 7.9 shows typical cases that use the simply supported beam of Fig. 7.8 on spring supports for which, in the decompositions, we would and would not encounter a zero diagonal element (see Example 7.13).

Assume that a zero diagonal element d_{ii} is encountered in the Gauss elimination. To be able to proceed with the solution, it is necessary to interchange the ith row with another row, say the jth row, where $j > i$. The new diagonal element should not be zero, and to increase solution accuracy, it should be large (see Section 7.5). This row interchange corresponds to a rearranging of the equations, where it should be noted that the row interchange results in the coefficient matrix no longer being symmetric. On the other hand, symmetry would be preserved if we were to interchange not only the ith and jth rows but also the corresponding columns to obtain a nonzero diagonal element in row i, which is not always possible (see Example 10.4). In effect, the interchange of columns and rows corresponds to a rearranging of the associated constraint problems in such a way that these have nonzero eigenvalues.

The remedy of row interchanges assumes that it can be arranged that the new diagonal element is nonzero. In fact, this will always be the case, unless the matrix has a zero eigenvalue of multiplicity m and $i = n - m + 1$. In this case the matrix is singular and d_{ii} is zero, but all other elements of the last m rows of the upper triangular factor of the coefficient matrix are also zero, and the factorization of the matrix has already been completed. In other words, since the number of rows that are made up of zero elements only is equal to the multiplicity m of the zero eigenvalue, the last $m - 1$ matrices $\mathbf{L}_{n-1}^{-1}, \mathbf{L}_{n-2}^{-1}, \ldots, \mathbf{L}_{n-m+1}^{-1}$ in (7.10) cannot and need not be calculated.

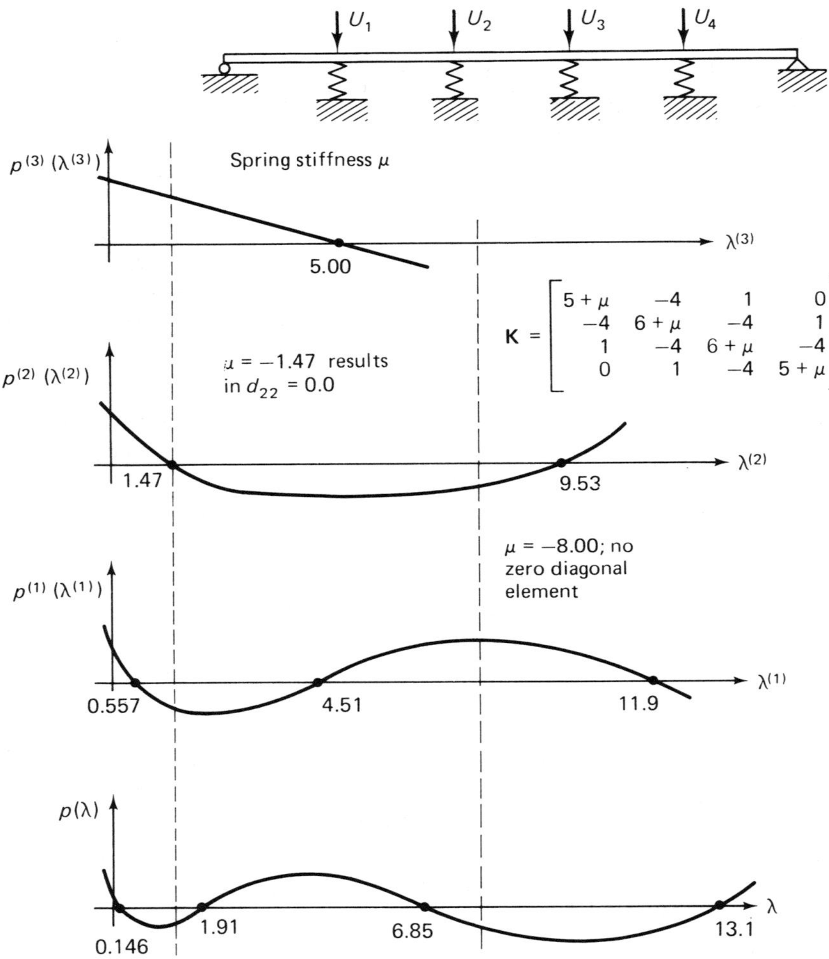

Figure 7.9. Simply supported beam on spring supports; negative spring stiffness can result in solution difficulties.

We shall therefore be able to solve for m linearly independent solutions by assuming appropriate values for the m last entries of the solution vector.

Of particular interest is the case $\mathbf{K}$ representing the stiffness matrix of an unrestrained finite element assemblage. The interchange of rows and corresponding columns during the Gauss elimination always results, in this case, in associated constraint problems with positive definite stiffness matrices, until the elements in the last rows of the reduced stiffness matrix [i.e., $\mathbf{S}$ in (7.10)] all become zero. The number of rows that are made up of zero elements is equal to the number of rigid body modes in the system.

Example 7.12

Consider the beam element in Fig. 7.10(a). The stiffness matrix of the element is

$$\mathbf{K} = \begin{bmatrix} 12 & -6 & -12 & -6 \\ -6 & 4 & 6 & 2 \\ -12 & 6 & 12 & 6 \\ -6 & 2 & 6 & 4 \end{bmatrix}$$

Show that Gauss elimination results into the third and fourth row to consist of zero elements only, and evaluate formally the rigid body mode displacements.

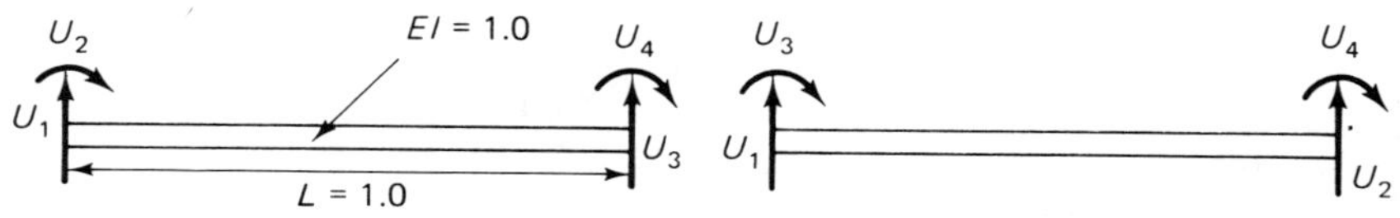

(a) Initial degree of freedom numbering

(b) Degree of freedom numbering that requires column and row interchange

Figure 7.10. Beam element with two rigid body modes.

Using the procedure in (7.10), we expect to arrive at a matrix $\mathbf{S}$ with its last two rows consisting of zero elements only, because the beam element has two rigid body modes, corresponding to vertical translation and rotation. We have

$$\mathbf{L}_1^{-1} = \begin{bmatrix} 1 & & & \\ \frac{1}{2} & 1 & 0 & \\ 1 & 0 & 1 & \\ \frac{1}{2} & 0 & 0 & 1 \end{bmatrix}$$

Hence

$$\mathbf{L}_1^{-1}\mathbf{K} = \begin{bmatrix} 12 & -6 & -12 & -6 \\ 0 & 1 & 0 & -1 \\ 0 & 0 & 0 & 0 \\ 0 & -1 & 0 & 1 \end{bmatrix}$$

Then

$$\mathbf{L}_2^{-1} = \begin{bmatrix} 1 & & & \\ 0 & 1 & & \\ 0 & 0 & 1 & \\ 0 & 1 & 0 & 1 \end{bmatrix}$$

and

$$\mathbf{S} = \mathbf{L}_2^{-1}\mathbf{L}_1^{-1}\mathbf{K} = \begin{bmatrix} 12 & -6 & -12 & -6 \\ 0 & 1 & 0 & -1 \\ 0 & 0 & 0 & 0 \\ 0 & 0 & 0 & 0 \end{bmatrix} \tag{a}$$

Therefore, as expected, the last two rows in **S** consist of zero elements, and $\mathbf{L}_3^{-1}$ cannot and need not be calculated. We should also note that if the numbering of the degrees of freedom of the beam element were initially as in Fig. 7.10(b), we would need to interchange rows and columns 2 with rows and columns 3 in order to be able to continue with the triangularization. This, however, is equivalent to using the degree of freedom numbering that we were concerned with in the first place, i.e., the numbering in Fig. 7.10(a).

Using the matrix **S** in (a), we can now formally evaluate the rigid body mode displacements of the beam, i.e., solve the equations $\mathbf{KU} = \mathbf{0}$ to obtain two linearly independent solutions for **U**.

First assume that $U_4 = 1$ and $U_3 = 0$; then using **S**, we obtain

$$\begin{aligned} 12U_1 - 6U_2 &= 6 \\ U_2 &= 1 \end{aligned}$$

Hence

$$U_1 = 1, \qquad U_2 = 1, \qquad U_3 = 0, \qquad U_4 = 1$$

Then assume that $U_4 = 0$, $U_3 = 1$, to obtain

$$\begin{aligned} 12U_1 - 6U_2 &= 12 \\ U_2 &= 0 \end{aligned}$$

Hence

$$U_1 = 1, \qquad U_2 = 0, \qquad U_3 = 1, \qquad U_4 = 0$$

It should be noted that the rigid body displacement vectors are not unique, but instead the two-dimensional subspace spanned by any two linearly independent rigid body mode vectors is unique. Therefore, any rigid body displacement vector can be written

$$\begin{bmatrix} U_1 \\ U_2 \\ U_3 \\ U_4 \end{bmatrix} = \alpha_1 \begin{bmatrix} 1 \\ 1 \\ 0 \\ 1 \end{bmatrix} + \alpha_2 \begin{bmatrix} 1 \\ 0 \\ 1 \\ 0 \end{bmatrix}$$

where α_1 and α_2 are constants.

In the above discussion we investigated the possibility of obtaining the solution to a system of simultaneous equations by referring to the eigenvalues of the coefficient matrix and of the matrices derived from it by deleting the last rows and columns. However, we should also realize that the Gauss elimination determines implicitly the rank of the coefficient matrix; i.e., the dimension of the row and column space of the matrix is obtained (see Section 2.2). Namely, assuming that interchanges are performed in the Gauss elimination on a general coefficient matrix, the final number of nonzero rows in the reduced matrix is equal to the rank of the matrix. This follows because at each stage of the Gauss elimination a multiple of one row is subtracted from another row. Therefore, if the last i rows have only zero elements, it follows that only $n - i$ rows were linearly independent. Considering an element stiffness matrix, we therefore conclude that a stiffness matrix of order n, in which i rigid body modes have not been eliminated, has rank $n - i$. We should note that the determination of the rank of a matrix by Gauss elimination is applicable to indefinite matrices but that interchanges must be carried out.

Example 7.13

Use the Gauss elimination procedure to factorize the matrix $\mathbf{K} - \lambda_1^{(2)}\mathbf{I}$ into an upper and lower triangular matrix, where $\mathbf{K}$ is the stiffness matrix considered in Example 7.2 and $\lambda_1^{(2)}$ is the smallest eigenvalue of the eigenproblem corresponding to the second associated constraint problem. Thus determine the rank of $\mathbf{K} - \lambda_1^{(2)}\mathbf{I}$.

We have

$$\mathbf{K} = \begin{bmatrix} 5 & -4 & 1 & 0 \\ -4 & 6 & -4 & 1 \\ 1 & -4 & 6 & -4 \\ 0 & 1 & -4 & 5 \end{bmatrix}; \quad \mathbf{K}^{(2)} = \begin{bmatrix} 5 & -4 \\ -4 & 6 \end{bmatrix}; \quad \lambda_1^{(2)} = 5.5 - \tfrac{1}{2}\sqrt{65}$$

Hence

$$\mathbf{K} - \lambda_1^{(2)}\mathbf{I} = \begin{bmatrix} 3.53 & -4 & 1 & 0 \\ -4 & 4.53 & -4 & 1 \\ 1 & -4 & 4.53 & -4 \\ 0 & 1 & -4 & 3.53 \end{bmatrix}$$

To set to zero all subdiagonal elements in the first column, we evaluate $\mathbf{L}_1^{-1}(\mathbf{K} - \lambda_1^{(2)}\mathbf{I})$, where

$$\mathbf{L}_1^{-1} = \begin{bmatrix} 1 & & & \\ 1.13 & 1 & & \\ -0.283 & & 1 & \\ 0 & & & 1 \end{bmatrix}$$

and hence

$$\mathbf{L}_1^{-1}(\mathbf{K} - \lambda_1^{(2)}\mathbf{I}) = \begin{bmatrix} 3.53 & -4 & 1 & 0 \\ 0 & 0 & -2.87 & 1 \\ 0 & -2.87 & 4.25 & -4 \\ 0 & 1 & -4 & 3.53 \end{bmatrix} \tag{a}$$

To proceed with the Gauss factorization we need to interchange the second row with either the third or fourth row. Switching rows 2 and 3, we have

$$\bar{\mathbf{L}}_1^{-1}(\bar{\mathbf{K}} - \lambda_1^{(2)}\bar{\mathbf{I}}) = \begin{bmatrix} 3.53 & -4 & 1 & 0 \\ 0 & -2.87 & 4.25 & -4 \\ 0 & 0 & -2.87 & 1 \\ 0 & 1 & -4 & 3.53 \end{bmatrix}$$

where

$$\bar{\mathbf{K}} = \begin{bmatrix} 5 & -4 & 1 & 0 \\ 1 & -4 & 6 & -4 \\ -4 & 6 & -4 & 1 \\ 0 & 1 & -4 & 5 \end{bmatrix}; \qquad \bar{\mathbf{L}}_1^{-1} = \begin{bmatrix} 1 & & & \\ -0.283 & 1 & & \\ 1.13 & & 1 & \\ 0 & & & 1 \end{bmatrix};$$

$$\bar{\mathbf{I}} = \begin{bmatrix} 1 & 0 & 0 & 0 \\ 0 & 0 & 1 & 0 \\ 0 & 1 & 0 & 0 \\ 0 & 0 & 0 & 1 \end{bmatrix}$$

We can now complete the factorization, using

$$\mathbf{L}_2^{-1} = \begin{bmatrix} 1 & & & \\ & 1 & & \\ & 0 & 1 & \\ & 0.349 & & 1 \end{bmatrix}$$

We obtain

$$\mathbf{L}_2^{-1}\bar{\mathbf{L}}_1^{-1}(\bar{\mathbf{K}} - \lambda_1^{(2)}\bar{\mathbf{I}}) = \begin{bmatrix} 3.53 & -4 & 1 & 0 \\ 0 & -2.87 & 4.25 & -4 \\ 0 & 0 & -2.87 & 1 \\ 0 & 0 & -2.52 & 2.14 \end{bmatrix}$$

and

$$\mathbf{L}_3^{-1} = \begin{bmatrix} 1 & & & \\ & 1 & & \\ & & 1 & \\ & & -0.878 & 1 \end{bmatrix}$$

Hence, defining $\mathbf{S} = \mathbf{L}_3^{-1}\mathbf{L}_2^{-1}\bar{\mathbf{L}}_1^{-1}(\bar{\mathbf{K}} - \lambda_1^{(2)}\bar{\mathbf{I}})$, we have

$$\mathbf{S} = \begin{bmatrix} 3.53 & -4 & 1 & 0 \\ & -2.87 & 4.25 & -4 \\ & & -2.87 & 1 \\ & & & 1.26 \end{bmatrix}$$

Furthermore, we have $(\bar{\mathbf{K}} - \lambda_1^{(2)}\bar{\mathbf{I}}) = \mathbf{L}\mathbf{S}$, where

$$\mathbf{L} = \begin{bmatrix} 1 & & & \\ 0.283 & 1 & & \\ -1.13 & 0 & 1 & \\ 0 & -0.349 & 0.878 & 1 \end{bmatrix}$$

We should note that the matrix $\bar{\mathbf{K}} - \lambda_1^{(2)}\bar{\mathbf{I}}$ is nonsymmetric, and therefore we cannot write $\mathbf{S} = \mathbf{D}\mathbf{L}^T$. However, if we would have interchanged not only rows but also the corresponding columns, the coefficient matrix would have remained symmetric. To determine the rank of $\mathbf{K} - \lambda_1^{(2)}\mathbf{I}$, we note that no zero rows have been obtained in $\mathbf{S}$; hence the rank of $\mathbf{K} - \lambda_1^{(2)}\mathbf{I}$ is four.

Now, interchanging the second and third rows and columns of the matrix in (a), we obtain

$$\bar{\mathbf{L}}_1^{-1}(\hat{\mathbf{K}} - \lambda_1^{(2)}\mathbf{I}) = \begin{bmatrix} 3.53 & 1 & -4 & 0 \\ 0 & 4.25 & -2.87 & -4 \\ 0 & -2.87 & 0 & 1 \\ 0 & -4 & 1 & 3.53 \end{bmatrix}$$

where

$$\hat{\mathbf{K}} = \begin{bmatrix} 5 & 1 & -4 & 0 \\ 1 & 6 & -4 & -4 \\ -4 & -4 & 6 & 1 \\ 0 & -4 & 1 & 5 \end{bmatrix}; \qquad \bar{\mathbf{L}}_1^{-1} = \begin{bmatrix} 1 & & & \\ -0.283 & 1 & & \\ 1.13 & & 1 & \\ 0 & & & 1 \end{bmatrix}$$

In the next step we use

$$\mathbf{L}_2^{-1} = \begin{bmatrix} 1 & & & \\ & 1 & & \\ & 0.675 & 1 & \\ & 0.942 & & 1 \end{bmatrix}$$

and obtain

$$\mathbf{L}_2^{-1}\bar{\mathbf{L}}_1^{-1}(\hat{\mathbf{K}} - \lambda_1^{(2)}\mathbf{I}) = \begin{bmatrix} 3.53 & 1 & -4 & 0 \\ 0 & 4.25 & -2.87 & -4 \\ 0 & 0 & -1.94 & -1.70 \\ 0 & 0 & -1.70 & -0.24 \end{bmatrix}$$

The factorization is completed using

$$\mathbf{L}_3^{-1} = \begin{bmatrix} 1 & & & \\ & 1 & & \\ & & 1 & \\ & & -0.878 & 1 \end{bmatrix}$$

$$\mathbf{L}_3^{-1}\mathbf{L}_2^{-1}\bar{\mathbf{L}}_1^{-1}(\hat{\mathbf{K}} - \lambda_1^{(2)}\mathbf{I}) = \begin{bmatrix} 3.53 & 1 & -4 & 0 \\ & 4.25 & -2.87 & -4 \\ & & -1.94 & -1.70 \\ & & & 1.26 \end{bmatrix}$$

Hence we have $(\hat{\mathbf{K}} - \lambda_1^{(2)}\mathbf{I}) = \mathbf{LS}$, where

$$\mathbf{L} = \begin{bmatrix} 1 & & & \\ 0.283 & 1 & & \\ -1.13 & -0.675 & 1 & \\ 0 & -0.942 & 0.878 & 1 \end{bmatrix}; \qquad \mathbf{S} = \begin{bmatrix} 3.53 & 1 & -4 & 0 \\ & 4.25 & -2.87 & -4 \\ & & -1.94 & -1.70 \\ & & & 1.26 \end{bmatrix}$$

and, as may be checked, $\mathbf{S} = \mathbf{DL}^T$.

So far, we have implicitly assumed in the discussion that all calculations are performed in exact arithmetic, although in some example solutions, the calculations have been performed to an accuracy of only a few digits. In general, using the digital computer in practical calculations, diagonal elements exactly equal to zero are hardly ever generated in the Gauss triangular factorization of a stiffness matrix. Instead, owing to the finite precision arithmetic of the computer, diagonal elements which are small in relation to the other diagonal elements and which cause large multipliers would nearly always be calculated. Therefore, a breakdown of Gauss elimination as considered above will hardly ever occur. However, if small diagonal elements are used, the errors in the factorization can be large, and this is really what we are interested in. The estimation of errors in the solution of simultaneous equations is discussed further in Section 7.5.

7.3 DIRECT SOLUTIONS USING ORTHOGONAL MATRICES

In Section 7.2.2 we have been writing Gauss elimination as premultiplication of $\mathbf{K}$ by matrices $\mathbf{L}_i^{-1}$, $i = 1, 2, \ldots, n-1$, where n is the order of $\mathbf{K}$ and $\mathbf{L}_i^{-1}$ was defined in (7.11). In fact, each matrix $\mathbf{L}_i^{-1}$ is a product of elementary matrices

$$\mathbf{L}_i^{-1} = \mathbf{L}_{n,i}^{-1}\mathbf{L}_{n-1,i}^{-1} \ldots \mathbf{L}_{i+1,i}^{-1} \tag{7.44}$$

so that (7.10) becomes

$$\mathbf{L}_{n,n-1}^{-1} \ldots (\mathbf{L}_{n,2}^{-1}\mathbf{L}_{n-1,2}^{-1} \ldots \mathbf{L}_{3,2}^{-1})(\mathbf{L}_{n,1}^{-1}\mathbf{L}_{n-1,1}^{-1} \ldots \mathbf{L}_{2,1}^{-1})\mathbf{K} = \mathbf{S} \tag{7.45}$$

where

$$\mathbf{L}_{k,i}^{-1} = \begin{bmatrix} 1 & & & & & & \\ & \ddots & & & & & \\ & & 1 & & & & \\ & & 0 & \ddots & & & \\ & & \vdots & & & & \\ & & 0 & & & & \\ & & -l_{ki} & & \ddots & & \\ & & 0 & & & & \\ & & \vdots & & & \ddots & \\ & & 0 & & & & 1 \end{bmatrix}; \quad k = i+1, \ldots, n \tag{7.46}$$

and the Gauss multiplying factor $l_{i+j,i}$ was defined in (7.11). This definition showed that the reduction cannot be continued if $k_{ii}^{(i)}$ is zero, and numerical difficulties should be expected if $k_{ii}^{(i)}$ is small, or rather $l_{i+j,i}$ is large.

In practical finite element analysis using a direct solution scheme, the Gauss solution algorithm is nearly always used in one form or the other, because the procedure is cost-effective and numerically sufficiently stable. However, it should be realized that a direct solution can also be achieved using other techniques. The objective in this section is to briefly summarize two additional direct solution methods.

The basic function of each elementary matrix $\mathbf{L}_{j,i}^{-1}$ is to reduce an element of the stiffness matrix (considered in step i) to zero. Specifically, the elementary matrix $\mathbf{L}_{j,i}^{-1}$ reduces the element $k_{ji}^{(i)}$ to zero. As might be expected, there is no reason why instead of elementary matrices we could not use orthogonal matrices to reduce the element $k_{ji}^{(i)}$ to zero.[7] In Section 2.5 we defined rotation and reflection matrices as two different orthogonal matrices, and both can be used to reduce $\mathbf{K}$ into upper triangular form.

7.3.1 The Givens Factorization

In the Givens factorization, rotation matrices are used to reduce $\mathbf{K}$ into upper triangular form. In effect, each elementary matrix $\mathbf{L}_{j,i}^{-1}$ in (7.45) is replaced by a rotation matrix, so that we have

$$\mathbf{P}_G^T\mathbf{K} = \mathbf{S}_G \tag{7.47}$$

where

$$\mathbf{P}_G^T = \mathbf{P}_{n,n-1}^T \ldots (\mathbf{P}_{n,2}^T\mathbf{P}_{n-1,2}^T \ldots \mathbf{P}_{3,2}^T)(\mathbf{P}_{n,1}^T\mathbf{P}_{n-1,1}^T \ldots \mathbf{P}_{2,1}^T) \tag{7.48}$$

and $\mathbf{S}_G$ is an upper triangular matrix; the subscript G denotes that $\mathbf{S}_G$ is not equal to $\mathbf{S}$ in (7.10). The solution of the system of equations $\mathbf{KU} = \mathbf{R}$ in (7.1) is obtained using the decomposition of $\mathbf{K}$ established in (7.47),

$$\mathbf{K} = \mathbf{P}_G\mathbf{S}_G \tag{7.49}$$

Hence, analogously to the Gauss elimination solution, we need to reduce the load vector $\mathbf{R}$,

$$\mathbf{P}_G\mathbf{V} = \mathbf{R} \tag{7.50}$$

from which

$$\mathbf{V} = \mathbf{P}_G^T\mathbf{R} \tag{7.51}$$

and obtain the solution $\mathbf{U}$ by a back-substitution,

$$\mathbf{S}_G\mathbf{U} = \mathbf{V} \tag{7.52}$$

An important aspect is the evaluation of the rotation matrices $\mathbf{P}_{j,i}^T$, which must reduce to zero the elements in positions (j, i), $j > i$, of the coefficient matrix. A typical rotation matrix is given by

$$\mathbf{P}_{j,i}^T = \begin{bmatrix} 1 & & & & & & & & \\ & \ddots & & & & & & & \\ & & \cos\theta & & & & \sin\theta & & \\ & & & 1 & & & & & \\ & & & & \ddots & & & & \\ & & & & & 1 & & & \\ & & -\sin\theta & & & & \cos\theta & & \\ & & & & & & & 1 & \\ & & & & & & & & \ddots \end{bmatrix} \quad \begin{matrix} \\ \\ \leftarrow i\text{th} \\ \\ \\ \\ \leftarrow j\text{th row} \\ \\ \\ \end{matrix} \tag{7.53}$$

(with ith and jth columns marked above the $\cos\theta$ / $\sin\theta$ entries)

where θ is chosen in such way as to reduce to zero element (j, i).

Before deriving a typical matrix $P_{j,i}^T$, we should note the order in which the elements in the stiffness matrix are zeroed; namely, (7.48) shows that we use

$$i = 1, \qquad j = 2, 3, \ldots, n$$

and then

$$\begin{aligned} i &= 2, & j &= 3, 4, \ldots, n \\ &\vdots \\ i &= n-1, & j &= n \end{aligned}$$

Consider now the construction of the matrix $\mathbf{P}_{2,1}^T$, which is typical. The purpose of evaluating the matrix product $\tilde{\mathbf{K}} = \mathbf{P}_{2,1}^T\mathbf{K}$ is to reduce element (2, 1) in the original stiffness matrix to zero; i.e., we have

$$\tilde{\mathbf{K}} = \begin{bmatrix} \cos\theta & \sin\theta & & & & \\ -\sin\theta & \cos\theta & & & & \\ & & 1 & & & \\ & & & \ddots & & \\ & & & & & 1 \end{bmatrix} \begin{bmatrix} k_{11} & k_{12} & k_{13} & \ldots \\ k_{21} & k_{22} & k_{23} & \\ k_{31} & k_{32} & k_{33} & \\ \vdots & & & \end{bmatrix} \tag{7.54}$$

or

$$\tilde{\mathbf{K}} = \begin{bmatrix} \tilde{k}_{11} & \tilde{k}_{12} & \tilde{k}_{13} & \cdots \\ 0 & \tilde{k}_{22} & \tilde{k}_{23} & \\ \tilde{k}_{31} & \tilde{k}_{32} & \tilde{k}_{33} & \\ \cdot & & & \\ \cdot & & & \\ \cdot & & & \end{bmatrix}$$

The angle θ is chosen from the equation

$$-k_{11} \sin \theta + k_{21} \cos \theta = 0 \tag{7.55}$$

Hence

$$\cos \theta = \frac{k_{11}}{(k_{11}^2 + k_{21}^2)^{1/2}}; \qquad \sin \theta = \frac{k_{21}}{(k_{11}^2 + k_{21}^2)^{1/2}} \tag{7.56}$$

and if $k_{11}^2 + k_{21}^2 = 0$, no transformation is required; i.e., we use

$$\cos \theta = 1$$
$$\sin \theta = 0$$

It should be noted that it is not required that $k_{11} \neq 0.0$, as in the Gauss elimination procedure.

The premultiplication of the stiffness matrix $\mathbf{K}$ by the matrix $\mathbf{P}_{2,1}^T$ corresponds to linearly combining the first two rows of $\mathbf{K}$; i.e., for the elements of the resulting matrix $\tilde{\mathbf{K}}$, we have

$$\left.\begin{aligned} \tilde{k}_{1j} &= k_{1j} \cos \theta + k_{2j} \sin \theta \\ \tilde{k}_{2j} &= -k_{1j} \sin \theta + k_{2j} \cos \theta \end{aligned}\right\} j = 1, \ldots, n$$
$$\tilde{k}_{ij} = k_{ij} \qquad i = 3, \ldots, n; \quad j = 1, \ldots, n \tag{7.57}$$

which completes the operations with $\mathbf{P}_{2,1}^T$.

The elimination of the other subdiagonal elements, i.e., the evaluation of the other rotation matrices and the required matrix products, is carried out in an analogous way. We demonstrate the procedure in the following example.

Example 7.14

Use the Givens factorization to obtain the solution to the system of equations considered in Section 7.2.1, which rewritten are

$$\begin{bmatrix} 5 & -4 & 1 & 0 \\ -4 & 6 & -4 & 1 \\ 1 & -4 & 6 & -4 \\ 0 & 1 & -4 & 5 \end{bmatrix} \begin{bmatrix} U_1 \\ U_2 \\ U_3 \\ U_4 \end{bmatrix} = \begin{bmatrix} 0 \\ 1 \\ 0 \\ 0 \end{bmatrix}$$

Using the Givens method, we have for $i = 1, j = 2$,

$$\mathbf{P}_{2,1}^T = \begin{bmatrix} 0.780869 & -0.624695 & & \\ 0.624695 & 0.780869 & & \\ & & 1 & \\ & & & 1 \end{bmatrix}$$

Hence

$$\mathbf{P}_{2,1}^T\mathbf{K} = \begin{bmatrix} 6.403124 & -6.871646 & 3.279649 & -0.624695 \\ 0 & 2.186433 & -2.498780 & 0.780869 \\ 1 & -4 & 6 & -4 \\ 0 & 1 & -4 & 5 \end{bmatrix};$$

$$\mathbf{P}_{2,1}^T\mathbf{R} = \begin{bmatrix} -0.624695 \\ 0.780869 \\ 0 \\ 0 \end{bmatrix}$$

For the next step $i = 1, j = 3$, we use

$$\mathbf{P}_{3,1}^T = \begin{bmatrix} 0.988024 & 0 & 0.154303 & 0 \\ 0 & 1 & 0 & 0 \\ -0.154303 & 0 & 0.988024 & 0 \\ 0 & 0 & 0 & 1 \end{bmatrix}$$

Hence

$$\mathbf{P}_{3,1}^T\mathbf{P}_{2,1}^T\mathbf{K} = \begin{bmatrix} 6.480741 & -7.406561 & 4.166190 & -1.234427 \\ 0 & 2.186433 & -2.498780 & 0.780869 \\ 0 & -2.891776 & 5.422080 & -3.855702 \\ 0 & 1 & -4 & 5 \end{bmatrix}$$

$$\mathbf{P}_{3,1}^T\mathbf{P}_{2,1}^T\mathbf{R} = \begin{bmatrix} -0.617213 \\ 0.780869 \\ 0.096393 \\ 0 \end{bmatrix}$$

Writing down the results for the next steps:

$$i = 2, j = 3; \qquad \cos\theta = 0.603103, \sin\theta = -0.797664$$

$$\mathbf{P}_{3,2}^T\mathbf{P}_{3,1}^T\mathbf{P}_{2,1}^T\mathbf{K} = \begin{bmatrix} 6.480741 & -7.406561 & 4.166190 & -1.234427 \\ 0 & 3.625308 & -5.832017 & 3.547497 \\ 0 & 0 & 1.276885 & -1.702513 \\ 0 & 1 & -4 & 5 \end{bmatrix}$$

$$\mathbf{P}_{3,2}^T\mathbf{P}_{3,1}^T\mathbf{P}_{2,1}^T\mathbf{R} = \begin{bmatrix} -0.617213 \\ 0.394055 \\ 0.681005 \\ 0 \end{bmatrix}$$

$$i = 2, j = 4; \qquad \cos\theta = 0.963998, \sin\theta = 0.265908$$

$$\mathbf{P}_{4,2}^T\mathbf{P}_{3,2}^T\mathbf{P}_{3,1}^T\mathbf{P}_{2,1}^T\mathbf{K} = \begin{bmatrix} 6.480741 & -7.406561 & 4.166190 & -1.234427 \\ 0 & 3.760699 & -6.685687 & 4.748357 \\ 0 & 0 & 1.276885 & -1.7[\text{illegible}]513 \\ 0 & 0 & -2.305214 & 3.876950 \end{bmatrix}$$

$$\mathbf{P}_{4,2}^T\mathbf{P}_{3,2}^T\mathbf{P}_{3,1}^T\mathbf{P}_{2,1}^T\mathbf{R} = \begin{bmatrix} -0.617213 \\ 0.379869 \\ 0.681005 \\ -0.104782 \end{bmatrix}$$

In the last step we have

$$i = 3, j = 4; \qquad \cos\theta = 0.484544, \sin\theta = -0.874767$$

and obtain

$$\mathbf{P}_G^T\mathbf{K} = \begin{bmatrix} 6.480741 & -7.406561 & 4.166190 & -1.234427 \\ 0 & 3.760699 & -6.685687 & 4.748357 \\ 0 & 0 & 2.635231 & -4.216370 \\ 0 & 0 & 0 & 0.389249 \end{bmatrix} \tag{a}$$

$$\mathbf{P}_G^T\mathbf{R} = \begin{bmatrix} -0.617213 \\ 0.379869 \\ 0.421637 \\ 0.544949 \end{bmatrix} \tag{b}$$

The relations in (a) and (b) give $\mathbf{S}_G$ and $\mathbf{V}$, respectively. Using (7.52) for the back-substitution, we obtain

$$\mathbf{U} = \begin{bmatrix} 1.6 \\ 2.6 \\ 2.4 \\ 1.4 \end{bmatrix}$$

which is the solution already obtained in Section 7.2.1.

It should be noted that $\mathbf{S}_G = \mathbf{P}_G^T\mathbf{K}$ given in (a) is different from $\mathbf{S}$ obtained in Example 7.2, and that symmetry in the coefficient matrix is lost during the Givens reduction process.

7.3.2 The Householder Factorization

In the Householder reduction of $\mathbf{K}$ to upper triangular form, reflection matrices are used; i.e., we evaluate

$$\mathbf{P}_H^T\mathbf{K} = \mathbf{S}_H \tag{7.58}$$

where

$$\mathbf{P}_H^T = \mathbf{P}_{n-1}^T \ldots \mathbf{P}_2^T\mathbf{P}_1^T \tag{7.59}$$

and $\mathbf{S}_H$ is an upper triangular matrix. We shall prove in Example 7.15 that the upper triangular matrices $\mathbf{S}_G$ and $\mathbf{S}_H$ and orthogonal matrices $\mathbf{P}_G$ and $\mathbf{P}_H$ used in (7.47) and (7.58), respectively, are "essentially" the same. The matrices $\mathbf{P}_i^T$, $i = 1, \ldots, n-1$, are reflection matrices constructed in such a way that $\mathbf{P}_i^T$ reduces to zero the subdiagonal elements in column i of the current stiffness matrix. More specifically,

$$\mathbf{P}_i^T = \left[\begin{array}{c|c} \mathbf{I}_{i-1} & \mathbf{0} \\ \hline \mathbf{0} & \tilde{\mathbf{P}}_i \end{array}\right] \tag{7.60}$$

where $\mathbf{I}_{i-1}$ is the identity matrix of order $i-1$. Also,

$$\tilde{\mathbf{P}}_i = \mathbf{I} - \theta \mathbf{w}_i \mathbf{w}_i^T; \qquad \theta = \frac{2}{\mathbf{w}_i^T \mathbf{w}_i}$$

where $\tilde{\mathbf{P}}_i$ is a matrix and $\mathbf{w}_i$ is a vector of order $n - i + 1$. Since $\tilde{\mathbf{P}}_i$ is symmetric, we have that $\mathbf{P}_i$ is symmetric and $\mathbf{P}_i^T = \mathbf{P}_i$.

Once the factorization of $\mathbf{K}$ has been obtained, where $\mathbf{K} = \mathbf{P}_H \mathbf{S}_H$, the solution of the equations $\mathbf{KU} = \mathbf{R}$ is carried out as expressed in (7.50) to (7.52) (and replacing $\mathbf{S}_G$ by $\mathbf{S}_H$ and $\mathbf{P}_G$ by $\mathbf{P}_H$).

To demonstrate the reduction procedure of $\mathbf{K}$, consider the first step, i.e., the construction of $\mathbf{P}_1$, which is typical. Let

$$\tilde{\mathbf{K}} = \mathbf{P}_1 \mathbf{K} \tag{7.61}$$

and partition $\mathbf{K}$ as follows:

$$\mathbf{K} = [\mathbf{k}_1 \mid \mathbf{K}_1] \tag{7.62}$$

where $\mathbf{k}_1$ is the first column of $\mathbf{K}$. We also have

$$\mathbf{P}_1 = \mathbf{I} - \theta \mathbf{w}_1 \mathbf{w}_1^T \tag{7.63}$$

Since the requirement is that the subdiagonal elements in column 1 of $\tilde{\mathbf{K}}$ are zero, we have the condition that $\mathbf{P}_1\mathbf{k}_1$ is a vector with only the first element nonzero. This condition is fulfilled by choosing the vector $\mathbf{w}_1$ in the following way:

$$(\mathbf{I} - \theta \mathbf{w}_1 \mathbf{w}_1^T)\mathbf{k}_1 = \pm \|\mathbf{k}_1\|_2 \mathbf{e}_1 \tag{7.64}$$

where $\mathbf{e}_1$ is the n-dimensional unit vector. The relation (7.64) corresponds to a reflection of the vector $\mathbf{k}_1$ in the n-dimensional space in such a way that only the component corresponding to direction 1 is nonzero and is equal to the length of $\mathbf{k}_1$ (see Example 2.13).

To solve for $\mathbf{w}_1$, we obtain, from (7.64),

$$\mathbf{k}_1 - \bar{\theta} \mathbf{w}_1 = \pm \|\mathbf{k}_1\|_2 \mathbf{e}_1 \tag{7.65}$$

where $\bar{\theta}$ is a constant, $\bar{\theta} = \theta \mathbf{w}_1^T \mathbf{k}_1$. Since $\mathbf{w}_1$ determines only the direction of the reflection (see Fig 2.4), the vector may have arbitrary length and, for convenience, we set $\bar{\theta} = 1.0$. We thus obtain

$$\mathbf{w}_1 = \mathbf{k}_1 + \text{sign}\,(k_{11})\, \|\mathbf{k}_1\|_2 \mathbf{e}_1 \tag{7.66}$$

where the $+$ or $-$ sign in (7.65) is chosen with the objective that the absolute value of the first element in $\mathbf{w}_1$ is by $\|\mathbf{k}_1\|_2$ larger than the absolute value of k_{11}.

Considering now the calculation of $\tilde{\mathbf{K}}$ in (7.61), we should note that $\mathbf{P}_1$ is actually never evaluated explicitly, but we calculate

$$\mathbf{v}_1^T = \mathbf{w}_1^T \mathbf{K} \tag{7.67}$$

and then

$$\tilde{\mathbf{K}} = \mathbf{K} - \theta(\mathbf{w}_1 \mathbf{v}_1^T) \tag{7.68}$$

The other reflection matrices, $\mathbf{P}_2, \ldots, \mathbf{P}_{n-1}$, and required matrix products are obtained in a manner analogous to the evaluation of $\mathbf{P}_1$ and $\tilde{\mathbf{K}}$. We demonstrate the factorization in Example 7.16.

Example 7.15

Show that the upper triangular matrices $\mathbf{S}_G$ and $\mathbf{S}_H$ and orthogonal matrices $\mathbf{P}_G$ and $\mathbf{P}_H$ are "essentially" the same. The terminology "essentially" means that the rows in $\mathbf{S}_G$ and $\mathbf{S}_H$ and columns in $\mathbf{P}_G$ and $\mathbf{P}_H$ are the same apart from a multiplier of -1.

The relations in (7.49) and (7.58) give

$$\mathbf{K} = \mathbf{P}_G \mathbf{S}_G$$

and

$$\mathbf{K} = \mathbf{P}_H \mathbf{S}_H$$

Hence

$$\mathbf{P}_G \mathbf{S}_G = \mathbf{P}_H \mathbf{S}_H$$

or

$$\mathbf{P}_H^T \mathbf{P}_G = \mathbf{S}_H \mathbf{S}_G^{-1} = \mathbf{U} \tag{a}$$

where $\mathbf{U}$ must be an upper triangular matrix because $\mathbf{S}_H$ and $\mathbf{S}_G^{-1}$ are upper triangular matrices. We assumed in (a) that $\mathbf{S}_G$ is nonsingular, which is the case if $\mathbf{K}$ is positive definite. Using (a) we obtain

$$\mathbf{U}^T \mathbf{U} = \mathbf{I} \tag{b}$$

because $\mathbf{P}_H$ and $\mathbf{P}_G$ are orthogonal matrices; i.e.,

$$(\mathbf{P}_H^T \mathbf{P}_G)^T (\mathbf{P}_H^T \mathbf{P}_G) = \mathbf{I}$$

However, if we now equate elements of the matrices on either side of the relation in (b), we find that $\mathbf{U}$ must be a diagonal matrix with its diagonal elements $+1$ or -1. Using the relations in (a), we therefore conclude that $\mathbf{P}_H$ and $\mathbf{P}_G$ are equal except for a multiplier of -1 applied to their columns, and $\mathbf{S}_H$ and $\mathbf{S}_G$ are equal except for a multiplier of -1 applied to their rows.

Example 7.16

Use the Householder factorization to solve the equations considered in Example 7.14.

In the first step we have

$$\mathbf{w}_1 = \begin{bmatrix} 11.480741 \\ -4.000000 \\ 1.000000 \\ 0 \end{bmatrix}$$

and thus

$$\mathbf{P}_1^T\mathbf{K} = \begin{bmatrix} -6.480741 & 7.406561 & -4.166190 & 1.234427 \\ 0 & 2.025845 & -2.200050 & 0.569914 \\ 0 & -3.006461 & 5.550012 & -3.892478 \\ 0 & 1.000000 & -4.000000 & 5.000000 \end{bmatrix}$$

$$\mathbf{P}_1^T\mathbf{R} = \begin{bmatrix} 0.617213 \\ 0.784957 \\ 0.053761 \\ 0 \end{bmatrix}$$

In the second step we obtain

$$\mathbf{w}_2 = \begin{bmatrix} 5.786544 \\ -3.006461 \\ 1.000000 \end{bmatrix}$$

$$\mathbf{P}_2^T\mathbf{P}_1^T\mathbf{K} = \begin{bmatrix} -6.480741 & 7.406561 & -4.166190 & 1.234427 \\ 0 & -3.760699 & 6.685687 & -4.748357 \\ 0 & 0 & 0.933332 & -1.129313 \\ 0 & 0 & -2.464414 & 4.080924 \end{bmatrix}$$

$$\mathbf{P}_2^T\mathbf{P}_1^T\mathbf{R} = \begin{bmatrix} 0.617213 \\ -0.379869 \\ 0.658958 \\ -0.201299 \end{bmatrix}$$

And finally we have

$$\mathbf{w}_3 = \begin{bmatrix} 3.568564 \\ -2.464414 \end{bmatrix}$$

$$\mathbf{P}_3^T\mathbf{P}_2^T\mathbf{P}_1^T\mathbf{K} = \begin{bmatrix} -6.480741 & 7.406561 & -4.166190 & 1.234427 \\ 0 & -3.760699 & 6.685687 & -4.748357 \\ 0 & 0 & -2.635231 & 4.216370 \\ 0 & 0 & 0 & 0.389249 \end{bmatrix}$$

$$\mathbf{P}_3^T\mathbf{P}_2^T\mathbf{P}_1^T\mathbf{R} = \begin{bmatrix} 0.617213 \\ -0.379869 \\ -0.421637 \\ 0.544949 \end{bmatrix}$$

Now back-substituting, we obtain

$$\mathbf{U} = \begin{bmatrix} 1.6 \\ 2.6 \\ 2.4 \\ 1.4 \end{bmatrix}$$

As mentioned previously, the Givens and Householder factorizations are not frequently used in practice, primarily because the solution procedures are uneconomical when compared to the Gauss reduction scheme. During the factorization of a stiffness matrix, the symmetry and bandwidth properties are lost. An advantage of both the Givens and Householder factorizations is that they are numerically very stable and might be used effectively in the solution of ill-conditioned sets of equations. However, the main purpose of presenting the Givens and Householder factorization procedures was to introduce the use of rotation and reflection matrices, because we shall see in Chapter 11 that rotation and reflection matrices are used effectively in the solution of eigenproblems.

7.4 THE GAUSS–SEIDEL ITERATIVE SOLUTION

At this time practically all finite element large-scale computer programs use some form of Gauss elimination to solve the equilibrium equations $\mathbf{KU} = \mathbf{R}$. However, it is interesting to note that during the initial developments of the finite element method, iterative solution algorithms have been employed extensively, and much research is currently spent on improving various iterative solution schemes.[4-6,35-44] A basic disadvantage of an iterative solution is that the time of solution can only be estimated very approximately, because the number of iterations required for convergence depends on the condition number of the matrix $\mathbf{K}$ and whether effective acceleration factors are used. A direct solution using one of the techniques given in Section 7.2 is nearly always most effective. However, at least for instructive purposes, it is important to be familiar also with an iterative solution method that has been used extensively, namely, the Gauss–Seidel method. The objective in this section is to present briefly the basic formulas of the Gauss–Seidel iterative solution together with a physical interpretation of the procedure.

In the iterative solution of $\mathbf{KU} = \mathbf{R}$, it is necessary to use an initial estimate for the displacements $\mathbf{U}$, say $\mathbf{U}^{(1)}$, which, if no better value is known, may be a null vector. In the Gauss–Seidel iteration we then evaluate for $s = 1, 2, \ldots$:

$$U_i^{(s+1)} = k_{ii}^{-1} \left\{ R_i - \sum_{j=1}^{i-1} k_{ij} U_j^{(s+1)} - \sum_{j=i+1}^{n} k_{ij} U_j^{(s)} \right\} \tag{7.69}$$

where $U_i^{(s)}$ and R_i are the ith component of $\mathbf{U}$ and $\mathbf{R}$, and s indicates the cycle of iteration. Alternatively, we may write in matrix form,

$$\mathbf{U}^{(s+1)} = \mathbf{K}_D^{-1}\{\mathbf{R} - \mathbf{K}_L \mathbf{U}^{(s+1)} - \mathbf{K}_L^T \mathbf{U}^{(s)}\} \tag{7.70}$$

where $\mathbf{K}_D$ is a diagonal matrix, $\mathbf{K}_D = \text{diag}\,(k_{ii})$, and $\mathbf{K}_L$ is a lower triangular matrix with the elements k_{ij} such that:

$$\mathbf{K} = \mathbf{K}_L + \mathbf{K}_D + \mathbf{K}_L^T \tag{7.71}$$

The iteration is continued until the change in the current estimate of the displacement vector is small enough, i.e., until

$$\frac{\| \mathbf{U}^{(s+1)} - \mathbf{U}^{(s)} \|_2}{\| \mathbf{U}^{(s+1)} \|_2} < \epsilon \tag{7.72}$$

where ϵ is the convergence tolerance. The number of iterations depends on the "quality" of the starting vector $\mathbf{U}^{(1)}$ and on the conditioning of the matrix $\mathbf{K}$. But it is important to note that the iteration will always converge, provided that $\mathbf{K}$ is positive definite. Furthermore, the rate of convergence can be increased using overrelaxation, in which case the iteration is as follows:

$$\mathbf{U}^{(s+1)} = \mathbf{U}^{(s)} + \beta \mathbf{K}_D^{-1}\{\mathbf{R} - \mathbf{K}_L \mathbf{U}^{(s+1)} - \mathbf{K}_D \mathbf{U}^{(s)} - \mathbf{K}_L^T \mathbf{U}^{(s)}\} \tag{7.73}$$

where β is the overrelaxation factor. The optimum value of β depends on the matrix $\mathbf{K}$ but is usually between 1.3 and 1.9.

Example 7.17

Use the Gauss–Seidel iteration to solve the system of equations considered in Section 7.2.1.

The equations to be solved are

$$\begin{bmatrix} 5 & -4 & 1 & 0 \\ -4 & 6 & -4 & 1 \\ 1 & -4 & 6 & -4 \\ 0 & 1 & -4 & 5 \end{bmatrix} \begin{bmatrix} U_1 \\ U_2 \\ U_3 \\ U_4 \end{bmatrix} = \begin{bmatrix} 0 \\ 1 \\ 0 \\ 0 \end{bmatrix}$$

In the solution we use (7.73), which in this case becomes

$$\begin{bmatrix} U_1 \\ U_2 \\ U_3 \\ U_4 \end{bmatrix}^{(s+1)} = \begin{bmatrix} U_1 \\ U_2 \\ U_3 \\ U_4 \end{bmatrix}^{(s)} + \beta \begin{bmatrix} \frac{1}{5} & & & \\ & \frac{1}{6} & & \\ & & \frac{1}{6} & \\ & & & \frac{1}{5} \end{bmatrix}$$

$$\times \left\{ \begin{bmatrix} 0 \\ 1 \\ 0 \\ 0 \end{bmatrix} - \begin{bmatrix} 0 & & & \\ -4 & 0 & & \\ 1 & -4 & 0 & \\ 0 & 1 & -4 & 0 \end{bmatrix} \begin{bmatrix} U_1 \\ U_2 \\ U_3 \\ U_4 \end{bmatrix}^{(s+1)} - \begin{bmatrix} 5 & -4 & 1 & 0 \\ & 6 & -4 & 1 \\ & & 6 & -4 \\ & & & 5 \end{bmatrix} \begin{bmatrix} U_1 \\ U_2 \\ U_3 \\ U_4 \end{bmatrix}^{(s)} \right\}$$

We use as an initial guess

$$\mathbf{U}^{(1)} = \begin{bmatrix} 0 \\ 0 \\ 0 \\ 0 \end{bmatrix}$$

Consider first the solution without overrelaxation, i.e., $\beta = 1$. We obtain

$$\begin{bmatrix} U_1 \\ U_2 \\ U_3 \\ U_4 \end{bmatrix}^{(2)} = \begin{bmatrix} 0 \\ 0.167 \\ 0.111 \\ 0.0556 \end{bmatrix}; \qquad \begin{bmatrix} U_1 \\ U_2 \\ U_3 \\ U_4 \end{bmatrix}^{(3)} = \begin{bmatrix} 0.111 \\ 0.305 \\ 0.222 \\ 0.116 \end{bmatrix}$$

Using the convergence limit in (7.72) with $\epsilon = 0.001$, we have convergence after 104 iterations and

$$\begin{bmatrix} U_1 \\ U_2 \\ U_3 \\ U_4 \end{bmatrix}^{(104)} = \begin{bmatrix} 1.59 \\ 2.59 \\ 2.39 \\ 1.39 \end{bmatrix}$$

with the exact solution being

$$\begin{bmatrix} U_1 \\ U_2 \\ U_3 \\ U_4 \end{bmatrix} = \begin{bmatrix} 1.60 \\ 2.60 \\ 2.40 \\ 1.40 \end{bmatrix}$$

We now vary β and recalculate the solution. The following table gives the number of iterations required for convergence with $\epsilon = 0.001$ as a function of β:

β	1.0	1.1	1.2	1.3	1.4	1.5	1.6	1.7	1.8	1.9
Number of iterations	104	88	74	61	49	37	23	30	43	82

Hence, for this example, we find that the minimum number of iterations is required at $\beta = 1.6$.

It is instructive to identify the physical process that is followed in the solution procedure. For this purpose we note that on the right-hand side of (7.73) we evaluate an out-of-balance force corresponding to degree of freedom i,

$$Q_i^{(s)} = R_i - \sum_{j=1}^{i-1} k_{ij} U_j^{(s+1)} - \sum_{j=i}^{n} k_{ij} U_j^{(s)} \tag{7.74}$$

and then calculate an improved value for the corresponding displacement component, $U_i^{(s+1)}$, using

$$U_i^{(s+1)} = U_i^{(s)} + \beta k_{ii}^{-1} Q_i^{(s)} \tag{7.75}$$

where $i = 1, \ldots, n$. Assuming that $\beta = 1$, the correction to $U_i^{(s)}$ in (7.75) is calculated by applying the out-of-balance force $Q_i^{(s)}$ to the ith nodal degree of freedom, with all other nodal displacements kept fixed. The process is therefore identical to the moment-distribution procedure, which has been used extensively in hand calculation analysis of frames. However, faster convergence is achieved if the acceleration factor β is used.

Although direct solution methods are used almost exclusively in finite element analysis programs, it is important to recognize some advantages of the Gauss–Seidel method. The solution scheme might be used effectively in the problem areas of reanalysis and optimization,[45-50] although a form of Gauss elimination may still be more efficient. Namely, if in reanalysis a structure is changed only slightly, the previous solution is a good starting vector for the Gauss–Seidel iterative solution of the new structure.

Another advantage of the Gauss–Seidel method can be that the assembled stiffness matrix need not be formed and thus out-of-core solution is avoided, because all matrix multiplications can be carried out on the element level. For example, instead of calculating $\mathbf{K}_L\mathbf{U}^{(s+1)}$, we can evaluate $\sum_m \mathbf{K}_{Lm}\mathbf{U}^{(s+1)}$, where the summation goes over all finite elements and $\mathbf{K}_{Lm}$ is the contribution of the mth element to $\mathbf{K}_L$. The same procedure is generally used in the implementation of the central difference scheme for transient analysis (see Section 8.2.1).

7.5 SOLUTION ERRORS

In the preceding sections we presented various algorithms for the solution of $\mathbf{KU} = \mathbf{R}$. We applied the solution procedures for instructive purposes to some small problems, but in practical analysis the methods are employed on large systems of equations using the electronic digital computer. However, when using the computer it is important to note that the elements of the matrices can only be represented to a fixed number of digits, and this means that errors will be introduced in the solution.[1-10,51-59] The aim in this section is to discuss the solution errors that can occur in Gauss elimination, i.e., when any one of the algorithms described in Section 7.2 is employed, and to give guidelines for avoiding the introduction of large errors.

In order to identify the source of the errors, let us assume that we use a computer in which a number is represented using t digits in single precision. Then, to increase accuracy, double precision arithmetic may be specified, in which case each number is approximately represented using $2t$ digits. Table 7.1 gives the number t for some typical computers.

Considering the finite digit arithmetic computers, it is found that the t digits are used quite differently in different machines. However, most computers, in effect, perform the arithmetic operations and afterward "chop off" all digits beyond the number of digits carried. Therefore, for demonstration purposes we assume in this section that the computer at hand first adds,

TABLE 7.1 NUMBER OF DIGITS USED TO REPRESENT FLOATING-POINT NUMBER

Computer	Single Precision Arithmetic	Double Precision Arithmetic	t
UNIVAC	8	16	8
IBM	6	16	6
CDC	14	28	14

subtracts, multiplies, and divides two numbers exactly, and then to obtain the finite precision results, chops off all digits beyond the t digits used.

In order to demonstrate the finite precision arithmetic, assume that we want to solve the system of equations

$$\begin{bmatrix} 3.42521 & -3.42521 \\ -3.42521 & 101.2431 \end{bmatrix} \begin{bmatrix} U_1 \\ U_2 \end{bmatrix} = \begin{bmatrix} 1.3021 \\ 0.0 \end{bmatrix} \tag{7.76}$$

where **K** and **R** are given "exactly." The exact solution is (to 10 digits)

$$U_1 = 0.3934633449; \qquad U_2 = 0.0133114709 \tag{7.77}$$

Assume now that for a (hypothetical) computer at hand, $t = 3$; i.e., each number is represented to only 3 digits. In this case the solution to the equations in (7.76) would be obtained by first representing **K** and **R** using only the first three digits in each number, and then calculating the solution by always using only three-digit representations with chopping off of the additional digits. Employing the basic Gauss elimination algorithm (see Section 7.2.1), the solution would be as follows:

$$\begin{bmatrix} 3.42 & -3.42 \\ -3.42 & 101 \end{bmatrix} \begin{bmatrix} \hat{U}_1 \\ \hat{U}_2 \end{bmatrix} = \begin{bmatrix} 1.30 \\ 0 \end{bmatrix} \tag{7.78}$$

where the hats over U_1 and U_2 indicate that the solution of (7.78) is different from the solution of (7.76). Using that with chopping to three digits,

$$101 - \left(\frac{-3.42}{3.42}\right)(-3.42) = 101 - (-1)(-3.42) = 97.5$$

we obtain

$$\begin{bmatrix} 3.42 & -3.42 \\ 0.0 & 97.5 \end{bmatrix} \begin{bmatrix} \bar{U}_1 \\ \bar{U}_2 \end{bmatrix} = \begin{bmatrix} 1.30 \\ 1.30 \end{bmatrix} \tag{7.79}$$

where the bars over U_1 and U_2 indicate that we solve (7.78) approximately. Continuing to use three-digit chopping arithmetic, we have

$$\left.\begin{aligned} \bar{U}_2 &= \frac{1.30}{97.5} = 0.0133 \\ &\text{and} \\ \bar{U}_1 &= \frac{1}{3.42}[1.30 - (-3.42)(0.0133)] = \frac{1}{3.42}(1.34) \\ &= 0.391 \end{aligned}\right\} \tag{7.80}$$

Referring to the above example, we can identify two kinds of errors: a *truncation error* and *a round-off error*. The truncation error is the error arising because the exact matrix **K** and load vector **R** in (7.76) are only represented to three-digit precision, as given in (7.78). The round-off error is the error that arises in the solution of (7.78) because only three-digit arithmetic is used. Considering the situations in which each type of error would be large, we note that the truncation error can be large if the absolute magnitude of the elements in the matrix **K** including the diagonal elements varies by a large amount. The round-off error can be large if a small diagonal element d_{ii} is used that creates a large multiplier l_{ij}. The reason for the truncation and round-off errors to be large under the above conditions is that the basic operation in the factorization is a subtraction of a multiple of the pivot row from the rows below it. If in this operation numbers of widely different magnitudes are subtracted—that have, however, only been represented to a fixed number of digits—the errors in this basic operation can be relatively large. For general systems, matrix scaling may be used in some cases, in which the complete rows and columns of the matrix are scaled down or up to obtain matrix elements that are more equal in absolute magnitude.

As was pointed out above, the total error in the displacement solution is the sum of the truncation and the round-off errors. To identify the round-off errors and the truncation errors individually in the above example, we need to solve (7.78) exactly, in which case we obtain

$$\begin{bmatrix} 3.42 & -3.42 \\ 0 & 97.58 \end{bmatrix} \begin{bmatrix} \hat{U}_1 \\ \hat{U}_2 \end{bmatrix} = \begin{bmatrix} 1.30 \\ 1.30 \end{bmatrix} \tag{7.81}$$

and

$$\begin{aligned} \hat{U}_1 &= 0.3934393613 \\ \hat{U}_2 &= 0.0133224020 \end{aligned} \tag{7.82}$$

The error in the solution arising from initial truncation is therefore

$$\hat{\mathbf{r}} = \begin{bmatrix} U_1 \\ U_2 \end{bmatrix} - \begin{bmatrix} \hat{U}_1 \\ \hat{U}_2 \end{bmatrix} = \begin{bmatrix} 0.0000239836 \\ -0.0000109311 \end{bmatrix} \tag{7.83}$$

and the round-off error is

$$\bar{\mathbf{r}} = \begin{bmatrix} \hat{U}_1 \\ \hat{U}_2 \end{bmatrix} - \begin{bmatrix} \bar{U}_1 \\ \bar{U}_2 \end{bmatrix} = \begin{bmatrix} 0.0024393613 \\ 0.0000224020 \end{bmatrix} \tag{7.84}$$

The total error **r** is the sum of $\bar{\mathbf{r}}$ and $\hat{\mathbf{r}}$, or

$$\mathbf{r} = \begin{bmatrix} U_1 \\ U_2 \end{bmatrix} - \begin{bmatrix} \bar{U}_1 \\ \bar{U}_2 \end{bmatrix} = \begin{bmatrix} 0.0024633449 \\ 0.0000114709 \end{bmatrix} \tag{7.85}$$

In the above evaluation of the solution errors, we used the exact solutions to (7.76) and (7.81). In practical analyses, these exact solutions cannot be obtained; instead, double precision arithmetic could be employed to calculate close approximations to them.

Consider now that in a specific analysis the solution obtained to the equations $\mathbf{KU} = \mathbf{R}$ is $\bar{\mathbf{U}}$; i.e., because of truncation and round-off errors, $\bar{\mathbf{U}}$ is calculated instead of $\mathbf{U}$. It appears that the error in the solution can be obtained by evaluating a residual $\Delta\mathbf{R}$, where

$$\Delta\mathbf{R} = \mathbf{R} - \mathbf{K}\bar{\mathbf{U}} \tag{7.86}$$

In practice, $\Delta\mathbf{R}$ would be calculated using double precision arithmetic. Substituting $\mathbf{KU}$ for $\mathbf{R}$ into (7.86), we have for the solution error $\mathbf{r} = \mathbf{U} - \bar{\mathbf{U}}$,

$$\mathbf{r} = \mathbf{K}^{-1}\Delta\mathbf{R} \tag{7.87}$$

meaning that *although* $\Delta\mathbf{R}$ *may be small, the error in the solution may still be large*. On the other hand, for an accurate solution $\Delta\mathbf{R}$ must be small. Therefore, a small residual $\Delta\mathbf{R}$ is a necessary but not a sufficient condition for an accurate solution.

Example 7.18

Calculate $\Delta\mathbf{R}$ and $\mathbf{r}$ for the introductory example considered above.

Using the values for $\mathbf{R}$, $\mathbf{K}$, and $\bar{\mathbf{U}}$ given in (7.76) and (7.80), respectively, we obtain, using (7.86),

$$\Delta\mathbf{R} = \begin{bmatrix} 1.3021 \\ 0 \end{bmatrix} - \begin{bmatrix} 3.42521 & -3.42521 \\ -3.42521 & 101.2431 \end{bmatrix}\begin{bmatrix} 0.391 \\ 0.0133 \end{bmatrix}$$

or

$$\Delta\mathbf{R} = \begin{bmatrix} 0.00839818 \\ -0.00042520 \end{bmatrix}$$

Hence, using (7.87), we have

$$\mathbf{r} = \begin{bmatrix} 0.00253338 \\ 0.00008151 \end{bmatrix}$$

In this case, $\Delta\mathbf{R}$ and $\mathbf{r}$ are both relatively small, because $\mathbf{K}$ is well-conditioned.

Example 7.19

Consider the system of equations

$$\begin{bmatrix} 4.855 & -4 & 1 & 0 \\ -4 & 5.855 & -4 & 1 \\ 1 & -4 & 5.855 & -4 \\ 0 & 1 & -4 & 4.855 \end{bmatrix}\begin{bmatrix} U_1 \\ U_2 \\ U_3 \\ U_4 \end{bmatrix} = \begin{bmatrix} -1.59 \\ 1 \\ 1 \\ -1.64 \end{bmatrix}$$

Use six-digit arithmetic with chopping to calculate the solution.

Following the basic Gauss elimination process, we have

$$\begin{bmatrix} 4.85500 & -4 & 1 & 0 \\ 0 & 2.55944 & -3.17610 & 1 \\ 0 & -3.17610 & 5.64902 & -4 \\ 0 & 1 & -4 & 4.85500 \end{bmatrix} \begin{bmatrix} \bar{U}_1 \\ \bar{U}_2 \\ \bar{U}_3 \\ \bar{U}_4 \end{bmatrix} = \begin{bmatrix} -1.59000 \\ -0.309980 \\ 1.32719 \\ -1.64000 \end{bmatrix}$$

$$\begin{bmatrix} 4.85500 & -4 & 1 & 0 \\ 0 & 2.55944 & -3.17610 & 1 \\ 0 & 0 & 1.70771 & -2.75907 \\ 0 & 0 & -2.75907 & 4.46429 \end{bmatrix} \begin{bmatrix} \bar{U}_1 \\ \bar{U}_2 \\ \bar{U}_3 \\ \bar{U}_4 \end{bmatrix} = \begin{bmatrix} -1.59000 \\ -0.309980 \\ 0.942827 \\ -1.51888 \end{bmatrix}$$

$$\begin{bmatrix} 4.85500 & -4 & 1 & 0 \\ & 2.55944 & -3.17610 & 1 \\ & & 1.70771 & -2.75907 \\ & & & 0.006600 \end{bmatrix} \begin{bmatrix} \bar{U}_1 \\ \bar{U}_2 \\ \bar{U}_3 \\ \bar{U}_4 \end{bmatrix} = \begin{bmatrix} -1.59000 \\ -0.309980 \\ 0.942827 \\ 0.004390 \end{bmatrix}$$

The back-substitution yields

$$\bar{\mathbf{U}} = \begin{bmatrix} 0.686706 \\ 1.63768 \\ 1.62674 \\ 0.665151 \end{bmatrix}$$

The exact answer (to seven digits) is

$$\mathbf{U} = \begin{bmatrix} 0.7037247 \\ 1.6652256 \\ 1.6542831 \\ 0.6821567 \end{bmatrix}$$

Evaluating $\Delta\mathbf{R}$ as given in (7.86), we obtain

$$\Delta\mathbf{R} = \begin{bmatrix} -1.59 \\ 1 \\ 1 \\ -1.64 \end{bmatrix} - \begin{bmatrix} -1.59002237 \\ 0.99998340 \\ 0.99994470 \\ -1.639971895 \end{bmatrix} = \begin{bmatrix} 0.00002237 \\ 0.00001660 \\ 0.00005530 \\ -0.000028105 \end{bmatrix}$$

Also evaluating $\mathbf{r}$, we have

$$\mathbf{r} = \begin{bmatrix} 0.01702 \\ 0.02756 \\ 0.02754 \\ 0.01701 \end{bmatrix}$$

We therefore see that $\Delta\mathbf{R}$ is relatively much smaller than $\mathbf{r}$. Indeed, the displacement errors are of the order 1 to 2%, although the load errors seem to indicate an accurate solution of the equations.

Considering (7.87), we may expect that solution accuracy is difficult to obtain when the smallest eigenvalue of **K** is very small or nearly zero; i.e., the system can almost undergo rigid body motion. Namely, in that case the elements in $\mathbf{K}^{-1}$ are large and the solution errors may be large although $\mathbf{\Delta R}$ is small. Also, to substantiate this conclusion, we may realize that if λ_1 of **K** is small, the solution $\mathbf{KU} = \mathbf{R}$ may be thought of as one step of inverse iteration with a shift close to λ_1. But the analysis in Section 11.2.1 shows that in such a case the solution tends to have components of the corresponding eigenvector. These components now appear as solution errors.

To obtain more information on the solution errors, an analysis can be performed which shows that it is not only a small (near zero) eigenvalue λ_1 but a large ratio of the largest to the smallest eigenvalues of **K** that determines the solution errors. Namely, in the solution of $\mathbf{KU} = \mathbf{R}$, owing to truncation and round-off errors, we may assume that we in fact solve

$$(\mathbf{K} + \delta\mathbf{K})(\mathbf{U} + \delta\mathbf{U}) = \mathbf{R} \tag{7.88}$$

Assuming that $\delta\mathbf{K}\delta\mathbf{U}$ is small in relation to the other terms, we have approximately

$$\delta\mathbf{U} = -\mathbf{K}^{-1}\delta\mathbf{K}\mathbf{U} \tag{7.89}$$

or, taking norms,

$$\frac{\|\delta\mathbf{U}\|}{\|\mathbf{U}\|} \leq \text{cond}\,(\mathbf{K})\frac{\|\delta\mathbf{K}\|}{\|\mathbf{K}\|} \tag{7.90}$$

where cond (**K**) is the *condition number* of **K**,

$$\text{cond}\,(\mathbf{K}) = \frac{\lambda_n}{\lambda_1} \tag{7.91}$$

Therefore, *a large condition number means that solution errors are more likely.* Experience and tests have shown that in finite element analysis, round-off errors are not as serious as the initial truncation errors, and indeed it appears that for an error estimate only initial truncation errors must be considered.[59] To evaluate an estimate of the initial truncation errors, we can assume that for a t-digit precision computer,

$$\frac{\|\delta\mathbf{K}\|}{\|\mathbf{K}\|} = 10^{-t} \tag{7.92}$$

Also, assuming s-digit precision in the solution, we have

$$\frac{\|\delta\mathbf{U}\|}{\|\mathbf{U}\|} = 10^{-s} \tag{7.93}$$

Substituting (7.92) and (7.93) into (7.90), we obtain as an estimate of the number of accurate digits obtained in the solution,

$$s \geq t - \log_{10}[\text{cond}\,(\mathbf{K})] \tag{7.94}$$

Example 7.20

Calculate the condition number of the matrix **K** used in Example 7.19. Then estimate the accuracy that can be expected in the equation solution.

In this case we have

$$\lambda_1 = 0.000898$$

$$\lambda_4 = 12.9452$$

Hence

$$\text{cond}\,(\mathbf{K}) = 14415.6$$

and

$$\log_{10}\,[\text{cond}\,(\mathbf{K})] = 4.15883$$

Thus the number of accurate digits using six-digit arithmetic predicted using (7.94) are

$$s \geq 6 - 4.16$$

or one- to two-digit accuracy can be expected. Comparing this result with the results obtained in Example 7.19, we observe that, indeed, only one- to two-digit accuracy was obtained.

The condition number of **K** can in practice be evaluated approximately by calculating an upper bound for λ_n, say λ_n^u,

$$\lambda_n^u = ||\mathbf{K}|| \tag{7.95}$$

where any matrix norm may be used (see Example 7.21), and evaluating a lower bound for λ_1, say λ_1^l, using inverse iteration (see Section 11.2.1). We thus have

$$\text{cond}\,(\mathbf{K}) \doteq \frac{\lambda_n^u}{\lambda_1^l} \tag{7.96}$$

Example 7.21

Calculate an estimate on the condition number of the matrix **K** used in Example 7.19.

Here we have, using the ∞ norm (see Section 2.8),

$$||\mathbf{K}||_\infty = 14.855$$

and by inverse iteration we obtain $\lambda_1 = 0.0009$. Hence

$$\log_{10}\,[\text{cond}\,(\mathbf{K})] = 4.2176$$

and the conclusions reached in Example 7.20 are still valid.

Although much additional research is required in the area of finite element solution accuracy, the above considerations on round-off and truncation errors yield the following two important results:

(1) Both types of errors can be expected to be large if structures with widely varying stiffness are analyzed. Large stiffness differences may be due to different material moduli, or they may be the result of the

finite element modeling used, in which case a more effective model can frequently be chosen. This may be achieved by the use of finite elements that are nearly equal in size and have almost the same lengths in each dimension, the use of master–slave degrees of freedom (see Example 7.23) and relative degrees of freedom (see Example 7.24).

(2) Since truncation errors are most significant, to improve the solution accuracy it is necessary both to evaluate the stiffness matrix **K** and also the solution of **KU** = **R** in double precision. It is not sufficient (a) to evaluate **K** in single precision and then solve the equations in double precision (see Example 7.22), or (b) to evaluate **K** in single precision, solve the equations in single precision using a Gauss elimination procedure, and then iterate for an improvement in the solution employing, for example, the Gauss–Seidel method.

We demonstrate the two conclusions above by means of some simple examples.

Example 7.22

Consider the simple spring system in Fig. 7.11. Calculate the displacements when $k = 1$, $K = 10{,}000$ using four-digit arithmetic. The equilibrium equations of the system are

$$\begin{bmatrix} K & -K & 0 \\ -K & 2K & -K \\ 0 & -K & K+k \end{bmatrix} \begin{bmatrix} U_1 \\ U_2 \\ U_3 \end{bmatrix} = \begin{bmatrix} 1 \\ 0 \\ 1 \end{bmatrix}$$

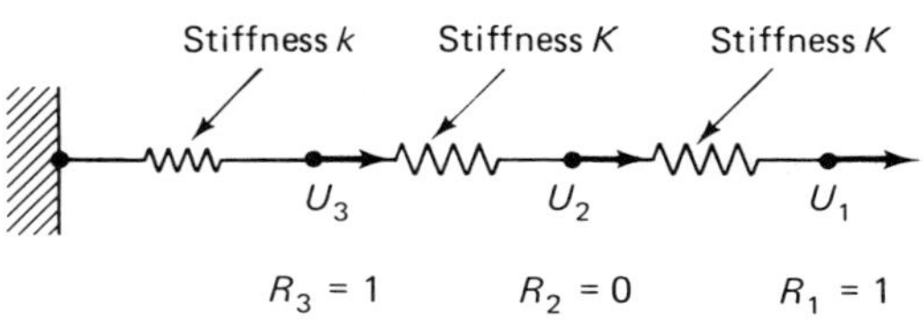

Figure 7.11. Simple spring system.

Substituting $K = 10{,}000$, $k = 1$, and using four-digit arithmetic, we have

$$\begin{bmatrix} 10{,}000 & -10{,}000 & 0 \\ -10{,}000 & 20{,}000 & -10{,}000 \\ 0 & -10{,}000 & 10{,}000 \end{bmatrix} \begin{bmatrix} U_1 \\ U_2 \\ U_3 \end{bmatrix} = \begin{bmatrix} 1 \\ 0 \\ 1 \end{bmatrix}$$

The triangularization of the coefficient matrix gives

$$\begin{bmatrix} 10{,}000 & -10{,}000 & 0 \\ 0 & 10{,}000 & -10{,}000 \\ 0 & 0 & 0 \end{bmatrix} \begin{bmatrix} U_1 \\ U_2 \\ U_3 \end{bmatrix} = \begin{bmatrix} 1.0 \\ 1.0 \\ 2.0 \end{bmatrix}$$

Hence a solution is not possible, because $d_{nn} = 0.0$.

To obtain a solution we may employ higher-digit arithmetic. In practice, this would mean that double precision arithmetic would be used, i.e., in this case, eight- instead of four-digit arithmetic.

Using eight-digit arithmetic (indeed five-digits would be sufficient), we obtain the exact solution as follows:

$$\begin{bmatrix} 10{,}000 & -10{,}000 & 0 \\ -10{,}000 & 20{,}000 & -10{,}000 \\ 0 & -10{,}000 & 10{,}001 \end{bmatrix} \begin{bmatrix} U_1 \\ U_2 \\ U_3 \end{bmatrix} = \begin{bmatrix} 1 \\ 0 \\ 1 \end{bmatrix}$$

$$\begin{bmatrix} 10{,}000 & -10{,}000 & 0 \\ 0 & 10{,}000 & -10{,}000 \\ 0 & 0 & 1 \end{bmatrix} \begin{bmatrix} U_1 \\ U_2 \\ U_3 \end{bmatrix} = \begin{bmatrix} 1.0 \\ 1.0 \\ 2.0 \end{bmatrix}$$

Hence

$$\mathbf{U} = \begin{bmatrix} 2.0002 \\ 2.0001 \\ 2.0 \end{bmatrix}$$

This example shows that a sufficient number of digits carried in the arithmetic may be vital for the solution not to break down.

Example 7.23

Use the master–slave solution procedure to analyze the system considered in Fig. 7.11.

The basic assumption in the master–slave analysis is

$$U_1 = U_2 = U_3$$

The equilibrium equation governing the system is thus

$$kU_1 = 2$$

Substituting for k, we obtain

$$U_1 = 2$$

and the complete solution is

$$\mathbf{U} = \begin{bmatrix} 2.0 \\ 2.0 \\ 2.0 \end{bmatrix}$$

This solution is approximate. However, comparing the solution with the exact result (see Example 7.22), we find that the main response is properly predicted.

Example 7.24

Use relative degrees of freedom to analyze the system in Fig. 7.11.

Using relative degrees of freedom, the displacement degrees of freedom defined are U_3, Δ_1, and Δ_2, where

$$U_2 = U_3 + \Delta_2$$
$$U_1 = U_2 + \Delta_1$$

or we have

$$\begin{bmatrix} U_1 \\ U_2 \\ U_3 \end{bmatrix} = \begin{bmatrix} 1 & 1 & 1 \\ 0 & 1 & 1 \\ 0 & 0 & 1 \end{bmatrix} \begin{bmatrix} \Delta_1 \\ \Delta_2 \\ U_3 \end{bmatrix} \tag{a}$$

The matrix relating the degrees of freedom Δ_1, Δ_2, and U_3 to the degrees of freedom U_1, U_2, and U_3 is the matrix $\mathbf{T}$. The equilibrium equations for the system using relative degrees of freedom are $(\mathbf{T}^T\mathbf{K}\mathbf{T})\mathbf{U} = \mathbf{T}^T\mathbf{R}$; i.e., the equilibrium equations are now

$$\begin{bmatrix} 10{,}000 & 0 & 0 \\ 0 & 10{,}000 & 0 \\ 0 & 0 & 1.0 \end{bmatrix} \begin{bmatrix} \Delta_1 \\ \Delta_2 \\ U_3 \end{bmatrix} = \begin{bmatrix} 1.0 \\ 1.0 \\ 2.0 \end{bmatrix} \tag{b}$$

with the solution

$$\Delta_1 = 0.0001$$
$$\Delta_2 = 0.0001$$
$$U_3 = 2.0$$

Hence we obtain

$$U_1 = 2.0002$$
$$U_2 = 2.0001$$
$$U_3 = 2.0000$$

Therefore, using four-digit arithmetic, we obtain the exact solution of the system if relative degrees of freedom are used (see Example 7.22). However, it should be noted that the equilibrium equations corresponding to the relative degrees of freedom would have to be formed directly, i.e., without the transformation used in this example.

REFERENCES

1. V. N. FADDEEVA, *Computational Methods of Linear Algebra*, Dover Publications, Inc., New York, N.Y., 1959.
2. G. E. FORSYTHE and C. B. MOLER, *Computer Solution of Linear Algebraic Systems*, Prentice-Hall, Inc., Englewood Cliffs, N.J., 1967.
3. V. V. KLYUYEV and N. I. KOKOVKIN-SCHERBAK, "On the Minimization of the Number of Arithmetic Operations for the Solution of Linear Algebraic Systems of Equations," Technical Report CS 24, Computer Science Department, Stanford University, 1965.
4. B. NOBLE, *Applied Linear Algebra*, Prentice-Hall, Inc., Englewood Cliffs, N.J., 1969.
5. C. E. FRÖBERG, *Introduction to Numerical Analysis*, Addison-Wesley Publishing Company, Inc., Reading, Mass., 1969.
6. S. H. CRANDALL, *Engineering Analysis*, McGraw-Hill Book Company, New York, N.Y., 1956.

7. J. H. WILKINSON, *The Algebraic Eigenvalue Problem*, Oxford University Press, Inc., London, 1965.
8. R. S. MARTIN, G. PETERS, and J. H. WILKINSON, "Symmetric Decomposition of a Positive Definite Matrix," *Numerische Mathematik*, Vol. 7, 1965, pp. 362–383.
9. C. LANCZOS, *Applied Analysis*, Prentice-Hall, Inc., Englewood Cliffs, N.J., 1956.
10. R. ZURMÜHL, *Matrizen*, Springer-Verlag, Berlin, 1964.
11. P. D. CROUT, "A Short Method for Evaluating Determinants and Solving Systems of Linear Equations with Real or Complex Coefficients," *A.I.E.E. Transactions*, Vol. 60, 1941, pp. 1235–1240.
12. E. L. WILSON, K. J. BATHE, and W. P. DOHERTY, "Direct Solution of Large Systems of Linear Equations," *Computers and Structures*, Vol. 4, pp. 363–372.
13. R. W. CLOUGH and J. PENZIEN, *Dynamics of Structures*, McGraw-Hill Book Company, New York, N.Y., 1975.
14. E. L. WILSON, "The Static Condensation Algorithm," *International Journal for Numerical Methods in Engineering*, Vol. 8, 1974, pp. 199–203.
15. E. L. WILSON, "Structural Analysis of Axisymmetric Solids," *A.I.A.A. Journal*, Vol. 3, 1965, pp. 2269–2274.
16. G. KRON, "Solving Highly Complex Elastic Structures in Easy Stages," *Journal of Applied Mechanics*, Vol. 22, 1955, pp. 235–244.
17. J. S. PRZEMIENIECKI, "Matrix Structural Analysis of Substructures," *A.I.A.A. Journal*, Vol. 1, 1963, pp. 138–147.
18. J. S. PRZEMIENIECKI, *Theory of Matrix Structural Analysis*, McGraw-Hill Book Company, New York, N.Y., 1968.
19. M. F. RUBINSTEIN, "Combined Analysis by Substructures and Recursion," *A.S.C.E., Journal of the Structural Division*, Vol. 93, No. ST2, Apr. 1967, pp. 231–235.
20. T. FURNIKE, "Computerized Multiple Level Substructuring Analysis," *Computers and Structures*, Vol. 2, 1972, pp. 1063–1073.
21. E. L. WILSON and H. H. DOVEY, "Static and Earthquake Analysis of Three-Dimensional Frame and Shear Wall Buildings," Report EERC 72-1, College of Engineering, University of California, Berkeley, 1972.
22. H. A. KAMEL, D. LIU, M. W. MCCABE, and V. PHILIPPOPOULOS, "Some Developments in the Analysis of Complex Ship Structures," *Advances in Computational Methods in Structural Mechanics and Design*, (J. T. Oden, R. W. Clough, Y. Yamamoto, eds.), University of Alabama Press, University of Alabama, Huntsville, Ala., 1972.
23. D. P. MONDKAR and G. H. POWELL, "Large Capacity Equation Solver for Structural Analysis," *Computers and Structures*, Vol. 4, 1974, pp. 699–728.
24. G. CANTIN, "An Equation Solver of Very Large Capacity," *International Journal for Numerical Methods in Engineering*, Vol. 3, 1971, pp. 379–388.
25. K. J. BATHE, E. L. WILSON, and F. E. PETERSON, "SAP IV—A Structural Analysis Program for Static and Dynamic Response of Linear Systems," Report EERC 73-11, College of Engineering University of California, Berkeley, June 1973, revised Apr. 1974.

26. K. J. Bathe, "ADINA—A Finite Element Program for Automatic Dynamic Incremental Nonlinear Analysis," Report 82448-1, Acoustics and Vibration Laboratory, Department of Mechanical Engineering, Massachusetts Institute of Technology, Cambridge, Mass., 1975.

27. EAC/EASE2—User Information Manual/Theoretical, Control Data Corporation Publication 84002700, Minneapolis, Minn., 1973.

28. W. Weaver, Jr., *Computer Programs for Structural Analysis*, Van Nostrand Reinhold Company, New York, 1967.

29. P. V. Marcal (ed.), "General Purpose Finite Element Computer Programs; Proceedings of Seminar A.S.M.E. Winter Annual Meeting, New York, Nov., 1970.

30. S. J. Fenves, N. Perrone, J. Robinson, and W. C. Schnobrich, *Numerical and Computer Methods in Structural Mechanics*, Academic Press, Inc., New York, N.Y., 1973.

31. W. Pilkey, K. Saczalski, and H. Schaeffer (eds.), *Structural Mechanics Computer Programs*, University Press of Virginia, Charlottesville, Va., 1974.

32. B. M. Irons, "A Frontal Solution Program for Finite Element Analysis," *International Journal for Numerical Methods in Engineering*, Vol. 2, 1970, pp. 5–32.

33. R. J. Melosh and R. M. Bamford, "Efficient Solution of Load-Deflection Equations," *A.S.C.E., Journal of the Structural Division*, Vol. 95, No. ST4, Apr. 1969, pp. 661–676.

34. T. K. Hellen, "A Frontal Solution for Finite Element Techniques," Central Electricity Generating Board, Report RD/B/N1459, Berkeley Nuclear Laboratories, Berkeley, Gloucestershire, 1969.

35. R. W. Clough and E. L. Wilson, "Stress Analysis of a Gravity Dam by the Finite Element Method," Proceedings, Symposium on the Use of Computers in Civil Engineering, Lisbon, Portugal, Oct. 1962.

36. V. B. Venkayya, "Iterative Method for the Analysis of Large Structural Systems," Air Force Flight Dynamics Laboratory, Wright-Patterson A.F.B., Ohio, Report AFFDL-TR-67-194, Apr. 1968.

37. R. V. Southwell, *Relaxation Methods in Theoretical Physics*, Oxford University Press, New York, 1946.

38. E. Stiefel, "On Some Relaxation Methods," *Zeitschrift für angewandte Mathematik und Physik*, Vol. 3, 1952, pp. 1–33.

39. M. F. Rubinstein and D. E. Wikholm, "Analysis by Group Iteration Using Substructures," *A.S.C.E., Journal of the Structural Division*, Vol. 94, No. ST2, Feb. 1968, pp. 363–375.

40. R. S. Varga, *Matrix Iterative Analysis*, Prentice-Hall, Inc., Englewood Cliffs, N.J., 1962.

41. R. Van Norton, "The Solution of Linear Equations by the Gauss–Seidel Method," in *Mathematical Methods for Digital Computers* (A. Ralston and H. S. Wilf, eds.), John Wiley & Sons, Inc., New York, N.Y., 1960.

42. F. BECKMAN, "The Solution of Linear Equations by the Conjugate Gradient Method," in *Mathematical Methods for Digital Computers*, Vol. 1 (A. Ralston and H. S. Wilf, eds.), John Wiley & Sons, Inc., New York, 1960.

43. J. K. REID, "On the Method of Conjugate Gradients for the Solution of Large Sparse Systems of Linear Equations," Conference on Large Sparse Sets of Linear Equations, St. Catherine's College, Oxford, Apr. 1970, pp. 231–254.

44. I. FRIED, "A Gradient Computational Procedure for the Solution of Large Problems Arising from the Finite Element Discretization Method," *International Journal for Numerical Methods in Engineering*, Vol. 2, 1970, pp. 477–494.

45. J. H. ARGYRIS, "The Matrix Analysis of Structures with Cutouts and Modifications," Proceedings, 9th International Congress of Applied Mechanics, Brussels, Vol. 67, 1956, pp. 131–142.

46. J. M. BENNETT, "Triangular Factors of Modified Matrices," *Numerische Mathematik*, Vol. 7, 1965, pp. 217–221.

47. J. SOBIESZCZANSKI, "Structural Modification by Perturbation Method," *A.S.C.E., Journal of the Structural Division*, Vol. 94, No. ST12, Dec. 1968, pp. 2799–2816.

48. D. KAVLIE and G. H. POWELL, "Efficient Reanalysis of Modified Structures," *A.S.C.E., Journal of the Structural Division*, Vol. 97, No. ST1, Jan. 1971, pp. 377–392 ("Discussion" by N. E. Wiberg, Oct. 1971).

49. J. H. ARGYRIS and J. R. ROY, "General Treatment of Structural Modifications," *A.S.C.E., Journal of the Structural Division*, Vol. 98, No. ST2, Feb. 1972, pp. 465–492.

50. B. MOHRAZ and R. N. WRIGHT, "Solving Topologically Modified Structures," paper presented at the National Symposium on Computerized Structural Analysis and Design, George Washington University, Washington, D.C., Mar. 1972.

51. J. H. WILKINSON, *Rounding Errors in Algebraic Processes*, Prentice-Hall, Inc., Englewood Cliffs, N.J., 1962.

52. J. VON NEUMANN and H. H. GOLDSTINE, "Numerical Inverting of Matrices of High Order," *Bulletin of the American Mathematical Society*, Vol. 53, 1947, pp. 1021–1099; and *Proceedings of the American Mathematical Society*, Vol. 2, 1951, pp. 188–202.

53. R. S. MARTIN, G. PETERS, and J. H. WILKINSON, "Iterative Refinement of the Solution of a Positive Definite System of Equations," *Numerische Mathematik*, Vol. 8, 1966, pp. 203–216.

54. J. H. WILKINSON, "The Solution of Ill-Conditioned Linear Equations," in *Mathematical Methods for Digital Computers*, Vol. 2 (A. Ralston and H. S. Wilf, eds.), John Wiley & Sons, Inc., New York, 1967.

55. F. L. BAUER, "Optimally Scaled Matrices," *Numerische Mathematik*, Vol. 5, No. 1, 1963, pp. 73–87.

56. B. M. IRONS, "Roundoff Criteria in Direct Stiffness Solutions," *A.I.A.A. Journal*, Vol. 6, No. 7, July 1968, pp. 1308–1312.

57. R. A. ROSANOFF, J. F. GLOUDEMAN, and S. LEVY, "Numerical Conditioning of Stiffness Matrix Formulations for Frame Structures," Proceedings, 2nd Conference on Matrix Methods in Structural Mechanics, AFFDL-TR-68-150, Wright-Patterson A.F.B., Ohio, 1968, pp. 1029–1060.

58. R. J. MELOSH, "Manipulation Errors in Finite Element Analysis," *Recent Advances in Matrix Methods of Structural Analysis and Design* (R. H. Gallagher, Y. Yamada, and J. T. Oden, eds.), University of Alabama Press, Huntsville, Ala., 1971.

59. J. R. ROY, "Numerical Error in Structural Solutions," *A.S.C.E., Journal of the Structural Division*, Vol. 97, No. ST4, Apr. 1971, pp. 1039–1054.

CHAPTER 8

SOLUTION OF EQUILIBRIUM EQUATIONS IN DYNAMIC ANALYSIS

8.1 INTRODUCTION

In Section 3.2.2 we derived the equations of equilibrium for a finite element system in motion, which, restating (3.76), were written

$$\mathbf{M}\ddot{\mathbf{U}} + \mathbf{C}\dot{\mathbf{U}} + \mathbf{K}\mathbf{U} = \mathbf{R} \tag{8.1}$$

where **M**, **C**, and **K** are the mass, damping, and stiffness matrices; **R** is the external load vector; and $\mathbf{U}$, $\dot{\mathbf{U}}$, and $\ddot{\mathbf{U}}$ are the displacement, velocity, and acceleration vectors of the finite element assemblage. It should be recalled that (8.1) was derived from considerations of statics at time t; i.e., (8.1) may be written

$$\mathbf{F}_I(t) + \mathbf{F}_D(t) + \mathbf{F}_E(t) = \mathbf{R}(t) \tag{8.2}$$

where $\mathbf{F}_I(t)$ are the inertia forces, $\mathbf{F}_I(t) = \mathbf{M}\ddot{\mathbf{U}}$, $\mathbf{F}_D(t)$ are the damping forces, $\mathbf{F}_D(t) = \mathbf{C}\dot{\mathbf{U}}$, and $\mathbf{F}_E(t)$ are the elastic forces, $\mathbf{F}_E(t) = \mathbf{K}\mathbf{U}$, all of them being time-dependent. Therefore, in dynamic analysis, in principle, static equilibrium at time t, which includes the effect of acceleration-dependent inertia forces and velocity-dependent damping forces, is considered. Vice versa, in static analysis the equations of motion in (8.1) are considered, with inertia and damping effects neglected.

The choice for a static or dynamic analysis (i.e., for including or neglecting velocity- and acceleration-dependent forces in the analysis) is usually decided by engineering judgment, the objective thereby being to reduce the analysis effort required. However, it should be realized that the assumptions of a static analysis should be justified, since otherwise the analysis results are meaningless. Indeed, in nonlinear analysis the assumption of neglecting inertia and damping forces may be so severe that a solution may be difficult or not possible to obtain.

Mathematically, (8.1) represents a system of linear differential equations of second order and, in principle, the solution to the equations can be obtained by standard procedures for the solution of differential equations with constant coefficients.[1-3] However, the procedures proposed for the solution of general systems of differential equations can become very expensive if the order of the matrices is large—unless specific advantage is taken of the special characteristics of the coefficient matrices **K**, **C**, and **M**. In practical finite element analysis, we are therefore mainly interested in a few effective methods and we will concentrate in the next sections on the presentation of those techniques.[4-19] The procedures that we will consider are divided into two methods of solution: direct integration and mode superposition. Although the two techniques may at first sight appear to be quite different, in fact, they are closely related, and the choice for one method or the other is determined only by their numerical effectiveness.

8.2 DIRECT INTEGRATION METHODS

In direct integration the equations in (8.1) *are integrated using a numerical step-by-step procedure, the term "direct" meaning that prior to the numerical integration, no transformation of the equations into a different form is carried out.* In essence, direct numerical integration is based on two ideas. First, instead of trying to satisfy (8.1) at any time t, it is aimed to satisfy (8.1) only at discrete time intervals Δt apart. This means that, basically, (static) equilibrium, which includes the effect of inertia and damping forces, is sought at discrete time points within the interval of solution. Therefore, it appears that all solution techniques employed in static analysis can probably also be used effectively in direct integration. The second idea on which a direct integration method is based is that a variation of displacements, velocities, and accelerations within each time interval Δt is assumed. As will be discussed in detail, it is the form of the assumption on the variation of displacements, velocities, and accelerations within each time interval that determines the accuracy, stability, and cost of the solution procedure.

In the following, assume that the displacement, velocity, and acceleration vectors at time 0, denoted by $\mathbf{U}_0$, $\dot{\mathbf{U}}_0$, and $\ddot{\mathbf{U}}_0$, respectively, are known, and let the solution to (8.1) be required from time 0 to time T. In the solution the time span under consideration, T, is subdivided into n equal time intervals Δt (i.e., $\Delta t = T/n$), and the integration scheme employed establishes an approximate solution at times $0, \Delta t, 2\,\Delta t, 3\,\Delta t, \ldots, t, t + \Delta t, \ldots, T$. Since an algorithm calculates the solution at the next required time from the solutions at the previous times considered, we derive the algorithms by assuming that the solutions at times $0, \Delta t, 2\,\Delta t, \ldots, t$ are known and that the solution at time $t+\Delta t$ is required next. The calculations performed to obtain the solution at time $t + \Delta t$ are typical for calculating the solution at time Δt later than considered so far, and thus establish the general algorithm which can be used to calculate the solution at all discrete time points.

In the following sections a few commonly used effective direct integration methods are presented. Considerations of accuracy, selection of time step size, and a discussion of the advantages of one method over the other are postponed until Chapter 9.

8.2.1 The Central Difference Method

If the equilibrium relation in (8.1) is regarded as a system of ordinary differential equations with constant coefficients, it follows that any convenient finite difference expressions to approximate the accelerations and velocities in terms of displacements can be used. Therefore, theoretically a large number of different finite difference expressions could be employed. However, the solution scheme should be effective, and it follows that only a few schemes need to be considered. One procedure that can be very effective in the solution of some problems is the central difference method,[1] in which it is assumed that

$$\ddot{\mathbf{U}}_t = \frac{1}{\Delta t^2}\{\mathbf{U}_{t-\Delta t} - 2\mathbf{U}_t + \mathbf{U}_{t+\Delta t}\} \tag{8.3}$$

The error in the expansion (8.3) is of order $(\Delta t)^2$, and to have the same order of error in the velocity expansion, we can use

$$\dot{\mathbf{U}}_t = \frac{1}{2\Delta t}(-\mathbf{U}_{t-\Delta t} + \mathbf{U}_{t+\Delta t}) \tag{8.4}$$

The displacement solution for time $t + \Delta t$ is obtained by considering (8.1) at time t, i.e.,

$$\mathbf{M}\ddot{\mathbf{U}}_t + \mathbf{C}\dot{\mathbf{U}}_t + \mathbf{K}\mathbf{U}_t = \mathbf{R}_t \tag{8.5}$$

Substituting the relations for $\ddot{\mathbf{U}}_t$ and $\dot{\mathbf{U}}_t$ in (8.3) and (8.4), respectively, into (8.5), we obtain

$$\left(\frac{1}{\Delta t^2}\mathbf{M} + \frac{1}{2\Delta t}\mathbf{C}\right)\mathbf{U}_{t+\Delta t} = \mathbf{R}_t - \left(\mathbf{K} - \frac{2}{\Delta t^2}\mathbf{M}\right)\mathbf{U}_t - \left(\frac{1}{\Delta t^2}\mathbf{M} - \frac{1}{2\Delta t}\mathbf{C}\right)\mathbf{U}_{t-\Delta t} \tag{8.6}$$

from which we can solve for $\mathbf{U}_{t+\Delta t}$. It should be noted that the solution of $\mathbf{U}_{t+\Delta t}$ is thus based on using the equilibrium conditions at time t; i.e. $\mathbf{U}_{t+\Delta t}$ is calculated by using (8.5). For this reason the integration procedure is called an *explicit integration method*, and it is noted that such integration schemes do not require a factorization of the (effective) stiffness matrix in the step-by-step solution. On the other hand, the Houbolt, Wilson, and Newmark methods, considered in the next sections, use the equilibrium conditions at time $t + \Delta t$ and are called *implicit integration methods.*

A second observation is that using the central difference method, the calculation of $\mathbf{U}_{t+\Delta t}$ involves $\mathbf{U}_t$ and $\mathbf{U}_{t-\Delta t}$. Therefore, to calculate the solution at time Δt, a special starting procedure must be used. Since $\mathbf{U}_0$, $\dot{\mathbf{U}}_0$, and $\ddot{\mathbf{U}}_0$ are known [note that with $\mathbf{U}_0$ and $\dot{\mathbf{U}}_0$ known, $\ddot{\mathbf{U}}_0$ can be calculated using (8.1) at time 0; see Example 8.1], the relations in (8.3) and (8.4) can be used to obtain $\mathbf{U}_{-\Delta t}$; i.e., we have

$$U_{-\Delta t}^{(i)} = U_0^{(i)} - \Delta t \dot{U}_0^{(i)} + \frac{\Delta t^2}{2} \ddot{U}_0^{(i)} \tag{8.7}$$

where the superscript (i) indicates the ith element of the vector considered. Table 8.1 summarizes the time integration scheme as it might be implemented in the computer.

TABLE 8.1 STEP-BY-STEP SOLUTION USING CENTRAL DIFFERENCE METHOD (GENERAL MASS AND DAMPING MATRICES)

A. *Initial Calculations:*

1. Form stiffness matrix $\mathbf{K}$, mass matrix $\mathbf{M}$, and damping matrix $\mathbf{C}$.
2. Initialize $\mathbf{U}_0$, $\dot{\mathbf{U}}_0$, and $\ddot{\mathbf{U}}_0$.
3. Select time step Δt, $\Delta t < \Delta t_{cr}$, and calculate integration constants:

$$a_0 = \frac{1}{\Delta t^2}; \qquad a_1 = \frac{1}{2\Delta t}; \qquad a_2 = 2a_0; \qquad a_3 = \frac{1}{a_2}$$

4. Calculate $\mathbf{U}_{-\Delta t} = \mathbf{U}_0 - \Delta t \dot{\mathbf{U}}_0 + a_3 \ddot{\mathbf{U}}_0$.
5. Form effective mass matrix $\hat{\mathbf{M}} = a_0 \mathbf{M} + a_1 \mathbf{C}$.
6. Triangularize $\hat{\mathbf{M}}$: $\hat{\mathbf{M}} = \mathbf{LDL}^T$.

B. *For Each Time Step:*

1. Calculate effective loads at time t:

$$\hat{\mathbf{R}}_t = \mathbf{R}_t - (\mathbf{K} - a_2\mathbf{M})\mathbf{U}_t - (a_0\mathbf{M} - a_1\mathbf{C})\mathbf{U}_{t-\Delta t}$$

2. Solve for displacements at time $t + \Delta t$:

$$\mathbf{LDL}^T \mathbf{U}_{t+\Delta t} = \hat{\mathbf{R}}_t$$

3. If required, evaluate accelerations and velocities at time t:

$$\ddot{\mathbf{U}}_t = a_0(\mathbf{U}_{t-\Delta t} - 2\mathbf{U}_t + \mathbf{U}_{t+\Delta t})$$
$$\dot{\mathbf{U}}_t = a_1(-\mathbf{U}_{t-\Delta t} + \mathbf{U}_{t+\Delta t})$$

Assume that the system has no physical damping; i.e., $\mathbf{C}$ is zero. In this case (8.6) reduces to

$$\left(\frac{1}{\Delta t^2}\mathbf{M}\right)\mathbf{U}_{t+\Delta t} = \hat{\mathbf{R}}_t \tag{8.8}$$

where

$$\hat{\mathbf{R}}_t = \mathbf{R}_t - \left(\mathbf{K} - \frac{2}{\Delta t^2}\mathbf{M}\right)\mathbf{U}_t - \left(\frac{1}{\Delta t^2}\mathbf{M}\right)\mathbf{U}_{t-\Delta t} \tag{8.9}$$

Therefore, if the mass matrix is diagonal, the system of equations in (8.1) can be solved without factorizing a matrix; i.e., only matrix multiplications are required to obtain the right-hand side effective load vector $\hat{\mathbf{R}}_t$, after which the displacement components are obtained, using

$$U_{t+\Delta t}^{(i)} = \hat{R}_t^{(i)}\left(\frac{\Delta t^2}{m_{ii}}\right) \tag{8.10}$$

where $U_{t+\Delta t}^{(i)}$ and $\hat{R}_t^{(i)}$ denote the ith components of the vectors $\mathbf{U}_{t+\Delta t}$ and $\hat{\mathbf{R}}_t$, respectively, and m_{ii} is the ith diagonal element of the mass matrix, and it is assumed that $m_{ii} > 0$.

If neither the stiffness nor the mass matrix of the element assemblage is to be triangularized, it is also not necessary to assemble $\mathbf{K}$ and $\mathbf{M}$. We have shown in Section 3.2.2 that

$$\mathbf{K} = \sum_i \mathbf{K}_i; \qquad \mathbf{M} = \sum_i \mathbf{M}_i \tag{8.11}$$

which means that $\mathbf{KU}_t$, $(2\mathbf{M}/\Delta t^2)\mathbf{U}_t$, and $(\mathbf{M}/\Delta t^2)\mathbf{U}_{t-\Delta t}$, as required in (8.9), can be calculated on the element level by summing the contributions from each element to the effective load vector. Hence $\hat{\mathbf{R}}_t$ is evaluated using

$$\hat{\mathbf{R}}_t = \mathbf{R}_t - \sum_i (\mathbf{K}_i\mathbf{U}_t) - \sum_i \frac{1}{\Delta t^2}\mathbf{M}_i(\mathbf{U}_{t-\Delta t} - 2\mathbf{U}_t) \tag{8.12}$$

where it is important to note that the products $\mathbf{K}_i\mathbf{U}_t$, and $\mathbf{M}_i(\mathbf{U}_{t-\Delta t} - 2\mathbf{U}_t)$ are evaluated using $\mathbf{K}_i$ and $\mathbf{M}_i$ in compacted form; i.e., $\mathbf{K}_i^e$ and $\mathbf{M}_i^e$ are employed (see Sections 3.2.1 and 6.2.3).

The advantage of using the central difference method in the form given in (8.10) and (8.12) now becomes apparent. Since no stiffness and mass matrices of the complete element assemblage need to be calculated, the solution can essentially be carried out on the element level and relatively little high-speed storage is required. The method becomes even more effective if element stiffness and mass matrices of subsequent elements are the same, because in that case it is only necessary to calculate or read from back-up storage the matrices corresponding to the first element in the series. Using the central difference scheme as given in (8.10) and (8.12), systems of very large order have been solved effectively.

Considering the shortcomings of the central difference method, it must be recognized that the effectiveness of the procedure depends on the use of a diagonal mass matrix and the neglect of general velocity-dependent damping forces. If only a diagonal damping matrix is included, the benefits of performing the solution on the element level are preserved. For practical purposes the disadvantage that only a diagonal mass matrix can be employed is usually not very serious, because good accuracy of solution can be obtained by using a fine-enough finite element discretization.

A second very important consideration in the use of the central difference scheme is that the integration method requires that the time step Δt is smaller than a critical value, Δt_{cr}, which can be calculated from the mass and stiffness properties of the complete element assemblage. More specifically, we will show in Chapter 9 that to obtain a valid solution,

$$\Delta t \leq \Delta t_{cr} = \frac{T_n}{\pi} \tag{8.13}$$

where T_n is the smallest period of the finite element assemblage and n is the order of the element system, respectively. The period T_n can be calculated using one of the techniques discussed in Chapters 11 and 12, or a lower bound on T_n may be evaluated using norms.

In the solution using (8.10), it was assumed that $m_{ii} > 0$ for all i. The relation in (8.13) states this requirement once more, because a zero diagonal element in a diagonal mass matrix means that the element assemblage has a zero period (see Section 10.2.4). In general, all diagonal elements of the mass matrix can be assumed to be larger than zero, in which case (8.13) gives a limit on the magnitude of the time step Δt that can be used in the integration. In the analysis of some problems (8.13) may not require an unduly small time step; but, in general, the time step that should be small enough for accuracy of integration may be many times larger than Δt_{cr} obtained from (8.13). Since the total cost of analysis is approximately inversely proportional to the magnitude of time step used, it follows that if the time step can be m times as large, the cost would be reduced by a factor of m.

We discuss the selection of the time step Δt for direct integration in Chapter 9. However, the reason for Δt to be artificially small in some analyses can already be explained using a simple argument. Assume that we consider the direct integration of the equilibrium equations in (8.1) when the order n of the matrices is relatively large, say n is 100, at least. The time step for the integration would be selected using (8.13). Assume that we now change the smallest diagonal element of the mass matrix to become very small and, in fact, nearly zero. As enumerated above, a diagonal element in the mass matrix cannot be exactly zero, because T_n would then be zero and the integration would not be possible. However, as the diagonal element in the mass matrix approaches zero, the smallest period of the system, and hence Δt_{cr}, approaches zero. Therefore, the reduction of one mass element necessitates a severe reduction in the time step size that can be used in the integration. On the other hand, since the order of the system is large, we would hardly expect that the dynamic response of the element assemblage changes very much when the smallest mass element is reduced, even to become zero. Hence the cost of analysis would be unduly large, only because of one very small mass element. The same condition is also reached when the stiffness of only one element is changed to become large.

Integration schemes that require the use of a time step Δt smaller than a critical time step Δt_{cr}, such as the central difference method, are said to be conditionally stable. If a time step is used larger than Δt_{cr}, the integration is unstable, meaning that any errors resulting from the numerical integration or round-off in the computer grow and make the response calculations worthless in most cases. The concept of stability of integration is very important, and we will discuss it further in Chapter 9. However, at this stage it is useful to consider the following example.

Example 8.1

Consider a simple system for which the governing equilibrium equations are

$$\begin{bmatrix} 2 & 0 \\ 0 & 1 \end{bmatrix}\begin{bmatrix} \ddot{U}_1 \\ \ddot{U}_2 \end{bmatrix} + \begin{bmatrix} 6 & -2 \\ -2 & 4 \end{bmatrix}\begin{bmatrix} U_1 \\ U_2 \end{bmatrix} = \begin{bmatrix} 0 \\ 10 \end{bmatrix} \tag{a}$$

The free vibration periods of the system are given in Example 8.7, where we find that $T_1 = 4.45$, $T_2 = 2.8$. Use the central difference method in direct integration with time steps (1) $\Delta t = T_2/10$ and (2) $\Delta t = 10T_2$ to calculate the response of the system for 12 steps. Assume that $\mathbf{U}_0 = \mathbf{0}$ and $\dot{\mathbf{U}}_0 = \mathbf{0}$.

The first step is to calculate $\ddot{\mathbf{U}}_0$ using the equations in (a) at time 0; i.e., we use

$$\begin{bmatrix} 2 & 0 \\ 0 & 1 \end{bmatrix}\ddot{\mathbf{U}}_0 + \begin{bmatrix} 6 & -2 \\ -2 & 4 \end{bmatrix}\begin{bmatrix} 0 \\ 0 \end{bmatrix} = \begin{bmatrix} 0 \\ 10 \end{bmatrix}$$

Hence

$$\ddot{\mathbf{U}}_0 = \begin{bmatrix} 0 \\ 10 \end{bmatrix}$$

Now we follow the calculations in Table 8.1.

Consider case (1), in which $\Delta t = 0.28$. We then have

$$a_0 = \frac{1}{(0.28)^2} = 12.8; \qquad a_1 = \frac{1}{(2)(0.28)} = 1.79$$

$$a_2 = 2a_0 = 25.5; \qquad a_3 = \frac{1}{a_2} = 0.0392$$

Hence

$$\mathbf{U}_{-\Delta t} = \begin{bmatrix} 0 \\ 0 \end{bmatrix} - 0.28\begin{bmatrix} 0 \\ 0 \end{bmatrix} + 0.0392\begin{bmatrix} 0 \\ 10 \end{bmatrix} = \begin{bmatrix} 0 \\ 0.392 \end{bmatrix}$$

$$\hat{\mathbf{M}} = 12.8\begin{bmatrix} 2 & 0 \\ 0 & 1 \end{bmatrix} + 1.79\begin{bmatrix} 0 & 0 \\ 0 & 0 \end{bmatrix}$$

$$= \begin{bmatrix} 25.5 & 0 \\ 0 & 12.8 \end{bmatrix}$$

The effective loads at time t are

$$\hat{\mathbf{R}}_t = \begin{bmatrix} 0 \\ 10 \end{bmatrix} + \begin{bmatrix} 45.0 & 2 \\ 2 & 21.5 \end{bmatrix} \mathbf{U}_t - \begin{bmatrix} 25.5 & 0 \\ 0 & 12.8 \end{bmatrix} \mathbf{U}_{t-\Delta t}$$

Hence we need to solve the following equations for each time step,

$$\begin{bmatrix} 25.5 & 0 \\ 0 & 12.8 \end{bmatrix} \mathbf{U}_{t+\Delta t} = \hat{\mathbf{R}}_t \tag{b}$$

It should be noted that the solution of the equations in (b) is trivial because the coefficient matrix is diagonal.

Calculating the solution to (b) for each time step, we obtain

Time	Δt	$2\Delta t$	$3\Delta t$	$4\Delta t$	$5\Delta t$	$6\Delta t$	$7\Delta t$	$8\Delta t$	$9\Delta t$	$10\Delta t$	$11\Delta t$	$12\Delta t$
$\mathbf{U}_t$	0 0.392	0.0307 1.45	0.168 2.83	0.487 4.14	1.02 5.02	1.70 5.26	2.40 4.90	2.91 4.17	3.07 3.37	2.77 2.78	2.04 2.54	1.02 2.60

The solution obtained is compared with the exact results in Example 8.7.

Consider now case (2), in which $\Delta t = 28$. Following through the same calculations, we find that

$$\mathbf{U}_{\Delta t} = \begin{bmatrix} 0 \\ 3.83 \times 10^3 \end{bmatrix}; \qquad \mathbf{U}_{2\Delta t} = \begin{bmatrix} 3.03 \times 10^6 \\ -1.21 \times 10^7 \end{bmatrix}$$

and the calculated displacements continue to increase. Since the time step Δt is about 6 times larger than T_1 and 10 times larger than T_2, we can certainly not expect accuracy in the numerical integration. But of particular interest is whether the calculated values decrease or increase. The increase in the values as observed in this example is a consequence of the time integration scheme not being stable. As pointed out above, the time step Δt must not be larger than Δt_{cr} for stability in the integration using the central difference method, where $\Delta t_{cr} = (1/\pi)T_2$. In this case the time step Δt is much larger, and the calculated response increases without bound. This is the typical phenomenon of instability. We shall see in Examples 8.2, 8.3, and 8.4 that the response predicted using $\Delta t = 28$ in the unconditionally stable Houbolt, Wilson θ, and Newmark methods is also very inaccurate, but does not increase.

We discussed above the main disadvantage of the central difference method: The scheme is only conditionally stable. Various other integration methods are also conditionally stable. Since the effective use of conditionally stable methods is restricted in structural analysis, we consider in the following sections commonly employed integration schemes which are unconditionally stable. The effectiveness of unconditionally stable integration schemes derives from the fact that to obtain accuracy in the integration, the time step Δt can be selected without a requirement such as (8.13), and in many cases Δt can be orders of magnitudes larger than (8.13) would allow. All integration methods discussed in the following are implicit; i.e., a triangularization of the stiffness matrix **K**, or rather an effective stiffness matrix, is required for solution.

8.2.2 The Houbolt Method

The Houbolt integration scheme is somewhat related to the previously discussed central difference method in that standard finite difference expressions are used to approximate the acceleration and velocity components in terms of the displacement components. The following finite difference expansions are employed in the Houbolt integration scheme:[11]

$$\ddot{\mathbf{U}}_{t+\Delta t} = \frac{1}{\Delta t^2}\{2\mathbf{U}_{t+\Delta t} - 5\mathbf{U}_t + 4\mathbf{U}_{t-\Delta t} - \mathbf{U}_{t-2\Delta t}\} \tag{8.14}$$

and

$$\dot{\mathbf{U}}_{t+\Delta t} = \frac{1}{6\Delta t}\{11\mathbf{U}_{t+\Delta t} - 18\mathbf{U}_t + 9\mathbf{U}_{t-\Delta t} - 2\mathbf{U}_{t-2\Delta t}\} \tag{8.15}$$

which are two backward-difference formulas with errors of order $(\Delta t)^2$.

In order to obtain the solution at time $t + \Delta t$, we now consider (8.1) at time $t + \Delta t$ (and not at time t as for the central difference method), which gives

$$\mathbf{M}\ddot{\mathbf{U}}_{t+\Delta t} + \mathbf{C}\dot{\mathbf{U}}_{t+\Delta t} + \mathbf{K}\mathbf{U}_{t+\Delta t} = \mathbf{R}_{t+\Delta t} \tag{8.16}$$

Substituting (8.14) and (8.15) into (8.16) and arranging all known vectors on the right-hand side, we obtain for the solution of $\mathbf{U}_{t+\Delta t}$,

$$\begin{aligned}\left(\frac{2}{\Delta t^2}\mathbf{M} + \frac{11}{6\Delta t}\mathbf{C} + \mathbf{K}\right)\mathbf{U}_{t+\Delta t} &= \mathbf{R}_{t+\Delta t} + \left(\frac{5}{\Delta t^2}\mathbf{M} + \frac{3}{\Delta t}\mathbf{C}\right)\mathbf{U}_t \\ &\quad - \left(\frac{4}{\Delta t^2}\mathbf{M} + \frac{3}{2\Delta t}\mathbf{C}\right)\mathbf{U}_{t-\Delta t} + \left(\frac{1}{\Delta t^2}\mathbf{M} + \frac{1}{3\Delta t}\mathbf{C}\right)\mathbf{U}_{t-2\Delta t}\end{aligned} \tag{8.17}$$

As shown in (8.17), the solution of $\mathbf{U}_{t+\Delta t}$ requires knowledge of $\mathbf{U}_t$, $\mathbf{U}_{t-\Delta t}$, and $\mathbf{U}_{t-2\Delta t}$. Although the knowledge of $\mathbf{U}_0$, $\dot{\mathbf{U}}_0$, and $\ddot{\mathbf{U}}_0$ is useful to start the Houbolt integration scheme, it is more accurate to calculate $\mathbf{U}_{\Delta t}$ and $\mathbf{U}_{2\Delta t}$ by some other means; i.e., we employ special starting procedures. One way of proceeding is to integrate (8.1) for the solution of $\mathbf{U}_{\Delta t}$ and $\mathbf{U}_{2\Delta t}$ using a different integration scheme, possibly a conditionally stable method such as the central difference scheme with a fraction of Δt as the time step (see Example 8.2). Table 8.2 summarizes the Houbolt integration procedure for use in a computer program.

A basic difference between the Houbolt method in Table 8.2 and the central difference scheme in Table 8.1 is the appearance of the stiffness matrix **K** as a factor to the required displacements $\mathbf{U}_{t+\Delta t}$. The term $\mathbf{K}\mathbf{U}_{t+\Delta t}$ appears because in (8.16) equilibrium is considered at time $t + \Delta t$ and not at time t as in the central difference method. The Houbolt method is, for this reason, an implicit integration scheme, whereas the central difference method was an explicit procedure. With regard to the time step Δt that can be used in the integration, there is no critical time-step limit, and Δt can in general be selected much larger than given in (8.13) for the central difference method.

A noteworthy point is that the step-by-step solution scheme based on the Houbolt method reduces directly to a static analysis, if mass and damping effects are neglected, whereas the central difference method solution in Table 8.1 could not be used. In other words, if $\mathbf{C} = \mathbf{0}$ and $\mathbf{M} = \mathbf{0}$, the solution method in Table 8.2 yields the static solution for time-dependent loads.

TABLE 8.2 Step-by-Step Solution Using Houbolt Integration Method

---

A. *Initial Calculations:*

1. Form stiffness matrix $\mathbf{K}$, mass matrix $\mathbf{M}$, and damping matrix $\mathbf{C}$.
2. Initialize $\mathbf{U}_0$, $\dot{\mathbf{U}}_0$, and $\ddot{\mathbf{U}}_0$.
3. Select time step Δt and calculate integration constants:

$$a_0 = \frac{2}{\Delta t^2}; \quad a_1 = \frac{11}{6\Delta t}; \quad a_2 = \frac{5}{\Delta t^2}; \quad a_3 = \frac{3}{\Delta t}; \quad a_4 = -2a_0;$$

$$a_5 = \frac{-a_3}{2}; \quad a_6 = \frac{a_0}{2}; \quad a_7 = \frac{a_3}{9}$$

4. Use special starting procedure to calculate $\mathbf{U}_{\Delta t}$ and $\mathbf{U}_{2\Delta t}$.
5. Calculate effective stiffness matrix $\hat{\mathbf{K}}$: $\hat{\mathbf{K}} = \mathbf{K} + a_0\mathbf{M} + a_1\mathbf{C}$.
6. Triangularize $\hat{\mathbf{K}}$: $\hat{\mathbf{K}} = \mathbf{LDL}^T$.

B. *For Each Time Step:*

1. Calculate effective loads at time $t + \Delta t$:

$$\hat{\mathbf{R}}_{t+\Delta t} = \mathbf{R}_{t+\Delta t} + \mathbf{M}(a_2\mathbf{U}_t + a_4\mathbf{U}_{t-\Delta t} + a_6\mathbf{U}_{t-2\Delta t}) + \mathbf{C}(a_3\mathbf{U}_t + a_5\mathbf{U}_{t-\Delta t} + a_7\mathbf{U}_{t-2\Delta t})$$

2. Solve for displacements at time $t + \Delta t$:

$$\mathbf{LDL}^T\mathbf{U}_{t+\Delta t} = \hat{\mathbf{R}}_{t+\Delta t}$$

3. If required, evaluate accelerations and velocities at time $t + \Delta t$:

$$\ddot{\mathbf{U}}_{t+\Delta t} = a_0\mathbf{U}_{t+\Delta t} - a_2\mathbf{U}_t - a_4\mathbf{U}_{t-\Delta t} - a_6\mathbf{U}_{t-2\Delta t}$$

$$\dot{\mathbf{U}}_{t+\Delta t} = a_1\mathbf{U}_{t+\Delta t} - a_3\mathbf{U}_t - a_5\mathbf{U}_{t-\Delta t} - a_7\mathbf{U}_{t-2\Delta t}$$

Example 8.2

Use the Houbolt direct integration scheme to calculate the response of the system considered in Example 8.1.

First we consider the case $\Delta t = 0.28$. We then have, following Table 8.2:

$$a_0 = 25.5; \quad a_1 = 6.55; \quad a_2 = 63.8; \quad a_3 = 10.7;$$

$$a_4 = -51.0; \quad a_5 = -5.36; \quad a_6 = 12.8; \quad a_7 = 1.19$$

To start the integration we need $\mathbf{U}_{\Delta t}$ and $\mathbf{U}_{2\Delta t}$. Let us here use simply the values calculated with the central difference method in Example 8.1, i.e.,

$$\mathbf{U}_{\Delta t} = \begin{bmatrix} 0.0 \\ 0.392 \end{bmatrix}; \qquad \mathbf{U}_{2\Delta t} = \begin{bmatrix} 0.0307 \\ 1.45 \end{bmatrix}$$

Next we calculate $\hat{\mathbf{K}}$ and obtain

$$\hat{\mathbf{K}} = \begin{bmatrix} 6 & -2 \\ -2 & 4 \end{bmatrix} + 25.5 \begin{bmatrix} 2 & 0 \\ 0 & 1 \end{bmatrix} = \begin{bmatrix} 57 & -2 \\ -2 & 29.5 \end{bmatrix}$$

For each time step we need $\hat{\mathbf{R}}_{t+\Delta t}$, which is in this case

$$\hat{\mathbf{R}}_{t+\Delta t} = \begin{bmatrix} 0 \\ 10 \end{bmatrix} + \begin{bmatrix} 2 & 0 \\ 0 & 1 \end{bmatrix} \{63.8\mathbf{U}_t - 51.0\mathbf{U}_{t-\Delta t} + 12.8\mathbf{U}_{t-2\Delta t}\}$$

Solving $\hat{\mathbf{K}}\mathbf{U}_{t+\Delta t} = \hat{\mathbf{R}}_{t+\Delta t}$ for 12 time steps, we obtain

Time	Δt	$2\Delta t$	$3\Delta t$	$4\Delta t$	$5\Delta t$	$6\Delta t$	$7\Delta t$	$8\Delta t$	$9\Delta t$	$10\Delta t$	$11\Delta t$	$12\Delta t$
$\mathbf{U}_t$	0 0.392	0.0307 1.45	0.167 2.80	0.461 4.08	0.923 5.02	1.50 5.43	2.11 5.31	2.60 4.77	2.86 4.01	2.80 3.24	2.40 2.63	1.72 2.28

The solution obtained is compared with the exact results in Example 8.7.

Next we consider the case $\Delta t = 28$ in order to observe the unconditional stability of the Houbolt operator. To start the integration we use the exact response at times Δt and $2\Delta t$ (see Example 8.7),

$$\mathbf{U}_{\Delta t} = \begin{bmatrix} 2.23 \\ 2.28 \end{bmatrix}; \qquad \mathbf{U}_{2\Delta t} = \begin{bmatrix} 2.98 \\ 3.17 \end{bmatrix}$$

It is interesting to compare $\hat{\mathbf{K}}$ with $\mathbf{K}$,

$$\hat{\mathbf{K}} = \begin{bmatrix} 6 & -2 \\ -2 & 4 \end{bmatrix} + 0.00255 \begin{bmatrix} 2 & 0 \\ 0 & 1 \end{bmatrix} = \begin{bmatrix} 6.0051 & -2.0000 \\ -2.0000 & 4.00255 \end{bmatrix}$$

where it is noted that $\hat{\mathbf{K}}$ is almost equal to $\mathbf{K}$. The displacement response over 12 time steps is given in the following table:

Time	Δt	$2\Delta t$	$3\Delta t$	$4\Delta t$	$5\Delta t$	$6\Delta t$	$7\Delta t$	$8\Delta t$	$9\Delta t$	$10\Delta t$	$11\Delta t$	$12\Delta t$
$\mathbf{U}_t$	2.23 2.28	2.98 3.17	1.00 3.00	1.00 3.00	1.00 3.00	1.00 3.00	1.00 3.00	1.00 3.00	1.00 3.00	1.00 3.00	1.00 3.00	1.00 3.00

It should be noted that the static solution is

$$\mathbf{U}_t = \begin{bmatrix} 1.0 \\ 3.0 \end{bmatrix}$$

Therefore, the displacement response very rapidly approaches the static solution.

8.2.3 The Wilson θ Method

The Wilson θ method is essentially an extension of the linear acceleration method, in which a linear variation of acceleration from time t to time $t + \Delta t$ is assumed. Referring to Fig. 8.1 in the Wilson θ method, the acceleration is assumed to be linear from time t to time $t + \theta\,\Delta t$, where $\theta \geq 1.0$.[14] When $\theta = 1.0$, the method reduces to the linear acceleration scheme, but we will

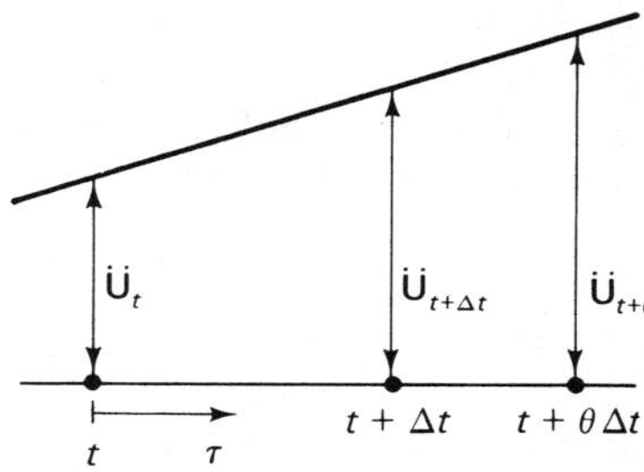

Figure 8.1. Linear acceleration assumption of Wilson θ method.

show in Chapter 9 that for unconditional stability we need to use $\theta \geq 1.37$, and usually we employ $\theta = 1.40$.

Let τ denote the increase in time, where $0 \leq \tau \leq \theta\Delta t$; then for the time interval t to $t + \theta\,\Delta t$, it is assumed that

$$\ddot{\mathbf{U}}_{t+\tau} = \ddot{\mathbf{U}}_t + \frac{\tau}{\theta\Delta t}(\ddot{\mathbf{U}}_{t+\theta\Delta t} - \ddot{\mathbf{U}}_t) \tag{8.18}$$

Integrating (8.18), we obtain

$$\dot{\mathbf{U}}_{t+\tau} = \dot{\mathbf{U}}_t + \ddot{\mathbf{U}}_t\tau + \frac{\tau^2}{2\theta\,\Delta t}(\ddot{\mathbf{U}}_{t+\theta\Delta t} - \ddot{\mathbf{U}}_t) \tag{8.19}$$

and

$$\mathbf{U}_{t+\tau} = \mathbf{U}_t + \dot{\mathbf{U}}_t\tau + \frac{1}{2}\ddot{\mathbf{U}}_t\tau^2 + \frac{1}{6\theta\,\Delta t}\tau^3(\ddot{\mathbf{U}}_{t+\theta\Delta t} - \ddot{\mathbf{U}}_t) \tag{8.20}$$

Using (8.19) and (8.20), we have, at time $t + \theta\,\Delta t$,

$$\dot{\mathbf{U}}_{t+\theta\Delta t} = \dot{\mathbf{U}}_t + \frac{\theta\,\Delta t}{2}(\ddot{\mathbf{U}}_{t+\theta\Delta t} + \ddot{\mathbf{U}}_t) \tag{8.21}$$

$$\mathbf{U}_{t+\theta\Delta t} = \mathbf{U}_t + \theta\,\Delta t\,\dot{\mathbf{U}}_t + \frac{\theta^2\,\Delta t^2}{6}(\ddot{\mathbf{U}}_{t+\theta\Delta t} + 2\ddot{\mathbf{U}}_t) \tag{8.22}$$

from which we can solve for $\ddot{\mathbf{U}}_{t+\theta\Delta t}$ and $\dot{\mathbf{U}}_{t+\theta\Delta t}$ in terms of $\mathbf{U}_{t+\theta\Delta t}$:

$$\ddot{\mathbf{U}}_{t+\theta\Delta t} = \frac{6}{\theta^2\,\Delta t^2}(\mathbf{U}_{t+\theta\Delta t} - \mathbf{U}_t) - \frac{6}{\theta\,\Delta t}\dot{\mathbf{U}}_t - 2\ddot{\mathbf{U}}_t \tag{8.23}$$

and

$$\dot{\mathbf{U}}_{t+\theta\Delta t} = \frac{3}{\theta\,\Delta t}(\mathbf{U}_{t+\theta\Delta t} - \mathbf{U}_t) - 2\dot{\mathbf{U}}_t - \frac{\theta\,\Delta t}{2}\ddot{\mathbf{U}}_t \tag{8.24}$$

To obtain the solution for the displacements, velocities, and accelerations at time $t + \Delta t$, the equilibrium equations (8.1) are considered at time $t + \theta\,\Delta t$. However, because the accelerations are assumed to vary linearly, a linearly projected load vector is used; i.e., the equation employed is

$$\mathbf{M}\ddot{\mathbf{U}}_{t+\theta\Delta t} + \mathbf{C}\dot{\mathbf{U}}_{t+\theta\Delta t} + \mathbf{K}\mathbf{U}_{t+\theta\Delta t} = \bar{\mathbf{R}}_{t+\theta\Delta t} \tag{8.25}$$

where

$$\bar{\mathbf{R}}_{t+\theta\Delta t} = \mathbf{R}_t + \theta(\mathbf{R}_{t+\Delta t} - \mathbf{R}_t) \tag{8.26}$$

Substituting (8.23) and (8.24) into (8.25), an equation is obtained from which $\mathbf{U}_{t+\theta\Delta t}$ can be solved. Then substituting $\mathbf{U}_{t+\theta\Delta t}$, into (8.23), we obtain $\ddot{\mathbf{U}}_{t+\Delta\theta t}$, which is used in (8.18), (8.19), and (8.20), all evaluated at $\tau = \Delta t$ to calculate $\ddot{\mathbf{U}}_{t+\Delta t}$, $\dot{\mathbf{U}}_{t+\Delta t}$, and $\mathbf{U}_{t+\Delta t}$. The complete algorithm used in the integration is given in Table 8.3.

TABLE 8.3 STEP-BY-STEP SOLUTION USING WILSON θ INTEGRATION METHOD

A. *Initial Calculations:*

1. Form stiffness matrix $\mathbf{K}$, mass matrix $\mathbf{M}$, and damping matrix $\mathbf{C}$.
2. Initialize $\mathbf{U}_0$, $\dot{\mathbf{U}}_0$, and $\ddot{\mathbf{U}}_0$.
3. Select time step Δt and calculate integration constants, $\theta = 1.4$ (usually):

$$a_0 = \frac{6}{(\theta\,\Delta t)^2}; \qquad a_1 = \frac{3}{\theta\,\Delta t}; \qquad a_2 = 2a_1; \qquad a_3 = \frac{\theta\,\Delta t}{2};$$

$$a_4 = \frac{a_0}{\theta}; \qquad a_5 = \frac{-a_2}{\theta}; \qquad a_6 = 1 - \frac{3}{\theta}; \qquad a_7 = \frac{\Delta t}{2};$$

$$a_8 = \frac{\Delta t^2}{6}$$

4. Form effective stiffness matrix $\hat{\mathbf{K}}$: $\hat{\mathbf{K}} = \mathbf{K} + a_0\mathbf{M} + a_1\mathbf{C}$.
5. Triangularize $\hat{\mathbf{K}}$: $\hat{\mathbf{K}} = \mathbf{L}\mathbf{D}\mathbf{L}^T$.

B. *For Each Time Step:*

1. Calculate effective loads at time $t + \Delta t$:

$$\hat{\mathbf{R}}_{t+\theta\Delta t} = \mathbf{R}_t + \theta(\mathbf{R}_{t+\Delta t} - \mathbf{R}_t) + \mathbf{M}(a_0\mathbf{U}_t + a_2\dot{\mathbf{U}}_t + 2\ddot{\mathbf{U}}_t) + \mathbf{C}(a_1\mathbf{U}_t + 2\dot{\mathbf{U}}_t + a_3\ddot{\mathbf{U}}_t)$$

2. Solve for displacements at time $t + \theta\,\Delta t$:

$$\mathbf{L}\mathbf{D}\mathbf{L}^T\mathbf{U}_{t+\theta\Delta t} = \hat{\mathbf{R}}_{t+\theta\Delta t}$$

3. Calculate displacements, velocities, and accelerations at time $t + \Delta t$:

$$\ddot{\mathbf{U}}_{t+\Delta t} = a_4(\mathbf{U}_{t+\theta\Delta t} - \mathbf{U}_t) + a_5\dot{\mathbf{U}}_t + a_6\ddot{\mathbf{U}}_t$$
$$\dot{\mathbf{U}}_{t+\Delta t} = \dot{\mathbf{U}}_t + a_7(\ddot{\mathbf{U}}_{t+\Delta t} + \ddot{\mathbf{U}}_t)$$
$$\mathbf{U}_{t+\Delta t} = \mathbf{U}_t + \Delta t\dot{\mathbf{U}}_t + a_8(\ddot{\mathbf{U}}_{t+\Delta t} + 2\ddot{\mathbf{U}}_t)$$

As pointed out earlier, the Wilson θ method is also an implicit integration method, because the stiffness matrix **K** is a coefficient matrix to the unknown displacement vector. It may also be noted that no special starting procedures are needed, since the displacements, velocities, and accelerations at time $t + \Delta t$ are expressed in terms of the same quantities at time t only. The stability and accuracy of the integration scheme is studied in Chapter 9.

Example 8.3

Calculate the displacement response of the system considered in Examples 8.1 and 8.2 using the Wilson θ method. Use $\theta = 1.4$.

First we consider the case $\Delta t = 0.28$. Following the steps of calculations in Table 8.3, we have

$$\mathbf{U}_0 = \begin{bmatrix} 0 \\ 0 \end{bmatrix}; \qquad \dot{\mathbf{U}}_0 = \begin{bmatrix} 0 \\ 0 \end{bmatrix}; \qquad \ddot{\mathbf{U}}_0 = \begin{bmatrix} 0 \\ 10 \end{bmatrix}$$

where $\ddot{\mathbf{U}}_0$ was evaluated in Example 8.1. Then

$$a_0 = 39.0; \quad a_1 = 7.65; \quad a_2 = 15.3; \quad a_3 = 0.196; \quad a_4 = 27.9;$$
$$a_5 = -10.9; \quad a_6 = -1.14; \quad a_7 = 0.14; \quad a_8 = 0.0131$$

and

$$\hat{\mathbf{K}} = \begin{bmatrix} 6 & -2 \\ -2 & 4 \end{bmatrix} + 39.0\begin{bmatrix} 2 & 0 \\ 0 & 1 \end{bmatrix} = \begin{bmatrix} 84.1 & -2 \\ -2 & 43.0 \end{bmatrix}$$

For each time step we need to evaluate

$$\hat{\mathbf{R}}_{t+\theta\Delta t} = \begin{bmatrix} 0 \\ 10 \end{bmatrix} + \begin{bmatrix} 0 \\ 0 \end{bmatrix} + \begin{bmatrix} 2 & 0 \\ 0 & 1 \end{bmatrix}\{39.0\mathbf{U}_t + 15.3\dot{\mathbf{U}}_t + 2\ddot{\mathbf{U}}_t\}$$

$$\hat{\mathbf{K}}\mathbf{U}_{t+\theta\Delta t} = \hat{\mathbf{R}}_{t+\theta\Delta t}$$

and then calculate

$$\ddot{\mathbf{U}}_{t+\Delta t} = 27.9(\mathbf{U}_{t+\theta\Delta t} - \mathbf{U}_t) - 10.9\dot{\mathbf{U}}_t - 1.14\ddot{\mathbf{U}}_t$$
$$\dot{\mathbf{U}}_{t+\Delta t} = \dot{\mathbf{U}}_t + 0.14(\ddot{\mathbf{U}}_{t+\Delta t} + \ddot{\mathbf{U}}_t)$$
$$\mathbf{U}_{t+\Delta t} = \mathbf{U}_t + 0.28\dot{\mathbf{U}}_t + 0.0131(\ddot{\mathbf{U}}_{t+\Delta t} + 2\ddot{\mathbf{U}}_t)$$

Time	Δt	$2\Delta t$	$3\Delta t$	$4\Delta t$	$5\Delta t$	$6\Delta t$	$7\Delta t$	$8\Delta t$	$9\Delta t$	$10\Delta t$	$11\Delta t$	$12\Delta t$
$\mathbf{U}_t$	0.00605	0.0525	0.196	0.490	0.952	1.54	2.16	2.67	2.92	2.82	2.33	1.54
	0.366	1.34	2.64	3.92	4.88	5.31	5.18	4.61	3.82	3.06	2.52	2.29

The solution obtained is compared with the exact results in Example 8.7.

Consider now the direct integration with a time step $\Delta t = 28$. In this case we have

$$\hat{\mathbf{K}} = \begin{bmatrix} 6 & -2 \\ -2 & 4 \end{bmatrix} + 0.00392\begin{bmatrix} 2 & 0 \\ 0 & 1 \end{bmatrix} = \begin{bmatrix} 6.00784 & -2 \\ -2 & 4.00392 \end{bmatrix}$$

where we note that $\hat{\mathbf{K}}$ is nearly equal to $\mathbf{K}$, as in the integration using the Houbolt method.

The displacement response obtained over 12 time steps is

Time	Δt	$2\Delta t$	$3\Delta t$	$4\Delta t$	$5\Delta t$	$6\Delta t$	$7\Delta t$	$8\Delta t$	$9\Delta t$	$10\Delta t$	$11\Delta t$	$12\Delta t$
$\mathbf{U}_t$	1.09 1123.	2.82 −834.	−2.61 674.	5.85 −519.	−4.47 406.	6.59 −308.	−4.38 24.2	5.97 −181.	−3.46 144.	4.92 −105.	−2.39 86.1	3.89 −60.8

Here it should be noted that the initial conditions on the accelerations, i.e., $\ddot{\mathbf{U}}_0 = \begin{bmatrix} 0 \\ 10 \end{bmatrix}$, cause the large initial displacement response. This response is damped out with increasing time. If we use $\ddot{\mathbf{U}} = \begin{bmatrix} 0 \\ 0 \end{bmatrix}$, the calculated displacement response is

Time	Δt	$2\Delta t$	$3\Delta t$	$4\Delta t$	$5\Delta t$	$6\Delta t$	$7\Delta t$	$8\Delta t$	$9\Delta t$	$10\Delta t$	$11\Delta t$	$12\Delta t$
$\mathbf{U}_t$	0.363 1.09	1.44 4.33	0.632 1.89	1.29 3.87	0.782 2.32	1.17 3.52	0.875 2.60	1.09 3.31	0.923 2.77	1.05 3.18	0.960 2.86	1.03 3.11

where it is observed that the static solution is approached (see Example 8.2).

8.2.4 The Newmark Method

The Newmark integration scheme can also be understood to be an extension of the linear acceleration method. The following assumptions are used:[17]

$$\dot{\mathbf{U}}_{t+\Delta t} = \dot{\mathbf{U}}_t + [(1 - \delta)\ddot{\mathbf{U}}_t + \delta\ddot{\mathbf{U}}_{t+\Delta t}]\,\Delta t \tag{8.27}$$

$$\mathbf{U}_{t+\Delta t} = \mathbf{U}_t + \dot{\mathbf{U}}_t\,\Delta t + [(\tfrac{1}{2} - \alpha)\ddot{\mathbf{U}}_t + \alpha\ddot{\mathbf{U}}_{t+\Delta t}]\,\Delta t^2 \tag{8.28}$$

where α and δ are parameters that can be determined to obtain integration accuracy and stability. When $\delta = \frac{1}{2}$ and $\alpha = \frac{1}{6}$, relations (8.27) and (8.28) correspond to the linear acceleration method (which is also obtained using $\theta = 1$ in the Wilson θ method). Newmark originally proposed as an unconditionally stable scheme the constant-average-acceleration method, in which case $\delta = \frac{1}{2}$ and $\alpha = \frac{1}{4}$ (see Fig. 8.2).

In addition to (8.27) and (8.28), for solution of the displacements, velocities, and accelerations at time $t + \Delta t$, the equilibrium equations (8.1) at time $t + \Delta t$ are also considered:

$$\mathbf{M}\ddot{\mathbf{U}}_{t+\Delta t} + \mathbf{C}\dot{\mathbf{U}}_{t+\Delta t} + \mathbf{K}\mathbf{U}_{t+\Delta t} = \mathbf{R}_{t+\Delta t} \tag{8.29}$$

Solving from (8.28) for $\ddot{\mathbf{U}}_{t+\Delta t}$ in terms of $\mathbf{U}_{t+\Delta t}$, and then substituting for $\ddot{\mathbf{U}}_{t+\Delta t}$ into (8.27), we obtain equations for $\ddot{\mathbf{U}}_{t+\Delta t}$ and $\dot{\mathbf{U}}_{t+\Delta t}$, each in terms of the unknown displacements $\mathbf{U}_{t+\Delta t}$ only. These two relations for $\dot{\mathbf{U}}_{t+\Delta t}$ and

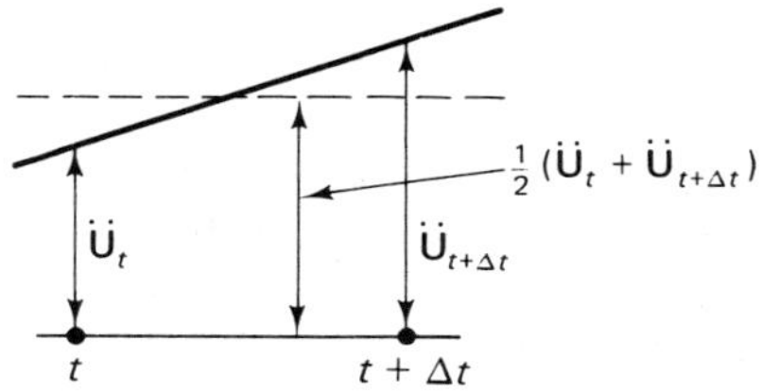

Figure 8.2. Newmark's constant-average-acceleration scheme.

$\ddot{\mathbf{U}}_{t+\Delta t}$ are substituted into (8.29) to solve for $\mathbf{U}_{t+\Delta t}$, after which, using (8.27) and (8.28), $\ddot{\mathbf{U}}_{t+\Delta t}$ and $\dot{\mathbf{U}}_{t+\Delta t}$ can also be calculated.

The complete algorithm using the Newmark integration scheme is given in Table 8.4. The close relationship between the computer implementation of the Newmark and the Wilson method should be noted, which makes it

TABLE 8.4 STEP-BY-STEP SOLUTION USING NEWMARK INTEGRATION METHOD

A. *Initial Calculations:*

1. Form stiffness matrix $\mathbf{K}$, mass matrix $\mathbf{M}$, and damping matrix $\mathbf{C}$.
2. Initialize $\mathbf{U}_0$, $\dot{\mathbf{U}}_0$, and $\ddot{\mathbf{U}}_0$.
3. Select time step size Δt, parameters α and δ, and calculate integration constants:

$$\delta \geq 0.50; \qquad \alpha \geq 0.25(0.5 + \delta)^2$$

$$a_0 = \frac{1}{\alpha\,\Delta t^2}; \qquad a_1 = \frac{\delta}{\alpha\,\Delta t}; \qquad a_2 = \frac{1}{\alpha\,\Delta t}; \qquad a_3 = \frac{1}{2\alpha} - 1;$$

$$a_4 = \frac{\delta}{\alpha} - 1; \qquad a_5 = \frac{\Delta t}{2}\left(\frac{\delta}{\alpha} - 2\right); \qquad a_6 = \Delta t(1 - \delta);$$

$$a_7 = \delta\,\Delta t$$

4. Form effective stiffness matrix $\hat{\mathbf{K}}$: $\hat{\mathbf{K}} = \mathbf{K} + a_0\mathbf{M} + a_1\mathbf{C}$.
5. Triangularize $\hat{\mathbf{K}}$: $\hat{\mathbf{K}} = \mathbf{L}\mathbf{D}\mathbf{L}^T$.

B. *For Each Time Step:*

1. Calculate effective loads at time $t + \Delta t$:

$$\hat{\mathbf{R}}_{t+\Delta t} = \mathbf{R}_{t+\Delta t} + \mathbf{M}(a_0\mathbf{U}_t + a_2\dot{\mathbf{U}}_t + a_3\ddot{\mathbf{U}}_t) + \mathbf{C}(a_1\mathbf{U}_t + a_4\dot{\mathbf{U}}_t + a_5\ddot{\mathbf{U}}_t)$$

2. Solve for displacements at time $t + \Delta t$:

$$\mathbf{L}\mathbf{D}\mathbf{L}^T\mathbf{U}_{t+\Delta t} = \hat{\mathbf{R}}_{t+\Delta t}$$

3. Calculate accelerations and velocities at time $t + \Delta t$:

$$\ddot{\mathbf{U}}_{t+\Delta t} = a_0(\mathbf{U}_{t+\Delta t} - \mathbf{U}_t) - a_2\dot{\mathbf{U}}_t - a_3\ddot{\mathbf{U}}_t$$

$$\dot{\mathbf{U}}_{t+\Delta t} = \dot{\mathbf{U}}_t + a_6\ddot{\mathbf{U}}_t + a_7\ddot{\mathbf{U}}_{t+\Delta t}$$

possible to conveniently use both integration schemes in one single computer program.[18,19] The stability and accuracy analysis of the Newmark method is presented in Chapter 9.

In the following discussion we consider first a simple example analysis using the Newmark method. Then, to indicate the general applicability of direct integration, we develop an integration scheme for the analysis of transient heat transfer problems.

Example 8.4

Calculate the displacement response of the system considered in Examples 8.1 to 8.3 using the Newmark method. Use $\alpha = 0.25$, $\delta = 0.5$.

Consider first the case $\Delta t = 0.28$. Following the steps of calculations given in Table 8.4, we have

$$\mathbf{U}_0 = \begin{bmatrix} 0 \\ 0 \end{bmatrix}; \qquad \dot{\mathbf{U}}_0 = \begin{bmatrix} 0 \\ 0 \end{bmatrix}; \qquad \ddot{\mathbf{U}}_0 = \begin{bmatrix} 0 \\ 10 \end{bmatrix}$$

The integration constants are

$$a_0 = 51.0; \qquad a_1 = 7.14; \qquad a_2 = 14.3; \qquad a_3 = 1.00;$$
$$a_4 = 1.00; \qquad a_5 = 0.00; \qquad a_6 = 0.14; \qquad a_7 = 0.14$$

Thus the effective stiffness matrix is

$$\hat{\mathbf{K}} = \begin{bmatrix} 6 & -2 \\ -2 & 4 \end{bmatrix} + 51.0\begin{bmatrix} 2 & 0 \\ 0 & 1 \end{bmatrix} = \begin{bmatrix} 108 & -2 \\ -2 & 55 \end{bmatrix}$$

For each time step we need to evaluate

$$\hat{\mathbf{R}}_{t+\Delta t} = \begin{bmatrix} 0 \\ 10 \end{bmatrix} + \begin{bmatrix} 2 & 0 \\ 0 & 1 \end{bmatrix}(51\mathbf{U}_t + 14.3\dot{\mathbf{U}}_t + 1.0\ddot{\mathbf{U}}_t)$$

Then

$$\hat{\mathbf{K}}\mathbf{U}_{t+\Delta t} = \hat{\mathbf{R}}_{t+\Delta t}$$

and

$$\ddot{\mathbf{U}}_{t+\Delta t} = 51.0(\mathbf{U}_{t+\Delta t} - \mathbf{U}_t) - 14.3\dot{\mathbf{U}}_t - 1.0\ddot{\mathbf{U}}_t$$
$$\dot{\mathbf{U}}_{t+\Delta t} = \dot{\mathbf{U}}_t + 0.14\ddot{\mathbf{U}}_t + 1.0\ddot{\mathbf{U}}_{t+\Delta t}$$

Performing the above calculations, we obtain

Time	Δt	$2\Delta t$	$3\Delta t$	$4\Delta t$	$5\Delta t$	$6\Delta t$	$7\Delta t$	$8\Delta t$	$9\Delta t$	$10\Delta t$	$11\Delta t$	$12\Delta t$
$\mathbf{U}_t$	0.00673 0.364	0.0504 1.35	0.189 2.68	0.485 4.00	0.961 4.95	1.58 5.34	2.23 5.13	2.76 4.48	3.00 3.64	2.85 2.90	2.28 2.44	1.40 2.31

The solution obtained is compared with the exact results in Example 8.7.

Next we consider the time step $\Delta t = 28.0$. In this case we have

$$\hat{\mathbf{K}} = \begin{bmatrix} 6 & -2 \\ -2 & 4 \end{bmatrix} + 0.0051\begin{bmatrix} 2 & 0 \\ 0 & 1 \end{bmatrix} = \begin{bmatrix} 6.0102 & -2.0000 \\ -2.0000 & 4.0051 \end{bmatrix}$$

Therefore, $\hat{\mathbf{K}}$ is, as in the integration using the Houbolt or Wilson θ method, nearly equal to $\mathbf{K}$.

Using the initial conditions $\ddot{\mathbf{U}}_0 = \begin{bmatrix} 0 \\ 10 \end{bmatrix}$, we obtain, as the displacement response,

Time	Δt	$2\Delta t$	$3\Delta t$	$4\Delta t$	$5\Delta t$	$6\Delta t$	$7\Delta t$	$8\Delta t$	$9\Delta t$	$10\Delta t$	$11\Delta t$	$12\Delta t$
$\mathbf{U}_t$	1.99 5.99	0.028 0.045	1.94 5.99	0.112 0.177	1.82 5.72	0.248 0.393	1.67 5.47	0.429 0.685	1.47 5.14	0.648 1.04	1.23 4.76	0.894 1.45

However, using as initial conditions $\ddot{\mathbf{U}}_0 = \begin{bmatrix} 0 \\ 0 \end{bmatrix}$, we obtain

Time	Δt	$2\Delta t$	$3\Delta t$	$4\Delta t$	$5\Delta t$	$6\Delta t$	$7\Delta t$	$8\Delta t$	$9\Delta t$	$10\Delta t$	$11\Delta t$	$12\Delta t$
$\mathbf{U}_t$	0.363 1.09	1.44 4.33	0.632 1.89	1.29 3.87	0.782 2.32	1.17 3.52	0.875 2.60	1.09 3.31	0.929 2.77	1.05 3.18	0.960 2.86	1.03 2.11

and hence the static solution is approached (see Examples 8.2 and 8.3).

Example 8.5

Use the principles presented in this section to develop a step-by-step direct integration scheme for the solution of transient heat transfer problems. Use the heat flow equilibrium equations derived in Section 5.4 and assume that the temperature varies linearly over the time interval Δt.

The equilibrium equations obtained in Section 5.4 are given in relation (5.40) and are restated as follows:

$$\mathbf{C}\dot{\boldsymbol{\theta}} + \mathbf{K}\boldsymbol{\theta} = \mathbf{Q} \tag{a}$$

where $\mathbf{C}$ and $\mathbf{K}$ are the heat capacity and conductivity matrices, and $\boldsymbol{\theta}$ and $\mathbf{Q}$ are nodal point temperatures and nodal point heat flow inputs, respectively.

The assumption of a linear temperature variation over the time interval Δt gives

$$\boldsymbol{\theta}(\tau) = \boldsymbol{\theta}_t + \frac{\tau - t}{\Delta t}(\boldsymbol{\theta}_{t+\Delta t} - \boldsymbol{\theta}_t)$$

where

$$t \leq \tau \leq t + \Delta t$$

Hence we can use

$$\dot{\boldsymbol{\theta}}(\tau) = \frac{1}{\Delta t}(\boldsymbol{\theta}_{t+\Delta t} - \boldsymbol{\theta}_t)$$

and specifically at time $t + \Delta t$,

$$\dot{\boldsymbol{\theta}}_{t+\Delta t} = \frac{1}{\Delta t}(\boldsymbol{\theta}_{t+\Delta t} - \boldsymbol{\theta}_t) \tag{b}$$

In addition to (b) we also consider the heat flow equilibrium equation given in (a) at time $t + \Delta t$,

$$\mathbf{C}\dot{\boldsymbol{\theta}}_{t+\Delta t} + \mathbf{K}\boldsymbol{\theta}_{t+\Delta t} = \mathbf{Q}_{t+\Delta t} \tag{c}$$

Substituting for $\dot{\boldsymbol{\theta}}_{t+\Delta t}$ in (c) from the relation in (b), we obtain

$$\left(\frac{\mathbf{C}}{\Delta t} + \mathbf{K}\right)\boldsymbol{\theta}_{t+\Delta t} = \mathbf{Q}_{t+\Delta t} + \frac{\mathbf{C}}{\Delta t}\boldsymbol{\theta}_t \tag{d}$$

Therefore, if we know $\boldsymbol{\theta}_t$, we use (d) to calculate $\boldsymbol{\theta}_{t+\Delta t}$, and thus the relation in (d) establishes the step-by-step integration procedure. It should be noted that the $\mathbf{K}$ matrix appears in the coefficient matrix of $\boldsymbol{\theta}_{t+\Delta t}$; hence the integration method is implicit. We analyze the stability and accuracy characteristics of the method in Example 9.2.

8.3 MODE SUPERPOSITION

Tables 8.1 to 8.4, summarizing the direct integration schemes, show that if a diagonal mass matrix and no damping is assumed, the number of operations for one time step are somewhat larger than $2nm_{\mathbf{K}}$, where n and $m_{\mathbf{K}}$ are the order and half-bandwidth of the stiffness matrix considered, respectively. In the central difference method $2nm_{\mathbf{K}}$ operations are required for the product of the stiffness matrix times the displacement vector, and in the Houbolt, Wilson, and Newmark methods about $2nm_{\mathbf{K}}$ operations are required for the solution of the system equations in each time step. The initial triangular factorization of the effective stiffness matrix requires additional operations. Furthermore, if a consistent mass matrix is used, or a damping matrix is included in the analysis, an additional number of operations proportional to $nm_{\mathbf{K}}$ is required per time step in either case. Therefore, neglecting the operations for the initial calculations, a total number of about $\alpha nm_{\mathbf{K}}s$ operations are required in the complete integration, where α depends on the characteristics of the matrices used, $\alpha \geq 2$, and s is the number of time steps.

The above considerations show that the number of operations required in the direct integration are directly proportional to the number of time steps used in the analysis. Therefore, in general, the use of direct integration can be expected to be effective when the response for a relatively short duration (i.e., for a few time steps) is required. However, if the integration must be carried out for many time steps, it may be more effective to first transform the equilibrium equations in (8.1) into a form in which the step-by-step solution is less costly. In particular, since the number of operations required is directly proportional to the half-bandwidth $m_{\mathbf{K}}$ of the stiffness matrix, a reduction in $m_{\mathbf{K}}$ would decrease proportionally the cost of the step-by-step solution.

It is important at this stage to fully recognize what we have proposed to pursue. We recall that (8.1) are the equilibrium equations obtained when the finite element interpolation functions are used in the evaluation of the virtual work equation (3.17) (see Section 3.2.2). The resulting matrices **K**, **M**, and **C** have a bandwidth that is determined by the numbering of the finite element nodal points (see Section 6.2.3). Therefore, the topology of the finite element mesh determines the order and bandwidth of the system matrices. As pointed out in Section 6.2.3, in order to reduce the bandwidth of the system matrices, we may rearrange the nodal point numbering; however, there is a limit on the minimum bandwidth that can be obtained in this way, and we therefore set out to follow a different procedure.

8.3.1 Change of Basis to Modal Generalized Displacements

We propose to transform the equilibrium equations into a more effective form for direct integration by using the following transformation on the finite element displacements **U**:

$$\mathbf{U}(t) = \mathbf{P}\mathbf{X}(t) \tag{8.30}$$

where **P** is a square matrix and $\mathbf{X}(t)$ is a time-dependent vector of order n. The transformation matrix **P** is still unknown and will have to be determined. The components of **X** are referred to as *generalized displacements*. Substituting (8.30) into (8.1) and premultiplying by $\mathbf{P}^T$, we obtain

$$\tilde{\mathbf{M}}\ddot{\mathbf{X}}(t) + \tilde{\mathbf{C}}\dot{\mathbf{X}}(t) + \tilde{\mathbf{K}}\mathbf{X}(t) = \tilde{\mathbf{R}}(t) \tag{8.31}$$

where

$$\tilde{\mathbf{M}} = \mathbf{P}^T\mathbf{M}\mathbf{P}; \qquad \tilde{\mathbf{C}} = \mathbf{P}^T\mathbf{C}\mathbf{P}; \qquad \tilde{\mathbf{K}} = \mathbf{P}^T\mathbf{K}\mathbf{P}; \qquad \tilde{\mathbf{R}} = \mathbf{P}^T\mathbf{R} \tag{8.32}$$

It should be noted that this transformation is obtained by substituting (8.30) into (3.54) to express the element displacements in terms of the generalized displacements:

$$\mathbf{U}^{(m)}(x, y, z, t) = \mathbf{H}^{(m)}\mathbf{P}\mathbf{X}(t) \tag{8.33}$$

and then using (8.33) in the virtual work equation (3.58). Therefore, in essence, to obtain (8.31) from (8.1), a change of basis from the finite element displacement basis to a generalized displacement basis has been performed (see Section 2.5).

The objective of the transformation is to obtain new system stiffness, mass, and damping matrices, $\tilde{\mathbf{K}}$, $\tilde{\mathbf{M}}$, and $\tilde{\mathbf{C}}$, which have a smaller bandwidth than the original system matrices, and the transformation matrix **P** should be selected accordingly. In addition, it should be noted that **P** must be nonsingular (i.e., the rank of **P** must be n) in order to have a unique relation between any vectors **U** and **X** as expressed in (8.30).

In theory, there can be many different transformation matrices **P**, which would reduce the bandwidth of the system matrices. However, in practice, an effective transformation matrix is established using the displacement solutions of the free vibration equilibrium equations with damping neglected,

$$\mathbf{M}\ddot{\mathbf{U}} + \mathbf{K}\mathbf{U} = \mathbf{0} \tag{8.34}$$

The solution to (8.34) can be postulated to be of the form

$$\mathbf{U} = \boldsymbol{\phi} \sin \omega(t - t_0) \tag{8.35}$$

where $\boldsymbol{\phi}$ is a vector of order n, t the time variable, t_0 a time constant, and ω a constant identified to represent the frequency of vibration (rad/sec) of the vector $\boldsymbol{\phi}$.

Substituting (8.35) into (8.34), we obtain the generalized eigenproblem, from which $\boldsymbol{\phi}$ and ω must be determined:

$$\mathbf{K}\boldsymbol{\phi} = \omega^2 \mathbf{M}\boldsymbol{\phi} \tag{8.36}$$

The eigenproblem in (8.36) yields the n eigensolutions $(\omega_1^2, \boldsymbol{\phi}_1), (\omega_2^2, \boldsymbol{\phi}_2), \ldots, (\omega_n^2, \boldsymbol{\phi}_n)$, where the eigenvectors are **M**-orthonormalized (see Section 10.2.1); i.e.,

$$\boldsymbol{\phi}_i^T \mathbf{M} \boldsymbol{\phi}_j \begin{cases} = 1; & i = j \\ = 0; & i \neq j \end{cases} \tag{8.37}$$

and

$$0 \leq \omega_1^2 \leq \omega_2^2 \leq \omega_3^2 \cdots \leq \omega_n^2 \tag{8.38}$$

The vector $\boldsymbol{\phi}_i$ is called the ith-mode shape vector, and ω_i is the corresponding frequency of vibration (rad/sec). It should be emphasized that (8.34) is satisfied using any of the n displacement solutions $\boldsymbol{\phi}_i \sin \omega_i(t - t_0)$, $i = 1, 2, \ldots, n$. For a physical interpretation of ω_i and $\boldsymbol{\phi}_i$ see Example 8.6.

Defining a matrix $\boldsymbol{\Phi}$ whose columns are the eigenvectors $\boldsymbol{\phi}_i$ and a diagonal matrix $\boldsymbol{\Omega}^2$ which stores the eigenvalues ω_i^2 on its diagonal; i.e.,

$$\boldsymbol{\Phi} = [\boldsymbol{\phi}_1, \boldsymbol{\phi}_2, \cdots, \boldsymbol{\phi}_n]; \qquad \boldsymbol{\Omega}^2 = \begin{bmatrix} \omega_1^2 & & & & \\ & \omega_2^2 & & & \\ & & \cdot & & \\ & & & \cdot & \\ & & & & \cdot \\ & & & & & \omega_n^2 \end{bmatrix} \tag{8.39}$$

we can write the n solutions to (8.36) as

$$\mathbf{K}\boldsymbol{\Phi} = \mathbf{M}\boldsymbol{\Phi}\boldsymbol{\Omega}^2 \tag{8.40}$$

Since the eigenvectors are **M**-orthonormal, we have

$$\boldsymbol{\Phi}^T \mathbf{K} \boldsymbol{\Phi} = \boldsymbol{\Omega}^2; \qquad \boldsymbol{\Phi}^T \mathbf{M} \boldsymbol{\Phi} = \mathbf{I} \tag{8.41}$$

It is now apparent that the matrix $\boldsymbol{\Phi}$ would be a suitable transformation matrix **P** in (8.30). Using

$$\mathbf{U}(t) = \boldsymbol{\Phi}\mathbf{X}(t) \tag{8.42}$$

we obtain equilibrium equations that correspond to the modal generalized displacements

$$\ddot{\mathbf{X}}(t) + \boldsymbol{\Phi}^T \mathbf{C} \boldsymbol{\Phi} \dot{\mathbf{X}}(t) + \boldsymbol{\Omega}^2 \mathbf{X}(t) = \boldsymbol{\Phi}^T \mathbf{R}(t) \tag{8.43}$$

The initial conditions on $\mathbf{X}(t)$ are obtained using (8.42) and the **M**-orthonormality of $\mathbf{\Phi}$; i.e., at time 0 we have

$$\mathbf{X}_0 = \mathbf{\Phi}^T\mathbf{M}\mathbf{U}_0; \qquad \dot{\mathbf{X}}_0 = \mathbf{\Phi}^T\mathbf{M}\dot{\mathbf{U}}_0 \tag{8.44}$$

The equations in (8.43) show that if a damping matrix is not included in the analysis, the finite element equilibrium equations are decoupled when using in the transformation matrix **P** the free vibration mode shapes of the finite element system. Since the derivation of the damping matrix can in many cases not be carried out explicitly (see Section 3.2.2) but the damping effects can only be included approximately, it is reasonable to use a damping matrix that includes all required effects but at the same time allows an effective solution of the equilibrium equations. In many analyses damping effects are neglected altogether, and it is this case that we shall discuss first.

Example 8.6

Calculate the transformation matrix $\mathbf{\Phi}$ for the problem considered in Examples 8.1 to 8.4, and thus establish the decoupled equations of equilibrium in the basis of mode shape vectors.

For the system under consideration we have

$$\mathbf{K} = \begin{bmatrix} 6 & -2 \\ -2 & 4 \end{bmatrix}; \qquad \mathbf{M} = \begin{bmatrix} 2 & 0 \\ 0 & 1 \end{bmatrix}; \qquad \mathbf{R} = \begin{bmatrix} 0 \\ 10 \end{bmatrix}$$

The generalized eigenproblem to be solved is therefore

$$\begin{bmatrix} 6 & -2 \\ -2 & 4 \end{bmatrix} \boldsymbol{\phi} = \omega^2 \begin{bmatrix} 2 & 0 \\ 0 & 1 \end{bmatrix} \boldsymbol{\phi}$$

The solution is obtained by one of the methods given in Chapters 10 to 12. Here we simply give the two solutions without derivations:

$$\omega_1^2 = 2, \qquad \boldsymbol{\phi}_1 = \begin{bmatrix} \dfrac{1}{\sqrt{3}} \\[2ex] \dfrac{1}{\sqrt{3}} \end{bmatrix}$$

$$\omega_2^2 = 5, \qquad \boldsymbol{\phi}_2 = \begin{bmatrix} \dfrac{1}{2}\sqrt{\dfrac{2}{3}} \\[2ex] -\sqrt{\dfrac{2}{3}} \end{bmatrix}$$

Therefore, considering the free vibration equilibrium equations of the system

$$\begin{bmatrix} 2 & 0 \\ 0 & 1 \end{bmatrix} \ddot{\mathbf{U}}(t) + \begin{bmatrix} 6 & -2 \\ -2 & 4 \end{bmatrix} \mathbf{U}(t) = \mathbf{0} \tag{a}$$

the following two solutions are possible:

$$\mathbf{U}_1(t) = \begin{bmatrix} \dfrac{1}{\sqrt{3}} \\[2ex] \dfrac{1}{\sqrt{3}} \end{bmatrix} \sin\sqrt{2}\,(t - t_0^1) \quad \text{or} \quad \mathbf{U}_2(t) = \begin{bmatrix} \dfrac{1}{2}\sqrt{\dfrac{2}{3}} \\[2ex] -\sqrt{\dfrac{2}{3}} \end{bmatrix} \sin\sqrt{5}\,(t - t_0^2)$$

That the vectors $\mathbf{U}_1(t)$ and $\mathbf{U}_2(t)$ indeed satisfy the relation in (a) can be verified simply by substituting $\mathbf{U}_1$ or $\mathbf{U}_2$ into the equilibrium equations. The actual solution to the equations in (a) is of the form

$$\mathbf{U}(t) = \alpha \begin{bmatrix} \dfrac{1}{\sqrt{3}} \\ \dfrac{1}{\sqrt{3}} \end{bmatrix} \sin \sqrt{2}(t - t_0^1) + \beta \begin{bmatrix} \dfrac{1}{2}\sqrt{\dfrac{2}{3}} \\ -\sqrt{\dfrac{2}{3}} \end{bmatrix} \sin \sqrt{5}(t - t_0^2)$$

where α, β, t_0^1, and t_0^2 are determined by the initial conditions on $\mathbf{U}$ and $\dot{\mathbf{U}}$. In particular, if we impose initial conditions corresponding to α (or β) only, we have that the system vibrates in the corresponding eigenvector with frequency $\sqrt{2}$ rad/sec (or $\sqrt{5}$ rad/sec). The general procedure of solution for α, β, t_1^0, and t_2^0 is discussed in Section 8.3.2.

Having evaluated $(\omega_1^2, \boldsymbol{\phi}_1)$ and $(\omega_2^2, \boldsymbol{\phi}_2)$ we arrive at the following equilibrium equations in the basis of eigenvectors:

$$\ddot{\mathbf{X}}(t) + \begin{bmatrix} 2 & 0 \\ 0 & 5 \end{bmatrix} \mathbf{X}(t) = \begin{bmatrix} \dfrac{1}{\sqrt{3}} & \dfrac{1}{\sqrt{3}} \\ \dfrac{1}{2}\sqrt{\dfrac{2}{3}} & -\sqrt{\dfrac{2}{3}} \end{bmatrix} \begin{bmatrix} 0 \\ 10 \end{bmatrix}$$

or

$$\ddot{\mathbf{X}}(t) + \begin{bmatrix} 2 & 0 \\ 0 & 5 \end{bmatrix} \mathbf{X}(t) = \begin{bmatrix} \dfrac{10}{\sqrt{3}} \\ -10\sqrt{\dfrac{2}{3}} \end{bmatrix}$$

8.3.2 Analysis with Damping Neglected

If velocity-dependent damping effects are not included in the analysis, (8.43) reduces to

$$\ddot{\mathbf{X}}(t) + \boldsymbol{\Omega}^2\mathbf{X}(t) = \boldsymbol{\Phi}^T\mathbf{R}(t) \tag{8.45}$$

i.e., n individual equations of the form

$$\left.\begin{aligned} \ddot{x}_i(t) + \omega_i^2 x_i(t) &= r_i(t) \\ \text{where} \qquad r_i(t) &= \boldsymbol{\phi}_i^T\mathbf{R}(t) \end{aligned}\right\} i = 1, 2, \ldots, n \tag{8.46}$$

We note that the ith typical equation in (8.46) is the equilibrium equation of a single degree of freedom system with unit mass and stiffness ω_i^2. The initial conditions on the motion of this system are obtained from (8.44):

$$\begin{aligned} x_i|_{t=0} &= \boldsymbol{\phi}_i^T\mathbf{M}\mathbf{U}_0 \\ \dot{x}_i|_{t=0} &= \boldsymbol{\phi}_i^T\mathbf{M}\dot{\mathbf{U}}_0 \end{aligned} \tag{8.47}$$

The solution to each equation in (8.46) can be obtained using the integration algorithms in Tables 8.1 to 8.4 or can be calculated using the Duhamel integral:

$$x_i(t) = \frac{1}{\omega_i}\int_0^t r_i(\tau) \sin \omega_i(t - \tau)\, d\tau + \alpha_i \sin \omega_i t + \beta_i \cos \omega_i t \tag{8.48}$$

where α_i and β_i are determined from the initial conditions in (8.47). The

Duhamel integral in (8.48) must, in general, be evaluated numerically. In addition, it should be noted that various other integration methods could also be used in the solution of (8.46).

For the complete response, the solution to all n equations in (8.46), $i = 1, 2, \ldots, n$, must be calculated and then the finite element nodal point displacements are obtained by superposition of the response in each mode; i.e., using (8.42), we obtain

$$\mathbf{U}(t) = \sum_{i=1}^{n} \boldsymbol{\phi}_i x_i(t) \tag{8.49}$$

Therefore, in summary, the response analysis by mode superposition requires, first, the solution of the eigenvalues and eigenvectors of the problem in (8.36), then the solution of the decoupled equilibrium equations in (8.46), and, finally, the superposition of the response in each eigenvector as expressed in (8.49). In the analysis, the eigenvectors are the free vibration mode shapes of the finite element assemblage. As discussed earlier, the choice between mode superposition analysis and direct integration described in Section 8.2 is merely one of numerical effectiveness. The solutions obtained using either procedure are identical within the numerical errors of the time integration schemes used [if the same time integration methods are used in direct integration and the solution of (8.46), the same numerical errors are present] and the round-off errors in the computer.

Example 8.7

Use mode superposition to calculate the displacement response of the system considered in Examples 8.1 to 8.4 and 8.6. (1) Calculate the exact response by integrating each of the two decoupled equilibrium equations exactly. (2) Use the Newmark method with time step $\Delta t = 0.28$ for the time integration.

We established the decoupled equilibrium equations of the system under consideration in Example 8.6; i.e., the two equilibrium equations to be solved are

$$\begin{aligned} \ddot{x}_1 + 2x_1 &= \frac{10}{\sqrt{3}} \\ \ddot{x}_2 + 5x_2 &= -10\sqrt{\frac{2}{3}} \end{aligned} \tag{a}$$

The initial conditions on the system are $\mathbf{U}|_{t=0} = \mathbf{0}$, $\dot{\mathbf{U}}|_{t=0} = \mathbf{0}$, and hence, using (8.47), we have

$$\begin{aligned} x_1|_{t=0} &= 0 \qquad \dot{x}_1|_{t=0} = 0 \\ x_2|_{t=0} &= 0 \qquad \dot{x}_2|_{t=0} = 0 \end{aligned} \tag{b}$$

Also, to obtain $\mathbf{U}$ we need to use the relation in (8.42), which, using the eigenvectors calculated in Example 8.6, gives

$$\mathbf{U}(t) = \begin{bmatrix} \dfrac{1}{\sqrt{3}} & \dfrac{1}{2}\sqrt{\dfrac{2}{3}} \\ \dfrac{1}{\sqrt{3}} & -\sqrt{\dfrac{2}{3}} \end{bmatrix} \mathbf{X}(t) \tag{c}$$

The exact solutions to the equations in (a) and (b) are

$$\begin{aligned} x_1 &= \frac{5}{\sqrt{3}}(1 - \cos\sqrt{2}\,t) \\ x_2 &= 2\sqrt{\frac{2}{3}}(-1 + \cos\sqrt{5}\,t) \end{aligned} \tag{d}$$

Hence using (c) we have

$$\mathbf{U}(t) = \begin{bmatrix} \frac{1}{\sqrt{3}} & \frac{1}{2}\sqrt{\frac{2}{3}} \\ \frac{1}{\sqrt{3}} & -\sqrt{\frac{2}{3}} \end{bmatrix} \begin{bmatrix} \frac{5}{\sqrt{3}}(1 - \cos\sqrt{2}\,t) \\ 2\sqrt{\frac{2}{3}}(-1 + \cos\sqrt{5}\,t) \end{bmatrix} \tag{e}$$

Evaluating the displacements from (e) for times $\Delta t, 2\Delta t, \ldots, 12\Delta t$, where $\Delta t = 0.28$, we obtain

Time	Δt	$2\Delta t$	$3\Delta t$	$4\Delta t$	$5\Delta t$	$6\Delta t$	$7\Delta t$	$8\Delta t$	$9\Delta t$	$10\Delta t$	$11\Delta t$	$12\Delta t$
$\mathbf{U}_t$	0.003	0.038	0.176	0.486	0.996	1.66	2.338	2.861	3.052	2.806	2.131	1.157
	0.382	1.41	2.78	4.09	5.00	5.29	4.986	4.277	3.457	2.806	2.484	2.489

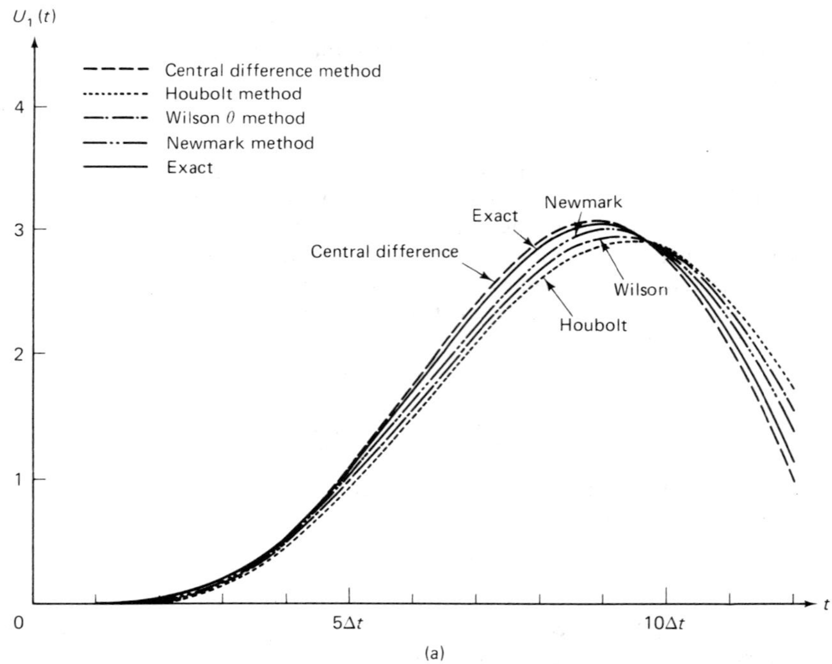

Figure 8.3. Displacement response of system considered in Examples 8.1, 8.2, 8.3, 8.4, and 8.7.

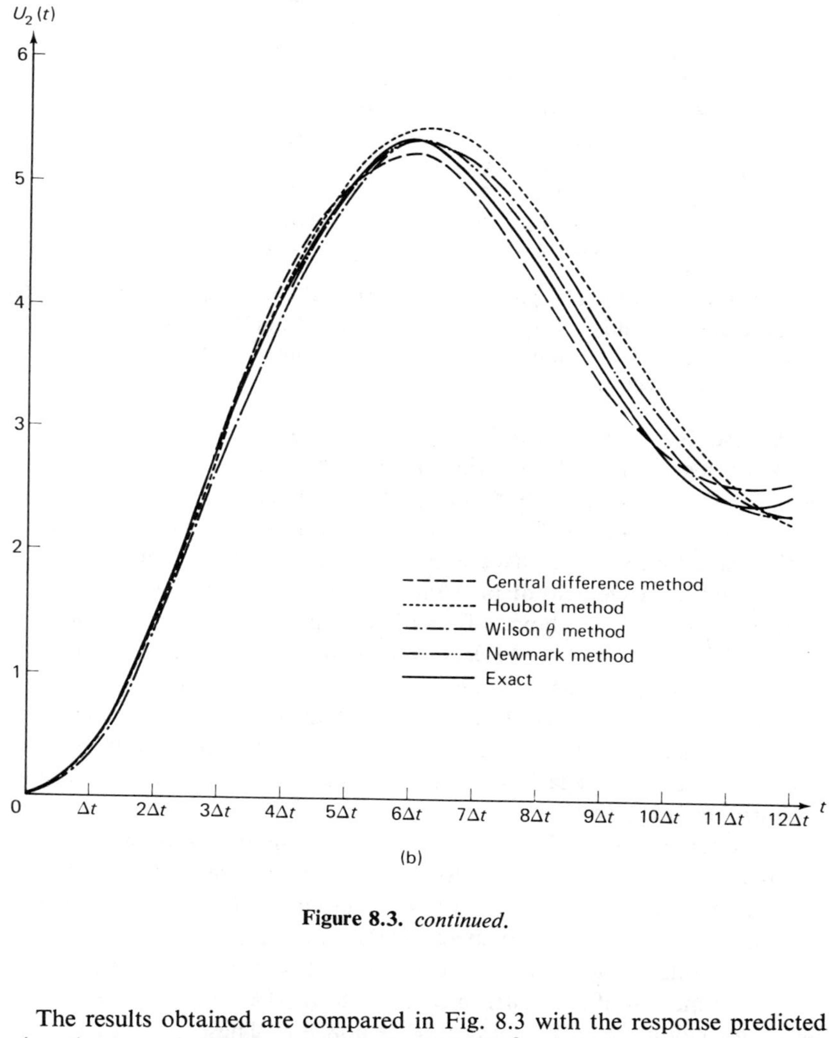

(b)

Figure 8.3. *continued.*

The results obtained are compared in Fig. 8.3 with the response predicted using the central difference, Houbolt, Wilson θ, and Newmark methods in Examples 8.1 to 8.4, respectively. The discussion in Chapter 9 will show that the time step Δt selected for the direct integrations is relatively large, and with this in mind it may be noted that the direct integration schemes predict a fair approximation to the exact response of the system.

Instead of evaluating the exact response as given in (d), we could use a numerical integration scheme to solve the equations in (a). Here we employ the Newmark method and obtain:

Time	Δt	$2\Delta t$	$3\Delta t$	$4\Delta t$	$5\Delta t$	$6\Delta t$
$x_1(t)$	0.2258	0.8199	1.807	2.379	4.123	5.064
$x_2(t)$	−0.3046	−0.7920	−2.1239	−2.939	−3.258	−2.632

	$7\Delta t$	$8\Delta t$	$9\Delta t$	$10\Delta t$	$11\Delta t$	$12\Delta t$
	5.579	5.774	5.521	4.855	3.866	2.773
	−2.161	−1.156	−0.3307	−0.004083	−0.2482	−1.088

The solution for $U_1(t)$ and $U_2(t)$ is now evaluated by substituting for $\mathbf{X}(t)$ in the relation given in (c). As expected, it is found that the displacement response thus predicted is the same as the response obtained when using the Newmark method in direct integration.

As discussed so far, the only difference between a mode superposition and a direct integration analysis is that prior to the time integration, a change of basis is carried out, namely from the finite element coordinate basis to the basis of eigenvectors of the generalized eigenproblem $\mathbf{K}\boldsymbol{\phi} = \omega^2\mathbf{M}\boldsymbol{\phi}$. Since mathematically the same space is spanned by the n eigenvectors as by the n nodal point finite element displacements, the same solution must be obtained in both analyses. The choice of whether to use direct integration or mode superposition will therefore be decided by considerations of effectiveness only. However, this choice can only be made once an important additional aspect of mode superposition, as used in practical analysis, has been presented. In fact, it is this aspect that makes mode superposition analysis of some structures feasible with regard to cost, whereas direct integration would be prohibitively expensive.

Consider the decoupled equilibrium equations in (8.46). We note that if $r_i(t)$ is zero and the initial conditions on x_i and $\dot{x}_i$ are zero, then x_i is zero at all time t. These specific conditions would be realized, for example, if all finite element nodal point displacements and velocities are zero at time 0 and the applied loading is of the form $\mathbf{R}(t) = \mathbf{M}\boldsymbol{\phi}_j f(t)$, where $f(t)$ is an arbitrary function of t. In such a case, since $\boldsymbol{\phi}_i^T\mathbf{M}\boldsymbol{\phi}_j = \delta_{ij}$ (δ_{ij} = Kronecker delta), we would have that only $x_j(t)$ is nonzero. These are rather stringent conditions, and in general analysis can hardly be expected to apply exactly to many of the n equations in (8.46), because the loading is in general arbitrary. However, in addition to the fact that the loading may be nearly orthogonal to $\boldsymbol{\phi}_i$, it is in particular the frequency content of the loading that determines whether the ith equation in (8.46) will contribute significantly to the response. Namely, the response $x_i(t)$ is relatively large if the excitation frequency lies near ω_i. Furthermore, additional transient response of a finite element system is determined by those modes for which nonzero initial conditions are calculated in (8.47).

To demonstrate these basic considerations, we present the following example.

Example 8.8

Consider a one degree of freedom system with the equilibrium equation

$$\ddot{x}(t) + \omega^2 x(t) = R \sin pt$$

and initial conditions

$$x|_{t=0} = 0, \qquad \dot{x}|_{t=0} = 1 \tag{a}$$

Use the Duhamel integral to calculate the displacement response.

We note that the system is subjected to a periodic force input and a nonzero initial velocity. Using the relation in (8.48), we obtain

$$x(t) = \frac{R}{\omega} \int_0^t \sin p\tau \sin \omega(t - \tau)\, d\tau + \alpha \sin \omega t + \beta \cos \omega t$$

Evaluating the integral, we obtain

$$x(t) = \frac{R/\omega^2}{1 - p^2/\omega^2} \sin pt + \alpha \sin \omega t + \beta \cos \omega t \tag{b}$$

We now need to use the initial conditions to evaluate α and β. The solution at time $t = 0$ is

$$x|_{t=0} = \beta$$

$$\dot{x}|_{t=0} = \frac{Rp/\omega^2}{1 - p^2/\omega^2} + \alpha\omega$$

Using the conditions in (a), we obtain

$$\beta = 0; \qquad \alpha = \frac{1}{\omega} - \frac{Rp/\omega^3}{1 - p^2/\omega^2}$$

Substituting for α and β into (b), we thus have

$$x(t) = \frac{R/\omega^2}{1 - p^2/\omega^2} \sin pt + \left(\frac{1}{\omega} - \frac{Rp/\omega^3}{1 - p^2/\omega^2}\right) \sin \omega t$$

It should be noted that if $p = \omega$, in which case we have resonance conditions, infinite displacement response would be obtained. In practice, the presence of damping prevents infinite displacement response, but the displacements can become very large.

Considering the analysis of a multiple degree of freedom system, we therefore find that the response in a vibration mode is large if the excitation frequency is close to the natural frequency of vibration.

It is for the above reasons that, as shown by experience, for many types of practical loading conditions, only a fraction of the total number of decoupled equations need be considered, in order to obtain a good approximate solution to the actual response of the system. Most frequently, only the first p equilibrium equations need be used; i.e., we need to include in the analysis the equations (8.46) for $i = 1, 2, \ldots, p$, where $p \ll n$, in order to obtain a good approximate solution. This means that we also need to solve only for the lowest p eigenvalues and corresponding eigenvectors of the problem in (8.36), and we only sum in (8.49) the response in the first p modes.

It is primarily for the fact that in a mode superposition analysis only a few modes may need to be considered that the mode superposition procedure can be much more effective than direct integration. However, it also follows that the effectiveness of mode superposition depends on the number of modes that must be included in the analysis. In general, the structure considered and the spatial distribution and frequency content of the loading determine the number of modes to be used. For earthquake loading, in some cases only the 10 lowest modes need be considered, although the order of the system n may be larger than 1000. On the other hand, for blast or shock loading, many more modes need generally be included, and p may be as large as $2n/3$. Finally, in vibration excitation analysis, only a few intermediate frequencies may be excited, such as all frequencies between the lower and upper frequency limits ω_l and ω_u, respectively.

Considering the problem of selecting the number of modes to be included in the mode superposition analysis, it should always be kept in mind that an approximate solution to the dynamic equilibrium equations in (8.1) is sought. Therefore, if not enough modes are considered, the equations in (8.1) are not solved accurately enough. But this means, in effect, that equilibrium, including inertia forces, is not satisfied for the approximate response calculated. Denoting by $\mathbf{U}^p$ the response predicted by mode superposition when p modes are considered, an indication of the accuracy of analysis at any time t is obtained by calculating the error measure ϵ^p, defined as

$$\epsilon^p(t) = \frac{\|\mathbf{R}(t) - [\mathbf{M}\ddot{\mathbf{U}}^p(t) + \mathbf{K}\mathbf{U}^p(t)]\|_2}{\|\mathbf{R}(t)\|_2} \tag{8.50}$$

If a good approximate solution of the system equilibrium equations in (8.1) has been obtained, $\epsilon^p(t)$ will be small at any time t. But it must be noted that $\mathbf{U}^p(t)$ must have been obtained by an accurate calculation of the response in each of the p modes considered, because this way the only error is due to not including enough modes in the analysis.

It should be noted that the error measure ϵ^p calculated in (8.50) determines how well equilibrium including inertia forces in satisfied, and is a measure of the nodal point loads not balanced by inertia and elastic nodal point forces [see (8.2)]. Alternatively, we may say that ϵ^p is a measure of that part of the external load vector that has not been included in the mode superposition analysis, because not all mode shape vectors have been used. It is here instructive to note that in a direct integration analysis, ϵ^p is always zero at times 0, Δt, 2 Δt, . . . [except when using the Wilson θ method because (8.25) is solved rather than (8.29), but these errors can be expected to be small]. In summary, therefore, assuming that the decoupled equations in (8.46) have been solved accurately, the errors in a mode superposition analysis using $p < n$ are due to the fact that not enough modes have been used, whereas the errors in a direct integration analysis arise because too large a time step is employed.

One further important aspect should be mentioned in this context. So far, we have only been concerned with the exact solution of the finite element system equilibrium equations in (8.1). However, what we really aim to obtain is a good approximation to the actual exact response of the structure under consideration. We showed in Section 5.3 that under certain conditions the finite element analysis can be understood to be a Ritz analysis. Based on the discussion in Sections 5.3 and 10.3.2, it can be concluded that in such a case in finite element analysis, upper bounds to the "exact" frequencies of the actual structure are obtained. In general, the finite element analysis approximates the lowest exact frequencies best, and little or no accuracy can be expected in approximating the higher frequencies and mode shapes. Therefore, there is usually little justification for including the response in the mode shapes corresponding to the high frequencies in the analysis. In fact, the finite element mesh should be chosen such that all important exact frequencies and vibration mode shapes are well approximated, and then the solution need only be calculated including the response in these modes. However, this can be achieved precisely using mode superposition analysis by considering only the important modes of the finite element system.

From the above discussion it appears that the mode superposition procedure has an inherent advantage over direct integration in that the response corresponding to the higher probably inaccurate frequencies of the finite element system is not included in the analysis. However, assuming that in finite element analysis all important frequencies are predicted accurately, little response is calculated in the higher modes of the system (which are presumably not important), and the inclusion of the finite element system high-frequency response will not seriously affect the accuracy of the solution. In addition, we will discuss in Chapter 9 that in direct integration, advantage can be taken also of integrating accurately only the first p equations in (8.46) and neglecting the high-frequency response of the finite element system. This is achieved by using an unconditionally stable direct integration method and selecting an appropriate integration time step Δt, which, in general, is much larger than the integration step used with a conditionally stable integration scheme.

8.3.3 Analysis with Damping Included

The general form of the equilibrium equations of the finite element system in the basis of eigenvectors $\boldsymbol{\phi}_i$, $i = 1, \ldots, n$ was given in (8.43), which shows that provided that damping effects are neglected, the equilibrium equations decouple and the time integration can be carried out individually for each equation. Considering the analysis of systems in which damping effects cannot be neglected, we still would like to deal with decoupled equilibrium equations in (8.43), merely to be able to use essentially the same computational procedure whether damping effects are included or neglected. In general, the damping matrix **C** cannot be constructed from element damping matrices, such as the mass and stiffness matrices of the element assemblage, and its purpose is to approximate the overall energy dissipation during the system response. The mode superposition analysis is particularly effective if it can be assumed that damping is proportional, in which case

$$\boldsymbol{\phi}_i^T \mathbf{C} \boldsymbol{\phi}_j = 2\omega_i \xi_i \delta_{ij} \tag{8.51}$$

where ξ_i is a modal damping parameter and δ_{ij} is the Kronecker delta ($\delta_{ij} = 1$ for $i = j$, $\delta_{ij} = 0$ for $i \neq j$). Therefore, using (8.51) it is assumed that the eigenvectors $\boldsymbol{\phi}_i$, $i = 1, 2, \ldots, n$, are also **C**-orthogonal and the equations in (8.43) reduce to n equations of the form

$$\ddot{x}_i(t) + 2\omega_i \xi_i \dot{x}_i(t) + \omega_i^2 x_i(t) = r_i(t) \tag{8.52}$$

where $r_i(t)$ and the initial conditions on $x_i(t)$ have already been defined in (8.46) and (8.47). We note that (8.52) is the equilibrium equation governing motion of the single degree of freedom system considered in (8.46) when ξ_i is the damping ratio.[4]

If the relation in (8.51) is used to account for damping effects, the procedure of solution of the finite element equilibrium equations in (8.43) is the same as in the case when damping is neglected (see Section 8.3.2) except that the response in each mode is obtained by solving (8.52). This response can be calculated using an integration scheme such as those given in Tables 8.1 to 8.4 or by evaluating the Duhamel integral to obtain

$$\begin{aligned} x_i(t) = \frac{1}{\bar{\omega}_i} \int_0^t r_i(\tau) e^{-\xi_i \omega_i (t-\tau)} \sin \bar{\omega}_i (t - \tau)\, d\tau \\ + e^{-\xi_i \omega_i t} \{\alpha_i \sin \bar{\omega}_i t + \beta_i \cos \bar{\omega}_i t\} \end{aligned} \tag{8.53}$$

where

$$\bar{\omega}_i = \omega_i \sqrt{1 - \xi_i^2} \tag{8.54}$$

and α_i and β_i are calculated using the initial conditions in (8.47).

In considering the implications of using (8.51) to take account of damping effects, the following observations are made. Firstly, the assumption in (8.51) means that the total damping in the structure is the sum of individual damping in each mode. The damping in one mode could be observed, for example, by imposing initial conditions corresponding to that mode only (i.e., $\mathbf{U}_0 = \boldsymbol{\phi}_i$ for mode i) and measuring the amplitude decay during the free damped vibration. In fact, the ability to measure values for the damping ratios ξ_i, and thus approximate in many cases in a realistic manner the damping behavior of the complete structural system, is an important consideration. A second observation relating to the mode superposition analysis is that for the numerical solution of the finite element equilibrium equations in (8.1) using the decoupled equations in (8.52), we do not calculate the damping matrix **C**, but only the stiffness and mass matrices **K** and **M**.

As discussed, damping effects can readily be taken into account in mode superposition analysis provided that (8.51) is satisfied. However, assume that it would be numerically more effective to use direct step-by-step integration and that realistic damping ratios ξ_i, $i = 1, \ldots, p$ are known. In that case, it is necessary to evaluate the matrix **C** explicitly, which when substituted into (8.51) yields the established damping ratios ξ_i. If $p = 2$, Rayleigh damping can be assumed, which is of the form

$$\mathbf{C} = \alpha\mathbf{M} + \beta\mathbf{K} \tag{8.55}$$

where α and β are constants to be determined from two given damping ratios that correspond to two unequal frequencies of vibration.

Example 8.9

Assume that for a multiple degree of freedom system $\omega_1 = 2$ and $\omega_2 = 3$, and that in those two modes we require 2% and 10% critical damping, respectively; i.e., we require $\xi_1 = 0.02$ and $\xi_2 = 0.10$. Establish the constants α and β for Rayleigh damping in order that a direct step-by-step integration can be carried out.

In Rayleigh damping we have

$$\mathbf{C} = \alpha\mathbf{M} + \beta\mathbf{K} \tag{a}$$

But using the relation in (8.51) we obtain using (a)

$$\boldsymbol{\phi}_i^T(\alpha\mathbf{M} + \beta\mathbf{K})\boldsymbol{\phi}_i = 2\omega_i\xi_i$$

or

$$\alpha + \beta\omega_i^2 = 2\omega_i\xi_i \tag{b}$$

Using this relation for ω_1, ξ_1 and ω_2, ξ_2, we obtain two equations for α and β:

$$\begin{aligned} \alpha + 4\beta &= 0.08 \\ \alpha + 9\beta &= 0.60 \end{aligned} \tag{c}$$

The solution of (c) is $\alpha = -0.336$ and $\beta = 0.104$. Thus the damping matrix to be used is

$$\mathbf{C} = -0.336\mathbf{M} + 0.104\mathbf{K} \tag{d}$$

With the damping matrix given, we can now establish the damping ratio that is specified at any value of ω_i, when the Rayleigh damping matrix in (d) is used. Namely, the relation in (b) gives

$$\xi_i = \frac{-0.336 + 0.104\omega_i^2}{2\omega_i}$$

for all values of ω_i.

In actual analysis it may well be that the damping ratios are known for many more than two frequencies. In that case two average values, say $\bar{\xi}_1$ and $\bar{\xi}_2$, are used to evaluate α and β. Consider the following example.

Example 8.10

Assume that the approximate damping to be specified for a multiple degree of freedom system is as follows:

$$\xi_1 = 0.002; \quad \omega_1 = 2; \quad \xi_2 = 0.03; \quad \omega_2 = 3$$
$$\xi_3 = 0.04; \quad \omega_3 = 7; \quad \xi_4 = 0.10; \quad \omega_4 = 15$$
$$\xi_5 = 0.14; \quad \omega_5 = 19$$

Choose appropriate Rayleigh damping parameters α and β.

As in Example 8.9, we determine α and β from the relation

$$\alpha + \beta\omega_i^2 = 2\omega_i\xi_i \tag{a}$$

However, only two pairs of values $(\bar{\xi}_1, \bar{\omega}_1)$ and $(\bar{\xi}_2, \bar{\omega}_2)$ determine α and β. Considering the spacing of the frequencies, we use

$$\begin{aligned} \bar{\xi}_1 &= 0.03; \quad \bar{\omega}_1 = 4 \\ \bar{\xi}_2 &= 0.12; \quad \bar{\omega}_2 = 17 \end{aligned} \tag{b}$$

For the values in (b) we obtain, using (a),

$$\alpha + 16\beta = 0.24$$
$$\alpha + 289\beta = 4.08$$

Hence $\alpha = 0.01498$, $\beta = 0.01405$, and we obtain

$$\mathbf{C} = 0.01498\mathbf{M} + 0.01405\mathbf{K} \tag{c}$$

We can now calculate which actual damping ratios are employed when

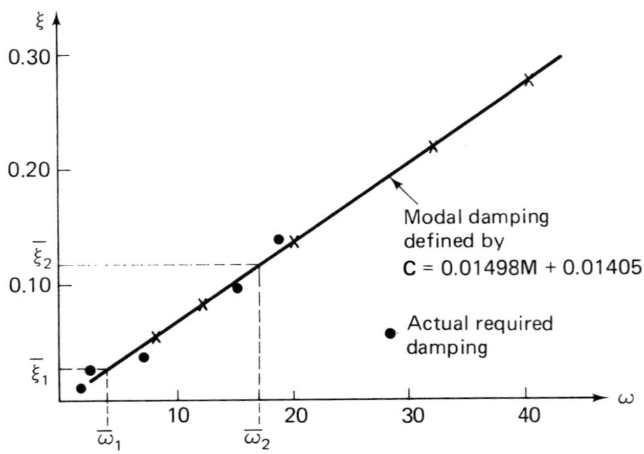

Figure 8.4. Damping as a function of frequency.

the damping matrix **C** in (c) is used. Namely, from (a) we obtain

$$\xi_i = \frac{0.01498 + 0.01405\omega_i^2}{2\omega_i}$$

Figure 8.4 shows the relation of ξ_i as a function of ω_i.

The procedure of calculating α and β in Examples 8.9 and 8.10 may suggest the use of a more complicated damping matrix if more than only two damping ratios are used to establish **C**. Assume that the p damping ratios ξ_i, $i = 1, 2, \ldots, p$ are used to define **C**. Then a damping matrix that satisfies the relation in (8.51) is obtained using the Caughey series,[20]

$$\mathbf{C} = \mathbf{M} \sum_{k=0}^{p-1} a_k [\mathbf{M}^{-1} \ \mathbf{K}]^k \tag{8.56}$$

where the coefficients a_k, $k = 1, \ldots, p$, are calculated from the p simultaneous equations

$$\xi_i = \frac{1}{2}\left(\frac{a_0}{\omega_i} + a_1\omega_i + a_2\omega_i^3 + \ldots + a_{p-1}\omega_i^{2p-3}\right) \tag{8.57}$$

We should note that with $p = 2$, (8.56) reduces to Rayleigh damping, as presented in (8.55). An important observation is that if $p > 2$, the damping matrix **C** in (8.56) is, in general, a full matrix. Similarly, a second procedure to construct a damping matrix yields also, in general, a full damping matrix.[21] Since the cost of analysis is increased by a very significant amount if the damping matrix is not banded, in most practical analyses using direct integration, Rayleigh damping is assumed. A disadvantage of Rayleigh damping is that the higher modes are considerably more damped than the lower modes, for which the Rayleigh constants have been selected (see Example 8.10).

In practice, reasonable Rayleigh coefficients in the analysis of a specific structure may often be selected using available information on the damping characteristics of a typical similar structure; i.e., approximately the same α and β values are used in the analysis of similar structures. The magnitude of the Rayleigh coefficients is to a large extent determined by the energy dissipation characteristics of the structural materials.

In the above discussion we assumed that the damping characteristics of the structure can be represented appropriately using proportional damping, either in a mode superposition analysis or in a direct integration procedure. In many analyses, the assumption of proportional damping [i.e., that (8.51) is satisfied] is adequate. However, in the analysis of structures with widely varying material properties, nonproportional damping may need to be used. For example, in the analysis of foundation-structure interaction problems, significantly more damping may be observed in the foundation than in the surface structure. In this case it may be reasonable to assign in the construction of the damping matrix different Rayleigh coefficients α and β to different parts of the structure, which results into a damping matrix that does not satisfy the relation in (8.51). Another case of nonproportional damping is encountered when concentrated dampers corresponding to specific degrees of freedom (e.g., at the support points of a structure) are specified.

The solution of the finite element system equilibrium equations with nonproportional damping can be obtained using the direct integration algorithms in Tables 8.1 to 8.4 without modifications, because the property of the damping matrix did not enter the derivation of the solution procedures. On the other hand considering mode superposition analysis using the free vibration mode shapes with damping neglected as base vectors, we find that $\mathbf{\Phi}^T\mathbf{C}\mathbf{\Phi}$ in (8.43) is in the case of nonproportional damping a full matrix. In other words, the equilibrium equations in the basis of mode shape vectors are no longer decoupled. But, if it can be assumed that the primary response of the system is still contained in the subspace spanned by $\boldsymbol{\phi}_1, \ldots, \boldsymbol{\phi}_p$, it is only necessary to consider the first p equations in (8.43). Assuming that the coupling in the damping matrix $\mathbf{\Phi}^T\mathbf{C}\mathbf{\Phi}$ between x_i, $i = 1, \ldots, p$, and x_i, $i = p + 1, \ldots, n$, can be neglected, the first p equations in (8.43) decouple from the equations $(p + 1)$ to n and can be solved by direct integration using the algorithms in Table 8.1 to 8.4. (See Example 8.11). In an alternative analysis procedure, the decoupling of the finite element equilibrium equations is achieved by solving a quadratic eigenproblem, in which case complex frequencies and vibration mode shapes are calculated.

Example 8.11

Consider the solution of the equilibrium equation

$$\begin{bmatrix} \frac{1}{2} & & \\ & 1 & \\ & & \frac{1}{2} \end{bmatrix}\ddot{\mathbf{U}} + \begin{bmatrix} 0.1 & & \\ & 0 & \\ & & 0.5 \end{bmatrix}\dot{\mathbf{U}} + \begin{bmatrix} 2 & -1 & 0 \\ -1 & 4 & -1 \\ 0 & -1 & 2 \end{bmatrix}\mathbf{U} = \mathbf{R}(t) \qquad \text{(a)}$$

The free vibration mode shapes with damping neglected and corresponding frequencies of vibration are calculated in Example 10.4 and are

$$\mathbf{\Phi} = \begin{bmatrix} \sqrt{\frac{1}{2}} & 1 & -\sqrt{\frac{1}{2}} \\ \sqrt{\frac{1}{2}} & 0 & \sqrt{\frac{1}{2}} \\ \sqrt{\frac{1}{2}} & -1 & -\sqrt{\frac{1}{2}} \end{bmatrix}; \qquad \mathbf{\Omega}^2 = \begin{bmatrix} 2 & & \\ & 4 & \\ & & 6 \end{bmatrix}$$

Transform the equilibrium equations in (a) to equilibrium relations in the mode shape basis.

Using $\mathbf{U} = \mathbf{\Phi}\mathbf{X}$, we obtain corresponding to (8.43) the equilibrium relations

$$\ddot{\mathbf{X}}(t) + \begin{bmatrix} 0.3 & -0.2\sqrt{2} & -0.3 \\ -0.2\sqrt{2} & 0.6 & 0.2\sqrt{2} \\ -0.3 & 0.2\sqrt{2} & 0.3 \end{bmatrix} \dot{\mathbf{X}}(t) + \begin{bmatrix} 2 & & \\ & 4 & \\ & & 6 \end{bmatrix} \mathbf{X}(t)$$

$$= \begin{bmatrix} \sqrt{\frac{1}{2}} & \sqrt{\frac{1}{2}} & \sqrt{\frac{1}{2}} \\ 1 & 0 & -1 \\ -\sqrt{\frac{1}{2}} & \sqrt{\frac{1}{2}} & -\sqrt{\frac{1}{2}} \end{bmatrix} \mathbf{R}(t) \qquad \text{(b)}$$

If it were now known that because of the specific loading applied, the primary response lies in the first mode only, we may obtain an approximate response by solving only

$$\ddot{x}_1(t) + 0.3\dot{x}_1(t) + 2x_1(t) = \begin{bmatrix} \frac{1}{\sqrt{2}} & \frac{1}{\sqrt{2}} & \frac{1}{\sqrt{2}} \end{bmatrix} \mathbf{R}(t) \qquad \text{(c)}$$

and then calculating

$$\mathbf{U}(t) = \begin{bmatrix} \frac{1}{\sqrt{2}} \\ \frac{1}{\sqrt{2}} \\ \frac{1}{\sqrt{2}} \end{bmatrix} x_1(t)$$

However, it should be noted that because $\mathbf{\Phi}^T\mathbf{C}\mathbf{\Phi}$ in (b) is full, the solution of $x_1(t)$ from (c) does not give the actual response in the first mode, because the damping coupling has been neglected.

REFERENCES

1. L. Collatz, *The Numerical Treatment of Differential Equations*, Springer-Verlag, New York, N.Y., 1966.
2. S. H. Crandall, *Engineering Analysis*, McGraw-Hill Book Company, New York, 1956.
3. C. E. Fröberg, *Introduction to Numerical Analysis*, Addison-Wesley Publishing Company, Inc., Reading, Mass., 1969.
4. R. W. Clough and J. Penzien, *Dynamics of Structures*, McGraw-Hill Book Company, New York, N.Y., 1975.
5. J. M. Biggs, *Introduction to Structural Dynamics*, McGraw-Hill Book Company, New York, N.Y., 1964.
6. W. C. Hurty and M. F. Rubinstein, *Dynamics of Structures*, Prentice-Hall, Inc., Englewood Cliffs, N.J., 1964.
7. M. F. Rubinstein, *Structural Systems—Statics, Dynamics and Stability*, Prentice-Hall, Inc., Englewood Cliffs, N.J., 1970.

8. S. H. CRANDALL, D. C. KARNOPP, E. F. KURTZ, and D. C. PRIDMORE-BROWN, *Dynamics of Mechanical and Electromechanical Systems*, McGraw-Hill Book Company, New York, N.Y., 1967.

9. R. W. CLOUGH, "Analysis of Structural Vibrations and Dynamic Response," *Recent Advances in Matrix Methods of Structural Analysis and Design*, (R. H. Gallagher, Y. Yamada, and J. T. Oden, eds.), University of Alabama Press, Huntsville, Ala., 1971.

10. R. W. CLOUGH and K. J. BATHE, "Finite Element Analysis of Dynamic Response," *Advances in Computational Methods in Structural Mechanics and Design*, (J. T. Oden, R. W. Clough, and Y. Yamamoto, eds.), University of Alabama Press, Huntsville, Ala., 1972.

11. J. C. HOUBOLT, "A Recurrence Matrix Solution for the Dynamic Response of Elastic Aircraft," *Journal of Aeronautical Science*, Vol. 17, 1950, pp. 540–550.

12. E. L. WILSON, "A Computer Program for the Dynamic Stress Analysis of Underground Structures," Report UC SESM 68-1, Department of Civil Engineering, University of California, Berkeley, 1968.

13. E. L. WILSON, "Elastic Dynamic Response of Axisymmetric Structures," Report UC SESM 69-2, Department of Civil Engineering, University of California, Berkeley, 1969.

14. E. L. WILSON, I. FARHOOMAND, and K. J. BATHE, "Nonlinear Dynamic Analysis of Complex Structures," *International Journal of Earthquake Engineering and Structural Dynamics*, Vol. 1, 1973, pp. 241–252.

15. K. J. BATHE, E. L. WILSON, and F. E. PETERSON, "SAP IV—A Structural Analysis Program for Static and Dynamic Response of Linear Systems," Report EERC 73-11, College of Engineering, University of California, Berkeley, June 1973, revised Apr. 1974.

16. EAC/EASE2 Dynamics—User Information Manual/Theoretical, Control Data Corporation (in press).

17. N. M. NEWMARK, "A Method of Computation for Structural Dynamics," *A.S.C.E., Journal of Engineering Mechanics Division*, Vol. 85, 1959, pp. 67–94.

18. K. J. BATHE, "ADINA—A Finite Element Program for Automatic Dynamic Incremental Nonlinear Analysis," Report 82448-1, Acoustics and Vibration Laboratory, Department of Mechanical Engineering, Massachusetts Institute of Technology, Cambridge, Mass., 1975.

19. K. J. BATHE, H. OZDEMIR, and E. L. WILSON, "Static and Dynamic Geometric and Material Nonlinear Analysis," Report UC SESM 74-4, College of Engineering, University of California, Berkeley, Feb. 1974.

20. T. K. CAUGHEY, "Classical Normal Modes in Damped Linear Systems," *Journal of Applied Mechanics*, Vol. 27, 1960, pp. 269–271.

21. E. L. WILSON and J. PENZIEN, "Evaluation of Orthogonal Damping Matrices," *International Journal for Numerical Methods in Engineering*, Vol. 4, No. 1, Jan. 1972, pp. 5–10.

CHAPTER 9

ANALYSIS OF DIRECT INTEGRATION METHODS

9.1 INTRODUCTION

In Chapter 8 we presented the two principal procedures used for the solution of the dynamic equilibrium equations

$$\mathbf{M}\ddot{\mathbf{U}}(t) + \mathbf{C}\dot{\mathbf{U}}(t) + \mathbf{K}\mathbf{U}(t) = \mathbf{R}(t) \tag{9.1}$$

where the matrices and vectors have been defined in Section 8.1. The two procedures were mode superposition and direct integration. The integration schemes considered were the central difference method, the Houbolt method, the Wilson θ method, and the Newmark integration procedure (see Tables 8.1 to 8.4). We stated that using the central difference scheme, a time step Δt smaller than a critical time step Δt_{cr} has to be used; but when employing the other three integration schemes, a similar time step limitation is not applicable.

An important observation was that the cost of a direct integration analysis (i.e., the number of operations required) is directly proportional to the number of time steps required for solution. It follows that the selection of an appropriate time step in direct integration is of much importance. On one hand, the time step must be small enough to obtain accuracy in the solution; but, on the other hand, the time step must not be smaller than necessary, because this would mean that the solution is more costly than actually required. The aim in this chapter is to discuss in detail the problem of selecting an appropriate time step Δt for direct integration. The two fundamental concepts to be considered are those of stability and accuracy of the integration schemes. The analysis of the stability and accuracy characteristics of the integration methods results into guidelines for the selection of an appropriate time step.

A first fundamental observation for the analysis of a direct integration method is the relation between mode superposition and direct integration. We pointed out in Chapter 8 that, in essence, using either procedure the solution is obtained by numerical integration. However, in the mode superposition analysis, a change of basis from the finite element coordinate basis to the basis of eigenvectors of the generalized eigenproblem,

$$\mathbf{K}\boldsymbol{\phi} = \omega^2\mathbf{M}\boldsymbol{\phi} \tag{9.2}$$

is performed prior to the time integration. Writing

$$\mathbf{U}(t) = \boldsymbol{\Phi}\mathbf{X}(t) \tag{9.3}$$

where the columns in $\boldsymbol{\Phi}$ are the $\mathbf{M}$-orthonormalized eigenvectors (free vibration modes) $\boldsymbol{\phi}_1, \ldots, \boldsymbol{\phi}_n$, and substituting for $\mathbf{U}(t)$ into (9.1), we obtain

$$\ddot{\mathbf{X}}(t) + \boldsymbol{\Delta}\dot{\mathbf{X}}(t) + \boldsymbol{\Omega}^2\mathbf{X}(t) = \boldsymbol{\Phi}^T\mathbf{R}(t) \tag{9.4}$$

where $\boldsymbol{\Omega}^2$ is a diagonal matrix listing the eigenvalues of (9.2) (free vibration frequencies squared) $\omega_1^2, \ldots, \omega_n^2$. Assuming that the damping is proportional, $\boldsymbol{\Delta}$ is a diagonal matrix, $\boldsymbol{\Delta} = \text{diag}(2\omega_i\xi_i)$, where ξ_i is the damping ratio in the ith mode.

The equation in (9.4) consists of n uncoupled equations, which can be solved, for example, using the Duhamel integral. Alternatively, one of the numerical integration schemes discussed as direct integration procedures may be used. Because the periods of vibration T_i, $i = 1, \ldots, n$, are known, where $T_i = 2\pi/\omega_i$, we can choose in the numerical integration of each equation in (9.4) an appropriate time step that assures a required level of accuracy. On the other hand, if all n equations in (9.4) are integrated using the same time step Δt, then the mode superposition analysis is completely equivalent to a direct integration analysis, in which the same integration scheme and the same time step Δt are used. In other words, the solution of the finite element system equilibrium equations would be identical using either procedure. Therefore, *to study the accuracy of direct integration, we may focus attention on the integration of the equations in* (9.4) *with a common time step* Δt *instead of considering* (9.1). *In this way, the variables to be considered in the stability and accuracy analysis of the direct integration method are only* Δt, ω_i, *and* ξ_i, $i = 1, \ldots, n$, *and not all elements of the stiffness, mass, and damping matrices. Furthermore, because all n equations in* (9.4) *are similar, we only need to study the integration of one typical row in* (9.4), *which may be written*

$$\ddot{x} + 2\xi\omega\dot{x} + \omega^2 x = r \tag{9.5}$$

and is the equilibrium equation governing motion of a single degree of freedom system with free vibration period T, damping ratio ξ, and applied load r.

It may be mentioned here that the above procedure of changing basis, i.e., using the transformation in (9.3), is also used in the convergence analysis of eigenvalue and eigenvector solution methods (see Chapter 11). The reason for carrying out the transformation is the same as above; namely, many fewer variables need to be considered in the analysis.

Considering the solution characteristics of a direct integration method, the problem is, therefore, to estimate the integration errors in the solution of (9.5) as a function of $\Delta t/T$, ξ, and r. Various procedures could be followed[1-23]. However, in the following discussion we employ a relatively simple and constructive procedure in which the first step is to evaluate an approximation and load operator which relates explicitly the unknown required variables at time $t + \Delta t$ to previously calculated quantities.

9.2 DIRECT INTEGRATION APPROXIMATION AND LOAD OPERATORS

In analogy to the derivation of the direct integration methods (see Section 8.2), assume that we have obtained the required solution for the discrete times $0, \Delta t, 2\Delta t, 3\Delta t, \ldots, t - \Delta t, t$, and that the solution for time $t + \Delta t$ is required next. Then for the specific integration method considered, we aim to establish the following recursive relationship:

$$\hat{\mathbf{X}}_{t+\Delta t} = \mathbf{A}\hat{\mathbf{X}}_t + \mathbf{L}r_{t+\nu} \tag{9.6}$$

where $\hat{\mathbf{X}}_{t+\Delta t}$ and $\hat{\mathbf{X}}_t$ are vectors storing the solution quantities (e.g., displacements, velocities, etc.) and $r_{t+\nu}$ is the load at time $t + \nu$. We will see that ν may be 0, Δt, or $\theta\,\Delta t$ for the integration methods considered. The matrix $\mathbf{A}$ and vector $\mathbf{L}$ are the *integration approximation and load operators*, respectively. Each quantity in (9.6) depends on the specific integration scheme employed. However, before deriving the matrices and vectors corresponding to the different integration procedures, we note that (9.6) can be used to calculate the solution at any time $t + n\,\Delta t$; namely, applying (9.6) recursively, we obtain

$$\begin{aligned}\hat{\mathbf{X}}_{t+n\Delta t} = \mathbf{A}^n\hat{\mathbf{X}}_t + \mathbf{A}^{n-1}\mathbf{L}r_{t+\nu} + \mathbf{A}^{n-2}\mathbf{L}r_{t+\Delta t+\nu} + \cdots \\ + \mathbf{A}\mathbf{L}r_{t+(n-2)\Delta t+\nu} + \mathbf{L}r_{t+(n-1)\Delta t+\nu}\end{aligned} \tag{9.7}$$

It is this relation that we will use for the study of the stability and accuracy of the integration methods. In the following sections we derive the operators $\mathbf{A}$ and $\mathbf{L}$ for the different integration methods considered, where we refer to the presentations in Sections 8.2.1 to 8.2.4.

9.2.1 The Central Difference Method

In the central difference integration scheme we use (8.3) and (8.4) to approximate the acceleration and velocity at time t, respectively. The equilibrium equation (9.5) is considered at time t; i.e., we use

$$\ddot{x}_t + 2\xi\omega\dot{x}_t + \omega^2 x_t = r_t \tag{9.8}$$

$$\ddot{x}_t = \frac{1}{\Delta t^2}\{x_{t-\Delta t} - 2x_t + x_{t+\Delta t}\} \tag{9.9}$$

$$\dot{x}_t = \frac{1}{2\Delta t}\{-x_{t-\Delta t} + x_{t+\Delta t}\} \tag{9.10}$$

Substituting (9.9) and (9.10) into (9.8) and solving for $x_{t+\Delta t}$, we obtain

$$x_{t+\Delta t} = \frac{2 - \omega^2 \Delta t^2}{1 + \xi\omega\Delta t}x_t - \frac{1 - \xi\omega\,\Delta t}{1 + \xi\omega\,\Delta t}x_{t-\Delta t} + \frac{\Delta t^2}{1 + \xi\omega\,\Delta t}r_t \tag{9.11}$$

The solution (9.11) can now be written in the form (9.6); i.e., we have

$$\begin{bmatrix} x_{t+\Delta t} \\ x_t \end{bmatrix} = \mathbf{A}\begin{bmatrix} x_t \\ x_{t-\Delta t} \end{bmatrix} + \mathbf{L}r_t \tag{9.12}$$

where

$$\mathbf{A} = \begin{bmatrix} \dfrac{2 - \omega^2\Delta t^2}{1 + \xi\omega\,\Delta t} & -\dfrac{1 - \xi\omega\,\Delta t}{1 + \xi\omega\,\Delta t} \\ 1 & 0 \end{bmatrix} \tag{9.13}$$

and

$$\mathbf{L} = \begin{bmatrix} \dfrac{\Delta t^2}{1 + \xi\omega\,\Delta t} \\ 0 \end{bmatrix} \tag{9.14}$$

9.2.2 The Houbolt Method

In the Houbolt integration scheme the equilibrium equation (9.5) is considered at time $t + \Delta t$ and two backward difference formulas are used for the acceleration and velocity at time $t + \Delta t$; i.e., we use

$$\ddot{x}_{t+\Delta t} + 2\xi\omega\dot{x}_{t+\Delta t} + \omega^2 x_{t+\Delta t} = r_{t+\Delta t} \tag{9.15}$$

$$\ddot{x}_{t+\Delta t} = \frac{1}{\Delta t^2}(2x_{t+\Delta t} - 5x_t + 4x_{t-\Delta t} - x_{t-2\Delta t}) \tag{9.16}$$

$$\dot{x}_{t+\Delta t} = \frac{1}{6\Delta t}(11x_{t+\Delta t} - 18x_t + 9x_{t-\Delta t} - 2x_{t-2\Delta t}) \tag{9.17}$$

Substituting (9.16) and (9.17) into (9.15), we can establish the relation

$$\begin{bmatrix} x_{t+\Delta t} \\ x_t \\ x_{t-\Delta t} \end{bmatrix} = \mathbf{A}\begin{bmatrix} x_t \\ x_{t-\Delta t} \\ x_{t-2\Delta t} \end{bmatrix} + \mathbf{L}r_{t+\Delta t} \tag{9.18}$$

where

$$\mathbf{A} = \begin{bmatrix} \frac{5\beta}{\omega^2\,\Delta t^2} + 6\kappa & -\left(\frac{4\beta}{\omega^2\,\Delta t^2} + 3\kappa\right) & \frac{\beta}{\omega^2\,\Delta t^2} + \frac{2\kappa}{3} \\ 1 & 0 & 0 \\ 0 & 1 & 0 \end{bmatrix} \tag{9.19}$$

$$\beta = \left(\frac{2}{\omega^2\,\Delta t^2} + \frac{11\xi}{3\omega\,\Delta t} + 1\right)^{-1}; \qquad \kappa = \frac{\xi\beta}{\omega\,\Delta t} \tag{9.20}$$

and

$$\mathbf{L} = \begin{bmatrix} \frac{\beta}{\omega^2} \\ 0 \\ 0 \end{bmatrix} \tag{9.21}$$

9.2.3 The Wilson θ Method

The basic assumption in the Wilson θ method is that the acceleration varies linearly over the time interval from t to $t + \theta\,\Delta t$, where $\theta \geq 1$ and is determined to obtain optimum stability and accuracy characteristics. Let τ denote the increase in time from time t, where $0 \leq \tau \leq \theta\,\Delta t$; then for the time interval t to $t + \theta\,\Delta t$, we have

$$\ddot{x}_{t+\tau} = \ddot{x}_t + (\ddot{x}_{t+\Delta t} - \ddot{x}_t)\frac{\tau}{\Delta t} \tag{9.22}$$

$$\dot{x}_{t+\tau} = \dot{x}_t + \ddot{x}_t\tau + (\ddot{x}_{t+\Delta t} - \ddot{x}_t)\frac{\tau^2}{2\Delta t} \tag{9.23}$$

$$x_{t+\tau} = x_t + \dot{x}_t\tau + \tfrac{1}{2}\ddot{x}_t\tau^2 + (\ddot{x}_{t+\Delta t} - \ddot{x}_t)\frac{\tau^3}{6\Delta t} \tag{9.24}$$

At time $t + \Delta t$ we have

$$\dot{x}_{t+\Delta t} = \dot{x}_t + (\ddot{x}_{t+\Delta t} + \ddot{x}_t)\frac{\Delta t}{2} \tag{9.25}$$

$$x_{t+\Delta t} = x_t + \dot{x}_t\,\Delta t + (2\ddot{x}_t + \ddot{x}_{t+\Delta t})\frac{\Delta t^2}{6} \tag{9.26}$$

Furthermore, in the Wilson θ method the equilibrium equation (9.5) is considered at time $t + \theta\,\Delta t$ [with a projected load, see (8.25)], which gives

$$\ddot{x}_{t+\theta\Delta t} + 2\xi\omega\dot{x}_{t+\theta\Delta t} + \omega^2 x_{t+\theta\Delta t} = r_{t+\theta\Delta t} \tag{9.27}$$

Using (9.22), (9.23), and (9.24) at time $\tau = \theta\,\Delta t$ to substitute into (9.27), an equation is obtained with $\ddot{x}_{t+\Delta t}$ as the only unknown. Solving for $\ddot{x}_{t+\Delta t}$ and substituting into (9.25) and (9.26), the following relationship of the form (9.6) is established:

$$\begin{bmatrix} \ddot{x}_{t+\Delta t} \\ \dot{x}_{t+\Delta t} \\ x_{t+\Delta t} \end{bmatrix} = \mathbf{A} \begin{bmatrix} \ddot{x}_t \\ \dot{x}_t \\ x_t \end{bmatrix} + \mathbf{L} r_{t+\theta\Delta t} \tag{9.28}$$

where

$$\mathbf{A} = \begin{bmatrix} 1 - \frac{\beta\theta^2}{3} - \frac{1}{\theta} - \kappa\theta & \frac{1}{\Delta t}(-\beta\theta - 2\kappa) & \frac{1}{\Delta t^2}(-\beta) \\ \Delta t\left(1 - \frac{1}{2\theta} - \frac{\beta\theta^2}{6} - \frac{\kappa\theta}{2}\right) & 1 - \frac{\beta\theta}{2} - \kappa & \frac{1}{\Delta t}\left(-\frac{\beta}{2}\right) \\ \Delta t^2\left(\frac{1}{2} - \frac{1}{6\theta} - \frac{\beta\theta^2}{18} - \frac{\kappa\theta}{6}\right) & \Delta t\left(1 - \frac{\beta\theta}{6} - \frac{\kappa}{3}\right) & 1 - \frac{\beta}{6} \end{bmatrix} \tag{9.29}$$

$$\beta = \left(\frac{\theta}{\omega^2\,\Delta t^2} + \frac{\xi\theta^2}{\omega\,\Delta t} + \frac{\theta^3}{6}\right)^{-1}; \qquad \kappa = \frac{\xi\beta}{\omega\,\Delta t} \tag{9.30}$$

and

$$\mathbf{L} = \begin{bmatrix} \frac{\beta}{\omega^2\,\Delta t^2} \\ \frac{\beta}{2\omega^2\,\Delta t} \\ \frac{\beta}{6\omega^2} \end{bmatrix} \tag{9.31}$$

9.2.4 The Newmark Method

In the Newmark integration scheme the equilibrium equation (9.5) is considered at time $t + \Delta t$; i.e., we use

$$\ddot{x}_{t+\Delta t} + 2\xi\omega\dot{x}_{t+\Delta t} + \omega^2 x_{t+\Delta t} = r_{t+\Delta t} \tag{9.32}$$

and the following expansions are employed for the velocity and displacement at time $t + \Delta t$:

$$\dot{x}_{t+\Delta t} = \dot{x}_t + [(1 - \delta)\ddot{x}_t + \delta\ddot{x}_{t+\Delta t}]\,\Delta t \tag{9.33}$$

$$x_{t+\Delta t} = x_t + \dot{x}_t\,\Delta t + [(\tfrac{1}{2} - \alpha)\ddot{x}_t + \alpha\ddot{x}_{t+\Delta t}]\,\Delta t^2 \tag{9.34}$$

where δ and α are parameters to be chosen to obtain optimum stability and accuracy. Newmark proposed as an unconditionally stable scheme the constant-average-acceleration method, in which case $\delta = \frac{1}{2}$ and $\alpha = \frac{1}{4}$.

Substituting for $\dot{x}_{t+\Delta t}$ and $x_{t+\Delta t}$ into (9.32), we can solve for $\ddot{x}_{t+\Delta t}$ and then use (9.33) and (9.34) to calculate $\dot{x}_{t+\Delta t}$ and $x_{t+\Delta t}$. We thus can establish the relation

$$\begin{bmatrix} \ddot{x}_{t+\Delta t} \\ \dot{x}_{t+\Delta t} \\ x_{t+\Delta t} \end{bmatrix} = \mathbf{A} \begin{bmatrix} \ddot{x}_t \\ \dot{x}_t \\ x_t \end{bmatrix} + \mathbf{L}r_{t+\Delta t} \tag{9.35}$$

where

$$\mathbf{A} = \begin{bmatrix} -(\frac{1}{2} - \alpha)\beta - 2(1 - \delta)\kappa & \frac{1}{\Delta t}(-\beta - 2\kappa) & \frac{1}{\Delta t^2}(-\beta) \\ \Delta t[1 - \delta - (\frac{1}{2} - \alpha)\delta\beta - 2(1 - \delta)\delta\kappa] & 1 - \beta\delta - 2\delta\kappa & \frac{1}{\Delta t}(-\beta\delta) \\ \Delta t^2[\frac{1}{2} - \alpha - (\frac{1}{2} - \alpha)\alpha\beta - 2(1 - \delta)\alpha\kappa] & \Delta t(1 - \alpha\beta - 2\alpha\kappa) & (1 - \alpha\beta) \end{bmatrix} \tag{9.36}$$

$$\beta = \left(\frac{1}{\omega^2\,\Delta t^2} + \frac{2\xi\delta}{\omega\,\Delta t} + \alpha\right)^{-1}; \qquad \kappa = \frac{\xi\beta}{\omega\,\Delta t} \tag{9.37}$$

and

$$\mathbf{L} = \begin{bmatrix} \dfrac{\beta}{\omega^2\,\Delta t^2} \\ \dfrac{\beta\delta}{\omega^2\,\Delta t} \\ \dfrac{\alpha\beta}{\omega^2} \end{bmatrix} \tag{9.38}$$

We should note the close relationship between the Newmark operators in (9.36) and (9.38) and the Wilson θ method operators in (9.29) and (9.31). That is, using $\delta = \frac{1}{2}$, $\alpha = \frac{1}{6}$, and $\theta = 1.0$, the same approximation and load operators are obtained in both methods. This should be expected because for the above parameters both methods assume a linear variation of acceleration over the time interval t to $t + \Delta t$.

9.3 STABILITY ANALYSIS

The aim in the numerical integration of the finite element system equilibrium equations is to evaluate a good approximation to the actual dynamic response of the structure under consideration. In order to predict the dynamic response of the structure accurately, it would seem that all system equilibrium equations in (9.1) must be integrated to high precision, and this means that all n equations of the form (9.5) need be integrated accurately. Since in direct integration the same time step is used for each equation of the form (9.5), Δt would have to be selected corresponding to the smallest period in the system, which may mean that the time step is very small indeed. As an estimate of Δt thus required, it appears that if the smallest period is T_n, Δt would have to be about $T_n/10$. However, we discussed in Section 8.3.2 that in many analyses the primary response lies in only some modes of vibration and that for this reason, only some mode shapes are considered in mode superposition analysis. In addition, it was pointed out that in many analyses there is little justification to include the response predicted in the higher modes, because the frequencies and mode shapes of the finite element mesh can only be crude approximations to the "exact" quantities. Therefore, the finite element idealization has to be chosen in such a way that the lowest p frequencies and mode shapes of the structure are predicted accurately, where p is determined by the distribution and frequency content of the loading.

Considering direct integration, it can therefore be concluded that in many analyses we are only interested in integrating the first p equations of the n equations in (9.4) accurately. This means that we would be able to revise Δt to be about $T_p/10$, i.e., T_p/T_n times larger than our first estimate. In practical analysis the ratio T_p/T_n can be very large, say of the order 1000, meaning that the analysis would be much more effective using $\Delta t = T_p/10$. However, assuming that we select a time step Δt of magnitude $T_p/10$, we realize that in the direct integration also the response in the higher modes is automatically integrated with the same time step. Since we cannot possibly integrate accurately the response in those modes for which Δt is larger than half the natural period T, an important question is: What "response" is predicted in the numerical integration of (9.5) when $\Delta t/T$ is large? This is, in essence, the question of stability of an integration scheme. Stability of an integration method means that the initial conditions for the equations with a large value $\Delta t/T$ must not be amplified artificially and thus make the integration of the lower modes worthless. Stability also means that any errors in the displacements, velocities, and accelerations at time t, which may be due to round-off in the computer, do not grow in the integration. Stability is assured if the time step is small enough to integrate accurately the response in the highest frequency component. But this may require a very small time step, and, as was pointed out above, the accurate integration of the high-frequency response predicted by the finite element assemblage is in many cases not justified and therefore not necessary.

The stability of an integration method is determined by examining the behavior of the numerical solution for arbitrary initial conditions. Therefore, we consider the integration of (9.5) when no load is specified; i.e., $r = 0$. The solution for prescribed initial conditions only as obtained from (9.7) is, hence,

$$\hat{\mathbf{X}}_{t+n\Delta t} = \mathbf{A}^n \hat{\mathbf{X}}_t \tag{9.39}$$

Considering the stability of integration methods, we have procedures that are unconditionally stable and that are only conditionally stable. *An integration method is unconditionally stable if the solution for any initial conditions does not grow without bound for any time step Δt, in particular when $\Delta t/T$ is large. The method is only conditionally stable if the above only holds provided that $\Delta t/T$ is smaller than a certain value, usually called the stability limit.*

To investigate the stability of an integration method we may conveniently use the relation in (9.39), where the approximation operators corresponding to some popular integration schemes have been presented in Section 9.2.

For the analysis we use the spectral decomposition of $\mathbf{A}$ and have

$$\mathbf{A}^n = \mathbf{P}\mathbf{J}^n\mathbf{P}^{-1} \tag{9.40}$$

where $\mathbf{P}$ is the matrix of eigenvectors of $\mathbf{A}$ and $\mathbf{J}$ is the Jordan form of $\mathbf{A}$ with eigenvalues λ_i of $\mathbf{A}$ on its diagonal.[3] We considered in Section 2.6 the case of $\mathbf{A}$ being symmetric (see Example 2.20); however, the approximation operator is in general nonsymmetric and we therefore must use the more general decomposition $\mathbf{A} = \mathbf{PJP}^{-1}$. Let $\rho(\mathbf{A})$ be the spectral radius of $\mathbf{A}$ defined as

$$\rho(\mathbf{A}) = \max |\lambda_i|; \qquad i = 1, 2, \ldots \tag{9.41}$$

then $\mathbf{J}^n$ *is bounded for* $n \longrightarrow \infty$ *if and only if* $\rho(\mathbf{A}) \leq 1$. *This is our stability criterion.* Furthermore, $\mathbf{J}^n \longrightarrow 0$ if $\rho(\mathbf{A}) < 1$, and the smaller $\rho(\mathbf{A})$, the more rapid is the convergence.

Since the stability of an integration method depends only on the eigenvalues of the approximation operator, it may be convenient to apply a similarity transformation on $\mathbf{A}$ before evaluating the eigenvalues. In the case of the Newmark method and Wilson θ method we apply the similarity transformation $\mathbf{D}^{-1}\mathbf{AD}$, where $\mathbf{D}$ is a diagonal matrix with $d_{ii} = (\Delta t)^i$. As might be expected, we then find that the spectral radii and therefore the stability of the integration methods depend only on the time ratio $\Delta t/T$, the damping ratio ξ, and the integration parameters used. Therefore, for given $\Delta t/T$ and ξ, it is possible in the Wilson θ method and in the Newmark method to vary the parameters θ and α, δ, repectively, to obtain optimum stability and accuracy characteristics.

Consider as a simple example the stability analysis of the central difference method.

Example 9.1

Analyze the central difference method for its integration stability. Consider the case $\xi = 0.0$.

We need to calculate the spectral radius of the approximation operator given in (9.13) when $\xi = 0$. The eigenvalue problem $\mathbf{Au} = \lambda\mathbf{u}$ to be solved is

$$\begin{bmatrix} 2 - \omega^2\,\Delta t^2 & -1 \\ 1 & 0 \end{bmatrix}\mathbf{u} = \lambda\mathbf{u} \tag{a}$$

The eigenvalues are the roots of the characteristic polynomial $p(\lambda)$ (see Section 2.6) defined as

$$p(\lambda) = (2 - \omega^2\,\Delta t^2 - \lambda)(-\lambda) + 1$$

Hence

$$\lambda_1 = \frac{2 - \omega^2\,\Delta t^2}{2} + \sqrt{\frac{(2 - \omega^2\,\Delta t^2)^2}{4} - 1}$$

$$\lambda_2 = \frac{2 - \omega^2\,\Delta t^2}{2} - \sqrt{\frac{(2 - \omega^2\,\Delta t^2)^2}{4} - 1}$$

For stability we need that the absolute value of λ_1 and λ_2 be smaller or equal to 1; i.e., the spectral radius $\rho(\mathbf{A})$ of the matrix $\mathbf{A}$ in (a) must satisfy the condition $\rho(\mathbf{A}) \leq 1$. Figure 9.1 shows $\rho(\mathbf{A})$ as a function of $\Delta t/T$, where $\omega = 2\pi/T$. The spectral radius is smaller than 1 for $\Delta t/T < 1/\pi$, and hence the central difference method is stable provided that $\Delta t \leq \Delta t_{cr}$, where $\Delta t_{cr} = T_n/\pi$.

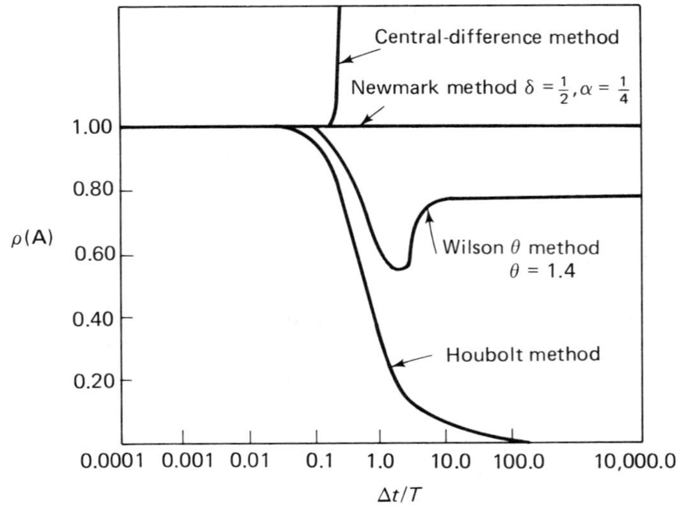

Figure 9.1. Spectral radii of approximation operators, case $\xi = 0.0$.

By the same procedure as employed in Example 9.1, the Wilson θ method, Newmark, and Houbolt methods can be analyzed for stability using the corresponding approximation operators. Assuming no physical damping (i.e., $\xi = 0.0$) Fig 9.1 shows the stability characteristics of the methods considered in Section 8.2. It is noted that the central difference method is only conditionally stable as evaluated in Example 9.1, and that the Newmark, Wilson, and Houbolt methods are unconditionally stable. For the case $\xi > 0$, the stability analysis would include ξ as an additional variable. However, in general, small values of ξ do not change the overall stability characteristics of an integration scheme.

In order to evaluate an optimum value of θ for the Wilson θ method, the variation of the spectral radius of the approximation operator as a function of θ must be calculated as shown in Fig. 9.2. It is seen that unconditional stability is obtained when $\theta \geq 1.37$. This information must be supplemented by the accuracy of the method in order to arrive at the optimum value of θ (see Section 9.4).

Considering the Newmark method, the two parameters α and δ can be varied to obtain optimum stability and accuracy. The integration scheme is unconditionally stable provided that $\delta \geq 0.5$ and $\alpha \geq 0.25(\delta + 0.5)^2$. We will see in the next section that the method corresponding to $\delta = 0.5$ and $\alpha = 0.25$ has the most desirable accuracy characteristics.

Although only the most widely used integration schemes have been discussed, the above considerations already show that a large number of

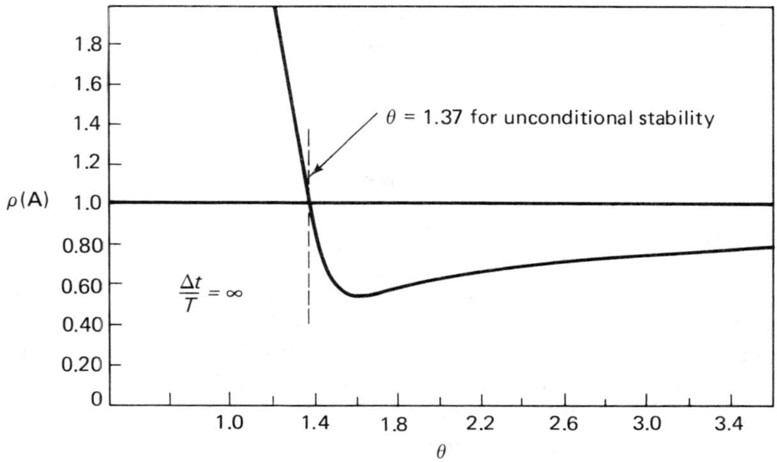

Figure 9.2. Spectral radius $\rho(\mathbf{A})$ as a function of θ in the Wilson θ method.

integration methods are available.[1-23] Additional schemes not mentioned include, for example, the stiffly stable methods,[18-23] and further research is being carried out to obtain improved integration procedures. With the large number of integration schemes available, the analyst has to make a choice as to which method to use. This choice is largely determined by the accuracy characteristics of the method, i.e., the accuracy that can be obtained in the integration for a given time step Δt.

9.4 ACCURACY ANALYSIS

The decision as to which integration operator to use in a practical analysis is governed by the cost of solution, which in turn is determined by the number of time steps required in the integration. If a conditionally stable algorithm such as the central difference method is employed, the time step size, and hence the number of time steps for a given time range considered, is determined by the critical time step Δt_{cr}, and not much choice is available. However, using an unconditionally stable operator, the time step has to be chosen to yield an accurate and effective solution. Because the direct integration of the equilibrium equations in (9.1) is equivalent to integrating simultaneously all n decoupled equations of the form (9.5), we can study the integration accuracy obtained in the solution of (9.1) by assessing the accuracy obtained in the integration of (9.5) as a function of $\Delta t/T$, ξ, and r. The solution to (9.5) was given in (9.7), and it is this equation that we use to assess the integration errors.

Various studies of the integration accuracies of the Newmark, Wilson, Houbolt, and central difference methods have been reported,[8-15] and because the accuracy of a specific procedure depends on a number of factors, in the following we only summarize some important solution characteristics. For this purpose we consider the solution of the initial value problem defined by

$$\left.\begin{aligned} &\ddot{x} + \omega^2 x = 0 \\ \text{and} \qquad & x_0 = 1.0; \qquad \dot{x}_0 = 0.0; \qquad \ddot{x}_0 = -\omega^2 \end{aligned}\right\} \tag{9.42}$$

for which the exact solution is $x = \cos \omega t$. For a complete analysis we would have to consider also (1) the initial value problem corresponding to $x_0 = 0.0$, $\dot{x}_0 = \omega$, $\ddot{x}_0 = 0$ with the exact solution $x = \sin \omega t$, and (2) the solution for a general loading condition. In addition, the influence of the damping parameter ξ would need to be investigated. However, the significant solution characteristics can already be demonstrated by considering only the numerical solution of the problem in (9.42).

The Newmark and Wilson methods can be used directly with the initial conditions given in (9.42). However, in the Houbolt method, the initial conditions are only defined by initial displacements, and in the following study the exact displacement values for $x_{\Delta t}$ and $x_{2\Delta t}$ obtained using the solution $x = \cos \omega t$ have been employed.

The numerical solution of (9.42) using the different integration methods shows that the errors in the integrations can be measured in terms of period elongation and amplitude decay. Figure 9.3 shows the percentage period elongations and amplitude decays in the implicit integration schemes discussed as a function of $\Delta t/T$. The relationships have been obtained by evaluating (9.7) using the computer and comparing the exact solution $x = \cos \omega t$ with the numerical solutions. It should be noted that (9.7) only yields the solution at the discrete time points Δt apart. To obtain maximum displacements, as required in Fig. 9.3, the relations in (9.23), (9.24) and (9.33), (9.34), respectively, have been employed in the Wilson θ and Newmark methods, and an interpolating polynomial of order 3, which fits the displacements at the discrete time points Δt apart, has been used in the Houbolt method.

The curves in Fig. 9.3 show that, in general, the numerical integrations using any one of the methods are accurate when $\Delta t/T$ is smaller than about 0.01. However, when the time step-to-period ratio is larger, the various integration methods exhibit quite different characteristics. Notably, for a given ratio $\Delta t/T$, the Wilson θ method with $\theta = 1.4$ introduces less amplitude decay and period elongation than the Houbolt method, and the Newmark constant-average-acceleration method introduces only period elongation and no amplitude decay.

The characteristics of the integration errors exhibited in Fig. 9.3 are used in the discussion of the simultaneous integration of all n equations of form (9.5), as required in the solution of (9.4) and thus in the solution of (9.1). We observe that the equations for which the time step-to-period ratio is small

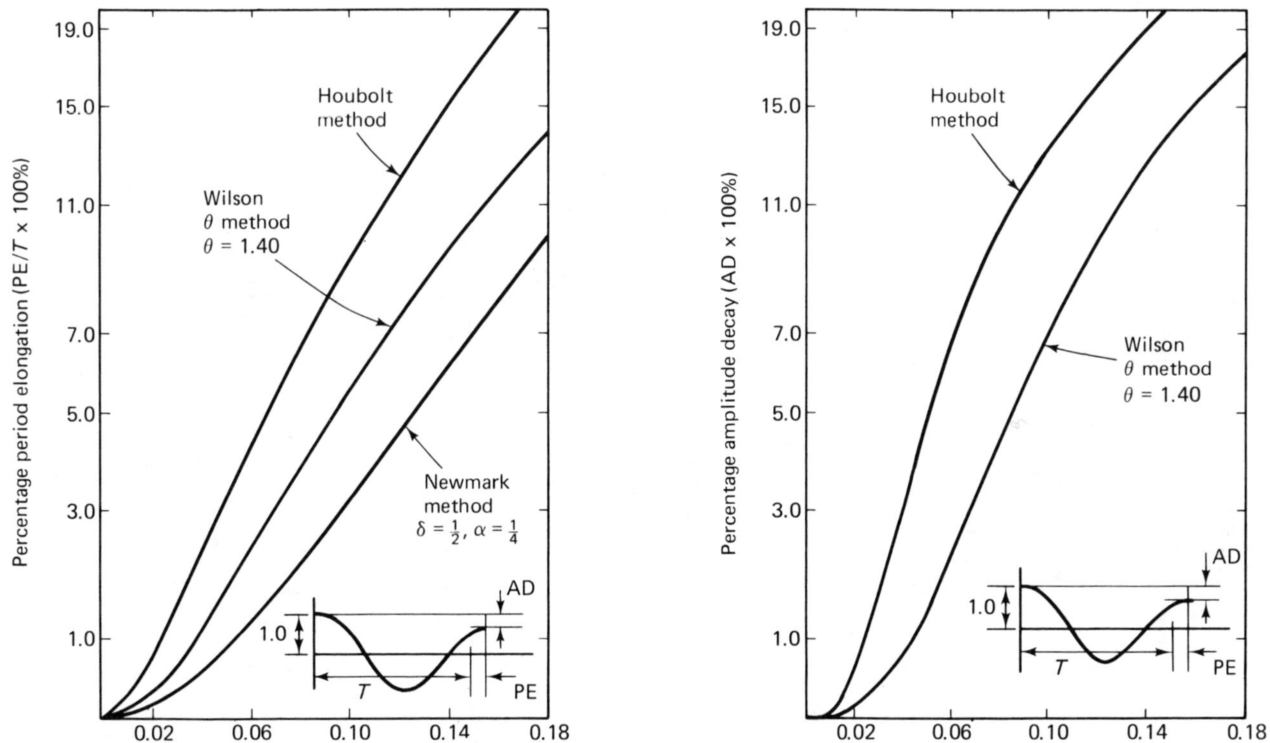

Figure 9.3. Percentage period elongations and amplitude decays.

are integrated accurately, but that the response in the equations for which $\Delta t/T$ is large is not obtained with any precision.

An important problem is the choice of an appropriate time step Δt. Using the central difference method, the time step has to be chosen such that Δt is smaller than Δt_{cr} evaluated in Example 9.1, which means that Δt will, in general, be small enough to obtain accuracy in the integration of practically all n equations in (9.4). However, using one of the unconditionally stable schemes, we realize that the time step Δt can be much larger and should only be small enough that the response in all modes which significantly contribute to the total structural response is calculated accurately. The other modal response components are not evaluated accurately, but the errors are unimportant because the response measured in those components is negligible. However, it need be noted again that this applies only if the integration scheme is unconditionally stable, i.e., if it is assumed that the amplitudes in the components which are negligible do not grow.

Using one of the unconditionally stable algorithms the time step Δt can be chosen by referring to the integration errors displayed in Fig. 9.3. Since one important phenomenon is the amplitude decay observed in the numerical integration of the modes for which $\Delta t/T$ is large, consider the following demonstrative case. Assume that using the Wilson method with $\theta = 1.4$, a time step is selected which gives $\Delta t/T_1 = 0.01$, where T_1 is the fundamental period of a six degree of freedom system and $T_{i+1} = T_i/10$, $i = 1, \ldots, 5$. Let the initial conditions in each mode be those given in (9.42), and let the integration be performed for 100 time steps. Figure 9.4 shows the response calculated in the fundamental and higher modes. It is observed that the amplitude decay caused by the numerical integration errors effectively "filters" the high mode response out of the solution. The same effect is obtained using the Houbolt method, whereas when the Newmark constant-

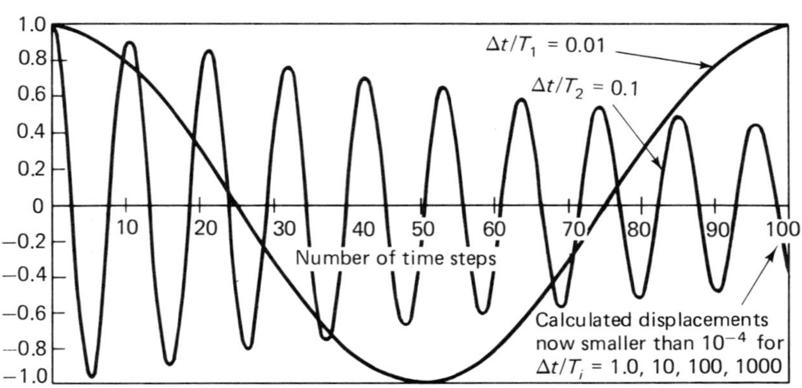

Figure 9.4. Displacement response predicted with increasing $\Delta t/T$ ratio; Wilson method, $\theta = 1.4$.

average-acceleration scheme is employed, which does not introduce amplitude decay, the high frequency response is retained in the solution. In order to obtain amplitude decay using the Newmark method, it is necessary to employ $\delta > 0.5$ and correspondingly $\alpha = \frac{1}{4}(\delta + 0.5)^2$.

It should be noted that the above considerations may be considered to be rather theoretical, because in a practical analysis the initial conditions on the second to sixth mode should be negligible for the above integration time step to be realistic. However, integration accuracy cannot possibly be obtained in the response of the modes for which the time step-to-period ratio is large, and it might be best to simply suppress any response calculations in these modes. On the other hand, since the response in those modes for which the time step-to-period ratio is large is supposed to be negligible, the filtering should not actually be required, and because of the smallest numerical errors, the Newmark method appears to be most effective. However, in most practical analyses, little difference in the computational effort using either the Newmark or Wilson θ method is observed, because the errors resulting into the period elongations require about the same time step using either procedure. On the other hand, the Houbolt method has the disadvantage that special starting procedures must be used.

It should be realized that the stability and accuracy analyses given in this chapter assume that linear finite element analysis is carried out. In particular, the stability limits are only applicable in linear analysis, whereas many more considerations are required in nonlinear response calculations.[24-30]

In order to indicate the general applicability of the stability and accuracy analysis procedure given in this chapter, we give the following example, in which the direct integration of the heat transfer equilibrium equations is considered.

Example 9.2

Use the stability and accuracy analysis procedure discussed in this chapter to evaluate the stability and accuracy characteristics of the integration algorithm proposed in Example 8.5 for heat transfer analysis.

The heat flow equilibrium equations considered are

$$\mathbf{C}\dot{\boldsymbol{\theta}} + \mathbf{K}\boldsymbol{\theta} = \mathbf{Q} \tag{a}$$

where $\mathbf{C}$ and $\mathbf{K}$ are the heat capacity and conductivity matrices, respectively, and $\mathbf{Q}$ and $\boldsymbol{\theta}$ are vectors of nodal point heat flow input and temperatures.

To analyze the step-by-step integration algorithm given in Example 8.5, we proceed in an analogous manner to the analysis of the integration schemes used in dynamic response calculations. The first step is to represent the heat flow equilibrium equations in (a) in the basis of thermal mode vectors. These vectors are obtained from the solution of the eigenproblem

$$\mathbf{K}\boldsymbol{\phi} = \lambda\mathbf{C}\boldsymbol{\phi} \tag{b}$$

Similar to the eigenproblem considered in dynamic analysis, the problem in (b) has n solutions, which can be written

$$\mathbf{K\Phi} = \mathbf{C\Phi\Lambda}$$

where $\mathbf{\Phi}$ is a matrix storing the thermal mode vectors, $\mathbf{\Phi} = [\boldsymbol{\phi}_1, \ldots, \boldsymbol{\phi}_n]$ and $\mathbf{\Lambda}$ is a diagonal matrix listing the eigenvalues or thermal frequencies, $\mathbf{\Lambda} = \text{diag}(\lambda_i)$. The basic properties of the eigenvectors are the orthogonality conditions,

$$\mathbf{\Phi}^T\mathbf{K\Phi} = \mathbf{\Lambda}; \qquad \mathbf{\Phi}^T\mathbf{C\Phi} = \mathbf{I} \tag{c}$$

Therefore, if we define generalized temperatures $\mathbf{X}(t)$,

$$\boldsymbol{\theta} = \mathbf{\Phi X} \tag{d}$$

substitute for $\boldsymbol{\theta}$ into (a), and use the conditions in (c), we obtain

$$\dot{\mathbf{X}} + \mathbf{\Lambda X} = \mathbf{\Phi}^T\mathbf{Q} \tag{e}$$

The stability and accuracy of an integration scheme can now be investigated by studying the integration algorithm as applied to a typical equation of (e), which may be written

$$\dot{x} + \lambda x = q \tag{f}$$

where

$$q = \boldsymbol{\phi}_i^T\mathbf{Q}$$

The basic approximation used in the integration algorithm proposed in Example 8.5 was

$$\dot{x}_{t+\Delta t} = \frac{1}{\Delta t}(x_{t+\Delta t} - x_t) \tag{g}$$

Using (f) at time $t + \Delta t$ and substituting for $\dot{x}_{t+\Delta t}$ from (g), we obtain

$$\frac{1}{\Delta t}(x_{t+\Delta t} - x_t) + \lambda x_{t+\Delta t} = q_{t+\Delta t}$$

which may be written in the form (9.6); i.e.,

$$x_{t+\Delta t} = Ax_t + Lq_{t+\Delta t} \tag{h}$$

where

$$A = \frac{1}{1 + \lambda\,\Delta t} \tag{i}$$

$$L = \frac{\Delta t}{1 + \lambda\,\Delta t}$$

It should be noted that (h) could also have been obtained by using the transformation in (d) directly in equation (d) of Example 8.5.

The recursion relation in (h) can be used to study the stability and accuracy of the integration scheme. The solution at time $t + n\,\Delta t$ is

$$x_{t+n\Delta t} = A^n x_t + A^{n-1}Lq_{t+\Delta t} + \ldots + Lq_{t+n\Delta t} \tag{j}$$

The integration method is unconditionally stable if A^n is bounded as $n \longrightarrow \infty$ for any time step Δt. This means that A in (i) shall be smaller than or equal to 1, and we need

$$1 + \lambda\,\Delta t \geq 1$$

However, the thermal frequency $\lambda \geq 0$, and hence the integration algorithm is unconditionally stable.

The accuracy of the integration algorithm clearly depends on $\lambda\ \Delta t$ and can be studied conveniently using the relation in (j).

REFERENCES

1. L. COLLATZ, *The Numerical Treatment of Differential Equations*, Springer-Verlag, New York, 1966.
2. S. H. CRANDALL, *Engineering Analysis*, McGraw-Hill Book Company, New York, 1956.
3. C. E. FRÖBERG, *Introduction to Numerical Analysis*, Addison-Wesley Publishing Company, Inc., Reading, Mass., 1969.
4. P. E. LAX and R. D. RICHTMYER, "Survey of the Stability of Finite Difference Equations," *Communications in Pure Applied Mathematics*, Vol. 9, 1956, pp. 267–293.
5. P. HENRICI, *Error Propagation for Difference Methods*, John Wiley & Sons, Inc., New York, 1963.
6. G. DAHLQUIST, "A Special Stability Criterion for Linear Multistep Methods," *BIT*, Vol. 3, 1963, pp. 27–43.
7. J. W. LEECH, P. T. HSU, and E. W. MACK, "Stability of a Finite Difference Method for Solving Matrix Equations," *A.I.A.A. Journal*, Vol. 3, 1965, pp. 2172–2173.
8. D. E. JOHNSON, "A Proof of the Stability of the Houbolt Method," *A.I.A.A. Journal*, Vol. 4, 1966, pp. 1450–1451.
9. G. P. DESTEFANO, "Causes of Instabilities in Numerical Integration Techniques," *International Journal of Computational Mathematics*, Vol. 2, 1968, pp. 123–142.
10. R. E. NICKELL, "On the Stability of Approximation Operators in Problems of Structural Dynamics," *International Journal of Solids and Structures*, Vol. 7, 1971, pp. 301–319.
11. G. L. GOUDREAU and R. L. TAYLOR, "Evaluation of Numerical Integration Methods in Elastodynamics," *Computer Methods in Applied Mechanics and Engineering*, Vol. 2, No. 1, 1973, pp. 69–97.
12. K. J. BATHE and E. L. WILSON, "Stability and Accuracy Analysis of Direct Integration Methods," *International Journal of Earthquake Engineering and Structural Dynamics*, Vol. 1, 1973, pp. 283–291.
13. R. D. KRIEG and S. W. KEY, "Transient Shell Response by Numerical Time Integration," *Advances in Computational Methods in Structural Mechanics and Design* (J. T. Oden, R. W. Clough, and Y. Yamamoto, eds.), University of Alabama Press, Huntsville, Ala., 1972.
14. R. D. KRIEG, "Unconditional Stability in Numerical Time Integration Methods," *Trans. A.S.M.E., Journal of Applied Mechanics*, June 1973, pp. 417–421.
15. R. E. NICKELL, "Direct Integration in Structural Dynamics," *A.S.C.E., Journal of Engineering Mechanics Division*, Vol. 99, 1973, pp. 303–317.

16. M. GERADIN, "A Classification and Discussion of Integration Operators for Transient Structural Response," *A.I.A.A.*, Aerospace Science Meeting, Washington, D.C., 1974.

17. J. H. ARGYRIS and D. W. SCHARPF, "Finite Elements in Time and Space," *The Aeronautical Journal of the Royal Aeronautical Society*, 73, 1969, pp. 1041–1044.

18. C. W. GEAR, "Numerical Integration of Stiff Ordinary Differential Equations," Report 221, Department of Computer Sciences, University of Illinois, Urbana–Champaign, January 1967.

19. C. W. GEAR, "The Automatic Integration of Stiff Ordinary Differential Equations," *Information Processing*, Vol. 68, 1969, pp. 187–193.

20. I. W. SANDBERG and H. SCHICKMAN, "Numerical Integration of Systems of Stiff Nonlinear Differential Equations," *Bell Systems Technical Journal*, Vol. 47, 1968, pp. 511–527.

21. W. LINIGER and R. A. WILLOUGHBY, "Efficient Integration Methods for Stiff Systems of Ordinary Differential Equations," *S.I.A.M., Journal of Numerical Analysis*, Vol. 7, No. 1, 1970, pp. 47–66.

22. M. K. JAIN and V. K. SRIVASTAVA, "Optimal Stiffly Stable Methods for Ordinary Differential Equations," Report 402, Department of Computer Sciences, University of Illinois, Urbana–Champaign, June 1970.

23. C. DILL and C. W. GEAR, "A Graphical Search for Stiffly Stable Methods for Ordinary Differential Equations," *Journal of the Association for Computing Machinery*, Vol. 18, 1971, pp. 75–79.

24. J. A. STRICKLIN, J. E. MARTINEZ, J. R. TILLERSON, J. H. HONG, and W. E. HAISLER, "Nonlinear Dynamic Analysis of Shells of Revolution by the Matrix Displacement Method," *A.I.A.A. Journal*, Vol. 9, No. 4, 1971.

25. G. E. WEEKS, "Temporal Operators for Nonlinear Structural Dynamics Problems," *A.S.C.E., Journal of Engineering Mechanics Division*, Vol. 98, No. EM5, Proc. Paper 9260, Oct. 1972, pp. 1087–1104.

26. T. BELYTSCHKO and B. J. HSIEH, "Nonlinear Transient Analysis of Shells and Solids of Revolution by Convected Elements," A.I.A.A. paper No. 73-359, AIAA/ASME/SAE 14th Structures, Structural Dynamics and Materials Conference, Williamsburg, Va., Mar. 1973.

27. J. H. ARGYRIS, P. C. DUNNE, and T. ANGELOPOULOS, "Nonlinear Oscillations Using the Finite Element Technique," *Computer Methods in Applied Mechanics and Engineering*, Vol. 2, 1973, pp. 203–250.

28. J. F. MCNAMARA and P. V. MARCAL, "Incremental Stiffness Method for Finite Element Analysis of the Nonlinear Dynamic Problem," *Numerical and Computer Methods in Structural Mechanics* (S. J. Fenves, N. Perrone, J. Robinson, and W. C. Schnobrich, eds.), Academic Press, Inc., New York, 1973.

29. K. J. BATHE, H. OZDEMIR, and E. L. WILSON, "Static and Dynamic Geometric and Material Nonlinear Analysis," Report UC SESM 74-4, Department of Civil Engineering, University of California, 1974.

30. K. J. BATHE, "An Assessment of Current Finite Element Analysis of Nonlinear Problems in Solid Mechanics," Proceedings, Symposium on the Numerical Solution of Partial Differential Equations, May 1975, Academic Press, Inc., 1976.

CHAPTER 10

PRELIMINARIES TO THE SOLUTION OF EIGENPROBLEMS

10.1 INTRODUCTION

In various sections of the preceding chapters we encountered eigenproblems and the statement of their solutions. We did not at that time discuss how to obtain the required eigenvalues and eigenvectors. It is the purpose of this and the next chapters to describe the actual solution procedures used to solve the eigenproblems of interest. Before presenting the algorithms, we discuss in this chapter some important basic considerations for the solution of eigenproblems.[1-7]

First, let us briefly summarize the eigenproblems that we want to solve. The simplest problem encountered is the standard eigenproblem,

$$\mathbf{K}\boldsymbol{\phi} = \lambda\boldsymbol{\phi} \tag{10.1}$$

where $\mathbf{K}$ is the stiffness matrix of a single finite element or of an element assemblage. We recall that $\mathbf{K}$ has order n and half-bandwidth $m_{\mathbf{K}}$ (i.e., the total bandwidth is $2m_{\mathbf{K}} + 1$) and that $\mathbf{K}$ is positive semidefinite or positive definite. There are n eigenvalues and corresponding eigenvectors satisfying (10.1). The ith eigenpair is denoted as $(\lambda_i, \boldsymbol{\phi}_i)$, where the eigenvalues are ordered according to their magnitudes:

$$0 \leq \lambda_1 \leq \lambda_2 \leq \ldots \leq \lambda_{n-1} \leq \lambda_n \tag{10.2}$$

The solution for p eigenpairs can be written

$$\mathbf{K}\boldsymbol{\Phi} = \boldsymbol{\Phi}\boldsymbol{\Lambda} \tag{10.3}$$

where $\mathbf{\Phi}$ is an $n \times p$ matrix with its columns equal to the p eigenvectors and $\mathbf{\Lambda}$ is a $p \times p$ diagonal matrix listing the corresponding eigenvalues. As an example, (10.3) may represent the solution to the lowest p eigenvalues and corresponding eigenvectors of $\mathbf{K}$, in which case $\mathbf{\Phi} = [\boldsymbol{\phi}_1, \ldots, \boldsymbol{\phi}_p]$ and $\mathbf{\Lambda} =$ diag (λ_i), $i = 1, \ldots, p$. We recall that if $\mathbf{K}$ is positive definite, $\lambda_i > 0$, $i = 1, \ldots, n$, and if $\mathbf{K}$ is positive semidefinite, $\lambda_i \geq 0$, $i = 1, \ldots, n$, where the number of zero eigenvalues is equal to the number of rigid body modes in the system.

The solution of the eigenvalue problem in (10.1) is, for example, sought in the evaluation of an element stiffness matrix or in the calculation of the condition number of a structure stiffness matrix. We discussed in Section 3.2.5 that the representation of the element stiffness matrix in its canonical form (i.e., in the eigenvector basis) is used to evaluate the effectiveness of the element. In this case all eigenvalues and vectors of $\mathbf{K}$ must be calculated. On the other hand, to evaluate the condition number of a stiffness matrix, only the smallest and largest eigenvalues are required (see Section 7.5).

Before proceeding to the generalized eigenproblems, we should mention that other standard eigenproblems may also need to be solved. For example, we may require the eigenvalues of the mass matrix $\mathbf{M}$, in which case $\mathbf{M}$ replaces $\mathbf{K}$ in (10.1). Similarly, we may want to solve for the eigenvalues of a conductivity or heat capacity matrix in heat flow analysis (see Section 5.4).

A very frequently considered eigenproblem is the one to be solved in vibration mode superposition analysis (see Section 8.3). In this case we consider the generalized eigenproblem

$$\mathbf{K}\boldsymbol{\phi} = \lambda \mathbf{M}\boldsymbol{\phi} \tag{10.4}$$

where $\mathbf{K}$ and $\mathbf{M}$ are, respectively, the stiffness matrix and mass matrix of the finite element assemblage. The eigenvalues λ_i and eigenvectors $\boldsymbol{\phi}_i$ are the free vibration frequencies (rad/sec) squared, ω_i^2, and corresponding mode shape vectors, respectively. The properties of $\mathbf{K}$ are as discussed above. The mass matrix may be banded, in which case its half-bandwidth $m_{\mathbf{M}}$ is equal to $m_{\mathbf{K}}$, or $\mathbf{M}$ may be diagonal with $m_{ii} \geq 0$; i.e., some diagonal elements may possibly be zero. A banded mass matrix, obtained in a consistent mass analysis, is always positive definite, whereas a lumped mass matrix is positive definite only if all diagonal elements are larger than zero. In general, a diagonal mass matrix is positive semidefinite.

In analogy to (10.3), the solution for p eigenvalues and corresponding eigenvectors of (10.4) can be written

$$\mathbf{K}\mathbf{\Phi} = \mathbf{M}\mathbf{\Phi}\mathbf{\Lambda} \tag{10.5}$$

where the columns in $\mathbf{\Phi}$ are the eigenvectors and $\mathbf{\Lambda}$ is a diagonal matrix listing the corresponding eigenvalues.

We should note that the generalized eigenproblem in (10.4) reduces to the standard eigenproblem in (10.1) if **M** is an identity matrix. In other words, the eigenvalues and eigenvectors in (10.3) can also be thought of as frequencies squared and vibration mode shapes of the system when unit mass is specified at each degree of freedom. Corresponding to the possible eigenvalues in the solution of (10.1), the generalized eigenproblem in (10.4) has eigenvalues $\lambda_i \geq 0$, $i = 1, \ldots, n$, where the number of zero eigenvalues is again equal to the number of rigid body modes in the system.

Two other generalized eigenproblems should be mentioned briefly. The first problem is solved in linearized buckling analysis, in which case we consider[8]

$$\mathbf{K}_G \boldsymbol{\phi} = \lambda \mathbf{K} \boldsymbol{\phi} \tag{10.6}$$

where $\mathbf{K}_G$ is the nonlinear strain (geometric or initial stress) stiffness matrix and **K** is the linear strain stiffness matrix. The nonlinear strain stiffness matrix is banded with the same bandwidth as **K** and is, in general, indefinite. The eigenvalues of (10.6) can be positive and negative and they give the buckling loads with the largest eigenvalue defining the smallest critical load.

A third generalized eigenproblem is encountered in heat transfer analysis, where we consider the equation

$$\mathbf{K} \boldsymbol{\phi} = \lambda \mathbf{C} \boldsymbol{\phi} \tag{10.7}$$

where **K** is the heat conductivity matrix and **C** is the heat capacity matrix. The eigenvalues and eigenvectors are the thermal frequencies and mode shapes, respectively. The solution of (10.7) is required in heat transfer analysis using mode superposition. The matrices **K** and **C** in (10.7) are positive definite or positive semidefinite, so that the eigenvalues of (10.7) are $\lambda_i \geq 0$, $i = 1, \ldots, n$.

In this and the next chapter we discuss the solution of the eigenproblems $\mathbf{K}\boldsymbol{\phi} = \lambda\boldsymbol{\phi}$ and $\mathbf{K}\boldsymbol{\phi} = \lambda\mathbf{M}\boldsymbol{\phi}$ in (10.1) and (10.4). These eigenproblems are encountered frequently in practice. However, it should be realized that all algorithms to be presented are also applicable to the solution of other eigenproblems, provided they are of the same form and the matrices satisfy the appropriate conditions of positive definiteness, semidefiniteness, and so on. For example, to solve the problem in (10.7), the mass matrix **M** need simply be replaced by the heat capacity matrix **C**, and the matrix **K** is the heat conductivity matrix.

Considering the actual computer solution of the required eigenproblems, we recall that in the introduction to equation solution procedures in static analysis (see Section 7.1), we observed the importance of using effective calculation procedures. This is even more so in eigensystem calculations, because the solution of eigenvalues and corresponding eigenvectors requires, in general, much more computer effort than the solution of static equilibrium equations. A particularly important consideration is that the solution algorithms must be stable, which is more difficult to achieve than in static analysis.

A variety of eigensystem solution methods have been developed and are reported in the literature.[1-7] Most of the techniques have been devised for rather general matrices. However, in finite element analysis we are concerned with the solution of the specific eigenproblems summarized above, in which each of the matrices has specific properties, such as being banded, positive definite, and so on. The eigensystem solution algorithms should take advantage of these properties in order to make a more economical solution possible.

The objective in this chapter is to lay the foundation for a thorough understanding of effective eigensolution methods. This is accomplished by first discussing the properties of the matrices, eigenvalues, and eigenvectors of the problems of interest, and then presenting some approximate solution techniques. The actual solution methods recommended for use are presented in Chapters 11 and 12.

10.2 FUNDAMENTAL FACTS USED IN THE SOLUTION OF EIGENSYSTEMS

Before the working of any eigensystem solution procedure can be properly studied, it is necessary first to thoroughly understand the different properties of the matrices and eigenvalues and eigenvectors considered.[1-7] In particular, we will find that all solution methods are, in essence, based on these fundamental properties. We therefore want to summarize in this section the important properties of the matrices and their eigensystems, although some of the material has already been presented in other sections of the book. As pointed out in Section 10.1, we consider the eigenproblem $\mathbf{K}\boldsymbol{\phi} = \lambda\mathbf{M}\boldsymbol{\phi}$, which reduces to $\mathbf{K}\boldsymbol{\phi} = \lambda\boldsymbol{\phi}$ when $\mathbf{M} = \mathbf{I}$, but the observations made are equally applicable to other eigenproblems of interest.

10.2.1 Properties of the Eigenvectors

It was stated that the solution of the generalized eigenproblem $\mathbf{K}\boldsymbol{\phi} = \lambda\mathbf{M}\boldsymbol{\phi}$ yields n eigenvalues $\lambda_1, \ldots, \lambda_n$, ordered as shown in (10.2), and corresponding eigenvectors $\boldsymbol{\phi}_1, \ldots, \boldsymbol{\phi}_n$. Each eigenpair $(\lambda_i, \boldsymbol{\phi}_i)$ satisfies (10.4); i.e., we have

$$\mathbf{K}\boldsymbol{\phi}_i = \lambda_i\mathbf{M}\boldsymbol{\phi}_i; \qquad i = 1, \ldots, n \tag{10.8}$$

The significance of (10.8) should be well understood. The equation says that if we establish a vector $\lambda_i\mathbf{M}\boldsymbol{\phi}_i$ and use it as a load vector $\mathbf{R}$ in the equation $\mathbf{KU} = \mathbf{R}$, then $\mathbf{U} = \boldsymbol{\phi}_i$. This thought may immediately suggest the use of static solution algorithms for the calculation of an eigenvector. We will see later that the $\mathbf{LDL}^T$ decomposition algorithm is indeed an important part of eigensolution procedures.

The equation in (10.8) also shows that an eigenvector is only defined within a multiple of itself; i.e., we also have

$$\mathbf{K}(\alpha\boldsymbol{\phi}_i) = \lambda_i\mathbf{M}(\alpha\boldsymbol{\phi}_i) \tag{10.9}$$

where α is a nonzero constant. Therefore, with $\boldsymbol{\phi}_i$ being an eigenvector, $\alpha\boldsymbol{\phi}_i$ is also an eigenvector, and we say that an eigenvector is only defined by its direction in the n-dimensional space considered. However, in our discussion we refer to *the* eigenvectors $\boldsymbol{\phi}_i$ as satisfying (10.8) and also the relation $\boldsymbol{\phi}_i^T\mathbf{M}\boldsymbol{\phi}_i = 1$, which fixes the lengths of the eigenvectors, i.e., the absolute magnitude of the elements in each eigenvector. However, we may note that the eigenvectors are still only defined within a multiplier of -1.

An important relation which the eigenvectors satisfy is that of **M**-orthonormality; i.e., we have

$$\boldsymbol{\phi}_i^T\mathbf{M}\boldsymbol{\phi}_j = \delta_{ij} \tag{10.10}$$

where δ_{ij} is the Kronecker delta. This relation follows from the orthonormality of the eigenvectors of standard eigenproblems (see Section 2.6) and is discussed further in Section 10.2.5. Premultiplying (10.8) by $\boldsymbol{\phi}_j$ transposed and using the condition in (10.10), we obtain

$$\boldsymbol{\phi}_i^T\mathbf{K}\boldsymbol{\phi}_j = \lambda_i\delta_{ij} \tag{10.11}$$

meaning that the eigenvectors are also **K**-orthogonal. When using the relations in (10.10) and (10.11), it should be kept in mind that the **M**- and **K**-orthogonality follow from (10.8), and that (10.8) is the basic equation to be satisfied. In other words, if we believe that we have an eigenvector and eigenvalue, then as a check we should substitute into (10.8) (see Example 10.3).

So far we have made no mention of multiple eigenvalues and corresponding eigenvectors. It is important to realize that in this case the eigenvectors are not unique but that we can always choose a set of **M**-orthonormal eigenvectors, which span the subspace that corresponds to a multiple eigenvalue (see Section 2.6). In other words, assume that λ_i has multiplicity m (i.e., $\lambda_i = \lambda_{i+1} = \ldots = \lambda_{i+m-1}$), then we can choose m eigenvectors $\boldsymbol{\phi}_i, \ldots, \boldsymbol{\phi}_{i+m-1}$, which span the m-dimensional subspace corresponding to the eigenvalues of magnitude λ_i and which satisfy the orthogonality relations in (10.10) and (10.11). However, the eigenvectors are not unique; instead, the eigenspace corresponding to λ_i is unique. We demonstrate the results by means of some examples.

Example 10.1

The stiffness matrix and mass matrix of a two-degree-of-freedom system are

$$\mathbf{K} = \begin{bmatrix} 5 & -2 \\ -2 & 2 \end{bmatrix}; \qquad \mathbf{M} = \begin{bmatrix} \frac{5}{4} & 0 \\ 0 & \frac{1}{5} \end{bmatrix}$$

It is believed that the two eigenpairs of the problem $\mathbf{K}\boldsymbol{\phi} = \lambda\boldsymbol{\phi}$ are

$$(d_1, \mathbf{v}_1) = \left(1, \begin{bmatrix} \dfrac{1}{\sqrt{5}} \\ \dfrac{2}{\sqrt{5}} \end{bmatrix}\right); \qquad (d_2, \mathbf{v}_2) = \left(6, \begin{bmatrix} \dfrac{2}{\sqrt{5}} \\ -\dfrac{1}{\sqrt{5}} \end{bmatrix}\right) \tag{a}$$

and the two eigenpairs of the problem $\mathbf{K}\boldsymbol{\phi} = \lambda\mathbf{M}\boldsymbol{\phi}$ are

$$(g_1, \mathbf{w}_1) = \left(2, \begin{bmatrix} \frac{4}{5} \\ 1 \end{bmatrix}\right); \qquad (g_2, \mathbf{w}_2) = \left(12, \begin{bmatrix} \frac{2}{5} \\ -2 \end{bmatrix}\right) \tag{b}$$

Verify that we indeed have in (a) and (b) the eigensolutions of the problems $\mathbf{K}\boldsymbol{\phi} = \lambda\boldsymbol{\phi}$ and $\mathbf{K}\boldsymbol{\phi} = \lambda\mathbf{M}\boldsymbol{\phi}$, respectively.

Consider first the problem $\mathbf{K}\boldsymbol{\phi} = \lambda\boldsymbol{\phi}$. The values given in (a) are indeed the eigensolution if they satisfy the relation in (10.8) with $\mathbf{M} = \mathbf{I}$ and, to fix the lengths of the vectors, the orthonormality relations in (10.10) with $\mathbf{M} = \mathbf{I}$. Substituting into (10.3), which expresses the relation in (10.8) for all eigenpairs, we have

$$\begin{bmatrix} 5 & -2 \\ -2 & 2 \end{bmatrix} \begin{bmatrix} \frac{1}{\sqrt{5}} & \frac{2}{\sqrt{5}} \\ \frac{2}{\sqrt{5}} & -\frac{1}{\sqrt{5}} \end{bmatrix} = \begin{bmatrix} \frac{1}{\sqrt{5}} & \frac{2}{\sqrt{5}} \\ \frac{2}{\sqrt{5}} & -\frac{1}{\sqrt{5}} \end{bmatrix} \begin{bmatrix} 1 & 0 \\ 0 & 6 \end{bmatrix}$$

or

$$\begin{bmatrix} \frac{1}{\sqrt{5}} & \frac{12}{\sqrt{5}} \\ \frac{2}{\sqrt{5}} & -\frac{6}{\sqrt{5}} \end{bmatrix} = \begin{bmatrix} \frac{1}{\sqrt{5}} & \frac{12}{\sqrt{5}} \\ \frac{2}{\sqrt{5}} & -\frac{6}{\sqrt{5}} \end{bmatrix}$$

Evaluating (10.10) we obtain

$$\mathbf{v}_1^T\mathbf{v}_1 = \begin{bmatrix} \frac{1}{\sqrt{5}} & \frac{2}{\sqrt{5}} \end{bmatrix} \begin{bmatrix} \frac{1}{\sqrt{5}} \\ \frac{2}{\sqrt{5}} \end{bmatrix} = 1$$

$$\mathbf{v}_2^T\mathbf{v}_1 = \begin{bmatrix} \frac{2}{\sqrt{5}} & -\frac{1}{\sqrt{5}} \end{bmatrix} \begin{bmatrix} \frac{1}{\sqrt{5}} \\ \frac{2}{\sqrt{5}} \end{bmatrix} = 0$$

$$\mathbf{v}_2^T\mathbf{v}_2 = \begin{bmatrix} \frac{2}{\sqrt{5}} & -\frac{1}{\sqrt{5}} \end{bmatrix} \begin{bmatrix} \frac{2}{\sqrt{5}} \\ -\frac{1}{\sqrt{5}} \end{bmatrix} = 1$$

$$\mathbf{v}_1^T\mathbf{v}_2 = \mathbf{v}_2^T\mathbf{v}_1 = 0$$

Therefore, the relations in (10.3) and (10.10) are satisfied and we have $\lambda_1 = d_1$, $\lambda_2 = d_2$, $\boldsymbol{\phi}_1 = \mathbf{v}_1$, and $\boldsymbol{\phi}_2 = \mathbf{v}_2$.

To check whether we have in (b) the eigensolution to $\mathbf{K}\boldsymbol{\phi} = \lambda\mathbf{M}\boldsymbol{\phi}$, we proceed in an analogous way. Substituting into (10.5) we have

$$\begin{bmatrix} 5 & -2 \\ -2 & 2 \end{bmatrix} \begin{bmatrix} \frac{4}{5} & \frac{2}{5} \\ 1 & -2 \end{bmatrix} = \begin{bmatrix} \frac{5}{4} & 0 \\ 0 & \frac{1}{5} \end{bmatrix} \begin{bmatrix} \frac{4}{5} & \frac{2}{5} \\ 1 & -2 \end{bmatrix} \begin{bmatrix} 2 & 0 \\ 0 & 12 \end{bmatrix}$$

or

$$\begin{bmatrix} 2 & 6 \\ \frac{2}{5} & -\frac{24}{5} \end{bmatrix} = \begin{bmatrix} 2 & 6 \\ \frac{2}{5} & -\frac{24}{5} \end{bmatrix}$$

and evaluating (10.10) we obtain

$$\mathbf{w}_1^T\mathbf{M}\mathbf{w}_1 = [\tfrac{4}{5} \quad 1]\begin{bmatrix} \frac{5}{4} & 0 \\ 0 & \frac{1}{5} \end{bmatrix}\begin{bmatrix} \frac{4}{5} \\ 1 \end{bmatrix} = 1$$

$$\mathbf{w}_2^T\mathbf{M}\mathbf{w}_1 = [\tfrac{2}{5} \quad -2]\begin{bmatrix} \frac{5}{4} & 0 \\ 0 & \frac{1}{5} \end{bmatrix}\begin{bmatrix} \frac{4}{5} \\ 1 \end{bmatrix} = 0$$

$$\mathbf{w}_2^T\mathbf{M}\mathbf{w}_2 = [\tfrac{2}{5} \quad -2]\begin{bmatrix} \frac{5}{4} & 0 \\ 0 & \frac{1}{5} \end{bmatrix}\begin{bmatrix} \frac{2}{5} \\ -2 \end{bmatrix} = 1$$

$$\mathbf{w}_1^T\mathbf{M}\mathbf{w}_2 = \mathbf{w}_2^T\mathbf{M}\mathbf{w}_1 = 0$$

Hence the relations in (10.5) and (10.10) are satisfied and we have

$$\lambda_1 = g_1, \lambda_2 = g_2, \boldsymbol{\phi}_1 = \mathbf{w}_1, \text{ and } \boldsymbol{\phi}_2 = \mathbf{w}_2.$$

Example 10.2

Consider the eigenproblem

$$\mathbf{K}\boldsymbol{\phi} = \lambda\boldsymbol{\phi} \qquad \text{with } \mathbf{K} = \begin{bmatrix} 2 & & \\ & 2 & \\ & & 3 \end{bmatrix}$$

and show that the eigenvectors corresponding to the multiple eigenvalue are not unique.

The eigenvalues of $\mathbf{K}$ are $\lambda_1 = 2$, $\lambda_2 = 2$, and $\lambda_3 = 3$, and a set of eigenvectors is

$$\boldsymbol{\phi}_1 = \begin{bmatrix} 1 \\ 0 \\ 0 \end{bmatrix}; \qquad \boldsymbol{\phi}_2 = \begin{bmatrix} 0 \\ 1 \\ 0 \end{bmatrix}; \qquad \boldsymbol{\phi}_3 = \begin{bmatrix} 0 \\ 0 \\ 1 \end{bmatrix} \tag{a}$$

where $\boldsymbol{\phi}_3$ is unique. These values could be checked as in Example 10.1. However, any linear combinations of $\boldsymbol{\phi}_1$ and $\boldsymbol{\phi}_2$ given in (a) that satisfy the orthonormality conditions in (10.10) with $\mathbf{M} = \mathbf{I}$ would also be eigenvectors. For example, we could use

$$\boldsymbol{\phi}_1 = \begin{bmatrix} \frac{1}{\sqrt{2}} \\ \frac{1}{\sqrt{2}} \\ 0 \end{bmatrix} \quad \text{and} \quad \boldsymbol{\phi}_2 = \begin{bmatrix} \frac{1}{\sqrt{2}} \\ -\frac{1}{\sqrt{2}} \\ 0 \end{bmatrix}$$

That $\boldsymbol{\phi}_1$ and $\boldsymbol{\phi}_2$ are indeed eigenvectors corresponding to $\lambda_1 = \lambda_2 = 2$ can again be checked as in Example 10.1. It should be noted that any eigenvectors $\boldsymbol{\phi}_1$ and $\boldsymbol{\phi}_2$ provide a basis for the unique two-dimensional eigenspace that corresponds to λ_1 and λ_2.

The solution to (10.4) for all p required eigenvalues and corresponding eigenvectors was established in (10.5). Using the relations in (10.10) and (10.11), we may now write

$$\boldsymbol{\Phi}^T\mathbf{K}\boldsymbol{\Phi} = \boldsymbol{\Lambda} \tag{10.12}$$

and

$$\boldsymbol{\Phi}^T\mathbf{M}\boldsymbol{\Phi} = \mathbf{I} \tag{10.13}$$

where the p columns of $\mathbf{\Phi}$ are the eigenvectors. *It is very important to note that* (10.12) *and* (10.13) *are conditions which the eigenvectors must satisfy, but that if the* **M**- *and* **K**-*orthonormality are satisfied, the p vectors need not necessarily be eigenvectors, unless* $p = n$. In other words, assume that $\mathbf{X}$ stores p vectors, $p < n$, and that $\mathbf{X}^T\mathbf{K}\mathbf{X} = \mathbf{D}$ and $\mathbf{X}^T\mathbf{M}\mathbf{X} = \mathbf{I}$; then the vectors in $\mathbf{X}$ and the diagonal elements in $\mathbf{D}$ may or may not be eigenvectors and eigenvalues of (10.4). However, if $p = n$, then $\mathbf{X} = \mathbf{\Phi}$ and $\mathbf{D} = \mathbf{\Lambda}$, because only the eigenvectors span the complete n-dimensional space and diagonalize the matrices $\mathbf{K}$ and $\mathbf{M}$. The above aspect is very important and is frequently not observed in practice (see Section 10.3.2). To underline the concept we present the following example.

Example 10.3

Consider the eigenproblem $\mathbf{K}\boldsymbol{\phi} = \lambda\mathbf{M}\boldsymbol{\phi}$, where

$$\mathbf{K} = \begin{bmatrix} 2 & -1 & 0 \\ -1 & 4 & -1 \\ 0 & -1 & 2 \end{bmatrix}; \qquad \mathbf{M} = \begin{bmatrix} \frac{1}{2} & 0 & 0 \\ 0 & 1 & 0 \\ 0 & 0 & \frac{1}{2} \end{bmatrix}$$

and the two vectors

$$\mathbf{v}_1 = \begin{bmatrix} 1 \\ \dfrac{1}{\sqrt{2}} \\ 0 \end{bmatrix}; \qquad \mathbf{v}_2 = \begin{bmatrix} 1 \\ -\dfrac{1}{\sqrt{2}} \\ 0 \end{bmatrix}$$

Show that the vectors $\mathbf{v}_1$ and $\mathbf{v}_2$ satisfy the orthogonality relations in (10.12) and (10.13) [i.e., the relations in (10.11) and (10.10)] but that they are not eigenvectors.

For the check we let $\mathbf{v}_1$ and $\mathbf{v}_2$ be the columns in $\mathbf{\Phi}$, and we evaluate (10.12) and (10.13). Thus we obtain

$$\begin{bmatrix} 1 & \dfrac{1}{\sqrt{2}} & 0 \\ 1 & -\dfrac{1}{\sqrt{2}} & 0 \end{bmatrix} \begin{bmatrix} 2 & -1 & 0 \\ -1 & 4 & -1 \\ 0 & -1 & 2 \end{bmatrix} \begin{bmatrix} 1 & 1 \\ \dfrac{1}{\sqrt{2}} & -\dfrac{1}{\sqrt{2}} \\ 0 & 0 \end{bmatrix} = \begin{bmatrix} (4 - \sqrt{2}) & 0 \\ 0 & (4 + \sqrt{2}) \end{bmatrix} \tag{a}$$

and

$$\begin{bmatrix} 1 & \dfrac{1}{\sqrt{2}} & 0 \\ 1 & -\dfrac{1}{\sqrt{2}} & 0 \end{bmatrix} \begin{bmatrix} \frac{1}{2} & & \\ & 1 & \\ & & \frac{1}{2} \end{bmatrix} \begin{bmatrix} 1 & 1 \\ \dfrac{1}{\sqrt{2}} & -\dfrac{1}{\sqrt{2}} \\ 0 & 0 \end{bmatrix} = \begin{bmatrix} 1 & 0 \\ 0 & 1 \end{bmatrix}$$

Hence the orthogonality relations are satisfied. To show that $\mathbf{v}_1$ and $\mathbf{v}_2$ are not eigenvectors, we employ (10.8). For example,

$$\mathbf{K}\mathbf{v}_1 = \begin{bmatrix} 2 - \dfrac{1}{\sqrt{2}} \\ -1 + \dfrac{4}{\sqrt{2}} \\ -\dfrac{1}{\sqrt{2}} \end{bmatrix}; \qquad \mathbf{M}\mathbf{v}_1 = \begin{bmatrix} \dfrac{1}{2} \\ \dfrac{1}{\sqrt{2}} \\ 0 \end{bmatrix}$$

However, the vector $\mathbf{K}\mathbf{v}_1$ cannot be equal to the vector $\alpha\mathbf{M}\mathbf{v}_1$, where α is a scalar; i.e., $\mathbf{K}\mathbf{v}_1$ is not parallel to $\mathbf{M}\mathbf{v}_1$, and therefore $\mathbf{v}_1$ is not an eigenvector. Similarly, $\mathbf{v}_2$ is not an eigenvector and the values $(4 - \sqrt{2})$ and $(4 + \sqrt{2})$ calculated in (a) are not eigenvalues. The actual eigenvalues and corresponding eigenvectors are given in Example 10.4.

In the preceding presentation we considered the properties of the eigenvectors of the problem $\mathbf{K}\boldsymbol{\phi} = \lambda\mathbf{M}\boldsymbol{\phi}$, and we should now briefly comment on the properties of the eigenvectors calculated in the solution of the other eigenvalue problems of interest. The comment is simple: the orthogonality relations discussed here hold equally for the eigenvectors of the problems encountered in buckling analysis and heat transfer analysis. That is, we also have in buckling analysis, using the notation in (10.6),

$$\left.\begin{aligned} \boldsymbol{\phi}_i^T\mathbf{K}\boldsymbol{\phi}_j &= \delta_{ij}; \qquad & \boldsymbol{\phi}_i^T\mathbf{K}_G\boldsymbol{\phi}_j &= \lambda_i\delta_{ij} \\ \boldsymbol{\Phi}^T\mathbf{K}\boldsymbol{\Phi} &= \mathbf{I}; \qquad & \boldsymbol{\Phi}^T\mathbf{K}_G\boldsymbol{\Phi} &= \boldsymbol{\Lambda} \end{aligned}\right\} \tag{10.14}$$

and in heat transfer analysis, using the notation in (10.7), we have

$$\left.\begin{aligned} \boldsymbol{\phi}_i^T\mathbf{C}\boldsymbol{\phi}_j &= \delta_{ij}; \qquad & \boldsymbol{\phi}_i^T\mathbf{K}\boldsymbol{\phi}_j &= \lambda_i\delta_{ij} \\ \boldsymbol{\Phi}^T\mathbf{C}\boldsymbol{\Phi} &= \mathbf{I}; \qquad & \boldsymbol{\Phi}^T\mathbf{K}\boldsymbol{\Phi} &= \boldsymbol{\Lambda} \end{aligned}\right\} \tag{10.15}$$

As for the eigenproblem $\mathbf{K}\boldsymbol{\phi} = \lambda\mathbf{M}\boldsymbol{\phi}$, the proof of the relations in (10.14) and (10.15) depends on the fact that the generalized eigenproblems can be transformed to a standard form. We discuss this matter further in Section 10.2.5.

10.2.2 The Characteristic Polynomials of the Eigenproblem $\mathbf{K}\boldsymbol{\phi} = \lambda\mathbf{M}\boldsymbol{\phi}$ and of Its Associated Constraint Problems

An important property of the eigenvalues of the problem $\mathbf{K}\boldsymbol{\phi} = \lambda\mathbf{M}\boldsymbol{\phi}$ is that they are the roots of the characteristic polynomial,

$$p(\lambda) = \det(\mathbf{K} - \lambda\mathbf{M}) \tag{10.16}$$

We can show that this property derives from the basic relation in (10.8). Rewriting (10.8) in the form

$$(\mathbf{K} - \lambda_i\mathbf{M})\boldsymbol{\phi}_i = \mathbf{0} \tag{10.17}$$

we observe that (10.8) can only be satisfied for nontrivial $\boldsymbol{\phi}_i$ (i.e., $\boldsymbol{\phi}_i$ not being equal to a null vector) provided that the matrix $\mathbf{K} - \lambda_i\mathbf{M}$ is singular. This means that if we factorize $\mathbf{K} - \lambda_i\mathbf{M}$ into a unit lower triangular matrix $\mathbf{L}$ and an upper triangular matrix $\mathbf{S}$ using Gauss elimination, we have that $s_{nn} = 0.0$. However, since

$$p(\lambda_i) = \det \mathbf{LS} = \prod_{i=1}^{n} s_{ii} \tag{10.18}$$

it follows that $p(\lambda_i) = 0.0$. Furthermore, if λ_i has multiplicity m we also have $s_{n-1,n-1} = \ldots = s_{n-m+1,n-m+1} = 0.0$. We should note that in the factorization of $\mathbf{K} - \lambda_i\mathbf{M}$, interchanges may be needed, in which case the factorization of $\mathbf{K} - \lambda_i\mathbf{M}$ with its rows and possibly its columns interchanged is obtained [each row and each column interchange then introduces a sign change in the determinant which must be taken into account (see Section 1.8)]. If no interchanges are carried out or row and corresponding column interchanges are performed, which in practice is nearly always possible (but see Example 10.4 for a case where it is not possible), the coefficient matrix remains symmetric. In this case we can write for (10.18),

$$p(\lambda_i) = \det \mathbf{LDL}^T = \prod_{i=1}^{n} d_{ii} \tag{10.19}$$

where $\mathbf{LDL}^T$ is the factorization of $\mathbf{K} - \lambda_i\mathbf{M}$ or the matrix derived from it by interchanging rows and corresponding columns, i.e., using a different ordering for the system degrees of freedom (see Section 7.2.5). The condition $s_{nn} = 0.0$ is now $d_{nn} = 0.0$, and in case λ_i has multiplicity m, the last m elements of $\mathbf{D}$ are zero.

In Section 7.2.5 we discussed the Sturm sequence property of the characteristic polynomials of the constraint problems associated with the problem $\mathbf{K}\boldsymbol{\phi} = \lambda\boldsymbol{\phi}$. The same properties that we observed in that discussion are applicable also to the characteristic polynomials of the constraint problems associated with the problem $\mathbf{K}\boldsymbol{\phi} = \lambda\mathbf{M}\boldsymbol{\phi}$. The proof follows from the fact that the generalized eigenproblem $\mathbf{K}\boldsymbol{\phi} = \lambda\mathbf{M}\boldsymbol{\phi}$ can be transformed to a standard eigenproblem, for which the Sturm sequence property of the characteristic polynomials holds. Referring the proof to Section 10.2.5, Example 10.11, let us summarize the important result.

The eigenproblem of the rth associated constraint problem corresponding to $\mathbf{K}\boldsymbol{\phi} = \lambda\mathbf{M}\boldsymbol{\phi}$ is given by

$$\mathbf{K}^{(r)}\boldsymbol{\phi}^{(r)} = \lambda^{(r)}\mathbf{M}^{(r)}\boldsymbol{\phi}^{(r)} \tag{10.20}$$

where all matrices are of order $n - r$, and $\mathbf{K}^{(r)}$ and $\mathbf{M}^{(r)}$ are obtained by deleting from $\mathbf{K}$ and $\mathbf{M}$ the last r rows and columns. The characteristic polynomial of the rth associated constraint problem is

$$p^{(r)}(\lambda^{(r)}) = \det (\mathbf{K}^{(r)} - \lambda^{(r)}\mathbf{M}^{(r)}) \tag{10.21}$$

and as for the special case $\mathbf{M} = \mathbf{I}$, we have that the eigenvalues of the $(r + 1)$st constraint problem separate those of the rth constraint problem; i.e., as stated in (7.38), we again have

$$\lambda_1^{(r)} \leq \lambda_1^{(r+1)} \leq \lambda_2^{(r)} \leq \lambda_2^{(r+1)} \ldots \leq \lambda_{n-r-1}^{(r)} \leq \lambda_{n-r-1}^{(r+1)} \leq \lambda_{n-r}^{(r)} \tag{10.22}$$

Example 10.4

Consider the eigenproblem $\mathbf{K}\boldsymbol{\phi} = \lambda\mathbf{M}\boldsymbol{\phi}$, where

$$\mathbf{K} = \begin{bmatrix} 2 & -1 & 0 \\ -1 & 4 & -1 \\ 0 & -1 & 2 \end{bmatrix}; \qquad \mathbf{M} = \begin{bmatrix} \frac{1}{2} & & \\ & 1 & \\ & & \frac{1}{2} \end{bmatrix}$$

(a) Calculate the eigenvalues using the characteristic polynomial as defined in (10.16).
(b) Solve for the eigenvectors $\boldsymbol{\phi}_i$, $i = 1, 2, 3$, by using the relation in (10.17) and the **M**-orthonormality condition of the eigenvectors.
(c) Calculate the eigenvalues of the associated constraint problems and show that the eigenvalue separation property given in (10.22) holds.

Using (10.16) we obtain the characteristic polynomial

$$p(\lambda) = (2 - \tfrac{1}{2}\lambda)(4 - \lambda)(2 - \tfrac{1}{2}\lambda) - (-1)(-1)(2 - \tfrac{1}{2}\lambda) - (-1)(-1)(2 - \tfrac{1}{2}\lambda)$$

Hence

$$p(\lambda) = -\tfrac{1}{4}\lambda^3 + 3\lambda^2 - 11\lambda + 12$$

and we have

$$\lambda_1 = 2; \qquad \lambda_2 = 4; \qquad \lambda_3 = 6$$

To obtain the corresponding eigenvectors, we use the relation in (10.17). For λ_1 we have

$$\begin{bmatrix} 1 & -1 & 0 \\ -1 & 2 & -1 \\ 0 & -1 & 1 \end{bmatrix} \boldsymbol{\phi}_1 = \mathbf{0} \tag{a}$$

The coefficient matrix $\mathbf{K} - \lambda_1\mathbf{M}$ in (a) can be factorized into $\mathbf{LDL}^T$ without interchanges. Using the procedure described in Section 7.2.2, we obtain

$$\begin{bmatrix} 1 & & \\ -1 & 1 & \\ 0 & -1 & 1 \end{bmatrix} \begin{bmatrix} 1 & & \\ & 1 & \\ & & 0 \end{bmatrix} \begin{bmatrix} 1 & -1 & 0 \\ & 1 & -1 \\ & & 1 \end{bmatrix} \boldsymbol{\phi}_1 = \mathbf{0} \tag{b}$$

We note that $d_{33} = 0.0$. To evaluate $\boldsymbol{\phi}_1$, we obtain from (b),

$$\begin{bmatrix} 1 & -1 & 0 \\ & 1 & -1 \\ & & 0 \end{bmatrix} \boldsymbol{\phi}_1 = \mathbf{0}$$

Using also that $\boldsymbol{\phi}_1^T\mathbf{M}\boldsymbol{\phi}_1 = 1$ we have

$$\boldsymbol{\phi}_1^T = \begin{bmatrix} \frac{1}{\sqrt{2}} & \frac{1}{\sqrt{2}} & \frac{1}{\sqrt{2}} \end{bmatrix}$$

To obtain $\boldsymbol{\phi}_2$ and $\boldsymbol{\phi}_3$, we proceed in an analogous way. Evaluating $\mathbf{K} - \lambda_2\mathbf{M}$, we obtain from (10.17),

$$\begin{bmatrix} 0 & -1 & 0 \\ -1 & 0 & -1 \\ 0 & -1 & 0 \end{bmatrix} \boldsymbol{\phi}_2 = \mathbf{0}$$

In this case we cannot factorize the coefficient matrix preserving symmetry; i.e., we need to interchange the first and second rows only (and not the corresponding columns). This row interchange results into the relation

$$\begin{bmatrix} -1 & 0 & -1 \\ 0 & -1 & 0 \\ 0 & -1 & 0 \end{bmatrix} \boldsymbol{\phi}_2 = \mathbf{0}$$

Factorizing the coefficient matrix into a lower unit triangular matrix **L** and an upper triangular matrix **S**, we obtain

$$\begin{bmatrix} 1 & & \\ 0 & 1 & \\ 0 & 1 & 1 \end{bmatrix} \begin{bmatrix} -1 & 0 & -1 \\ & -1 & 0 \\ & & 0 \end{bmatrix} \boldsymbol{\phi}_2 = \mathbf{0}$$

and hence $s_{33} = 0.0$. To solve for $\boldsymbol{\phi}_2$, we use

$$\begin{bmatrix} -1 & 0 & -1 \\ & -1 & 0 \\ & & 0 \end{bmatrix} \boldsymbol{\phi}_2 = \mathbf{0}$$

and that $\boldsymbol{\phi}_2^T \mathbf{M} \boldsymbol{\phi}_2 = 1$. Thus we obtain

$$\boldsymbol{\phi}_2^T = [-1 \quad 0 \quad 1]$$

To calculate $\boldsymbol{\phi}_3$, we evaluate $\mathbf{K} - \lambda_3 \mathbf{M}$ and have

$$\begin{bmatrix} -1 & -1 & 0 \\ -1 & -2 & -1 \\ 0 & -1 & -1 \end{bmatrix} \boldsymbol{\phi}_3 = \mathbf{0}$$

The coefficient matrix can be factorized into $\mathbf{LDL}^T$ without interchanges; i.e., we have

$$\begin{bmatrix} 1 & & \\ 1 & 1 & \\ 0 & 1 & 1 \end{bmatrix} \begin{bmatrix} -1 & & \\ & -1 & \\ & & 0 \end{bmatrix} \begin{bmatrix} 1 & 1 & 0 \\ & 1 & 1 \\ & & 1 \end{bmatrix} \boldsymbol{\phi}_3 = \mathbf{0}$$

We note that $d_{33} = 0.0$. To calculate $\boldsymbol{\phi}_3$ we use

$$\begin{bmatrix} -1 & -1 & 0 \\ 0 & -1 & -1 \\ 0 & 0 & 0 \end{bmatrix} \boldsymbol{\phi}_3 = \mathbf{0}$$

and $\boldsymbol{\phi}_3^T \mathbf{M} \boldsymbol{\phi}_3 = 1$. Hence

$$\boldsymbol{\phi}_3^T = \begin{bmatrix} \frac{1}{\sqrt{2}} & -\frac{1}{\sqrt{2}} & \frac{1}{\sqrt{2}} \end{bmatrix}$$

The eigenvalues of the first associated constraint problem are obtained from the solution of

$$\begin{bmatrix} 2 & -1 \\ -1 & 4 \end{bmatrix} \boldsymbol{\phi}^{(1)} = \lambda^{(1)} \begin{bmatrix} \frac{1}{2} & 0 \\ 0 & 1 \end{bmatrix} \boldsymbol{\phi}^{(1)}$$

Hence

$$p^{(1)}(\lambda^{(1)}) = \tfrac{1}{2} \lambda^{(1)^2} - 4\lambda^{(1)} + 7$$

and

$$\lambda_1^{(1)} = 4 - \sqrt{2}; \qquad \lambda_2^{(1)} = 4 + \sqrt{2}$$

Also $\lambda_1^{(2)} = 4$, and hence the eigenvalue separation property given in (10.22) is in this case:

1. For the eigenvalues of the first and second associated constraint problems,

$$4 - \sqrt{2} \leq 4 \leq 4 + \sqrt{2}$$

2. For the eigenvalues of $\mathbf{K\phi} = \lambda\mathbf{M\phi}$ and the first associated constraint problem,

$$2 \leq (4 - \sqrt{2}) \leq 4 \leq (4 + \sqrt{2}) \leq 6$$

An important fact that follows from the property of the separation of eigenvalues as expressed in (10.22) is the following. Assume that we can factorize the matrix $\mathbf{K} - \mu\mathbf{M}$ into $\mathbf{LDL}^T$; i.e., none of the associated constraint problems has a zero eigenvalue. For simplicity of discussion let us first assume that all eigenvalues are distinct; i.e., there are no multiple eigenvalues. *The important fact is that in the decomposition of* $\mathbf{K} - \mu\mathbf{M}$, *the number of negative elements in* $\mathbf{D}$ *is equal to the number of eigenvalues smaller than* μ. Conversely, if $\lambda_i < \mu < \lambda_{i+1}$, there are exactly i negative diagonal elements in $\mathbf{D}$. The proof is obtained using the separation property in (10.22), and is relatively easily outlined by the following considerations. Referring to Fig. 10.1, assume that in the sketch of the characteristic polynomials we connect by straight lines all eigenvalues $\lambda_1^{(r)}$, $r = 0, 1, \ldots,$ with $\lambda_1^{(0)} = \lambda_1$, and call the resulting curve C_1. Similarly, we establish curves $C_2, C_3, \ldots$ as indicated in Fig. 10.1. Consider now that $\lambda_i < \mu < \lambda_{i+1}$ and draw a vertical line corresponding to μ in the figure of the characteristic polynomials; i.e., this line establishes where μ lies in relation to the eigenvalues of the associated constraint problems. We note that the line corresponding to μ must cross the curves $C_1, \ldots, C_i$ and because of the eigenvalue separation property cannot cross the curves $C_{i+1}, \ldots, C_n$. However, since

$$p^{(r)}(\mu) = \prod_{i=1}^{n-r} d_{ii} \tag{10.23}$$

and since each crossing of μ with an envelope C_k corresponds to a negative element appearing in $\mathbf{D}$, we have exactly i negative elements in $\mathbf{D}$.

The above considerations also hold in the case of multiple eigenvalues; i.e., in Fig. 10.1 we would merely have that some eigenvalues are equal, but the argument given above would not change.

The property that the number of negative elements in $\mathbf{D}$ is equal to the eigenvalues smaller than μ can be used directly in the solution of eigenvalues (see Section 11.5). Namely, by assuming a shift μ and checking whether μ is smaller or larger than the required eigenvalue, we can successively reduce the interval in which the eigenvalue must lie. We demonstrate the solution procedure in the following example.

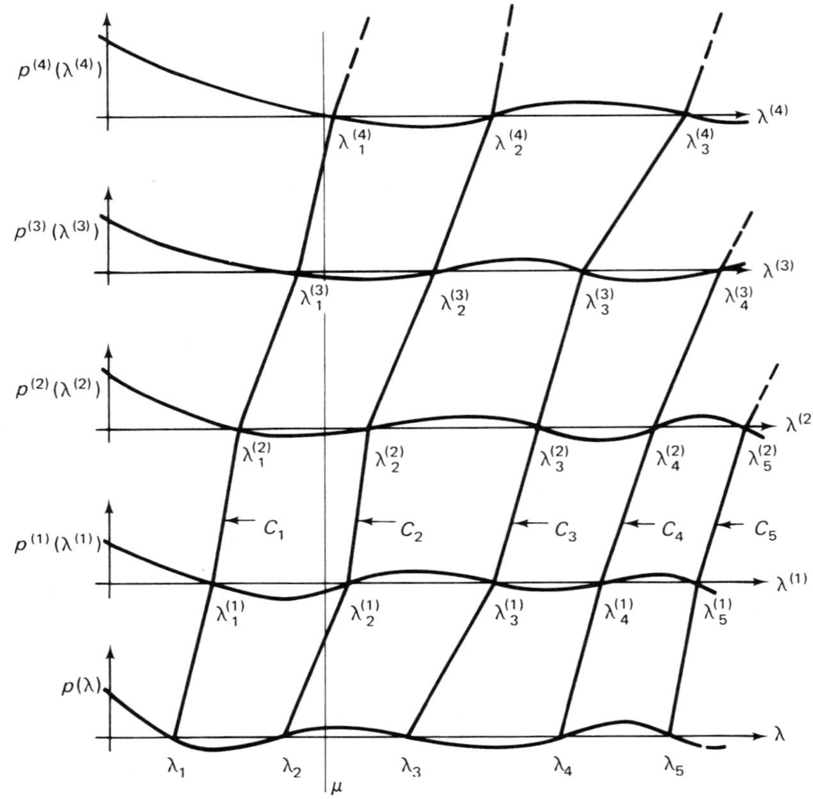

Figure 10.1. Construction of curves C_i for the characteristic polynomials of the problem $\mathbf{K}\boldsymbol{\phi} = \lambda\mathbf{M}\boldsymbol{\phi}$ and of the associated constraint problems.

Example 10.5

Use the fact that the number of negative elements in $\mathbf{D}$, where $\mathbf{LDL}^T = \mathbf{K} - \mu\mathbf{M}$, is equal to the number of eigenvalues smaller than μ in order to calculate λ_2 of $\mathbf{K}\boldsymbol{\phi} = \lambda\mathbf{M}\boldsymbol{\phi}$, where

$$\mathbf{K} = \begin{bmatrix} 2 & -1 & 0 \\ -1 & 4 & -1 \\ 0 & -1 & 2 \end{bmatrix}; \qquad \mathbf{M} = \begin{bmatrix} \frac{1}{2} & & \\ & 1 & \\ & & \frac{1}{2} \end{bmatrix}$$

The three eigenvalues of the problem have already been calculated in Example 10.4. We now proceed in the following systematic steps:

1. Let us assume $\mu = 1$ and evaluate $\mathbf{LDL}^T$ of $\mathbf{K} - \mu\mathbf{M}$,

$$\mathbf{K} - \mu\mathbf{M} = \begin{bmatrix} \frac{3}{2} & -1 & 0 \\ -1 & 3 & -1 \\ 0 & -1 & \frac{3}{2} \end{bmatrix}$$

Hence

$$\mathbf{LDL}^T = \begin{bmatrix} 1 & & \\ -\frac{2}{3} & 1 & \\ 0 & -\frac{3}{7} & 1 \end{bmatrix} \begin{bmatrix} \frac{3}{2} & & \\ & \frac{7}{3} & \\ & & \frac{15}{14} \end{bmatrix} \begin{bmatrix} 1 & -\frac{2}{3} & 0 \\ & 1 & -\frac{3}{7} \\ & & 1 \end{bmatrix}$$

Since all elements in $\mathbf{D}$ are larger than zero, we have $\lambda_1 > 1$.

2. We now try $\mu = 8$, where

$$\mathbf{K} - \mu\mathbf{M} = \begin{bmatrix} -2 & -1 & 0 \\ -1 & -4 & -1 \\ 0 & -1 & -2 \end{bmatrix}$$

and

$$\mathbf{LDL}^T = \begin{bmatrix} 1 & & \\ \frac{1}{2} & 1 & \\ 0 & \frac{2}{7} & 1 \end{bmatrix} \begin{bmatrix} -2 & & \\ & -\frac{7}{2} & \\ & & -\frac{12}{7} \end{bmatrix} \begin{bmatrix} 1 & \frac{1}{2} & 0 \\ & 1 & \frac{2}{7} \\ & & 1 \end{bmatrix}$$

Since all three diagonal elements are smaller than zero, it follows that $\lambda_3 < 8$.

3. The next estimate μ should logically lie between 1 and 8; we choose $\mu = 5$, for which

$$\mathbf{K} - \mu\mathbf{M} = \begin{bmatrix} -\frac{1}{2} & -1 & 0 \\ -1 & -1 & -1 \\ 0 & -1 & -\frac{1}{2} \end{bmatrix}$$

and

$$\mathbf{LDL}^T = \begin{bmatrix} 1 & & \\ 2 & 1 & \\ 0 & -1 & 1 \end{bmatrix} \begin{bmatrix} -\frac{1}{2} & & \\ & +1 & \\ & & -\frac{3}{2} \end{bmatrix} \begin{bmatrix} 1 & 2 & 0 \\ & 1 & -1 \\ & & 1 \end{bmatrix}$$

Since two negative elements are in $\mathbf{D}$, we have that $\lambda_2 < 5$.

4. The next estimate must lie between 1 and 5. Let us use $\mu = 3$, in which case

$$\mathbf{K} - \mu\mathbf{M} = \begin{bmatrix} \frac{1}{2} & -1 & 0 \\ -1 & 1 & -1 \\ 0 & -1 & \frac{1}{2} \end{bmatrix}$$

and

$$\mathbf{LDL}^T = \begin{bmatrix} 1 & & \\ -2 & 1 & \\ 0 & 1 & 1 \end{bmatrix} \begin{bmatrix} \frac{1}{2} & & \\ & -1 & \\ & & \frac{3}{2} \end{bmatrix} \begin{bmatrix} 1 & -2 & 0 \\ & 1 & 1 \\ & & 1 \end{bmatrix}$$

Hence $\lambda_2 > 3$, because there is only one negative element in $\mathbf{D}$.

The pattern of the solution procedure has now been established. So far we know that $3 < \lambda_2 < 5$. In order to obtain a closer estimate on λ_2 we would continue choosing a shift μ in the interval 3 to 5 and investigate whether the new shift is smaller or larger than λ_2. By always choosing an appropriate new shift the required eigenvalue can be determined very accurately (see Section 11.5). It should be noted that we did not need to use interchanges in the factorizations of $\mathbf{K} - \mu\mathbf{M}$ carried out above.

10.2.3 Shifting

An important procedure that is used extensively in the solution of eigenvalues and eigenvectors is shifting. The purpose of shifting is to accelerate the calculations of the required eigensystem. In the solution of $\mathbf{K}\boldsymbol{\phi} = \lambda\mathbf{M}\boldsymbol{\phi}$, we perform a shift ρ on $\mathbf{K}$ by calculating

$$\hat{\mathbf{K}} = \mathbf{K} - \rho\mathbf{M} \tag{10.24}$$

and we then consider the eigenproblem

$$\hat{\mathbf{K}}\boldsymbol{\psi} = \mu\mathbf{M}\boldsymbol{\psi} \tag{10.25}$$

To identify how the eigenvalues and eigenvectors of $\mathbf{K}\boldsymbol{\phi} = \lambda\mathbf{M}\boldsymbol{\phi}$ are related to those of the problem $\hat{\mathbf{K}}\boldsymbol{\psi} = \mu\mathbf{M}\boldsymbol{\psi}$ we rewrite (10.25) in the form

$$\mathbf{K}\boldsymbol{\psi} = \gamma\mathbf{M}\boldsymbol{\psi} \tag{10.26}$$

where $\gamma = \rho + \mu$. However, (10.26) is, in fact, the eigenproblem $\mathbf{K}\boldsymbol{\phi} = \lambda\mathbf{M}\boldsymbol{\phi}$, and since the solution of this problem is unique, we have

$$\lambda_i = \rho + \mu_i; \qquad \boldsymbol{\phi}_i = \boldsymbol{\psi}_i \tag{10.27}$$

In other words, the eigenvectors of $\hat{\mathbf{K}}\boldsymbol{\psi} = \mu\mathbf{M}\boldsymbol{\psi}$ *are the same as the eigenvectors of* $\mathbf{K}\boldsymbol{\phi} = \lambda\mathbf{M}\boldsymbol{\phi}$*, but the eigenvalues have been decreased by* ρ. A frequent application of shifting occurs in the calculation of rigid body modes when an algorithm is to be used that is not designed explicitly to calculate zero eigenvalues. We illustrate such an application in the following example.

Example 10.6

Consider the eigenproblem

$$\begin{bmatrix} 3 & -3 \\ -3 & 3 \end{bmatrix}\boldsymbol{\phi} = \lambda\begin{bmatrix} 2 & 1 \\ 1 & 2 \end{bmatrix}\boldsymbol{\phi} \tag{a}$$

Calculate the eigenvalues and eigenvectors. Then impose a shift $\rho = -2$ and solve again for the eigenvalues and corresponding eigenvectors.

To calculate the eigenvalues we use the characteristic polynomial

$$p(\lambda) = \det(\mathbf{K} - \lambda\mathbf{M}) = 3\lambda^2 - 18\lambda$$

and thus obtain $\lambda_1 = 0$, $\lambda_2 = 6$. To calculate $\boldsymbol{\phi}_1$ and $\boldsymbol{\phi}_2$ we use the relation in (10.17) and the mass orthonormality condition $\boldsymbol{\phi}_i^T\mathbf{M}\boldsymbol{\phi}_i = 1$. We have

$$\begin{bmatrix} 3 & -3 \\ -3 & 3 \end{bmatrix}\boldsymbol{\phi}_1 = \mathbf{0}; \qquad \text{hence } \boldsymbol{\phi}_1 = \begin{bmatrix} \dfrac{1}{\sqrt{6}} \\ \dfrac{1}{\sqrt{6}} \end{bmatrix} \tag{b}$$

and

$$\begin{bmatrix} -9 & -9 \\ -9 & -9 \end{bmatrix}\boldsymbol{\phi}_2 = \mathbf{0}; \qquad \text{hence } \boldsymbol{\phi}_2 = \begin{bmatrix} \dfrac{1}{\sqrt{2}} \\ -\dfrac{1}{\sqrt{2}} \end{bmatrix} \tag{c}$$

Imposing a shift of $\rho = -2$, we obtain the problem

$$\begin{bmatrix} 7 & -1 \\ -1 & 7 \end{bmatrix} \boldsymbol{\phi} = \lambda \begin{bmatrix} 2 & 1 \\ 1 & 2 \end{bmatrix} \boldsymbol{\phi} \tag{d}$$

Proceeding as before, we have

$$p(\lambda) = \lambda^2 - 10\lambda + 16$$

and obtain as the roots $\lambda_1 = 2$, $\lambda_2 = 8$. Hence the eigenvalues have increased by 2; i.e., they have decreased by ρ.

The eigenvectors would be calculated using (10.17). However, we note that this relation again yields the equations in (b) and (c), and therefore the eigenvectors of the problem in (d) are those of the problem in (a).

An important observation resulting from the above discussion is that, in principle, we only need solution algorithms to calculate the eigenvalues and corresponding eigenvectors of the problem $\mathbf{K}\boldsymbol{\phi} = \lambda\mathbf{M}\boldsymbol{\phi}$ when all eigenvalues are larger than zero. This follows, because if rigid body modes are present, we may always operate on a shifted stiffness matrix that renders all eigenvalues positive.

Extensive applications of shifting are given in Chapters 11 and 12, where the various eigensystem solution algorithms are discussed.

10.2.4 Effect of Zero Mass

Considering a lumped mass analysis, we pointed out that the mass matrix $\mathbf{M}$ is diagonal with positive and possibly some zero diagonal elements. If all elements m_{ii} are larger than zero, the eigenvalues λ_i cannot be obtained without the use of an eigenvalue solution algorithm as described in Chapter 11. However, if $\mathbf{M}$ has some zero diagonal elements, say r diagonal elements in $\mathbf{M}$ are zero, we can immediately say that the problem $\mathbf{K}\boldsymbol{\phi} = \lambda\mathbf{M}\boldsymbol{\phi}$ has the eigenvalues $\lambda_n = \lambda_{n-1} = \ldots = \lambda_{n-r+1} = \infty$, and can also construct the corresponding eigenvectors by inspection.

To obtain the above result, let us recall the fundamental objective in an eigensolution. It is important to remember that all we require is a vector $\boldsymbol{\phi}$ and scalar λ that satisfy the equation

$$\mathbf{K}\boldsymbol{\phi} = \lambda\mathbf{M}\boldsymbol{\phi} \tag{10.4}$$

where $\boldsymbol{\phi}$ is nontrivial; i.e., $\boldsymbol{\phi}$ is a vector with at least one element in it nonzero. In other words, if we have a vector $\boldsymbol{\phi}$ and scalar λ that satisfy (10.4), then λ and $\boldsymbol{\phi}$ are an eigenvalue λ_i and eigenvector $\boldsymbol{\phi}_i$, respectively, where it should be noted that it does not matter how $\boldsymbol{\phi}$ and λ have been obtained. If, for example, we can guess $\boldsymbol{\phi}$ and λ, we should certainly take advantage of it. This may be the case when rigid body modes are present in the structural element assemblage. Thus, if we know that the element assemblage can undergo a rigid body mode, we have $\lambda_1 = 0$ and need to seek $\boldsymbol{\phi}_1$ to satisfy the equation $\mathbf{K}\boldsymbol{\phi}_1 = \mathbf{0}$. In general, the solution of $\boldsymbol{\phi}_1$ must be obtained using an equation solver, but in a simple finite element assemblage, we may be able to identify $\boldsymbol{\phi}_1$ by inspection.

In the case of r zero diagonal elements in a diagonal mass matrix $\mathbf{M}$, we can always immediately establish r eigenvalues and corresponding eigenvectors. Namely, rewriting the eigenproblem in (10.4) in the form

$$\mathbf{M}\boldsymbol{\phi} = \mu\mathbf{K}\boldsymbol{\phi} \tag{10.28}$$

where $\mu = \lambda^{-1}$, we find that if $m_{kk} = 0.0$, we have an eigenpair $(\mu_i, \boldsymbol{\phi}_i) = (0.0, \mathbf{e}_k)$; i.e.,

$$\boldsymbol{\phi}_i^T = [0 \quad 0 \quad \dots \quad 0 \quad \underset{\uparrow\, k\text{th element}}{1} \quad 0 \quad \dots \quad 0]; \qquad \mu_i = 0.0 \tag{10.29}$$

That $\boldsymbol{\phi}_i$ and μ_i in (10.29) are indeed an eigenvector and eigenvalue of (10.28) is verified by simply substituting into (10.28) and noting that $(\mu_i, \boldsymbol{\phi}_i)$ is a nontrivial solution. Since $\mu = \lambda^{-1}$, we therefore found that an eigenpair of $\mathbf{K}\boldsymbol{\phi} = \lambda\mathbf{M}\boldsymbol{\phi}$ is given by $(\lambda_i, \boldsymbol{\phi}_i) = (\infty, \mathbf{e}_k)$. Considering the case of r zero diagonal elements in $\mathbf{M}$, it follows that there are r infinite eigenvalues, and the corresponding eigenvectors can be taken as unit vectors with the 1 in a location corresponding to a zero mass element in $\mathbf{M}$. We should note that since λ_n is then an eigenvalue of multiplicity r, the corresponding eigenvectors are not unique (see Section 10.2.1). In addition we note that the length of an eigenvector cannot be fixed using the condition of $\mathbf{M}$-orthonormality. We demonstrate how we establish the eigenvalues and eigenvectors by means of a small example.

Example 10.7

Consider the eigenproblem

$$\begin{bmatrix} 2 & -1 & & \\ -1 & 2 & -1 & \\ & -1 & 2 & -1 \\ & & -1 & 1 \end{bmatrix} \boldsymbol{\phi} = \lambda \begin{bmatrix} 0 & & & \\ & 2 & & \\ & & 0 & \\ & & & 1 \end{bmatrix} \boldsymbol{\phi}$$

Establish λ_3, λ_4 and $\boldsymbol{\phi}_3$, $\boldsymbol{\phi}_4$.

There are two zero diagonal elements in $\mathbf{M}$; hence $\lambda_3 = \infty$, $\lambda_4 = \infty$. As corresponding eigenvectors we can use

$$\boldsymbol{\phi}_3 = \begin{bmatrix} 1 \\ 0 \\ 0 \\ 0 \end{bmatrix}; \qquad \boldsymbol{\phi}_4 = \begin{bmatrix} 0 \\ 0 \\ 1 \\ 0 \end{bmatrix} \tag{a}$$

Alternatively, any linear combination of $\boldsymbol{\phi}_3$ and $\boldsymbol{\phi}_4$ given in (a) would represent an eigenvector. We should note that $\boldsymbol{\phi}_i^T\mathbf{M}\boldsymbol{\phi}_i = 0$ for $i = 3, 4$, and therefore the magnitude of the elements in $\boldsymbol{\phi}_i$ cannot be fixed using the $\mathbf{M}$-orthonormality condition.

10.2.5 Transformation of the Generalized Eigenproblem $\mathbf{K}\boldsymbol{\phi} = \lambda\mathbf{M}\boldsymbol{\phi}$ to a Standard Form

The most common eigenproblems that are encountered in general scientific analysis are standard eigenproblems, and most other eigenproblems can be reduced to a standard form. For this reason, the solution of standard eigenproblems has attracted much attention in numerical analysis, and many solution algorithms are available. The purpose of this section is to show how the eigenproblem $\mathbf{K}\boldsymbol{\phi} = \lambda\mathbf{M}\boldsymbol{\phi}$ can be reduced to a standard form. The implications of the transformation are twofold. First, because the transformation is possible, use can be made of the various solution algorithms available for standard eigenproblems. We will see that the effectiveness of the eigensolution procedure employed depends to a large degree on the decision whether or not to carry out a transformation to a standard form. Second, *if a generalized eigenproblem can be written in standard form, the properties of the eigenvalues, eigenvectors, and characteristic polynomials of the generalized eigenproblem can be deduced from the properties of the corresponding quantities of the standard eigenproblem.* Realizing that the properties of standard eigenproblems are more easily assessed, it is to a large extent for the second reason that the transformation to a standard eigenproblem is important to be studied. Indeed, after having presented the transformation procedures, we will show how the properties of the eigenvectors (see Section 10.2.1) and the properties of the characteristic polynomials (see Section 10.2.2) of the problem $\mathbf{K}\boldsymbol{\phi} = \lambda\mathbf{M}\boldsymbol{\phi}$ are derived from the corresponding properties of the standard eigenproblem.

In the following we assume that $\mathbf{M}$ is positive definite. This is the case when $\mathbf{M}$ is diagonal with $m_{ii} > 0$, $i = 1, \ldots, n$, or $\mathbf{M}$ is banded, as in a consistent mass analysis. If $\mathbf{M}$ is diagonal with some zero diagonal elements, we first need to perform static condensation on the massless degrees of freedom as described in Section 10.3.1. Assuming that $\mathbf{M}$ is positive definite, we can transform the generalized eigenproblem $\mathbf{K}\boldsymbol{\phi} = \lambda\mathbf{M}\boldsymbol{\phi}$ given in (10.4) by using a decomposition of $\mathbf{M}$ of the form

$$\mathbf{M} = \mathbf{S}\mathbf{S}^T \tag{10.30}$$

where $\mathbf{S}$ is any nonsingular matrix. Namely, substituting for $\mathbf{M}$ into (10.4), we have

$$\mathbf{K}\boldsymbol{\phi} = \lambda\mathbf{S}\mathbf{S}^T\boldsymbol{\phi} \tag{10.31}$$

Premultiplying both sides of (10.31) by $\mathbf{S}^{-1}$ and defining a vector

$$\tilde{\boldsymbol{\phi}} = \mathbf{S}^T\boldsymbol{\phi} \tag{10.32}$$

we obtain the standard eigenproblem

$$\tilde{\mathbf{K}}\tilde{\boldsymbol{\phi}} = \lambda\tilde{\boldsymbol{\phi}} \tag{10.33}$$

where

$$\tilde{\mathbf{K}} = \mathbf{S}^{-1}\mathbf{K}\mathbf{S}^{-T} \tag{10.34}$$

One of two decompositions of $\mathbf{M}$ is used in general: the Cholesky factorization or the spectral decomposition of $\mathbf{M}$. The Cholesky factorization of $\mathbf{M}$ is obtained as described in Section 7.2.4 and yields $\mathbf{M} = \tilde{\mathbf{L}}_{\mathbf{M}}\tilde{\mathbf{L}}_{\mathbf{M}}^T$. In (10.30) to (10.34) we therefore have

$$\mathbf{S} = \tilde{\mathbf{L}}_{\mathbf{M}} \tag{10.35}$$

The spectral decomposition of $\mathbf{M}$ requires solution of the complete eigensystem of $\mathbf{M}$. Denoting the matrix of orthonormal eigenvectors by $\mathbf{R}$ and the diagonal matrix of eigenvalues by $\mathbf{D}^2$, we have

$$\mathbf{M} = \mathbf{R}\mathbf{D}^2\mathbf{R}^T \tag{10.36}$$

and we use in (10.30) to (10.34),

$$\mathbf{S} = \mathbf{R}\mathbf{D} \tag{10.37}$$

It should be noted that when $\mathbf{M}$ is diagonal, the matrices $\mathbf{S}$ in (10.35) and (10.37) are the same, but when $\mathbf{M}$ is banded, they are different.

Considering the effectiveness of the solution of the required eigenvalues and eigenvectors of (10.33), it is most important that $\tilde{\mathbf{K}}$ has the same bandwidth as $\mathbf{K}$ when $\mathbf{M}$ is diagonal. However, in case $\mathbf{M}$ is banded, $\tilde{\mathbf{K}}$ in (10.33) is in general a full matrix, which makes the transformation ineffective in almost all large-order finite element analyses. This will become more apparent in Chapters 11 and 12 when various eigensystem solution algorithms are discussed.

Comparing the Cholesky factorization and the spectral decomposition of $\mathbf{M}$, it may be noted that the use of the Cholesky factors is in general computationally more efficient than the use of the spectral decomposition because fewer operations are involved in calculating $\tilde{\mathbf{L}}_{\mathbf{M}}$ than $\mathbf{R}$ and $\mathbf{D}$. However, the spectral decomposition of $\mathbf{M}$ may yield a more accurate solution of $\mathbf{K}\boldsymbol{\phi} = \lambda\mathbf{M}\boldsymbol{\phi}$. Assume that $\mathbf{M}$ is ill-conditioned with respect to inversion; then the transformation process to the standard eigenproblem is also ill-conditioned. In that case it is important to employ the more stable transformation procedure. Using the Cholesky factorization of $\mathbf{M}$ without pivoting, we find that $\tilde{\mathbf{L}}_{\mathbf{M}}^{-1}$ has large elements in many locations because of the coupling in $\mathbf{M}$ and $\tilde{\mathbf{L}}_{\mathbf{M}}^{-1}$. Consequently, $\tilde{\mathbf{K}}$ is calculated with little precision, and the lowest eigenvalues and corresponding eigenvectors are determined inaccurately.

On the other hand, using the spectral decomposition of $\mathbf{M}$, good accuracy may be obtained in the elements of $\mathbf{R}$ and $\mathbf{D}^2$, although some elements in $\mathbf{D}^2$ are small in relation to the other elements. The ill-conditioning of $\mathbf{M}$ is now only concentrated in the small elements of $\mathbf{D}^2$, and considering $\tilde{\mathbf{K}}$, only those rows and columns which correspond to the small elements in $\mathbf{D}$ have large elements, and the eigenvalues of normal size are more likely to be preserved accurately.

Consider the following examples of transforming the generalized eigenvalue problem $\mathbf{K}\boldsymbol{\phi} = \lambda\mathbf{M}\boldsymbol{\phi}$ to a standard form.

Example 10.8

Consider the problem $\mathbf{K}\boldsymbol{\phi} = \lambda\mathbf{M}\boldsymbol{\phi}$, where

$$\mathbf{K} = \begin{bmatrix} 3 & -1 & 0 \\ -1 & 2 & -1 \\ 0 & -1 & 1 \end{bmatrix}; \qquad \mathbf{M} = \begin{bmatrix} 2 & 1 & 0 \\ 1 & 3 & 1 \\ 0 & 1 & 2 \end{bmatrix}$$

Use the Cholesky factorization of $\mathbf{M}$ to calculate the matrix $\tilde{\mathbf{K}}$ of a corresponding standard eigenproblem.

We first calculate the $\mathbf{LDL}^T$ decomposition of $\mathbf{M}$,

$$\mathbf{M} = \begin{bmatrix} 1 & & \\ \frac{1}{2} & 1 & \\ 0 & \frac{2}{5} & 1 \end{bmatrix} \begin{bmatrix} 2 & & \\ & \frac{5}{2} & \\ & & \frac{8}{5} \end{bmatrix} \begin{bmatrix} 1 & \frac{1}{2} & 0 \\ & 1 & \frac{2}{5} \\ & & 1 \end{bmatrix}$$

Hence the Cholesky factor of $\mathbf{M}$ is (see Section 7.2.4)

$$\tilde{\mathbf{L}}_{\mathbf{M}} = \begin{bmatrix} \sqrt{2} & & \\ \dfrac{1}{\sqrt{2}} & \sqrt{\dfrac{5}{2}} & \\ 0 & \sqrt{\dfrac{2}{5}} & \sqrt{\dfrac{8}{5}} \end{bmatrix}$$

and

$$\tilde{\mathbf{L}}_{\mathbf{M}}^{-1} = \begin{bmatrix} \dfrac{1}{\sqrt{2}} & & \\ -\dfrac{1}{\sqrt{10}} & \sqrt{\dfrac{2}{5}} & \\ \dfrac{1}{\sqrt{40}} & -\dfrac{1}{\sqrt{10}} & \sqrt{\dfrac{5}{8}} \end{bmatrix}$$

The matrix of the standard eigenproblem $\tilde{\mathbf{K}} = \tilde{\mathbf{L}}_{\mathbf{M}}^{-1}\mathbf{K}\tilde{\mathbf{L}}_{\mathbf{M}}^{-T}$ is in this case,

$$\tilde{\mathbf{K}} = \begin{bmatrix} \dfrac{3}{2} & -\dfrac{\sqrt{5}}{2} & \sqrt{\dfrac{5}{16}} \\ -\dfrac{\sqrt{5}}{2} & \dfrac{3}{2} & -\dfrac{5}{4} \\ \sqrt{\dfrac{5}{16}} & -\dfrac{5}{4} & \dfrac{3}{2} \end{bmatrix}$$

Example 10.9

Consider the generalized eigenproblem of Example 10.8. Use the spectral decomposition of $\mathbf{M}$ to calculate the matrix $\tilde{\mathbf{K}}$ of a corresponding standard eigenproblem.

The eigenvalues and corresponding eigenvectors of the problem $\mathbf{M}\boldsymbol{\phi} = \lambda\boldsymbol{\phi}$

can be calculated as shown in Example 10.4. We obtain $\lambda_1 = 1, \lambda_2 = 2, \lambda_3 = 4$, and

$$\boldsymbol{\phi}_1 = \begin{bmatrix} \frac{1}{\sqrt{3}} \\ -\frac{1}{\sqrt{3}} \\ \frac{1}{\sqrt{3}} \end{bmatrix}; \qquad \boldsymbol{\phi}_2 = \begin{bmatrix} \frac{1}{\sqrt{2}} \\ 0 \\ -\frac{1}{\sqrt{2}} \end{bmatrix}; \qquad \boldsymbol{\phi}_3 = \begin{bmatrix} \frac{1}{\sqrt{6}} \\ \frac{2}{\sqrt{6}} \\ \frac{1}{\sqrt{6}} \end{bmatrix}$$

Hence the decomposition $\mathbf{M} = \mathbf{R}\mathbf{D}^2\mathbf{R}^T$ is

$$\mathbf{M} = \begin{bmatrix} \frac{1}{\sqrt{3}} & \frac{1}{\sqrt{2}} & \frac{1}{\sqrt{6}} \\ -\frac{1}{\sqrt{3}} & 0 & \frac{2}{\sqrt{6}} \\ \frac{1}{\sqrt{3}} & -\frac{1}{\sqrt{2}} & \frac{1}{\sqrt{6}} \end{bmatrix} \begin{bmatrix} 1 & & \\ & 2 & \\ & & 4 \end{bmatrix} \begin{bmatrix} \frac{1}{\sqrt{3}} & -\frac{1}{\sqrt{3}} & \frac{1}{\sqrt{3}} \\ \frac{1}{\sqrt{2}} & 0 & -\frac{1}{\sqrt{2}} \\ \frac{1}{\sqrt{6}} & \frac{2}{\sqrt{6}} & \frac{1}{\sqrt{6}} \end{bmatrix}$$

Noting that $\mathbf{S} = \mathbf{R}\mathbf{D}$ and $\mathbf{S}^{-1} = \mathbf{D}^{-1}\mathbf{R}^T$ because $\mathbf{R}\mathbf{R}^T = \mathbf{I}$, we obtain

$$\mathbf{S}^{-1} = \begin{bmatrix} \frac{1}{\sqrt{3}} & -\frac{1}{\sqrt{3}} & \frac{1}{\sqrt{3}} \\ \frac{1}{2} & 0 & -\frac{1}{2} \\ \frac{1}{2\sqrt{6}} & \frac{1}{\sqrt{6}} & \frac{1}{2\sqrt{6}} \end{bmatrix}$$

The matrix of the standard eigenproblem is $\tilde{\mathbf{K}} = \mathbf{S}^{-1}\mathbf{K}\mathbf{S}^{-T}$; i.e.,

$$\tilde{\mathbf{K}} = \begin{bmatrix} \frac{10}{3} & \frac{1}{\sqrt{3}} & -\frac{1}{3\sqrt{2}} \\ \frac{1}{\sqrt{3}} & 1 & \frac{1}{2\sqrt{6}} \\ -\frac{1}{3\sqrt{2}} & \frac{1}{2\sqrt{6}} & \frac{1}{6} \end{bmatrix}$$

We should note that the matrix $\tilde{\mathbf{K}}$ obtained here is different from the matrix $\tilde{\mathbf{K}}$ derived in Example 10.8.

In the above discussion we considered only the factorization of $\mathbf{M}$ into $\mathbf{M} = \mathbf{S}\mathbf{S}^T$ and then the transformation of $\mathbf{K}\boldsymbol{\phi} = \lambda\mathbf{M}\boldsymbol{\phi}$ into the form given in (10.33). We pointed out that this transformation can yield inaccurate results if $\mathbf{M}$ is ill-conditioned. In such a case it seems natural to avoid the decomposition of $\mathbf{M}$ and instead use a factorization of $\mathbf{K}$. Rewriting $\mathbf{K}\boldsymbol{\phi} = \lambda\mathbf{M}\boldsymbol{\phi}$ in the form $\mathbf{M}\boldsymbol{\phi} = (1/\lambda)\mathbf{K}\boldsymbol{\phi}$, we can use an analogous procedure to obtain the eigenproblem

$$\tilde{\mathbf{M}}\tilde{\boldsymbol{\phi}} = \frac{1}{\lambda}\tilde{\boldsymbol{\phi}} \tag{10.38}$$

where

$$\tilde{\mathbf{M}} = \mathbf{S}^{-1}\mathbf{M}\mathbf{S}^{-T} \tag{10.39}$$

$$\mathbf{K} = \mathbf{S}\mathbf{S}^{T} \tag{10.40}$$

$$\tilde{\boldsymbol{\phi}} = \mathbf{S}^{T}\boldsymbol{\phi} \tag{10.41}$$

and $\mathbf{S}$ is obtained using the Cholesky factor or spectral decomposition of $\mathbf{K}$. If $\mathbf{K}$ is well-conditioned, the transformation is also well-conditioned. However, since $\mathbf{K}$ is always banded, we have that $\tilde{\mathbf{M}}$ is always a full matrix, and the transformation is inefficient.

As we pointed out earlier, the possibility of actually solving a generalized eigenproblem by first transforming it into a standard form is only one reason why we considered the above transformations. The second reason is that the properties of the eigensolution of the generalized eigenproblem can be deduced from the solution of the corresponding standard eigenproblem. Specifically, we can derive the orthogonality properties of the eigenvectors as given in (10.10) and (10.11), and the Sturm sequence property of the characteristic polynomials of the eigenproblem $\mathbf{K}\boldsymbol{\phi} = \lambda\mathbf{M}\boldsymbol{\phi}$ and of its associated constraint problems as given in (10.22). In both cases no fundamentally new concepts need be proved; instead; the corresponding properties of the standard eigenproblem, which is obtained from the generalized eigenproblem, are used. We give the proofs in the following examples as an application of the transformation of a generalized eigenproblem to a standard form.

Example 10.10

Show that the eigenvectors of the problem $\mathbf{K}\boldsymbol{\phi} = \lambda\mathbf{M}\boldsymbol{\phi}$ are $\mathbf{M}$- and $\mathbf{K}$-orthogonal, and discuss the orthogonality of the eigenvectors of the problems $\mathbf{K}_G\boldsymbol{\phi} = \lambda\mathbf{K}\boldsymbol{\phi}$ given in (10.6) and $\mathbf{K}\boldsymbol{\phi} = \lambda\mathbf{C}\boldsymbol{\phi}$ given in (10.7).

The eigenvector orthogonality is proved by transforming the generalized eigenproblem to a standard form and using the fact that the eigenvectors of a standard eigenproblem with a symmetric matrix are orthogonal. Consider first the problem $\mathbf{K}\boldsymbol{\phi} = \lambda\mathbf{M}\boldsymbol{\phi}$ and assume that $\mathbf{M}$ is positive definite. Then we can use the transformation in (10.30) to (10.34) to obtain, as an equivalent eigenproblem,

$$\tilde{\mathbf{K}}\tilde{\boldsymbol{\phi}} = \lambda\tilde{\boldsymbol{\phi}}$$

where

$$\mathbf{M} = \mathbf{S}\mathbf{S}^{T}; \qquad \tilde{\mathbf{K}} = \mathbf{S}^{-1}\mathbf{K}\mathbf{S}^{-T}; \qquad \tilde{\boldsymbol{\phi}} = \mathbf{S}^{T}\boldsymbol{\phi}$$

But since the eigenvectors $\tilde{\boldsymbol{\phi}}_i$ of the problem $\tilde{\mathbf{K}}\tilde{\boldsymbol{\phi}} = \lambda\tilde{\boldsymbol{\phi}}$ have the properties

$$\tilde{\boldsymbol{\phi}}_i^T\tilde{\boldsymbol{\phi}}_j = \delta_{ij}; \qquad \tilde{\boldsymbol{\phi}}_i^T\tilde{\mathbf{K}}\tilde{\boldsymbol{\phi}}_j = \lambda_i\delta_{ij}$$

we have, substituting $\tilde{\boldsymbol{\phi}}_i = \mathbf{S}^T\boldsymbol{\phi}_i$, $\tilde{\boldsymbol{\phi}}_j = \mathbf{S}^T\boldsymbol{\phi}_j$,

$$\boldsymbol{\phi}_i^T\mathbf{M}\boldsymbol{\phi}_j = \delta_{ij}; \qquad \boldsymbol{\phi}_i^T\mathbf{K}\boldsymbol{\phi}_j = \lambda_i\delta_{ij} \tag{a}$$

In case $\mathbf{M}$ is not positive definite, we consider the eigenproblem $\mathbf{M}\boldsymbol{\phi} = (1/\lambda)\mathbf{K}\boldsymbol{\phi}$ (with $\mathbf{K}$ positive definite or a shift must be imposed; see Section 10.2.3).

We now use the transformation

$$\tilde{\mathbf{M}}\tilde{\boldsymbol{\phi}} = \left(\frac{1}{\lambda}\right)\tilde{\boldsymbol{\phi}}$$

where

$$\mathbf{K} = \mathbf{S}\mathbf{S}^T; \qquad \tilde{\mathbf{M}} = \mathbf{S}^{-1}\mathbf{M}\mathbf{S}^{-T}; \qquad \tilde{\boldsymbol{\phi}} = \mathbf{S}^T\boldsymbol{\phi}$$

and the properties

$$\tilde{\boldsymbol{\phi}}_i^T\tilde{\boldsymbol{\phi}}_j = \delta_{ij}; \qquad \tilde{\boldsymbol{\phi}}_i^T\tilde{\mathbf{M}}\tilde{\boldsymbol{\phi}}_j = \left(\frac{1}{\lambda_i}\right)\delta_{ij}$$

Substituting for $\tilde{\boldsymbol{\phi}}_i$ and $\tilde{\boldsymbol{\phi}}_j$, we obtain

$$\boldsymbol{\phi}_i^T\mathbf{K}\boldsymbol{\phi}_j = \delta_{ij}; \qquad \boldsymbol{\phi}_i^T\mathbf{M}\boldsymbol{\phi}_j = \left(\frac{1}{\lambda_i}\right)\delta_{ij} \tag{b}$$

with the eigenvectors now being **K**- orthonormalized, because the problem $\mathbf{M}\boldsymbol{\phi} = (1/\lambda)\mathbf{K}\boldsymbol{\phi}$ was considered. To obtain the same vectors as in the problem $\mathbf{K}\boldsymbol{\phi} = \lambda\mathbf{M}\boldsymbol{\phi}$, we need to multiply the eigenvectors $\boldsymbol{\phi}_i$ of the problem $\mathbf{M}\boldsymbol{\phi} = (1/\lambda)\mathbf{K}\boldsymbol{\phi}$ by the factors $\sqrt{\lambda_i}$, $i = 1, \ldots, n$.

Considering these proofs, we note that we arranged the eigenproblems in such a way that the matrix associated with the eigenvalue; i.e., the matrix on the right-hand side of the eigenvalue problem is positive definite. This is necessary to be able to carry out the transformation of the generalized eigenproblem to a standard form and thus derive the eigenvector orthogonality properties in (a) and (b). However, considering the problems $\mathbf{K}_G\boldsymbol{\phi} = \lambda\mathbf{K}\boldsymbol{\phi}$ and $\mathbf{K}\boldsymbol{\phi} = \lambda\mathbf{C}\boldsymbol{\phi}$ given in (10.6) and (10.7), we can proceed in a similar manner. The results would be the eigenvector orthogonality properties given in (10.14) and (10.15).

Example 10.11

Prove the Sturm sequence property of the characteristic polynomials of the problem $\mathbf{K}\boldsymbol{\phi} = \lambda\mathbf{M}\boldsymbol{\phi}$ and the associated constraint problems. Demonstrate the proof for the following matrices:

$$\mathbf{K} = \begin{bmatrix} 3 & -1 & \\ -1 & 2 & -1 \\ & -1 & 1 \end{bmatrix}; \qquad \mathbf{M} = \begin{bmatrix} 4 & 4 & \\ 4 & 8 & 4 \\ & 4 & 8 \end{bmatrix} \tag{a}$$

The proof that we consider here is based on the transformation of the eigenproblems $\mathbf{K}\boldsymbol{\phi} = \lambda\mathbf{M}\boldsymbol{\phi}$ and $\mathbf{K}^{(r)}\boldsymbol{\phi}^{(r)} = \lambda^{(r)}\mathbf{M}^{(r)}\boldsymbol{\phi}^{(r)}$ to standard eigenproblems, for which the characteristic polynomials are known to form a Sturm sequence.

As in Example 10.10, we assume first that **M** is positive definite. In this case we can transform the problem $\mathbf{K}\boldsymbol{\phi} = \lambda\mathbf{M}\boldsymbol{\phi}$ into the form

$$\tilde{\mathbf{K}}\tilde{\boldsymbol{\phi}} = \lambda\tilde{\boldsymbol{\phi}}$$

where

$$\tilde{\mathbf{K}} = \tilde{\mathbf{L}}_{\mathbf{M}}^{-1}\mathbf{K}\tilde{\mathbf{L}}_{\mathbf{M}}^{-T}; \qquad \mathbf{M} = \tilde{\mathbf{L}}_{\mathbf{M}}\tilde{\mathbf{L}}_{\mathbf{M}}^T; \qquad \tilde{\boldsymbol{\phi}} = \tilde{\mathbf{L}}_{\mathbf{M}}^T\boldsymbol{\phi}$$

and $\tilde{\mathbf{L}}_{\mathbf{M}}$ is the Cholesky factor of **M**.

Considering the eigenproblems $\tilde{\mathbf{K}}\tilde{\boldsymbol{\phi}} = \lambda\tilde{\boldsymbol{\phi}}$ and $\tilde{\mathbf{K}}^{(r)}\tilde{\boldsymbol{\phi}}^{(r)} = \lambda^{(r)}\tilde{\boldsymbol{\phi}}^{(r)}$, $r = 1, \ldots, n - 1$, we know that the characteristic polynomials form a Sturm sequence.

On the other hand, if we consider the eigenproblem $\mathbf{K}\boldsymbol{\phi} = \lambda\mathbf{M}\boldsymbol{\phi}$ and the eigenproblems of its associated constraint problems, i.e., $\mathbf{K}^{(r)}\boldsymbol{\phi}^{(r)} = \lambda^{(r)}\mathbf{M}^{(r)}\boldsymbol{\phi}^{(r)}$, we note that the problems $\tilde{\mathbf{K}}^{(r)}\tilde{\boldsymbol{\phi}}^{(r)} = \lambda^{(r)}\tilde{\boldsymbol{\phi}}^{(r)}$ and $\mathbf{K}^{(r)}\boldsymbol{\phi}^{(r)} = \lambda^{(r)}\mathbf{M}^{(r)}\boldsymbol{\phi}^{(r)}$ have the same eigenvalues. Namely, $\tilde{\mathbf{K}}^{(r)}\tilde{\boldsymbol{\phi}}^{(r)} = \lambda^{(r)}\tilde{\boldsymbol{\phi}}^{(r)}$ is a standard eigenproblem corresponding to $\mathbf{K}^{(r)}\boldsymbol{\phi}^{(r)} = \lambda^{(r)}\mathbf{M}^{(r)}\boldsymbol{\phi}^{(r)}$; i.e.,

$$\tilde{\mathbf{K}}^{(r)} = \tilde{\mathbf{L}}_{\mathbf{M}}^{(r)-1}\mathbf{K}^{(r)}\tilde{\mathbf{L}}_{\mathbf{M}}^{(r)-T}; \qquad \mathbf{M}^{(r)} = \tilde{\mathbf{L}}_{\mathbf{M}}^{(r)}\tilde{\mathbf{L}}_{\mathbf{M}}^{(r)T}; \qquad \tilde{\boldsymbol{\phi}}^{(r)} = \tilde{\mathbf{L}}_{\mathbf{M}}^{(r)T}\boldsymbol{\phi}^{(r)} \tag{b}$$

Hence the Sturm sequence property also holds for the characteristic polynomials of $\mathbf{K}\boldsymbol{\phi} = \lambda\mathbf{M}\boldsymbol{\phi}$ and the associated constraint problems.

For the example to be considered we have

$$\tilde{\mathbf{L}}_{\mathbf{M}} = \begin{bmatrix} 2 & 0 & 0 \\ 2 & 2 & 0 \\ 0 & 2 & 2 \end{bmatrix} \tag{c}$$

Hence

$$\tilde{\mathbf{K}} = \begin{bmatrix} \frac{1}{2} & 0 & 0 \\ -\frac{1}{2} & \frac{1}{2} & 0 \\ \frac{1}{2} & -\frac{1}{2} & \frac{1}{2} \end{bmatrix} \begin{bmatrix} 3 & -1 & 0 \\ -1 & 2 & -1 \\ 0 & -1 & 1 \end{bmatrix} \begin{bmatrix} \frac{1}{2} & -\frac{1}{2} & \frac{1}{2} \\ 0 & \frac{1}{2} & -\frac{1}{2} \\ 0 & 0 & \frac{1}{2} \end{bmatrix} = \begin{bmatrix} \frac{3}{4} & -1 & 1 \\ -1 & \frac{7}{4} & -2 \\ 1 & -2 & \frac{5}{2} \end{bmatrix} \tag{d}$$

Using $\tilde{\mathbf{K}}$ in (d) to obtain $\tilde{\mathbf{K}}^{(1)}$ and $\tilde{\mathbf{K}}^{(2)}$, we have

$$\tilde{\mathbf{K}}^{(1)} = \begin{bmatrix} \frac{3}{4} & -1 \\ -1 & \frac{7}{4} \end{bmatrix}; \qquad \tilde{\mathbf{K}}^{(2)} = [\tfrac{3}{4}]$$

On the other hand, we can obtain the same matrices $\tilde{\mathbf{K}}^{(1)}$ and $\tilde{\mathbf{K}}^{(2)}$ using the relations in (b),

$$\tilde{\mathbf{K}}^{(1)} = \tilde{\mathbf{L}}_{\mathbf{M}}^{(1)-1}\mathbf{K}^{(1)}\tilde{\mathbf{L}}_{\mathbf{M}}^{(1)-T}; \qquad \tilde{\mathbf{K}}^{(2)} = \tilde{\mathbf{L}}_{\mathbf{M}}^{(2)-1}\mathbf{K}^{(2)}\tilde{\mathbf{L}}_{\mathbf{M}}^{(2)-T}$$

where $\mathbf{K}^{(r)}$ and $\mathbf{M}^{(r)}$ (to calculate $\mathbf{L}_{\mathbf{M}}^{(r)}$) are obtained from $\mathbf{K}$ and $\mathbf{M}$ in (a).

In the preceding discussion we assumed that $\mathbf{M}$ is positive definite. If $\mathbf{M}$ is positive semidefinite, we can consider the problem $\mathbf{M}\boldsymbol{\phi} = (1/\lambda)\mathbf{K}\boldsymbol{\phi}$ instead, in which $\mathbf{K}$ is positive definite (this may mean that a shift has to be imposed; see Section 10.2.3) and thus show that the Sturm sequence property still holds.

It may be noted that it follows from the above discussion that the characteristic polynomials of the eigenproblems $\mathbf{K}_G\boldsymbol{\phi} = \lambda\mathbf{K}\boldsymbol{\phi}$ and $\mathbf{K}\boldsymbol{\phi} = \lambda\mathbf{C}\boldsymbol{\phi}$ given in (10.6) and (10.7) and of their associated constraint problems also form a Sturm sequence.

10.3 APPROXIMATE SOLUTION TECHNIQUES

It is apparent from the nature of a dynamic problem that a dynamic response calculation must be substantially more costly than a static analysis. Namely, whereas in a static analysis the solution is obtained in one step, in dynamics the solution is required at a number of discrete time points over the time interval considered. Indeed, we found that in a direct step-by-step integration solution, an equation of statics, which includes the effects of inertia and damping forces, is considered at the end of each discrete time step (see Section 8.2).

Considering a mode superposition analysis, the main computational effort is spent in the calculation of the required frequencies and mode shapes, which also requires considerably more effort than a static analysis. It is therefore natural that much attention has been directed toward effective algorithms for the calculation of the required eigensystem in the problem $\mathbf{K}\boldsymbol{\phi} = \lambda\mathbf{M}\boldsymbol{\phi}$. In fact, because the "exact" solution of the required eigenvalues and corresponding eigenvectors can be prohibitively expensive when the order of the system is large and a "conventional" technique is used, approximate techniques of solution have been developed. The purpose of this section is to present the major approximate methods that have been designed and are currently still in use.[8-26]

The approximate solution techniques have primarily been developed to calculate the lowest eigenvalues and corresponding eigenvectors in the problem $\mathbf{K}\boldsymbol{\phi} = \lambda\mathbf{M}\boldsymbol{\phi}$, when the order of the system is large. Most programs use exact solution techniques in the analysis of small-order systems. However, the problem of calculating the few lowest eigenpairs of relatively large-order systems is very important and is encountered in all branches of structural engineering and in particular in earthquake response analysis. In the following sections we present three major techniques, all of which, in fact, can be considered to be a Ritz analysis. *The aim in the presentation is not to advocate the implementation of any one of these methods, but rather to describe their practical use, the limitations, and the assumptions employed. Moreover, the relationships between the approximate techniques are described, and in Chapter 12 we will, in fact, find that all approximate techniques considered here can be understood to be one iteration step in the subspace iteration algorithm* (*see Section 12.3*).

10.3.1 Static Condensation

We encountered the procedure of static condensation already in the solution of static equilibrium equations, where we showed that static condensation is, in fact, an application of Gauss elimination (see Section 7.2.4). In static condensation we eliminated those degrees of freedom that are not required to appear in the global finite element assemblage. For example, the displacement degrees of freedom at the internal nodes of a finite element can be condensed out, because they do not take part in imposing interelement continuity. We mentioned already in Section 7.2.4 that the name "static condensation" was actually coined in dynamic analysis.

The basic assumption of static condensation in the calculation of frequencies and mode shapes is that the mass of the structure can be lumped at only some specific degrees of freedom without much effect on the accuracy of the frequencies and mode shapes of interest. In case of a lumped mass matrix with some zero diagonal elements, some of the mass lumping has already been carried out. However, additional mass lumping is in general required. Typically, we may have that the ratio of mass degrees of freedom to the total number of degrees of freedom is somewhere between $\frac{1}{2}$ and $\frac{1}{10}$. The more mass lumping is performed, the less computer effort is required in the solution, however, the more probable it is also that the required frequencies and mode shapes are not predicted accurately. We shall have more to say about this later.

Assume that the mass lumping has been carried out. By partitioning $\mathbf{K}$ and $\mathbf{M}$, we can then write the eigenproblem $\mathbf{K}\boldsymbol{\phi} = \lambda\mathbf{M}\boldsymbol{\phi}$ in the form

$$\begin{bmatrix} \mathbf{K}_{aa} & \mathbf{K}_{ac} \\ \mathbf{K}_{ca} & \mathbf{K}_{cc} \end{bmatrix} \begin{bmatrix} \boldsymbol{\phi}_a \\ \boldsymbol{\phi}_c \end{bmatrix} = \lambda \begin{bmatrix} \mathbf{M}_a & \mathbf{0} \\ \mathbf{0} & \mathbf{0} \end{bmatrix} \begin{bmatrix} \boldsymbol{\phi}_a \\ \boldsymbol{\phi}_c \end{bmatrix} \tag{10.42}$$

where $\boldsymbol{\phi}_a$ and $\boldsymbol{\phi}_c$ are the displacements at the mass and the massless degrees of freedom, respectively, and $\mathbf{M}_a$ is a diagonal mass matrix. The relation in (10.42) gives the condition

$$\mathbf{K}_{ca}\boldsymbol{\phi}_a + \mathbf{K}_{cc}\boldsymbol{\phi}_c = \mathbf{0} \tag{10.43}$$

which can be used to eliminate $\boldsymbol{\phi}_c$. From (10.43) we obtain

$$\boldsymbol{\phi}_c = -\mathbf{K}_{cc}^{-1}\mathbf{K}_{ca}\boldsymbol{\phi}_a \tag{10.44}$$

and substituting into (10.42), we obtain the reduced eigenproblem

$$\mathbf{K}_a\boldsymbol{\phi}_a = \lambda\mathbf{M}_a\boldsymbol{\phi}_a \tag{10.45}$$

where

$$\mathbf{K}_a = \mathbf{K}_{aa} - \mathbf{K}_{ac}\mathbf{K}_{cc}^{-1}\mathbf{K}_{ca} \tag{10.46}$$

The solution of the generalized eigenproblem in (10.45) is in most cases obtained by transforming the problem first into a standard form as described in Section 10.2.5. Since $\mathbf{M}_a$ is a diagonal mass matrix with all its diagonal elements positive and probably not small, the transformation is in general well-conditioned.

The analogy to the use of static condensation in static analysis should be noted. Namely, realizing that the right-hand side of (10.42) may be understood to be a load vector $\mathbf{R}$, where

$$\mathbf{R} = \begin{bmatrix} \lambda\mathbf{M}_a\boldsymbol{\phi}_a \\ \mathbf{0} \end{bmatrix} \tag{10.47}$$

we can use Gauss elimination on the massless degrees of freedom in the same way as we do on the degrees of freedom associated with the interior nodes of an element or a substructure (see Section 7.2.4).

One important aspect should be observed when comparing the static condensation procedure on the massless degrees of freedom in (10.42) to (10.46) on the one side, with Gauss elimination or static condensation in static analysis on the other side. Considering (10.47), we find that the loads at the $\boldsymbol{\phi}_a$ degrees of freedom depend on the eigenvalue (free vibration frequency squared) and eigenvector (mode shape displacements). This means that in (10.45) a further reduction of the number of degrees of freedom to be considered is not possible. This is a basic difference to static condensation as applied in static analysis, where the loads are given explicitly and their effect can be carried over to the remaining degrees of freedom.

Example 10.12

Use static condensation to calculate the eigenvalues and eigenvectors of the problem $\mathbf{K}\boldsymbol{\phi} = \lambda\mathbf{M}\boldsymbol{\phi}$, where

$$\mathbf{K} = \begin{bmatrix} 2 & -1 & 0 & 0 \\ -1 & 2 & -1 & 0 \\ 0 & -1 & 2 & -1 \\ 0 & 0 & -1 & 1 \end{bmatrix}; \qquad \mathbf{M} = \begin{bmatrix} 0 & & & \\ & 2 & & \\ & & 0 & \\ & & & 1 \end{bmatrix}$$

First we rearrange columns and rows to obtain the form given in (10.42), which is

$$\begin{bmatrix} 2 & 0 & -1 & -1 \\ 0 & 1 & 0 & -1 \\ -1 & 0 & 2 & 0 \\ -1 & -1 & 0 & 2 \end{bmatrix} \begin{bmatrix} \boldsymbol{\phi}_a \\ \boldsymbol{\phi}_c \end{bmatrix} = \lambda \begin{bmatrix} 2 & & & \\ & 1 & & \\ & & 0 & \\ & & & 0 \end{bmatrix} \begin{bmatrix} \boldsymbol{\phi}_a \\ \boldsymbol{\phi}_c \end{bmatrix}$$

Hence $\mathbf{K}_a$ given in (10.46) is, in this case,

$$\mathbf{K}_a = \begin{bmatrix} 2 & 0 \\ 0 & 1 \end{bmatrix} - \begin{bmatrix} -1 & -1 \\ 0 & -1 \end{bmatrix} \begin{bmatrix} \frac{1}{2} & 0 \\ 0 & \frac{1}{2} \end{bmatrix} \begin{bmatrix} -1 & 0 \\ -1 & -1 \end{bmatrix} = \begin{bmatrix} 1 & -\frac{1}{2} \\ -\frac{1}{2} & \frac{1}{2} \end{bmatrix}$$

The eigenproblem $\mathbf{K}_a\boldsymbol{\phi}_a = \lambda\mathbf{M}_a\boldsymbol{\phi}_a$ is, therefore,

$$\begin{bmatrix} 1 & -\frac{1}{2} \\ -\frac{1}{2} & \frac{1}{2} \end{bmatrix} \boldsymbol{\phi}_a = \lambda \begin{bmatrix} 2 & \\ & 1 \end{bmatrix} \boldsymbol{\phi}_a$$

and we have

$$\det(\mathbf{K}_a - \lambda\mathbf{M}_a) = 2\lambda^2 - 2\lambda + \tfrac{1}{4}$$

Hence

$$\lambda_1 = \frac{1}{2} - \frac{\sqrt{2}}{4}; \qquad \lambda_2 = \frac{1}{2} + \frac{\sqrt{2}}{4}$$

The corresponding eigenvectors are calculated using

$$(\mathbf{K}_a - \lambda_i\mathbf{M}_a)\boldsymbol{\phi}_{a_i} = \mathbf{0}; \qquad \boldsymbol{\phi}_{a_i}^T\mathbf{M}_a\boldsymbol{\phi}_{a_i} = 1$$

Hence

$$\boldsymbol{\phi}_{a_1} = \begin{bmatrix} \frac{1}{2} \\ \frac{\sqrt{2}}{2} \end{bmatrix}; \qquad \boldsymbol{\phi}_{a_2} = \begin{bmatrix} -\frac{1}{2} \\ \frac{\sqrt{2}}{2} \end{bmatrix}$$

Using (10.44), we obtain

$$\boldsymbol{\phi}_{c_1} = -\begin{bmatrix} \frac{1}{2} & 0 \\ 0 & \frac{1}{2} \end{bmatrix} \begin{bmatrix} -1 & 0 \\ -1 & -1 \end{bmatrix} \begin{bmatrix} \frac{1}{2} \\ \frac{\sqrt{2}}{2} \end{bmatrix} = \begin{bmatrix} \frac{1}{4} \\ \frac{1+\sqrt{2}}{4} \end{bmatrix}$$

$$\boldsymbol{\phi}_{c_2} = -\begin{bmatrix} \frac{1}{2} & 0 \\ 0 & \frac{1}{2} \end{bmatrix}\begin{bmatrix} -1 & 0 \\ -1 & -1 \end{bmatrix}\begin{bmatrix} -\frac{1}{2} \\ \frac{\sqrt{2}}{2} \end{bmatrix} = \begin{bmatrix} -\frac{1}{4} \\ \frac{-1+\sqrt{2}}{4} \end{bmatrix}$$

Therefore, the solution to the eigenproblem $\mathbf{K}\boldsymbol{\phi} = \lambda\mathbf{M}\boldsymbol{\phi}$ is

$$\lambda_1 = \frac{1}{2} - \frac{\sqrt{2}}{4}; \qquad \boldsymbol{\phi}_1 = \begin{bmatrix} \frac{1}{4} \\ \frac{1}{2} \\ \frac{1+\sqrt{2}}{4} \\ \frac{\sqrt{2}}{2} \end{bmatrix}$$

$$\lambda_2 = \frac{1}{2} + \frac{\sqrt{2}}{4}; \qquad \boldsymbol{\phi}_2 = \begin{bmatrix} -\frac{1}{4} \\ -\frac{1}{2} \\ \frac{-1+\sqrt{2}}{4} \\ \frac{\sqrt{2}}{2} \end{bmatrix}$$

$$\lambda_3 = \infty; \qquad \boldsymbol{\phi}_3 = \begin{bmatrix} 1 \\ 0 \\ 0 \\ 0 \end{bmatrix}$$

$$\lambda_4 = \infty; \qquad \boldsymbol{\phi}_4 = \begin{bmatrix} 0 \\ 0 \\ 1 \\ 0 \end{bmatrix}$$

In the above discussion we gave the formal matrix equations for carrying out static condensation. The main computational effort is in calculating $\mathbf{K}_a$ given in (10.46), where it should be noted that in practice a formal inversion of $\mathbf{K}_{cc}$ is not performed. Instead, $\mathbf{K}_a$ can be obtained conveniently using the Cholesky factor $\tilde{\mathbf{L}}_c$ of $\mathbf{K}_{cc}$. If we factorize $\mathbf{K}_{cc}$,

$$\mathbf{K}_{cc} = \tilde{\mathbf{L}}_c\tilde{\mathbf{L}}_c^T \tag{10.48}$$

we can calculate $\mathbf{K}_a$ in the following way:

$$\mathbf{K}_a = \mathbf{K}_{aa} - \mathbf{Y}^T\mathbf{Y} \tag{10.49}$$

where $\mathbf{Y}$ is solved from

$$\tilde{\mathbf{L}}_c\mathbf{Y} = \mathbf{K}_{ca} \tag{10.50}$$

As was pointed out earlier, the above procedure is, in fact, Gauss elimination of the massless degrees of freedom, i.e., elimination of those degrees of freedom at which no external forces (mass effects) are acting. Therefore, an alternative procedure to the one given in (10.42) to (10.50) is to directly use Gauss elimination on the $\boldsymbol{\phi}_c$ degrees of freedom without partitioning $\mathbf{K}$ into the submatrices $\mathbf{K}_{aa}$, $\mathbf{K}_{cc}$, $\mathbf{K}_{ac}$, and $\mathbf{K}_{ca}$, because Gauss elimination can be performed in any order (see Example 7.1, Section 7.2.1). However, the bandwidth of the stiffness matrix will then, in general, increase during the reduction process, and problems of storage need be considered.

For the solution of the eigenproblem $\mathbf{K}_a\boldsymbol{\phi}_a = \lambda\mathbf{M}_a\boldsymbol{\phi}_a$, it is important to note that $\mathbf{K}_a$ is, in general, a full matrix, and the solution is relatively expensive unless the order of the matrices is small.

Instead of calculating the matrix $\mathbf{K}_a$, it may be preferable to evaluate the flexibility matrix $\mathbf{F}_a = \mathbf{K}_a^{-1}$, which is obtained using

$$\begin{bmatrix} \mathbf{K}_{aa} & \mathbf{K}_{ac} \\ \mathbf{K}_{ca} & \mathbf{K}_{cc} \end{bmatrix} \begin{bmatrix} \mathbf{F}_a \\ \mathbf{F}_c \end{bmatrix} = \begin{bmatrix} \mathbf{I} \\ \mathbf{0} \end{bmatrix} \tag{10.51}$$

where $\mathbf{I}$ is a unit matrix of the same order as $\mathbf{K}_{aa}$. Therefore, in (10.51) we solve for the displacements of the structure when unit loads are applied in turn at the mass degrees of freedom. Although the degrees of freedom have been partitioned in (10.51), there is no need for it in this analysis (see Example 10.13). Having solved for $\mathbf{F}_a$, we now consider instead of (10.45) the eigenproblem

$$\left(\frac{1}{\lambda}\right)\boldsymbol{\phi}_a = \mathbf{F}_a\mathbf{M}_a\boldsymbol{\phi}_a \tag{10.52}$$

Although this eigenproblem is of a slightly different form than the generalized problem $\mathbf{K}\boldsymbol{\phi} = \lambda\mathbf{M}\boldsymbol{\phi}$, the transformation to a standard problem proceeds in much the same way (see Section 10.2.5). For the transformation we define

$$\tilde{\boldsymbol{\phi}}_a = \mathbf{M}_a^{1/2}\boldsymbol{\phi}_a \tag{10.53}$$

where $\mathbf{M}_a^{1/2}$ is a diagonal matrix with its ith diagonal element equal to the root of the ith diagonal element of $\mathbf{M}_a$. Premultiplying both sides of (10.52) by $\mathbf{M}_a^{1/2}$ and substituting the relation in (10.53), we obtain

$$\tilde{\mathbf{F}}_a\tilde{\boldsymbol{\phi}}_a = \left(\frac{1}{\lambda}\right)\tilde{\boldsymbol{\phi}}_a \tag{10.54}$$

$$\tilde{\mathbf{F}}_a = \mathbf{M}_a^{1/2}\mathbf{F}_a\mathbf{M}_a^{1/2} \tag{10.55}$$

Once the displacements $\boldsymbol{\phi}_a$ have been calculated, we obtain the complete displacement vector using

$$\begin{bmatrix} \boldsymbol{\phi}_a \\ \boldsymbol{\phi}_c \end{bmatrix} = \begin{bmatrix} \mathbf{I} \\ \mathbf{F}_c\mathbf{K}_a \end{bmatrix}\boldsymbol{\phi}_a \tag{10.56}$$

where $\mathbf{F}_c$ was calculated in (10.51). The relation in (10.56) is arrived at by realizing that the forces applied at the mass degrees of freedom to impose $\boldsymbol{\phi}_a$ are $\mathbf{K}_a\boldsymbol{\phi}_a$. Using (10.51), the corresponding displacements at all degrees of freedom are given in (10.56).

Example 10.13

Use the procedure given in (10.51) to (10.56) to calculate the eigenvalues and eigenvectors of the problem $\mathbf{K}\boldsymbol{\phi} = \lambda\mathbf{M}\boldsymbol{\phi}$ considered in Example 10.12.

The first step is to solve the equations

$$\begin{bmatrix} 2 & -1 & 0 & 0 \\ -1 & 2 & -1 & 0 \\ 0 & -1 & 2 & -1 \\ 0 & 0 & -1 & 1 \end{bmatrix} [\mathbf{v}_1 \quad \mathbf{v}_2] = \begin{bmatrix} 0 & 0 \\ 1 & 0 \\ 0 & 0 \\ 0 & 1 \end{bmatrix} \tag{a}$$

where we did not interchange rows and columns in $\mathbf{K}$ in order to obtain the form in (10.51).

For the solution of the equations in (a), we use the $\mathbf{LDL}^T$ decomposition of $\mathbf{K}$, where

$$\mathbf{L} = \begin{bmatrix} 1 & & & \\ -\frac{1}{2} & 1 & & \\ 0 & -\frac{2}{3} & 1 & \\ 0 & 0 & -\frac{3}{4} & 1 \end{bmatrix}; \qquad \mathbf{D} = \begin{bmatrix} 2 & & & \\ & \frac{3}{2} & & \\ & & \frac{4}{3} & \\ & & & \frac{1}{4} \end{bmatrix}$$

Hence we obtain

$$\mathbf{v}_1^T = [1 \quad 2 \quad 2 \quad 2]; \qquad \mathbf{v}_2^T = [1 \quad 2 \quad 3 \quad 4]$$

and hence

$$\mathbf{F}_a = \begin{bmatrix} 2 & 2 \\ 2 & 4 \end{bmatrix}; \qquad \mathbf{F}_c = \begin{bmatrix} 1 & 1 \\ 2 & 3 \end{bmatrix}$$

$$\tilde{\mathbf{F}}_a = \begin{bmatrix} \sqrt{2} & 0 \\ 0 & 1 \end{bmatrix} \begin{bmatrix} 2 & 2 \\ 2 & 4 \end{bmatrix} \begin{bmatrix} \sqrt{2} & 0 \\ 0 & 1 \end{bmatrix} = \begin{bmatrix} 4 & 2\sqrt{2} \\ 2\sqrt{2} & 4 \end{bmatrix}$$

The solution of the eigenproblem

$$\tilde{\mathbf{F}}_a\tilde{\boldsymbol{\phi}}_a = \mu\tilde{\boldsymbol{\phi}}_a$$

gives

$$\begin{aligned} \mu_1 &= 4 - 2\sqrt{2}; \qquad \tilde{\boldsymbol{\phi}}_{a_1} = \begin{bmatrix} -\dfrac{1}{\sqrt{2}} \\ \dfrac{1}{\sqrt{2}} \end{bmatrix} \\ \mu_2 &= 4 + 2\sqrt{2}; \qquad \tilde{\boldsymbol{\phi}}_{a_2} = \begin{bmatrix} \dfrac{1}{\sqrt{2}} \\ \dfrac{1}{\sqrt{2}} \end{bmatrix} \end{aligned} \tag{b}$$

Since $\boldsymbol{\phi}_a = \mathbf{M}_a^{-1/2}\tilde{\boldsymbol{\phi}}_a$, we have

$$\boldsymbol{\phi}_{a_1} = \begin{bmatrix} \frac{1}{\sqrt{2}} & 0 \\ 0 & 1 \end{bmatrix} \begin{bmatrix} -\frac{1}{\sqrt{2}} \\ \frac{1}{\sqrt{2}} \end{bmatrix} = \begin{bmatrix} -\frac{1}{2} \\ \frac{1}{\sqrt{2}} \end{bmatrix}; \qquad \boldsymbol{\phi}_{a_2} = \begin{bmatrix} \frac{1}{2} \\ \frac{1}{\sqrt{2}} \end{bmatrix} \tag{c}$$

The vectors $\boldsymbol{\phi}_{c_1}$ and $\boldsymbol{\phi}_{c_2}$ are calculated using (10.56); hence

$$\boldsymbol{\phi}_{c_1} = \begin{bmatrix} -\frac{1}{4} \\ \frac{-1+\sqrt{2}}{4} \end{bmatrix}; \qquad \boldsymbol{\phi}_{c_2} = \begin{bmatrix} \frac{1}{4} \\ \frac{1+\sqrt{2}}{4} \end{bmatrix} \tag{d}$$

Since $\mu = 1/\lambda$ we realize that in (b), (c), and (d) we have the same solution as obtained in Example 10.12.

Considering the different procedures of eliminating the massless degrees of freedom, the results of the eigensystem analysis are the same irrespective of the procedure followed, i.e., whether $\mathbf{K}_a$ or $\mathbf{F}_a$ is established and whether the eigenproblem in (10.45) or in (10.52) is solved. The basic assumption in the analysis is that resulting from mass lumping. As we discussed in Section 10.2.4, each zero mass corresponds to an infinite frequency in the system. Therefore, in approximating the equation $\mathbf{K}\boldsymbol{\phi} = \lambda\mathbf{M}\boldsymbol{\phi}$ by the equation in (10.42), we replace, in fact, some of the frequencies of $\mathbf{K}\boldsymbol{\phi} = \lambda\mathbf{M}\boldsymbol{\phi}$ by infinite frequencies and assume that the lowest frequencies solved from either equation are not much different. The accuracy with which the lowest frequencies of $\mathbf{K}\boldsymbol{\phi} = \lambda\mathbf{M}\boldsymbol{\phi}$ are approximated by solving $\mathbf{K}_a\boldsymbol{\phi}_a = \lambda\mathbf{M}_a\boldsymbol{\phi}_a$ depends on the specific mass lumping chosen and may be adequate or crude indeed. In general, more accuracy can be expected if more mass degrees of freedom are included. However, realizing that the static condensation results in $\mathbf{K}_a$ to have a larger bandwidth than $\mathbf{K}$ (and $\mathbf{F}_a$ is certainly full), the computational effort required in the solution of the reduced eigenproblem increases rapidly as the order of $\mathbf{K}_a$ becomes large (see Section 11.3). On the other hand, if sufficient mass degrees of freedom for accuracy of solution are selected, we may no longer want to calculate the complete eigensystem of $\mathbf{K}_a\boldsymbol{\phi}_a = \lambda\mathbf{M}_a\boldsymbol{\phi}_a$ but only the smallest eigenvalues and corresponding vectors. However, in this case we may just as well consider the problem $\mathbf{K}\boldsymbol{\phi} = \lambda\mathbf{M}\boldsymbol{\phi}$ without mass lumping and solve directly only for the eigenvalues and vectors of interest using one of the algorithms described in Chapter 12.

In summary, the main shortcoming of the mass lumping procedure followed by static condensation is that the accuracy of solution depends to a large degree on the experience of the analyst to distribute the mass appropriately, and that the solution accuracy is actually not assessed. We consider the following example to show the approximation that can typically result.

Example 10.14

In Example 10.4 we calculated the eigensystem of the problem $\mathbf{K}\boldsymbol{\phi} = \lambda\mathbf{M}\boldsymbol{\phi}$, where $\mathbf{K}$ and $\mathbf{M}$ are given in the example. To evaluate an approximation to the smallest eigenvalue and corresponding eigenvector, consider instead the following eigenproblem, in which the mass is lumped:

$$\begin{bmatrix} 2 & -1 & 0 \\ -1 & 4 & -1 \\ 0 & -1 & 2 \end{bmatrix}\boldsymbol{\phi} = \lambda\begin{bmatrix} 0 & & \\ & 2 & \\ & & 0 \end{bmatrix}\boldsymbol{\phi} \tag{a}$$

Using the procedure given in (10.51) to (10.56), we obtain

$$\mathbf{F}_a = \tfrac{1}{3}; \qquad \mathbf{F}_c = \begin{bmatrix} \frac{1}{6} \\ \frac{1}{6} \end{bmatrix}$$

Hence $\lambda_1 = \frac{3}{2}$, $\boldsymbol{\phi}_{a_1} = 1/\sqrt{2}$, and

$$\boldsymbol{\phi}_{c_1} = \begin{bmatrix} \dfrac{1}{2\sqrt{2}} \\[2mm] \dfrac{1}{2\sqrt{2}} \end{bmatrix}$$

The solution of the eigenproblem in (a) for the smallest eigenvalue and corresponding eigenvector is hence

$$\lambda_1 = \tfrac{3}{2}; \qquad \boldsymbol{\phi}_1 = \begin{bmatrix} \dfrac{1}{2\sqrt{2}} \\[2mm] \dfrac{1}{\sqrt{2}} \\[2mm] \dfrac{1}{2\sqrt{2}} \end{bmatrix}$$

whereas the solution of the original problem (see Example 10.4) is

$$\lambda_1 = 2; \qquad \boldsymbol{\phi}_1 = \begin{bmatrix} \dfrac{1}{\sqrt{2}} \\[2mm] \dfrac{1}{\sqrt{2}} \\[2mm] \dfrac{1}{\sqrt{2}} \end{bmatrix}$$

It should be noted that using the mass lumping procedure, the eigenvalues can be smaller—as in this example—or larger than the eigenvalues of the original system.

10.3.2 Rayleigh–Ritz Analysis

A most general technique for finding approximations to the lowest eigenvalues and corresponding eigenvectors of the problem $\mathbf{K}\boldsymbol{\phi} = \lambda\mathbf{M}\boldsymbol{\phi}$ is the Rayleigh–Ritz analysis. *The static condensation procedure in Section 10.3.1, the component mode synthesis described in the next section, and various other methods can be understood to be Ritz analyses.* As we will see, the techniques differ only in the choice of the Ritz basis vectors assumed in the analysis. In the following we first present the Rayleigh–Ritz analysis procedure in general and then show how other techniques relate to it.

The eigenproblem that we consider is

$$\mathbf{K}\boldsymbol{\phi} = \lambda\mathbf{M}\boldsymbol{\phi} \tag{10.4}$$

where we now first assume for clarity of presentation that **K** and **M** are both positive definite, which assures that the eigenvalues are all positive; i.e., $\lambda_1 > 0$. As we pointed out in Section 10.2.3, **K** can be assumed positive definite, because a shift can always be introduced to obtain a shifted stiffness matrix which satisfies this condition. As for the mass matrix, we now assume that **M** is a consistent mass matrix or a lumped mass matrix with no zero diagonal elements, which is a condition that we shall later relax.

Consider first the Rayleigh minimum principle, which states that

$$\lambda_1 = \min \rho(\boldsymbol{\phi}) \tag{10.57}$$

where the minimum is taken over all possible vectors $\boldsymbol{\phi}$, and $\rho(\boldsymbol{\phi})$ is the Rayleigh quotient

$$\rho(\boldsymbol{\phi}) = \frac{\boldsymbol{\phi}^T\mathbf{K}\boldsymbol{\phi}}{\boldsymbol{\phi}^T\mathbf{M}\boldsymbol{\phi}} \tag{10.58}$$

This Rayleigh quotient is obtained from the Rayleigh quotient of the standard eigenvalue problem $\tilde{\mathbf{K}}\tilde{\boldsymbol{\phi}} = \lambda\tilde{\boldsymbol{\phi}}$ (see Sections 2.7 and 10.2.5). Since both **K** and **M** are positive definite, $\rho(\boldsymbol{\phi})$ has finite values for all $\boldsymbol{\phi}$. Referring to Section 2.7 the bounds on the Rayleigh quotient are

$$0 < \lambda_1 \leq \rho(\boldsymbol{\phi}) \leq \lambda_n < \infty \tag{10.59}$$

In the Ritz analysis we consider a set of vectors $\bar{\boldsymbol{\phi}}$, which are linear combinations of the Ritz basis vectors $\boldsymbol{\psi}_i$, $i = 1, \ldots, q$; i.e., a typical vector is given by

$$\bar{\boldsymbol{\phi}} = \sum_{i=1}^{q} x_i \boldsymbol{\psi}_i \tag{10.60}$$

where the x_i are the Ritz coordinates. Since $\bar{\boldsymbol{\phi}}$ is a linear combination of the Ritz basis vectors, $\bar{\boldsymbol{\phi}}$ cannot be any arbitrary vector, but instead lies in the subspace spanned by the Ritz basis vectors, which we call V_q (see Sections 2.2 and 12.3). It should be noted that the vectors $\boldsymbol{\psi}_i$, $i = 1, \ldots, q$, must be linearly independent; therefore, the subspace V_q has dimension q. Also, denoting the n-dimensional space in which the matrices **K** and **M** are defined by V_n, we have that V_q is contained in V_n.

In the Rayleigh–Ritz analysis we aim to determine the specific vectors $\bar{\boldsymbol{\phi}}_i$, $i = 1, \ldots, q$, which, with the constraint of lying in the subspace spanned by the Ritz basis vectors, "best" approximate the required eigenvectors. For this purpose we invoke the Rayleigh minimum principle. The use of this principle determines in what sense the solution does "best" approximate the eigenvectors sought, an aspect that we shall point out during the presentation of the solution procedure.

To invoke the Rayleigh minimum principle on $\bar{\boldsymbol{\phi}}$, we first evaluate the Rayleigh quotient,

$$\rho(\bar{\boldsymbol{\phi}}) = \frac{\sum_{j=1}^{q} \sum_{i=1}^{q} x_i x_j \tilde{k}_{ij}}{\sum_{j=1}^{q} \sum_{i=1}^{q} x_i x_j \tilde{m}_{ij}} = \frac{\tilde{k}}{\tilde{m}} \tag{10.61}$$

where

$$\tilde{k}_{ij} = \boldsymbol{\psi}_i^T \mathbf{K} \boldsymbol{\psi}_j \tag{10.62}$$

$$\tilde{m}_{ij} = \boldsymbol{\psi}_i^T \mathbf{M} \boldsymbol{\psi}_j \tag{10.63}$$

The necessary condition for a minimum of $\rho(\bar{\boldsymbol{\phi}})$ given in (10.61) is $\partial \rho(\bar{\boldsymbol{\phi}})/\partial x_i = 0$, $i = 1, \ldots, q$, because the x_i are the only variables. However,

$$\frac{\partial \rho(\bar{\boldsymbol{\phi}})}{\partial x_i} = \frac{2\tilde{m} \sum_{j=1}^{q} x_j \tilde{k}_{ij} - 2\tilde{k} \sum_{j=1}^{q} x_j \tilde{m}_{ij}}{\tilde{m}^2} \tag{10.64}$$

and using $\rho = \tilde{k}/\tilde{m}$, the condition for a minimum of $\rho(\bar{\boldsymbol{\phi}})$ is

$$\sum_{j=1}^{q} x_j(\tilde{k}_{ij} - \rho \tilde{m}_{ij}) = 0 \qquad \text{for } i = 1, \ldots, q \tag{10.65}$$

In actual analysis we write the q equations in (10.65) in matrix form, thus obtaining the eigenproblem

$$\tilde{\mathbf{K}}\mathbf{x} = \rho \tilde{\mathbf{M}}\mathbf{x} \tag{10.66}$$

where $\tilde{\mathbf{K}}$ and $\tilde{\mathbf{M}}$ are $q \times q$ matrices with typical elements defined in (10.62) and (10.63), respectively, and $\mathbf{x}$ is a vector of the Ritz coordinates sought:

$$\mathbf{x}^T = [x_1 \quad x_2 \quad \ldots \quad x_q] \tag{10.67}$$

The solution to (10.66) yields q eigenvalues $\rho_1, \ldots, \rho_q$, which are approximations to $\lambda_1, \ldots, \lambda_q$, and q eigenvectors,

$$\begin{aligned} \mathbf{x}_1^T &= [x_1^1 \quad x_2^1 \quad \ldots \quad x_q^1] \\ \mathbf{x}_2^T &= [x_1^2 \quad x_2^2 \quad \cdots \quad x_q^2] \\ &\vdots \\ \mathbf{x}_q^T &= [x_1^q \quad x_2^q \quad \ldots \quad x_q^q] \end{aligned} \tag{10.68}$$

The eigenvectors $\mathbf{x}_i$ are used to evaluate the vectors $\bar{\boldsymbol{\phi}}_1, \ldots, \bar{\boldsymbol{\phi}}_q$, which are approximations to the eigenvectors $\boldsymbol{\phi}_1, \ldots, \boldsymbol{\phi}_q$. Using (10.68) and (10.60), we have

$$\bar{\boldsymbol{\phi}}_i = \sum_{j=1}^{q} x_j^i \boldsymbol{\psi}_j; \qquad i = 1, \ldots, q \tag{10.69}$$

An important feature of the eigenvalue approximations calculated in the analysis is that they are upper bound approximations to the eigenvalues of interest; i.e.,

$$\lambda_1 \leq \rho_1; \quad \lambda_2 \leq \rho_2; \quad \lambda_3 \leq \rho_3; \ldots; \quad \lambda_q \leq \rho_q \leq \lambda_n \tag{10.70}$$

meaning that since $\mathbf{K}$ and $\mathbf{M}$ are assumed to be positive definite, $\tilde{\mathbf{K}}$ and $\tilde{\mathbf{M}}$ are also positive definite matrices.

The proof of the inequality in (10.70) shows the actual mechanism that is used to obtain the eigenvalue approximations ρ_i. To calculate ρ_1 we search for the minimum of $\rho(\bar{\boldsymbol{\phi}})$ that can be reached by linearly combining all available Ritz basis vectors. The inequality $\lambda_1 \leq \rho_1$ follows from the Rayleigh minimum principle in (10.57) and because V_q is contained in the n-dimensional space V_n, in which $\mathbf{K}$ and $\mathbf{M}$ are defined.

The condition that is employed to obtain ρ_2 is typical of the mechanism used to calculate the approximations to the higher eigenvalues. First we observe that for the eigenvalue problem $\mathbf{K}\boldsymbol{\phi} = \lambda\mathbf{M}\boldsymbol{\phi}$, we have

$$\lambda_2 = \min \rho(\boldsymbol{\phi}) \tag{10.71}$$

where the minimum is now taken over all possible vectors $\boldsymbol{\phi}$ in V_n that satisfy the orthogonality condition (see Section 2.7)

$$\boldsymbol{\phi}^T\mathbf{M}\boldsymbol{\phi}_1 = 0 \tag{10.72}$$

Considering the approximate eigenvectors $\bar{\boldsymbol{\phi}}_i$ obtained in the Rayleigh–Ritz analysis, we observe that

$$\bar{\boldsymbol{\phi}}_i^T\mathbf{M}\bar{\boldsymbol{\phi}}_j = \delta_{ij} \tag{10.73}$$

where δ_{ij} = Kronecker delta, and that, therefore, in the above Rayleigh–Ritz analysis we obtained ρ_2 by evaluating

$$\rho_2 = \min \rho(\bar{\boldsymbol{\phi}}) \tag{10.74}$$

where the minimum was taken over all possible vectors $\bar{\boldsymbol{\phi}}$ in V_q that satisfy the orthogonality condition

$$\bar{\boldsymbol{\phi}}^T\mathbf{M}\bar{\boldsymbol{\phi}}_1 = 0 \tag{10.75}$$

To show that $\lambda_2 \leq \rho_2$, we consider an auxiliary problem; i.e., assume that we evaluate

$$\tilde{\rho}_2 = \min \rho(\bar{\boldsymbol{\phi}}) \tag{10.76}$$

where the minimum is taken over all vectors $\bar{\boldsymbol{\phi}}$ that satisfy the condition

$$\bar{\boldsymbol{\phi}}^T\mathbf{M}\boldsymbol{\phi}_1 = 0 \tag{10.77}$$

The problem defined in (10.76) and (10.77) is the same as the problem in (10.71) and (10.72), except that in the latter case the minimum is taken over all $\boldsymbol{\phi}$, whereas in the problem of (10.76) and (10.77) we consider all vectors $\bar{\boldsymbol{\phi}}$ in V_q. Then since V_q is contained in V_n, we have $\lambda_2 \leq \tilde{\rho}_2$. On the other hand, $\tilde{\rho}_2 \leq \rho_2$, because the most severe constraint on $\bar{\boldsymbol{\phi}}$ in (10.77) is $\bar{\boldsymbol{\phi}}_1$. Therefore, we have

$$\lambda_2 \leq \tilde{\rho}_2 \leq \rho_2 \tag{10.78}$$

The basis for the calculation of $\bar{\boldsymbol{\phi}}_2$ and hence ρ_2 is that the minimum of $\rho(\bar{\boldsymbol{\phi}})$ is sought with the orthogonality condition in (10.75) on $\bar{\boldsymbol{\phi}}_1$. Similarly, to obtain ρ_i and $\bar{\boldsymbol{\phi}}_i$, we in fact minimize $\rho(\bar{\boldsymbol{\phi}})$ with the orthogonality conditions $\bar{\boldsymbol{\phi}}^T\mathbf{M}\bar{\boldsymbol{\phi}}_j = 0$ for $j = 1, \ldots, i - 1$. Accordingly, the inequality on ρ_i in (10.70) can be proved in an analogous manner to the procedure used above for ρ_2, but all $(i - 1)$ constraint equations need to be satisfied.

The observation that $(i - 1)$ constraint equations need to be fulfilled in the evaluation of ρ_i also indicates that we can expect less accuracy in the approximation of the higher eigenvalues than in the approximation of the lower eigenvalues, for which less constraints are imposed. This is generally also observed in actual analysis.

Considering the procedure in practical dynamic analysis, the Ritz basis functions may be calculated from a static solution in which q load patters are specified in $\mathbf{R}$; i.e., we consider

$$\mathbf{K}\boldsymbol{\Psi} = \mathbf{R} \tag{10.79}$$

where $\boldsymbol{\Psi}$ is an $n \times q$ matrix storing the Ritz basis vectors; i.e., $\boldsymbol{\Psi} = [\boldsymbol{\psi}_1, \ldots, \boldsymbol{\psi}_q]$. The analysis is continued by evaluating the projections of $\mathbf{K}$ and $\mathbf{M}$ onto the subspace V_q spanned by the vectors $\boldsymbol{\psi}_i$, $i = 1, \ldots, q$; i.e., we calculate

$$\tilde{\mathbf{K}} = \boldsymbol{\Psi}^T\mathbf{K}\boldsymbol{\Psi} \tag{10.80}$$

and

$$\tilde{\mathbf{M}} = \boldsymbol{\Psi}^T\mathbf{M}\boldsymbol{\Psi} \tag{10.81}$$

where because of (10.79) we have

$$\tilde{\mathbf{K}} = \boldsymbol{\Psi}^T\mathbf{R} \tag{10.82}$$

Next we solve the eigenproblem $\tilde{\mathbf{K}}\mathbf{x} = \rho\tilde{\mathbf{M}}\mathbf{x}$, the solution of which can be written

$$\tilde{\mathbf{K}}\mathbf{X} = \tilde{\mathbf{M}}\mathbf{X}\boldsymbol{\rho} \tag{10.83}$$

where $\boldsymbol{\rho}$ is a diagonal matrix listing the eigenvalue approximations ρ_i, $\boldsymbol{\rho} = \text{diag}\,(\rho_i)$, and $\mathbf{X}$ is a matrix storing the $\tilde{\mathbf{M}}$-orthonormal eigenvectors $\mathbf{x}_1, \ldots, \mathbf{x}_q$. The approximations to the eigenvectors of the problem $\mathbf{K}\boldsymbol{\phi} = \lambda\mathbf{M}\boldsymbol{\phi}$ are then

$$\bar{\boldsymbol{\Phi}} = \boldsymbol{\Psi}\mathbf{X} \tag{10.84}$$

So far we assumed that the mass matrix of the finite element system is positive definite; i.e., $\mathbf{M}$ is not a diagonal mass matrix with some zero diagonal elements. The reason for this assumption was to avoid the case $\bar{\boldsymbol{\phi}}^T\mathbf{M}\bar{\boldsymbol{\phi}}$ equal to zero in the calculation of the Rayleigh quotient, in which case $\rho(\bar{\boldsymbol{\phi}})$ gives an infinite eigenvalue. However, the Rayleigh–Ritz analysis can be carried out as described above when $\mathbf{M}$ is a diagonal matrix with some zero diagonal elements, provided the Ritz basis vectors are selected to lie in the subspace that corresponds to the finite eigenvalues. In addition, the

Ritz basis vectors must be linearly independent when considering only the mass degrees of freedom, in order to obtain a positive definite matrix $\tilde{\mathbf{M}}$. One way of achieving this in practice is to excite in each of the load vectors in $\mathbf{R}$ in (10.79) different mass degrees of freedom (see Section 12.3.3 and Example 10.16).

Of particular interest are the errors that we may expect in the solution. *Although we have shown that an eigenvalue calculated from the Ritz analysis is an upper bound on the corresponding exact eigenvalue of the system, we did not establish anything about the actual error in the eigenvalue.* This error depends on the Ritz basis vectors used, because the vectors $\bar{\boldsymbol{\phi}}$ are linear combinations of the Ritz basis vectors $\boldsymbol{\psi}_i$, $i = 1, \ldots, q$. We can only obtain good results if the vectors $\boldsymbol{\psi}_i$ span a subspace V_q that is close to the least dominant subspace of $\mathbf{K}$ and $\mathbf{M}$ spanned by $\boldsymbol{\phi}_1, \ldots, \boldsymbol{\phi}_q$. It should be noted that this does not mean that the Ritz basis vectors should each be close to an eigenvector sought, but rather that linear combinations of the Ritz basis vectors can establish good approximations of the required eigenvectors of $\mathbf{K}\boldsymbol{\phi} = \lambda\mathbf{M}\boldsymbol{\phi}$. We further discuss the selection of good Ritz basis vectors and the approximations involved in the analysis in Section 12.3, when we present the subspace iteration method, because this method uses the Ritz analysis technique.

To demonstrate the Rayleigh–Ritz analysis procedure, consider the following examples.

Example 10.15

Obtain approximate solutions to the eigenproblem $\mathbf{K}\boldsymbol{\phi} = \lambda\mathbf{M}\boldsymbol{\phi}$ considered in Example 10.4, where

$$\mathbf{K} = \begin{bmatrix} 2 & -1 & 0 \\ -1 & 4 & -1 \\ 0 & -1 & 2 \end{bmatrix}; \qquad \mathbf{M} = \begin{bmatrix} \frac{1}{2} & & \\ & 1 & \\ & & \frac{1}{2} \end{bmatrix}$$

The exact eigenvalues are $\lambda_1 = 2$, $\lambda_2 = 4$, $\lambda_3 = 6$.

1. Use the following load vectors to generate the Ritz basis vectors

$$\mathbf{R} = \begin{bmatrix} 1 & 0 \\ 0 & 0 \\ 0 & 1 \end{bmatrix}$$

2. Then use a different set of load vectors to generate the Ritz basis vectors

$$\mathbf{R} = \begin{bmatrix} 1 & 0 \\ 1 & 1 \\ 1 & 0 \end{bmatrix}$$

In the Ritz analysis we employ the relations in (10.79) to (10.84), and obtain, in case (1),

$$\begin{bmatrix} 2 & -1 & 0 \\ -1 & 4 & -1 \\ 0 & -1 & 2 \end{bmatrix} \mathbf{\Psi} = \begin{bmatrix} 1 & 0 \\ 0 & 0 \\ 0 & 1 \end{bmatrix}$$

Hence

$$\mathbf{\Psi} = \begin{bmatrix} \frac{7}{12} & \frac{1}{12} \\ \frac{1}{6} & \frac{1}{6} \\ \frac{1}{12} & \frac{7}{12} \end{bmatrix}$$

and

$$\tilde{\mathbf{K}} = \frac{1}{12}\begin{bmatrix} 7 & 1 \\ 1 & 7 \end{bmatrix}; \qquad \tilde{\mathbf{M}} = \frac{1}{144}\begin{bmatrix} 29 & 11 \\ 11 & 29 \end{bmatrix}$$

The solution of the eigenproblem $\tilde{\mathbf{K}}\mathbf{x} = \rho\tilde{\mathbf{M}}\mathbf{x}$ is

$$(\rho_1, \mathbf{x}_1) = \left(2.40043, \begin{bmatrix} 1.34183 \\ 1.34183 \end{bmatrix}\right); \qquad (\rho_2, \mathbf{x}_2) = \left(4.00320, \begin{bmatrix} 2.00080 \\ -2.00080 \end{bmatrix}\right)$$

Hence we have as eigenvalue approximations $\rho_1 = 2.40$, $\rho_2 = 4.00$, and evaluating

$$\bar{\mathbf{\Phi}} = \begin{bmatrix} \frac{7}{12} & \frac{1}{12} \\ \frac{1}{6} & \frac{1}{6} \\ \frac{1}{12} & \frac{7}{12} \end{bmatrix} \begin{bmatrix} 1.342 & 2.000 \\ 1.342 & -2.000 \end{bmatrix} = \begin{bmatrix} 0.895 & 1.00 \\ 0.447 & 0 \\ 0.895 & -1.00 \end{bmatrix}$$

we have

$$\bar{\boldsymbol{\phi}}_1 = \begin{bmatrix} 0.895 \\ 0.447 \\ 0.895 \end{bmatrix}; \qquad \bar{\boldsymbol{\phi}}_2 = \begin{bmatrix} 1.00 \\ 0.00 \\ -1.00 \end{bmatrix}$$

Next we assume the load vectors in (2) and solve

$$\begin{bmatrix} 2 & -1 & 0 \\ -1 & 4 & -1 \\ 0 & -1 & 2 \end{bmatrix} \mathbf{\Psi} = \begin{bmatrix} 1 & 0 \\ 1 & 1 \\ 1 & 0 \end{bmatrix}$$

Hence

$$\mathbf{\Psi} = \begin{bmatrix} \frac{5}{6} & \frac{1}{6} \\ \frac{2}{3} & \frac{1}{3} \\ \frac{5}{6} & \frac{1}{6} \end{bmatrix}$$

and

$$\tilde{\mathbf{K}} = \begin{bmatrix} \frac{7}{3} & \frac{2}{3} \\ \frac{2}{3} & \frac{1}{3} \end{bmatrix}; \qquad \tilde{\mathbf{M}} = \frac{1}{36}\begin{bmatrix} 41 & 13 \\ 13 & 5 \end{bmatrix}$$

The solution of the eigenproblem $\tilde{\mathbf{K}}\mathbf{x} = \rho\tilde{\mathbf{M}}\mathbf{x}$ gives

$$(\rho_1, \mathbf{x}_1) = \left(2.0000, \begin{bmatrix} 0.70711 \\ 0.70711 \end{bmatrix}\right); \qquad (\rho_2, \mathbf{x}_2) = \left(6.0000, \begin{bmatrix} -2.1213 \\ 6.3640 \end{bmatrix}\right)$$

Hence we have as eigenvalue approximations $\rho_1 = 2.00$, $\rho_2 = 6.00$, and evaluating

$$\bar{\mathbf{\Phi}} = \begin{bmatrix} \frac{5}{6} & \frac{1}{6} \\ \frac{2}{3} & \frac{1}{3} \\ \frac{5}{6} & \frac{1}{6} \end{bmatrix} \begin{bmatrix} 0.70711 & -2.1213 \\ 0.70711 & 6.3640 \end{bmatrix} = \begin{bmatrix} 0.70711 & -0.70708 \\ 0.70711 & 0.70713 \\ 0.70711 & -0.70708 \end{bmatrix}$$

we have

$$\bar{\boldsymbol{\phi}}_1 = \begin{bmatrix} 0.70711 \\ 0.70711 \\ 0.70711 \end{bmatrix}; \qquad \bar{\boldsymbol{\phi}}_2 = \begin{bmatrix} -0.70708 \\ 0.70713 \\ -0.70708 \end{bmatrix}$$

Comparing the results with the exact solution, it is interesting to note that in case (1) $\rho_1 > \lambda_1$ and $\rho_2 = \lambda_2$, whereas in case (2), $\rho_1 = \lambda_1$ and $\rho_2 = \lambda_3$. In both cases we did not obtain good approximations to the lowest two eigenvalues, and it is clearly demonstrated that the results depend completely on the initial Ritz basis vectors chosen.

Example 10.16

Use the Rayleigh–Ritz analysis to calculate an approximation to λ_1 and $\boldsymbol{\phi}_1$ of the eigenproblem considered in Example 10.12.

We note that in this case **M** is positive semidefinite. Therefore, to carry out the Ritz analysis we need to choose a load vector in **R** that excites at least one mass. Assume that we use

$$\mathbf{R}^T = [0 \quad 1 \quad 0 \quad 0]$$

Then the solution of (10.79) yields (see Example 10.13)

$$\boldsymbol{\Psi}^T = [1 \quad 2 \quad 2 \quad 2]$$

Hence

$$\tilde{\mathbf{K}} = [2]; \qquad \tilde{\mathbf{M}} = [12]$$

$$\rho_1 = \frac{1}{6}; \qquad \mathbf{x}_1 = \left[\frac{1}{2\sqrt{3}}\right]$$

and

$$\bar{\boldsymbol{\phi}}_1^T = \left[\frac{1}{2\sqrt{3}} \quad \frac{1}{\sqrt{3}} \quad \frac{1}{\sqrt{3}} \quad \frac{1}{\sqrt{3}}\right]$$

Hence we have as expected that $\rho_1 > \lambda_1$.

The Ritz analysis procedure presented above is a very general tool, and, as pointed out earlier, various analysis methods known under different names can actually be shown to be Ritz analyses. In Section 10.3.3 we present the component mode synthesis as a Ritz analysis. In the following we briefly want to show that the technique of static condensation as described in Section 10.3.1 is, in fact, also a Ritz analysis.

In the static condensation analysis we assumed that all mass can be lumped at q degrees of freedom. Therefore, as an approximation to the eigenproblem $\mathbf{K}\boldsymbol{\phi} = \lambda\mathbf{M}\boldsymbol{\phi}$, we obtained the following problem:

$$\begin{bmatrix} \mathbf{K}_{aa} & \mathbf{K}_{ac} \\ \mathbf{K}_{ca} & \mathbf{K}_{cc} \end{bmatrix} \begin{bmatrix} \boldsymbol{\phi}_a \\ \boldsymbol{\phi}_c \end{bmatrix} = \lambda \begin{bmatrix} \mathbf{M}_a & \mathbf{0} \\ \mathbf{0} & \mathbf{0} \end{bmatrix} \begin{bmatrix} \boldsymbol{\phi}_a \\ \boldsymbol{\phi}_c \end{bmatrix} \tag{10.42}$$

with q finite and $(n - q)$ infinite eigenvalues, which correspond to the massless degrees of freedom (see Section 10.2.4). To calculate the finite eigenvalues, we used static condensation on the massless degrees of freedom and arrived at the eigenproblem

$$\mathbf{K}_a \boldsymbol{\phi}_a = \lambda \mathbf{M}_a \boldsymbol{\phi}_a \tag{10.45}$$

where $\mathbf{K}_a$ is defined in (10.46). However, this solution is actually a Ritz analysis of the lumped mass model considered in (10.42). The Ritz basis vectors are the displacement patterns associated with the $\boldsymbol{\phi}_a$ degrees of freedom when the $\boldsymbol{\phi}_c$ degrees of freedom are released. Solving the equations

$$\begin{bmatrix} \mathbf{K}_{aa} & \mathbf{K}_{ac} \\ \mathbf{K}_{ca} & \mathbf{K}_{cc} \end{bmatrix} \begin{bmatrix} \mathbf{F}_a \\ \mathbf{F}_c \end{bmatrix} = \begin{bmatrix} \mathbf{I} \\ \mathbf{0} \end{bmatrix} \tag{10.51}$$

in which $\mathbf{F}_a = \mathbf{K}_a^{-1}$, we have that the Ritz basis vectors to be used in (10.80), (10.81), and (10.84) are

$$\boldsymbol{\Psi} = \begin{bmatrix} \mathbf{I} \\ \mathbf{F}_c \mathbf{K}_a \end{bmatrix} \tag{10.85}$$

To verify that a Ritz analysis with the base vectors in (10.85) yields in fact (10.45), we evaluate (10.80) and (10.81). Substituting for $\boldsymbol{\Psi}$ and $\mathbf{K}$ in (10.80), we obtain

$$\tilde{\mathbf{K}} = [\mathbf{I} \quad (\mathbf{F}_c \mathbf{K}_a)^T] \begin{bmatrix} \mathbf{K}_{aa} & \mathbf{K}_{ac} \\ \mathbf{K}_{ca} & \mathbf{K}_{cc} \end{bmatrix} \begin{bmatrix} \mathbf{I} \\ \mathbf{F}_c \mathbf{K}_a \end{bmatrix} \tag{10.86}$$

which, using (10.51), reduces to

$$\tilde{\mathbf{K}} = \mathbf{K}_a \tag{10.87}$$

Similarly, substituting for $\boldsymbol{\Psi}$ and $\mathbf{M}$ in (10.81), we have

$$\tilde{\mathbf{M}} = [\mathbf{I} \quad (\mathbf{F}_c \mathbf{K}_a)^T] \begin{bmatrix} \mathbf{M}_a & \mathbf{0} \\ \mathbf{0} & \mathbf{0} \end{bmatrix} \begin{bmatrix} \mathbf{I} \\ \mathbf{F}_c \mathbf{K}_a \end{bmatrix} \tag{10.88}$$

or

$$\tilde{\mathbf{M}} = \mathbf{M}_a \tag{10.89}$$

Hence, in the static condensation we actually perform a Ritz analysis of the lumped mass model. It should be noted that in the analysis we calculate the q finite eigenvalues exactly (i.e., $\rho_i = \lambda_i$ for $i = 1, \ldots, q$), because the Ritz basis vectors span the q-dimensional subspace corresponding to the finite eigenvalues. In practice, the evaluation of the vectors $\boldsymbol{\Psi}$ in (10.85) is not necessary (and would be costly), and instead the Ritz analysis is better carried out using

$$\boldsymbol{\Psi} = \begin{bmatrix} \mathbf{F}_a \\ \mathbf{F}_c \end{bmatrix} \tag{10.90}$$

Since the vectors in (10.90) span the same subspace as the vectors in (10.85), the same eigenvalues and eigenvectors are calculated employing either set of base vectors. Specifically, using (10.90) we obtain in the Ritz analysis the reduced eigenproblem

$$\mathbf{F}_a\mathbf{x} = \lambda\mathbf{F}_a\mathbf{M}_a\mathbf{F}_a\mathbf{x} \tag{10.91}$$

To show that this eigenproblem is indeed equivalent to the problem in (10.45), we premultiply both sides in (10.91) by $\mathbf{K}_a$ and use the transformation $\mathbf{x} = \mathbf{K}_a\tilde{\mathbf{x}}$, giving $\mathbf{K}_a\tilde{\mathbf{x}} = \lambda\mathbf{M}_a\tilde{\mathbf{x}}$, i.e., the problem in (10.45).

Example 10.17

Use the Ritz analysis procedure to perform static condensation of the massless degrees of freedom in the problem $\mathbf{K}\boldsymbol{\phi} = \lambda\mathbf{M}\boldsymbol{\phi}$ considered in Example 10.12.

We first need to evaluate the Ritz basis vectors given in (10.90). This has been done in Example 10.13, where we found that

$$\mathbf{F}_a = \begin{bmatrix} 2 & 2 \\ 2 & 4 \end{bmatrix}; \qquad \mathbf{F}_c = \begin{bmatrix} 1 & 1 \\ 2 & 3 \end{bmatrix}$$

The Ritz reduction given in (10.91) thus yields the eigenproblem

$$\begin{bmatrix} 2 & 2 \\ 2 & 4 \end{bmatrix}\mathbf{x} = \lambda\begin{bmatrix} 12 & 16 \\ 16 & 24 \end{bmatrix}\mathbf{x}$$

10.3.3 Component Mode Synthesis

As for the static condensation procedure, the component mode synthesis is, in fact, a Ritz analysis, and the method might have been presented in the previous section as a specific application. However, as was repeatedly pointed out, the most important aspect in Ritz analysis is the selection of appropriate Ritz basis vectors, because the results can only be as good as the Ritz basis vectors allow them to be. The specific scheme used in the component mode synthesis is of particular interest, which is the reason we want to devote a separate section to the discussion of the method.

The component mode synthesis has been developed to a large extent as a natural consequence of the analysis procedure followed in practice when large and complex structures are analyzed. The general practical procedure is that different groups perform the analyses of different components of the structure under consideration. For example, in a reactor analysis, one group may analyze a main pipe and another group a piping system attached to it. In a first preliminary analysis, both groups work separately and model the effects of the other components on the specific component that they consider in an approximate manner. For example, in the analysis of the two piping systems referred to above, the group analyzing the side branch may assume full fixity at the point of intersection with the main pipe, and the group analyzing the main pipe may introduce a concentrated spring and mass to allow for the side branch. The advantage of considering separately the components of the structure is primarily one of time scheduling; i.e., the separate groups can work on the analyses and designs of the components at the same time. It is primarily for this reason that the component mode synthesis is very appealing in the analysis and design of large structural systems.

Assume that the preliminary analyses of the components have been carried out and that the complete structure shall now be analyzed. It is at this stage that the component mode synthesis is a natural procedure to use. Namely, with the mode shape characteristics of each component known, it appears natural to use this information in estimating the frequencies and mode shapes of the complete structure. The specific procedure may vary according to the specific information available, but, in essence, the mode shapes of the components are used in a Rayleigh–Ritz analysis to calculate approximate mode shapes and frequencies of the complete structure.

Consider for illustration that each component structure was obtained by fixing all its boundary degrees of freedom, and denote the stiffness matrices of the component structures by $\mathbf{K}_{\mathrm{I}}, \mathbf{K}_{\mathrm{II}}, \ldots, \mathbf{K}_{\mathrm{M}}$. Assume that component structure L is only coupled to component structures $L-1$ and $L+1$; then we can write for the stiffness matrix of the complete structure,

$$\mathbf{K} = \begin{bmatrix} \mathbf{K}_{\mathrm{I}} & \mathbf{K}_{\mathrm{I,II}} & & \\ \mathbf{K}_{\mathrm{I,II}}^T & \mathbf{K}_{\mathrm{II}} & \mathbf{K}_{\mathrm{II,III}} & \\ & & & \\ & & & \\ & & \mathbf{K}_{\mathrm{M-1,M}}^T & \mathbf{K}_{\mathrm{M}} \end{bmatrix} \tag{10.92}$$

where $\mathbf{K}_{L-1,L}$ represents the stiffness coupling between the component structures $L-1$ and L. Using an analogous notation for the mass matrices, we also have

$$\mathbf{M} = \begin{bmatrix} \mathbf{M}_{\mathrm{I}} & \mathbf{M}_{\mathrm{I,II}} & & \\ \mathbf{M}_{\mathrm{I,II}}^T & \mathbf{M}_{\mathrm{II}} & \mathbf{M}_{\mathrm{II,III}} & \\ & & & \\ & & & \\ & & \mathbf{M}_{\mathrm{M-1,M}}^T & \mathbf{M}_{\mathrm{M}} \end{bmatrix} \tag{10.93}$$

where the mass coupling matrices $\mathbf{M}_{L-1,L}$ are zero in case of a lumped mass matrix. Assume that the lowest eigenvalues and corresponding eigenvectors of each component structure have been calculated; i.e., we have for each component structure,

$$\left.\begin{aligned} \mathbf{K}_{\mathrm{I}}\mathbf{\Phi}_{\mathrm{I}} &= \mathbf{M}_{\mathrm{I}}\mathbf{\Phi}_{\mathrm{I}}\mathbf{\Lambda}_{\mathrm{I}} \\ \mathbf{K}_{\mathrm{II}}\mathbf{\Phi}_{\mathrm{II}} &= \mathbf{M}_{\mathrm{II}}\mathbf{\Phi}_{\mathrm{II}}\mathbf{\Lambda}_{\mathrm{II}} \\ &\;\;\vdots \\ \mathbf{K}_{\mathrm{M}}\mathbf{\Phi}_{\mathrm{M}} &= \mathbf{M}_{\mathrm{M}}\mathbf{\Phi}_{\mathrm{M}}\mathbf{\Lambda}_{\mathrm{M}} \end{aligned}\right\} \tag{10.94}$$

where $\mathbf{\Phi}_L$ and $\mathbf{\Lambda}_L$ are the matrices of calculated eigenvectors and eigenvalues of the Lth component structure.

In a component mode synthesis, approximate mode shapes and frequencies are obtained by performing a Rayleigh–Ritz analysis with the following assumed loads on the right-hand side of (10.79),

$$\mathbf{R} = \begin{bmatrix} \mathbf{\Phi}_{\text{I}} & \mathbf{0} & \mathbf{0} & \cdots \\ \mathbf{0} & \mathbf{I}_{\text{I,II}} & \mathbf{0} & \\ \mathbf{\Phi}_{\text{II}} & \mathbf{0} & \mathbf{0} & \cdots \\ \mathbf{0} & \mathbf{0} & \mathbf{I}_{\text{II,III}} & \\ \cdot & \cdot & \cdot & \\ \cdot & \cdot & \cdot & \cdots \\ \cdot & \cdot & \cdot & \\ \mathbf{\Phi}_{\text{M}} & \mathbf{0} & & \end{bmatrix} \tag{10.95}$$

where $\mathbf{I}_{L-1,L}$ is a unit matrix of order equal to the connection degrees of freedom between component structures $L - 1$ and L. The unit matrices correspond to loads that are applied to the connection degrees of freedom of the component structures. Since in the derivation of the mode shape matrices used in (10.95), the component structures were fixed at their boundaries, the unit loads have the effect of releasing these connection degrees of freedom. If, on the other hand, the connection degrees of freedom have been included in the analysis of the component structures, we may dispense with the unit matrices in $\mathbf{R}$.

An important consideration is the accuracy that can be expected in the above component mode synthesis. Since a Ritz analysis is performed, all accuracy considerations discussed in Section 10.3.2 are directly applicable; i.e., the analysis yields upper bounds to the exact eigenvalues of the problem $\mathbf{K}\boldsymbol{\phi} = \lambda\mathbf{M}\boldsymbol{\phi}$. However, the actual accuracy achieved in the solution is not known, although it can be evaluated, for example, as described in Section 10.4. The fact that the solution accuracy is highly dependent on the vectors used in $\mathbf{R}$ (i.e., the Ritz basis vectors) is, as in all Ritz analyses, the main defect of the method. However, in practice, reasonable accuracy can often be obtained because the eigenvectors corresponding to the smallest eigenvalues of the component structures are used in $\mathbf{R}$. We demonstrate the analysis procedure in the following example.

Example 10.18

Consider the eigenproblem $\mathbf{K}\boldsymbol{\phi} = \lambda\mathbf{M}\boldsymbol{\phi}$, where

$$\mathbf{K} = \begin{bmatrix} 2 & -1 & & & \\ -1 & 2 & -1 & & \\ & -1 & 2 & -1 & \\ & & -1 & 2 & -1 \\ & & & -1 & 1 \end{bmatrix}; \qquad \mathbf{M} = \begin{bmatrix} 1 & & & & \\ & 1 & & & \\ & & 1 & & \\ & & & 1 & \\ & & & & \frac{1}{2} \end{bmatrix}$$

Use the substructure eigenproblems indicated by the dashed lines in **K** and **M** to establish the load matrix given in (10.95) for a component mode synthesis analysis. Then calculate eigenvalue and eigenvector approximations.

Here we have for substructure I,

$$\mathbf{K}_{\text{I}} = \begin{bmatrix} 2 & -1 \\ -1 & 2 \end{bmatrix}; \qquad \mathbf{M}_{\text{I}} = \begin{bmatrix} 1 & 0 \\ 0 & 1 \end{bmatrix}$$

with the eigensolution

$$\lambda_1 = 1, \quad \lambda_2 = 3; \qquad \boldsymbol{\phi}_1 = \begin{bmatrix} \frac{\sqrt{2}}{2} \\ \frac{\sqrt{2}}{2} \end{bmatrix}, \quad \boldsymbol{\phi}_2 = \begin{bmatrix} -\frac{\sqrt{2}}{2} \\ \frac{\sqrt{2}}{2} \end{bmatrix}$$

and for substructure II,

$$\mathbf{K}_{\text{II}} = \begin{bmatrix} 2 & -1 \\ -1 & 1 \end{bmatrix}; \qquad \mathbf{M}_{\text{II}} = \begin{bmatrix} 1 & 0 \\ 0 & \frac{1}{2} \end{bmatrix}$$

with the eigensolution

$$\lambda_1 = 2 - \sqrt{2}, \quad \lambda_2 = 2 + \sqrt{2}; \qquad \boldsymbol{\phi}_1 = \begin{bmatrix} \frac{\sqrt{2}}{2} \\ 1 \end{bmatrix}, \quad \boldsymbol{\phi}_2 = \begin{bmatrix} -\frac{\sqrt{2}}{2} \\ 1 \end{bmatrix}$$

Thus we have for the matrix **R** in (10.95),

$$\mathbf{R} = \begin{bmatrix} \frac{\sqrt{2}}{2} & -\frac{\sqrt{2}}{2} & 0 \\ \frac{\sqrt{2}}{2} & \frac{\sqrt{2}}{2} & 0 \\ 0 & 0 & 1 \\ \frac{\sqrt{2}}{2} & -\frac{\sqrt{2}}{2} & 0 \\ 1 & 1 & 0 \end{bmatrix}$$

Now performing the Ritz analysis as given in (10.79) to (10.84), we obtain

$$\tilde{\mathbf{K}} = \begin{bmatrix} 22.40 & 5.328 & 7.243 \\ 5.328 & 2.257 & 1.586 \\ 7.243 & 1.586 & 3 \end{bmatrix}$$

$$\tilde{\mathbf{M}} = \begin{bmatrix} 222.4 & 50.69 & 77.69 \\ 50.69 & 11.94 & 17.59 \\ 77.69 & 17.59 & 27.5 \end{bmatrix}$$

and hence

$$\boldsymbol{\rho} = \begin{bmatrix} 0.098 & & \\ & 2.83 & \\ & & 1.82 \end{bmatrix}$$

$$\bar{\boldsymbol{\Phi}} = \begin{bmatrix} 0.207 & -0.773 & 0.00690 \\ 0.181 & 0.0984 & -0.0655 \\ 0.509 & 1.47 & 0.443 \\ 0.594 & -0.385 & -0.166 \\ 0.655 & 0.574 & -0.978 \end{bmatrix}$$

The exact eigenvalues are

$$\lambda_1 = 0.098; \quad \lambda_2 = 0.824; \quad \lambda_3 = 2.00; \quad \lambda_4 = 3.18; \quad \lambda_5 = 3.90$$

and hence we note that we obtained in ρ_1 a good approximation to λ_1, but ρ_2 and ρ_3 do not represent approximations to eigenvalues.

10.4 SOLUTION ERRORS

An important part of an eigenvalue and vector solution is to estimate the accuracy with which the required eigensystem has been calculated. Since an eigensystem solution is necessarily iterative, the solution is terminated once convergence within the prescribed tolerances has been obtained. Although a good approximation to the required eigensystem should have been obtained at convergence, no definite conclusion can in most cases be drawn about the actual accuracy of the calculated values. Furthermore, in case one of the approximate solution techniques outlined in Section 10.3 is used, an estimate on the actual solution accuracy obtained is even more important.

In order to identify the accuracy that has been obtained in an eigensolution, we recall that the equation to be solved is

$$\mathbf{K}\boldsymbol{\phi} = \lambda\mathbf{M}\boldsymbol{\phi} \tag{10.96}$$

Assume that using any one solution procedure we obtained an approximation $\bar{\lambda}$ and $\bar{\boldsymbol{\phi}}$ to an eigenpair. It is important to note that at this stage it does not matter how the values $\bar{\lambda}$ and $\bar{\boldsymbol{\phi}}$ have been evaluated and indeed they may have been "guessed." Then, similar to the error analysis employed in the solution of the static equilibrium equations (see Section 7.5), we can evaluate a residual vector, $\mathbf{r}$, which gives important information about the accuracy with which $\bar{\lambda}$ and $\bar{\boldsymbol{\phi}}$ approximate the eigenpair, where

$$\mathbf{r} = \mathbf{K}\bar{\boldsymbol{\phi}} - \bar{\lambda}\mathbf{M}\bar{\boldsymbol{\phi}} \tag{10.97}$$

Let us assume first that $\mathbf{M} = \mathbf{I}$. In that case we can write, using the relations in (10.12) and (10.13),

$$\mathbf{r} = \boldsymbol{\Phi}(\boldsymbol{\Lambda} - \bar{\lambda}\mathbf{I})\boldsymbol{\Phi}^T\bar{\boldsymbol{\phi}} \tag{10.98}$$

or because $\bar{\lambda}$ is not equal but only close to an eigenvalue, we have

$$\bar{\boldsymbol{\phi}} = \boldsymbol{\Phi}(\boldsymbol{\Lambda} - \bar{\lambda}\mathbf{I})^{-1}\boldsymbol{\Phi}^T\mathbf{r} \tag{10.99}$$

Hence, because $||\bar{\boldsymbol{\phi}}||_2 = 1$, taking norms we obtain

$$1 \le ||(\boldsymbol{\Lambda} - \bar{\lambda}\mathbf{I})^{-1}||_2||\mathbf{r}||_2 \tag{10.100}$$

But since

$$||(\boldsymbol{\Lambda} - \bar{\lambda}\mathbf{I})^{-1}||_2 = \max_i \frac{1}{|\lambda_i - \bar{\lambda}|}$$

we have

$$\min_i |\lambda_i - \bar{\lambda}| \le ||\mathbf{r}||_2 \tag{10.101}$$

Therefore, a conclusive statement can be made about the accuracy with which $\bar{\lambda}$ approximates an eigenvalue λ_i by evaluating $\|\mathbf{r}\|_2$ as expressed in (10.101). This is quite different from the information that could be obtained from the evaluation of the residual vector **r** *in the solution of the static equilibrium equations.*

Although the relation in (10.101) establishes that $\bar{\lambda}$ is close to an eigenvalue provided that $\|\mathbf{r}\|_2$ is small, it should be recognized that the relation does not tell which eigenvalue is approximated. In fact, to identify which specific eigenvalue has been approximated, it is necessary to use the Sturm sequence property (see Section 10.2.2 and the following example).

Example 10.19

Consider the eigenproblem $\mathbf{K}\boldsymbol{\phi} = \lambda\boldsymbol{\phi}$, where

$$\mathbf{K} = \begin{bmatrix} 3 & -1 & 0 \\ -1 & 2 & -1 \\ 0 & -1 & 3 \end{bmatrix}$$

The eigensolution is $\lambda_1 = 1$, $\lambda_2 = 3$, $\lambda_3 = 4$, and

$$\boldsymbol{\phi}_1 = \frac{1}{\sqrt{6}}\begin{bmatrix} 1 \\ 2 \\ 1 \end{bmatrix}; \qquad \boldsymbol{\phi}_2 = \frac{1}{\sqrt{2}}\begin{bmatrix} 1 \\ 0 \\ -1 \end{bmatrix}; \qquad \boldsymbol{\phi}_3 = \frac{1}{\sqrt{3}}\begin{bmatrix} 1 \\ -1 \\ 1 \end{bmatrix}$$

Assume that we calculated

$$\bar{\lambda} = 3.1 \quad \text{and} \quad \bar{\boldsymbol{\phi}} = \begin{bmatrix} 0.7 \\ 0.1414 \\ -0.7 \end{bmatrix}$$

as approximations to λ_2 and $\boldsymbol{\phi}_2$. Apply the error bound relation in (10.101).

We have in this case

$$\mathbf{r} = \begin{bmatrix} 3 & -1 & 0 \\ -1 & 2 & -1 \\ 0 & -1 & 3 \end{bmatrix}\begin{bmatrix} 0.7 \\ 0.1414 \\ -0.7 \end{bmatrix} - 3.1\begin{bmatrix} 1 & & \\ & 1 & \\ & & 1 \end{bmatrix}\begin{bmatrix} 0.7 \\ 0.1414 \\ -0.7 \end{bmatrix}$$

Hence

$$\mathbf{r} = \begin{bmatrix} -0.2114 \\ -0.1555 \\ -0.2114 \end{bmatrix}; \qquad \|\mathbf{r}\|_2 = 0.3370$$

The relation in (10.101) now gives

$$|\lambda_2 - \bar{\lambda}| \leq 0.3370$$

which is indeed true because $\bar{\lambda} - \lambda_2 = 0.1$.

Assume now that we only have calculated $\bar{\lambda}$ and $\bar{\boldsymbol{\phi}}$ and do not know which eigenvalue and eigenvector they approximate. In this case we can use the relation in (10.101) to establish bounds on the unknown exact eigenvalue in order to apply Sturm sequence checks (see Section 10.2.2).

For the example considered here we have

$$2.7630 \leq \lambda_i \leq 3.4370$$

Let us use as a lower bound 2.7 and as an upper bound 3.5. The $\mathbf{LDL}^T$ triangular factorization of $\mathbf{K} - \mu\mathbf{I}$ gives, at $\mu = 2.7$,

$$\begin{bmatrix} 0.3 & -1 & 0 \\ -1 & -0.7 & -1 \\ 0 & -1 & 0.3 \end{bmatrix} = \begin{bmatrix} 1 & & \\ -3.333 & 1 & \\ 0 & 0.2479 & 1 \end{bmatrix} \begin{bmatrix} 0.3 & & \\ & -4.033 & \\ & & 0.3248 \end{bmatrix} \begin{bmatrix} 1 & -3.333 & 0 \\ & 1 & 0.2479 \\ & & 1 \end{bmatrix} \qquad \text{(a)}$$

and at $\mu = 3.5$,

$$\begin{bmatrix} -0.5 & -1 & 0 \\ -1 & -1.5 & -1 \\ 0 & -1 & -0.5 \end{bmatrix} = \begin{bmatrix} 1 & & \\ 2 & 1 & \\ 0 & -2 & 1 \end{bmatrix} \begin{bmatrix} -0.5 & & \\ & 0.5 & \\ & & -2.5 \end{bmatrix} \begin{bmatrix} 1 & 2 & 0 \\ & 1 & -2 \\ & & 1 \end{bmatrix} \qquad \text{(b)}$$

But there is one negative element in $\mathbf{D}$ of (a) and two negative elements in $\mathbf{D}$ of (b); hence we can conclude that $2.7 < \lambda_2 < 3.5$. Furthermore, it follows that $\bar{\lambda}$ and $\bar{\boldsymbol{\phi}}$ are approximations to λ_2 and $\boldsymbol{\phi}_2$.

Considering now the accuracy with which $\bar{\boldsymbol{\phi}}$ approximates an eigenvector, an analysis equivalent to the one above not only requires the evaluation of $\|\mathbf{r}\|_2$, but also the spacing between the individual eigenvalues is needed. In actual analysis this spacing is only known approximately, because the eigenvalues have only been evaluated to a specific accuracy.

Assume that $\bar{\lambda}$ and $\bar{\boldsymbol{\phi}}$ have been calculated, where $\|\bar{\boldsymbol{\phi}}\|_2 = 1$, and that $\bar{\lambda}$ approximates the eigenvalues λ_i, $i = p, \ldots, q$. For the error analysis we also assume that the eigenvalues λ_i for all i but $i \neq p, \ldots, q$ are known (although we would need to use here the calculated eigenvalues). The final result of the accuracy analysis is that if $|\lambda_i - \bar{\lambda}| \leq \|\mathbf{r}\|_2$ for $i = p, \ldots, q$ and $|\lambda_i - \bar{\lambda}| \geq s$ for all i, $i \neq p, \ldots, q$, then there is a vector $\tilde{\boldsymbol{\phi}} = \alpha_p\boldsymbol{\phi}_p + \ldots + \alpha_q\boldsymbol{\phi}_q$, for which $\|\bar{\boldsymbol{\phi}} - \tilde{\boldsymbol{\phi}}\|_2 \leq \|\mathbf{r}\|_2/s$ (see Example 10.20). Therefore, if $\bar{\lambda}$ is an approximation to a single eigenvalue λ_i, the corresponding vector $\bar{\boldsymbol{\phi}}$ is an approximation to $\boldsymbol{\phi}_i$, where

$$\|\bar{\boldsymbol{\phi}} - \alpha_i\boldsymbol{\phi}_i\|_2 \leq \frac{\|\mathbf{r}\|_2}{s}; \qquad s = \min_{\substack{\text{all } j \\ j \neq i}} |\lambda_j - \bar{\lambda}| \qquad (10.102)$$

However, if $\bar{\lambda}$ is close to a number of eigenvalues $\lambda_p, \ldots, \lambda_q$, then the analysis only shows that the corresponding vector $\bar{\boldsymbol{\phi}}$ is close to a vector that lies in the subspace corresponding to $\boldsymbol{\phi}_p, \ldots, \boldsymbol{\phi}_q$. In practical analysis (i.e., mode superposition in dynamic response calculations) this is most likely all that is required, because the close eigenvalues may almost be dealt with as equal eigenvalues, in which case the calculated eigenvectors would also not be unique but lie in the subspace corresponding to the equal eigenvalues. In the following we first give the proof for the accuracy with which $\bar{\boldsymbol{\phi}}$ approximates an eigenvector and then demonstrate the results by means of examples.

Example 10.20

Assume that we have calculated $\bar{\lambda}$, $\bar{\boldsymbol{\phi}}$, with $\|\bar{\boldsymbol{\phi}}\|_2 = 1$, as eigenvalue and eigenvector approximations and that $\mathbf{K}\bar{\boldsymbol{\phi}} - \bar{\lambda}\bar{\boldsymbol{\phi}} = \mathbf{r}$. Consider the case in which $|\lambda_i - \bar{\lambda}| \leq \|\mathbf{r}\|_2$ for $i = 1, \ldots, q$ and $|\lambda_i - \bar{\lambda}| \geq s$ for $i = q + 1, \ldots, n$. Show that $\|\bar{\boldsymbol{\phi}} - \tilde{\boldsymbol{\phi}}\|_2 \leq \|\mathbf{r}\|_2/s$, where $\tilde{\boldsymbol{\phi}}$ is a vector in the subspace that corresponds to $\boldsymbol{\phi}_1, \ldots, \boldsymbol{\phi}_q$.

The calculated eigenvector approximation $\bar{\boldsymbol{\phi}}$ can be written as

$$\bar{\boldsymbol{\phi}} = \sum_{i=1}^{n} \alpha_i \boldsymbol{\phi}_i$$

Using $\tilde{\boldsymbol{\phi}} = \sum_{i=1}^{q} \alpha_i \boldsymbol{\phi}_i$ we have

$$\|\bar{\boldsymbol{\phi}} - \tilde{\boldsymbol{\phi}}\|_2 = \left\| \sum_{i=q+1}^{n} \alpha_i \boldsymbol{\phi}_i \right\|_2$$

or, because $\boldsymbol{\phi}_i^T \boldsymbol{\phi}_j = \delta_{ij}$,

$$\|\bar{\boldsymbol{\phi}} - \tilde{\boldsymbol{\phi}}\|_2 = \left\{ \sum_{i=q+1}^{n} \alpha_i^2 \right\}^{1/2} \tag{a}$$

But

$$\|\mathbf{r}\|_2 = \|\mathbf{K}\bar{\boldsymbol{\phi}} - \bar{\lambda}\bar{\boldsymbol{\phi}}\|_2$$

$$= \left\| \sum_{i=1}^{n} \alpha_i (\lambda_i - \bar{\lambda}) \boldsymbol{\phi}_i \right\|_2$$

or

$$\|\mathbf{r}\|_2 = \left\{ \sum_{i=1}^{n} \alpha_i^2 (\lambda_i - \bar{\lambda})^2 \right\}^{1/2}$$

which gives

$$\|\mathbf{r}\|_2 \geq s \left\{ \sum_{i=q+1}^{n} \alpha_i^2 \right\}^{1/2} \tag{b}$$

Hence combining (a) and (b), we obtain

$$\|\bar{\boldsymbol{\phi}} - \tilde{\boldsymbol{\phi}}\|_2 \leq \frac{\|\mathbf{r}\|_2}{s}$$

Example 10.21

Consider the eigenproblem of Example 10.19. Assume that λ_1 and λ_3 are known (i.e., $\lambda_1 = 1$, $\lambda_3 = 4$) and that $\bar{\lambda}$ and $\bar{\boldsymbol{\phi}}$ given in Example 10.19 have been evaluated. (In actual analysis we would only have approximations to λ_1 and λ_3, and all error bound calculations would be approximate.) Estimate the accuracy with which $\bar{\boldsymbol{\phi}}$ approximates $\boldsymbol{\phi}_2$.

For the estimate we use the relation in (10.102). In this case we have

$$\|\bar{\boldsymbol{\phi}} - \alpha_2 \boldsymbol{\phi}_2\|_2 \leq \frac{0.3370}{s}$$

with

$$s = \min_{i=1,3} |\lambda_i - \bar{\lambda}|$$

Hence, since $\bar{\lambda} = 3.1$, we have $s = 0.9$ and

$$\|\bar{\boldsymbol{\phi}} - \alpha_2\boldsymbol{\phi}_2\|_2 \leq 0.3744$$

Evaluating $\|\bar{\boldsymbol{\phi}} - \boldsymbol{\phi}_2\|_2$ exactly, we have

$$\|\bar{\boldsymbol{\phi}} - \boldsymbol{\phi}_2\|_2 = \left\{\left(0.7 - \frac{1}{\sqrt{2}}\right)^2 + (0.1414 - 0)^2 + \left(-0.7 + \frac{1}{\sqrt{2}}\right)^2\right\}^{1/2} = 0.1418$$

Example 10.22

Consider the eigenproblem $\mathbf{K}\boldsymbol{\phi} = \lambda\boldsymbol{\phi}$, where

$$\mathbf{K} = \begin{bmatrix} 100 & -1 \\ -1 & 100 \end{bmatrix}$$

The eigenvalues and eigenvectors of the problem are

$$\lambda_1 = 99, \quad \boldsymbol{\phi}_1 = \frac{1}{\sqrt{2}}\begin{bmatrix} 1 \\ 1 \end{bmatrix}; \qquad \lambda_2 = 101, \quad \boldsymbol{\phi}_2 = \frac{1}{\sqrt{2}}\begin{bmatrix} 1 \\ -1 \end{bmatrix}$$

Assume that we have calculated eigenvalue and eigenvector approximations $\bar{\lambda} = 100$, $\bar{\boldsymbol{\phi}} = \begin{bmatrix} 1 \\ 0 \end{bmatrix}$. Evaluate $\mathbf{r}$ and thus establish the relations given in (10.101) and (10.102).

First we calculate $\mathbf{r}$ as given in (10.97),

$$\mathbf{r} = \begin{bmatrix} 100 & -1 \\ -1 & 100 \end{bmatrix}\begin{bmatrix} 1 \\ 0 \end{bmatrix} - 100\begin{bmatrix} 1 \\ 0 \end{bmatrix} = \begin{bmatrix} 0 \\ -1 \end{bmatrix}$$

Hence $\|\mathbf{r}\|_2 = 1$ and (10.101) yields

$$\min_i |\lambda_i - \bar{\lambda}| \leq 1 \qquad \text{(a)}$$

Therefore, we can conclude that an eigenvalue has been approximated with about 1% or less error. Since we know λ_1 and λ_2, we can compare $\bar{\lambda}$ with λ_1 or λ_2 and find that (a) does indeed hold.

Considering now the eigenvector approximation $\bar{\boldsymbol{\phi}}$, we note that $\bar{\boldsymbol{\phi}}$ does not approximate either $\boldsymbol{\phi}_1$ or $\boldsymbol{\phi}_2$. This is also reflected by evaluating the relation (10.102). Assuming that $\bar{\boldsymbol{\phi}}$ is an approximation to $\boldsymbol{\phi}_1$, which gives $s = 1$, we have

$$\|\bar{\boldsymbol{\phi}} - \alpha_1\boldsymbol{\phi}_1\|_2 \leq 1$$

Similarly, assuming that $\bar{\boldsymbol{\phi}}$ is an approximation to $\boldsymbol{\phi}_2$, we obtain

$$\|\bar{\boldsymbol{\phi}} - \alpha_2\boldsymbol{\phi}_2\|_2 \leq 1$$

and in both cases the bound obtained is very large (note that $\|\boldsymbol{\phi}_1\|_2 = 1$ and $\|\boldsymbol{\phi}_2\|_2 = 1$), indicating that $\bar{\boldsymbol{\phi}}$ does not approximate an eigenvector.

So far we dealt with the accuracy of an eigensolution obtained for the problem $\mathbf{K}\boldsymbol{\phi} = \lambda\boldsymbol{\phi}$. In order to estimate the accuracy obtained in the solution of a generalized eigenproblem $\mathbf{K}\boldsymbol{\phi} = \lambda\mathbf{M}\boldsymbol{\phi}$ we can use the results obtained by transforming the generalized eigenproblem to the standard form. Using the information given in Section 10.2.5, we can write the generalized problem

$\mathbf{K}\boldsymbol{\phi} = \lambda\mathbf{M}\boldsymbol{\phi}$ in the form $\tilde{\mathbf{K}}\tilde{\boldsymbol{\phi}} = \lambda\tilde{\boldsymbol{\phi}}$, where $\tilde{\boldsymbol{\phi}} = \mathbf{S}^T\boldsymbol{\phi}$. Assume that we calculated as an approximation to λ_i and $\boldsymbol{\phi}_i$ the values $\bar{\lambda}$ and $\bar{\boldsymbol{\phi}}$. Then, in analogy to the calculations performed above, we can evaluate an error vector, $\mathbf{r}_\mathbf{M}$, where

$$\mathbf{r}_\mathbf{M} = \mathbf{K}\bar{\boldsymbol{\phi}} - \bar{\lambda}\mathbf{M}\bar{\boldsymbol{\phi}} \tag{10.103}$$

In order to relate the error vector in (10.103) to the error vector that corresponds to the standard eigenproblem, we use that $\mathbf{M} = \mathbf{S}\mathbf{S}^T$, and then

$$\mathbf{r} = \tilde{\mathbf{K}}\tilde{\boldsymbol{\phi}} - \bar{\lambda}\tilde{\boldsymbol{\phi}} \tag{10.104}$$

where $\mathbf{r} = \mathbf{S}^{-1}\mathbf{r}_\mathbf{M}$, $\tilde{\boldsymbol{\phi}} = \mathbf{S}^T\bar{\boldsymbol{\phi}}$, and $\tilde{\mathbf{K}} = \mathbf{S}^{-1}\mathbf{K}\mathbf{S}^{-T}$. It is the vector $\mathbf{S}^{-1}\mathbf{r}_\mathbf{M}$ that we would need to use, therefore, to calculate an actual error bound, as given in (10.101). It should be noted that these error bound calculations would require the factorization of $\mathbf{M}$ into $\mathbf{S}\mathbf{S}^T$, where it is assumed that $\mathbf{M}$ is positive definite.

Although exact error bound calculations can be carried out as discussed above, in practical finite element analysis it is expedient to evaluate for the calculated eigenvalue and vector approximations $\bar{\lambda}$ and $\bar{\boldsymbol{\phi}}$ the following error measure (see Tables 12.1 and 12.3):

$$\epsilon = \frac{\|\mathbf{K}\bar{\boldsymbol{\phi}} - \bar{\lambda}\mathbf{M}\bar{\boldsymbol{\phi}}\|_2}{\|\mathbf{K}\bar{\boldsymbol{\phi}}\|_2} \tag{10.105}$$

Since, physically, $\mathbf{K}\bar{\boldsymbol{\phi}}$ represents the elastic nodal point forces and $\bar{\lambda}\mathbf{M}\bar{\boldsymbol{\phi}}$ represents the inertia nodal point forces when the finite element assemblage is vibrating in the mode $\bar{\boldsymbol{\phi}}$, we evaluate in (10.105) the norm of out-of-balance nodal point forces divided by the norm of elastic nodal point forces. This quantity should be small if $\bar{\lambda}$ and $\bar{\boldsymbol{\phi}}$ are an accurate solution of an eigenpair.

If $\mathbf{M} = \mathbf{I}$, it should be noted that we can write

$$\bar{\lambda}\epsilon = \|\mathbf{r}\|_2 \tag{10.106}$$

and hence

$$\epsilon \geq \min_i \frac{|\lambda_i - \bar{\lambda}|}{\bar{\lambda}} \tag{10.107}$$

Example 10.23

Consider the eigenproblem $\mathbf{K}\boldsymbol{\phi} = \lambda\mathbf{M}\boldsymbol{\phi}$, where

$$\mathbf{K} = \begin{bmatrix} 10 & -10 \\ -10 & 100 \end{bmatrix}; \qquad \mathbf{M} = \begin{bmatrix} 2 & 1 \\ 1 & 4 \end{bmatrix}$$

The exact eigenvalues and eigenvectors to 12-digit precision are

$$\lambda_1 = 3.863385512876; \qquad \boldsymbol{\phi}_1 = \begin{bmatrix} 0.640776011246 \\ 0.105070337503 \end{bmatrix}$$

$$\lambda_2 = 33.279471629982; \qquad \boldsymbol{\phi}_2 = \begin{bmatrix} -0.401041986380 \\ 0.524093989558 \end{bmatrix}$$

Assume that $\bar{\boldsymbol{\phi}} = (\boldsymbol{\phi}_1 + \delta\boldsymbol{\phi}_2)c$, where c is such that $\bar{\boldsymbol{\phi}}^T\mathbf{M}\bar{\boldsymbol{\phi}} = 1$ and $\delta = 10^{-1}, 10^{-3}$, and 10^{-6}. For each value of δ, evaluate $\bar{\lambda}$ as the Rayleigh quotient of $\bar{\boldsymbol{\phi}}$, and calculate the exact error bound on $|\lambda_1 - \bar{\lambda}|$ and the error value ϵ given in (10.105).

The following table summarizes the results obtained. The equations used to evaluate the error measures are given in (10.103) to (10.105). The results in the table show that for each value of δ, the relation in (10.101) is satisfied and that ϵ is also small for an accurate solution.

δ	10^{-1}	10^{-3}	10^{-6}
$\bar{\boldsymbol{\phi}}$	0.597690792656 0.156698194481	0.640374649073 0.105594378695	0.640775610204 0.105070861597
$\bar{\boldsymbol{\phi}}^T\mathbf{K}\bar{\boldsymbol{\phi}}$	4.154633890275	3.863414928932	3.863385512905
$\bar{\lambda}$	4.154633890275	3.863414928932	3.863385512905
$\mathbf{r}_M$	−1.207470493734 4.605630581124	−0.008218153965 0.049838803226	−0.000008177422 0.000049870085
$\mathbf{r}$	1.634419466242 1.411679295681	0.021106743617 0.015042545327	0.000021152364 0.000015049775
$\|\lambda_1 - \bar{\lambda}\|$	0.291248377399	0.000029416056	0.000000000029
$\|\|\mathbf{r}\|\|_2$	2.159667897036	0.025918580132	0.000025959936
ϵ	0.447113235813	0.007458208660	0.000007491764

REFERENCES

1. B. NOBLE, *Applied Linear Algebra*, Prentice-Hall, Inc., Englewood Cliffs, N.J., 1969.
2. J. H. WILKINSON, *The Algebraic Eigenvalue Problem*, Clarendon Press, Oxford, 1965.
3. R. ZURMÜHL, *Matrizen*, Springer-Verlag, Berlin, 1964.
4. H. R. SCHWARZ, H. RUTISHAUSER, and E. STIEFEL, *Matrizen-Numerik*, B. G. Teubner, Stuttgart, 1972.
5. C. LANCZOS, *Applied Analysis*, Prentice-Hall, Inc., Englewood Cliffs, N.J., 1956.
6. L. COLLATZ, *The Numerical Treatment of Differential Equations*, Springer-Verlag, New York, 1966.
7. S. H. CRANDALL, *Engineering Analysis*, McGraw-Hill Book Company, New York, 1956.

8. K. J. Bathe, "Solution Methods for Large Generalized Eigenvalue Problems in Structural Engineering," Report UC SESM 71-20, Civil Engineering Department, University of California, Berkeley, California.

9. E. L. Wilson, "Earthquake Analysis of Reactor Structures," A.S.M.E. First National Congress on Pressure Vessels and Piping Components, San Francisco, May 1971.

10. B. M. Irons, "Eigenvalue Economisers in Vibration Problems," *Journal of the Royal Aeronautical Society*, Vol. 67, 1963, p. 526.

11. B. M. Irons, "Structural Eigenvalue Problems: Elimination of Unwanted Variables," *A.I.A.A., Journal*, Vol. 3, No. 5, July 1965.

12. R. J. Guyan, "Reduction of Stiffness and Mass Matrices," *A.I.A.A. Journal*, Vol. 3, No. 2, 1965.

13. R. Uhrig, "Reduction of the Number of Unknowns in the Displacement Method Applied to Kinetic Problems," *Journal of Sound and Vibration*, Vol. 4, No. 2, 1966, pp. 149–155.

14. R. G. Anderson, B. M. Irons, and O. C. Zienkiewicz, "Vibration and Stability of Plates Using Finite Elements," *International Journal of Solids and Structures*, Vol. 4, 1968, pp. 1031–1055.

15. J. N. Ramsden and J. R. Stoker, "Mass Condensation; a Semi-automatic Method for Reducing the Size of Vibration Problems," *International Journal for Numerical Methods in Engineering*, Vol. 1, 1969, pp. 333–349.

16. R. W. Clough, "Analysis of Structural Vibrations and Dynamic Response," in *Recent Advances in Matrix Methods of Structural Analysis and Design* (R. H. Gallagher, Y. Yamada, and J. T. Oden, eds.), University of Alabama Press, University of Alabama, Huntsville, Ala., 1971.

17. G. C. Wright and G. A. Miles, "An Economical Method for Determining the Smallest Eigenvalues of Large Linear Systems," *International Journal for Numerical Methods in Engineering*, Vol. 3, 1971, pp. 25–33.

18. M. Geradin, "Error Bounds for Eigenvalue Analysis by Elimination of Variables," *Journal of Sound and Vibration*, Vol. 19, 1971, pp. 111–132.

19. W. L. Craver, Jr., and D. M. Egle, "A Method for Selection of Significant Terms in the Assumed Solution in a Rayleigh–Ritz Analysis," *Journal of Sound and Vibration*, Vol. 22, 1972, pp. 133–142.

20. W. C. Hurty, "Dynamic Analysis of Structural Systems by Component Modal Synthesis," Report 32-530, Jet Propulsion Laboratory Report, Pasadena, Calif., 1964.

21. W. C. Hurty, "Dynamic Analysis of Structural Systems Using Component Modes," *A.I.A.A. Journal*, Vol. 3, 1965, p. 678.

22. R. L. Goldman, "Vibration Analysis by Dynamic Partitioning, *A.I.A.A. Journal*, Vol. 7, No. 6, 1969.

23. S. Hon, "Review of Modal Synthesis Techniques and a New Approach," *Shock and Vibration Bulletin*, Vol. 40, No. 4, 1969.

24. R. L. BAJAN, C. C. FENG, and T. J. JASZLICS, "Vibration Analysis of Complex Structural Systems by Modal Substitution," *Shock and Vibration Bulletin*, Vol. 39, No. 3, 1969.

25. R. R. CRAIG and M. C. C. BAMPTON, "Coupling of Substructures for Dynamic Analysis," *A.I.A.A. Journal*, Vol. 6, No. 7, 1968.

26. R. L. SMITH, "Substructuring Techniques to Analyze Complex Structures," Graduate Student Report, Structural Engineering and Structural Mechanics, University of California, Department of Civil Engineering, Berkeley, California, 1974.

CHAPTER 11

SOLUTION METHODS FOR EIGENPROBLEMS

11.1 INTRODUCTION

In Chapter 10 we discussed the basic facts that are used in eigenvalue and vector solutions and some techniques for the calculation of approximations to the required eigensystem. The purpose of this and the next chapter is to present effective eigensolution techniques. The methods considered here are based on the fundamental aspects discussed in Chapter 10. Therefore, for a thorough understanding of the solution techniques to be presented in this chapter, it is necessary to be very familiar with the material discussed in Chapter 10. In addition, we also employ the notation that was defined in Chapter 10.

As before, we concentrate on the solution of the eigenproblem

$$\mathbf{K}\boldsymbol{\phi} = \lambda\mathbf{M}\boldsymbol{\phi} \tag{11.1}$$

and, in particular, on the calculation of the lowest eigenvalues $\lambda_1, \ldots, \lambda_q$ and corresponding eigenvectors $\boldsymbol{\phi}_1, \ldots, \boldsymbol{\phi}_q$. *The solution methods under consideration can be subdivided into four groups, corresponding to which basic property is used as the basis of the solution algorithm.*[1-11]

The vector iteration methods make up the first group, in which the basic property used is that

$$\mathbf{K}\boldsymbol{\phi}_i = \lambda_i\mathbf{M}\boldsymbol{\phi}_i \tag{11.2}$$

The transformation methods constitute the second group, using that

$$\boldsymbol{\Phi}^T\mathbf{K}\boldsymbol{\Phi} = \boldsymbol{\Lambda} \tag{11.3}$$

and

$$\mathbf{\Phi}^T\mathbf{M}\mathbf{\Phi} = \mathbf{I} \tag{11.4}$$

where $\mathbf{\Phi} = [\boldsymbol{\phi}_1, \ldots, \boldsymbol{\phi}_n]$ and $\mathbf{\Lambda} = \text{diag}(\lambda_1), i = 1, \ldots, n$. The solution methods of the third group are polynomial iteration techniques which operate on the fact that

$$p(\lambda_i) = 0 \tag{11.5}$$

where

$$p(\lambda) = \det(\mathbf{K} - \lambda\mathbf{M}) \tag{11.6}$$

The solution methods of the fourth group employ the Sturm sequence property of the characteristic polynomials

$$p(\lambda) = \det(\mathbf{K} - \lambda\mathbf{M}) \tag{11.7}$$

and

$$p^{(r)}(\lambda^{(r)}) = \det(\mathbf{K}^{(r)} - \lambda^{(r)}\mathbf{M}^{(r)}); \qquad r = 1, \ldots, n-1 \tag{11.8}$$

where $p^{(r)}(\lambda^{(r)})$ is the characteristic polynomial of the rth associated constraint problem corresponding to $\mathbf{K}\boldsymbol{\phi} = \lambda\mathbf{M}\boldsymbol{\phi}$.

In addition to using an algorithm that is based on the basic properties summarized above, techniques are under investigation which are based on minimizing the Rayleigh quotient[12-14] (see Section 2.6).

A number of solution algorithms have been developed within each of the above four groups of solution methods.[1-10] However, for an effective calculation of the required eigensystem of $\mathbf{K}\boldsymbol{\phi} = \lambda\mathbf{M}\boldsymbol{\phi}$, only a few techniques need to be considered. Corresponding to the four groups above, we present in Sections 11.2 to 11.5 those techniques that are most important in finite element analysis. In addition to those techniques that can be classified to fall into one of the four groups, we discuss in Chapter 12 the determinant search technique and the subspace iteration method, both of which use a combination of the fundamental properties given in (11.2) to (11.8).

Before presenting the solution techniques of interest, a few basic additional points should be noted. It is important to realize that *all solution methods must be iterative in nature, because, basically, solving the eigenvalue problem* $\mathbf{K}\boldsymbol{\phi} = \lambda\mathbf{M}\boldsymbol{\phi}$ *is equivalent to calculating the roots of the polynomial* $p(\lambda)$, *which has order equal to the order of* **K** *and* **M**. Since there are for the general case no explicit formulas available for the calculation of the roots of $p(\lambda)$ when the order of p is larger than 4, an iterative solution method has to be used. However, before iteration is started, we may choose to transform the matrices **K** and **M** into a form that allows a more economical solution of the required eigensystem (see Section 11.3.3).

Although iteration is needed in the solution of an eigenpair $(\lambda_i, \boldsymbol{\phi}_i)$, it should be noted that once one member of the eigenpair has been calculated, we can obtain the other member without further iteration. Assume that λ_i

has been evaluated by iteration; then we can obtain $\boldsymbol{\phi}_i$ using (11.2); i.e., $\boldsymbol{\phi}_i$ is calculated by solving

$$(\mathbf{K} - \lambda_i\mathbf{M})\boldsymbol{\phi}_i = \mathbf{0} \tag{11.9}$$

On the other hand, if we have evaluated $\boldsymbol{\phi}_i$ by iteration, we can obtain the required eigenvalue from the Rayleigh quotient; i.e., using (11.3) and (11.4), we have

$$\lambda_i = \boldsymbol{\phi}_i^T\mathbf{K}\boldsymbol{\phi}_i; \qquad \boldsymbol{\phi}_i^T\mathbf{M}\boldsymbol{\phi}_i = 1 \tag{11.10}$$

Therefore, when considering the design of an effective solution method, a basic question is whether we should first solve for the eigenvalue λ_i and then calculate the eigenvector $\boldsymbol{\phi}_i$, or vice versa, or whether it is most economical to solve for both λ_i and $\boldsymbol{\phi}_i$ simultaneously. The answer to this question depends on the solution requirements and the properties of the matrices **K** and **M**, i.e., such factors as the number of required eigenpairs, the order of **K** and **M**, the bandwidth of **K**, and whether **M** is banded.

The above discussion already leads us to realize that there is no single algorithm that always provides an efficient solution. Indeed, an algorithm that is effective in the solution of a specific problem may be totally inadequate for the solution of another problem. Therefore, an important aspect in the following sections is to give guidelines for the selection of the appropriate solution method corresponding to a given problem.

The effectiveness of a solution method depends largely on two factors: firstly, the possibility of a reliable use of the procedure and, secondly, the cost of solution. The solution cost is essentially determined by the number of high-speed storage operations required and the possibility of efficient use of back-up storage in the solution of out-of-core systems. The solution costs using the different algorithms under consideration are discussed in the following sections of this and the next chapter.

It is most important that a solution method can be used in a reliable manner. First, for well-defined structural stiffness and mass matrices, the solution to the problem $\mathbf{K}\boldsymbol{\phi} = \lambda\mathbf{M}\boldsymbol{\phi}$ must be obtained to the required precision without solution breakdown. However, almost equally important is a second aspect, namely that meaningful information is generated at an early stage of solution in case **K** and **M** are not well defined, for example, **K** is indefinite as a result of data input error. It may also be helpful if the solution method identifies at an early stage of solution a badly conditioned system. The importance of these aspects and practical considerations are pointed out when we discuss in Section 12.4 the selection of an appropriate solution routine corresponding to a given problem.

11.2 VECTOR ITERATION METHODS

As has been pointed out already, in the solution of an eigenvector or an eigenvalue we do need to use iteration. In Section 11.1 we classed the solution methods according to the basic relation upon which they operate. In the vector iteration methods the basic relation considered is[1-9]

$$\mathbf{K}\boldsymbol{\phi} = \lambda \mathbf{M}\boldsymbol{\phi} \tag{11.1}$$

The aim is to satisfy the equation in (11.1) by directly operating upon it. Consider that we assume a vector for $\boldsymbol{\phi}$, say $\mathbf{x}_1$, and assume a value for λ, say $\lambda = 1$. We can then evaluate the right-hand side of (11.1); i.e., we may calculate

$$\mathbf{R}_1 = (1)\mathbf{M}\mathbf{x}_1 \tag{11.11}$$

Since $\mathbf{x}_1$ is an arbitrarily assumed vector, we do not have, in general, that $\mathbf{K}\mathbf{x}_1 = \mathbf{R}_1$. If we had that $\mathbf{K}\mathbf{x}_1 = \mathbf{R}_1$, then $\mathbf{x}_1$ would be an eigenvector and, except for trivial cases, our assumptions would have been extremely lucky. Instead, we have an equilibrium equation as encountered in static analysis (see Section 7.2), which we may write

$$\mathbf{K}\mathbf{x}_2 = \mathbf{R}_1, \qquad \mathbf{x}_2 \neq \mathbf{x}_1 \tag{11.12}$$

where $\mathbf{x}_2$ is the displacement solution corresponding to the applied forces $\mathbf{R}_1$. Since we know that we have to use iteration to solve for an eigenvector, we may now intuitively feel that $\mathbf{x}_2$ may be a better approximation to an eigenvector than was $\mathbf{x}_1$. This is indeed the case, and by repeating the cycle we obtain an increasingly better approximation to an eigenvector.

The procedure described above is the basis of inverse iteration. We will see that other vector iteration techniques work in a similar way. Specifically, in forward iteration, the iterative cycle is reversed; i.e., in the first step we evaluate $\mathbf{R}_1 = \mathbf{K}\mathbf{x}_1$ and then obtain the improved approximation, $\mathbf{x}_2$, to an eigenvector by solving $\mathbf{M}\mathbf{x}_2 = \mathbf{R}_1$.

The basic steps of the vector iteration schemes and of the other solution methods that we consider later can be introduced using intuition. Namely, we need to satisfy one of the basic relations summarized in Section 11.1 and try to do so by some iterative cycle. However, the real justification for using any one of the methods derives from the fact that they do work and that they can be used economically.

11.2.1 Inverse Iteration

The technique of inverse iteration is very effectively used to calculate an eigenvector, and at the same time the corresponding eigenvalue can also be evaluated. Inverse iteration is employed in various important iteration procedures, including the determinant search and subspace iteration methods described in Sections 12.2 and 12.3. It is therefore important that we discuss the method in detail.

In this section we assume that $\mathbf{K}$ is positive definite, whereas $\mathbf{M}$ may be a diagonal mass matrix with or without zero diagonal elements or may be a banded mass matrix. If $\mathbf{K}$ is only positive semidefinite, a shift should be used prior to the iteration (see Section 11.2.3).

In the following we first consider the basic equations used in inverse iteration and then present a more effective form of the technique. In the solution we assume a starting iteration vector $\mathbf{x}_1$ and then evaluate in each iteration step $k = 1, 2, \ldots$:

$$\mathbf{K}\bar{\mathbf{x}}_{k+1} = \mathbf{M}\mathbf{x}_k \tag{11.13}$$

and

$$\mathbf{x}_{k+1} = \frac{\bar{\mathbf{x}}_{k+1}}{(\bar{\mathbf{x}}_{k+1}^T \mathbf{M} \bar{\mathbf{x}}_{k+1})^{1/2}} \tag{11.14}$$

where provided that $\mathbf{x}_1$ is not $\mathbf{M}$-orthogonal to $\boldsymbol{\phi}_1$, meaning that $\mathbf{x}_1^T \mathbf{M} \boldsymbol{\phi}_1 \neq 0$, we have that

$$\mathbf{x}_{k+1} \longrightarrow \boldsymbol{\phi}_1 \qquad \text{as } k \longrightarrow \infty$$

The basic step in the iteration is the solution of the equations in (11.13) in which we evaluate a vector $\bar{\mathbf{x}}_{k+1}$ with a direction closer to an eigenvector than the previous iteration vector $\mathbf{x}_k$. The calculation in (11.14) merely assures that the $\mathbf{M}$-weighted length of the new iteration vector $\mathbf{x}_{k+1}$ is unity; i.e., we want that $\mathbf{x}_{k+1}$ satisfies the mass orthonormality relation

$$\mathbf{x}_{k+1}^T \mathbf{M} \mathbf{x}_{k+1} = 1 \tag{11.15}$$

Substituting for $\mathbf{x}_{k+1}$ from (11.14) into (11.15), we find that (11.15) is indeed satisfied. If the scaling in (11.14) is not included in the iteration, the elements of the iteration vectors grow (or decrease) in each step and the iteration vectors do not converge to $\boldsymbol{\phi}_1$ but to a multiple of it. We illustrate the procedure by means of the following example.

Example 11.1

Consider the eigenproblem $\mathbf{K}\boldsymbol{\phi} = \lambda \mathbf{M} \boldsymbol{\phi}$, where

$$\mathbf{K} = \begin{bmatrix} 2 & -1 & 0 & 0 \\ -1 & 2 & -1 & 0 \\ 0 & -1 & 2 & -1 \\ 0 & 0 & -1 & 1 \end{bmatrix}; \qquad \mathbf{M} = \begin{bmatrix} 0 & & & \\ & 2 & & \\ & & 0 & \\ & & & 1 \end{bmatrix}$$

The eigenvalues and corresponding vectors of the problem have been evaluated in Examples 10.12 and 10.13. Use two steps of inverse iteration to evaluate an approximation to $\boldsymbol{\phi}_1$.

The first step is to decompose $\mathbf{K}$ into $\mathbf{L}\mathbf{D}\mathbf{L}^T$ in order to be able to solve the equations in (11.13). We obtained the triangular factors of $\mathbf{K}$ in Example 10.13.

As starting iteration vector we need a vector that is not orthogonal to $\boldsymbol{\phi}_1$. Since we do not know $\boldsymbol{\phi}_1$, we cannot make sure that $\boldsymbol{\phi}_1^T \mathbf{M} \mathbf{x}_1 \neq 0$, but we want to pick a vector that is not likely to be orthogonal to $\boldsymbol{\phi}_1$. Experience has shown that a good starting vector is in many cases a unit full vector (but see Example 11.6 for a case in which a unit full vector is a bad choice). In this example we use

$$\mathbf{x}_1^T = [1 \quad 1 \quad 1 \quad 1]$$

and then obtain, for $k = 1$,

$$\begin{bmatrix} 2 & -1 & 0 & 0 \\ -1 & 2 & -1 & 0 \\ 0 & -1 & 2 & -1 \\ 0 & 0 & -1 & 1 \end{bmatrix} \bar{\mathbf{x}}_2 = \begin{bmatrix} 0 & & & \\ & 2 & & \\ & & 0 & \\ & & & 1 \end{bmatrix} \begin{bmatrix} 1 \\ 1 \\ 1 \\ 1 \end{bmatrix}$$

Hence

$$\bar{\mathbf{x}}_2 = \begin{bmatrix} 3 \\ 6 \\ 7 \\ 8 \end{bmatrix}; \qquad \bar{\mathbf{x}}_2^T \mathbf{M} \bar{\mathbf{x}}_2 = 136$$

and

$$\mathbf{x}_2 = \frac{1}{\sqrt{136}} \begin{bmatrix} 3 \\ 6 \\ 7 \\ 8 \end{bmatrix}$$

Note that the zero diagonal elements in **M** do not introduce solution difficulties. Proceeding to the next iteration, $k = 2$, we use

$$\begin{bmatrix} 2 & -1 & 0 & 0 \\ -1 & 2 & -1 & 0 \\ 0 & -1 & 2 & -1 \\ 0 & 0 & -1 & 1 \end{bmatrix} \bar{\mathbf{x}}_3 = \begin{bmatrix} 0 & & & \\ & 2 & & \\ & & 0 & \\ & & & 1 \end{bmatrix} \begin{bmatrix} \frac{3}{\sqrt{136}} \\ \frac{6}{\sqrt{136}} \\ \frac{7}{\sqrt{136}} \\ \frac{8}{\sqrt{136}} \end{bmatrix}$$

Hence

$$\bar{\mathbf{x}}_3 = \frac{1}{\sqrt{136}} \begin{bmatrix} 20 \\ 40 \\ 48 \\ 56 \end{bmatrix}; \qquad \bar{\mathbf{x}}_3^T \mathbf{M} \bar{\mathbf{x}}_3 = \frac{6336}{136}$$

and

$$\mathbf{x}_3 = \frac{1}{\sqrt{6336}} \begin{bmatrix} 20 \\ 40 \\ 48 \\ 56 \end{bmatrix}$$

Comparing $\mathbf{x}_3$ with the exact solution (see Example 10.12) we have

$$\mathbf{x}_3 = \begin{bmatrix} 0.251 \\ 0.503 \\ 0.603 \\ 0.704 \end{bmatrix} \quad \text{and} \quad \boldsymbol{\phi}_1 = \begin{bmatrix} 0.250 \\ 0.500 \\ 0.602 \\ 0.707 \end{bmatrix}$$

Hence with only two iterations we have already obtained a fair approximation to $\boldsymbol{\phi}_1$.

The relations in (11.13) and (11.14) state the basic inverse iteration algorithm. However, in actual computer implementation it is more effective to iterate as follows. Assuming that $\mathbf{y}_1 = \mathbf{M}\mathbf{x}_1$, we evaluate for $k = 1, 2, \ldots,$

$$\mathbf{K}\bar{\mathbf{x}}_{k+1} = \mathbf{y}_k \tag{11.16}$$

$$\bar{\mathbf{y}}_{k+1} = \mathbf{M}\bar{\mathbf{x}}_{k+1} \tag{11.17}$$

$$\rho(\bar{\mathbf{x}}_{k+1}) = \frac{\bar{\mathbf{x}}_{k+1}^T \mathbf{y}_k}{\bar{\mathbf{x}}_{k+1}^T \bar{\mathbf{y}}_{k+1}} \tag{11.18}$$

$$\mathbf{y}_{k+1} = \frac{\bar{\mathbf{y}}_{k+1}}{(\bar{\mathbf{x}}_{k+1}^T \bar{\mathbf{y}}_{k+1})^{1/2}} \tag{11.19}$$

where, provided that $\mathbf{y}_1^T \boldsymbol{\phi}_1 \neq 0$,

$$\mathbf{y}_{k+1} \longrightarrow \mathbf{M}\boldsymbol{\phi}_1 \quad \text{and} \quad \rho(\bar{\mathbf{x}}_{k+1}) \longrightarrow \lambda_1 \qquad \text{as } k \longrightarrow \infty$$

It should be noted that we essentially dispense in (11.16) to (11.19) with the calculation of the matrix product $\mathbf{M}\mathbf{x}_k$ in (11.13) by iterating on $\mathbf{y}_k$ rather than on $\mathbf{x}_k$. But the value of $\bar{\mathbf{y}}_{k+1}$ is evaluated in either procedure; i.e., $\bar{\mathbf{y}}_{k+1}$ must be calculated in (11.14) and is evaluated in (11.17). Using the second iteration procedure, we obtain in (11.18) an approximation to the eigenvalue λ_1 given by the Rayleigh quotient $\rho(\bar{\mathbf{x}}_{k+1})$. It is this approximation to λ_1 which is conveniently used to determine convergence in the iteration. Denoting the current approximation to λ_1 by $\lambda_1^{(k+1)}$ [i.e., $\lambda_1^{(k+1)} = \rho(\bar{\mathbf{x}}_{k+1})$], we have convergence when

$$\frac{|\lambda_1^{(k+1)} - \lambda_1^{(k)}|}{\lambda_1^{(k+1)}} \leq \text{tol} \tag{11.20}$$

where tol should be 10^{-2s} or smaller when the eigenvalue λ_1 is required to $2s$-digit accuracy. The eigenvector will then be accurate to about s or more digits [see (11.33)]. Let l be the last iteration; then we have

$$\lambda_1 \doteq \rho(\bar{\mathbf{x}}_{l+1}) \tag{11.21}$$

and

$$\boldsymbol{\phi}_1 \doteq \frac{\bar{\mathbf{x}}_{l+1}}{(\bar{\mathbf{x}}_{l+1}^T \bar{\mathbf{y}}_{l+1})^{1/2}} \tag{11.22}$$

Example 11.2

Use the inverse iteration procedure given in (11.16) to (11.19) to evaluate an approximation to λ_1 and $\boldsymbol{\phi}_1$ of the eigenproblem $\mathbf{K}\boldsymbol{\phi} = \lambda\mathbf{M}\boldsymbol{\phi}$ considered in Example 11.1. Use tol $= 10^{-6}$ (i.e., $s = 3$) in (11.20), in order to measure convergence.

As in Example 11.1 we start the iteration with

$$\mathbf{x}_1 = \begin{bmatrix} 1 \\ 1 \\ 1 \\ 1 \end{bmatrix}$$

Proceeding as given in (11.16) to (11.19), we obtain for $k = 1$,

$$\mathbf{y}_1 = \begin{bmatrix} 0 \\ 2 \\ 0 \\ 1 \end{bmatrix}; \qquad \bar{\mathbf{x}}_2 = \begin{bmatrix} 3 \\ 6 \\ 7 \\ 8 \end{bmatrix}; \qquad \bar{\mathbf{y}}_2 = \begin{bmatrix} 0 \\ 12 \\ 0 \\ 8 \end{bmatrix}$$

$$\rho(\bar{\mathbf{x}}_2) = \frac{\bar{\mathbf{x}}_2^T \mathbf{y}_1}{\bar{\mathbf{x}}_2^T \bar{\mathbf{y}}_2} = 0.1470588$$

and

$$\mathbf{y}_2 = \begin{bmatrix} 0.0 \\ 1.02899 \\ 0.0 \\ 0.68599 \end{bmatrix}$$

TABLE 11.1

k	$\bar{\mathbf{x}}_{k+1}$	$\bar{\mathbf{y}}_{k+1}$	$\rho(\bar{\mathbf{x}}_{k+1})$	$\dfrac{\lvert \lambda_1^{(k+1)} - \lambda_1^{(k)} \rvert}{\lambda_1^{(k+1)}}$	$\mathbf{y}_k$
1	3 6 7 8	0 12 0 8	0.1470588	—	0 1.02899 0 0.68599
2	1.71499 3.42997 4.11597 4.80196	0 6.85994 0 4.80196	0.1464646	0.004056795132	0 1.00504 0 0.70353
3	1.70856 3.41713 4.12066 4.82418	0 6.83426 0 4.82418	0.1464471	0.000119538581	0 1.00087 0 0.70649
4	1.70736 3.41472 4.12121 4.82771	0 6.82944 0 4.82771	0.1464466	0.000003518989	0 1.00015 0 0.70700
5	1.70715 3.41430 4.12130 4.82830	0 6.82860 0 4.82830	0.1464466	0.000000103589	0 1.00003 0 0.70709

The next iterations are carried out in the same way. The results are summarized in Table 11.1. It is seen that after five iterations, convergence has been achieved. It should be noted that the Rayleigh quotient $\rho(\bar{\mathbf{x}}_{k+1})$ converges much faster than the vector $\bar{\mathbf{x}}_{k+1}$ (see Example 11.3) and converges from above to λ_1. Using (11.21) and (11.22), we have

$$\lambda_1 \doteq 0.146447; \qquad \boldsymbol{\phi}_1 \doteq \begin{bmatrix} 0.25001 \\ 0.50001 \\ 0.60355 \\ 0.70709 \end{bmatrix}$$

In the above discussion we have merely stated the iteration scheme and its convergence. We then applied the method in two examples, but did not formally prove convergence. In the following we derive the convergence properties, because we believe that the proof is very instructive.

The first step in the proof of convergence and of the convergence rate given here is similar to the procedure used in the analysis of direct integration methods (see Chapter 9). The fundamental equation used in inverse iteration is the relation in (11.13). Neglecting the scaling of the elements in the iteration vector, we basically use for $k = 1, 2, \ldots,$

$$\mathbf{K}\mathbf{x}_{k+1} = \mathbf{M}\mathbf{x}_k \tag{11.23}$$

where we stated that $\mathbf{x}_{k+1}$ will now converge to a multiple of $\boldsymbol{\phi}_1$. To show convergence it is convenient (as in the analysis of direct integration procedures) to change basis from the finite element coordinate basis to the basis of eigenvectors; namely, we can write for any iteration vector $\mathbf{x}_k$,

$$\mathbf{x}_k = \boldsymbol{\Phi}\mathbf{z}_k \tag{11.24}$$

where $\boldsymbol{\Phi}$ is the matrix of eigenvectors $\boldsymbol{\Phi} = [\boldsymbol{\phi}_1, \ldots, \boldsymbol{\phi}_n]$. It should be realized that because $\boldsymbol{\Phi}$ is nonsingular, there is a unique vector $\mathbf{z}_k$ for any vector $\mathbf{x}_k$. Substituting for $\mathbf{x}_k$ and $\mathbf{x}_{k+1}$ from (11.24) into (11.23), premultiplying by $\boldsymbol{\Phi}^T$ and using the orthogonality relations $\boldsymbol{\Phi}^T\mathbf{K}\boldsymbol{\Phi} = \boldsymbol{\Lambda}$ and $\boldsymbol{\Phi}^T\mathbf{M}\boldsymbol{\Phi} = \mathbf{I}$, we obtain

$$\boldsymbol{\Lambda}\mathbf{z}_{k+1} = \mathbf{z}_k \tag{11.25}$$

where $\boldsymbol{\Lambda} = \operatorname{diag}(\lambda_i)$. Comparing (11.25) with (11.23) we find that the iterations are of the same form with $\mathbf{K} = \boldsymbol{\Lambda}$ and $\mathbf{M} = \mathbf{I}$. We may wonder why the transformation in (11.24) is used, since $\boldsymbol{\Phi}$ is unknown. However, we should realize that the transformation is only employed to investigate the convergence behavior of inverse iteration. Namely, because *in theory* (11.25) is equivalent to (11.23), the convergence properties of (11.25) are also those of (11.23). But the convergence characteristics of (11.25) are relatively easy to investigate, since the eigenvalues are the diagonal elements of $\boldsymbol{\Lambda}$ and the eigenvectors are the unit vectors $\mathbf{e}_i$, where

$$\mathbf{e}_i^T = [0 \ldots 0 \quad \underset{\substack{\uparrow \\ i\text{th location}}}{1} \quad 0 \ldots 0] \tag{11.26}$$

In the presentation of the inverse iteration algorithms given in (11.13) to (11.14) and (11.16) to (11.22), we stated that the starting iteration vector $\mathbf{x}_1$ must not be **M**-orthogonal to $\boldsymbol{\phi}_1$. Equivalently, in (11.25) the iteration vector $\mathbf{z}_1$ must not be orthogonal to $\mathbf{e}_1$. Assume that we use

$$\mathbf{z}_1^T = [1 \quad 1 \quad 1 \dots 1] \tag{11.27}$$

We discuss the effect of this assumption in Section 11.2.6. Then using (11.25) for $k = 1, \dots, l$, we obtain

$$\mathbf{z}_{l+1}^T = \begin{bmatrix} \frac{1}{\lambda_1^l} & \frac{1}{\lambda_2^l} & \cdots & \frac{1}{\lambda_n^l} \end{bmatrix} \tag{11.28}$$

Let us first assume that $\lambda_1 < \lambda_2$. To show that $\mathbf{z}_{l+1}$ converges to a multiple of $\mathbf{e}_1$ as $l \longrightarrow \infty$, we multiply $\mathbf{z}_{l+1}$ in (11.28) by λ_1^l to obtain

$$\bar{\mathbf{z}}_{l+1} = \begin{bmatrix} 1 \\ (\lambda_1/\lambda_2)^l \\ \cdot \\ \cdot \\ \cdot \\ (\lambda_1/\lambda_n)^l \end{bmatrix} \tag{11.29}$$

and observe that $\bar{\mathbf{z}}_{l+1}$ converges to $\mathbf{e}_1$ as $l \longrightarrow \infty$. Hence $\mathbf{z}_{l+1}$ converges to a multiple of $\mathbf{e}_1$ as $l \longrightarrow \infty$.

To evaluate the order and rate of convergence, we use the convergence definition given in Section 2.8. For the iteration here under consideration we obtain

$$\lim_{l \to \infty} \frac{\| \bar{\mathbf{z}}_{l+1} - \mathbf{e}_1 \|_2}{\| \bar{\mathbf{z}}_l - \mathbf{e}_1 \|_2} = \frac{\lambda_1}{\lambda_2} \tag{11.30}$$

Hence convergence is linear and the rate of convergence is λ_1/λ_2. This convergence rate is also shown in the iteration vector $\bar{\mathbf{z}}_{l+1}$ in (11.29); i.e., those elements in the iteration vector that should tend to zero do so with at least the ratio λ_1/λ_2 in each additional iteration. Thus if $\lambda_2 > \lambda_1$, it is the relative magnitude of λ_1 to λ_2 that determines how fast the iteration vector converges to the eigenvector $\boldsymbol{\phi}_1$.

In the above discussion we assumed that $\lambda_1 < \lambda_2$. Let us now consider the case of a multiple eigenvalue, namely $\lambda_1 = \lambda_2 = \dots = \lambda_m$. Then we have in (11.29),

$$\bar{\mathbf{z}}_{l+1}^T = \begin{bmatrix} 1 & 1 & \dots & 1 & \left(\frac{\lambda_1}{\lambda_{m+1}}\right)^l & \cdots & \left(\frac{\lambda_1}{\lambda_n}\right)^l \end{bmatrix} \tag{11.31}$$

and the convergence rate of the iteration vector is λ_1/λ_{m+1}. Therefore, in general, the rate of convergence of the iteration vector in inverse iteration is given by the ratio of λ_1 to the next distinct eigenvalue.

In the iteration given in (11.16) to (11.22), we obtain an approximation to the eigenvalue λ_1 by evaluating the Rayleigh quotient. Corresponding to

(11.18) the Rayleigh quotient calculated in the iteration of (11.25) would be

$$\rho(\mathbf{z}_{k+1}) = \frac{\mathbf{z}_{k+1}^T \mathbf{z}_k}{\mathbf{z}_{k+1}^T \mathbf{z}_{k+1}} \tag{11.32}$$

Assume that we consider the last iteration in which $k = l$. Then substituting for $\mathbf{z}_l$ and $\mathbf{z}_{l+1}$ from (11.28) into (11.32), we obtain

$$\rho(\mathbf{z}_{l+1}) = \frac{\lambda_1 \sum_{i=1}^{n} (\lambda_1/\lambda_i)^{2l-1}}{\sum_{i=1}^{n} (\lambda_1/\lambda_i)^{2l}} \tag{11.33}$$

Hence we have for λ_1 being a simple or multiple eigenvalue,

$$\rho(\mathbf{z}_{l+1}) \longrightarrow \lambda_1 \qquad \text{as } l \longrightarrow \infty$$

Also, convergence is linear with the rate equal to $(\lambda_1/\lambda_{m+1})^2$, where λ_{m+1} is defined as in (11.31). This convergence rate substantiates the observation that if an eigenvector is known with an error ϵ, then the Rayleigh quotient yields an approximation to the corresponding eigenvalue with error ϵ^2 (see Section 2.7).

Before demonstrating the results by means of a small example, it should be recalled that we assumed in the above analysis a full unit starting iteration vector as given in (11.27). The convergence properties derived hold for any starting iteration vector that is not orthogonal to the eigenvector of interest, but the convergence rates can in many practical analyses only be observed as the number of iterations becomes large. The same observation also holds for any of the other convergence analyses that are presented in the following sections. We discuss this observation with other important practical aspects in Section 11.2.6.

Example 11.3

For the problem considered in Example 11.2, calculate the ultimate convergence rates of the iteration vector and the Rayleigh quotient. Compare the ultimate convergence rates with those actually observed in the inverse iteration carried out in Example 11.2.

For the evaluation of the theoretical convergence rates, we need λ_1 and λ_2. We calculated the eigenvalues in Example 10.12 and found

$$\lambda_1 = \frac{1}{2} - \frac{\sqrt{2}}{4}$$

$$\lambda_2 = \frac{1}{2} + \frac{\sqrt{2}}{4}$$

Hence the ultimate convergence rate of the iteration vector is

$$\frac{\lambda_1}{\lambda_2} = 0.17$$

and the ultimate convergence rate of the Rayleigh quotient is

$$\left(\frac{\lambda_1}{\lambda_2}\right)^2 = 0.029$$

The actual vector convergence obtained is observed by evaluating the ratio r_{k+1}, $k = 1, 2, \ldots$, where

$$r_{k+1} = \frac{\|\mathbf{x}_{k+1} - \boldsymbol{\phi}_1\|_2}{\|\mathbf{x}_k - \boldsymbol{\phi}_1\|_2}$$

and we assume that $\boldsymbol{\phi}_1$ is obtained in the last iteration [see (11.22)].

For the iteration in Example 11.2, we thus obtain

$$r_2 = 0.026083; \qquad r_3 = 0.170559; \qquad r_4 = 0.167134; \qquad r_5 = 0.144251$$

Ignoring r_2 because the iteration just started, we see that the theoretical and actual convergence rates compare quite well.

Similarly, the actual convergence of the Rayleigh quotient calculated in Example 11.2 is observed by evaluating

$$\epsilon_{k+1} = \frac{|\rho(\bar{\mathbf{x}}_{k+1}) - \lambda_1|}{|\rho(\bar{\mathbf{x}}_k) - \lambda_1|}$$

where we use the converged value of the Rayleigh quotient for λ_1. In the iteration of Example 11.2 we have

$$\epsilon_3 = 0.028768; \qquad \epsilon_4 = 0.027778; \qquad \epsilon_5 = 0$$

Hence, we see that the theoretical and observed convergence rates agree again quite well in this solution.

11.2.2 Forward Iteration

The method of forward iteration is complementary to the inverse iteration technique, in that the method yields the eigenvector corresponding to the largest eigenvalue. Whereas we assumed in inverse iteration that $\mathbf{K}$ be positive definite, we assume in this section that $\mathbf{M}$ is positive definite; otherwise, a shift must be used (see Section 11.2.3). Having chosen a starting iteration vector $\mathbf{x}_1$, in forward iteration we evaluate, for $k = 1, 2, \ldots,$

$$\mathbf{M}\bar{\mathbf{x}}_{k+1} = \mathbf{K}\mathbf{x}_k \tag{11.34}$$

and

$$\mathbf{x}_{k+1} = \frac{\bar{\mathbf{x}}_{k+1}}{(\bar{\mathbf{x}}_{k+1}^T\mathbf{M}\bar{\mathbf{x}}_{k+1})^{1/2}} \tag{11.35}$$

where provided that $\mathbf{x}_1$ is not $\mathbf{M}$-orthogonal to $\boldsymbol{\phi}_n$, we have that

$$\mathbf{x}_{k+1} \longrightarrow \boldsymbol{\phi}_n \qquad \text{as } k \longrightarrow \infty$$

The analogy to inverse iteration should be noted; the only difference is that we solve (11.34) rather than (11.13) to obtain an improved eigenvector. This means, in practice, that in the inverse iteration we need to triangularize the matrix $\mathbf{K}$ and in the forward iteration we decompose $\mathbf{M}$.

A more effective forward iteration procedure than in (11.34) and (11.35) would be obtained by using equations that are analogous to those in (11.16) to (11.22). Assuming that $\mathbf{y}_1 = \mathbf{K}\mathbf{x}_1$, we evaluate for $k = 1, 2, \ldots,$

$$\mathbf{M}\bar{\mathbf{x}}_{k+1} = \mathbf{y}_k \tag{11.36}$$

$$\bar{\mathbf{y}}_{k+1} = \mathbf{K}\bar{\mathbf{x}}_{k+1} \tag{11.37}$$

$$\rho(\bar{\mathbf{x}}_{k+1}) = \frac{\bar{\mathbf{x}}_{k+1}^T \bar{\mathbf{y}}_{k+1}}{\bar{\mathbf{x}}_{k+1}^T \mathbf{y}_k} \tag{11.38}$$

$$\mathbf{y}_{k+1} = \frac{\bar{\mathbf{y}}_{k+1}}{(\bar{\mathbf{x}}_{k+1}^T \mathbf{y}_k)^{1/2}} \tag{11.39}$$

where provided that $\boldsymbol{\phi}_n^T \mathbf{y}_1 \neq 0$,

$$\mathbf{y}_{k+1} \longrightarrow \mathbf{K}\boldsymbol{\phi}_n \quad \text{and} \quad \rho(\bar{\mathbf{x}}_{k+1}) \longrightarrow \lambda_n \qquad \text{as } k \longrightarrow \infty$$

Convergence in the iteration could again be measured as given in (11.20), and denoting by l the last iteration, we have

$$\lambda_n \doteq \rho(\bar{\mathbf{x}}_{l+1}) \tag{11.40}$$

and

$$\boldsymbol{\phi}_n \doteq \frac{\bar{\mathbf{x}}_{l+1}}{(\bar{\mathbf{x}}_{l+1}^T \mathbf{y}_l)^{1/2}} \tag{11.41}$$

Considering the analysis of convergence of the iteration vector to $\boldsymbol{\phi}_n$, it can be carried out following the same procedure that was used in the evaluation of the convergence characteristics of inverse iteration. Alternatively, we may use the results that we obtained in the analysis of inverse iteration. Namely, assume that we write the eigenproblem $\mathbf{K}\boldsymbol{\phi} = \lambda\mathbf{M}\boldsymbol{\phi}$ in the form $\mathbf{M}\boldsymbol{\phi} = \lambda^{-1}\mathbf{K}\boldsymbol{\phi}$; then using inverse iteration to solve for an eigenvector and corresponding eigenvalue is equivalent to performing forward iteration on the problem $\mathbf{K}\boldsymbol{\phi} = \lambda\mathbf{M}\boldsymbol{\phi}$. But since we converge in the inverse iteration of (11.16) to (11.22) to the smallest eigenvalue and corresponding eigenvector, and since for the problem $\mathbf{M}\boldsymbol{\phi} = \lambda^{-1}\mathbf{K}\boldsymbol{\phi}$, this eigenvalue is λ_n^{-1}, where λ_n is the largest eigenvalue of $\mathbf{K}\boldsymbol{\phi} = \lambda\mathbf{M}\boldsymbol{\phi}$, we converge in the forward iteration of (11.36) to (11.41) to λ_n and $\boldsymbol{\phi}_n$ and the convergence rate of the iteration vector is λ_{n-1}/λ_n. We should note that the Rayleigh quotient evaluated in (11.38) is $(\bar{\mathbf{x}}_{k+1}^T\mathbf{K}\bar{\mathbf{x}}_{k+1})/(\bar{\mathbf{x}}_{k+1}^T\mathbf{M}\bar{\mathbf{x}}_{k+1})$, i.e., just the inverse of the Rayleigh quotient to calculate an approximation to λ_n^{-1} in the problem $\mathbf{M}\boldsymbol{\phi} = \lambda^{-1}\mathbf{K}\boldsymbol{\phi}$.

We demonstrate the iteration and convergence in the following example.

Example 11.4

Use forward iteration as given in (11.36) to (11.41) with tol 10^{-6} in (11.20) to evaluate λ_4 and $\boldsymbol{\phi}_4$ of the eigenproblem $\mathbf{K}\boldsymbol{\phi} = \lambda\mathbf{M}\boldsymbol{\phi}$, where

$$\mathbf{K} = \begin{bmatrix} 5 & -4 & 1 & 0 \\ -4 & 6 & -4 & 1 \\ 1 & -4 & 6 & -4 \\ 0 & 1 & -4 & 5 \end{bmatrix}; \qquad \mathbf{M} = \begin{bmatrix} 2 & & & \\ & 2 & & \\ & & 1 & \\ & & & 1 \end{bmatrix}$$

The physical problem considered in this example is the free vibration characteristics of the simply supported beam shown in Fig. 7.1 with the above mass matrix.

Starting the iteration with

$$\mathbf{x}_1 = \begin{bmatrix} 1 \\ 1 \\ 1 \\ 1 \end{bmatrix}$$

we calculate in the inverse iteration the values summarized in Table 11.2.

TABLE 11.2

k	$\bar{\mathbf{x}}_{k+1}$	$\bar{\mathbf{y}}_{k+1}$	$\rho(\bar{\mathbf{x}}_{k+1})$	$\mathbf{y}_{k+1}$	$\dfrac{\lvert \lambda_1^{(k+1)} - \lambda_1^{(k)} \rvert}{\lambda_1^{(k+1)}}$
1	1	6	5.93333	2.1909	—
	−0.5	−1		−0.3652	
	−1	−11		−4.0166	
	2	13.5		4.9295	
2	1.0955	2.1909	8.57886	0.3345	0.3084
	−0.1826	15.5188		2.3694	
	−4.0166	−41.9921		−6.4112	
	4.9295	40.5315		6.1882	
3	0.1672	−10.3137	10.15966	−1.1372	0.1556
	1.1847	38.2720		4.2198	
	−6.4112	−67.7914		−7.4748	
	6.1882	57.7704		6.3696	
4	−0.5686	−18.7569	10.55204	−1.8219	0.03719
	2.1099	51.2010		4.9733	
	−7.4745	−79.3332		−7.7059	
	6.3696	63.8557		6.2025	
5	−0.9110	−22.2074	10.62367	−2.0995	0.006743
	2.4867	55.5901		5.2556	
	−7.7059	−89.9033		−7.7433	
	6.2025	64.3230		6.0812	
6	−1.0498	−23.5033	10.63595	−2.2115	0.001154
	2.6278	57.0203		5.3651	
	−7.7433	−82.3457		−7.7480	
	6.0812	64.0072		6.0225	
7	−1.1057	−24.0068	10.63802	−2.2570	0.0001954
	2.6826	57.5326		5.4089	
	−7.7479	−82.4138		−7.7481	
	6.0225	63.7869		5.9969	
8	−1.1285	−24.2083	10.63838	−2.2756	0.00003304
	2.7044	57.7298		5.4267	
	−7.7481	−82.4222		−7.7478	
	5.9969	68.6811		5.9861	
9	−1.1378	−24.2902	10.63844	−2.2833	0.000005584
	2.7133	57.8086		5.4340	
	−7.7478	−82.4224		−7.7476	
	5.9861	63.6351		5.9816	
10	−1.1416	−24.3237	10.63845	−2.2864	0.0000009437
	2.7170	57.8405		5.4369	
	−7.7476	−82.4219		−7.7476	
	5.9816	63.6157		5.9898	

Hence, we need 10 iterations for a convergence tolerance of 10^{-6} in (11.20), and we then use, as given in (11.40) and (11.41),

$$\lambda_4 \doteq 10.63845; \qquad \boldsymbol{\phi}_4 \doteq \begin{bmatrix} -0.10731 \\ 0.25539 \\ -0.72827 \\ 0.56227 \end{bmatrix}$$

11.2.3 Shifting in Vector Iteration

The convergence analysis of inverse iteration in Section 11.2.1 showed that assuming $\lambda_1 < \lambda_2$, the iteration vector converges with a rate λ_1/λ_2 to the eigenvector $\boldsymbol{\phi}_1$. Therefore, depending on the magnitude of λ_1 and λ_2, the convergence rate can be arbitrarily low, say $\lambda_1/\lambda_2 = 0.99999$, or can be very high, say $\lambda_1/\lambda_2 = 0.01$. Similarly, in forward iteration the convergence rate can be low or high. Therefore, a natural question must be how to improve the convergence rate in the vector iterations. We show in this section that the convergence rate can be much improved by shifting. In addition, a shift can be used to obtain convergence to an eigenpair other than $(\lambda_1, \boldsymbol{\phi}_1)$ and $(\lambda_n, \boldsymbol{\phi}_n)$ in inverse and forward iterations, respectively, and a shift is used effectively in inverse iteration when $\mathbf{K}$ is positive semidefinite and in forward iteration when $\mathbf{M}$ is diagonal with some zero diagonal elements (see Example 11.6).

Assume that a shift μ is applied as described in Section 10.2.3; then we consider the eigenproblem

$$(\mathbf{K} - \mu\mathbf{M})\boldsymbol{\phi} = \eta\mathbf{M}\boldsymbol{\phi} \tag{11.42}$$

where the eigenvalues of the original problem $\mathbf{K}\boldsymbol{\phi} = \lambda\mathbf{M}\boldsymbol{\phi}$ and of the problem in (11.42) are related by $\eta_i = \lambda_i - \mu$, $i = 1, \ldots, n$. To analyze the convergence properties of inverse and forward iteration when applied to the problem in (11.42), we follow in all respects the procedure used in Section 11.2.1. The first step is to consider the problem in the basis of eigenvectors $\boldsymbol{\Phi}$. Using the transformation

$$\boldsymbol{\phi} = \boldsymbol{\Phi}\boldsymbol{\psi} \tag{11.43}$$

we obtain for the convergence analysis the equivalent eigenproblem

$$(\boldsymbol{\Lambda} - \mu\mathbf{I})\boldsymbol{\psi} = \eta\boldsymbol{\psi} \tag{11.44}$$

Consider first inverse iteration and assume that all eigenvalues are distinct. In that case we obtain, using the notation in Section 11.2.1,

$$\mathbf{z}_{l+1}^T = \begin{bmatrix} \dfrac{1}{(\lambda_1 - \mu)^l} & \dfrac{1}{(\lambda_2 - \mu)^l} & \cdots & \dfrac{1}{(\lambda_n - \mu)^l} \end{bmatrix} \tag{11.45}$$

where it is assumed that all $(\lambda_i - \mu)$ are nonzero, but they may be positive or negative. Assume that $(\lambda_i - \mu)$ is smallest in absolute magnitude when $i = j$; then multiplying $\mathbf{z}_{l+1}$ by $(\lambda_j - \mu)^l$, we obtain

$$\bar{\mathbf{z}}_{l+1} = \begin{bmatrix} \left(\dfrac{\lambda_j - \mu}{\lambda_1 - \mu}\right)^l \\ \cdot \\ \cdot \\ \cdot \\ \left(\dfrac{\lambda_j - \mu}{\lambda_{j-1} - \mu}\right)^l \\ 1 \\ \left(\dfrac{\lambda_j - \mu}{\lambda_{j+1} - \mu}\right)^l \\ \cdot \\ \cdot \\ \cdot \\ \left(\dfrac{\lambda_j - \mu}{\lambda_n - \mu}\right)^l \end{bmatrix} \tag{11.46}$$

where $|(\lambda_j - \mu)/(\lambda_p - \mu)| < 1$ for all $p \neq j$. Hence, in the iteration we have $\bar{\mathbf{z}}_{l+1} \longrightarrow \mathbf{e}_j$, meaning that in inverse iteration to solve (11.42), the iteration vector converges to $\boldsymbol{\phi}_j$. Furthermore, we obtain $\lambda_j = \eta_j + \mu$. The convergence rate in the iteration is determined by the element $(\lambda_j - \mu)/(\lambda_p - \mu)$, which is largest in absolute magnitude, $p \neq j$; i.e., the convergence rate, r, is

$$r = \max_{p \neq j} \left| \frac{\lambda_j - \mu}{\lambda_p - \mu} \right| \tag{11.47}$$

Since λ_j is nearest μ, the convergence rate of the iteration vector in (11.42) to the eigenvector $\boldsymbol{\phi}_j$ is either

$$\left| \frac{\lambda_j - \mu}{\lambda_{j-1} - \mu} \right| \quad \text{or} \quad \left| \frac{\lambda_j - \mu}{\lambda_{j+1} - \mu} \right|$$

whichever is larger. The convergence rate for a typical case is shown in Fig. 11.1.

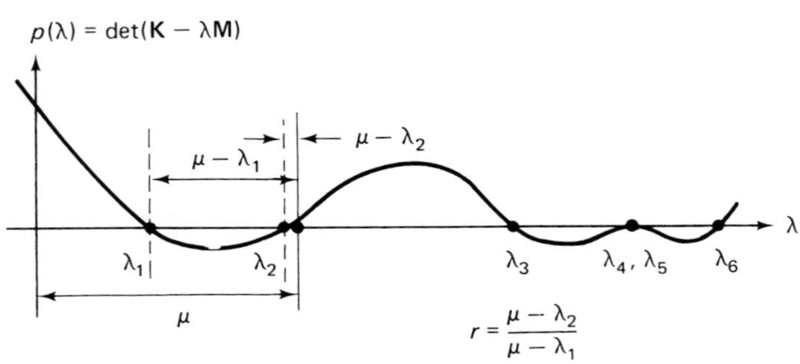

Figure 11.1. Example of vector convergence rate r in inverse iteration.

Using the results of the above convergence analysis and of the analysis of inverse iteration without shifting (see Section 11.2.1), two additional conclusions are reached. Firstly, we observe that the convergence rate of the Rayleigh quotient, which for μ nearest to λ_j converges to $(\lambda_j - \mu)$, is

$$\left|\frac{\lambda_j - \mu}{\lambda_{j-1} - \mu}\right|^2 \quad \text{or} \quad \left|\frac{\lambda_j - \mu}{\lambda_{j+1} - \mu}\right|^2$$

whichever is larger.

The second observation concerns the case λ_j being a multiple eigenvalue. The analysis in Section 11.2.1 and the conclusions above show that if $\lambda_j = \lambda_{j+1} = \ldots = \lambda_{j+m-1}$, the rate of convergence of the iteration vector is

$$\max_{p \neq j, j+1, \ldots, j+m-1} \left|\frac{\lambda_j - \mu}{\lambda_p - \mu}\right|$$

and convergence occurs to a vector in the subspace corresponding to λ_j.

The important point in inverse iteration with shifting is that by choosing a shift near enough the specific eigenvalue of interest, we can, in theory, have a convergence rate that is as high as required; i.e., we would only need to make $|\lambda_j - \mu|$ small enough in relation to $|\lambda_p - \mu|$ defined above. However, in an actual solution scheme the difficulty is to find an appropriate μ, for which we consider various methods in the next sections.

Example 11.5

Use inverse iteration as given in (11.16) to (11.22) in order to calculate $(\lambda_1, \boldsymbol{\phi}_1)$ of the problem $\mathbf{K}\boldsymbol{\phi} = \lambda\mathbf{M}\boldsymbol{\phi}$, where $\mathbf{K}$ and $\mathbf{M}$ are given in Example 11.4. Then impose the shift $\mu = 10$ and show that in the inverse iteration convergence occurs toward λ_4 and $\boldsymbol{\phi}_4$.

Using inverse iteration on $\mathbf{K}\boldsymbol{\phi} = \lambda\mathbf{M}\boldsymbol{\phi}$ as in Example 11.2 gives convergence after three iterations with a tolerance of 10^{-6},

$$\lambda_1 \doteq 0.09654; \qquad \boldsymbol{\phi}_1 \doteq \begin{bmatrix} 0.3126 \\ 0.4955 \\ 0.4791 \\ 0.2898 \end{bmatrix}$$

Now imposing a shift of $\mu = 10$, we obtain

$$\mathbf{K} - \mu\mathbf{M} = \begin{bmatrix} -15 & -4 & 1 & 0 \\ -4 & -14 & -4 & 1 \\ 1 & -4 & -4 & -4 \\ 0 & 1 & -4 & -5 \end{bmatrix}$$

Using inverse iteration on the problem $(\mathbf{K} - \mu\mathbf{M})\boldsymbol{\phi} = \eta\mathbf{M}\boldsymbol{\phi}$, we obtain convergence after six iterations with

$$\rho(\bar{\mathbf{x}}_7) = 0.6385; \qquad \mathbf{x}_7 = \begin{bmatrix} -0.1076 \\ 0.2556 \\ -0.7283 \\ 0.5620 \end{bmatrix}$$

Since we imposed a shift, we do know that $\mu + \rho(\bar{\mathbf{x}}_7)$ is an approximation to an eigenvalue and $\mathbf{x}_7$ is an approximation to the corresponding eigenvector. But we do not know which eigenpair has been approximated. By comparing $\mathbf{x}_7$ with the results obtained in Example 11.4 we find that we have

$$\lambda_4 \doteq \mu + \rho(\mathbf{x}_7) = 10.6385; \qquad \boldsymbol{\phi}_4 \doteq \mathbf{x}_7$$

Example 11.6

Consider the unsupported beam element depicted in Fig. 7.10. Show that the usual inverse iteration algorithm of calculating λ_1 and $\boldsymbol{\phi}_1$ does not work, but that after imposing a shift the standard algorithm can again be applied.

The first step in the inverse iteration defined in (11.16) is, in this case with $\mathbf{M} = \mathbf{I}$,

$$\begin{bmatrix} 12 & -6 & -12 & -6 \\ -6 & 4 & 6 & 2 \\ -12 & 6 & 12 & 6 \\ -6 & 2 & 6 & 4 \end{bmatrix} \bar{\mathbf{x}}_2 = \begin{bmatrix} 1 \\ 1 \\ 1 \\ 1 \end{bmatrix} \tag{a}$$

Using Gauss elimination to solve the equations, we arrive at

$$\begin{bmatrix} 12 & -6 & -12 & -6 \\ & 1 & 0 & -1 \\ & & 0 & 0 \\ & & & 0 \end{bmatrix} \bar{\mathbf{x}}_2 = \begin{bmatrix} 1 \\ \frac{3}{2} \\ 2 \\ \frac{7}{2} \end{bmatrix}$$

and hence the equations in (a) have no solution. They only have a solution if the right-hand side [i.e., $\mathbf{x}_1$ in (11.16)] is a null vector. There would be no difficulty in modifying the solution procedure when a singular coefficient matrix is encountered, and the advantage would be that the eigenvector would be calculated in one iteration. On the other hand, if we impose a shift, we can use the standard iteration procedure, and stability problems are avoided in the calculation of other eigenvalues and eigenvectors. Assume that we use $\mu = -6$ so that all λ_i are positive. Then we have

$$\mathbf{K} - \mu\mathbf{I} = \begin{bmatrix} 18 & -6 & -12 & -6 \\ -6 & 10 & 6 & 2 \\ -12 & 6 & 18 & 6 \\ -6 & 2 & 6 & 10 \end{bmatrix}$$

The inverse iteration can now be performed in the standard manner using a full unit starting iteration vector. Convergence is achieved after five iterations to a tolerance of 10^{-6}, and we have

$$\rho(\bar{\mathbf{x}}_6) = 6.000000; \qquad \mathbf{x}_6 = \begin{bmatrix} 0.73784 \\ 0.42165 \\ 0.31625 \\ 0.42165 \end{bmatrix}$$

Hence taking account of the shift, we have

$$\lambda_1 \doteq 0.0; \qquad \boldsymbol{\phi}_1 \doteq \mathbf{x}_6$$

We showed above that the rate of convergence in inverse iteration can be greatly increased by shifting. We may now wonder whether the convergence rate in forward iteration can be increased in a similar way. In analogy to the convergence proof of inverse iteration with a shift, we can generalize the convergence analysis of forward iteration when a shift μ is used. The final result is that the iteration vector converges to the eigenvector $\boldsymbol{\phi}_j$ which corresponds to the largest eigenvalue $|\lambda_j - \mu|$ of the problem in (11.42), where

$$|\lambda_j - \mu| = \max_{\text{all } i} |\lambda_i - \mu| \tag{11.48}$$

The convergence rate of the iteration vector is given by

$$r = \max_{p \neq j} \left| \frac{\lambda_p - \mu}{\lambda_j - \mu} \right| \tag{11.49}$$

which, in fact, is the ratio of the second largest eigenvalue to the largest eigenvalue (both measured in absolute values) of the problem $(\mathbf{K} - \mu\mathbf{M})\boldsymbol{\phi} = \eta\mathbf{M}\boldsymbol{\phi}$. In case λ_j is a multiple eigenvalue, say $\lambda_j = \lambda_{j+1} = \ldots = \lambda_{j+m-1}$, the iteration vector converges to a vector in the subspace corresponding to λ_j, and the rate of convergence is

$$\max_{p \neq j, j+1, \ldots, j+m-1} \left| \frac{\lambda_p - \mu}{\lambda_j - \mu} \right|$$

The main difference between the convergence rate in (11.47) and (11.49) is that, in (11.47), λ_p is in the denominator, whereas in (11.49), λ_p is in the numerator. This limits the convergence rate in forward iteration and by means of shifting convergence can be obtained only to the eigenpair $(\lambda_n, \boldsymbol{\phi}_n)$ or to the eigenpair $(\lambda_1, \boldsymbol{\phi}_1)$. To achieve the highest convergence rates to $\boldsymbol{\phi}_n$ and $\boldsymbol{\phi}_1$, we need to choose $\mu = (\lambda_1 + \lambda_{n-1})/2$ and $\mu = (\lambda_2 + \lambda_n)/2$, respectively, and have the corresponding convergence rates

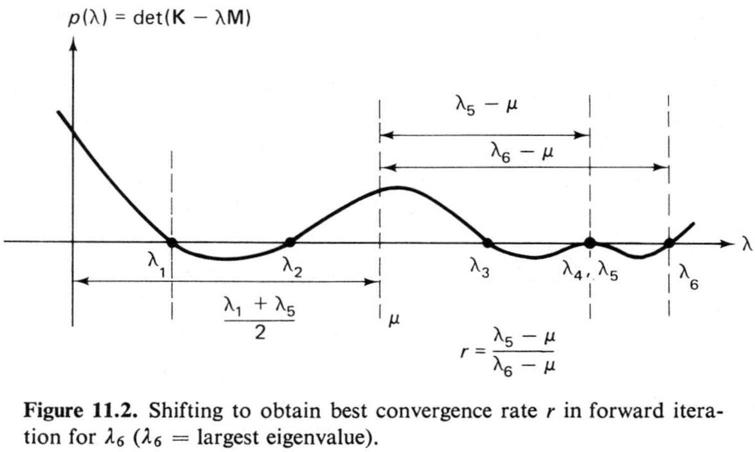

Figure 11.2. Shifting to obtain best convergence rate r in forward iteration for λ_6 (λ_6 = largest eigenvalue).

$$\left| \frac{\lambda_{n-1} - \dfrac{\lambda_1 + \lambda_{n-1}}{2}}{\lambda_n - \dfrac{\lambda_1 + \lambda_{n-1}}{2}} \right| \quad \text{and} \quad \left| \frac{\lambda_2 - \dfrac{\lambda_2 + \lambda_n}{2}}{\lambda_1 - \dfrac{\lambda_2 + \lambda_n}{2}} \right|$$

(see Fig. 11.2). Therefore, a much higher convergence rate can be obtained with shifting in inverse iteration than using forward iteration. For this reason and because a shift can be chosen to converge to any eigenpair, inverse iteration is much more important in practical analysis, and in the algorithms presented later, we always use inverse iteration whenever vector iteration is required.

11.2.4 Rayleigh Quotient Iteration

We discussed in Section 11.2.3 that the convergence rate in inverse iteration can be much improved by shifting. In practice, the difficulty lies in choosing an appropriate shift. One possibility is to use as shift value the Rayleigh quotient calculated in (11.18), which is an approximation to the eigenvalue sought. If a new shift using (11.18) is evaluated in each iteration, we have the Rayleigh quotient iteration.[15-18] In this procedure we assume a starting iteration vector $\mathbf{x}_1$, hence $\mathbf{y}_1 = \mathbf{M}\mathbf{x}_1$, a starting shift, $\rho(\bar{\mathbf{x}}_1)$, which is usually zero, and then evaluate for $k = 1, 2, \ldots$:

$$(\mathbf{K} - \rho(\bar{\mathbf{x}}_k)\mathbf{M})\bar{\mathbf{x}}_{k+1} = \mathbf{y}_k \tag{11.50}$$

$$\bar{\mathbf{y}}_{k+1} = \mathbf{M}\bar{\mathbf{x}}_{k+1} \tag{11.51}$$

$$\rho(\bar{\mathbf{x}}_{k+1}) = \frac{\bar{\mathbf{x}}_{k+1}^T \mathbf{y}_k}{\bar{\mathbf{x}}_{k+1}^T \bar{\mathbf{y}}_{k+1}} + \rho(\bar{\mathbf{x}}_k) \tag{11.52}$$

$$\mathbf{y}_{k+1} = \frac{\bar{\mathbf{y}}_{k+1}}{(\bar{\mathbf{x}}_{k+1}^T \bar{\mathbf{y}}_{k+1})^{1/2}} \tag{11.53}$$

where now

$$\mathbf{y}_{k+1} \longrightarrow \mathbf{M}\boldsymbol{\phi}_i \quad \text{and} \quad \rho(\bar{\mathbf{x}}_{k+1}) \longrightarrow \lambda_i \qquad \text{as } k \longrightarrow \infty$$

The eigenvalue λ_i and corresponding eigenvector $\boldsymbol{\phi}_i$ to which the iteration converges depend on the starting iteration vector $\mathbf{x}_1$ and the initial shift $\rho(\bar{\mathbf{x}}_1)$. If $\mathbf{x}_1$ has strong components of an eigenvector, say $\boldsymbol{\phi}_k$, and $\rho(\bar{\mathbf{x}}_2)$ provides a sufficiently close shift to the corresponding eigenvalue λ_k, then the iteration converges to the eigenpair $(\lambda_k, \boldsymbol{\phi}_k)$, and the ultimate order of convergence for both λ_k and $\boldsymbol{\phi}_k$ is cubic. Otherwise, convergence will occur to another eigenpair (disregarding a special case that has a very small probability), but *convergence will always be cubic.*[17] This excellent convergence behavior is a most important observation. We may intuitively explain it by the fact that in inverse iteration the vector converges linearly, and with an error of order ϵ in the vector the Rayleigh quotient predicts the eigenvalue with an error of order ϵ^2. Since the eigenvalue approximation used as a shift has a direct effect on the approximation to be obtained for the eigenvector, and vice versa, it seems probable that in Rayleigh quotient iteration, the order of convergence is cubic for both the eigenvalue and eigenvector.

To analyze the convergence characteristics of Rayleigh quotient iteration we may proceed in the same way as in the analysis of inverse iteration; i.e., we consider the iteration in the basis of the eigenvectors. In this case we use the transformation in (11.24) and write the two basic equations of Rayleigh quotient iteration [i.e., (11.50) and (11.52), respectively] in the following form:

$$(\mathbf{\Lambda} - \rho(\mathbf{z}_k)\mathbf{I})\mathbf{z}_{k+1} = \mathbf{z}_k \tag{11.54}$$

and

$$\rho(\mathbf{z}_{k+1}) = \frac{\mathbf{z}_{k+1}^T \mathbf{z}_k}{\mathbf{z}_{k+1}^T \mathbf{z}_{k+1}} + \rho(\mathbf{z}_k) \tag{11.55}$$

where the length normalization of the iteration vector has been omitted.

To consider the convergence characteristics of the iteration vector, let us perform an approximate convergence analysis that gives insight into the working of the algorithm. Assume that the current iteration vector $\mathbf{z}_l$ is already close to the eigenvector $\mathbf{e}_1$; i.e., we have

$$\mathbf{z}_l^T = [1 \quad o(\epsilon) \quad o(\epsilon) \ldots o(\epsilon)] \tag{11.56}$$

where $o(\epsilon)$ denotes "of order ϵ" and $\epsilon \ll 1$. We then obtain

$$\rho(\mathbf{z}_l) = \lambda_1 + o(\epsilon^2) \tag{11.57}$$

Solving from (11.54) for $\mathbf{z}_{l+1}$, we thus have

$$\mathbf{z}_{l+1}^T = \left[\frac{1}{o(\epsilon^2)} \quad \frac{o(\epsilon)}{\lambda_2 - \lambda_1} \cdots \frac{o(\epsilon)}{\lambda_n - \lambda_1}\right] \tag{11.58}$$

In order to assess the convergence of the iteration vector we normalize to 1 the first component of $\mathbf{z}_{l+1}$, to obtain

$$\bar{\mathbf{z}}_{l+1}^T = [1 \quad o(\epsilon^3) \quad o(\epsilon^3) \quad \ldots \quad o(\epsilon^3)] \tag{11.59}$$

Hence the elements that in $\mathbf{z}_l$ have been of order ϵ are now of order ϵ^3, indicating cubic convergence.

Consider the following example to demonstrate the characteristics of Rayleigh quotient iteration.

Example 11.7

Perform the Rayleigh quotient iteration on the problem $\mathbf{\Lambda}\boldsymbol{\phi} = \lambda\boldsymbol{\phi}$, where

$$\mathbf{\Lambda} = \begin{bmatrix} 2 & 0 \\ 0 & 6 \end{bmatrix}$$

Use as starting iteration vectors $\mathbf{x}_1$ the vectors

$$(1) \quad \mathbf{x}_1 = \begin{bmatrix} 1 \\ 1 \end{bmatrix}; \qquad (2) \quad \mathbf{x}_1 = \begin{bmatrix} 1 \\ 0.1 \end{bmatrix}$$

Using the relations given in (11.50) to (11.53) [with $\rho(\bar{\mathbf{x}}_1) = 0.0$], we obtain, in case (1),

$$\bar{\mathbf{x}}_2 = \begin{bmatrix} 0.500 \\ 0.166667 \end{bmatrix}; \qquad \rho(\bar{\mathbf{x}}_2) = 2.40$$

$$\mathbf{y}_2 = \begin{bmatrix} 0.94868 \\ 0.31623 \end{bmatrix}$$

$$\bar{\mathbf{x}}_3 = \begin{bmatrix} -2.37171 \\ 0.08784 \end{bmatrix}; \qquad \rho(\bar{\mathbf{x}}_3) = 2.00548$$

$$\mathbf{y}_3 = \begin{bmatrix} -0.99931 \\ 0.03701 \end{bmatrix}$$

$$\bar{\mathbf{x}}_4 = \begin{bmatrix} 182.37496 \\ 0.00927 \end{bmatrix}; \qquad \rho(\bar{\mathbf{x}}_4) = 2.000000$$

and

$$\mathbf{y}_4 = \begin{bmatrix} 1.0000 \\ 0.00005 \end{bmatrix}$$

Hence we see that in three steps of iteration we obtained a good approximation to the required eigenvalue and eigenvector.

In case (2) we have

$$\bar{\mathbf{x}}_2 = \begin{bmatrix} 0.50000 \\ 0.016667 \end{bmatrix}; \qquad \rho(\bar{\mathbf{x}}_2) = 2.00444$$

$$\mathbf{y}_2 = \begin{bmatrix} 0.99944 \\ 0.033315 \end{bmatrix}$$

and then

$$\bar{\mathbf{x}}_3 = \begin{bmatrix} -225.125 \\ 0.00834 \end{bmatrix}; \qquad \rho(\bar{\mathbf{x}}_3) = 2.000001$$

$$\mathbf{y}_3 = \begin{bmatrix} -1.00000 \\ 0.000037 \end{bmatrix}$$

We observe that in this case two iterations are sufficient to obtain a good approximation to the required eigenvalue and eigenvector, because the starting iteration vector was already closer to the required eigenvector.

As was pointed out in the preceding discussion, the Rayleigh quotient iteration can, in principle, converge to any eigenpair. Therefore, if we are interested in the smallest p eigenvalues and corresponding eigenvectors, we need to supplement the Rayleigh quotient iterations by another technique to assure convergence to one of the eigenpairs sought. For example, to calculate the smallest eigenvalue and corresponding eigenvector, we may first use the inverse iteration in (11.16) to (11.19) without shifting to obtain an iteration vector that is a good approximation of $\boldsymbol{\phi}_1$, and only then start with Rayleigh quotient iteration.[22] However, the difficulty lies in assessing how many inverse iterations must be performed before Rayleigh quotient shifting can be started and yet convergence to $\boldsymbol{\phi}_1$ and λ_1 is achieved. Unfortunately, this question can in general not be resolved, and it is necessary to use the Sturm sequence property to make sure that the required eigenvalue and corresponding eigenvector have indeed been calculated (see Section 11.5).

11.2.5 Matrix Deflation and Gram–Schmidt Orthogonalization

In Sections 11.2.1 to 11.2.4 we discussed how an eigenvalue and corresponding eigenvector can be calculated using vector iteration. The basic inverse iteration technique converges to λ_1 and $\boldsymbol{\phi}_1$ (see Section 11.2.1), the basic forward iteration can be used to calculate λ_n and $\boldsymbol{\phi}_n$ (see Section 11.2.2), but the methods can also be employed with shifting to calculate other eigenvalues and corresponding eigenvectors (see Section 11.2.3). Assume now that we have calculated a specific eigenpair, say $(\lambda_k, \boldsymbol{\phi}_k)$, using either method and that we require the solution of another eigenpair. To assure that we do not converge again to λ_k and $\boldsymbol{\phi}_k$, we need to deflate either the matrices or the iteration vectors.[1-10,19,20]

Matrix deflation has been applied extensively in the solution of standard eigenproblems. The problem may be $\mathbf{K}\boldsymbol{\phi} = \lambda\boldsymbol{\phi}$, i.e., when $\mathbf{M}$ is the identity matrix in $\mathbf{K}\boldsymbol{\phi} = \lambda\mathbf{M}\boldsymbol{\phi}$, or may be $\tilde{\mathbf{K}}\tilde{\boldsymbol{\phi}} = \lambda\tilde{\boldsymbol{\phi}}$, which is obtained by transforming the generalized eigenproblem into a standard form (see Section 10.2.5). We recall that this transformation is effective when $\mathbf{M}$ is diagonal and all diagonal elements are larger than zero, because in such a case $\tilde{\mathbf{K}}$ has the same bandwidth as $\mathbf{K}$.

Consider the deflation of $\mathbf{K}\boldsymbol{\phi} = \lambda\boldsymbol{\phi}$, because the deflation of $\tilde{\mathbf{K}}\tilde{\boldsymbol{\phi}} = \lambda\tilde{\boldsymbol{\phi}}$ would be obtained in the same way. A stable matrix deflation can be carried out by finding an orthogonal matrix $\mathbf{P}$ whose first column is the calculated eigenvector $\boldsymbol{\phi}_k$.

Writing $\mathbf{P}$ as

$$\mathbf{P} = [\boldsymbol{\phi}_k, \mathbf{p}_2, \ldots, \mathbf{p}_n] \tag{11.60}$$

we need to have $\boldsymbol{\phi}_k^T\mathbf{p}_i = 0$ for $i = 2, \ldots, n$. It then follows that

$$\mathbf{P}^T\mathbf{K}\mathbf{P} = \begin{bmatrix} \lambda_k & \mathbf{0} \\ \mathbf{0} & \mathbf{K}_1 \end{bmatrix} \tag{11.61}$$

because $\boldsymbol{\phi}_k^T\boldsymbol{\phi}_k = 1$. The important point is that $\mathbf{P}^T\mathbf{K}\mathbf{P}$ has the same eigenvalues as $\mathbf{K}$, and therefore $\mathbf{K}_1$ must have all eigenvalues of $\mathbf{K}$ except λ_k. In addition, denoting the eigenvectors of $\mathbf{P}^T\mathbf{K}\mathbf{P}$ by $\bar{\boldsymbol{\phi}}_i$, we have that

$$\boldsymbol{\phi}_i = \mathbf{P}\bar{\boldsymbol{\phi}}_i \tag{11.62}$$

It is important to note that the matrix $\mathbf{P}$ is not unique and that various techniques can be used to construct an appropriate transformation matrix. Since $\mathbf{K}$ is banded we would like to have that the transformation does not destroy the bandform, and a very effective procedure can be used.[21,22]

From this discussion it follows that once a second required eigenpair using $\mathbf{K}_1$ has been evaluated, the process of deflation could be repeated by working with $\mathbf{K}_1$ rather than with $\mathbf{K}$. Therefore, we may continue deflating until all required eigenvalues and eigenvectors have been calculated. The disadvantage of matrix deflation is that the eigenvectors have to be calculated to very high precision to avoid the accumulation of errors introduced in the deflation process.

Instead of matrix deflation, we may deflate the iteration vector in order to converge to an eigenpair other than $(\lambda_k, \boldsymbol{\phi}_k)$. The basis of vector deflation is that in order for an iteration vector to converge in forward or inverse iteration to a required eigenvector, the iteration vector must not be orthogonal to it. Hence, conversely, if the iteration vector is orthogonalized to the eigenvectors already calculated, we eliminate the possibility that the iteration converges to any one of them, and, as we will see, convergence occurs instead to another eigenvector.

A particular vector orthogonalization procedure that is employed extensively is the Gram–Schmidt method. The procedure can be used in the solution of the generalized eigenproblem $\mathbf{K}\boldsymbol{\phi} = \lambda\mathbf{M}\boldsymbol{\phi}$, where $\mathbf{M}$ can take the different forms that we encounter in finite element analysis.

In order to consider a general case, assume that we have calculated in inverse iteration the eigenvectors $\boldsymbol{\phi}_1, \boldsymbol{\phi}_2, \ldots, \boldsymbol{\phi}_m$, and that we want to $\mathbf{M}$-orthogonalize $\mathbf{x}_1$ to these eigenvectors. In Gram–Schmidt orthogonalization a vector $\tilde{\mathbf{x}}_1$, which is $\mathbf{M}$-orthogonal to the eigenvectors $\boldsymbol{\phi}_i$, $i = 1, \ldots, m$, is calculated using

$$\tilde{\mathbf{x}}_1 = \mathbf{x}_1 - \sum_{i=1}^{m} \alpha_i \boldsymbol{\phi}_i \tag{11.63}$$

where the coefficients α_i are obtained using the conditions that $\boldsymbol{\phi}_i^T\mathbf{M}\tilde{\mathbf{x}}_1 = 0$, $i = 1, \ldots, m$, and $\boldsymbol{\phi}_i^T\mathbf{M}\boldsymbol{\phi}_j = \delta_{ij}$. Premultiplying both sides of (11.63) by $\boldsymbol{\phi}_i^T\mathbf{M}$, we therefore obtain

$$\alpha_i = \boldsymbol{\phi}_i^T\mathbf{M}\mathbf{x}_1; \qquad i = 1, \ldots, m \tag{11.64}$$

In the inverse iteration we would now use $\tilde{\mathbf{x}}_1$ as the starting iteration vector instead of $\mathbf{x}_1$ and provided that $\mathbf{x}_1^T\mathbf{M}\boldsymbol{\phi}_{m+1} \neq 0$, convergence occurs (at least in theory, see Section 11.2.6) to $\boldsymbol{\phi}_{m+1}$ and λ_{m+1}.

To prove the convergence given above, we consider as before the iteration process in the basis of eigenvectors; i.e., we analyze the iteration given in (11.25) when the Gram–Schmidt orthogonalization is included. In this case the eigenvectors corresponding to the smallest eigenvalues are $\mathbf{e}_i$, $i = 1, \ldots, m$. Carrying out the deflation of the starting iteration vector $\mathbf{z}_1$ in (11.27), we obtain

$$\tilde{\mathbf{z}}_1 = \mathbf{z}_1 - \sum_{i=1}^{m} \alpha_i \mathbf{e}_i \tag{11.65}$$

with

$$\begin{aligned}\alpha_i &= \mathbf{e}_i^T\mathbf{z}_1 \\ &= 1; \qquad i = 1, \ldots, m\end{aligned} \tag{11.66}$$

Hence

$$\tilde{\mathbf{z}}_1^T = [0 \quad \ldots \quad 0 \quad \underset{\text{element } m+1}{1} \quad \ldots \quad 1] \tag{11.67}$$

Using now $\tilde{\mathbf{z}}_1$ as the starting iteration vector and performing the convergence analysis as discussed in Section 11.2.1, we find that if $\lambda_{m+2} > \lambda_{m+1}$, we have that $\tilde{\mathbf{z}}_{l+1} \longrightarrow \mathbf{e}_{m+1}$, as was required to prove. Furthermore, we find that the rate of convergence of the eigenvector is $\lambda_{m+1}/\lambda_{m+2}$, and in case λ_{m+1} is a multiple eigenvalue, the rate of convergence is given by the ratio of λ_{m+1} to the next distinct eigenvalue.

Although so far we have discussed Gram–Schmidt orthogonalization in connection with vector inverse iteration, it should be realized that the orthogonalization procedure can be used equally well in the other vector iteration methods. All convergence considerations discussed in the presentation of inverse iteration, forward iteration, and Rayleigh quotient iteration are also applicable when Gram–Schmidt orthogonalization is included if it is taken into account that convergence to the eigenvectors already calculated is not possible.

Example 11.8

Calculate, using Gram–Schmidt orthogonalization, an appropriate starting iteration vector for the solution of the problem $\mathbf{K}\boldsymbol{\phi} = \lambda\mathbf{M}\boldsymbol{\phi}$, where $\mathbf{K}$ and $\mathbf{M}$ are given in Example 11.4. Assume that the eigenpairs $(\lambda_1, \boldsymbol{\phi}_1)$ and $(\lambda_4, \boldsymbol{\phi}_4)$ are known as obtained in Example 11.5 and that convergence to another eigenpair is sought.

To determine an appropriate starting iteration vector, we want to deflate the unit full vector of the iteration vectors $\boldsymbol{\phi}_1$ and $\boldsymbol{\phi}_4$; i.e., (11.63) reads

$$\tilde{\mathbf{x}}_1 = \begin{bmatrix} 1 \\ 1 \\ 1 \\ 1 \end{bmatrix} - \alpha_1\boldsymbol{\phi}_1 - \alpha_4\boldsymbol{\phi}_4$$

where α_1 and α_4 are obtained using (11.64):

$$\alpha_1 = \boldsymbol{\phi}_1^T\mathbf{M}\mathbf{x}_1; \qquad \alpha_4 = \boldsymbol{\phi}_4^T\mathbf{M}\mathbf{x}_1$$

Substituting for $\mathbf{M}$, $\boldsymbol{\phi}_1$, and $\boldsymbol{\phi}_4$, we obtain

$$\alpha_1 = 2.385; \qquad \alpha_4 = 0.1299$$

Then, to a few-digit accuracy,

$$\tilde{\mathbf{x}}_1 = \begin{bmatrix} 0.2683 \\ -0.2149 \\ -0.04812 \\ 0.2358 \end{bmatrix}$$

11.2.6 Some Practical Considerations Concerning Vector Iterations

So far we have discussed the theory used in vector iteration techniques. However, for a proper computer implementation of the methods, it is important to interpret the results obtained concerning the iteration procedures. Of particular importance are practical convergence and stability considerations when any one of the techniques is used.

A first important point is that the convergence rates of the iterations may turn out when measured in practice to be rather theoretical, Namely, we assumed that the starting iteration vector $\mathbf{z}_1$ in (11.27) is a unit full vector, which corresponds to a vector $\mathbf{x}_1 = \sum_{i=1}^{n} \boldsymbol{\phi}_i$. This means that the starting iteration vector is equally strong in each of the eigenvectors $\boldsymbol{\phi}_i$. We chose this starting iteration vector to identify easily the theoretical convergence rate with which the iteration vector approaches the required eigenvector. However, in practice it is hardly possible to pick $\mathbf{x}_1 = \sum_{i=1}^{n} \boldsymbol{\phi}_i$ as the starting iteration vector, and instead we have

$$\mathbf{x}_1 = \sum_{i=1}^{n} \alpha_i \boldsymbol{\phi}_i \tag{11.68}$$

where the α_i are arbitrary constants. This vector $\mathbf{x}_1$ corresponds to the following vector in the basis of eigenvectors:

$$\mathbf{z}_1 = \begin{bmatrix} \alpha_1 \\ \cdot \\ \cdot \\ \cdot \\ \alpha_n \end{bmatrix} \tag{11.69}$$

To identify the effect of the constants α_i, consider as an example the convergence analysis of inverse iteration without shifting when the starting vector in (11.68) is used and $\lambda_2 > \lambda_1$. The conclusions reached will be equally applicable to other vector iteration methods. As before, we consider the iteration in the basis of eigenvectors $\boldsymbol{\Phi}$ and require $\alpha_1 \neq 0$ in order to have $\mathbf{x}_1^T \mathbf{M} \boldsymbol{\phi}_1 \neq 0$. After l inverse iterations we now have instead of (11.29)

$$\bar{\mathbf{z}}_{l+1} = \begin{bmatrix} 1 \\ \beta_2 \left(\frac{\lambda_1}{\lambda_2}\right)^l \\ \cdot \\ \cdot \\ \cdot \\ \beta_n \left(\frac{\lambda_1}{\lambda_n}\right)^l \end{bmatrix} \tag{11.70}$$

$$\beta_i = \frac{\alpha_i}{\alpha_1}; \qquad i = 2, \ldots, n \tag{11.71}$$

Therefore, the iteration vector obtained now has the multipliers β_i in its last $(n - 1)$ components. In the iteration the ith component is still decreasing with each iteration, as in (11.29) by the factor λ_1/λ_i, $i = 2, \ldots, n$, and the rate of convergence is λ_1/λ_2, as was already derived in Section 11.2.1. However, in practical analysis the unknown coefficients β_i may produce the result that the theoretical convergence rate is not observed for many iterations. In practice, therefore, not only the order and rate of convergence, but equally important the "quality" of the starting iteration vector determines the number of iterations required for convergence. Furthermore, it is important to use a high-enough convergence tolerance to prevent premature acceptance of the iteration vector as an approximation to the required eigenvector.

Together with the vector iterations, we may use a matrix deflation procedure or Gram–Schmidt vector orthogonalization to obtain convergence to an eigenpair not already calculated (see Section 11.2.5). We mentioned already that for matrix deflation, the eigenvectors have to be evaluated to relatively high precision to preserve stability. Considering the Gram–Schmidt orthogonalization, the method must also be used with care. Namely, if the technique is employed in inverse or forward iteration without shifting, it is necessary to calculate the eigenvectors to high precision in order that Gram–Schmidt orthogonalization will work. In addition, the iteration vector should be orthogonalized in each iteration to the eigenvectors already calculated.

It is useful to draw an important conclusion. We pointed out earlier in the presentation of the vector iteration techniques that it is difficult (and indeed theory shows impossible) to *assure* convergence to a specific (but arbitrarily selected) eigenvalue and corresponding eigenvector. The discussion concerning practical aspects in this section substantiates those observations, and it is concluded that the vector iteration procedures and the Gram–Schmidt orthogonalization process must be employed with care if a specific eigenvalue and corresponding eigenvector are required. We will see in Chapter 12 that, in fact, both techniques are best employed and are used very effectively in conjunction with other solution strategies.

11.3 TRANSFORMATION METHODS

We pointed out in Section 11.1 that the transformation methods comprise a group of eigensystem solution procedures that employ the basic properties of the eigenvectors in the matrix $\mathbf{\Phi}$,

$$\mathbf{\Phi}^T\mathbf{K}\mathbf{\Phi} = \mathbf{\Lambda} \tag{11.3}$$

and

$$\mathbf{\Phi}^T\mathbf{M}\mathbf{\Phi} = \mathbf{I} \tag{11.4}$$

Since the matrix $\mathbf{\Phi}$, of order $n \times n$, which diagonalizes $\mathbf{K}$ and $\mathbf{M}$ in the way given in (11.3) and (11.4) is unique, we can try to construct it by iteration.[1-9] The basic scheme is to reduce $\mathbf{K}$ and $\mathbf{M}$ into diagonal form using successive pre- and postmultiplication by matrices $\mathbf{P}_k^T$ and $\mathbf{P}_k$, respectively, where $k = 1, 2, \ldots$. Specifically, if we define $\mathbf{K}_1 = \mathbf{K}$ and $\mathbf{M}_1 = \mathbf{M}$, we form

$$\left.\begin{aligned} \mathbf{K}_2 &= \mathbf{P}_1^T\mathbf{K}_1\mathbf{P}_1 \\ \mathbf{K}_3 &= \mathbf{P}_2^T\mathbf{K}_2\mathbf{P}_2 \\ &\;\vdots \\ \mathbf{K}_{k+1} &= \mathbf{P}_k^T\mathbf{K}_k\mathbf{P}_k \\ &\;\vdots \end{aligned}\right\} \tag{11.72}$$

Similarly,

$$\left.\begin{aligned} \mathbf{M}_2 &= \mathbf{P}_1^T\mathbf{M}_1\mathbf{P}_1 \\ \mathbf{M}_3 &= \mathbf{P}_2^T\mathbf{M}_2\mathbf{P}_2 \\ &\;\vdots \\ \mathbf{M}_{k+1} &= \mathbf{P}_k^T\mathbf{M}_k\mathbf{P}_k \\ &\;\vdots \end{aligned}\right\} \tag{11.73}$$

where the matrices $\mathbf{P}_k$ are selected to bring $\mathbf{K}_k$ and $\mathbf{M}_k$ closer to diagonal form. Then for a proper procedure we apparently need to have

$$\mathbf{K}_{k+1} \longrightarrow \mathbf{\Lambda} \quad \text{and} \quad \mathbf{M}_{k+1} \longrightarrow \mathbf{I} \qquad \text{as } k \longrightarrow \infty$$

in which case, with l being the last iteration,

$$\mathbf{\Phi} = \mathbf{P}_1\mathbf{P}_2 \ldots \mathbf{P}_l \tag{11.74}$$

In practice, it is not necessary that $\mathbf{M}_{k+1}$ converges to $\mathbf{I}$ and $\mathbf{K}_{k+1}$ to $\mathbf{\Lambda}$, but they only need to converge to diagonal form. Namely, if

$$\mathbf{K}_{k+1} \longrightarrow \text{diag}\,(K_r) \quad \text{and} \quad \mathbf{M}_{k+1} \longrightarrow \text{diag}\,(M_r) \qquad \text{as } k \longrightarrow \infty$$

then with l indicating the last iteration and disregarding that the eigenvalues and eigenvectors may not be in the usual order,

$$\mathbf{\Lambda} = \text{diag}\left(\frac{K_r^{(l+1)}}{M_r^{(l+1)}}\right) \tag{11.75}$$

and

$$\mathbf{\Phi} = \mathbf{P}_1\mathbf{P}_2 \ldots \mathbf{P}_l\,\text{diag}\left(\frac{1}{\sqrt{M_r^{(l+1)}}}\right) \tag{11.76}$$

Using the basic idea described above, a number of different iteration methods have been proposed. We shall discuss in the next sections only the Jacobi and the Householder–QR methods, which are believed to be most effective in finite element analysis. However, before presenting the techniques in detail we should point out one important aspect. In the above introduction it was implied that iteration is started with pre- and postmultiplication by $\mathbf{P}_1^T$ and $\mathbf{P}_1$, respectively, which is indeed the case in the Jacobi solution methods.

However, it should be noted that, alternatively, we may first aim to transform the eigenvalue problem $\mathbf{K}\boldsymbol{\phi} = \lambda\mathbf{M}\boldsymbol{\phi}$ into a form that is more economical to use in the iteration. In particular, when $\mathbf{M} = \mathbf{I}$, the first m transformations in (11.72) may be used to reduce $\mathbf{K}$ into tridiagonal form without iteration, after which the matrices $\mathbf{P}_i$, $i = m + 1, \ldots, l$, are applied in an iterative manner to bring $\mathbf{K}_{m+1}$ into diagonal form. In such a case the first matrices $\mathbf{P}_1, \ldots, \mathbf{P}_m$ may be of different form than the later applied matrices $\mathbf{P}_{m+1}, \ldots, \mathbf{P}_l$. An application of this procedure is the Householder–QR method, in which Householder matrices are used to first transform $\mathbf{K}$ into tridiagonal form and then rotation matrices are employed in the QR transformations. The same solution strategy can also be used to solve the generalized eigenproblem $\mathbf{K}\boldsymbol{\phi} = \lambda\mathbf{M}\boldsymbol{\phi}$, $\mathbf{M} \neq \mathbf{I}$, provided that the problem is first transformed into the standard form.

11.3.1 The Jacobi Method

The basic Jacobi solution method has been developed for the solution of standard eigenproblems (i.e., $\mathbf{M}$ being the identity matrix), and we consider it in this section. The method was proposed over a century ago and has been used extensively.[1-9, 23-27] A major advantage of the procedure is its simplicity and stability. Since the eigenvector properties in (11.3) and (11.4) (with $\mathbf{M} = \mathbf{I}$) are applicable to all symmetric matrices $\mathbf{K}$ with no restriction on the eigenvalues, the Jacobi method can be used to calculate negative, zero, or positive eigenvalues.

Considering the standard eigenproblem $\mathbf{K}\boldsymbol{\phi} = \lambda\boldsymbol{\phi}$, the kth iteration step defined in (11.72) reduces to

$$\mathbf{K}_{k+1} = \mathbf{P}_k^T\mathbf{K}_k\mathbf{P}_k \tag{11.77}$$

where $\mathbf{P}_k$ is an orthogonal matrix; i.e., (11.73) gives

$$\mathbf{P}_k^T\mathbf{P}_k = \mathbf{I} \tag{11.78}$$

In the Jacobi solution the matrix $\mathbf{P}_k$ is a rotation matrix which is selected in such way that an off-diagonal element in $\mathbf{K}_k$ is zeroed. If element (i, j) is to be reduced to zero, the corresponding orthogonal matrix $\mathbf{P}_k$ is

$$\mathbf{P}_k = \begin{bmatrix} 1 & & & & & & \\ & \ddots & & & & & \\ & & \cos\theta & & -\sin\theta & & \\ & & & \ddots & & & \\ & & \sin\theta & & \cos\theta & & \\ & & & & & \ddots & \\ & & & & & & 1 \end{bmatrix} \begin{matrix} \\ \\ \leftarrow i\text{th} \\ \\ \leftarrow j\text{th row} \\ \\ \\ \end{matrix} \tag{11.79}$$

(with $\cos\theta$, $\sin\theta$ in the ith column and $-\sin\theta$, $\cos\theta$ in the jth column) where θ is selected from the condition that element (i, j) in $\mathbf{K}_{k+1}$ be zero. Denoting element (i, j) in $\mathbf{K}_k$ by $k_{ij}^{(k)}$, we use

$$\tan 2\theta = \frac{2k_{ij}^{(k)}}{k_{ii}^{(k)} - k_{jj}^{(k)}} \qquad \text{for } k_{ii}^{(k)} \neq k_{jj}^{(k)} \tag{11.80}$$

and

$$\theta = \frac{\pi}{4} \qquad \text{for } k_{ii}^{(k)} = k_{jj}^{(k)} \tag{11.81}$$

It should be noted that the numerical evaluation of $\mathbf{K}_{k+1}$ in (11.77) requires only the linear combination of two rows and two columns. In addition, advantage should also be taken of the fact that $\mathbf{K}_k$ is symmetric for all k; i.e., we should only work on the upper (or lower) triangular part of the matrix, including its diagonal elements.

An important point to emphasize is that although the transformation in (11.77) reduces an off-diagonal element in $\mathbf{K}_k$ to zero, this element will again become nonzero during the transformations that follow. Therefore, for the design of an actual algorithm, we have to decide which element to reduce to zero. One choice is to always zero the largest off-diagonal element in $\mathbf{K}_k$. However, the search for the largest element is time consuming, and it may be preferable to simply carry out the Jacobi transformations systematically, row-by-row or column-by-column, which is known as the *cyclic Jacobi procedure*. Running once over all off-diagonal elements is one *sweep*. The disadvantage of this procedure is that regardless of its size, an off-diagonal element is always zeroed; i.e., the element may already be nearly zero and a rotation is still applied.

A procedure that has been used very effectively is the *threshold Jacobi method*, in which the off-diagonal elements are tested sequentially, namely row-by-row (or column-by-column), and a rotation is only applied if the element is larger than the threshold for that sweep. To define an appropriate threshold we note that, physically, in the diagonalization of $\mathbf{K}$ we want to reduce the coupling between the degrees of freedom i and j. A measure of this coupling is given by $(k_{ij}^2/k_{ii}k_{jj})^{1/2}$, and it is this factor that can be used effectively in deciding whether to apply a rotation. In addition to having a realistic threshold tolerance, it is also necessary to measure convergence. As described above, $\mathbf{K}_{k+1} \longrightarrow \mathbf{\Lambda}$ as $k \longrightarrow \infty$, but in the numerical computations we seek only a close-enough approximation to the eigenvalues and corresponding eigenvectors. Let l be the last iteration; i.e., we have, to the precision required,

$$\mathbf{K}_{l+1} \doteq \mathbf{\Lambda} \tag{11.82}$$

Then we say that convergence to a tolerance s has been achieved provided that

$$\frac{|k_{ii}^{(l+1)} - k_{ii}^{(l)}|}{k_{ii}^{(l+1)}} \leq 10^{-s}; \qquad i = 1, \ldots, n \tag{11.83}$$

and

$$\left[\frac{(k_{ij}^{(l+1)})^2}{k_{ii}^{(l+1)}k_{jj}^{(l+1)}}\right]^{1/2} \leq 10^{-s}; \qquad \text{all } i, j;\ i < j \tag{11.84}$$

The relation in (11.83) has to be satisfied because the element $k_{ii}^{(l+1)}$ is the current approximation to an eigenvalue, and the relation states that the current and last approximations to the eigenvalues did not change in the first s digits. This convergence measure is essentially the same as the one used in vector iteration in (11.20). The relation in (11.84) assures that the off-diagonal elements are indeed small.

Having discussed the main aspects of the iteration, we may now summarize the actual solution procedure. The following steps have been used in a threshold Jacobi iteration:

(1) Initialize the threshold for the sweep. Typically the threshold used for sweep m may be 10^{-2m}.
(2) For all (i, j) with $i < j$, calculate the coupling factor $[(k_{ij}^{(k)})^2/k_{ii}^{(k)}k_{jj}^{(k)}]^{1/2}$ and apply a transformation if the factor is larger than the current threshold.
(3) Use (11.83) to check for convergence. If the relation in (11.83) is not satisfied, continue with the next sweep; i.e., go to (1). If (11.83) is satisfied, check if (11.84) is also satisfied; if "yes," the iteration converged; if "no," continue with the next sweep.

So far we have stated the algorithm but we have not shown that convergence will indeed always occur. The proof of convergence has been given elsewhere[23-27] and will not be repeated here, because little additional insight into the working of the solution procedure would be gained. However, one important point should be noted—that convergence is quadratic once the off-diagonal elements are small. Since rapid convergence is obtained once the off-diagonal elements are small, little extra cost is involved in solving for the eigensystem to high accuracy when an approximate solution has been obtained. In practical solutions we use $m = 2$ and $s = 12$, and about six sweeps are required for solution of the eigensystem to high accuracy. A program used is given in the next section, when we discuss the solution of the generalized eigenproblem $\mathbf{K}\boldsymbol{\phi} = \lambda\mathbf{M}\boldsymbol{\phi}$.

Example 11.9

Calculate the eigensystem of the matrix $\mathbf{K}$, where

$$\mathbf{K} = \begin{bmatrix} 5 & -4 & 1 & 0 \\ -4 & 6 & -4 & 1 \\ 1 & -4 & 6 & -4 \\ 0 & 1 & -4 & 5 \end{bmatrix}$$

Use the threshold Jacobi iteration described above.

To demonstrate the solution algorithm we give one sweep in detail and then the results obtained in the next sweeps.

For sweep 1 we have as a threshold 10^{-2}. We therefore obtain the following results. For $i = 1, j = 2$:

$$\cos\theta = 0.7497; \qquad \sin\theta = 0.6618$$

and thus

$$\mathbf{P}_1 = \begin{bmatrix} 0.7497 & -0.6618 & 0 & 0 \\ 0.6618 & 0.7497 & 0 & 0 \\ 0 & 0 & 1 & 0 \\ 0 & 0 & 0 & 1 \end{bmatrix}$$

$$\mathbf{P}_1^T\mathbf{K}\mathbf{P}_1 = \begin{bmatrix} 1.469 & 0 & -1.898 & 0.6618 \\ 0 & 9.531 & -3.661 & 0.7497 \\ -1.898 & -3.661 & 6 & -4 \\ 0.6618 & 0.7497 & -4 & 5 \end{bmatrix}$$

For $i = 1, j = 3$:

$$\cos\theta = 0.9398; \qquad \sin\theta = 0.3416$$

$$\mathbf{P}_2 = \begin{bmatrix} 0.9398 & 0 & -0.3416 & 0 \\ 0 & 1 & 0 & 0 \\ 0.3416 & 0 & 0.9398 & 0 \\ 0 & 0 & 0 & 1 \end{bmatrix}$$

$$\mathbf{P}_2^T\mathbf{P}_1^T\mathbf{K}\mathbf{P}_1\mathbf{P}_2 = \begin{bmatrix} 0.7792 & -1.250 & 0 & -0.7444 \\ -1.250 & 9.531 & -3.440 & 0.7497 \\ 0 & -3.440 & 6.690 & -3.986 \\ -0.7444 & 0.7497 & -3.986 & 5 \end{bmatrix}$$

$$\mathbf{P}_1\mathbf{P}_2 = \begin{bmatrix} 0.7046 & -0.6618 & -0.2561 & 0 \\ 0.6220 & 0.7497 & -0.2261 & 0 \\ 0.3416 & 0 & 0.9398 & 0 \\ 0 & 0 & 0 & 1 \end{bmatrix}$$

For $i = 1, j = 4$:

$$\cos\theta = 0.9857; \qquad \sin\theta = 0.1687$$

$$\mathbf{P}_3 = \begin{bmatrix} 0.9857 & 0 & 0 & -0.1687 \\ 0 & 1 & 0 & 0 \\ 0 & 0 & 1 & 0 \\ 0.1687 & 0 & 0 & 0.9857 \end{bmatrix}$$

$$\mathbf{P}_3^T\mathbf{P}_2^T\mathbf{P}_1^T\mathbf{K}\mathbf{P}_1\mathbf{P}_2\mathbf{P}_3 = \begin{bmatrix} 0.6518 & -1.106 & -0.6725 & 0 \\ -1.106 & 9.531 & -3.440 & 0.9499 \\ -0.6725 & -3.440 & 6.690 & -3.928 \\ 0 & 0.9499 & -3.928 & 5.127 \end{bmatrix}$$

$$\mathbf{P}_1\mathbf{P}_2\mathbf{P}_3 = \begin{bmatrix} 0.6945 & -0.6618 & -0.2561 & -0.1189 \\ 0.6131 & 0.7497 & -0.2261 & -0.1050 \\ 0.3367 & 0 & 0.9398 & -0.0576 \\ 0.1687 & 0 & 0 & 0.9857 \end{bmatrix}$$

For $i = 2, j = 3$:

$$\cos\theta = 0.8312; \qquad \sin\theta = -0.5560$$

$$\mathbf{P}_4 = \begin{bmatrix} 1 & 0 & 0 & 0 \\ 0 & 0.8312 & 0.5560 & 0 \\ 0 & -0.5560 & 0.8312 & 0 \\ 0 & 0 & 0 & 1 \end{bmatrix}$$

$$\mathbf{P}_4^T\mathbf{P}_3^T\mathbf{P}_2^T\mathbf{P}_1^T\mathbf{K}\mathbf{P}_1\mathbf{P}_2\mathbf{P}_3\mathbf{P}_4 = \begin{bmatrix} 0.6518 & 0.5453 & -1.174 & 0 \\ -0.5453 & 11.83 & 0 & 2.974 \\ -1.174 & 0 & 4.388 & -2.737 \\ 0 & 2.974 & -2.737 & 5.127 \end{bmatrix}$$

$$\mathbf{P}_1\mathbf{P}_2\mathbf{P}_3\mathbf{P}_4 = \begin{bmatrix} 0.6945 & -0.4077 & -0.5808 & -0.1189 \\ 0.6131 & 0.7488 & 0.2289 & -0.1050 \\ 0.3367 & -0.5226 & 0.7812 & -0.0576 \\ 0.1682 & 0 & 0 & 0.9857 \end{bmatrix}$$

For $i = 2, j = 4$:

$$\cos\theta = 0.9349; \qquad \sin\theta = 0.3549$$

$$\mathbf{P}_5 = \begin{bmatrix} 1 & 0 & 0 & 0 \\ 0 & 0.9349 & 0 & -0.3549 \\ 0 & 0 & 1 & 0 \\ 0 & 0.3549 & 0 & 0.9349 \end{bmatrix}$$

$$\mathbf{P}_5^T\mathbf{P}_4^T\mathbf{P}_3^T\mathbf{P}_2^T\mathbf{P}_1^T\mathbf{K}\mathbf{P}_1\mathbf{P}_2\mathbf{P}_3\mathbf{P}_4\mathbf{P}_5 = \begin{bmatrix} 0.6518 & -0.5098 & -1.174 & 0.1935 \\ -0.5098 & 12.96 & 0.9713 & 0 \\ -1.174 & -0.9713 & 4.388 & -2.559 \\ 0.1935 & 0 & -2.559 & 3.999 \end{bmatrix}$$

$$\mathbf{P}_1\mathbf{P}_2\mathbf{P}_3\mathbf{P}_4\mathbf{P}_5 = \begin{bmatrix} 0.6945 & -0.4233 & -0.5808 & 0.0335 \\ 0.6131 & 0.6628 & 0.2289 & -0.3639 \\ 0.3367 & 0.5090 & 0.7812 & 0.1316 \\ 0.1687 & 0.3498 & 0 & 0.9213 \end{bmatrix}$$

To complete the sweep, we zero element (3, 4), using

$$\cos\theta = 0.7335; \qquad \sin\theta = -0.6797$$

$$\mathbf{P}_6 = \begin{bmatrix} 1 & 0 & 0 & 0 \\ 0 & 1 & 0 & 0 \\ 0 & 0 & 0.7335 & 0.6797 \\ 0 & 0 & -0.6797 & 0.7335 \end{bmatrix}$$

and hence the approximations obtained for $\mathbf{\Lambda}$ and $\mathbf{\Phi}$ are

$$\mathbf{\Lambda} = \mathbf{P}_6^T \ldots \mathbf{P}_1^T \mathbf{K} \mathbf{P}_1 \ldots \mathbf{P}_6$$

i.e.,

$$\mathbf{\Lambda} \doteq \begin{bmatrix} 0.6518 & -0.5098 & -0.9926 & -0.6560 \\ -0.5098 & 12.96 & -0.7124 & -0.6602 \\ -0.9926 & -0.7124 & 6.7596 & 0 \\ -0.6560 & -0.6602 & 0 & 1.6272 \end{bmatrix}$$

and

$$\mathbf{\Phi} \doteq \mathbf{P}_1 \ldots \mathbf{P}_6$$

i.e.,

$$\mathbf{\Phi} \doteq \begin{bmatrix} 0.6945 & -0.4233 & -0.4488 & -0.3702 \\ 0.6131 & 0.6628 & 0.4152 & -0.1113 \\ 0.3367 & -0.5090 & 0.4835 & 0.6275 \\ 0.1687 & 0.3498 & -0.6264 & 0.6759 \end{bmatrix}$$

After the second sweep we obtain

$$\mathbf{\Lambda} \doteq \begin{bmatrix} 0.1563 & -0.3635 & 0.0063 & -0.0176 \\ -0.3635 & 13.08 & -0.0020 & 0 \\ 0.0063 & -0.0020 & 6.845 & 0 \\ -0.0176 & 0 & 0 & 1.910 \end{bmatrix}$$

$$\mathbf{\Phi} \doteq \begin{bmatrix} 0.3875 & -0.3612 & -0.6017 & -0.5978 \\ 0.5884 & 0.6184 & 0.3710 & -0.3657 \\ 0.6148 & -0.5843 & 0.3714 & 0.3777 \\ 0.3546 & 0.3816 & -0.6020 & 0.6052 \end{bmatrix}$$

And after the third sweep we have

$$\mathbf{\Lambda} \doteq \begin{bmatrix} 0.1459 & & & \\ & 13.09 & & \\ & & 6.854 & \\ & & & 1.910 \end{bmatrix}$$

$$\mathbf{\Phi} \doteq \begin{bmatrix} 0.3717 & -0.3717 & -0.6015 & -0.6015 \\ 0.6015 & 0.6015 & 0.3717 & -0.3717 \\ 0.6015 & -0.6015 & 0.3717 & 0.3717 \\ 0.3717 & 0.3717 & -0.6015 & 0.6015 \end{bmatrix}$$

The approximation for $\mathbf{\Lambda}$ is diagonal to the precision given and we can use

$$\lambda_1 \doteq 0.1459; \qquad \boldsymbol{\phi}_1 \doteq \begin{bmatrix} 0.3717 \\ 0.6015 \\ 0.6015 \\ 0.3717 \end{bmatrix}$$

$$\lambda_2 \doteq 1.910; \qquad \boldsymbol{\phi}_2 \doteq \begin{bmatrix} -0.6015 \\ -0.3717 \\ 0.3717 \\ 0.6015 \end{bmatrix}$$

$$\lambda_3 \doteq 6.854; \qquad \boldsymbol{\phi}_3 \doteq \begin{bmatrix} -0.6015 \\ 0.3717 \\ 0.3717 \\ -0.6015 \end{bmatrix}$$

$$\lambda_4 \doteq 13.09; \qquad \boldsymbol{\phi}_4 \doteq \begin{bmatrix} -0.3717 \\ 0.6015 \\ -0.6015 \\ 0.3717 \end{bmatrix}$$

It should be noted that the eigenvalues and eigenvectors did not appear in the usual order in the approximation for $\boldsymbol{\Lambda}$ and $\boldsymbol{\Phi}$.

In the following example we demonstrate the quadratic convergence when the off-diagonal elements are already small.

Example 11.10

Consider the Jacobi solution of the eigenproblem $\mathbf{K}\boldsymbol{\phi} = \lambda\boldsymbol{\phi}$, where

$$\mathbf{K} = \begin{bmatrix} k_{11} & o(\epsilon) & o(\epsilon) \\ o(\epsilon) & k_{22} & o(\epsilon) \\ o(\epsilon) & o(\epsilon) & k_{33} \end{bmatrix}$$

The symbol $o(\epsilon)$ signifies "of order ϵ," where $\epsilon \ll k_{ii}$, $i = 1, 2, 3$. Show that after one complete sweep, all off-diagonal elements are of order ϵ^2, meaning that convergence is quadratic.

Since the rotations to be applied are small, we make the assumption that $\sin\theta = \theta$ and $\cos\theta = 1$. Hence the relation in (11.80) gives

$$\theta = \frac{k_{ij}^{(k)}}{k_{ii}^{(k)} - k_{jj}^{(k)}}$$

In one sweep we need to set to zero, in succession, once all off-diagonal elements. Using $\mathbf{K}_1 = \mathbf{K}$ we obtain $\mathbf{K}_2$ by zeroing element (1, 2) in $\mathbf{K}_1$,

$$\mathbf{K}_2 = \mathbf{P}_1^T\mathbf{K}_1\mathbf{P}_1$$

where

$$\mathbf{P}_1 = \begin{bmatrix} 1 & \dfrac{-o(\epsilon)}{k_{11} - k_{22}} & 0 \\ \dfrac{o(\epsilon)}{k_{11} - k_{22}} & 1 & 0 \\ 0 & 0 & 1 \end{bmatrix}$$

Hence

$$\mathbf{K}_2 = \begin{bmatrix} k_{11} + o(\epsilon^2) & 0 & o(\epsilon) \\ 0 & k_{22} + o(\epsilon^2) & o(\epsilon) \\ o(\epsilon) & o(\epsilon) & k_{33} \end{bmatrix}$$

Similarly, we zero element (1, 3) in $\mathbf{K}_2$ to obtain $\mathbf{K}_3$,

$$\mathbf{K}_3 = \begin{bmatrix} k_{11} + o(\epsilon^2) & o(\epsilon^2) & 0 \\ o(\epsilon^2) & k_{22} + o(\epsilon^2) & o(\epsilon) \\ 0 & o(\epsilon) & k_{33} + o(\epsilon^2) \end{bmatrix}$$

Finally, we zero element (2, 3) in $\mathbf{K}_3$ and have

$$\mathbf{K}_4 = \begin{bmatrix} k_{11} + o(\epsilon^2) & o(\epsilon^2) & o(\epsilon^2) \\ o(\epsilon^2) & k_{22} + o(\epsilon^2) & 0 \\ o(\epsilon^2) & 0 & k_{33} + o(\epsilon^2) \end{bmatrix}$$

with all off-diagonal elements at least $o(\epsilon^2)$.

11.3.2 The Generalized Jacobi Method

In the previous section we discussed the solution of the standard eigenproblem $\mathbf{K}\boldsymbol{\phi} = \lambda\boldsymbol{\phi}$ using the conventional Jacobi rotation matrices in order to reduce $\mathbf{K}$ into diagonal form. To solve the generalized problem $\mathbf{K}\boldsymbol{\phi} = \lambda\mathbf{M}\boldsymbol{\phi}$, $\mathbf{M} \neq \mathbf{I}$, using the standard Jacobi method it would be necessary to first transform the problem into a standard form. However, this transformation can be dispensed with by using a generalized Jacobi solution method, which operates directly on $\mathbf{K}$ and $\mathbf{M}$.[11,28] The algorithm proceeds as summarized in (11.72) to (11.76) and is a natural extension of the standard Jacobi solution scheme; i.e., the generalized method reduces to the scheme presented for the problem $\mathbf{K}\boldsymbol{\phi} = \lambda\boldsymbol{\phi}$ when $\mathbf{M}$ is an identity matrix.

Referring to the discussion in the previous section, in the generalized Jacobi iteration we use the following matrix $\mathbf{P}_k$:

$$\mathbf{P}_k = \begin{bmatrix} 1 & & & & & & \\ & \ddots & & & & & \\ & & 1 & & \alpha & & \\ & & & \ddots & & & \\ & & \gamma & & 1 & & \\ & & & & & \ddots & \\ & & & & & & 1 \end{bmatrix} \begin{matrix} \\ \\ \leftarrow i\text{th} \\ \\ \leftarrow j\text{th row} \\ \\ \\ \end{matrix} \qquad (i\text{th}, j\text{th column}) \tag{11.85}$$

where the constants α and γ are selected in such a way as to reduce to zero simultaneously elements (i, j) in $\mathbf{K}_k$ and $\mathbf{M}_k$. Therefore, the values of α and γ are a function of the elements $k_{ij}^{(k)}$, $k_{ii}^{(k)}$, $k_{jj}^{(k)}$, $m_{ij}^{(k)}$, $m_{ii}^{(k)}$, and $m_{jj}^{(k)}$, where the superscript (k) indicates that the kth iteration is considered. Performing the multiplications $\mathbf{P}_k^T\mathbf{K}_k\mathbf{P}_k$ and $\mathbf{P}_k^T\mathbf{M}_k\mathbf{P}_k$ and using the condition that $k_{ij}^{(k+1)}$ and $m_{ij}^{(k+1)}$ shall be zero, we obtain the following two equations for α and γ:

$$\alpha k_{ii}^{(k)} + (1 + \alpha\gamma)k_{ij}^{(k)} + \gamma k_{jj}^{(k)} = 0 \tag{11.86}$$

and

$$\alpha m_{ii}^{(k)} + (1 + \alpha\gamma)m_{ij}^{(k)} + \gamma m_{jj}^{(k)} = 0 \tag{11.87}$$

If

$$\frac{k_{ii}^{(k)}}{m_{ii}^{(k)}} = \frac{k_{jj}^{(k)}}{m_{jj}^{(k)}} = \frac{k_{ij}^{(k)}}{m_{ij}^{(k)}}$$

(i.e., the submatrices considered are scalar multiples, which may be regarded to be a trivial case), we use $\alpha = 0$ and $\gamma = -k_{ij}^{(k)}/k_{jj}^{(k)}$. In general, to solve for α and γ from (11.86) and and (11.87) we define

$$\left.\begin{aligned} \bar{k}_{ii}^{(k)} &= k_{ii}^{(k)}m_{ij}^{(k)} - m_{ii}^{(k)}k_{ij}^{(k)} \\ \bar{k}_{jj}^{(k)} &= k_{jj}^{(k)}m_{ij}^{(k)} - m_{jj}^{(k)}k_{ij}^{(k)} \\ \bar{k}^{(k)} &= k_{ii}^{(k)}m_{jj}^{(k)} - k_{jj}^{(k)}m_{ii}^{(k)} \end{aligned}\right\} \tag{11.88}$$

and

$$\gamma = -\frac{\bar{k}_{ii}^{(k)}}{x}; \qquad \alpha = \frac{\bar{k}_{jj}^{(k)}}{x} \tag{11.89}$$

The value of x to obtain α and γ is then to be determined using

$$x = \frac{\bar{k}^{(k)}}{2} + \operatorname{sign}(\bar{k}^{(k)})\sqrt{\left(\frac{\bar{k}^{(k)}}{2}\right)^2 + \bar{k}_{ii}^{(k)}\bar{k}_{jj}^{(k)}} \tag{11.90}$$

The above relations for α and γ are used and have primarily been developed for the case when $\mathbf{M}$ is a positive definite full or banded mass matrix. In that case (and, in fact, also under less restrictive conditions) we have that

$$\left(\frac{\bar{k}^{(k)}}{2}\right)^2 + \bar{k}_{ii}^{(k)}\bar{k}_{jj}^{(k)} > 0,$$

and hence x is always nonzero. In addition, also, $\det \mathbf{P}_k \neq 0$, which indeed is the necessary condition for the algorithm to work.

The generalized Jacobi solution procedure has been used a great deal in the subspace iteration method (see Section 12.3) and when a consistent mass idealization is employed. However, other situations may arise as well. Assume that $\mathbf{M}$ is a diagonal mass matrix, $\mathbf{M} \neq \mathbf{I}$, and $m_{ii} > 0$, in which case we employ in (11.88)

$$\bar{k}_{ii}^{(k)} = -m_{ii}^{(k)}k_{ij}^{(k)}; \qquad \bar{k}_{jj}^{(k)} = -m_{jj}^{(k)}k_{ij}^{(k)} \tag{11.91}$$

and otherwise (11.85) to (11.90) are used as before. However, if $\mathbf{M} = \mathbf{I}$, the relation in (11.87) yields that $\alpha = -\gamma$, and we recognize that $\mathbf{P}_k$ in (11.85) is a multiple of the rotation matrix defined in (11.79) (see Example 11.11).

TABLE 11.3 SUMMARY OF GENERALIZED JACOBI SOLUTION

Operation	Calculation	Number of Operations	Required Storage
Calculation of coupling factors	$\dfrac{k_{ij}^{(k)^2}}{k_{ii}^{(k)}k_{jj}^{(k)}}, \quad \dfrac{m_{ij}^{(k)^2}}{m_{ii}^{(k)}m_{jj}^{(k)}}$	6	
Transformation to zero elements (i, j)	$\bar{k}_{ii}^{(k)} = k_{ii}^{(k)}m_{ij}^{(k)} - m_{ii}^{(k)}k_{ij}^{(k)}$ $\bar{k}_{jj}^{(k)} = k_{jj}^{(k)}m_{ij}^{(k)} - m_{jj}^{(k)}k_{ij}^{(k)}$ $\bar{k}^{(k)} = k_{ii}^{(k)}m_{jj}^{(k)} - k_{jj}^{(k)}m_{ii}^{(k)}$ $x = \dfrac{\bar{k}^{(k)}}{2} + (\text{sign } \bar{k}^{(k)})\sqrt{\left[\left(\dfrac{\bar{k}^{(k)}}{2}\right)^2 + \bar{k}_{ii}^{(k)}\bar{k}_{jj}^{(k)}\right]}$ $\gamma = -\dfrac{\bar{k}_{ii}^{(k)}}{x}, \quad \alpha = \dfrac{\bar{k}_{jj}^{(k)}}{x}$ $\mathbf{K}_{k+1} = \mathbf{P}_k^T\mathbf{K}_k\mathbf{P}_k, \quad \mathbf{M}_{k+1} = \mathbf{P}_k^T\mathbf{M}_k\mathbf{P}_k$	$4n + 12$	Using symmetry of matrices $n(n + 2)$
Calculation of eigenvectors	$(\mathbf{P}_1 \ldots \mathbf{P}_{k-1})\mathbf{P}_k$	$2n$	n^2
	Total for one sweep	$3n^3 + 6n^2$	$2n^2 + 2n$

In addition, it should be mentioned that the solution procedure can be adapted to solve the problem $\mathbf{K}\boldsymbol{\phi} = \lambda\mathbf{M}\boldsymbol{\phi}$, when $\mathbf{M}$ is a diagonal matrix with some zero diagonal elements.

The complete solution process is analogous to the Jacobi iteration in the solution of the problem $\mathbf{K}\boldsymbol{\phi} = \lambda\boldsymbol{\phi}$, which was presented in the preceding section. The differences lie in that now a mass coupling factor $[(m_{ij}^{(k)})^2/m_{ii}^{(k)}m_{jj}^{(k)}]^{1/2}$ must also be calculated, unless $\mathbf{M}$ is diagonal, and the transformation is applied to $\mathbf{K}_k$ and $\mathbf{M}_k$. Convergence is measured by comparing successive eigenvalue approximations and by testing if all off-diagonal elements are small enough; i.e., with l being the last iteration, convergence has been achieved if

$$\frac{|\lambda_i^{(l+1)} - \lambda_i^{(l)}|}{\lambda_i^{(l+1)}} \leq 10^{-s}; \qquad i = 1, \ldots, n \tag{11.92}$$

where

$$\lambda_i^{(l)} = \frac{k_{ii}^{(l)}}{m_{ii}^{(l)}}; \qquad \lambda_i^{(l+1)} = \frac{k_{ii}^{(l+1)}}{m_{ii}^{(l+1)}} \tag{11.93}$$

and

$$\left[\frac{(k_{ij}^{(l+1)})^2}{k_{ii}^{(l+1)}k_{jj}^{(l+1)}}\right]^{1/2} \leq 10^{-s}; \qquad \left[\frac{(m_{ij}^{(l+1)})^2}{m_{ii}^{(l+1)}m_{jj}^{(l+1)}}\right]^{1/2} \leq 10^{-s}; \qquad \text{all } i, j; \qquad i < j \tag{11.94}$$

where 10^{-s} is the convergence tolerance.

Table 11.3 summarizes the solution procedure for the case $\mathbf{M}$ full (or banded) and positive definite. The relations given in Table 11.3 are employed directly in the Subroutine JACOBI, which is presented at the end of this section. We should note that Table 11.3 also gives an operation count of the solution process and the storage requirements. The total number of operations in one sweep as given in the table are an upper bound because it is assumed that both matrices are full and that all off-diagonal elements are zeroed; i.e., the threshold tolerance is never passed. Considering the number of sweeps required for solution, the same experience as with the solution of standard eigenproblems holds; i.e., with $m = 2$ and $s = 12$ in the iteration (see Section 11.3.1) about six sweeps are required for solution of the eigensystem to high accuracy.

The following examples demonstrate some of the characteristics of the generalized Jacobi solution algorithm.

Example 11.11

Prove that the generalized Jacobi method reduces to the standard technique when $\mathbf{M} = \mathbf{I}$.

For the proof we need only consider the calculation of the transformation matrices that would be used to zero a typical off-diagonal element. We want to show that the transformation matrices obtained in the standard and generalized Jacobi methods are multiples of each other, namely in that case we could, by proper scaling, obtain the standard iteration scheme. Since each step of iteration consists of applying a rotation in the (i, j)th plane, we can without loss of generality consider the solution of the problem

$$\begin{bmatrix} k_{11} & k_{12} \\ k_{12} & k_{22} \end{bmatrix} \boldsymbol{\phi} = \lambda \boldsymbol{\phi}$$

Using (11.88) to (11.90) we thus obtain

$$\alpha = -\gamma; \qquad \mathbf{P}_1 = \begin{bmatrix} 1 & -\gamma \\ \gamma & 1 \end{bmatrix} \tag{a}$$

and

$$\gamma = \frac{-k_{11} + k_{22} \pm \sqrt{(k_{11} - k_{22})^2 + 4k_{12}^2}}{2k_{12}}$$

On the other hand, in the standard Jacobi solution we use

$$\mathbf{P}_1 = \begin{bmatrix} \cos\theta & -\sin\theta \\ \sin\theta & \cos\theta \end{bmatrix}$$

which may be written as

$$\mathbf{P}_1 = \cos\theta \begin{bmatrix} 1 & -\tan\theta \\ \tan\theta & 1 \end{bmatrix} \tag{b}$$

Thus, $\mathbf{P}_1$ in (b) would be a multiple of $\mathbf{P}_1$ in (a) if $\tan\theta = \gamma$. In the standard Jacobi method we obtain $\tan 2\theta$ using (11.80). In this case, we have

$$\tan 2\theta = \frac{2k_{12}}{k_{11} - k_{22}} \tag{c}$$

We also have, by simple trigonometry,

$$\tan 2\theta = \frac{2\tan\theta}{1 - \tan^2\theta} \tag{d}$$

Using (c) and (d) we can solve for $\tan\theta$ to be used in (b) and obtain

$$\tan\theta = \frac{-k_{11} + k_{22} \pm \sqrt{(k_{11} - k_{22})^2 + 4k_{12}^2}}{2k_{12}}$$

Hence $\gamma = \tan\theta$ and the generalized Jacobi iteration is equivalent to the standard method when $\mathbf{M} = \mathbf{I}$.

Example 11.12

Use the generalized Jacobi method to calculate the eigensystem of the problem $\mathbf{K}\boldsymbol{\phi} = \lambda\mathbf{M}\boldsymbol{\phi}$.

(1) In the first case let

$$\mathbf{K} = \begin{bmatrix} 1 & -1 \\ -1 & 1 \end{bmatrix}; \qquad \mathbf{M} = \begin{bmatrix} 2 & 1 \\ 1 & 2 \end{bmatrix}$$

We note that $\mathbf{K}$ is singular and hence we expect a zero eigenvalue.

(2) Then let

$$\mathbf{K} = \begin{bmatrix} 2 & 1 \\ 1 & 2 \end{bmatrix}; \qquad \mathbf{M} = \begin{bmatrix} 2 & 0 \\ 0 & 0 \end{bmatrix}$$

in which case we have an infinite eigenvalue.

For solution, we use the relations in (11.85) to (11.90). Considering the problem in case (1), we obtain

$$\bar{k}_{11}^{(1)} = 3; \qquad \bar{k}_{22}^{(1)} = 3; \qquad \bar{k}^{(1)} = 0$$

$$x = 3; \qquad \gamma = -1; \qquad \alpha = 1$$

$$\mathbf{P}_1 = \begin{bmatrix} 1 & 1 \\ -1 & 1 \end{bmatrix}$$

Hence

$$\mathbf{P}_1^T\mathbf{K}\mathbf{P}_1 = \begin{bmatrix} 4 & 0 \\ 0 & 0 \end{bmatrix}; \qquad \mathbf{P}_1^T\mathbf{M}\mathbf{P}_1 = \begin{bmatrix} 2 & 0 \\ 0 & 6 \end{bmatrix}$$

To obtain $\mathbf{\Lambda}$ and $\mathbf{\Phi}$ we use (11.75) and (11.76) and arrange the columns in the matrices in the appropriate order. Hence

$$\mathbf{\Lambda} = \begin{bmatrix} 0 & \\ & 2 \end{bmatrix}; \qquad \mathbf{\Phi} = \begin{bmatrix} \frac{1}{\sqrt{6}} & \frac{1}{\sqrt{2}} \\ \frac{1}{\sqrt{6}} & -\frac{1}{\sqrt{2}} \end{bmatrix}$$

Now consider the problem in case (2). Here we have

$$\bar{k}_{11}^{(1)} = -2; \qquad \bar{k}_{22}^{(1)} = 0; \qquad \bar{k}^{(1)} = -4$$

$$x = -4; \qquad \alpha = 0; \qquad \gamma = -\tfrac{1}{2}$$

$$\mathbf{P}_1 = \begin{bmatrix} 1 & 0 \\ -\frac{1}{2} & 1 \end{bmatrix}$$

Hence

$$\mathbf{P}_1^T\mathbf{K}\mathbf{P}_1 = \begin{bmatrix} \frac{3}{2} & 0 \\ 0 & 2 \end{bmatrix}; \qquad \mathbf{P}_1^T\mathbf{M}\mathbf{P}_1 = \begin{bmatrix} 2 & 0 \\ 0 & 0 \end{bmatrix}$$

and

$$\mathbf{\Lambda} = \begin{bmatrix} \frac{3}{4} & \\ & \infty \end{bmatrix}; \qquad \boldsymbol{\phi}_1 = \begin{bmatrix} \frac{1}{\sqrt{2}} \\ -\frac{1}{2\sqrt{2}} \end{bmatrix}$$

The above discussion of the generalized Jacobi solution method has already indicated in some way the advantages of the solution technique. Firstly, the transformation of the generalized eigenproblem to the standard form is avoided. This is particularly advantageous when (1) the matrices are ill-conditioned, and (2) the off-diagonal elements in $\mathbf{K}$ and $\mathbf{M}$ are already small or, equivalently, when there are only a few nonzero off-diagonal elements. In the first case the direct solution of $\mathbf{K}\boldsymbol{\phi} = \lambda\mathbf{M}\boldsymbol{\phi}$ avoids the solution of a standard eigenproblem of a matrix with very large and small elements (see Section 10.2.5). In the second case the eigenproblem is already nearly solved, because the zeroing of small or only a few off-diagonal elements in $\mathbf{K}$ and $\mathbf{M}$ will not result into a large change of the diagonal elements of the matrices, the ratios of which are the eigenvalues. In addition fast convergence can be expected when the off-diagonal elements are small (see Section 11.3.1). We will see that this case arises in the subspace iteration method described in Section 12.3, which is one reason why the generalized Jacobi method is used effectively in that technique.

It should be noted that the Jacobi solution methods solve simultaneously for all eigenvalues and corresponding eigenvectors. However, in finite element analysis we require in most cases only some eigenpairs, and the use of a Jacobi solution procedure can be very inefficient, in particular when the order of **K** and **M** is large. In such cases much more effective solution methods are available, which only solve for the specific eigenvalues and eigenvectors that are actually required. However, the generalized Jacobi solution method presented in this section can be used very effectively as part of those solution strategies (see Section 12.3). When the order of the matrices **K** and **M** is relatively small, the solution of the eigenproblem is not very expensive, and the Jacobi iteration may also be attractive because of its simplicity and elegance of solution.

Subroutine JACOBI: Program JACOBI is used to calculate all eigenvalues and corresponding eigenvectors of the generalized eigenproblem $\mathbf{K}\boldsymbol{\phi} = \lambda\mathbf{M}\boldsymbol{\phi}$. The argument variables and use of the subroutine are defined using comment cards in the program.

```
      SUBROUTINE JACOBI (A,B,X,EIGV,D,N,RTOL,NSMAX,IFPR,IOUT)
C .....................................................................
C .                                                                   .
C .   P R O G R A M                                                   .
C .        TO SOLVE THE GENERALIZED EIGENPROBLEM USING THE            .
C .        GENERALIZED JACOBI ITERATION                               .
C .                                                                   .
C . - - INPUT VARIABLES - -                                           .
C .        A(N,N)    = STIFFNESS MATRIX (ASSUMED POSITIVE DEFINITE)   .
C .        B(N,N)    = MASS MATRIX (ASSUMED POSITIVE DEFINITE)        .
C .        X(N,N)    = MATRIX STORING EIGENVECTORS ON SOLUTION EXIT   .
C .        EIGV(N)   = VECTOR STORING EIGENVALUES ON SOLUTION EXIT    .
C .        D(N)      = WORKING VECTOR                                 .
C .        N         = ORDER OF MATRICES A AND B                      .
C .        RTOL      = CONVERGENCE TOLERANCE (USUALLY SET TO 10.**-12).
C .        NSMAX     = MAXIMUM NUMBER OF SWEEPS ALLOWED               .
C .                                  (USUALLY SET TO 15)              .
C .        IFPR      = FLAG FOR PRINTING DURING ITERATION             .
C .            EQ.0    NO PRINTING                                    .
C .            EQ.1    INTERMEDIATE RESULTS ARE PRINTED               .
C .        IOUT      = OUTPUT DEVICE NUMBER                           .
C .                                                                   .
C . - - OUTPUT - -                                                    .
C .        A(N,N)    = DIAGONALIZED STIFFNESS MATRIX                  .
C .        B(N,N)    = DIGONALIZED MASS MATRIX                        .
C .        X(N,N)    = EIGENVECTORS STORED COLUMNWISE                 .
C .        EIGV(N)   = EIGENVALUES                                    .
C .                                                                   .
C .....................................................................
      IMPLICIT REAL*8 (A-H,O-Z)
      ABS(X)=DABS(X)
      SQRT(X)=DSQRT(X)
C .....................................................................
C .        THIS PROGRAM IS USED IN SINGLE PRECISION ARITHMETIC ON     .
C .        CDC EQUIPMENT AND DOUBLE PRECISION ARITHMETIC ON IBM       .
C .        OR UNIVAC MACHINES .ACTIVATE,DEACTIVATE OR ADJUST ABOVE    .
C .        CARDS FOR SINGLE OR DOUBLE PRECISION ARITHMETIC
C .....................................................................
      DIMENSION A(N,N),B(N,N),X(N,N),EIGV(N),D(N)
C
C     INITIALIZE EIGENVALUE AND EIGENVECTOR MATRICES
C
      DO 10 I=1,N
      IF (A(I,I).GT.0. .AND. B(I,I).GT.0.) GO TO 4
      WRITE (IOUT,2020)
```

```
      STOP
    4 D(I)=A(I,I)/B(I,I)
   10 EIGV(I)=D(I)
      DO 30 I=1,N
      DO 20 J=1,N
   20 X(I,J)=0.
   30 X(I,I)=1.
      IF(N.EQ.1) RETURN
C
C     INITIALIZE SWEEP COUNTER AND BEGIN ITERATION
C
      NSWEEP=0
      NR=N-1
   40 NSWEEP=NSWEEP+1
      IF(IFPR.EQ.1) WRITE(IOUT,2000) NSWEEP
C
C     CHECK IF PRESENT OFF-DIAGONAL ELEMENT IS LARGE ENOUGH TO REQUIRE ZEROING
C
      EPS=(.01**NSWEEP)**2
      DO 210 J=1,NR
      JJ=J+1
      DO 210 K=JJ,N
      EPTOLA=(A(J,K)*A(J,K))/(A(J,J)*A(K,K))
      EPTOLB=(B(J,K)*B(J,K))/(B(J,J)*B(K,K))
      IF((EPTOLA.LT.EPS).AND.(EPTOLB.LT.EPS))GO TO 210
C
C     IF ZEROING IS REQUIRED, CALCULATE THE ROTATION MATRIX ELEMENTS CA AND CG
C
      AKK=A(K,K)*B(J,K)-B(K,K)*A(J,K)
      AJJ=A(J,J)*B(J,K)-B(J,J)*A(J,K)
      AB=A(J,J)*B(K,K)-A(K,K)*B(J,J)
      CHECK=(AB*AB+4.*AKK*AJJ)/4.
      IF(CHECK) 50,60,60
   50 WRITE(IOUT,2020)
      STOP
   60 SQCH=SQRT(CHECK)
      D1=AB/2.+SQCH
      D2=AB/2.-SQCH
      DEN=D1
      IF(ABS(D2).GT.ABS(D1)) DEN=D2
      IF(DEN) 80,70,80
   70 CA=0.
      CG=-A(J,K)/A(K,K)
      GO TO 90
   80 CA=AKK/DEN
      CG=-AJJ/DEN
C
C     PERFORM THE GENERALIZED ROTATION TO ZERO THE PRESENT OFF-DIAGONAL ELEMENT
C
   90 IF(N-2) 100,190,100
  100 JP1=J+1
      JM1=J-1
      KP1=K+1
      KM1=K-1
      IF(JM1-1) 130,110,110
  110 DO 120 I=1,JM1
      AJ=A(I,J)
      BJ=B(I,J)
      AK=A(I,K)
      BK=B(I,K)
      A(I,J)=AJ+CG*AK
      B(I,J)=BJ+CG*BK
      A(I,K)=AK+CA*AJ
  120 B(I,K)=BK+CA*BJ
  130 IF(KP1-N) 140,140,160
  140 DO 150 I=KP1,N
      AJ=A(J,I)
      BJ=B(J,I)
      AK=A(K,I)
      BK=B(K,I)
      A(J,I)=AJ+CG*AK
      B(J,I)=BJ+CG*BK
      A(K,I)=AK+CA*AJ
  150 B(K,I)=BK+CA*BJ
  160 IF(JP1-KM1) 170,170,190
  170 DO 180 I=JP1,KM1
      AJ=A(J,I)
      BJ=B(J,I)
      AK=A(I,K)
      BK=B(I,K)
      A(J,I)=AJ+CG*AK
      B(J,I)=BJ+CG*BK
      A(I,K)=AK+CA*AJ
  180 B(I,K)=BK+CA*BJ
  190 AK=A(K,K)
      BK=B(K,K)
      A(K,K)=AK+2.*CA*A(J,K)+CA*CA*A(J,J)
      B(K,K)=BK+2.*CA*B(J,K)+CA*CA*B(J,J)
      A(J,J)=A(J,J)+2.*CG*A(J,K)+CG*CG*AK
      B(J,J)=B(J,J)+2.*CG*B(J,K)+CG*CG*BK
      A(J,K)=0.
      B(J,K)=0.
C
```

```
C     UPDATE THE EIGENVECTOR MATRIX AFTER EACH ROTATION
C
      DO 200 I=1,N
      XJ=X(I,J)
      XK=X(I,K)
      X(I,J)=XJ+CG*XK
  200 X(I,K)=XK+CA*XJ
  210 CONTINUE
C
C     UPDATE THE EIGENVALUES AFTER EACH SWEEP
C
      DO 220 I=1,N
      IF (A(I,I).GT.0. .AND. B(I,I).GT.0.) GO TO 220
      WRITE(IOUT,2020)
      STOP
  220 EIGV(I)=A(I,I)/B(I,I)
      IF(IFPR.EQ.0) GO TO 230
      WRITE(IOUT,2030)
      WRITE(IOUT,2010) (EIGV(I),I=1,N)
C
C     CHECK FOR CONVERGENCE
C
  230 DO 240 I=1,N
      TOL=RTOL*D(I)
      DIF=ABS(EIGV(I)-D(I))
      IF(DIF.GT.TOL) GO TO 280
  240 CONTINUE
C
C     CHECK ALL OFF-DIAGONAL ELEMENTS TO SEE IF ANOTHER SWEEP IS REQUIRED
C
      EPS=RTOL**2
      DO 250 J=1,NR
      JJ=J+1
      DO 250 K=JJ,N
      EPSA=(A(J,K)*A(J,K))/(A(J,J)*A(K,K))
      EPSB=(B(J,K)*B(J,K))/(B(J,J)*B(K,K))
      IF((EPSA.LT.EPS).AND.(EPSB.LT.EPS))GO TO 250
      GO TO 280
  250 CONTINUE
C
C     FILL OUT BOTTOM TRIANGLE OF RESULTANT MATRICES AND SCALE EIGENVECTORS
C
  255 DO 260 I=1,N
      DO 260 J=1,N
      A(J,I)=A(I,J)
  260 B(J,I)=B(I,J)
      DO 270 J=1,N
      BB=SQRT(B(J,J))
      DO 270 K=1,N
  270 X(K,J)=X(K,J)/BB
      RETURN
C
C     UPDATE  D  MATRIX AND START NEW SWEEP, IF ALLOWED
C
  280 DO 290 I=1,N
  290 D(I)=EIGV(I)
      IF(NSWEEP.LT.NSMAX) GO TO 40
      GO TO 255
 2000 FORMAT(27H0SWEEP NUMBER IN *JACOBI* = ,I4)
 2010 FORMAT(1H0,6E20.12)
 2020 FORMAT (25H0*** ERROR  SOLUTION STOP /
     1        30H MATRICES NOT POSITIVE DEFINITE)
 2030 FORMAT(36H0CURRENT EIGENVALUES IN *JACOBI* ARE,/)
      END
```

11.3.3 The Householder-QR-Inverse Iteration Solution

Another most important transformation solution technique is the Householder-QR-inverse iteration (HQRI) method, although this procedure is restricted to the solution of the standard eigenproblem.[1,17,18,29-36] Therefore, if the generalized eigenproblem $\mathbf{K}\boldsymbol{\phi} = \lambda\mathbf{M}\boldsymbol{\phi}$ is considered, it must first be transformed into the standard form before the HQRI solution technique can be used. As pointed out in Section 10.2.5, this transformation is only effective in some cases.

In the following discussion, we consider the problem $\mathbf{K}\boldsymbol{\phi} = \lambda\boldsymbol{\phi}$, in which

$\mathbf{K}$ may have zero (and could also have negative) eigenvalues. Therefore, it is not necessary to impose a shift prior to applying the HQRI algorithm in order to solve for only positive eigenvalues (see Section 10.2.3). The name "HQRI solution method" stands for the following three solution steps:

1. Householder transformations are employed to reduce the matrix $\mathbf{K}$ into tridiagonal form.
2. QR iteration yields all eigenvalues.
3. Using inverse iteration, the required eigenvectors of the tridiagonal matrix are calculated and transformed to obtain the eigenvectors of $\mathbf{K}$.

A basic difference to the Jacobi solution method is therefore that the matrix is first transformed without iteration into a tridiagonal form. This matrix can then be used effectively in the QR iterative solution, in which all eigenvalues are calculated. Finally, only those eigenvectors that are actually requested are evaluated. We will note that unless many eigenvectors must be calculated, the transformation of $\mathbf{K}$ into tridiagonal form requires most of the numerical operations. In the following we consider in detail the three distinct steps carried out in the HQRI solution.

The Householder reduction: The Householder reduction to tridiagonal form involves $(n - 2)$ transformations of the form (11.72); i.e., using $\mathbf{K}_1 = \mathbf{K}$, we calculate

$$\mathbf{K}_{k+1} = \mathbf{P}_k^T \mathbf{K}_k \mathbf{P}_k; \qquad k = 1, \ldots, n-2 \tag{11.95}$$

where the $\mathbf{P}_k$ are Householder transformation matrices (see Sections 2.5 and 7.3.2):

$$\mathbf{P}_k = \mathbf{I} - \theta \mathbf{w}_k \mathbf{w}_k^T \tag{11.96}$$

$$\theta = \frac{2}{\mathbf{w}_k^T \mathbf{w}_k} \tag{11.97}$$

To show how the vector $\mathbf{w}_k$ which defines the matrix $\mathbf{P}_k$ is calculated, we consider the case $k = 1$, which is typical. First, we partition $\mathbf{K}_1$, $\mathbf{P}_1$, and $\mathbf{w}_1$ into submatrices as follows:

$$\left.\begin{aligned} \mathbf{P}_1 &= \left[\begin{array}{c|c} 1 & \mathbf{0} \\ \hline \mathbf{0} & \bar{\mathbf{P}}_1 \end{array}\right]; \qquad \mathbf{w}_1 = \left[\begin{array}{c} 0 \\ \hline \bar{\mathbf{w}}_1 \end{array}\right] \\ \mathbf{K}_1 &= \left[\begin{array}{c|c} k_{11} & \mathbf{k}_1^T \\ \hline \mathbf{k}_1 & \mathbf{K}_{11} \end{array}\right] \end{aligned}\right\} \tag{11.98}$$

where $\mathbf{K}_{11}$, $\bar{\mathbf{P}}_1$, and $\bar{\mathbf{w}}_1$ are of order $(n - 1)$. In the general case of step k, we have corresponding matrices of order $(n - k)$. Performing the multiplication in (11.95), we obtain, using the notation of (11.98),

$$\mathbf{K}_2 = \left[\begin{array}{c|c} k_{11} & \mathbf{k}_1^T \bar{\mathbf{P}}_1 \\ \hline \bar{\mathbf{P}}_1^T \mathbf{k}_1 & \bar{\mathbf{P}}_1^T \mathbf{K}_{11} \bar{\mathbf{P}}_1 \end{array}\right] \tag{11.99}$$

The condition is now that the first column and row of $\mathbf{K}_2$ be in tridiagonal form; i.e., we want that $\mathbf{K}_2$ is of the form

$$\mathbf{K}_2 = \left[\begin{array}{c|cccc} k_{11} & \text{x} & 0 & \cdots & 0 \\ \hline \text{x} & & & & \\ 0 & & & & \\ \cdot & & \bar{\mathbf{K}}_2 & & \\ \cdot & & & & \\ \cdot & & & & \\ 0 & & & & \end{array}\right] \tag{11.100}$$

where x indicates a nonzero value, and

$$\bar{\mathbf{K}}_2 = \bar{\mathbf{P}}_1^T \mathbf{K}_{11} \bar{\mathbf{P}}_1 \tag{11.101}$$

The form of $\mathbf{K}_2$ in (1.100) is achieved by realizing that $\bar{\mathbf{P}}_1$ is a reflection matrix (see Section 2.5). Therefore, we can use $\bar{\mathbf{P}}_1$ to reflect the vector $\mathbf{k}_1$ of $\mathbf{K}_1$ in (11.98) into a vector that has only its first component nonzero. Since the length of the new vector must be the length of $\mathbf{k}_1$, we determine $\bar{\mathbf{w}}_1$ from the condition

$$(\mathbf{I} - \theta \bar{\mathbf{w}}_1 \bar{\mathbf{w}}_1^T)\mathbf{k}_1 = \pm \|\mathbf{k}_1\|_2 \mathbf{e}_1 \tag{11.102}$$

where $\mathbf{e}_1$ is a unit vector of dimension $(n-1)$; i.e., $\mathbf{e}_1^T = [1 \quad 0 \quad 0 \ldots 0]$, and the $+$ or $-$ sign can be selected to obtain best numerical stability. Noting that we only need to solve for a multiple of $\bar{\mathbf{w}}_1$ [i.e., only the direction of the vector is important (see Section 2.5)], we obtain from (11.102) as a suitable value for $\bar{\mathbf{w}}_1$,

$$\bar{\mathbf{w}}_1 = \mathbf{k}_1 + \text{sign}\,(k_{21}) \|\mathbf{k}_1\|_2 \mathbf{e}_1 \tag{11.103}$$

where k_{21} is element (2, 1) of $\mathbf{K}_1$.

With $\bar{\mathbf{w}}_1$ defined in (11.103), the first Householder transformation, $k = 1$ in (11.95), can be carried out. In the next step, $k = 2$, we can consider the matrix $\bar{\mathbf{K}}_2$ in (11.100) in the same way as we considered $\mathbf{K}_1$ in (11.98) to (11.103), because the reduction of the first column and row of $\bar{\mathbf{K}}_2$ does not affect the first column and row in $\mathbf{K}_2$. Thus the general algorithm for the transformation of $\mathbf{K}$ into tridiagonal form is established. We demonstrate the procedure in the following example.

Example 11.13

Use Householder transformation matrices to reduce $\mathbf{K}$ to tridiagonal form, where

$$\mathbf{K} = \begin{bmatrix} 5 & -4 & 1 & 0 \\ -4 & 6 & -4 & 1 \\ 1 & -4 & 6 & -4 \\ 0 & 1 & -4 & 5 \end{bmatrix}$$

Here we have using (11.95) to (11.103) to reduce column 1,

$$\bar{\mathbf{w}}_1 = \begin{bmatrix} -4 \\ 1 \\ 0 \end{bmatrix} - 4.123 \begin{bmatrix} 1 \\ 0 \\ 0 \end{bmatrix} = \begin{bmatrix} -8.1231 \\ 1 \\ 0 \end{bmatrix}$$

Hence

$$\mathbf{w}_1 = \begin{bmatrix} 0 \\ -8.1231 \\ 1 \\ 0 \end{bmatrix}; \qquad \theta_1 = 0.0298575$$

$$\mathbf{P}_1 = \begin{bmatrix} 1 & 0 & 0 & 0 \\ 0 & -0.9701 & 0.2425 & 0 \\ 0 & 0.2425 & 0.9701 & 0 \\ 0 & 0 & 0 & 1 \end{bmatrix}$$

and

$$\mathbf{K}_2 = \begin{bmatrix} 5 & 4.1231 & 0 & 0 \\ 4.1231 & 7.8823 & 3.5294 & -1.9403 \\ 0 & 3.5294 & 4.1177 & -3.6380 \\ 0 & -1.9403 & -3.6380 & 5 \end{bmatrix}$$

Next we reduce column 2,

$$\bar{\mathbf{w}}_2 = \begin{bmatrix} 3.5294 \\ -1.9403 \end{bmatrix} + 4.0276 \begin{bmatrix} 1 \\ 0 \end{bmatrix} = \begin{bmatrix} 7.5570 \\ -1.9403 \end{bmatrix}$$

$$\mathbf{w}_2 = \begin{bmatrix} 0 \\ 0 \\ 7.5570 \\ -1.9403 \end{bmatrix}; \qquad \theta_2 = 0.0328553$$

$$\mathbf{P}_2 = \begin{bmatrix} 1 & 0 & 0 & 0 \\ 0 & 1 & 0 & 0 \\ 0 & 0 & -0.8763 & 0.4817 \\ 0 & 0 & 0.4817 & 0.8763 \end{bmatrix}$$

Hence

$$\mathbf{K}_3 = \begin{bmatrix} 5 & 4.1231 & 0 & 0 \\ 4.1231 & 7.8823 & -4.0276 & 0 \\ 0 & -4.0276 & 7.3941 & 2.3219 \\ 0 & 0 & 2.3219 & 1.7236 \end{bmatrix}$$

Some important numerical aspects should be noted. First, the reduced matrices $\mathbf{K}_2, \mathbf{K}_3, \ldots, \mathbf{K}_{n-1}$ are symmetric. This means that in the reduction we need only store the lower symmetric part of $\mathbf{K}$. Furthermore, to store the

$\bar{\mathbf{w}}_k$, $k = 1, 2, \ldots, n-2$, we can use the storage locations below the sub-diagonal elements in the currently being reduced matrix.

A disadvantage of the Householder transformations is that the bandwidth is increased in the unreduced part of $\mathbf{K}_{k+1}$. Hence, in the reduction, essentially no advantage can be taken of the bandedness of $\mathbf{K}$.

An important aspect in the transformation is the evaluation of the matrix product $\bar{\mathbf{P}}_1^T\mathbf{K}_{11}\bar{\mathbf{P}}_1$ and the similar products required in the next steps. In the most general case a triple matrix product with matrices of order n requires $2n^3$ operations, and if this many operations were required, the Householder reduction would be quite uneconomical. However, by taking advantage of the special nature of the matrix $\bar{\mathbf{P}}_1$, we can evaluate the product $\bar{\mathbf{P}}_1^T\mathbf{K}_{11}\bar{\mathbf{P}}_1$ by calculating

$$\left.\begin{aligned} \mathbf{v}_1 &= \mathbf{K}_{11}\bar{\mathbf{w}}_1 \\ \mathbf{p}_1^T &= \theta_1\mathbf{v}_1^T \\ \beta_1 &= \mathbf{p}_1^T\bar{\mathbf{w}}_1 \\ \mathbf{q}_1 &= \mathbf{p}_1 - \theta_1\beta_1\bar{\mathbf{w}}_1 \end{aligned}\right\} \tag{11.104}$$

and then

$$\bar{\mathbf{P}}_1^T\mathbf{K}_{11}\bar{\mathbf{P}}_1 = \mathbf{K}_{11} - \bar{\mathbf{w}}_1\mathbf{p}_1^T - \mathbf{q}_1\bar{\mathbf{w}}_1^T \tag{11.105}$$

which requires only about $3m^2 + 3m$ operations, where m is the order of $\bar{\mathbf{P}}_1$ and $\mathbf{K}_{11}$ (i.e., $m = n - 1$ in this case). Hence the multiplication $\bar{\mathbf{P}}_1^T\mathbf{K}_{11}\bar{\mathbf{P}}_1$ requires a number of operations of the order m-squared rather than m-cubed, which is a very significant reduction. We demonstrate the procedure given in (11.104) and (11.105) by reworking Example 11.13.

Example 11.14

Use the relations given in (11.104) and (11.105) to reduce the matrix $\mathbf{K}$ of Example 11.13 into tridiagonal form.

Here we obtain for the reduction of column 1 using $\bar{\mathbf{w}}_1$ and θ_1 calculated in Example 11.13,

$$\mathbf{v}_1 = \begin{bmatrix} 6 & -4 & 1 \\ -4 & 6 & -4 \\ 1 & -4 & 5 \end{bmatrix}\bar{\mathbf{w}}_1 = \begin{bmatrix} -52.738 \\ 38.4924 \\ -12.1231 \end{bmatrix}$$

$$\mathbf{p}_1^T = [-1.5746 \quad 1.1493 \quad -0.36197]; \qquad \beta_1 = 13.9403; \qquad \mathbf{q}_1 = \begin{bmatrix} 1.8064 \\ 0.7331 \\ -0.3620 \end{bmatrix}$$

$$\bar{\mathbf{P}}_1^T\mathbf{K}_{11}\bar{\mathbf{P}}_1 = \begin{bmatrix} 6 & -4 & 1 \\ -4 & 6 & -4 \\ 1 & -4 & 5 \end{bmatrix} - \begin{bmatrix} 12.7910 & -9.335 & 2.9403 \\ -1.5746 & 1.1493 & -0.3620 \\ 0 & 0 & 0 \end{bmatrix}$$

$$- \begin{bmatrix} -14.6734 & 1.8064 & 0 \\ -5.9548 & 0.73307 & 0 \\ 2.9403 & -0.3620 & 0 \end{bmatrix}$$

or

$$\bar{\mathbf{P}}_1^T\mathbf{K}_{11}\bar{\mathbf{P}}_1 = \begin{bmatrix} 7.8823 & 3.5294 & -1.9403 \\ 3.5294 & 4.1177 & -3.6380 \\ -1.9403 & -3.6380 & 5 \end{bmatrix}$$

and hence

$$\mathbf{K}_2 = \begin{bmatrix} 5 & 4.1231 & 0 & 0 \\ 4.1231 & 7.8823 & 3.5294 & -1.9403 \\ 0 & 3.5294 & 4.1177 & -3.6380 \\ 0 & -1.9403 & -3.6380 & 5 \end{bmatrix}$$

Next we reduce the second column

$$\mathbf{v}_2 = \begin{bmatrix} 4.1177 & -3.6380 \\ -3.6380 & 5 \end{bmatrix} \bar{\mathbf{w}}_2 = \begin{bmatrix} 38.1759 \\ -37.1941 \end{bmatrix}$$

$$\mathbf{p}_2^T = [1.2543 \quad -1.2220]; \qquad \beta_2 = 11.8497; \qquad \mathbf{q}_2 = \begin{bmatrix} -1.6878 \\ -0.4666 \end{bmatrix}$$

$$\bar{\mathbf{P}}_2^T\mathbf{K}_{22}\bar{\mathbf{P}}_2 = \begin{bmatrix} 4.1177 & -3.6380 \\ -3.6380 & 5 \end{bmatrix} - \begin{bmatrix} 9.4786 & -9.2348 \\ -2.4337 & 2.3711 \end{bmatrix}$$
$$- \begin{bmatrix} -12.7550 & 3.2749 \\ -3.5263 & 0.90538 \end{bmatrix}$$
$$= \begin{bmatrix} 7.3941 & 2.3219 \\ 2.3219 & 1.7236 \end{bmatrix}$$

and hence

$$\mathbf{K}_3 = \begin{bmatrix} 5 & 4.1231 & 0 & 0 \\ 4.1231 & 7.9823 & -4.0276 & 0 \\ 0 & -4.0276 & 7.3941 & 2.3219 \\ 0 & 0 & 2.3219 & 1.7236 \end{bmatrix}$$

The QR iteration: In the HQRI solution procedure, the QR iteration is applied to the tridiagonal matrix obtained by the Householder transformation of **K**. However, it should be realized that the QR iteration could be applied to the original matrix **K** as well, and that the transformation of **K** into tridiagonal form prior to the iteration is merely carried out to improve the efficiency of solution. In the following we therefore consider first how the iteration is applied to a general symmetric matrix **K**.

The name "QR iteration" derives from the notation used in the algorithm. Namely, the basic step in the iteration is to decompose **K** into the form

$$\mathbf{K} = \mathbf{QR} \tag{11.106}$$

where **Q** is an orthogonal and **R** is an upper triangular matrix. We then form

$$\mathbf{RQ} = \mathbf{Q}^T\mathbf{KQ} \tag{11.107}$$

Therefore, by calculating **RQ** we in fact carry out a transformation of the form (11.72).

The factorization in (11.106) could be obtained by applying the Gram–Schmidt process to the columns of $\mathbf{K}$. In practice, it is more effective to reduce $\mathbf{K}$ into upper triangular form using Jacobi rotation matrices; i.e., we evaluate (see Section 7.3.1)

$$\mathbf{P}_{n,n-1}^T \cdots \mathbf{P}_{3,1}^T \mathbf{P}_{2,1}^T \mathbf{K} = \mathbf{R} \tag{11.108}$$

where the matrix $\mathbf{P}_{j,i}^T$ is given in (7.53) and is selected to zero element (j, i). Using (11.108), we have, corresponding to (11.106),

$$\mathbf{Q} = \mathbf{P}_{2,1}\mathbf{P}_{3,1} \cdots \mathbf{P}_{n,n-1} \tag{11.109}$$

The QR iteration algorithm is obtained by repeating the process given in (11.106) and (11.107). Using the notation $\mathbf{K}_1 = \mathbf{K}$, we form

$$\mathbf{K}_k = \mathbf{Q}_k\mathbf{R}_k \tag{11.110}$$

and then

$$\mathbf{K}_{k+1} = \mathbf{R}_k\mathbf{Q}_k \tag{11.111}$$

where, then, disregarding that eigenvalues and eigenvectors may not be in the usual order,

$$\mathbf{K}_{k+1} \longrightarrow \mathbf{\Lambda} \quad \text{and} \quad \mathbf{Q}_1 \cdots \mathbf{Q}_{k-1}\mathbf{Q}_k \longrightarrow \mathbf{\Phi} \qquad \text{as } k \longrightarrow \infty$$

We demonstrate the iteration process in the following example.

Example 11.15

Use the QR iteration with $\mathbf{Q}$ obtained as a product of Jacobi rotation matrices to calculate the eigensystem of $\mathbf{K}$, where

$$\mathbf{K} = \begin{bmatrix} 5 & -4 & 1 & 0 \\ -4 & 6 & -4 & 1 \\ 1 & -4 & 6 & -4 \\ 0 & 1 & -4 & 5 \end{bmatrix}$$

In the solution we want to employ the Jacobi rotation matrices to establish the matrices $\mathbf{Q}_k$. Using the matrix $\mathbf{P}_{j,i}^T$ given in (7.53) to reduce to zero element (j, i) in the current matrix, we have

$$\sin\theta = \frac{\bar{k}_{ji}}{\sqrt{\bar{k}_{ii}^2 + \bar{k}_{ji}^2}}$$

$$\cos\theta = \frac{\bar{k}_{ii}}{\sqrt{\bar{k}_{ii}^2 + \bar{k}_{ji}^2}}$$

where the bar indicates that the elements of the current matrix are used.

Proceeding as given in (11.108), we obtain for element (2, 1),

$$\sin\theta = -0.6247; \qquad \cos\theta = 0.7809$$

$$\mathbf{P}_{2,1} = \begin{bmatrix} 0.7809 & 0.6247 & 0 & 0 \\ -0.6247 & 0.7809 & 0 & 0 \\ 0 & 0 & 1 & 0 \\ 0 & 0 & 0 & 1 \end{bmatrix}$$

and

$$\mathbf{P}_{2,1}^T\mathbf{K} = \begin{bmatrix} 6.403 & -6.872 & 3.280 & -0.6247 \\ 0 & 2.186 & -2.499 & 0.7809 \\ 1 & -4 & 6 & -4 \\ 0 & 1 & -4 & 5 \end{bmatrix}$$

Next we obtain zero element (3, 1),

$$\mathbf{P}_{3,1}^T\mathbf{P}_{2,1}^T\mathbf{K} = \begin{bmatrix} 6.481 & -7.407 & 4.166 & -1.234 \\ 0 & 2.186 & -2.499 & 0.7809 \\ 0 & -2.892 & 5.422 & -3.856 \\ 0 & 1 & -4 & 5 \end{bmatrix}$$

Noting that element (4, 1) is zero already, we continue with the factorization by zeroing element (3, 2),

$$\mathbf{P}_{3,2}^T\mathbf{P}_{3,1}^T\mathbf{P}_{2,1}^T\mathbf{K} = \begin{bmatrix} 6.481 & -7.407 & 4.166 & -1.234 \\ 0 & 3.625 & -5.832 & 3.546 \\ 0 & 0 & 1.277 & -1.703 \\ 0 & 1 & -4 & 5 \end{bmatrix}$$

Proceeding in the same manner, we obtain

$$\mathbf{P}_{4,2}^T\mathbf{P}_{3,2}^T\mathbf{P}_{3,1}^T\mathbf{P}_{2,1}^T\mathbf{K} = \begin{bmatrix} 6.481 & -7.407 & 4.166 & -1.234 \\ 0 & 3.761 & -6.686 & 4.748 \\ 0 & 0 & 1.277 & -1.703 \\ 0 & 0 & -2.305 & 3.877 \end{bmatrix}$$

and finally

$$\mathbf{R}_1 = \mathbf{P}_{4,3}^T\mathbf{P}_{4,2}^T\mathbf{P}_{3,2}^T\mathbf{P}_{3,1}^T\mathbf{P}_{2,1}^T\mathbf{K}$$

$$\mathbf{R}_1 = \begin{bmatrix} 6.481 & -7.407 & 4.166 & -1.234 \\ 0 & 3.761 & -6.686 & 4.748 \\ 0 & 0 & 2.635 & -4.216 \\ 0 & 0 & 0 & 0.3892 \end{bmatrix}.$$

Also, we have

$$\mathbf{Q}_1 = \mathbf{P}_{2,1}\mathbf{P}_{3,1}\mathbf{P}_{3,2}\mathbf{P}_{4,2}\mathbf{P}_{4,3}$$

$$\mathbf{Q}_1 = \begin{bmatrix} 0.7715 & 0.4558 & 0.3162 & 0.3114 \\ -0.6172 & 0.3799 & 0.4216 & 0.5449 \\ 0.1543 & -0.7597 & 0.1054 & 0.6228 \\ 0 & 0.2659 & -0.8433 & 0.4671 \end{bmatrix}$$

The first iteration step of QR is completed by calculating

$$\mathbf{K}_2 = \mathbf{R}_1\mathbf{Q}_1$$

to obtain

$$\mathbf{K}_2 = \begin{bmatrix} 10.21 & -3.353 & 0.4066 & 0 \\ -3.353 & 7.771 & -3.123 & 0.1035 \\ 0.4066 & -3.123 & 3.833 & -0.3282 \\ 0 & 0.1035 & -0.3282 & 0.1818 \end{bmatrix}$$

The following results are obtained in the next steps of QR. For $k = 2$:

$$\mathbf{R}_2 = \begin{bmatrix} 10.76 & -5.723 & 1.504 & -0.0446 \\ 0 & 6.974 & -4.163 & 0.2284 \\ 0 & 0 & 2.265 & -0.2752 \\ 0 & 0 & 0 & 0.1471 \end{bmatrix}$$

$$\mathbf{Q}_1\mathbf{Q}_2 = \begin{bmatrix} 0.6024 & 0.4943 & 0.5084 & 0.3665 \\ -0.6885 & -0.0257 & 0.4099 & 0.5978 \\ 0.3873 & -0.6409 & -0.2715 & 0.6046 \\ -0.1147 & 0.5867 & 0.7070 & 0.3779 \end{bmatrix}$$

$$\mathbf{K}_3 = \mathbf{R}_2\mathbf{Q}_2 = \begin{bmatrix} 12.05 & -2.331 & 0.0856 & 0 \\ -2.331 & 7.726 & -0.9483 & 0.0022 \\ 0.0856 & -0.9483 & 2.0740 & -0.0173 \\ 0 & 0.0022 & -0.0173 & 0.1461 \end{bmatrix}$$

For $k = 3$:

$$\mathbf{R}_3 = \begin{bmatrix} 12.28 & -3.761 & 0.2785 & -0.0005 \\ 0 & 7.202 & -1.173 & 0.0044 \\ 0 & 0 & 1.938 & -0.0182 \\ 0 & 0 & 0 & 0.1459 \end{bmatrix}$$

$$\mathbf{Q}_1\mathbf{Q}_2\mathbf{Q}_3 = \begin{bmatrix} 0.5011 & 0.5302 & 0.5743 & 0.3713 \\ -0.6682 & -0.2076 & 0.3860 & 0.6012 \\ 0.5000 & -0.5157 & -0.3492 & 0.6018 \\ -0.2290 & 0.6401 & -0.6319 & 0.3722 \end{bmatrix}$$

$$\mathbf{K}_4 = \mathbf{R}_3\mathbf{Q}_3 = \begin{bmatrix} 12.77 & -1.375 & 0.0135 & 0 \\ -1.375 & 7.162 & -0.2481 & 0 \\ 0.0135 & -0.2481 & 1.922 & -0.0013 \\ 0 & 0 & -0.0013 & 0.1459 \end{bmatrix}$$

And after nine iterations, we have:

$$\mathbf{R}_9 = \begin{bmatrix} 13.09 & -0.0869 & 0 & 0 \\ 0 & 6.854 & -0.0005 & 0 \\ 0 & 0 & 1.910 & 0 \\ 0 & 0 & 0 & 0.1459 \end{bmatrix}$$

$$\mathbf{Q}_1\mathbf{Q}_2\mathbf{Q}_3\mathbf{Q}_4\mathbf{Q}_5\mathbf{Q}_6\mathbf{Q}_7\mathbf{Q}_8\mathbf{Q}_9 = \begin{bmatrix} 0.3746 & 0.5997 & 0.6015 & 0.3717 \\ -0.6033 & -0.3689 & 0.3718 & 0.6015 \\ 0.5997 & -0.3746 & -0.3717 & 0.6015 \\ -0.3689 & 0.6033 & -0.6015 & 0.3718 \end{bmatrix}$$

$$\mathbf{K}_{10} = \begin{bmatrix} 13.09 & -0.0298 & 0 & 0 \\ -0.0298 & 6.8542 & -0.0001 & 0 \\ 0 & -0.0001 & 1.910 & 0 \\ 0 & 0 & 0 & 0.1459 \end{bmatrix}$$

Thus we have, after nine steps of QR iteration,

$$\lambda_1 \doteq 0.1459; \qquad \boldsymbol{\phi}_1 \doteq \begin{bmatrix} 0.3717 \\ 0.6015 \\ 0.6015 \\ 0.3718 \end{bmatrix}$$

$$\lambda_2 \doteq 1.910; \qquad \boldsymbol{\phi}_2 \doteq \begin{bmatrix} 0.6015 \\ 0.3718 \\ -0.3717 \\ -0.6015 \end{bmatrix}$$

$$\lambda_3 \doteq 6.854; \qquad \boldsymbol{\phi}_3 \doteq \begin{bmatrix} 0.5997 \\ -0.3689 \\ -0.3746 \\ 0.6033 \end{bmatrix}$$

$$\lambda_4 \doteq 13.09; \qquad \boldsymbol{\phi}_4 \doteq \begin{bmatrix} 0.3746 \\ -0.6033 \\ 0.5997 \\ -0.3689 \end{bmatrix}$$

These results can be compared with the results obtained using the Jacobi method in Example 11.9. It is interesting to note that in the above solution, λ_1 and $\boldsymbol{\phi}_1$ converged first and indeed were well predicted after only three QR iterations. This is a consequence of the fact that QR iteration is closely related to inverse iteration (see Example 11.16), in which the lowest eigenvalues and corresponding vectors converge first.

Although the QR iteration may look similar to the Jacobi solution procedure, the method is, in fact, completely different. This may be observed by studying the convergence characteristics of the QR solution procedure, because it is then found that *the QR method is intimately related to inverse iteration.* In Example 11.16 we compare QR and inverse iteration, where it is assumed that the matrix **K** is nonsingular. As we may recall, this assumption is necessary for inverse iteration and can always be satisfied by using a shift (see Section 10.2.3).

Example 11.16

Show the theoretical relationship between QR and inverse iteration.

In the QR method, we obtain after l steps of iteration,

$$\mathbf{K}_{l+1} = \mathbf{Q}_l^T \mathbf{Q}_{l-1}^T \ldots \mathbf{Q}_1^T \mathbf{K}_1 \mathbf{Q}_1 \ldots \mathbf{Q}_{l-1} \mathbf{Q}_l$$

or

$$\mathbf{K}_{l+1} = \mathbf{P}_l^T \mathbf{K}_1 \mathbf{P}_l; \qquad \mathbf{P}_l = \mathbf{Q}_1 \ldots \mathbf{Q}_l$$

Let us define

$$\mathbf{S}_l = \mathbf{R}_l \ldots \mathbf{R}_1$$

Then we have

$$\begin{aligned} \mathbf{P}_l \mathbf{S}_l &= \mathbf{P}_{l-1} \mathbf{Q}_l \mathbf{R}_l \mathbf{S}_{l-1} \\ &= \mathbf{P}_{l-1} \mathbf{K}_l \mathbf{S}_{l-1} \end{aligned}$$

If we note that

$$\mathbf{K}_1 \mathbf{P}_{l-1} = \mathbf{P}_{l-1} \mathbf{K}_l$$

we get

$$\mathbf{P}_l \mathbf{S}_l = \mathbf{K}_1 \mathbf{P}_{l-1} \mathbf{S}_{l-1}$$

In an analogous manner we obtain $\mathbf{P}_{l-1} \mathbf{S}_{l-1} = \mathbf{K}_1 \mathbf{P}_{l-2} \mathbf{S}_{l-2}$, and so on, and thus conclude that

$$\mathbf{P}_l \mathbf{S}_l = \mathbf{K}^l \tag{a}$$

Assuming that $\mathbf{K}$ is nonsingular, we have from (a)

$$\mathbf{P}_l = \mathbf{K}^{-l} \mathbf{S}_l^T$$

or equating columns on both sides,

$$\mathbf{P}_l \mathbf{E} = \mathbf{K}^{-l} \mathbf{S}_l^T \mathbf{E} \tag{b}$$

where $\mathbf{E}$ consists of the last p columns of $\mathbf{I}$.

Consider now inverse iteration on p vectors. This iteration process can be written

$$\mathbf{K} \mathbf{X}_k = \mathbf{X}_{k-1} \mathbf{L}_k; \qquad k = 1, 2, \ldots$$

where $\mathbf{L}_k$ is a lower triangular matrix chosen so that $\mathbf{X}_k^T \mathbf{X}_k = \mathbf{I}$. The matrix $\mathbf{L}_k$ can be determined using the Gram–Schmidt process on the iteration vectors. Hence after l steps we have

$$\mathbf{X}_l = \mathbf{K}^{-l} \mathbf{X}_0 \bar{\mathbf{L}}_l; \qquad \bar{\mathbf{L}}_l = \mathbf{L}_1 \ldots \mathbf{L}_l \tag{c}$$

On the other hand, the relation in (b) can be written

$$\mathbf{P}_l \mathbf{E} = \mathbf{K}^{-l} \mathbf{E} \bar{\mathbf{S}}_l \tag{d}$$

where $\bar{\mathbf{S}}_l$ consists of the last p columns and rows of $\mathbf{S}_l^T$. Using that $\mathbf{X}_l^T \mathbf{X}_l = \mathbf{I}$ and $(\mathbf{P}_l \mathbf{E})^T (\mathbf{P}_l \mathbf{E}) = \mathbf{I}$, we obtain, from (c) and (d), respectively,

$$\bar{\mathbf{L}}_l^{-T} \bar{\mathbf{L}}_l^{-1} = \mathbf{X}_0^T \mathbf{K}^{-2l} \mathbf{X}_0 \tag{e}$$

and

$$\bar{\mathbf{S}}_l^{-T} \bar{\mathbf{S}}_l^{-1} = \mathbf{E}^T \mathbf{K}^{-2l} \mathbf{E} \tag{f}$$

The equations in (e) and (f) can now be used to show the relationship between inverse iteration and the QR solution procedure. Namely, if we choose $\mathbf{X}_0 = \mathbf{E}$, we find from (e) and (f) that $\bar{\mathbf{L}}_l = \bar{\mathbf{S}}_l$, because these matrices are the Cholesky factors of the same positive definite matrix. However, referring then to (c) and (d), we can conclude that the inverse iteration yields vectors $\mathbf{X}_l$ which are the last p columns in $\mathbf{P}_l$ of the QR solution.

The relationship between the QR solution method and simple inverse iteration suggests that an acceleration of convergence in the QR iteration described in (11.110) and (11.111) should be possible. This is indeed the case and, in practice, QR iteration is used with shifting; i.e., instead of (11.110) and (11.111), the following decompositions are employed:

$$\mathbf{K}_k - \mu_k \mathbf{I} = \mathbf{Q}_k \mathbf{R}_k \tag{11.112}$$

$$\mathbf{K}_{k+1} = \mathbf{R}_k \mathbf{Q}_k + \mu_k \mathbf{I} \tag{11.113}$$

where then, as before,

$$\mathbf{K}_{k+1} \longrightarrow \mathbf{\Lambda} \quad \text{and} \quad \mathbf{Q}_1 \ldots \mathbf{Q}_{k-1}\mathbf{Q}_k \longrightarrow \mathbf{\Phi} \qquad \text{as } k \longrightarrow \infty$$

However, *if* μ_k *is element* (n, n) *of* $\mathbf{K}_k$, *the* QR *iteration corresponds to Rayleigh quotient iteration, ultimately giving cubic convergence.*

As pointed out earlier, in practice the QR iteration should be applied after reduction of **K** into tridiagonal form using Householder transformation matrices; i.e., the QR solution should be applied to the matrix $\mathbf{K}_{n-1}$ in (11.95), which we now call $\mathbf{T}_1$. When the matrix is tridiagonal, the QR process is very effective; i.e., by experience about $9n^2$ operations are required for solution of all eigenvalues. It is also not necessary to formally go through the procedure discussed above and demonstrated in Example 11.15; instead, we may use explicit formulas that relate the elements in $\mathbf{T}_{k+1}$ to the elements of $\mathbf{T}_k$, $k = 1, 2, \ldots$.[1, 36]

Calculation of eigenvectors: The eigenvalues are generally calculated to full machine precision, because convergence is very rapid in the QR iteration with shifting. Once the eigenvalues have been evaluated very accurately, we calculate only the required eigenvectors of the tridiagonal matrix $\mathbf{T}_1$ by simple inverse iteration with shifts equal to the corresponding eigenvalues. Two steps of inverse iteration starting with a full unit vector are usually sufficient. The eigenvectors of $\mathbf{T}_1$ then need to be transformed with the Householder transformations used to obtain the eigenvectors of **K**; i.e., denoting the ith eigenvector of $\mathbf{T}_1$ by $\boldsymbol{\psi}_i$, we have, using the transformation matrices $\mathbf{P}_k$ in (11.95),

$$\boldsymbol{\phi}_i = \mathbf{P}_1 \mathbf{P}_2 \ldots \mathbf{P}_{n-2} \boldsymbol{\psi}_i \tag{11.114}$$

With the three basic steps of the HQRI solution method described above, Table 11.4 summarizes the complete procedure and presents the high-speed storage needed and the number of operations required. It is noted that the greater part of the total number of operations is used for the Householder transformations in (11.95) and, in case many eigenvectors need be calculated, for the eigenvector transformations in (11.114). Therefore, it is seen that the calculation of the eigenvalues of $\mathbf{T}_1$ is not very expensive, but the preparation of **K** into a form that can be used effectively for the iteration process requires most of the numerical effort.

TABLE 11.4 SUMMARY OF HOUSEHOLDER-QR-INVERSE ITERATION SOLUTION

Operation	Calculation	Number of Operations	Required Storage
Householder transformations	$\mathbf{K}_{k+1} = \mathbf{P}_k^T \mathbf{K}_k \mathbf{P}_k;\ k = 1, 2, \ldots, n-2$ $\mathbf{K}_1 = \mathbf{K}$	$\frac{2}{3}n^3 + \frac{3}{2}n^2$	
QR iterations	$\mathbf{T}_{k+1} = \mathbf{Q}_k^T \mathbf{T}_k \mathbf{Q}_k;\ k = 1, 2, \ldots$ $\mathbf{T}_1 = \mathbf{K}_{n-1}$	$9n^2$	Using symmetry of matrix
Calculation of p eigenvectors	$(\mathbf{K}_{n-1} - \lambda_i \mathbf{I})\mathbf{x}_i^{(k+1)} = \mathbf{x}_i^{(k)};\ k = 1, 2, \quad i = 1, 2, \ldots, p$	$10pn$	$\frac{1}{2}[n(n+1)] + 6n$
Transformation of eigenvectors	$\boldsymbol{\phi}_i = \mathbf{P}_1 \ldots \mathbf{P}_{n-2}\mathbf{x}_i^{(3)};\quad i = 1, 2, \ldots, p$	$pn(n-1)$	
Total for all eigenvalues and p eigenvectors		$\frac{2}{3}n^3 + \frac{21}{2}n^2 + pn^2 + 9pn$	

It should be noted that Table 11.4 does not include the operations required for transforming a generalized eigenproblem into a standard form. If this transformation is carried out, the eigenvectors calculated in Table 11.4 must also be transformed to the eigenvectors of the generalized eigenproblem as discussed in Section 10.2.5.

11.4 POLYNOMIAL ITERATION TECHNIQUES

The close relationship between the calculation of the zeros of a polynomial and the evaluation of eigenvalues has been discussed in Section 10.2.2. Namely, defining the characteristic polynomial $p(\lambda)$, where

$$p(\lambda) = \det (\mathbf{K} - \lambda\mathbf{M}) \tag{11.6}$$

the zeros of $p(\lambda)$ are the eigenvalues of the eigenproblem $\mathbf{K}\boldsymbol{\phi} = \lambda\mathbf{M}\boldsymbol{\phi}$. To calculate the eigenvalues, we therefore may operate on $p(\lambda)$ to extract the zeros of the polynomial. A number of solution procedures have been proposed.[1-11] However, basically there are two strategies: explicit and implicit evaluation procedures, both of which may use the same basic iteration schemes.

In the discussion of the polynomial iteration schemes, we assume that the solution is carried out directly using **K** and **M** of the finite element assemblage, i.e., without transforming the problem into a different form. For example, if **M** is the identity matrix, we could transform **K** first into tridiagonal form as is done in the HQRI solution (see Section 11.3.3). In case $\mathbf{M} \neq \mathbf{I}$, we would need to transform the generalized eigenproblem into a standard form (see Section 10.2.5), before the Householder reduction into a tridiagonal matrix could be performed. Whichever problem we finally consider in the iterative solution of the required eigenvalues, the solution strategy would not be changed. However, if only a few eigenvalues are to be calculated, the direct solution using **K** and **M** is nearly always most effective (see Section 12.2).

It should be noted that using a polynomial iteration method, only the eigenvalues are calculated. The corresponding eigenvectors can then be obtained effectively by using inverse iteration with shifting; i.e., each required eigenvector is obtained by inverse iteration at a shift equal to the corresponding eigenvalue.

11.4.1 Explicit Polynomial Iteration

In the explicit polynomial iteration methods, the first step is to write $p(\lambda)$ in the form

$$p(\lambda) = a_0 + a_1\lambda + a_2\lambda^2 + \ldots + a_n\lambda^n \tag{11.115}$$

and evaluate the polynomial coefficients $a_0, a_1, \ldots, a_n$. The second step is to calculate the roots of the polynomial. We demonstrate the procedure by means of an example.

Example 11.17

Establish the coefficients of the characteristic polynomial of the problem $\mathbf{K}\boldsymbol{\phi} = \lambda\mathbf{M}\boldsymbol{\phi}$, where $\mathbf{K}$ and $\mathbf{M}$ are the matrices used in Example 11.4.

The problem is here to evaluate the expression

$$p(\lambda) = \det\begin{bmatrix} 5-2\lambda & -4 & 1 & 0 \\ -4 & 6-2\lambda & -4 & 1 \\ 1 & -4 & 6-\lambda & -4 \\ 0 & 1 & -4 & 5-\lambda \end{bmatrix}$$

Following the rules given in Section 1.8 for the evaluation of a determinant, we obtain

$$\begin{aligned} p(\lambda) = (5-2\lambda)\det&\begin{bmatrix} 6-2\lambda & -4 & 1 \\ -4 & 6-\lambda & -4 \\ 1 & -4 & 5-\lambda \end{bmatrix} \\ + (4)\det&\begin{bmatrix} -4 & -4 & 1 \\ 1 & 6-\lambda & -4 \\ 0 & -4 & 5-\lambda \end{bmatrix} \\ + (1)\det&\begin{bmatrix} -4 & 6-2\lambda & 1 \\ 1 & -4 & -4 \\ 0 & 1 & 5-\lambda \end{bmatrix} \end{aligned}$$

Hence

$$\begin{aligned} p(\lambda) = (5-2\lambda)\{&(6-2\lambda)[(6-\lambda)(5-\lambda)-16] \\ &+ 4[-4(5-\lambda)+4] + 16 - (6-\lambda)\} \\ &+ 4\{-4[(6-\lambda)(5-\lambda)-16] + 4(5-\lambda) - 4\} \\ &+ \{-4[(-4)(5-\lambda)+4] - (6-2\lambda)(5-\lambda) + 1\} \end{aligned}$$

and the expression finally reduces to

$$p(\lambda) = 4\lambda^4 - 66\lambda^3 + 276\lambda^2 - 285\lambda + 25$$

In the general case when n is large, we cannot evaluate the polynomial coefficients as easily as in the example above. The expansion of the determinant would require about n-factorial operations, which are far too many operations to make the method practical. However, other procedures have been developed; for example, the Newton identities may be used. Once the coefficients have been evaluated, it is necessary to employ a standard polynomial root finder, using, for example, a Newton iteration or secant iteration to evaluate the required eigenvalues.

Although the procedure seems most natural to use, one difficulty has caused the method to be almost completely abandoned for the solution of eigenvalue problems. A basic defect of the method is that small errors in the coefficients $a_0, \ldots, a_n$ cause large errors in the roots of the polynomial. But small errors are almost unavoidable, owing to round-off in the computer.

Therefore, an explicit evaluation of the coefficients $a_0, \ldots, a_n$ from $\mathbf{K}$ and $\mathbf{M}$ with subsequent solution of the required eigenvalues is not effective in general analysis.

11.4.2 Implicit Polynomial Iteration

In an implicit polynomial iteration solution we evaluate the value of $p(\lambda)$ directly without calculating first the coefficients $a_0, \ldots, a_n$ in (11.115). The value of $p(\lambda)$ can be obtained effectively by decomposing $\mathbf{K} - \lambda\mathbf{M}$ into a lower unit triangular matrix $\mathbf{L}$ and an upper triangular matrix $\mathbf{S}$; i.e., we have

$$\mathbf{K} - \lambda\mathbf{M} = \mathbf{LS} \tag{11.116}$$

where then

$$\det(\mathbf{K} - \lambda\mathbf{M}) = \prod_{i=1}^{n} s_{ii} \tag{11.117}$$

The decomposition of $\mathbf{K} - \lambda\mathbf{M}$ is obtained as discussed in Section 7.2, but, as pointed out in Section 7.2.5, may require interchanges when $\lambda > \lambda_1$. In case row and corresponding column interchanges are carried out, the coefficient matrix remains symmetric and, in effect, the degree of freedom numbering has then merely been rearranged. In other words, the stiffness and mass matrices of the finite element system actually used correspond to a different degree of freedom numbering than originally specified. On the other hand, if row interchanges only are employed, a nonsymmetric coefficient matrix is obtained. However, in either case it is important to note that the required row and column interchanges could have been carried out prior to the Gauss elimination process to obtain an "effective" coefficient matrix $\mathbf{K} - \lambda\mathbf{M}$ that is considered in (11.116), and no more interchanges would then be needed. Each row or column interchange introduced merely effects a change in sign of the determinant. In practice, we do not know the actual row and column interchanges that will be required, but the realization that all interchanges could have been carried out prior to the factorization shows that we can always use the Gauss elimination procedure given in (7.10) to (7.14), provided that we admit that the "effective" initial coefficient matrix may be nonsymmetric. Consider the following example.

Example 11.18

Use Gauss elimination to evaluate $p(\lambda) = \det(\mathbf{K} - \lambda\mathbf{M})$, where

$$\mathbf{K} = \begin{bmatrix} 2 & -1 & 0 \\ -1 & 4 & -1 \\ 0 & -1 & 2 \end{bmatrix}; \qquad \mathbf{M} = \begin{bmatrix} 1 & & \\ & 1 & \\ & & \frac{1}{2} \end{bmatrix}; \qquad \lambda = 2$$

In this case we have

$$\mathbf{K} - \lambda\mathbf{M} = \begin{bmatrix} 0 & -1 & 0 \\ -1 & 2 & -1 \\ 0 & -1 & 1 \end{bmatrix}$$

Since the first diagonal element is zero, we need to use interchanges. Assume that we interchange the first and second rows (and not the corresponding columns); then we effectively factorize $\bar{\mathbf{K}} - \lambda\bar{\mathbf{M}}$, where

$$\bar{\mathbf{K}} = \begin{bmatrix} -1 & 4 & -1 \\ 2 & -1 & 0 \\ 0 & -1 & 2 \end{bmatrix}; \qquad \bar{\mathbf{M}} = \begin{bmatrix} 0 & 1 & 0 \\ 1 & 0 & 0 \\ 0 & 0 & \frac{1}{2} \end{bmatrix}$$

The factorization of $\bar{\mathbf{K}} - \lambda\bar{\mathbf{M}}$ is now obtained in the usual way (see Examples 10.4 and 10.5),

$$\bar{\mathbf{K}} - \lambda\bar{\mathbf{M}} = \begin{bmatrix} 1 & & \\ 0 & 1 & \\ 0 & 1 & 1 \end{bmatrix}\begin{bmatrix} -1 & 2 & -1 \\ & -1 & 0 \\ & & 1 \end{bmatrix}$$

Hence

$$\det(\bar{\mathbf{K}} - \lambda\bar{\mathbf{M}}) = (-1)(-1)(1) = 1$$

and taking account of the fact that the row interchange changed the sign of the determinant (see Section 1.8), we have

$$\det(\mathbf{K} - \lambda\mathbf{M}) = -1$$

As pointed out above, if in the Gauss elimination no interchanges have been carried out or each row interchange was accompanied by a corresponding column interchange, the coefficient matrix $\mathbf{K} - \lambda\mathbf{M}$ in (11.116) is symmetric. In this case we have as in Section (7.2.2) that $\mathbf{S} = \mathbf{D}\mathbf{L}^T$ and hence

$$\det(\mathbf{K} - \lambda\mathbf{M}) = \prod_{i=1}^{n} d_{ii} \tag{11.118}$$

In the determinant search solution method (see Section 12.2) we only use a factorization if it can be completed without interchanges and therefore always employ (11.118). In this case, one polynomial evaluation requires about $\frac{1}{2}nm_{\mathbf{K}}^2$ operations, where n is the order of $\mathbf{K}$ and $\mathbf{M}$ and $m_{\mathbf{K}}$ is the half-bandwidth of $\mathbf{K}$.

With a scheme available for an accurate evaluation of $p(\lambda)$, we can now use a number of iteration schemes to calculate a root of the polynomial. One commonly used and simple technique is the secant method, in which a linear interpolation is employed; i.e., let $\mu_{k-1} < \mu_k$, then we iterate using

$$\mu_{k+1} = \mu_k - \frac{p(\mu_k)}{p(\mu_k) - p(\mu_{k-1})}(\mu_k - \mu_{k-1}) \tag{11.119}$$

where μ_k is the kth iterate (see Fig. 11.3). We may note that the secant method in (11.119) is an approximation to the Newton iteration,

$$\mu_{k+1} = \mu_k - \frac{p(\mu_k)}{p'(\mu_k)} \tag{11.120}$$

where $p'(\mu_k)$ is approximated using

$$p'(\mu_k) \doteq \frac{p(\mu_k) - p(\mu_{k-1})}{\mu_k - \mu_{k-1}} \tag{11.121}$$

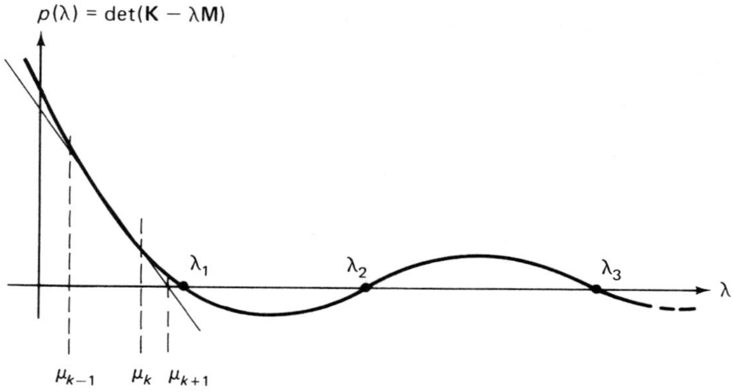

Figure 11.3. Secant iteration for calculation of λ_1.

An actual scheme to accurately evaluate $p'(\mu_k)$ has been tested but was not found to be effective.[11]

Another technique commonly used in the solution of complex eigenvalue problems is Muller's method, in which case a quadratic interpolation is employed. A disadvantage of Muller's method in the solution of $\mathbf{K}\boldsymbol{\phi} = \lambda\mathbf{M}\boldsymbol{\phi}$ is that although the starting values μ_k, μ_{k-1}, and μ_{k-2} are real, the calculated value μ_{k+1} may be complex.

It should be noted that so far we did not discuss to which eigenvalue any one of the iteration strategies converges. This depends on the starting iteration values. Using μ_{k-1} and μ_k both smaller than λ_1, the Newton and secant iterations converge montonically to λ_1, as shown in Fig. 11.3, and the ultimate order of convergence is quadratic and linear, respectively. In effect, convergence is achieved in the iterations because $p''(\lambda) > 0$ for $\lambda < \lambda_1$, which is always the case for any order and bandwidth of **K** and **M**. However, convergence cannot be guaranteed for arbitrary starting values μ_{k-1} and μ_k. Therefore, a solution algorithm to calculate a set of required eigenvalues using the secant iteration must be designed with care to assure convergence to the actually required roots.

Consider the following example of a secant iteration.

Example 11.19

Use the secant iteration to calculate λ_1 of the eigenproblem $\mathbf{K}\boldsymbol{\phi} = \lambda\mathbf{M}\boldsymbol{\phi}$, where

$$\mathbf{K} = \begin{bmatrix} 2 & -1 & 0 \\ -1 & 4 & -1 \\ 0 & -1 & 2 \end{bmatrix}; \qquad \mathbf{M} = \begin{bmatrix} \frac{1}{2} & & \\ & 1 & \\ & & \frac{1}{2} \end{bmatrix}$$

For the secant iteration we need two starting values for μ_1 and μ_2 that are lower bounds to λ_1. Let $\mu_1 = -1$ and $\mu_2 = 0$ Then we have

$$p(-1) = \det \begin{bmatrix} \frac{5}{2} & -1 & 0 \\ -1 & 5 & -1 \\ 0 & -1 & \frac{5}{2} \end{bmatrix}$$

$$= \det \begin{bmatrix} 1 & & \\ -\frac{2}{5} & 1 & \\ 0 & -\frac{5}{23} & 1 \end{bmatrix} \begin{bmatrix} \frac{5}{2} & & \\ & \frac{23}{5} & \\ & & \frac{105}{46} \end{bmatrix} \begin{bmatrix} 1 & -\frac{2}{5} & 0 \\ & 1 & -\frac{5}{23} \\ & & 1 \end{bmatrix}$$

Hence

$$p(-1) = (\tfrac{5}{2})(\tfrac{23}{5})(\tfrac{105}{46}) = 26.25$$

Similarly,

$$p(0) = \det \begin{bmatrix} 2 & -1 & 0 \\ -1 & 4 & -1 \\ 0 & -1 & 2 \end{bmatrix} = 12$$

Now using (11.119) we obtain as the next shift,

$$\mu_3 = 0 - \frac{12}{12 - 26.25}[0 - (-1)]$$

Hence

$$\mu_3 = 0.8421$$

Continuing in the same way, we obtain

$$p(0.8421) = 4.7150$$
$$\mu_4 = 1.3871$$
$$p(1.3871) = 1.8467$$
$$\mu_5 = 1.7380$$
$$p(1.7380) = 0.63136$$
$$\mu_6 = 1.9203$$
$$p(1.9203) = 0.16899$$
$$\mu_7 = 1.9870$$
$$p(1.9870) = 0.026347$$
$$\mu_8 = 1.9993$$

Hence after six iterations we have as an approximation to λ_1, $\lambda_1 \doteq 1.9993$.

11.5 METHODS BASED ON THE STURM SEQUENCE PROPERTY

In Section 10.2.2 we discussed the Sturm sequence property of the characteristic polynomials of the problem $\mathbf{K}\boldsymbol{\phi} = \lambda\mathbf{M}\boldsymbol{\phi}$ and of its associated constraint problems. The main result was the following. Assume that for a shift μ_k, the Gauss factorization of $\mathbf{K} - \mu_k\mathbf{M}$ into $\mathbf{LDL}^T$ can be obtained. Then the number of negative elements in **D** is equal to the number of eigenvalues smaller than μ_k. This result can directly be used to construct an algorithm for the calculation of eigenvalues and corresponding eigenvectors.[1,11,37-40] As in the discussion of the polynomial iteration methods, we assume in the following that the solution is performed directly using **K** and **M**, although the same strategies could be used after the generalized eigenproblem has been transformed into a different form. Also, as in Section 11.4, the solution method to be presented only solves for the eigenvalues, and the corresponding eigenvectors would be calculated using inverse iteration with shifting (see Section 11.2.3).

Consider that we want to solve for all eigenvalues between λ_l and λ_u, where λ_l and λ_u are the lower and upper limits, respectively. For example, we may have a case such as depicted in Fig. 11.4 or λ_l may be zero, in which

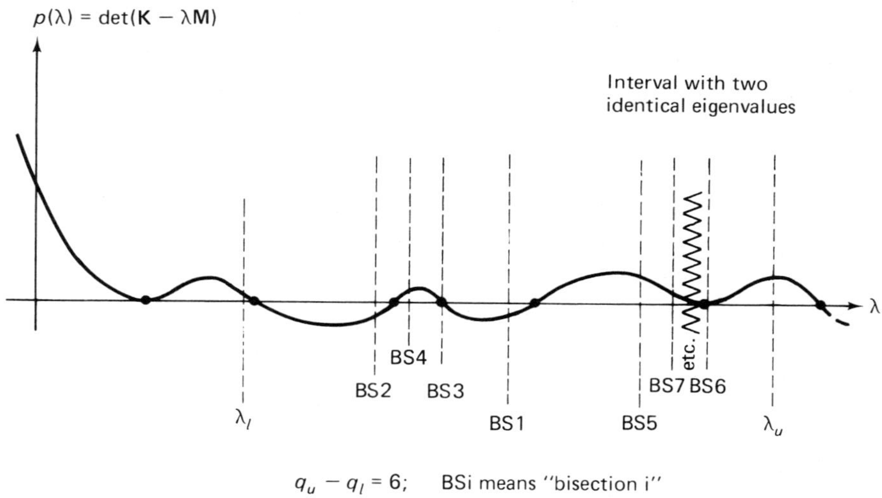

Figure 11.4. Use of Sturm sequence property to isolate eigenvalues.

case we would need to solve for all eigenvalues up to the value λ_u. The basis of the solution procedure is the triangular factorization of $\mathbf{K} - \mu_k\mathbf{M}$, where μ_k is selected in such way as to obtain from the positive or negative signs of the diagonal elements in the factorization meaningful information about the unknown and required eigenvalues. The following solution procedure, known as the *bisection method*, may be used (refer to Fig. 11.4 for a typical example):

1. Factorize $\mathbf{K} - \lambda_l\mathbf{M}$ and hence find how many eigenvalues, say q_l, are smaller than λ_l.

2. Apply the Sturm sequence check at $\mathbf{K} - \lambda_u \mathbf{M}$ and hence find how many eigenvalues, say q_u, are smaller than λ_u. There are thus $q_u - q_l$ eigenvalues between λ_u and λ_l.
3. Use a simple scheme of bisection to identify intervals in which the individual eigenvalues lie. In this process, those intervals in which more than one eigenvalue are known to lie are successively bisected and the Sturm sequence check is carried out until all eigenvalues are isolated.
4. Calculate the eigenvalues to the accuracy required and then obtain the corresponding eigenvectors by inverse iteration.

To obtain the eigenvalues accurately in step 4, the method of bisection is abandoned and a more efficient procedure is generally used. For example, the secant iteration presented in Section 11.4.2 can be employed once the eigenvalues have been isolated (see Example 11.20).

The above technique to calculate the required eigenvalues is straightforward. However, the method can be quite inefficient, because each iteration in step 3 requires a triangular factorization and because many iterations may be needed. The number of iterations can be particularly high, when multiple eigenvalues (see Fig. 11.4) or eigenvalue clusters must be evaluated, in which case additional strategies to accelerate the process must be included. In general, the method is effective if only a few factorizations are needed to identify intervals of individual eigenvalues, and the smallest required eigenvalue is much larger than λ_1.

Example 11.20

Use the bisection method followed by secant iteration to calculate λ_2 of the problem $\mathbf{K}\boldsymbol{\phi} = \lambda \mathbf{M}\boldsymbol{\phi}$, where

$$\mathbf{K} = \begin{bmatrix} 2 & -1 & 0 \\ -1 & 4 & -1 \\ 0 & -1 & 2 \end{bmatrix}; \qquad \mathbf{M} = \begin{bmatrix} \frac{1}{2} & & \\ & 1 & \\ & & \frac{1}{2} \end{bmatrix}$$

We already considered this problem in Example 10.5, where we isolated the eigenvalues using the bisection technique. Specifically, we found that

$$\lambda_1 < 3 < \lambda_2 < 5 < \lambda_3$$

Hence we can start the secant iteration with $\mu_1 = 3$, $\mu_2 = 5$, and using the results of Example 10.5:

$$p(\mu_1) = -\tfrac{3}{4}; \qquad p(\mu_2) = \tfrac{3}{4}$$

Using (11.119), we obtain

$$\mu_3 = 5 - \frac{\frac{3}{4}}{(\frac{3}{4}) - (-\frac{3}{4})}(5 - 3)$$

or

$$\mu_3 = 4$$

Next we need to evaluate

$$p(\mu_3) = \det(\mathbf{K} - \mu_3\mathbf{M})$$

and we find that $p(\mu_3) = 0.0$. Hence $\lambda_2 = \mu_3 = 4$. Therefore, in this case one step of secant iteration was sufficient to calculate λ_2. But it may be noted that to evaluate $p(\mu_3)$ by Gauss factorization of $\mathbf{K} - \mu_3\mathbf{M}$, a row interchange is needed (see Example 10.4).

It is important to note that we assumed in the above presentation that the factorization of $\mathbf{K} - \mu_k\mathbf{M}$ into $\mathbf{LDL}^T$ can be obtained. However, we discussed in Sections 7.2.5 and 11.4.2 that if $\mu_k > \lambda_1$, interchanges may be required. If each row interchange is accompanied by a corresponding column interchange, the coefficient matrix remains symmetric and the interchanges correspond merely to a rearranging of the degree of freedom numbering. Hence the $\mathbf{LDL}^T$ factorization of a coefficient matrix that corresponds to a different degree of freedom numbering is obtained. However, if row interchanges only are used, a nonsymmetric coefficient matrix is obtained. In this case the basic algorithm described above can still be used, but proper account has to be taken of the effect of each row interchange on the number of negative pivot elements that will be encountered.

The important point is that in the bisection algorithm, the accuracy with which the eigenvalues are calculated depends on the possibility of obtaining an accurate factorization of $\mathbf{K} - \mu_k\mathbf{M}$. Therefore, it is necessary in practical calculations to allow interchanges in the Gauss elimination, and it has been found that it is most effective to work with row interchanges only. There is no need to immediately start row interchanges, except when multiplier growth is encountered (see Section 7.5). However, the possible need for row interchanges means that more high-speed storage must be allocated, because the half-bandwidth may increase from $m_{\mathbf{K}}$ to $2m_{\mathbf{K}}$; and if row interchanges are carried out, the number of operations increases also.

The bisection method has two major disadvantages: (1) it is necessary to complete the factorization of $\mathbf{K} - \mu_k\mathbf{M}$ with as much accuracy as possible, although the coefficient matrix may be ill-conditioned (indeed, this may still be the case even when row interchanges are included), and (2) convergence can be very slow when a cluster of eigenvalues has to be solved. However, the Sturm sequence property can be employed in conjunction with other solution strategies, and it is in such cases that the property can be extremely useful. In particular, the Sturm sequence property is employed in the determinant search algorithm (see Section 12.2), in which the factorization of $\mathbf{K} - \mu_k\mathbf{M}$ is carried out without interchanges but is completed only provided that no instability occurs. If the factorization proves to be unstable, a different μ_k is selected. This is possible because the final accuracy with which the eigenvalues and eigenvectors are calculated does not depend on the specific shift μ_k used in the solution.

REFERENCES

1. J. H. WILKINSON, *The Algebraic Eigenvalue Problem*, Oxford University Press, Inc., London, 1965.
2. V. N. FADDEEVA, *Computational Methods of Linear Algebra*, Dover Publications, Inc., New York, 1959.
3. B. NOBLE, *Applied Linear Algebra*, Prentice-Hall, Inc., Englewood Cliffs, N.J., 1969.
4. C. E. FRÖBERG, *Introduction to Numerical Analysis*, Addison-Wesley Publishing Company, Inc., Reading, Mass., 1969.
5. C. LANCZOS, *Applied Analysis*, Prentice-Hall, Inc., Englewood Cliffs, N.J., 1956.
6. S. H. CRANDALL, *Engineering Analysis*, McGraw-Hill Book Company, New York, 1956.
7. R. ZURMÜHL, *Matrizen*, Springer-Verlag, Berlin, 1964.
8. H. R. SCHWARZ, H. RUTISHAUSER, and E. STIEFEL, *Matrizen-Numerik*, B. G. Teubner, Stuttgart, 1972.
9. A. S. HOUSEHOLDER, *Principles of Numerical Analysis*, McGraw-Hill Book Company, New York, 1963.
10. G. PETERS and J. H. WILKINSON, "Ax = λBx and the Generalized Eigenproblem," *S.I.A.M., Journal of Numerical Analysis*, Vol. 7, 1970, pp. 479–492.
11. K. J. BATHE, "Solution Methods for Large Generalized Eigenvalue Problems in Structural Engineering," Report UC SESM 71-20, Civil Engineering Department, University of California, Berkeley, California.
12. W. W. BRADBARY and R. FLETCHER, "New Iterative Methods for Solution of the Eigenproblem Arising in Structural Dynamics," Proceedings, 2nd Conference on Matrix Methods in Structural Mechanics, Wright-Patterson A.F.B., Ohio, 1968.
13. R. FOX and M. KAPOOR, "A Minimization Method for the Solution of the Eigenproblem Arising in Structural Dynamics," Proceedings, 2nd Conference on Matrix Methods in Structural Mechanics, Wright-Patterson A.F.B., Ohio, 1968.
14. I. FRIED, "Optimal Gradient Minimization Scheme for Finite Element Eigenproblems," *Journal of Sound and Vibration*, Vol. 20, No. 3, 1972, pp. 333–342.
15. S. H. CRANDALL, "Iterative Procedures Related to Relaxation Methods for Eigenvalue Problems," *Proceedings, Royal Society*, London, Vol. A207, 1951, p. 416.
16. A. M. OSTROWSKI, "On the Convergence of the Rayleigh Quotient Iteration for the Computation of the Characteristic Roots and Vectors, Parts I–VI," *Archives for Rational Mechanics and Analysis*, Vols. 1–4, 1958–1959.
17. B. N. PARLETT and W. KAHAN, "On the Convergence of a Practical QR Algorithm," Proceedings, I.F.I.P. Congress, 1968.
18. W. G. POOLE, Jr., "A Geometric Convergence Theory for the Eigenvalues of a Symmetric Tridiagonal Matrix," Technical Report CS 43, Computer Science Department, Stanford University, California, 1966.

19. H. WIELANDT, "Bestimmung höherer Eigenwerte durch gebrochene Iteration," Bericht B44/J/37 der Aerodynamischen Versuchsanstalt, Göttingen, 1944.

20. E. E. OSBORNE, "On Acceleration and Matrix Deflation Processes Used with the Power Method," *Journal of the Society for Industrial and Applied Mathematics,* Vol. 6, 1958, pp. 279–287.

21. H. RUTISHAUSER, "Deflation bei Bandmatrizen," *Z.A.M.P.*, Vol. 10, 1959, pp. 314–319.

22. C. A. FELIPPA, "BANEIG—Eigenvalue Routine for Symmetric Band Matrices," Computer Programming Series, SESM, Department of Civil Engineering, University of California, Berkeley, California, 1966.

23. C. G. J. JACOBI, "Über ein leichtes Verfahren die in der Theorie der Säculärstörungen vorkommenden Gleichungen numerisch aufzulösen," *Crelle's Journal*, Vol. 30, 1846, pp. 51–94.

24. H. H. GOLDSTINE, F. J. MURRAY, and J. von NEUMANN, "The Jacobi Method for Real Symmetric Matrices," *Journal of the Association of Computing Machines*, Vol. 6, 1959, pp. 59–96.

25. A. SCHÖNHAGE, "Zur Konvergenz des Jacobi-Verfahrens," *Numerische Mathematik*, Vol. 3, 1961, pp. 374–380.

26. G. E. FORSYTHE and P. HENRICI, "The Cyclic Jacobi Method for Computing the Principal Values of a Complex Matrix," *Transactions of the American Mathematical Society*, Vol. 94, 1960, pp. 1–23.

27. J. H. WILKINSON, "Note on the Quadratic Convergence of the Cyclic Jacobi Process," *Numerische Mathematik*, Vol. 4, 1962, pp. 296–300.

28. S. FALK and P. LANGEMEYER, "Das Jacobische Rotationsverfahren für Reellsymmetrische Matrizenpaare," *Elektronische Datenverarbeitung*, 1960, pp. 30–34.

29. R. S. MARTIN, C. REINSCH, and J. H. WILKINSON, "Householder's Tridiagonalization of a Symmetric Matrix," *Numerische Mathematik*, Vol. 11, 1968, pp. 181–195.

30. J. G. F. FRANCIS, "The QR Transformation, Parts I and II," *Computer Journal*, Vol. 4, 1961–1962, pp. 265, 332.

31. V. N. KUBLANOVSKAYA, "On Some Algorithms for the Solution of the Complete Problem of Proper Values," *Journal of Computer Mathematics and Mathematical Physics*, Vol. 1, 1961.

32. B. N. PARLETT, "Convergence of the QR Algorithm," *Numerische Mathematik*, Vol. 7, 1965, p. 187; Vol. 10, 1967, p. 163.

33. B. N. PARLETT, "Global Convergence of the Basic QR Algorithm on Hessenberg Matrices," *Mathematical Computation*, Vol. 22, 1968, p. 803.

34. J. H. WILKINSON, "The QR Algorithm for Real Symmetric Matrices with Multiple Eigenvalues," *Computer Journal*, Vol. 8, 1965, p. 85.

35. W. KAHAN and J. VARAH, "Two Working Algorithms for the Eigenvalues of a Symmetric Tridiagonal Matrix," Technical Report CS 43, Computer Science Department, Stanford University, California, 1966.

36. J. M. ORTEGA and H. F. KAISER, "The LL^T and QR Methods for Symmetric Tridiagonal Matrices," *Computer Journal*, Vol. 6, 1963, pp. 99–101.

37. J. W. GIVENS, "A Method of Computing Eigenvalues and Eigenvectors Suggested by Classical Results on Symmetric Matrices," *National Bureau of Standards Applied Mathematics Series 29*, Government Printing Office, Washington, D.C., L953, pp. 117–122.

38. W. BARTH, R. S. MARTIN, and J. H. WILKINSON, "Calculation of the Eigenvalues of a Symmetric Tridiagonal Matrix by the Method of Bisection," *Numerische Mathematik*, Vol. 9, 1967, pp. 386–393.

39. K. K. GUPTA, "Vibration of Frames and Other Structures with Banded Stiffness Matrix," *International Journal for Numerical Methods in Engineering*, Vol. 2, 1970, pp. 221–228.

40. K. K. GUPTA, "Solution of Eigenvalue Problems by the Sturm Sequence Method," *International Journal for Numerical Methods in Engineering*, Vol. 4, 1972, pp. 379–404.

CHAPTER 12

SOLUTION OF LARGE EIGENPROBLEMS

12.1 INTRODUCTION

In Chapter 11 we described individually four basic groups of solution techniques to calculate either an eigenvalue or an eigenvector, or all eigenvalues and eigenvectors simultaneously. The solution methods described are based on some fundamental facts as summarized in Section 11.1. Considering the effectiveness of the solution procedures, none of the methods is always most efficient, but the solution technique to be used should be selected according to the specific problem to be solved.

As was discussed in Section 8.3, the efficient solution of the required eigenvalues and eigenvectors is of much importance in mode superposition analysis of large finite element systems. Indeed, because procedures for the exact solution of large eigenproblems have been expensive for a long time, approximate solution techniques have been and are still used. Some commonly employed approximate solution methods have been discussed in Section 10.3. The final conclusion was that the approximate techniques must be used with care and cannot be recommended for general use.

Since it is very important to employ an optimum algorithm for the calculation of the required eigenvalues and eigenvectors in the problem $\mathbf{K}\boldsymbol{\phi} = \lambda\mathbf{M}\boldsymbol{\phi}$ when the order of $\mathbf{K}$ and $\mathbf{M}$ is large, it appears natural to try and combine the basic solution methods presented in Sections 11.2 to 11.5 in an effort to design more effective techniques. The solution methods to be designed should take specific advantage of the solution requirements and exploit the advantages of each basic solution procedure when taking account of the specific characteristics of **K** and **M**. In the following, we describe in detail two solution methods, the determinant search technique and the subspace iteration method, which are believed to be very effective in finite element analysis.[1-4] The techniques have both been extensively used in general purpose finite element analysis programs.[5-8] We assume in the description of the algorithms that the material covered in the preceding sections is known, and we make extensive reference to it.

Before presenting the solution techniques, we should define what we mean by a large eigenproblem. As we discussed in Section 8.3, in finite element analysis we are mainly interested in the smallest eigenvalues. Correspondingly, we define an eigenproblem as large if it is much cheaper to solve for only the required eigenvalues and eigenvectors instead of simply calculating all. In general, the system will be large if the high-speed core storage of a reasonably sized computer is too small for using an in-core Householder-QR-inverse iteration method or the generalized Jacobi solution technique. In that case it is most probably more effective to employ the determinant search or subspace iteration methods described in the following sections, because out-of-core solutions using transformation methods that calculate all eigenvalues require much use of back-up storage and are expensive. However, even for in-core solutions, the determinant search and subspace iteration techniques are in most cases more effective than the HQRI or Jacobi solution methods, because they calculate directly (i.e., without a transformation to the standard eigenproblem) only those eigenvalues and corresponding eigenvectors that are actually required. In addition, during the solution process optimum advantage is taken of the special characteristics of the matrices **K** and **M**.

The determinant search technique and the subspace iteration method have both been developed for the solution of the p lowest eigenvalues and corresponding eigenvectors, but by using shift strategies, the techniques can be modified to calculate the largest eigenvalues and corresponding eigenvectors as required in buckling analysis. In essence, the determinant search method is effectively used for in-core solution of small banded systems, whereas the subspace iteration technique is most efficient in the analysis of large banded systems.

12.2 THE DETERMINANT SEARCH METHOD

The fundamental techniques used in the determinant search solution method are: an iteration with the characteristic polynomial $p(\lambda)$, *the use of the Sturm sequence property of the characteristic polynomials* $p(\lambda)$ *and* $p^{(r)}(\lambda^{(r)})$, *and vector inverse iteration*. These techniques have been described in Sections 11.4.2, 11.5, and 11.2.3, respectively, where also the advantages and disadvantages of each procedure have been discussed. *The specific aim in the development of the determinant search method was to combine and extend these techniques in an optimum manner for the solution of the smallest eigenvalues and corresponding eigenvectors in the problem* $\mathbf{K}\boldsymbol{\phi} = \lambda\mathbf{M}\boldsymbol{\phi}$. It should be noted that each of the fundamental techniques can be used directly for the solution of the eigenproblem $\mathbf{K}\boldsymbol{\phi} = \lambda\mathbf{M}\boldsymbol{\phi}$; i.e., no transformation to the standard eigenproblem is required. This is a very important point, because difficulties that may arise from ill-conditioning or loss of bandwidth due to the transformation are avoided.

12.2.1 Preliminary Considerations

We recall that the iteration on the polynomial $p(\lambda)$ or the use of the Sturm sequence property are basic techniques to calculate an eigenvalue, whereas inverse iteration is a basic technique for evaluating an eigenvector. Also, once an eigenvalue or an eigenvector has been calculated, the other member of the eigenpair can be evaluated without further iteration, i.e.:

1. Given λ_i, the eigenvector $\boldsymbol{\phi}_i$ is obtained by solving $(\mathbf{K} - \lambda_i\mathbf{M})\boldsymbol{\phi}_i = \mathbf{0}$.
2. Given $\boldsymbol{\phi}_i$, the eigenvalue λ_i is obtained from the Rayleigh quotient $\lambda_i = \boldsymbol{\phi}_i^T\mathbf{K}\boldsymbol{\phi}_i/\boldsymbol{\phi}_i^T\mathbf{M}\boldsymbol{\phi}_i$.

If, however, only approximations to $\boldsymbol{\phi}_i$ or λ_i are known, we need further iteration to evaluate one member of the eigenpair $(\lambda_i, \boldsymbol{\phi}_i)$ accurately, after which the other member can be obtained as in 1 or 2 above.

Assume that we have a shift μ, for example, obtained by polynomial iteration, where $|\lambda_i - \mu|/|\lambda_j - \mu|$ is small for all $j, j \neq i$. In this case we can use inverse iteration with shift μ to calculate $\boldsymbol{\phi}_i$, as discussed in Section 11.2.3. In the iteration we may use Rayleigh quotient shifting to accelerate convergence, in which case we would obtain λ_i and $\boldsymbol{\phi}_i$ simultaneously. On the other hand, assume that we have a vector $\boldsymbol{\phi}$, which is a close approximation to the eigenvector $\boldsymbol{\phi}_i$. In such case we may use the Rayleigh quotient to obtain a shift near λ_i and then proceed again with inverse iteration and possibly Rayleigh quotient shifting. But it is important to note that a Rayleigh quotient shift can only be applied once the iteration vector is a close enough approximation to the eigenvector sought, because only in such a case will convergence occur to the required eigenpair (see Section 11.2.4).

Basically, therefore, the problem reduces to finding an approximation to either the next required eigenvalue or the corresponding eigenvector. Once an acceptable approximation has been obtained, the eigenpair is evaluated accurately by vector inverse iteration or Rayleigh quotient iteration. However, the choice of whether to find an approximation to the unknown eigenvalue or the corresponding eigenvector or to both simultaneously depends on the relative cost of the evaluations. Considering vector iteration as the basic technique to obtain an eigenvector, we discussed in Section 11.2 the difficulties of obtaining by simple inverse iteration (without shifting) an approximation to an eigenvector $\boldsymbol{\phi}_i$, where $i > 1$. In the vector iteration it is necessary to either deflate the matrix or the iteration vectors by the eigenpairs already calculated, which requires that the eigenpairs have been evaluated very accurately. Another disadvantage is that convergence can be very slow, namely, when eigenvalues are of nearly equal magnitude. *These difficulties suggest that it is more effective to calculate first an approximation to the unknown eigenvalue using a polynomial iteration scheme, and switch to inverse iteration only when a shift close to the required eigenvalue has been obtained. This is exactly the basic strategy followed in the determinant search solution method.*

12.2.2 The Solution Algorithm

Consider the iteration for the eigenpair $(\lambda_1, \boldsymbol{\phi}_1)$, which, as is pointed out later, is typical. Let μ_{k-1} and μ_k be two approximations to λ_1, where $\mu_{k-1} < \mu_k \leq \lambda_1$, as shown in Fig. 12.1. The first objective in the algorithm is to obtain as economically as possible a shift near λ_1. This is accomplished by using an accelerated secant iteration in which the next shift μ_{k+1} is calculated using

$$\mu_{k+1} = \mu_k - \eta \frac{p(\mu_k)}{p(\mu_k) - p(\mu_{k-1})}(\mu_k - \mu_{k-1}) \tag{12.1}$$

where η is a constant. When $\eta = 1$ this is the standard secant iteration described in Section 11.4.2, in which $\mu_{k+1} \longrightarrow \lambda_1$ as $k \longrightarrow \infty$, and convergence can be slow. However, since the objective is to merely obtain a shift near λ_1 the algorithm uses $\eta \geq 2$. For $\eta = 2$ it is known that $\mu_{k+1} \leq \mu_a$, where μ_a is the smallest stationary point of p (see Fig. 12.1). Thus the iteration with $\eta = 2$ can only jump over one root, which would be detected by a sign change in p. But it must be observed that in case λ_1 is a multiple root, the iteration can still be slow.

If the only information available were to be the value of p, it would be

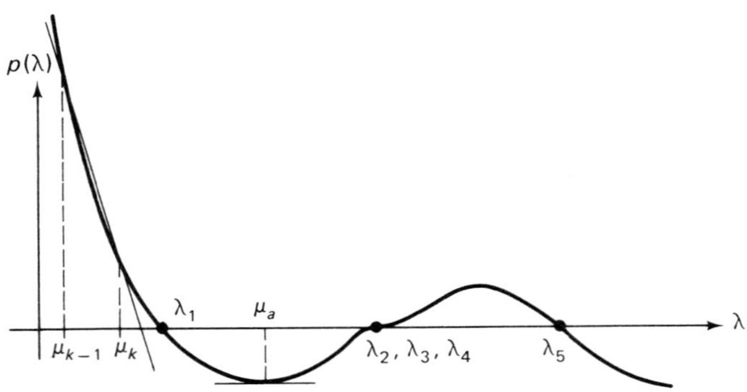

Figure 12.1. Characteristic polynomial $p(\lambda)$.

vitally important that the iteration not be able to jump over more than one root, since it would be impossible to detect when a number of roots had been passed. However, the algorithm also uses the Sturm sequence property, namely that the number of negative diagonal elements in the triangular factorization at μ_{k+1} is equal to the number of eigenvalues smaller than μ_{k+1} (see Section 11.5). Therefore, there is no restriction to $\eta = 2$, and the iteration can be accelerated further. In the polynomial iteration η is doubled after each step in which the iterates did not change in their two most significant digits, i.e., when the correction to the current shift is already relatively small. The iteration may thus jump over a single root, a multiple root, or into a cluster of roots, but this is always detected by counting the number of negative diagonal elements in the triangular factorization. Also, with the strategy adopted, the algorithm cannot jump far beyond the unknown roots.

Once it has been established that the secant iteration jumped over one or more unknown eigenvalues, vector inverse iteration with Gram–Schmidt orthogonalization is used to calculate the eigenvalues and eigenvectors accurately, as described in Section 11.2.5. Except in rare cases, no Rayleigh quotient shift is carried out, because once the iteration vector is an approximation to the eigenvector sought, convergence is usually rapid and another factorization would be inefficient. It should be noted that in this iteration, eigenvectors corresponding to eigenvalues larger than μ_{k+1} may be calculated if μ_{k+1} lies in an eigenvalue cluster. This is hardly a disadvantage, because these eigenvalues and eigenvectors are probably required anyway.

As the aim is merely to obtain a shift near the next unknown eigenvalue, it may happen that before any jump occurs, two successive iterates in (12.1) do not differ any more in their six most significant digits. The last iterate is then accepted as the place at which to switch to inverse iteration with Gram–Schmidt orthogonalization.

Of particular importance in the algorithm is the effective evaluation of $p(\lambda)$ (see Section 11.4.2). It should be noted that in the determinant search algorithm, the triangular factorization of $\mathbf{K} - \lambda\mathbf{M}$ into $\mathbf{LDL}^T$ is calculated using Gauss elimination without interchanges. Therefore, the stability of the decomposition is theoretically not guaranteed when $\lambda > \lambda_1$; i.e., multiplier growth could occur, which, however, would always be detected. It is now important to note that the algorithm only aims to find in the polynomial iteration a modest approximation to the required eigenvalue; therefore, if unacceptable multiplier growth does occur, λ is slightly changed in the algorithm, and another factorization is carried out. This is justified because in practice this condition is rarely reached. In summary, therefore, *the triangular factorization can be carried out without interchanges, because the evaluation of $p(\lambda)$ is only completed if no instability occurs, and the specific value used for the current shift does not affect the final accuracy of the eigenvalues and corresponding eigenvectors.*

The iteration described above for λ_1 and $\boldsymbol{\phi}_1$ is typical because the advantage of the one-sided approach to the eigenvalue is also obtained for any other eigenpair, say $(\lambda_{j+1}, \boldsymbol{\phi}_{j+1})$, by using instead of $p(\lambda)$ in (12.1) the deflated polynomial $p_j(\lambda)$ (Fig. 12.2), where

$$p_j(\lambda) = \frac{p(\lambda)}{\prod_{i=1}^{j} (\lambda - \lambda_i^{(l+1)})} \tag{12.2}$$

where $\lambda_i^{(l+1)}$ is the accepted approximation to the eigenvalue λ_i. Therefore, to iterate toward the next unknown eigenvalue, (12.2) is used to suppress all calculated eigenvalues from two previously obtained polynomial values which are the starting values in (12.1). It should be noted that in order not to divide in (12.2) by values close to zero, two λ values need be selected which are far enough from the calculated roots.

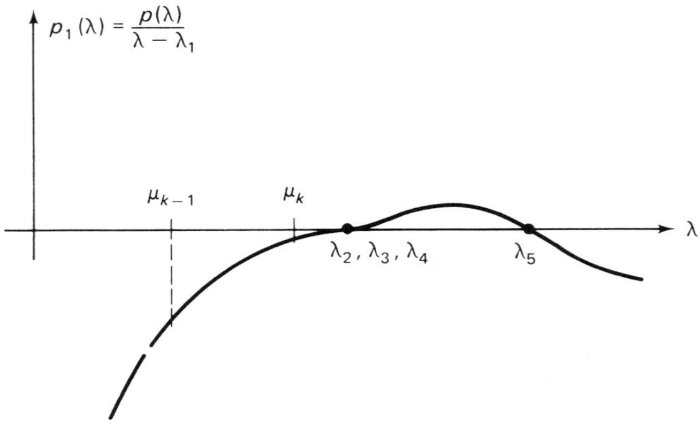

Figure 12.2. Deflated characteristic polynomial with λ_1 suppressed.

12.2.3 Final Remarks Concerning the Determinant Search Solution

A few practical details are worth mentioning:

1. A principal advantage of this solution procedure is that each eigenpair is essentially obtained independently from all those previously calculated. Eigenvalues and eigenvectors, therefore, need not be calculated to very high precision, as is the case when an explicit matrix deflation procedure is used (see Section 11.2.5). If not very high accuracy is required, the errors in earlier computed eigenpairs will not prevent attaining the required accuracy in the remaining eigenvalues and eigenvectors.

2. At a shift near the next required eigenvalue, Gram–Schmidt orthogonalization to all previously calculated eigenvectors is not necessary. Allowing for multiple eigenvalues, in the program developed, the iteration vector is only orthogonalized to the last six calculated eigenvectors.
3. It should be noted that in practical examples $p(\lambda)$ can be much larger than the overflow of the computer; therefore, scale factors which keep the implementation machine-independent must be introduced in the evaluation of $p(\lambda)$.[7]
4. To start the secant iteration, two lower bounds on λ_1 are needed. The first bound is $\mu_1 = 0.0$; the other bound is obtained from an approximation to $\boldsymbol{\phi}_1$ calculated by vector inverse iteration at λ equal to zero. Assume that $\mathbf{x}_k$ is the iteration vector after $(k - 1)$ inverse iterations; then we can use

$$\mu_2 = (1 - 0.01)\frac{\mathbf{x}_k^T\mathbf{K}\mathbf{x}_k}{\mathbf{x}_k^T\mathbf{M}\mathbf{x}_k} \tag{12.3}$$

   It may happen that μ_2 is larger than λ_1. However, this is detected by the Sturm sequence count in the factorization at μ_2. Let γ be the number of negative pivots in the triangular factorization. Then we divide μ_2 by $(\gamma + 1)$ until γ equals zero.
5. The algorithm is most efficiently used in an in-core solution routine. Because a few triangular factorizations are required in the calculation of each eigenpair, relatively much peripheral processor (tape and disc reading and writing) time must be expected in out-of-core solutions. Since the algorithm has been designed for the analysis of small banded systems, on reasonably sized computers, relatively large-order systems can be solved in high-speed storage.

In order to assess the effectiveness of the algorithm in the solution of a specific problem, Table 12.1 summarizes the steps in the solution procedure and gives the total number of operations used and the storage requirements. The number of iterations required for solution depends on the system considered, but, as assumed in Table 12.1, experience has shown that an average of about six secant steps and six inverse iterations are needed for the solution of one eigenpair. However, when solving for a cluster of eigenvalues, more iterations are probably required for the calculation of the first eigenpair, but then the next eigenpairs are obtained with only a few more iterations. Table 12.2 summarizes some approximate solution times used.

Assume that the determinant search algorithm has calculated all required eigenpairs. The solution is completed by calculating error estimates on the precision with which the eigenvalues and eigenvectors have been evaluated.

TABLE 12.1 SUMMARY OF DETERMINANT SEARCH SOLUTION

Operation	Calculation	Number of Operations: $m = m_{\mathbf{K}} = m_{\mathbf{M}}$	Number of Operations: $m = m_{\mathbf{K}},\ m_{\mathbf{M}} = 0$	$m = m_{\mathbf{K}} = m_{\mathbf{M}}$, $m = m_{\mathbf{K}},\ m_{\mathbf{M}} = 0$
Secant iteration	$\bar{\mathbf{K}} = \mathbf{K} - \mu_k \mathbf{M}$	$n(m+1)$	n	
	$\bar{\mathbf{K}} = \mathbf{LDL}^T$	$\frac{1}{2}nm^2 + \frac{3}{2}nm$	$\frac{1}{2}nm^2 + \frac{3}{2}nm$	
	$p(\mu_k) = \prod_{i=1}^{n} d_{ii}$	n	n	Using symmetry of matrices
Inverse iteration	$\bar{\mathbf{K}}\bar{\mathbf{x}}_{k+1} = \mathbf{y}_k$	$n(2m+1)$	$n(2m+1)$	
	$\bar{\mathbf{y}}_{k+1} = \mathbf{M}\bar{\mathbf{x}}_{k+1}$	$n(2m+1)$	n	
	$\rho(\bar{\mathbf{x}}_{k+1}) = \dfrac{\bar{\mathbf{x}}_{k+1}^T \mathbf{y}_k}{\bar{\mathbf{x}}_{k+1}^T \bar{\mathbf{y}}_{k+1}}$	$2n$	$2n$	$2n(m+1) + 9n$ $n(m+1) + 10n$
	$\mathbf{y}_{k+1} = \dfrac{\bar{\mathbf{y}}_{k+1} - \sum_{j=1}^{6} \alpha_{i-j} \bar{\boldsymbol{\phi}}_{i-j}}{(\bar{\mathbf{x}}_{k+1}^T \bar{\mathbf{y}}_{k+1})^{1/2}}$ where $\bar{\boldsymbol{\phi}}_j = \mathbf{M}\boldsymbol{\phi}_j$, $\alpha_j = \bar{\mathbf{x}}_{k+1}^T \bar{\boldsymbol{\phi}}_j$	$13n$	$13n$	(Algorithm has also been implemented as out-of core solver[7])
Error estimates	$\dfrac{\lVert \mathbf{K}\boldsymbol{\phi}_i^{(l+1)} - \lambda_i^{(l+1)} \mathbf{M}\boldsymbol{\phi}_i^{(l+1)} \rVert_2}{\lVert \mathbf{K}\boldsymbol{\phi}_i^{(l+1)} \rVert_2}$	$2nm + 4n$	$5nm + 2n$	
Total for p lowest eigenvalues and associated eigenvectors assuming six secant and six inverse iterations per eigenpair		$(3nm^2 + 41nm + 118n)p$	$(2nm^2 + 26nm + 116n)p$	

TABLE 12.2 SOLUTION TIMES USING DETERMINANT SEARCH ALGORITHM

System	System Order	Maximum Half Bandwidth m_K	Mass Matrix	Number of Eigenpairs	Computer Used	Central Processor Seconds
Plane Frame	297	29	Diagonal	3	CDC 6400	40
Dam	226	68	Banded	7	CDC 6600	71
Building	340	31	Diagonal	7	CDC 6600	20
Piping System	566	11	Diagonal	7	CDC 6600	11
Arch Dam	417	23	Diagonal	30	CDC 6600	98

The specific error quantities calculated have been presented in Section 10.4 and are, for each eigenpair,

$$\epsilon_i = \frac{\|\mathbf{K}\boldsymbol{\phi}_i^{(l+1)} - \lambda_i^{(l+1)}\mathbf{M}\boldsymbol{\phi}_i^{(l+1)}\|_2}{\|\mathbf{K}\boldsymbol{\phi}_i^{(l+1)}\|_2} \tag{12.4}$$

where $\boldsymbol{\phi}_i^{(l+1)}$ and $\lambda_i^{(l+1)}$ are the ith eigenvector and eigenvalue predicted in the last iteration.

As mentioned already, the determinant search method is used in a number of finite element computer programs, and actual subroutines have been published (see References 1, 5, and 7). However, the method is most effective as an in-core solver and considerable limitations may exist on the system size that can be considered. A more generally applicable solution procedure for large finite element systems is the subspace iteration method, which is presented next.

12.3 THE SUBSPACE ITERATION METHOD

The important idea in the determinant search method presented in the previous section was to combine polynomial iteration and vector inverse iteration in an optimum manner to obtain an effective algorithm for the solution of small banded systems. However, when the bandwidth of the matrices **K** and **M** increases, polynomial iterations are much more costly than vector inverse iterations. Figure 12.3 shows the number of inverse iterations that can be carried out for one polynomial iteration, when measured in terms of

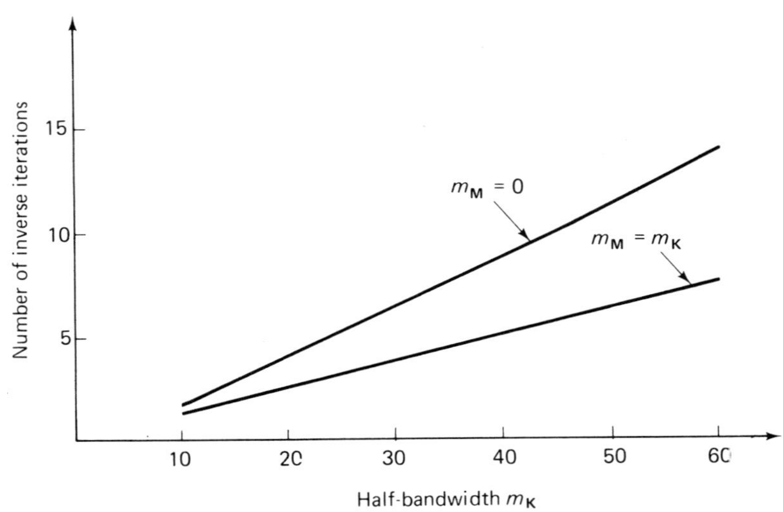

Figure 12.3. Number of inverse iterations per polynomial evaluation (Gram–Schmidt orthogonalization to last six eigenvectors).

high-speed storage operations. The relationship is shown as a function of the half-bandwidth m_K of the stiffness matrix, and it is assumed that in the vector inverse iterations, Gram–Schmidt orthogonalization is performed to the last six eigenvectors calculated. The figure shows that for a large bandwidth, many inverse iterations are equivalent to one polynomial evaluation. The cost of a polynomial evaluation when measured on vector inverse iterations increases still further when the operations have to be performed out-of-core, because a great deal more back-up storage reading and writing is needed in a polynomial evaluation than in vector inverse iteration. *These considerations suggest that for the solution of out-of-core systems and systems with large bandwidths, vector inverse iterations should be used primarily. In addition, if inverse iteration is performed simultaneously for all required eigenvectors, the solution of out-of-core systems is further enhanced, because the back-up storage reading and writing of the triangularized matrix and the iteration vectors is reduced. It is specifically with those ideas in mind that the subspace iteration method has been designed.*

In essence, the subspace iteration method consists of the following three steps:

(1) Establish q starting iteration vectors, $q > p$, where p is the number of eigenvalues and vectors to be calculated.
(2) Use simultaneous inverse iteration on the q vectors and Ritz analysis to extract the "best" eigenvalue and eigenvector approximations from the q iteration vectors.
(3) After iteration convergence, use the Sturm sequence check to verify that the required eigenvalues and corresponding eigenvectors have been calculated.

The solution procedure has been called subspace iteration method, because the iteration is equivalent to iterating with a q-dimensional subspace and should not be regarded as a simultaneous iteration with q individual iteration vectors. Specifically, as we discuss in the following sections, it must be noted that the selection of the starting iteration vectors in step (1) and the Sturm sequence check in step (3) are considered very important parts of the iteration procedure. Altogether, the subspace iteration method is largely based on various techniques that have been used earlier,[9-16] namely simultaneous vector iteration, Sturm sequence information (see Section 10.2.2), and Rayleigh-Ritz analysis (see Section 10.3.2), but most enlightening has been the work of Rutishauser.[17]

12.3.1 Preliminary Considerations

The basic objective in the subspace iteration method is to solve for the lowest p eigenvalues and corresponding eigenvectors satisfying

$$\mathbf{K\Phi} = \mathbf{M\Phi\Lambda} \tag{12.5}$$

where $\mathbf{\Lambda} = \text{diag}(\lambda_i)$ and $\mathbf{\Phi} = [\boldsymbol{\phi}_1, \ldots, \boldsymbol{\phi}_p]$.

In addition to the relation in (12.5), the eigenvectors also satisfy the orthogonality conditions

$$\mathbf{\Phi}^T\mathbf{K\Phi} = \mathbf{\Lambda}; \qquad \mathbf{\Phi}^T\mathbf{M\Phi} = \mathbf{I} \tag{12.6}$$

where $\mathbf{I}$ is a unit matrix of order p because $\mathbf{\Phi}$ stores only p eigenvectors. It is important to note that the relation in (12.5) is a necessary and sufficient condition for the vectors in $\mathbf{\Phi}$ to be eigenvectors but that the eigenvector orthogonality conditions in (12.6) are necessary but not sufficient. In other words, if we have p vectors that satisfy (12.6), $p < n$, then these vectors are not necessarily eigenvectors. However, if the p vectors satisfy (12.5), they are definitely eigenvectors, although we still need to make sure that they are indeed the p specific eigenvectors sought (see Section 10.2.1).

The essential idea of the subspace iteration method uses the fact that the eigenvectors in (12.5) form an $\mathbf{M}$-orthonormal basis of the p-dimensional least-dominant subspace of the operators $\mathbf{K}$ and $\mathbf{M}$, which we will now call E_∞ (see Section 2.2). In the solution the iteration with p linearly independent vectors can therefore be regarded as an iteration with a subspace. The starting iteration vectors span E_1 and iteration continues until, to sufficient accuracy, E_∞ is spanned. The fact that iteration is performed with a subspace has some important consequences. The total number of required iterations depends on how "close" E_1 is to E_∞, and not on how close each iteration vector is to an eigenvector. Hence the effectiveness of the algorithm lies in that it is much easier to establish a p-dimensional starting subspace which is close to E_∞ than to find p vectors that each are close to a required eigenvector. The specific algorithm used to establish the starting iteration vectors is described later. Also, *because iteration is performed with a subspace, convergence of the subspace is all that is required and not of individual iteration vectors to eigenvectors. In other words, if the iteration vectors are linear combinations of the required eigenvectors, the solution algorithm converges in one step.*

To demonstrate the essential idea, we first consider simultaneous vector inverse iteration on p vectors. Let $\mathbf{X}_1$ store the p starting iteration vectors, which span the starting subspace E_1. Simultaneous inverse iteration on the p vectors can be written

$$\mathbf{KX}_{k+1} = \mathbf{MX}_k; \qquad k = 1, 2, \ldots \tag{12.7}$$

where we now observe that the p iteration vectors in $\mathbf{X}_{k+1}$ span a p-dimensional subspace E_{k+1}, and the sequence of subspaces generated converges to E_∞, provided the starting vectors are not orthogonal to E_∞. This seems to contradict the fact that in this iteration each column in $\mathbf{X}_{k+1}$ is known to converge to the least dominant eigenvector unless the column is deficient in $\boldsymbol{\phi}_1$ (see Section 11.2.1). Actually, there is no contradiction. Although in exact arithmetic the vectors in $\mathbf{X}_{k+1}$ span $\mathbf{E}_{k+1}$, they do become more and more parallel and therefore a poorer and poorer basis for E_{k+1}. One way to preserve numerical stability is to generate orthogonal bases in the subspaces E_{k+1} using the Gram–Schmidt process (see Section 11.2.5). In this case we iterate for $k = 1, 2, \ldots$ as follows:

$$\mathbf{K}\bar{\mathbf{X}}_{k+1} = \mathbf{M}\mathbf{X}_k \tag{12.8}$$

$$\mathbf{X}_{k+1} = \bar{\mathbf{X}}_{k+1}\mathbf{R}_{k+1} \tag{12.9}$$

where $\mathbf{R}_{k+1}$ is an upper-triangular matrix chosen in such way that $\mathbf{X}_{k+1}^T\mathbf{M}\mathbf{X}_{k+1} = \mathbf{I}$. Then provided that the starting vectors in $\mathbf{X}_1$ are not deficient in the eigenvectors $\boldsymbol{\phi}_1, \boldsymbol{\phi}_2, \ldots, \boldsymbol{\phi}_p$, we have

$$\mathbf{X}_{k+1} \longrightarrow \boldsymbol{\Phi}; \qquad \mathbf{R}_{k+1} \longrightarrow \boldsymbol{\Lambda}$$

It is important to note that the iteration in (12.7) generates the same sequence of subspaces as the iteration in (12.8) and (12.9). However, the ith column in $\mathbf{X}_{k+1}$ of (12.8) and (12.9) converges linearly to $\boldsymbol{\phi}_i$ with the convergence rate equal to max $\{\lambda_{i-1}/\lambda_i, \lambda_i/\lambda_{i+1}\}$. To demonstrate the solution procedure, consider the following example.

Example 12.1

Consider the eigenproblem $\mathbf{K}\boldsymbol{\phi} = \lambda\mathbf{M}\boldsymbol{\phi}$, where

$$\mathbf{K} = \begin{bmatrix} 2 & -1 & 0 \\ -1 & 4 & -1 \\ 0 & -1 & 2 \end{bmatrix}; \qquad \mathbf{M} = \begin{bmatrix} \frac{1}{2} & & \\ & 1 & \\ & & \frac{1}{2} \end{bmatrix}$$

The two lowest eigenvalues and corresponding eigenvectors are (see Example 10.4)

$$\lambda_1 = 2, \quad \boldsymbol{\phi}_1 = \begin{bmatrix} \frac{1}{\sqrt{2}} \\ \frac{1}{\sqrt{2}} \\ \frac{1}{\sqrt{2}} \end{bmatrix}; \qquad \lambda_2 = 4, \quad \boldsymbol{\phi}_2 = \begin{bmatrix} -1 \\ 0 \\ 1 \end{bmatrix}$$

Use the simultaneous vector iteration with Gram–Schmidt orthogonalization given in (12.8) and (12.9) with starting vectors

$$\mathbf{X}_1 = \begin{bmatrix} 0 & 2 \\ 1 & 1 \\ 2 & 0 \end{bmatrix}$$

to calculate approximations to λ_1, $\boldsymbol{\phi}_1$ and λ_2, $\boldsymbol{\phi}_2$.

Solving for $\bar{\mathbf{X}}_2$ the relation $\mathbf{K}\bar{\mathbf{X}}_2 = \mathbf{M}\mathbf{X}_1$ gives

$$\bar{\mathbf{X}}_2 = \begin{bmatrix} 0.25 & 0.75 \\ 0.50 & 0.50 \\ 0.75 & 0.25 \end{bmatrix}$$

M-orthonormalizing $\bar{\mathbf{X}}_2$ gives

$$\mathbf{X}_2 = \begin{bmatrix} 0.3333 & 1.179 \\ 0.6667 & 0.2357 \\ 1.000 & -0.7071 \end{bmatrix}; \qquad \mathbf{R}_2 = \begin{bmatrix} 1.333 & -1.650 \\ 0 & 2.121 \end{bmatrix}$$

Proceeding similarly, we obtain the following results:

$$\mathbf{X}_3 = \begin{bmatrix} 0.5222 & 1.108 \\ 0.6963 & 0.1231 \\ 0.8704 & -0.8614 \end{bmatrix}; \qquad \mathbf{R}_3 = \begin{bmatrix} 2.089 & -0.9847 \\ 0 & 3.830 \end{bmatrix}$$

$$\mathbf{X}_4 = \begin{bmatrix} 0.6163 & 1.058 \\ 0.7044 & 0.0622 \\ 0.7924 & -0.9339 \end{bmatrix}; \qquad \mathbf{R}_4 = \begin{bmatrix} 2.023 & -0.5202 \\ 0 & 3.954 \end{bmatrix}$$

$$\mathbf{X}_5 = \begin{bmatrix} 0.6623 & 1.030 \\ 0.7064 & 0.0312 \\ 0.7506 & -0.9678 \end{bmatrix}; \qquad \mathbf{R}_5 = \begin{bmatrix} 2.006 & -0.2639 \\ 0 & 3.988 \end{bmatrix}$$

$$\mathbf{X}_6 = \begin{bmatrix} 0.6848 & 1.015 \\ 0.7069 & 0.0156 \\ 0.7290 & -0.9841 \end{bmatrix}; \qquad \mathbf{R}_6 = \begin{bmatrix} 2.001 & -0.1324 \\ 0 & 3.997 \end{bmatrix}$$

$$\mathbf{X}_7 = \begin{bmatrix} 0.6960 & 1.008 \\ 0.7071 & 0.0078 \\ 0.7181 & -0.9921 \end{bmatrix}; \qquad \mathbf{R}_7 = \begin{bmatrix} 2.000 & -0.0663 \\ 0 & 3.999 \end{bmatrix}$$

$$\mathbf{X}_8 = \begin{bmatrix} 0.7016 & 1.004 \\ 0.7071 & 0.0039 \\ 0.7126 & -0.9961 \end{bmatrix}; \qquad \mathbf{R}_8 = \begin{bmatrix} 2.000 & -0.0331 \\ 0 & 4.000 \end{bmatrix}$$

$$\mathbf{X}_9 = \begin{bmatrix} 0.7043 & 1.002 \\ 0.7071 & 0.0020 \\ 0.7099 & -0.9980 \end{bmatrix}; \qquad \mathbf{R}_9 = \begin{bmatrix} 2.000 & -0.0166 \\ 0 & 4.000 \end{bmatrix}$$

$$\mathbf{X}_{10} = \begin{bmatrix} 0.7057 & 1.001 \\ 0.7071 & 0.0010 \\ 0.7085 & -0.9990 \end{bmatrix}; \qquad \mathbf{R}_{10} = \begin{bmatrix} 2.000 & -0.0083 \\ 0 & 4.000 \end{bmatrix}$$

and after nine iterations we have

$$\boldsymbol{\phi}_1 \doteq \begin{bmatrix} 0.7057 \\ 0.7071 \\ 0.7085 \end{bmatrix}; \qquad \lambda_1 \doteq 2.000$$

$$\boldsymbol{\phi}_2 \doteq \begin{bmatrix} 1.001 \\ 0.0010 \\ -0.9990 \end{bmatrix}; \qquad \lambda_2 \doteq 4.000$$

It should be noted that although the vectors in $\mathbf{X}_1$ already span the space of $\boldsymbol{\phi}_1$ and $\boldsymbol{\phi}_2$, we need a relatively large number of iterations for convergence.

The solution in the preceding example demonstrates the iteration procedure in (12.8) and (12.9) and also shows the main deficiency of the method. Namely, each iteration vector is forced to converge to a different eigenvector by orthogonalizing the ith iteration vector to the $(i - 1)$ iteration vectors that have been orthogonalized already without allowing a more effective linear combination of the vectors to take place. In the preceding example, the iteration was started with two iteration vectors that were linear combinations of the required eigenvectors and this did not yield any advantage. In general, if the iteration vectors in $\mathbf{X}_{k+1}$ span E_∞ but are not eigenvectors (i.e., the vectors in $\mathbf{X}_{k+1}$ are linear combinations of the eigenvectors $\boldsymbol{\phi}_1, \ldots, \boldsymbol{\phi}_p$), then, although the subspace E_{k+1} already converged, many more iterations may be needed in order to turn the orthogonal basis of iteration vectors into the basis of eigenvectors.

12.3.2 Subspace Iteration

The following algorithm, which we call *subspace iteration*, finds an orthogonal basis of vectors in E_{k+1}, thus preserving numerical stability in the iteration of (12.7), and also calculates in one step the required eigenvectors when E_{k+1} converged to E_∞. This algorithm is the iteration used in the subspace iteration method, i.e. step 2 of the complete solution phase (see Section 12.3).

For $k = 1, 2, \ldots$, iterate from E_k to E_{k+1}:

$$\mathbf{K}\bar{\mathbf{X}}_{k+1} = \mathbf{M}\mathbf{X}_k \tag{12.10}$$

Find the projections of the operators $\mathbf{K}$ and $\mathbf{M}$ onto E_{k+1}:

$$\mathbf{K}_{k+1} = \bar{\mathbf{X}}_{k+1}^T\mathbf{K}\bar{\mathbf{X}}_{k+1} \tag{12.11}$$

$$\mathbf{M}_{k+1} = \bar{\mathbf{X}}_{k+1}^T\mathbf{M}\bar{\mathbf{X}}_{k+1} \tag{12.12}$$

Solve for the eigensystem of the projected operators:

$$\mathbf{K}_{k+1}\mathbf{Q}_{k+1} = \mathbf{M}_{k+1}\mathbf{Q}_{k+1}\boldsymbol{\Lambda}_{k+1} \tag{12.13}$$

Find an improved approximation to the eigenvectors:

$$\mathbf{X}_{k+1} = \bar{\mathbf{X}}_{k+1}\mathbf{Q}_{k+1} \tag{12.14}$$

Then, provided that the vectors in $\mathbf{X}_1$ are not orthogonal to one of the required eigenvectors, we have

$$\boldsymbol{\Lambda}_{k+1} \longrightarrow \boldsymbol{\Lambda} \quad \text{and} \quad \mathbf{X}_{k+1} \longrightarrow \boldsymbol{\Phi} \qquad \text{as } k \longrightarrow \infty$$

In the subspace iteration, it is implied that the iteration vectors are ordered in an appropriate way; i.e., the iteration vectors converging to $\boldsymbol{\phi}_1$, $\boldsymbol{\phi}_2, \ldots$ are stored as the first, second, ... columns of $\mathbf{X}_{k+1}$. We demonstrate the iteration procedure by calculating the solution to the problem considered in Example 12.1.

Example 12.2

Use the subspace iteration algorithm to solve the problem considered in Example 12.1.

Using the relations in (12.10) to (12.14) with $\mathbf{K}$, $\mathbf{M}$, and $\mathbf{X}_1$ given in Example 12.1, we obtain

$$\bar{\mathbf{X}}_2 = \frac{1}{4}\begin{bmatrix} 1 & 3 \\ 2 & 2 \\ 3 & 1 \end{bmatrix}$$

$$\mathbf{K}_2 = \frac{1}{4}\begin{bmatrix} 5 & 3 \\ 3 & 5 \end{bmatrix}; \qquad \mathbf{M}_2 = \frac{1}{16}\begin{bmatrix} 9 & 7 \\ 7 & 9 \end{bmatrix}$$

Hence

$$\mathbf{\Lambda}_2 = \begin{bmatrix} 2 & 0 \\ 0 & 4 \end{bmatrix}; \qquad \mathbf{Q}_2 = \begin{bmatrix} \frac{1}{\sqrt{2}} & 2 \\ \frac{1}{\sqrt{2}} & -2 \end{bmatrix}$$

and

$$\mathbf{X}_2 = \begin{bmatrix} \frac{1}{\sqrt{2}} & -1 \\ \frac{1}{\sqrt{2}} & 0 \\ \frac{1}{\sqrt{2}} & 1 \end{bmatrix}$$

Comparing the results with the solution obtained in Example 12.1, we observe that we have calculated the exact eigenvalues and eigenvectors in the first subspace iteration. This must be the case because the starting iteration vectors in $\mathbf{X}_1$ span the subspace defined by $\boldsymbol{\phi}_1$ and $\boldsymbol{\phi}_2$.

Considering the subspace iteration, a first observation is that $\mathbf{K}_{k+1}$ and $\mathbf{M}_{k+1}$ in (12.11) and (12.12), respectively, tend toward diagonal form as the number of iterations increases; i.e., $\mathbf{K}_{k+1}$ and $\mathbf{M}_{k+1}$ are diagonal matrices when the columns in $\mathbf{X}_{k+1}$ store multiples of eigenvectors. Hence it follows from the discussion in Section 11.3.2 that the generalized Jacobi method can be used very effectively for the solution of the eigenproblem in (12.13).

An important aspect is the convergence of the method. Assuming that in the iteration the vectors in $\mathbf{X}_{k+1}$ are ordered in such way that the ith diagonal element in $\mathbf{\Lambda}_{k+1}$ is larger than the $(i-1)$st element, $i = 2, \ldots, p$, then *the* ith *column in* $\mathbf{X}_{k+1}$ *converges linearly to* $\boldsymbol{\phi}_i$ *and the convergence rate is* λ_i/λ_{p+1}. Although this is an asymptotic convergence rate, it indicates that the smallest eigenvalues converge fastest. In addition, *a higher convergence rate can be obtained by using* q *iteration vectors, with* $q > p$. *However, using more iteration vectors will also increase the computer effort for one iteration. In practice, we have found that* $q = \min\{2p, p+8\}$ *is, in general, effective and we use this value*. Considering the convergence rate, it should be noted that multiple eigenvalues do not decrease the rate of convergence, provided that $\lambda_{q+1} > \lambda_p$.

As for the iteration schemes presented earlier, the theoretical convergence behavior can only be observed in practice when the iteration vectors are already relatively close to eigenvectors. However, in practice, we are very interested in knowing what happens already in the first few iterations when E_{k+1} is not yet "close" to E_∞. Indeed, the effectiveness of the algorithm lies to a large extent in that a few iterations can already give good approximations to the required eigenpairs, the reason being that one subspace iteration given in (12.10) to (12.14) is, in fact, a Ritz analysis, as described in Section 10.3.2. Therefore, *all characteristics of the Ritz analysis pertain also to the subspace iteration; i.e., the lowest eigenvalues are approximated best in the iteration and all eigenvalue approximations are upper bounds on the actual eigenvalues sought.* It follows that we may think of the subspace iteration as a repeated application of the Ritz analysis method of Section 10.3.2, in which the eigenvector approximations calculated in the previous iteration are used to form the right-hand side load vectors in the current iteration.

It is important to realize that using either one of the iteration procedures given in (12.7), (12.8) and (12.9), or (12.10) to (12.14), the same subspace E_{k+1} is spanned by the iteration vectors. Therefore, there is no need to iterate always as given in (12.10) to (12.14), but we may first use the simple inverse iteration in (12.7) or inverse iteration with Gram–Schmidt orthogonalization as given in (12.8) and (12.9), and finally use the subspace iteration scheme given in (12.10) to (12.14). The calculated results would be the same in theory as those obtained when subspace iterations only are performed. However, the difficulty then lies in deciding at what stage to orthogonalize the iteration vectors by using (12.11) to (12.14), because the iteration in (12.10) yields vectors that are more and more parallel. Also, the Gram–Schmidt orthogonalization is numerically not very stable. If the iteration vectors have become "too close to each other" because either the initial assumptions gave iteration vectors that were almost prarallel or by iterating without orthogonalization, it may be impossible to orthogonalize them because of finite precision arithmetic in the computer. Unfortunately, considering large finite element systems, the starting iteration vectors can in some cases be nearly parallel, although they span a subspace that is close to E_∞, and it is best to immediately orthogonalize the iteration vectors using the projections of $\mathbf{K}$ and $\mathbf{M}$ onto E_2. In addition, using the subspace iteration, we obtain in each iteration "best" approximations to the required eigenvalues and eigenvectors and can measure convergence in each iteration. It may also be noted that because of back-up storage reading and writing required, Gram–Schmidt orthogonalization can be ineffective in out-of-core solution.

12.3.3 Starting Iteration Vectors

The first step in the subspace iteration method is the selection of the starting iteration vectors in $\mathbf{X}_1$ [see (12.10)]. As pointed out earlier, if starting vectors are used that span the least dominant subspace, the iteration converges in one step. This is the case, for example, when there are only p nonzero masses in a diagonal mass matrix and the starting vectors are unit vectors $\mathbf{e}_i$ with the entries $+1$ corresponding to the mass degrees of freedom. In this case one subspace iteration is in effect a static condensation analysis. This follows from the discussion in Section 10.3.2 and because a subspace iteration embodies a Ritz analysis. Consider the following example.

Example 12.3

Use the subspace iteration to calculate the eigenpairs $(\lambda_1, \boldsymbol{\phi}_1)$ and $(\lambda_2, \boldsymbol{\phi}_2)$ of the problem $\mathbf{K}\boldsymbol{\phi} = \lambda\mathbf{M}\boldsymbol{\phi}$, where

$$\mathbf{K} = \begin{bmatrix} 2 & -1 & 0 & 0 \\ -1 & 2 & -1 & 0 \\ 0 & -1 & 2 & -1 \\ 0 & 0 & -1 & 1 \end{bmatrix}; \qquad \mathbf{M} = \begin{bmatrix} 0 & & & \\ & 2 & & \\ & & 0 & \\ & & & 1 \end{bmatrix}$$

As suggested above, we use as starting vectors the unit vectors $\mathbf{e}_2$ and $\mathbf{e}_4$. Following (12.10) to (12.14), we then obtain

$$\begin{bmatrix} 2 & -1 & 0 & 0 \\ -1 & 2 & -1 & 0 \\ 0 & -1 & 2 & -1 \\ 0 & 0 & -1 & 1 \end{bmatrix} \bar{\mathbf{X}}_2 = \begin{bmatrix} 0 & 0 \\ 2 & 0 \\ 0 & 0 \\ 0 & 1 \end{bmatrix}$$

$$\bar{\mathbf{X}}_2 = \begin{bmatrix} 2 & 1 \\ 4 & 2 \\ 4 & 3 \\ 4 & 4 \end{bmatrix}$$

and

$$\mathbf{K}_2 = 4\begin{bmatrix} 2 & 1 \\ 1 & 1 \end{bmatrix}; \qquad \mathbf{M}_2 = 8\begin{bmatrix} 6 & 4 \\ 4 & 3 \end{bmatrix}$$

Hence

$$\mathbf{\Lambda}_2 = \begin{bmatrix} \left(\frac{1}{2} - \frac{\sqrt{2}}{4}\right) & 0 \\ 0 & \left(\frac{1}{2} + \frac{\sqrt{2}}{4}\right) \end{bmatrix}; \qquad \mathbf{Q}_2 = \begin{bmatrix} \dfrac{1}{8 + 4\sqrt{2}} & \dfrac{1}{4\sqrt{2} - 8} \\ \dfrac{1}{4 + 4\sqrt{2}} & \dfrac{1}{4\sqrt{2} - 4} \end{bmatrix}$$

and

$$\mathbf{X}_2 = \begin{bmatrix} \frac{1}{4} & -\frac{1}{4} \\ \frac{1}{2} & -\frac{1}{2} \\ \dfrac{1 + \sqrt{2}}{4} & \dfrac{-1 + \sqrt{2}}{4} \\ \dfrac{\sqrt{2}}{2} & \dfrac{\sqrt{2}}{2} \end{bmatrix}$$

Comparing these results with the solution calculated in Example 10.12, we observe that we obtained the exact results with one subspace iteration.

A second case for which the subspace iteration can converge in one step arises when **K** and **M** are both diagonal. This is a rather trivial case, but it is considered for the development of an effective procedure for the selection of starting iteration vectors when general matrices are involved in the analysis. When **K** and **M** are diagonal, the iteration vectors should be unit vectors with the entries +1 corresponding to those degrees of freedom that have the smallest ratios k_{ii}/m_{ii}. These vectors are, in effect, the eigenvectors corresponding to the smallest eigenvalues, and this is why convergence is achieved in one step. We demonstrate the procedure in the following example.

Example 12.4

Construct starting iteration vectors for the solution of the lowest two eigenvalues by subspace iteration when considering the problem $\mathbf{K}\boldsymbol{\phi} = \lambda\mathbf{M}\boldsymbol{\phi}$, where

$$\mathbf{K} = \begin{bmatrix} 3 & & & \\ & 2 & & \\ & & 4 & \\ & & & 8 \end{bmatrix}; \qquad \mathbf{M} = \begin{bmatrix} 2 & & & \\ & 0 & & \\ & & 4 & \\ & & & 1 \end{bmatrix}$$

The ratios k_{ii}/m_{ii} are here $\frac{3}{2}$, ∞, 1, and 8 for $i = 1, \ldots, 4$, and indeed these are the eigenvalues of the problem. The starting vectors to be used are therefore

$$\mathbf{X}_1 = \left[\begin{array}{c:c} 0 & 1 \\ 0 & 0 \\ 1 & 0 \\ 0 & 0 \end{array}\right]$$

The vectors $\mathbf{X}_1$ are multiples of the required eigenvectors, and hence convergence occurs in the first step of iteration.

The two cases that we dealt with above involved rather special matrices; i.e., in the first case static condensation could be performed and in the second case the matrices **K** and **M** were diagonal. In both cases unit vectors $\mathbf{e}_i$ were employed, where $i = r_1, r_2, \ldots, r_p$, and the $r_j, j = 1, 2, \ldots, p$, corresponded to the p smallest values of k_{ii}/m_{ii} over all i. Using this notation we have $r_1 = 2$ and $r_2 = 4$ for Example 12.3 and $r_1 = 3$, $r_2 = 1$ for Example 12.4.

Although such specific matrices will hardly be encountered in general practical analysis, the results concerning the construction of the starting iteration vectors indicate how in general analysis effective starting iteration vectors may be selected. A general observation is that starting vectors that span the least dominant subspace E_∞ could be selected in the above cases, because the mass and stiffness properties have been lumped to a sufficient degree. In a general case, such lumping is not possible or would result in inaccurate stiffness and mass representations of the actual structure. However, although the matrices **K** and **M** do not have the exact same form used above, the discussion shows that the starting iteration vectors should be constructed to excite those degrees of freedom with which large mass and small stiffness are associated. Based on this observation, in essence, the following algorithm is being used effectively for the selection of the starting iteration vectors. The first column in $\mathbf{MX}_1$ is simply the diagonal of $\mathbf{M}$. This assures that all mass degrees of freedom are excited. The other columns in $\mathbf{MX}_1$ are unit vectors $\mathbf{e}_i$ with the entries $+1$ at the cordinates with the smallest ratios k_{ii}/m_{ii}. We note that except for the first iteration vector, this scheme would construct the same starting subspace as used in Examples 12.3 and 12.4. Since in the iteration we use q vectors, $q > p$, the first vector can be thought of as an additional vector. (In the analysis of large systems, an appropriate spacing between the unit entries in the starting vectors is also important and taken into account; see program SSPACE.)

The starting subspace described above is, in general, only an approximation to the actual subspace required, which we denoted as E_∞; however, the closer $\mathbf{K}$ and $\mathbf{M}$ are to the matrix forms used in Examples 12.3 and 12.4, the "better" is the starting subspace; i.e., the fewer iterations can be expected until convergence. In practice, the number of iterations required for convergence depends on the matrices $\mathbf{K}$ and $\mathbf{M}$ and on the accuracy required in the eigenvalues and eigenvectors. Experience has shown that with the starting subspace described above and $q = \min(2p, p + 8)$, of the order ten iterations are needed to calculate the largest eigenvalue λ_p to about six digit precision, with the smaller eigenvalues being more accurate.

The starting subspace above has been proved by experience to be quite effective. However, "better" starting vectors may be available in specific applications. For instance, in dynamic optimization, as the structure is modified in small steps, the eigensystem of the previous structure may be a good approximation to the eigensystem of the new structure. Similarly, if some eigenvectors have already been evaluated and we now want to solve for more eigenvectors, the already calculated eigenvectors would effectively be used in $\mathbf{X}_1$. If the eigenvectors of substructures have already been obtained, these may also be used in establishing the starting iteration vectors in $\mathbf{X}_1$ (see Section 10.3.3).

12.3.4 Convergence

In the subspace iteration it is necessary to measure convergence. Assume that in the iterations $(k - 1)$ and k, eigenvalue approximations $\lambda_i^{(k)}$ and $\lambda_i^{(k+1)}$, $i = 1, \ldots, p$, respectively, have been calculated. Convergence is measured as in inverse iteration (see Section 11.2.1); i.e., we require that

$$\frac{|\lambda_i^{(k+1)} - \lambda_i^{(k)}|}{\lambda_i^{(k+1)}} \leq \text{tol}; \qquad i = 1, \ldots, p \tag{12.15}$$

where tol should be 10^{-2s} when the eigenvalues shall be accurate to about

$2s$ digits. For example, if we iterate until all p ratios in (12.15) are smaller than 10^{-6}, we very likely have that λ_p has been approximated to about six-digit accuracy, and the smaller eigenvalues have been evaluated more accurately. Since the eigenvalue approximations are calculated using a Rayleigh quotient, the eigenvector approximations are accurate to only about s (or more) digits. It should be noted that the iteration is performed with q vectors, $q > p$, but convergence is measured only on the approximations predicted for the p smallest eigenvalues.

Another important aspect when using the subspace iteration technique is that of verifying that in fact the required eigenvalues and vectors have been calculated since the relations in (12.5) and (12.6) are satisfied by any eigenpairs. This verification is the third important phase of the subspace iteration method (see Section 12.3). As pointed out, the iteration in (12.10) to (12.14) converges in the limit to the eigenvectors $\boldsymbol{\phi}_1, \ldots, \boldsymbol{\phi}_p$, provided the starting iteration vectors in $\mathbf{X}_1$ are not orthogonal to any one of the required eigenvectors. The starting subspace described above has proved, by experience, to be very satisfactory in this regard, although there is no formal mathematical proof available that convergence will indeed always occur. However, once the convergence limit in (12.15) is satisfied, with s being at least equal to 3, we can make sure that the smallest eigenvalues and corresponding eigenvectors have indeed been calculated. For the check we use the Sturm sequence property of the characteristic polynomials of the problems $\mathbf{K}\boldsymbol{\phi} = \lambda\mathbf{M}\boldsymbol{\phi}$ and $\mathbf{K}^{(r)}\boldsymbol{\phi}^{(r)} = \lambda^{(r)}\mathbf{M}^{(r)}\boldsymbol{\phi}^{(r)}$ at a shift μ, where μ is just to the right of the calculated value for λ_p (see Fig. 12.4). The Sturm sequence property yields that in the Gauss factorization of $\mathbf{K} - \mu\mathbf{M}$ into $\mathbf{LDL}^T$, the number of negative elements in $\mathbf{D}$ is equal to the number of eigenvalues smaller than μ. Hence, in the case considered, we should have p negative elements in $\mathbf{D}$. However, in order to apply the Sturm sequence check, a meaningful μ must be used that takes account of the fact that we have obtained only approximations to the exact eigenvalues of the problem $\mathbf{K}\boldsymbol{\phi} = \lambda\mathbf{M}\boldsymbol{\phi}$. Let l be the last iteration, so that the calculated eigenvalues are $\lambda_1^{(l+1)}, \lambda_2^{(l+1)}, \ldots, \lambda_p^{(l+1)}$. A conservative esti-

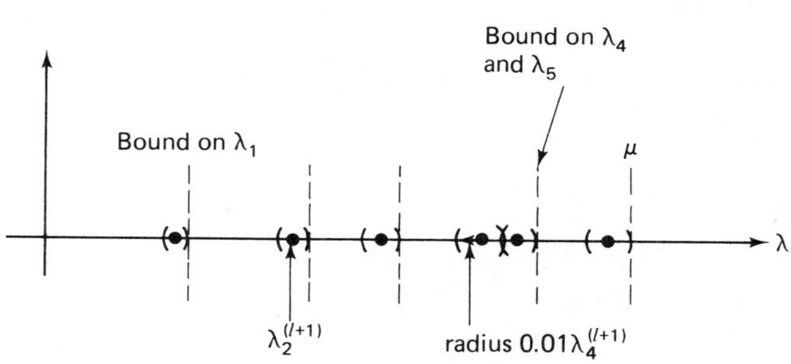

Figure 12.4. Bounds on eigenvalues to apply Sturm sequence check, $p = 6$.

mate for a region in which the exact eigenvalues of the problem $\mathbf{K}\boldsymbol{\phi} = \lambda\mathbf{M}\boldsymbol{\phi}$ lie is given by

$$0.99\lambda_i^{(l+1)} < \lambda_i < 1.01\lambda_i^{(l+1)} \tag{12.16}$$

where only the smallest eigenvalues that converged to a tolerance of 10^{-2s} should be included. The relation in (12.16) can be used to establish bounds on all exact eigenvalues, and hence a realistic Sturm sequence check can be applied.

12.3.5 Final Remarks Concerning the Subspace Iteration Method

The equations of subspace iteration have been presented in (12.10) to (12.14). However, in actual implementation, the solution can be performed more effectively as summarized in Table 12.3, which also gives the corresponding number of operations performed. Table 12.3 also shows that the solution is completed by calculating error estimates as in the determinant search solution (see Sections 12.2.3 and 10.4). It is noted that the subspace iteration method is very efficient in the analysis of systems with large bandwidth. Also, out-of-core solutions can be obtained effectively, because relatively little tape handling is necessary, and the lowest eigenvalues and corresponding vectors of very large systems can be solved.[1, 5-8] Table 12.4 summarizes some approximate solution times.

Considering the overall effectiveness of the subspace iteration method, it may be summarized that the method derives its efficiency from, firstly, the possibility of selecting a starting subspace that is relatively close to the required least dominant subspace, and, secondly, from the procedure of iterating from one subspace to the next and always calculating "best" eigenvalue and eigenvector approximations. The potential of the subspace iteration method lies in that a starting subspace, which is more effective than the one currently used, would further reduce the number of required iterations. In addition, there are promising possibilities for increasing the effectiveness of the method using shifting (such as employed in the determinant search method) and other acceleration strategies.[9]

The solution procedure is presented in a compact manner in the computer program given next.

SUBROUTINE SSPACE

Program SSPACE is an implementation of the subspace iteration method presented above for the solution of the smallest eigenvalues and corresponding eigenvectors of the generalized eigenproblem $\mathbf{K}\boldsymbol{\phi} = \lambda\mathbf{M}\boldsymbol{\phi}$. The argument variables and use of the subroutine are defined using comment cards in the program.

TABLE 12.3 Summary of Subspace Iteration Solution

Operation	Calculation	Number of Operations: $m = m_{\mathbf{K}} = m_{\mathbf{M}}$	Number of Operations: $m = m_{\mathbf{K}}$, $m_{\mathbf{M}} = 0$	Required Storage
Factorization of **K**	$\mathbf{K} = \mathbf{LDL}^T$	$\frac{1}{2}nm^2 + \frac{3}{2}nm$	$\frac{1}{2}nm^2 + \frac{3}{2}nm$	
Subspace iteration	$\mathbf{K}\bar{\mathbf{X}}_{k+1} = \mathbf{Y}_k$	$nq(2m+1)$	$nq(2m+1)$	
	$\mathbf{K}_{k+1} = \bar{\mathbf{X}}_{k+1}^T\mathbf{Y}_k$	$\frac{1}{2}nq(q+1)$	$\frac{1}{2}nq(q+1)$	
	$\bar{\mathbf{Y}}_{k+1} = \mathbf{M}\bar{\mathbf{X}}_{k+1}$	$nq(2m+1)$	nq	
	$\mathbf{M}_{k+1} = \bar{\mathbf{X}}_{k+1}^T\bar{\mathbf{Y}}_{k+1}$	$\frac{1}{2}nq(q+1)$	$\frac{1}{2}nq(q+1)$	Algorithm is effectively implemented as out-of-core solver
	$\mathbf{K}_{k+1}\mathbf{Q}_{k+1} = \mathbf{M}_{k+1}\mathbf{Q}_{k+1}\mathbf{\Omega}_{k+1}^2$	$O(q^3)$ neglected		
	$\mathbf{Y}_{k+1} = \bar{\mathbf{Y}}_{k+1}\mathbf{Q}_{k+1}$	nq^2	nq^2	
Sturm sequence check	$\bar{\mathbf{K}} = \mathbf{K} - \mu\mathbf{M}$	$n(m+1)$	n	
	$\bar{\mathbf{K}} = \mathbf{LDL}^T$	$\frac{1}{2}nm^2 + \frac{3}{2}nm$	$\frac{1}{2}nm^2 + \frac{3}{2}nm$	
Error estimates	$\dfrac{\lVert\mathbf{K}\boldsymbol{\phi}_i^{(l+1)} - \lambda_i^{(l+1)}\mathbf{M}\boldsymbol{\phi}_i^{(l+1)}\rVert_2}{\lVert\mathbf{K}\boldsymbol{\phi}_i^{(l+1)}\rVert_2}$	$2nm + 4n$	$5nm + 2n$	
Total for solution of p lowest eigenvalues and associated eigenvectors, assuming that ten iterations are required and $q = \min\{2p, p+8\}$		$nm^2 + nm(4+2p) + 4np + 20nq(2m + q + \frac{3}{2})$	$nm^2 + nm(3+5p) + 2np + 20nq(m + q + \frac{3}{2})$	

TABLE 12.4 Solution Times Using the Subspace Iteration Method

System	System Order	Maximum Half Bandwidth m_K	Mass Matrix	Number of Eigenpairs	Computer Used	Central Processor Seconds
3-Dimensional Building Frame	468	155	Diagonal	4	CDC 6400	160
Piping System	566	11	Diagonal	28	CDC 6600	142
Reactor Building with Foundation	1174	137	Diagonal	45	CDC 6600	890
Dam	2916	491	Diagonal	4	CDC 7600	495
Windtunnel	5952	215	Diagonal	10	CDC 7600	1000

```
      SUBROUTINE SSPACE (A,B,MAXA,R,EIGV,TT,W,AR,BR,VEC,D,RTOLV,BUP,BLO,
     1 BUPC,NN,NNM,NWK,NWM,NROOT,RTOL,NC,NNC,NITEM,IFSS,IFPR,NSTIF,IOUT)
C . . . . . . . . . . . . . . . . . . . . . . . . . . . . . . . . . . . .
C .                                                                     .
C .   P R O G R A M                                                     .
C .        TO SOLVE FOR THE SMALLEST EIGENVALUES AND CORRESPONDING      .
C .        EIGENVECTORS IN THE GENERALIZED EIGENPROBLEM USING THE       .
C .        SUBSPACE ITERATION METHOD                                    .
C .                                                                     .
C .  - - INPUT VARIABLES - -                                            .
C .        A(NWK)      = STIFFNESS MATRIX IN COMPACTED FORM (ASSUMED    .
C .                      POSITIVE DEFINITE)                             .
C .        B(NWM)      = MASS MATRIX IN COMPACTED FORM                  .
C .        MAXA(NNM)   = VECTOR CONTAINING ADDRESSES OF DIAGONAL        .
C .                      ELEMENTS OF STIFFNESS MATRIX A                 .
C .        R(NN,NC)    = EIGENVECTORS ON SOLUTION EXIT                  .
C .        EIGV(NC)    = EIGENVALUES ON SOLUTION EXIT                   .
C .        TT(NN)      = WORKING VECTOR                                 .
C .        W(NN)       = WORKING VECTOR                                 .
C .        AR(NNC)     = WORKING MATRIX STORING PROJECTION OF K         .
C .        BR(NNC)     = WORKING MATRIX STORING PROJECTION OF M         .
C .        VEC(NC,NC)= WORKING MATRIX                                   .
C .        D(NC)       = WORKING VECTOR                                 .
C .        RTOLV(NC)   = WORKING VECTOR                                 .
C .        BUP(NC)     = WORKING VECTOR                                 .
C .        BLO(NC)     = WORKING VECTOR                                 .
C .        BUPC(NC)    = WORKING VECTOR                                 .
C .        NN          = ORDER OF STIFFNESS AND MASS MATRICES           .
C .        NNM         = NN + 1                                         .
C .        NWK         = NUMBER OF ELEMENTS BELOW SKYLINE OF            .
C .                      STIFFNESS MATRIX                               .
C .        NWM         = NUMBER OF ELEMENTS BELOW SKYLINE OF            .
C .                      MASS MATRIX                                    .
C .                        I. E. NWM=NWK FOR CONSISTENT MASS MATRIX     .
C .                              NWM=NN  FOR LUMPED MASS MATRIX         .
C .        NROOT       = NUMBER OF REQUIRED EIGENVALUES AND EIGENVECTORS.
C .        RTOL        = CONVERGENCE TOLERANCE ON EIGENVALUES           .
C .                      ( 1.E-06 OR SMALLER )                          .
C .        NC          = NUMBER OF ITERATION VECTORS USED               .
C .                      (USUALLY SET TO MIN(2*NROOT, NROOT+8), BUT NC  .
C .                      CANNOT BE LARGER THAN THE NUMBER OF MASS       .
C .                      DEGREES OF FREEDOM)                            .
C .        NNC         = NC*(NC+1)/2 DIMENSION OF STORAGE VECTORS AR,BR .
C .        NITEM       = MAXIMUM NUMBER OF SUBSPACE ITERATIONS PERMITTED.
C .                      (USUALLY SET TO 16)                            .
C .                      THE PARAMETERS NC AND/OR NITEM MUST BE         .
C .                      INCREASED IF A SOLUTION HAS NOT CONVERGED      .
C .        IFSS        = FLAG FOR STURM SEQUENCE CHECK                  .
C .                        EQ.0  NO CHECK                               .
C .                        EQ.1  CHECK                                  .
C .        IFPR        = FLAG FOR PRINTING DURING ITERATION             .
C .                        EQ.0  NO PRINTING                            .
C .                        EQ.1  PRINT                                  .
C .        NSTIF       = SCRATCH FILE TO STORE STIFFNESS MATRIX         .
C .        IOUT        = OUTPUT PRINTING FILE                           .
C .                                                                     .
C .  - - OUTPUT - -                                                     .
C .        EIGV(NROOT) = EIGENVALUES                                    .
C .        R(NN,NROOT) = EIGENVECTORS                                   .
C .                                                                     .
C . . . . . . . . . . . . . . . . . . . . . . . . . . . . . . . . . . . .
      IMPLICIT REAL*8 (A-H,O-Z)
      ABS(X)=DABS(X)
      SQRT(X)=DSQRT(X)
C . . . . . . . . . . . . . . . . . . . . . . . . . . . . . . . . . . . .
C .        THIS PROGRAM IS USED IN SINGLE PRECISION ARITHMETIC ON       .
C .        CDC EQUIPMENT AND DOUBLE PRECISION ARITHMETIC ON IBM         .
C .        OR UNIVAC MACHINES .ACTIVATE,DEACTIVATE OR ADJUST ABOVE      .
C .        CARDS FOR SINGLE OR DOUBLE PRECISION ARITHMETIC              .
C . . . . . . . . . . . . . . . . . . . . . . . . . . . . . . . . . . . .
      DIMENSION A(NWK),B(NWM),R(NN,NC),TT(NN),W(NN),EIGV(NC),
     1          D(NC),VEC(NC,NC),AR(NNC),BR(NNC),RTOLV(NC),BUP(NC),
     2          BLO(NC),BUPC(NC)
      INTEGER MAXA(NNM)
C
C     SET TOLERANCE FOR JACOBI ITERATION
      TOLJ=0.000000000001
C
C     INITIALIZATION
C
      ICONV=0
      NSCH=0
      NSMAX=12
      N1=NC + 1
      NC1=NC - 1
      REWIND NSTIF
      WRITE (NSTIF) A
      DO 60 I=1,NC
   60 D(I)=0.
C
C     ESTABLISH STARTING ITERATION VECTORS
C
      ND=NN/NC
```

```
      IF (NWM.GT.NN) GO TO 4
      J=0
      DO 2 I=1,NN
      II=MAXA(I)
      R(I,1)=B(I)
      IF (B(I).GT.0) J=J+1
    2 W(I)=B(I)/A(II)
      IF (NC.LE.J) GO TO 16
      WRITE (IOUT,1007)
      STOP
    4 DO 10 I=1,NN
      II=MAXA(I)
      R(I,1)=B(II)
   10 W(I)=B(II)/A(II)
   16 DO 20 J=2,NC
      DO 20 I=1,NN
   20 R(I,J)=0.
C
      L=NN-ND
      DO 30 J=2,NC
      RT=0.
      DO 40 I=1,L
      IF (W(I).LT.RT) GO TO 40
      RT=W(I)
      IJ=I
   40 CONTINUE
      DO 50 I=L,NN
      IF (W(I).LE.RT) GO TO 50
      RT=W(I)
      IJ=I
   50 CONTINUE
      TT(J)=FLOAT(IJ)
      W(IJ)=0.
      L=L-ND
   30 R(IJ,J)=1.
C
      WRITE (IOUT,1008)
      WRITE (IOUT,1002) (TT(J),J=2,NC)
C
C     FACTORIZE MATRIX A INTO (L)*(D)*(L(T))
C
      ISH=0
      CALL DECOMP (A,MAXA,NN,ISH,IOUT)
C
C - - - S T A R T   O F   I T E R A T I O N   L O O P
C
      NITE=0
  100 NITE=NITE + 1
      IF (IFPR.EQ.0) GO TO 90
      WRITE (IOUT,1010) NITE
C
C     CALCULATE THE PROJECTIONS OF A AND B
C
   90 IJ=0
      DO 110 J=1,NC
      DO 120 K=1,NN
  120 TT(K)=R(K,J)
      CALL REDBAK (A,TT,MAXA,NN)
      DO 130 I=J,NC
      ART=0.
      DO 140 K=1,NN
  140 ART=ART + R(K,I)*TT(K)
      IJ=IJ + 1
  130 AR(IJ)=ART
      DO 150 K=1,NN
  150 R(K,J)=TT(K)
  110 CONTINUE
      IJ=0
      DO 160 J=1,NC
      CALL MULT (TT,B,R(1,J),MAXA,NN,NWM)
      DO 180 I=J,NC
      BRT=0.
      DO 190 K=1,NN
  190 BRT=BRT + R(K,I)*TT(K)
      IJ=IJ + 1
  180 BR(IJ)=BRT
      IF (ICONV.GT.0) GO TO 160
      DO 200 K=1,NN
  200 R(K,J)=TT(K)
  160 CONTINUE
C
C     SOLVE FOR EIGENSYSTEM OF SUBSPACE OPERATORS
C
      IF (IFPR.EQ.0) GO TO 320
      IND=1
  210 WRITE (IOUT,1020)
      II=1
      DO 300 I=1,NC
      ITEMP=II+NC-I
      WRITE(IOUT,1005) (AR(J),J=II,ITEMP)
  300 II=II + NC - I
      WRITE (IOUT,1030)
      II=1
```

```
      DO 310 I=1,NC
      ITEMP=II+NC-I
      WRITE(IOUT,1005) (BR(J),J=II,ITEMP)
  310 II=II + N1 - I
      IF(IND.EQ.2) GO TO 350
C
  320 CALL JACOBI (AR,BR,VEC,EIGV,W,NC,NNC,TOLJ,NSMAX,IFPR,IOUT)
C
      IF (IFPR.EQ.0) GO TO 350
      WRITE (IOUT,1040)
      IND=2
      GO TO 210
C
C     ARRANGE EIGENVALUES IN ASCENDING ORDER
C
  350 IS=0
      II=1
      DO 360 I=1,NC1
      ITEMP=II + N1 - I
      IF (EIGV(I+1).GE.EIGV(I)) GO TO 360
      IS=IS+1
      EIGVT=EIGV(I+1)
      EIGV(I+1)=EIGV(I)
      EIGV(I)=EIGVT
      BT=BR(ITEMP)
      BR(ITEMP)=BR(II)
      BR(II)=BT
      DO 370 K=1,NC
      RT=VEC(K,I+1)
      VEC(K,I+1)=VEC(K,I)
  370 VEC(K,I)=RT
  360 II=ITEMP
      IF (IS.GT.0) GO TO 350
      IF (IFPR.EQ.0) GO TO 375
      WRITE (IOUT,1035)
      WRITE (IOUT,1006) (EIGV(I),I=1,NC)
C
C     CALCULATE B TIMES APPROXIMATE EIGENVECTORS (ICONV.EQ.0)
C        OR      FINAL EIGENVECTOR APPROXIMATIONS (ICONV.GT.0)
C
  375 DO 420 I=1,NN
      DO 422 J=1,NC
  422 TT(J)=R(I,J)
      DO 424 K=1,NC
      RT=0.
      DO 430 L=1,NC
  430 RT=RT + TT(L)*VEC(L,K)
  424 R(I,K)=RT
  420 CONTINUE
      IF (ICONV.GT.0) GO TO 500
C
C     CHECK FOR CONVERGENCE OF EIGENVALUES
C
      DO 380 I=1,NC
      DIF=ABS(EIGV(I)-D(I))
  380 RTOLV(I)=DIF/EIGV(I)
      IF (IFPR.EQ.0) GO TO 385
      WRITE (IOUT,1050)
      WRITE (IOUT,1005) (RTOLV(I),I=1,NC)
C
  385 DO 390 I=1,NROOT
      IF (RTOLV(I).GT.RTOL) GO TO 400
  390 CONTINUE
      WRITE (IOUT,1060) RTOL
      ICONV=1
      GO TO 100
  400 IF (NITE.LT.NITEM) GO TO 410
      WRITE (IOUT,1070)
      ICONV=2
      IFSS=0
      GO TO 100
C
  410 DO 440 I=1,NC
  440 D(I)=EIGV(I)
      GO TO 100
C
C - - - E N D   O F   I T E R A T I O N   L O O P
C
  500 WRITE (IOUT,1100)
      WRITE (IOUT,1006) (EIGV(I),I=1,NROOT)
      WRITE (IOUT,1110)
      DO 530 J=1,NROOT
  530 WRITE (IOUT,1005) (R(K,J),K=1,NN)
C
C     CALCULATE AND PRINT ERROR NORMS
C
      REWIND NSTIF
      READ(NSTIF) A
C
      DO 580 L=1,NROOT
      RT=EIGV(L)
      CALL MULT(TT,A,R(1,L),MAXA,NN,NWK)
      VNORM=0.
```

```
      DO 590 I=1,NN
  590 VNORM=VNORM+TT(I)*TT(I)
      CALL MULT(W,B,R(1,L),MAXA,NN,NWM)
      WNORM=0.
      DO 600 I=1,NN
      TT(I)=TT(I) - RT*W(I)
  600 WNORM=WNORM + TT(I)*TT(I)
      VNORM=SQRT(VNORM)
      WNORM=SQRT(WNORM)
      D(L)=WNORM/VNORM
  580 CONTINUE
      WRITE(IOUT,1115)
      WRITE(IOUT,1006) (D(I),I=1,NROOT)
C
C     APPLY STURM SEQUENCE CHECK
C
      IF (IFSS.EQ.0) GO TO 700
      CALL SCHECK (EIGV,RTOLV,BUP,BLO,BUPC,D,NC,NEI,RTOL,SHIFT)
C
      WRITE (IOUT,1120) SHIFT
C
C     SHIFT MATRIX A
C
      REWIND NSTIF
      READ (NSTIF) A
      IF (NWM.GT.NN) GO TO 645
      DO 640 I=1,NN
      II=MAXA(I)
  640 A(II)=A(II) - B(I)*SHIFT
      GO TO 660
  645 DO 650 I=1,NWK
  650 A(I)=A(I) - B(I)*SHIFT
C
C     FACTORIZE SHIFTED MATRIX
C
  660 ISH=1
      CALL DECOMP (A,MAXA,NN,ISH,IOUT)
C
C     COUNT NUMBER OF NEGATIVE DIAGONAL ELEMENTS
C
      NSCH=0
      DO 664 I=1,NN
      II=MAXA(I)
      IF (A(II).LT.0.) NSCH=NSCH + 1
  664 CONTINUE
      IF (NSCH.EQ.NEI) GO TO 670
      NMIS=NSCH - NEI
      WRITE (IOUT,1130) NMIS
      GO TO 700
  670 WRITE (IOUT,1140) NSCH
  700 RETURN
C
 1002 FORMAT (1H0,10F10.0)
 1005 FORMAT (1H ,12E11.4)
 1006 FORMAT (1H0,6E22.14)
 1007 FORMAT (///63H STOP, NC IS LARGER THAN THE NUMBER OF MASS DEGREES
     1OF FREEDOM )
 1008 FORMAT ( ///,62H DEGREES OF FREEDOM EXCITED BY UNIT STARTING ITERA
     1TION VECTORS)
 1010 FORMAT (1H1,32HI T E R A T I O N   N U M B E R I4)
 1020 FORMAT (28H0PROJECTION OF A (MATRIX AR) )
 1030 FORMAT (28H0PROJECTION OF B (MATRIX BR) )
 1035 FORMAT (30H0EIGENVALUES OF AR-LAMBDA*BR  )
 1040 FORMAT (40H0AR AND BR AFTER JACOBI DIAGONALIZATION  )
 1050 FORMAT (43H0RELATIVE TOLERANCE REACHED ON EIGENVALUES  )
 1060 FORMAT (///,30H CONVERGENCE REACHED FOR RTOL  E10.4)
 1070 FORMAT (1H1,51H*** NO CONVERGENCE IN MAXIMUM NUMBER OF ITERATIONS
     1            9HPERMITTED/35H WE ACCEPT CURRENT ITERATION VALUES/
     2            42H THE STURM SEQUENCE CHECK IS NOT PERFORMED)
 1100 FORMAT (///,31H THE CALCULATED EIGENVALUES ARE  )
 1115 FORMAT (//1X,36HPRINT ERROR NORMS ON THE EIGENVALUES  )
 1110 FORMAT (//, 32H THE CALCULATED EIGENVECTORS ARE  //)
 1120 FORMAT  (///,23H CHECK APPLIED AT SHIFT   E22.14)
 1130 FORMAT (// 10H THERE ARE  I4,21H EIGENVALUES MISSING  )
 1140 FORMAT (// 20H WE FOUND THE LOWEST  I4,12H EIGENVALUES  )
C
      END
```

```
      SUBROUTINE DECOMP (A,MAXA,NN,ISH,IOUT)
C . . . . . . . . . . . . . . . . . . . . . . . . . . . . . . . . . . .
C .                                                                   .
C .   P R O G R A M                                                   .
C .        TO CALCULATE (L)*(D)*(L)(T) FACTORIZATION OF               .
C .        STIFFNESS MATRIX                                           .
C .                                                                   .
C . . . . . . . . . . . . . . . . . . . . . . . . . . . . . . . . . . .
C
      IMPLICIT REAL*8 (A-H,O-Z)
      ABS(X)=DABS(X)
      DIMENSION A(1),MAXA(1)
      IF (NN.EQ.1) RETURN
```

```
C
      DO 200 N=1,NN
      KN=MAXA(N)
      KL=KN + 1
      KU=MAXA(N+1) - 1
      KH=KU - KL
      IF (KH) 304,240,210
  210 K=N - KH
      IC=0
      KLT=KU
      DO 260 J=1,KH
      IC=IC + 1
      KLT=KLT - 1
      KI=MAXA(K)
      ND=MAXA(K+1) - KI - 1
      IF (ND) 260,260,270
  270 KK=MIN0(IC,ND)
      C=0.
      DO 280 L=1,KK
  280 C=C + A(KI+L)*A(KLT+L)
      A(KLT)=A(KLT) - C
  260 K=K + 1
  240 K=N
      B=0.
      DO 300 KK=KL,KU
      K=K - 1
      KI=MAXA(K)
      C=A(KK)/A(KI)
      IF (ABS(C).LT.1.E07) GO TO 290
      WRITE (IOUT,2010) N,C
      STOP
  290 B=B + C*A(KK)
  300 A(KK)=C
      A(KN)=A(KN) - B
  304 IF (A(KN)) 310,310,200
  310 IF (ISH.EQ.0) GO TO 320
      IF (A(KN).EQ.0.) A(KN)=-1.E-16
      GO TO 200
  320 WRITE(IOUT,2000) N,A(KN)
      STOP
  200 CONTINUE
C
      RETURN
 2000 FORMAT(//48H STOP - STIFFNESS MATRIX NOT POSITIVE DEFINITE   ,//
     1           32H NONPOSITIVE PIVOT FOR EQUATION  ,I4,//
     2           10H PIVOT =  ,E20.12 )
 2010 FORMAT (//47H STOP - STURM SEQUENCE CHECK FAILED BECAUSE OF
     135HMULTIPLIER GROWTH FOR COLUMN NUMBER,I4,//12H MULTIPLIER=,E20.8)
      END


      SUBROUTINE REDBAK (A,V,MAXA,NN)
C . . . . . . . . . . . . . . . . . . . . . . . . . . . . . . . . . . .
C .                                                                   .
C .   P R O G R A M                                                   .
C .        TO REDUCE AND BACK-SUBSTITUTE ITERATION VECTORS            .
C .                                                                   .
C . . . . . . . . . . . . . . . . . . . . . . . . . . . . . . . . . . .
C
      IMPLICIT REAL*8 (A-H,O-Z)
      DIMENSION A(1),V(1),MAXA(1)
C
      DO 400 N=1,NN
      KL=MAXA(N) + 1
      KU=MAXA(N+1) - 1
      IF (KU-KL) 400,410,410
  410 K=N
      C=0.
      DO 420 KK=KL,KU
      K=K - 1
  420 C=C + A(KK)*V(K)
      V(N)=V(N) - C
  400 CONTINUE
C
      DO 480 N=1,NN
      K=MAXA(N)
  480 V(N)=V(N)/A(K)
      IF (NN.EQ.1) RETURN
      N=NN
      DO 500 L=2,NN
      KL=MAXA(N) + 1
      KU=MAXA(N+1) - 1
      IF (KU-KL) 500,510,510
  510 K=N
      DO 520 KK=KL,KU
      K=K - 1
  520 V(K)=V(K) - A(KK)*V(N)
  500 N=N - 1
C
      RETURN
      END
```

```
      SUBROUTINE MULT (TT,B,RR,MAXA,NN,NWM)
C . . . . . . . . . . . . . . . . . . . . . . . . . . . . . . . . . . .
C .                                                                   .
C .   P R O G R A M                                                   .
C .        TO EVALUATE PRODUCT OF B TIMES RR AND STORE RESULT IN TT   .
C .                                                                   .
C . . . . . . . . . . . . . . . . . . . . . . . . . . . . . . . . . . .
C
      IMPLICIT REAL*8 (A-H,O-Z)
      DIMENSION TT(1),B(1),RR(1),MAXA(1)
C
      IF (NWM.GT.NN) GO TO 20
      DO 10 I=1,NN
   10 TT(I)=B(I)*RR(I)
      RETURN
C
   20 DO 40 I=1,NN
   40 TT(I)=0.
      DO 100 I=1,NN
      KL=MAXA(I)
      KU=MAXA(I+1) - 1
      II=I + 1
      CC=RR(I)
      DO 100 KK=KL,KU
      II=II - 1
  100 TT(II)=TT(II) + B(KK)*CC
      IF (NN.EQ.1) RETURN
      DO 200 I=2,NN
      KL=MAXA(I) + 1
      KU=MAXA(I+1) - 1
      IF (KU-KL) 200,210,210
  210 II=I
      AA=0.
      DO 220 KK=KL,KU
      II=II - 1
  220 AA=AA + B(KK)*RR(II)
      TT(I)=TT(I) + AA
  200 CONTINUE
C
      RETURN
      END
```

```
      SUBROUTINE SCHECK (EIGV,RTOLV,BUP,BLO,BUPC,NEIV,NC,NEI,RTOL,SHIFT)
C . . . . . . . . . . . . . . . . . . . . . . . . . . . . . . . . . . .
C .                                                                   .
C .   P R O G R A M                                                   .
C .        TO EVALUATE SHIFT FOR STURM SEQUENCE CHECK                 .
C .                                                                   .
C . . . . . . . . . . . . . . . . . . . . . . . . . . . . . . . . . . .
C
      IMPLICIT REAL*8 (A-H,O-Z)
      DIMENSION EIGV(NC),RTOLV(NC),BUP(NC),BLO(NC),BUPC(NC),NEIV(NC)
C
      FTOL=0.01
C
      DO 100 I=1,NC
      BUP(I)=EIGV(I)*(1. + FTOL)
  100 BLO(I)=EIGV(I)*(1. - FTOL)
      NROOT=0
      DO 120 I=1,NC
  120 IF (RTOLV(I).LT.RTOL) NROOT=NROOT+1
      IF (NROOT.GE.1) GO TO 200
      WRITE (6,1010)
      STOP
C
C       FIND UPPER BOUNDS ON EIGENVALUE CLUSTERS
C
  200 DO 240 I=1,NROOT
  240 NEIV(I)=1
      IF (NROOT.NE.1) GO TO 260
      BUPC(1)=BUP(1)
      LM=1
      L=1
      I=2
      GO TO 295
  260 L=1
      I=2
  270 IF (BUP(I-1).LE.BLO(I)) GO TO 280
      NEIV(L)=NEIV(L)+1
      I=I+1
      IF (I.LE.NROOT) GO TO 270
  280 BUPC(L)=BUP(I-1)
      IF (I.GT.NROOT) GO TO 290
      L=L+1
      I=I+1
      IF (I.LE.NROOT) GO TO 270
      BUPC(L)=BUP(I-1)
  290 LM=L
      IF (NROOT.EQ.NC) GO TO 300
```

```
  295 IF (BUP(I-1).LE.BLO(I)) GO TO 300
      IF (RTOLV(I).GT.RTOL) GO TO 300
      BUPC(L)=BUP(I)
      NEIV(L)=NEIV(L)+1
      NROOT=NROOT+1
      IF (NROOT.EQ.NC) GO TO 300
      I=I+1
      GO TO 295
C
C      FIND SHIFT
C
  300 WRITE (6,1020)
      WRITE (6,1005) (BUPC(I),I=1,LM)
      WRITE (6,1030)
      WRITE (6,1006) (NEIV(I),I=1,LM)
      LL=LM-1
      IF (LM.EQ.1) GO TO 310
  330 DO 320 I=1,LL
  320 NEIV(L)=NEIV(L)+NEIV(I)
      L=L-1
      LL=LL-1
      IF (L.NE.1) GO TO 330
  310 WRITE (6,1040)
      WRITE (6,1006) (NEIV(I),I=1,LM)
      L=0
      DO 340 I=1,LM
      L=L+1
      IF (NEIV(I).GE.NROOT) GO TO 350
  340 CONTINUE
  350 SHIFT=BUPC(L)
      NEI=NEIV(L)
C
      RETURN
C
 1005 FORMAT (1H0,6E22.14)
 1006 FORMAT (1H0,6I22)
 1010 FORMAT (37H0***ERROR   SOLUTION STOP IN *SCHECK*, / 12X,
     1        21HNO EIGENVALUES FOUND., / 1X)
 1020 FORMAT (///,37H UPPER BOUNDS ON EIGENVALUE CLUSTERS  )
 1030 FORMAT (34H0NO OF EIGENVALUES IN EACH CLUSTER  )
 1040 FORMAT (42H0NO OF EIGENVALUES LESS THAN UPPER BOUNDS   )
      END
```

```
      SUBROUTINE JACOBI (A,B,X,EIGV,D,N,NWA,RTOL,NSMAX,IFPR,IOUT)
C .................................................................
C .                                                               .
C .   P R O G R A M                                               .
C .        TO SOLVE THE GENERALIZED EIGENPROBLEM USING THE        .
C .        GENERALIZED JACOBI ITERATION                           .
C .................................................................
      IMPLICIT REAL*8(A-H,O-Z)
      ABS(X)=DABS(X)
      SQRT(X)=DSQRT(X)
      DIMENSION A(NWA),B(NWA),X(N,N),EIGV(N),D(N)
C
C     INITIALIZE EIGENVALUE AND EIGENVECTOR MATRICES
C
      N1=N + 1
      II=1
      DO 10 I=1,N
      IF(A(II).GT.0. .AND. B(II).GT.0.) GO TO 4
      WRITE(IOUT,2020) II,A(II),B(II)
      STOP
    4 D(I)=A(II)/B(II)
      EIGV(I)=D(I)
   10 II=II + N1 - I
      DO 30 I=1,N
      DO 20 J=1,N
   20 X(I,J)=0.
   30 X(I,I)=1.
      IF(N.EQ.1) RETURN
C
C     INITIALIZE SWEEP COUNTER AND BEGIN ITERATION
C
      NSWEEP=0
      NR=N-1
   40 NSWEEP=NSWEEP+1
      IF(IFPR.EQ.1)WRITE(IOUT,2000)NSWEEP
C
C     CHECK IF PRESENT OFF-DIAGONAL ELEMENT IS LARGE ENOUGH TO REQUIRE ZEROING
C
      EPS=(.01**NSWEEP)**2
      DO 210 J=1,NR
      JP1=J+1
      JM1=J-1
      LJK=JM1*N - JM1*J/2
      JJ=LJK + J
      DO 210 K=JP1,N
      KP1=K+1
```

```
      KM1=K-1
      JK=LJK + K
      KK=KM1*N - KM1*K/2 + K
      EPTOLA=(A(JK)*A(JK))/(A(JJ)*A(KK))
      EPTOLB=(B(JK)*B(JK))/(B(JJ)*B(KK))
      IF((EPTOLA.LT.EPS).AND.(EPTOLB.LT.EPS))GO TO 210
C
C     IF ZEROING IS REQUIRED, CALCULATE THE ROTATION MATRIX ELEMENTS CA AND CG
C
      AKK=A(KK)*B(JK)-B(KK)*A(JK)
      AJJ=A(JJ)*B(JK)-B(JJ)*A(JK)
      AB=A(JJ)*B(KK)-A(KK)*B(JJ)
      CHECK=(AB*AB+4.*AKK*AJJ)/4.
      IF(CHECK)50,60,60
   50 WRITE(IOUT,2020)
      STOP
   60 SQCH=SQRT(CHECK)
      D1=AB/2.+SQCH
      D2=AB/2.-SQCH
      DEN=D1
      IF(ABS(D2).GT.ABS(D1))DEN=D2
      IF(DEN)80,70,80
   70 CA=0.
      CG=-A(JK)/A(KK)
      GO TO 90
   80 CA=AKK/DEN
      CG=-AJJ/DEN
C
C     PERFORM THE GENERALIZED ROTATION TO ZERO THE PRESENT OFF-DIAGONAL ELEMENT
C
   90 IF(N-2)100,190,100
  100 IF(JM1-1)130,110,110
  110 DO 120 I=1,JM1
      IM1=I - 1
      IJ=IM1*N - IM1*I/2 + J
      IK=IM1*N - IM1*I/2 + K
      AJ=A(IJ)
      BJ=B(IJ)
      AK=A(IK)
      BK=B(IK)
      A(IJ)=AJ+CG*AK
      B(IJ)=BJ+CG*BK
      A(IK)=AK+CA*AJ
  120 B(IK)=BK+CA*BJ
  130 IF(KP1-N)140,140,160
  140 LJI=JM1*N - JM1*J/2
      LKI=KM1*N - KM1*K/2
      DO 150 I=KP1,N
      JI=LJI + I
      KI=LKI + I
      AJ=A(JI)
      BJ=B(JI)
      AK=A(KI)
      BK=B(KI)
      A(JI)=AJ+CG*AK
      B(JI)=BJ+CG*BK
      A(KI)=AK+CA*AJ
  150 B(KI)=BK+CA*BJ
  160 IF(JP1-KM1)170,170,190
  170 LJI=JM1*N - JM1*J/2
      DO 180 I=JP1,KM1
      JI=LJI + I
      IM1=I - 1
      IK=IM1*N - IM1*I/2 + K
      AJ=A(JI)
      BJ=B(JI)
      AK=A(IK)
      BK=B(IK)
      A(JI)=AJ+CG*AK
      B(JI)=BJ+CG*BK
      A(IK)=AK+CA*AJ
  180 B(IK)=BK+CA*BJ
  190 AK=A(KK)
      BK=B(KK)
      A(KK)=AK+2.*CA*A(JK)+CA*CA*A(JJ)
      B(KK)=BK+2.*CA*B(JK)+CA*CA*B(JJ)
      A(JJ)=A(JJ)+2.*CG*A(JK)+CG*CG*AK
      B(JJ)=B(JJ)+2.*CG*B(JK)+CG*CG*BK
      A(JK)=0.
      B(JK)=0.
C
C     UPDATE THE EIGENVECTOR MATRIX AFTER EACH ROTATION
C
      DO 200 I=1,N
      XJ=X(I,J)
      XK=X(I,K)
      X(I,J)=XJ+CG*XK
  200 X(I,K)=XK+CA*XJ
```

```
  210 CONTINUE
C
C     UPDATE THE EIGENVALUES AFTER EACH SWEEP
C
      II=1
      DO 220 I=1,N
      IF (A(II).GT.0. .AND. B(II).GT.0.) GO TO 215
      WRITE(IOUT,2020) II,A(II),B(II)
      STOP
  215 EIGV(I)=A(II)/B(II)
  220 II=II + N1 - I
      IF(IFPR.EQ.0)GO TO 230
      WRITE(IOUT,2030)
      WRITE(IOUT,2010) (EIGV(I),I=1,N)
C
C     CHECK FOR CONVERGENCE
C
  230 DO 240 I=1,N
      TOL=RTOL*D(I)
      DIF=ABS(EIGV(I)-D(I))
      IF(DIF.GT.TOL)GO TO 280
  240 CONTINUE
C
C     CHECK ALL OFF-DIAGONAL ELEMENTS TO SEE IF ANOTHER SWEEP IS REQUIRED
C
      EPS=RTOL**2
      DO 250 J=1,NR
      JM1=J-1
      JP1=J+1
      LJK=JM1*N - JM1*J/2
      JJ=LJK + J
      DO 250 K=JP1,N
      KM1=K-1
      JK=LJK + K
      KK=KM1*N - KM1*K/2 + K
      EPSA=(A(JK)*A(JK))/(A(JJ)*A(KK))
      EPSB=(B(JK)*B(JK))/(B(JJ)*B(KK))
      IF((EPSA.LT.EPS).AND.(EPSB.LT.EPS))GO TO 250
      GO TO 280
  250 CONTINUE
C
C     FILL OUT BOTTOM TRIANGLE OF RESULTANT MATRICES AND SCALE EIGENVECTORS
C
  255 II=1
      DO 275 I=1,N
      BB=SQRT(B(II))
      DO 270 K=1,N
  270 X(K,I)=X(K,I)/BB
  275 II=II + N1 - I
      RETURN
C
C     UPDATE  D  MATRIX AND START NEW SWEEP, IF ALLOWED
C
  280 DO 290 I=1,N
  290 D(I)=EIGV(I)
      IF(NSWEEP.LT.NSMAX)GO TO 40
      GO TO 255
 2000 FORMAT(27H0SWEEP NUMBER IN *JACOBI* = ,I4)
 2010 FORMAT(1H0,6E20.12)
 2020 FORMAT (25H0*** ERROR  SOLUTION STOP  /
     1         31H MATRICES NOT POSITIVE DEFINITE /
     2 4H II=,I4,6HA(II)=,E20.12,6HB(II)=,E20.12)
 2030 FORMAT(36H0CURRENT EIGENVALUES IN *JACOBI* ARE,/)
      END
```

12.4 SELECTION OF SOLUTION TECHNIQUE

In the selection of a solution method for an eigenproblem, it must be realized that there does not as yet exist a single algorithm that is always most effective. Which method is used most appropriately depends on the characteristics of **K** and **M** and primarily on the order and bandwidths of the matrices and the number of eigenvalues and eigenvectors required. For the most common eigenproblem, in which the lowest eigenvalues and corresponding eigenvectors will be calculated, the choice lies in practically all cases between the determinant search and subspace iteration methods. For this reason, the two techniques have been made available in general structural analysis programs.[5-8] In some special-purpose programs, the Householder-QR-inverse iteration and the generalized Jacobi iteration methods can also be used effectively (see Sections 11.3.2 and 11.3.3).

The appropriate solution technique for a given problem should be selected by considering the information given in Tables 11.3, 11.4, 12.1, and 12.3. The choice for a solution routine is governed by the number of operations needed for solution and the required high-speed storage. The Householder-QR-inverse iteration solution, the generalized Jacobi iteration, and the determinant search method are likely to be used on systems that can be solved in the high-speed storage of the computer. If the techniques are implemented in out-of-core solution routines, relatively much tape handling is necessary, but the high-speed storage requirements would be small.

The generalized Jacobi iteration is most efficient when the complete eigensystem is required and either not many off-diagonal elements are present or they are already small; i.e., the eigenproblem is already "nearly" solved. For this reason the technique is efficiently used for the solution of the eigensystem of $\mathbf{K}_{k+1}$ and $\mathbf{M}_{k+1}$ in (12.13) of the subspace iteration. When the order of the matrices is relatively small, the solution of the eigenproblem is not very expensive and the Jacobi iteration may also be attractive because of its simplicity and elegance of solution.

The Householder-QR-inverse iteration solution is most efficient when all eigenvalues and eigenvectors of a matrix are required and the matrix has a large bandwidth or is full. As was pointed out in Section 10.3.1, this solution requirement can arise after static condensation of the massless degrees of freedom. A full matrix is also obtained if the generalized problem with a banded mass matrix is transformed into the standard form (see Section 10.2.5).

Whether the procedure of static condensation and solution of the reduced eigenproblem is efficient depends on the original bandwidth of the stiffness matrix, the increase in bandwidth due to static condensation, the number of original and final degrees of freedom, and the number of required eigenvalues and eigenvectors. In most analyses mass is associated with about one-half or more of the degrees of freedom; therefore, if the order of the system is large, the static condensation still leads to a large-order system which may have lost the bandform. In this case a direct solution of the eigenproblem which takes full advantage of the banding characteristics and solves only for the required eigenvalues and eigenvectors is more efficient.

Considering the transformation to the standard eigenproblem, when the mass matrix is banded and the system is large, the transformation is almost always very inefficient.

The determinant search technique is very effectively used to calculate the lowest eigenvalues and corresponding eigenvectors of systems with small bandwidth. In the solution the eigenvalues and eigenvectors are calculated to high precision. If compacted storage is used, relatively large-order systems can also be solved in core (see Section 6.2.3). The use of a banded mass matrix increases the cost of solution relatively little. Note also that, depending on the bandwidth, to find the complete eigensystem, the determinant search method can be more efficient than the Householder-QR-inverse iteration solution.

The subspace iteration solution is very effective in the calculation of the lowest eigenvalues and corresponding eigenvectors of systems with large bandwidth and which are too large for the high-speed storage of the computer. It should be noted, however, that the eigensystem of the projected operators in (12.13) is calculated in high-speed storage, and that in the case when many vectors are calculated, this high-speed storage requirement may govern the problem size.

In addition to the total number of operations required for solution and the high-speed storage requirements, another consideration is important when selecting a solution routine. The determinant search and subspace iteration methods have the advantage that the procedures can detect at a relatively early stage whether the matrices $\mathbf{K}$ and $\mathbf{M}$ are well defined. In the use of a general analysis program, data input errors may cause $\mathbf{K}$ to be not positive definite, which would be detected using the subspace iteration method or determinant search technique when the matrix $\mathbf{K}$ is factorized into $\mathbf{LDL}^T$ at the beginning of the iteration. Namely, all elements in $\mathbf{D}$ must be larger than zero, and if negative elements are encountered, the solution would be stopped, with relatively little expense having occurred, because the actual iteration did not yet start. On the other hand, using the HQRI or Jacobi method, the solution may have progressed substantially before one of the diagonal elements becomes negative, indicating a data input error.

The most important eigenproblem in dynamic analysis is the solution of the lowest eigenvalues and corresponding eigenvectors. However, in some dynamic analyses, only eigenvalues within a specified intermediate range are of interest. If the order of the matrices is not large, a solution using the Householder-QR-inverse iteration technique, in which all eigenvalues are obtained, or a solution using the determinant search method, in which all eigenvalues below the maximum required are calculated, may be efficient. However, if only a very few eigenvalues and eigenvectors are needed, an algorithm based on a combination of the techniques used in the determinant search and subspace iteration methods would be very effective.

It should be pointed out that the above considerations for the choice of the appropriate solution algorithm are also applicable to eigenproblems other than $\mathbf{K}\boldsymbol{\phi} = \lambda\mathbf{M}\boldsymbol{\phi}$. Before actually presenting the various solution techniques, we summarized the eigenproblems encountered in finite element analysis and decided to concentrate on the solution of the required eigenvalues and corresponding eigenvectors in the problem $\mathbf{K}\boldsymbol{\phi} = \lambda\mathbf{M}\boldsymbol{\phi}$, where $\mathbf{K}$ is the stiffness matrix and $\mathbf{M}$ is the mass matrix of the system (see Section 10.1). The other eigenproblems would be obtained by using the appropriate matrices for $\mathbf{K}$ and $\mathbf{M}$. For example, in heat transfer analysis, $\mathbf{K}$ and $\mathbf{M}$ would be the conductivity and heat capacity matrices and the eigensystem solution would be analogous to the solution of the problem in dynamic analysis. In buckling analysis we have the problem $\mathbf{K}_G\boldsymbol{\phi} = \lambda\mathbf{K}\boldsymbol{\phi}$, where $\mathbf{K}$ and $\mathbf{K}_G$ are the linear and nonlinear strain stiffness matrices, and we require the largest eigenvalue and corresponding eigenvector. In this case the determinant search and subspace iteration methods can be used if prior to the solutions a shift is imposed that is larger than the largest eigenvalue, and the algorithms are modified to evaluate the smallest negative eigenvalue. Namely, the smallest negative eigenvalue of the shifted system added to the shift gives the required largest eigenvalue of $\mathbf{K}_G\boldsymbol{\phi} = \lambda\mathbf{K}\boldsymbol{\phi}$ (see Section 10.2.3).

In buckling analysis of large-order systems, the Householder-QR-inverse iteration and the generalized Jacobi iteration method are not efficient, and the determinant search or subspace iteration methods should be used. In this context it may be noted that for the solution of one eigenvalue only, both techniques follow virtually the same solution steps.

Finally, it should be noted once more that the solution methods presented in this chapter are believed to be most effective at this time. However, further improvements in eigensolution methods for large finite element systems must be anticipated and, indeed, are already under way.[7]

REFERENCES

1. K. J. BATHE, "Solution Methods of Large Generalized Eigenvalue Problems in Structural Engineering," Report UC SESM 71-20, Civil Engineering Department, University of California, Berkeley, 1971.
2. K. J. BATHE and E. L. WILSON, "Large Eigenvalue Problems in Dynamic Analysis," *A.S.C.E., Journal of Engineering Mechanics Division*, Vol. 98, 1972, pp. 1471–1485.
3. K. J. BATHE and E. L. WILSON, "Eigensolution of Large Structural Systems with Small Bandwidth," *A.S.C.E., Journal of Engineering Mechanics Division*, Vol. 99, 1973, pp. 467–479.
4. K. J. BATHE and E. L. WILSON, "Solution Methods for Eigenvalue Problems in Structural Mechanics," *International Journal for Numerical Methods in Engineering*, Vol. 6, 1973, pp. 213–226.
5. K. J. BATHE, E. L. WILSON, and F. E. PETERSON, "SAP IV—A Structural Analysis Program for Static and Dynamic Response of Linear Systems," Report EERC

73-11, College of Engineering, University of California, Berkeley, June 1973, revised Apr. 1974.

6. EAC/EASE2 Dynamics—User Information Manual/Theoretical, Control Data Corporation, Minneapolis, Minn. (in press).
7. K. J. Bathe, "ADINA—A Finite Element Program for Automatic Dynamic Incremental Nonlinear Analysis," Report 82448-1, Acoustics and Vibration Laboratory, Department of Mechanical Engineering, Massachusetts Institute of Technology, 1975.
8. K. J. Bathe, "An Assessment of Current Finite Element Analysis of Nonlinear Problems in Solid Mechanics," Proceedings, Symposium on the Numerical Solution of Partial Differential Equations, May 1975, Academic Press, 1976.
9. J. H. Wilkinson, *The Algebraic Eigenvalue Problem*, Clarendon Press, Oxford, 1965.
10. H. R. Schwarz, H. Rutishauser, and E. Stiefel, *Matrizen-Numerik*, B. G. Teubner, Stuttgart, 1972.
11. F. L. Bauer, "Das Verfahren der Treppen-Iteration und Verwandte Verfahren zur Lösung Algebraischer Eigenwertprobleme," *Z.A.M.P.*, Vol. 8, 1957, pp. 214–235.
12. O. E. Brönlund, "Eigenvalues of Large Matrices," Symposium on Finite Element Techniques at the Institut für Statik und Dynamik der Luft und Raumfahrtskonstruktionen, University of Stuttgart, Germany, 1969.
13. A. Jennings, "A Direct Iteration Method of Obtaining Latent Roots and Vectors of a Symmetric Matrix," *Proceedings of the Cambridge Philosophical Society*, Vol. 63, 1967, pp. 755–765.
14. A. Jennings and D. R. L. Orr, "Application of the Simultaneous Iteration Method to Undamped Vibration Problems," *International Journal for Numerical Methods in Engineering*, Vol. 3, 1971, pp. 13–24.
15. S. B. Dong, J. A. Wolf, and F. E. Peterson, "On a Direct-Iterative Eigensolution Technique," *International Journal for Numerical Methods in Engineering*, Vol. 4, 1972, pp. 155–162.
16. P. S. Jensen, "The Solution of Large Symmetric Eigenproblems by Sectioning," *S.I.A.M. Journal of Numerical Analysis*, Vol. 9, 1972, pp. 534–545.
17. H. Rutishauser, "Computational Aspects of F. L. Bauer's Simultaneous Iteration Method," *Numerische Mathematik*, Vol. 13, 1969, pp. 4–13.

INDEX

P

Q

R

S